Mathematical Analysis and Applications

Mathematical Analysis and Applications

Selected Topics

Edited by

Michael Ruzhansky
Hemen Dutta
Ravi P. Agarwal

Registered Office
John Wiley & Sons, Inc., 111 River Street, Hoboken, NJ 07030, USA

Editorial Office
111 River Street, Hoboken, NJ 07030, USA

For details of our global editorial offices, customer services, and more information about Wiley products visit us at www.wiley.com.

Wiley also publishes its books in a variety of electronic formats and by print-on-demand. Some content that appears in standard print versions of this book may not be available in other formats.

Library of Congress Cataloging-in-Publication Data:

Names: Ruzhansky, M. (Michael), editor. | Dutta, Hemen, 1981- editor. | Agarwal, Ravi P., editor.
Title: Mathematical analysis and applications : selected topics / edited by Michael Ruzhansky, Hemen Dutta, Ravi P. Agarwal.
Description: Hoboken, NJ : John Wiley & Sons, 2018. | Includes bibliographical references and index. |
Identifiers: LCCN 2017048922 (print) | LCCN 2017054738 (ebook) | ISBN 9781119414308 (pdf) | ISBN 9781119414339 (epub) | ISBN 9781119414346 (cloth)
Subjects: LCSH: Mathematical analysis.
Classification: LCC QA300 (ebook) | LCC QA300 .M225 2018 (print) | DDC 515–dc23
LC record available at https://lccn.loc.gov/2017048922

Cover Design: Wiley
Cover Image: © LoudRedCreative/Getty Images

Set in 10/12pt WarnockPro by SPi Global, Chennai, India

Printed in the United States of America

10 9 8 7 6 5 4 3 2 1

Contents

10 Stabilities and Instabilities of Rational Functional Equations and Euler–Lagrange–Jensen (a, b)-Sextic Functional Equations *341*
John Michael Rassias, Krishnan Ravi, and Beri V. Senthil Kumar

Preface

This book is designed for researchers, graduate students, educators, and practitioners with an interest in mathematical analysis in particular and in mathematics in general. The book aims to present theory, methods, and applications of the chosen topics under several chapters that have recent research importance and use. Emphasis is made to present the basic developments concerning each idea in full detail, and also contain the most recent advances made in the corresponding area of study. The text is presented in a self-contained manner, providing at least an idea of the proof of all results, and giving enough references to enable the interested reader to follow subsequent studies in a still developing field. There are 20 selected chapters in this book and they are organized as follows.

The chapter "Spaces of Asymptotically Developable Functions and Applications" presents the functional structure of the spaces of asymptotically developable functions in several complex variables. The authors also illustrate the notion of summability with some applications concerning singularly perturbed systems of ordinary differential equations and Pfaffian systems.

The chapter "Duality for Gaussian Processes from Random Signed Measures" proves a number of results for a general class of Gaussian processes. Two features are stressed, first the Gaussian processes are indexed by a general measure space; second, the authors "adjust" the associated reproducing kernel Hilbert spaces (RKHSs) to the measurable category. Among other things, this allows us to give a precise necessary and sufficient condition for equivalence of a pair of probability measures (in sample space), which determine the corresponding two Gaussian processes.

In the chapter "Many-body Wave Scattering Problems for Small Scatterers and Creating Materials with a Desired Refraction Coefficient," formulas are derived for solutions of many-body wave scattering problems by small impedance particles embedded in an inhomogeneous medium. The limiting case is considered when the size a of small particles tends to zero while their number tends to infinity at a suitable rate. Equations for the limiting effective (self-consistent) field in the medium are derived. The theory is based on a study of integral equations and asymptotic of their solutions as a tends to zero. The case of wave scattering by many small particles embedded in an inhomogeneous medium is also studied. Applications of this theory to creating materials with a desired refraction coefficient are given. A recipe is given for creating such materials by embedding into a given material many small impedance particles with prescribed boundary impedances.

The chapter "Generalized Convex Functions and their Applications" focuses on convex functions and their generalization. The definition of convex function along with some relevant properties of such functions is given first, followed by a discussion on a simple geometric property. Then the e-convex function is generalized and some of their properties are established. Moreover, a generalized s-convex function is presented in the second sense and the paper presents some new inequalities of generalized Hermite–Hadamard's type for the class of functions whose second local fractional derivatives of order α in absolute value at certain powers are generalized s-convex functions in the second sense. At the end, some applications to special means are also presented.

The chapter "Some Properties and Generalizations of the Catalan, Fuss, and Fuss–Catalan Numbers" presents an expository review and survey for analytic generalizations and properties of the Catalan numbers, the Fuss numbers, the Fuss–Catalan numbers, the Catalan function, the Catalan–Qi function, the q-Catalan–Qi numbers, and the Fuss–Catalan–Qi function.

The chapter "Trace Inequalities of Jensen Type for Selfadjoint Operators in Hilbert Spaces: A Survey of Recent Results" provides a survey of recent results for trace inequalities related to the celebrated Jensen's and Slater's inequalities and their reverses. Applications for various functions of interest such as power and logarithmic functions are also emphasized. Trace inequalities for bounded linear operators in complex Hilbert spaces play an important role in Physics, in general, and in Quantum Mechanics, in particular.

The chapter "Spectral Synthesis and its Applications" presents a survey about spectral analysis and spectral synthesis. The chapter recalls the most important classical results in the field and in some cases new proofs for them are given. It also presents the most recent results in discrete, nondiscrete, and spherical spectral synthesis together with some applications.

The chapter "Various Ulam–Hyers Stabilities of Euler–Lagrange–Jensen General $(a, b; k = a + b)$-Sextic Functional Equations" elucidates the historical development of well-known stabilities of various types of functional equations such as quintic, sextic, septic, and octic functional equations. It introduced a new generalized Euler–Lagrange–Jensen sextic functional equation, obtained its general solution and further investigated its various fundamental stabilities and instabilities by having employed the famous Hyers' direct method as well as the alternative fixed point method. The chapter is expected to be helpful to analyze the stability of various functional equations applied in the physical sciences.

The chapter "A Note on the Split Common Fixed-Point Problem and its Variant Forms" proposed new algorithms for solving the split common fixed point problem and its variant forms, and prove the convergence results of the proposed algorithms. The split common fixed point problems have found its applications in various branches of mathematics both pure and applied. It provides a unified structure to study a large number of nonlinear mappings.

The chapter "Stabilities and Instabilities of Rational Functional Equations and Euler–Lagrange–Jensen (a, b)-Sextic Functional Equations" comprises the growth, importance and relevance of functional equations in other fields. Its fundamental and basic results of various stabilities are presented. The stability results of various rational and Euler–Lagrange–Jensen sextic functional equations are investigated. Application and geometrical interpretation of rational functional equation are also illustrated for the readers to study similar problems.

The chapter "Attractor of the Generalized Contractive Iterated Function System" deals with the problems to construct the fractal sets of the iterated function system for certain finite collection of F-contraction mappings defined on metric spaces as well as b-metric spaces. A new iterated function system called generalized F-contractive iterated function system is defined. Further, a method is presented to construct new fractals; where the resulting fractals are often self-similar but more general.

The chapter "Regular and Rapid Variations and some Applications" presents an overview of recent results on regular and rapid variations of functions and sequences and some their applications in selection principles theory, game theory, and asymptotic analysis of solutions of differential equations.

The chapter "n-Inner Products, n-Norms, and Angles Between Two Subspaces" discusses the concepts of n-inner products and n-norms for any natural number n, which are generalizations of the concepts of inner products and norms. It presents some geometric aspects of n-normed spaces and n-inner product spaces, especially regarding the notion of orthogonality and angles between two subspaces of such a space.

The chapter "Proximal Fiber Bundles on Nerve Complexes" introduces proximal fiber bundles of nerve complexes. Briefly, a nerve complex is a collection of filled triangles (2-simplexes) that have nonempty intersection. Nerve complexes are important in the study of shapes with a number of recent applications that include the classification of object shapes in digital images. The focus of this chapter is on fiber bundles defined by projections on a set of fibers that are nerve complexes into a base space such as the set of descriptions of nerve complexes. Two forms of fiber bundles are introduced, namely, spatial and descriptive, including a descriptive form BreMiller–Sloyer sheaf on a Vietoris–Rips complex. The introduction to nerve complexes includes a recent extension of nerve complexes that includes nerve spokes. A nerve spoke is a collection of filled triangles that always includes filled triangle in a nerve complex. A natural transition from the study of fibers that are nerve complexes is in the form of projections of pairs of fibers onto a local nervous system complex, which is a collection of nerve complexes that are glued together with spokes common to the nerve fibers. A number of results are given for fiber bundles viewed in the context of proximity spaces.

The chapter "Approximation by Generalizations of Hybrid Baskakov Type Operators Preserving Exponential Functions" deals with the approximation properties of the certain Baskakov–Szász operators. It estimates the results for these hybrid Baskakov–Szász type operators for exponential test functions. It also estimates a quantitative asymptotic formula for such operators. Mathematica software is used to estimate the results.

The chapter "Well-Posed Minimization Problems via the Theory of Measures of Noncompactness" presents an analysis of the minimization problems for functionals defined, lower bounded and lower semicontinuous on a closed subset of a metric space. The main focus is on the well-posedness of minimization problems from the viewpoint of the theory of measures of noncompactness. The minimization problems for several functionals defined on some Banach spaces are also investigated as well. Thus, the chapter clarifies the role of the theory of measures of noncompactness in the general approach to the well-posedness of minimization problems.

The chapter "Some Recent Developments on Fixed Point Theory in Generalized Metric Spaces" discusses some important developments in the fixed point theory in metric spaces. Various advancements are explained in detail through useful and applicable results along with examples in generalized metric spaces.

The chapter "The Basel Problem with an Extension" presents some historical aspects to the famous Basel problem, which a number of brilliant mathematicians attempted, and which had remained unsolved for over 90 years. It was the genius Euler who provided a masterful solution and laid the foundations to the famous Riemann zeta function and the analysis of series. The chapter then investigates a related Euler sum and provides an explicit analytical representation, a closed form solution. The related Euler sum also represented in terms of logarithmic and hypergeometric functions. The integrals in question will be associated with the harmonic numbers of positive terms. A few examples of integrals provide an identity in terms of some special functions.

The chapter "Coupled Fixed Points and Coupled Coincidence Points via Fixed Point Theory" focuses on the study of the coupled fixed point and coupled coincidence point problems for single- and multi-valued operators. The study of this chapter is based on appropriate fixed point theorems in two types of generalized metric spaces. Some applications are also given to illustrate part of the abstract results presented in this chapter.

The chapter "The Corona Problem, Carleson Measures, and Applications" reviews the developments and generalizations of the Corona problem, the results on Carleson measures themselves and some applications of Carleson measures, in several different settings, starting from the disc in $\mathbb{C}$ (where the corona problem was originally set) arriving to the unit ball in $\mathbb{C}^n$, to bounded strongly pseudoconvex domains and even to domains in the quaternionic space. Both the corona problem and the Carleson measures still need investigation, as many open problems have not been solved yet. The open problems are also highlighted in this chapter. Carleson measures were introduced by Lennart Carleson in 1962 to solve an interpolation problem about bounded holomorphic functions called the corona problem.

The editors are grateful to the contributors for their timely contribution and co-operation throughout the editing process. The editors have benefited from the remarks and comments of several other experts on the topics covered in this book. The editors would also like to thank the book handling editors at Wiley and production staff members for their continuous support and help. Finally, the editors offer sincere thanks to all those who contributed directly or indirectly in completing this book project.

August 25, 2017

Michael Ruzhansky
London, UK

Hemen Dutta
Guwahati, India

Ravi P. Agarwal
Texas, USA

About the Editors

Michael Ruzhansky is a Professor at the Department of Mathematics, Imperial College London, UK. He has published over 100 research articles in leading international journals. He has also published 5 books and memoirs, and 9 edited volumes. His major research topics are related to pseudo-differential operators, harmonic analysis, functional analysis, partial differential equations, boundary value problems, and their applications. He is serving on the editorial board of many respected international journals and served as the President of the International Society of Analysis, Applications, and Computations (ISAAC) in the period 2009–2013.

Hemen Dutta is a Senior Assistant Professor of Mathematics at Gauhati University, India. He did his Master of Science (M.Sc.) in Mathematics, Post Graduate Diploma in Computer Application (PGDCA) and Ph.D. in Mathematics from Gauhati University, India. He received his M.Phil. in Mathematics from Madurai Kamaraj University, India. His research interest includes summability theory and functional analysis. He has to his credit more than 50 research papers and three books so far. He has delivered talks at foreign and national institutions. He has also organized a number of academic events. He is a member of several mathematical societies.

Ravi P. Agarwal is a Professor and the chair of the Department of Mathematics at Texas A&M University-Kingsville, USA. He has been actively involved in research as well as pedagogical activities for the last 45 years. Dr. Agarwal is the author or co-author of more than 1400 scientific papers and 40 research monographs. His major research interests include numerical analysis, differential and difference equations, inequalities, and fixed point theorems. Dr. Agarwal is the recipient of several notable honors and awards. He is on the editorial board of several journals in different capacities and also organized International Conferences.

List of Contributors

Mujahid Abbas
Department of Mathematics
Government College University
Katchery Road, Lahore 54000
Pakistan

and

Department of Mathematics
King Abdulaziz University
Jeddah 21589
Saudi Arabia

Józef Banaś
Department of Nonlinear Analysis
Rzeszów University of Technology
Aleja Powstańców Warszawy 8
35-959 Rzeszów
Poland

Dragan Djurčić
Faculty of Technical Sciences
Department of Mathematics
University of Kragujevac
34000 Čačak
Serbia

Silvestru Sever Dragomir
Mathematics Department
College of Engineering & Science
Victoria University
Melbourne 8001
Australia

and

DST-NRF Centre of Excellence in the
Mathematical and Statistical Sciences
School of Computer Science and
Applied Mathematics
University of the Witwatersrand
Johannesburg 2000
South Africa

Jorge Mozo Fernández
Departamento de Álgebra, Análisis
Matemático
Geometría y Topología
Facultad de Ciencias
Campus Miguel Delibes
Universidad de Valladolid
Paseo de Belén, 7, 47011 Valladolid
Spain

Hendra Gunawan
Department of Mathematics
Bandung Institute of Technology
Bandung 40132
Indonesia

Bai-Ni Guo
School of Mathematics and
Informatics
Henan Polytechnic University
Jiaozuo, Henan, 454010
China

Vijay Gupta
Department of Mathematics
Netaji Subhas Institute of Technology
Dwarka, New Delhi 110078
India

Palle E.T. Jorgensen
Department of Mathematics
The University of Iowa
Iowa City, IA 52242-1419
USA

Adem Kiliçman
Department of Mathematics
Faculty of Science
Putra University of Malaysia
43400 Serdang, Selangor
Malaysia

Ljubiša D.R. Kočinac
Faculty of Sciences and Mathematics
Department of Mathematics
University of Niš, 18000 Niš
Serbia

Somayya Komal
Theoretical and Computational
Science (TaCS) Centre, Department
of Mathematics, Faculty of Science
King Mongkut's University of
Technology Thonburi
Thung Khru, Bangkok 10140
Thailand

Poom Kumam
Theoretical and Computational
Science (TaCS) Centre, Department
of Mathematics, Faculty of Science
King Mongkut's University of
Technology Thonburi
Thung Khru, Bangkok 10140
Thailand

Beri V. Senthil Kumar
Department of Mathematics
C. Abdul Hakeem College of Engg.
and Tech.
Melvisharam 632 509, Tamil Nadu
India

Jelena V. Manojlović
Faculty of Sciences and Mathematics
Department of Mathematics
University of Niš
18000 Niš
Serbia

L.B. Mohammed
Department of Mathematics
Faculty of Science
Universiti Putra Malaysia
43400 Serdang, Selangor
Malaysia

Talat Nazir
Department of Mathematics
University of Jeddah
Jeddah 21589
Saudi Arabia

and

Department of Mathematics
COMSATS Institute of Information
Technology
Abbottabad 22060
Pakistan

Narasimman Pasupathi
Department of Mathematics
Thiruvalluvar University College of
Arts and Science
Tirupattur 635 901, Tamil Nadu
India

James F. Peters
Computational Intelligence
Laboratory
University of Manitoba
WPG, MB, R3T 5V6
Canada

and

Department of Mathematics
Faculty of Arts and Sciences
Adiyaman University
02040 Adiyaman
Turkey

Adrian Petruşel
Faculty of Mathematics and
Computer Science
Babeş-Bolyai University
400084 Cluj-Napoca
Romania

Gabriela Petruşel
Faculty of Business
Babeş-Bolyai University
400084 Cluj-Napoca
Romania

Feng Qi
Institute of Mathematics
Henan Polytechnic University
Jiaozuo, Henan, 454010
China

and

College of Mathematics
Inner Mongolia University for
Nationalities
Tongliao, Inner Mongolia, 028043
China

and

Department of Mathematics
College of Science
Tianjin Polytechnic University
Tianjin, 300387
China

Alexander G. Ramm
Department of Mathematics
Kansas State University
Manhattan, KS 66506-2602
USA

John Michael Rassias
Pedagogical Department E.E.,
Section of Mathematics and
Informatics
National and Capodistrian University
of Athens
Athens 15342
Greece

Krishnan Ravi
Department of Mathematics
Sacred Heart College
Tirupattur 635 601, Tamil Nadu
India

Wedad Saleh
Department of Mathematics, Faculty
of Science
Putra University of Malaysia
43400 Serdang, Selangor
Malaysia

Alberto Saracco
Dipartimento di Scienze
Matematiche, Fisiche e Informatiche
Universitá degli Studi di Parma,
43124
Italy

Anthony Sofo
Victoria University
Melbourne City, Victoria 8001
Australia

László Székelyhidi
Institute of Mathematics
University of Debrecen, H-4010
Debrecen
Hungary

and

Department of Mathematics
University of Botswana
Gaborone
Botswana

Feng Tian
Department of Mathematics
Hampton University
Hampton, VA 23668
USA

Sergio Alejandro Carrillo Torres
Escuela de Ciencias Exactas e
Ingeniería
Universidad Sergio Arboleda
Calle 74, 14-14, Bogotá
Colombia

Tomasz Zając
Department of Nonlinear Analysis
Rzeszów University of Technology
Aleja Powstańców Warszawy 8
35-959 Rzeszów
Poland

1

Spaces of Asymptotically Developable Functions and Applications

Sergio Alejandro Carrillo Torres[1] and Jorge Mozo Fernández[2]

[1]*Escuela de Ciencias Exactas e Ingeniería, Universidad Sergio Arboleda, Calle 74, 14-14, Bogotá, Colombia*
[2]*Departamento de Álgebra, Análisis Matemático, Geometría y Topología, Facultad de Ciencias, Campus Miguel Delibes, Universidad de Valladolid, Paseo de Belén, 7, 47011 Valladolid, Spain*

2010 AMS Subject Classification Primary 34E05, 34E15

1.1 Introduction and Some Notations

This chapter is a short review of some results concerning asymptotic expansions in several complex variables, and summability. Different notions of asymptotic expansions have been developed in literature in last decades, trying to generalize classical results regarding asymptotics in one variable, dating back to H. Poincaré, and further developed by Wasow [1], Ramis [2], Écalle [3], Balser [4], Braaksma [5, 6], and many others. In several variables, as main contributions, let us mention those of Gérard and Sibuya [7] in the 1970s, Majima [8, 9] in the 1980s, and more recently the notion of monomial asymptotic expansions and monomial summability of Canalis-Durand et al. [10]. It is also worth to mention here that the notion of composite asymptotic expansions, developed by Fruchard and Schäfke [11], has been very useful in the treatment of singularly perturbed linear differential equations.

We shall review mainly the notion of strong asymptotic expansions of Majima and monomial asymptotic expansions of Canalis-Durand, Mozo, and Schäfke, and clarifying some relations between them. These notions will be applied to several problems concerning summability of solutions of systems of ordinary differential equations ODEs and Pfaffian systems. Concerning Pfaffian systems, we will state some recent progresses of S. Carrillo, see [12] for complete details.

This chapter does not intend to be complete at all, the objective is only to present, in the opinion of the authors, some of the more relevant contributions concerning this wide theory. For the new results presented here, relevant precise references are given in the text. Such a survey, as we know, does not exist in the literature, so we think that it may be useful as a starting point for researchers in the area.

Mathematical Analysis and Applications: Selected Topics, First Edition.
Edited by Michael Ruzhansky, Hemen Dutta, and Ravi P. Agarwal.
© 2018 John Wiley & Sons, Inc. Published 2018 by John Wiley & Sons, Inc.

Some of the main notations used throughout the text will be the following:

$\mathbb{N}$ will denote the set of natural numbers including 0, and $\mathbb{N}^* = \mathbb{N} \backslash \{0\}$.

$\mathbf{a}, \mathbf{b}, \mathbf{z}, \ldots$ (boldface) will denote vectors: $\mathbf{a} = (a_1, a_2, \ldots, a_n)$, and so on.

If $\mathbf{a}, \mathbf{b} \in \mathbb{R}^n$, $\mathbf{a} \leq \mathbf{b}$ means that $\forall\, i, a_i \leq b_i$.

If $\mathbf{a}, \mathbf{b} \in \mathbb{R}^n$, $\mathbf{a} < \mathbf{b}$ means that $\forall\, i, a_i < b_i$.

$\mathbf{0} := (0, 0, \ldots, 0)$.

If U is an open set in $\mathbb{C}^n$, $\mathcal{O}(U)$ is the set of holomorphic functions on U. Similarly, $C^\infty(U)$ will denote the set of C^∞ functions on U, identifying $\mathbb{C}^n \cong \mathbb{R}^{2n}$.

$\mathbb{C}[[\mathbf{x}]]$ is the ring of formal power series in the variables $\mathbf{x}$.

$\mathbb{C}\{\mathbf{x}\}$ is the ring of convergent power series (at the origin) in the variables $\mathbf{x}$.

If $a < b$ and $R > 0$ (or $R = +\infty$), $V(a, b; R)$ will denote the sector of radius R and opening between the rays $\arg z = a$ and $\arg z = b$, that is, the set

$$\{z \in \mathbb{C}; \quad a < \arg z < b, \quad 0 < |z| < R\}.$$

Note that we are not restricted to the case $b - a \leq 2\pi$. If $b - a > 2\pi$, the sectors are to be understood in the Riemann surface of $\log z$, that is, the universal covering of $\mathbb{C} \backslash \{0\}$.

If $\mathbf{a} < \mathbf{b}$ and $\mathbf{R} > \mathbf{0}$, $V(\mathbf{a}, \mathbf{b}; \mathbf{R})$ will denote the polysector

$$V(a_1, b_1; R_1) \times V(a_2, b_2; R_2) \times \cdots \times V(a_n, b_n; R_n).$$

In general, multiindex notation will be freely used throughout the text. So, $\mathbf{x}^\mathbf{a}$ will denote $x_1^{a_1} x_2^{a_2} \cdots x_n^{a_n}$, $\mathbf{N}!^\mathbf{a}$ is $N_1!^{a_1} N_2!^{a_2} \cdots N_n!^{a_n}$, and so on.

1.2 Strong Asymptotic Expansions

The notion of asymptotics in one variable was introduced by Poincaré, trying to give a meaning to the notion of a sum for divergent series, that had been controversed and widely study since the times of Euler and Abel and was developed by different authors during the twentieth century, as Birkhoff, Wasow, Hukuhara, and Sibuya. Some good expositions with emphasis in the history, of the theory of divergent series, are due to Ramis [13, 14]. An important improvement was done at the end of the 1970s by Ramis with the introduction of the theory of *Gevrey summability*, generalizing Borel summability. The objective was to give a meaning to the formal power series appearing as solutions of systems of ODEs with irregular singular points. In other words, to define a *sum* (or several sums) for these series. It turns out that not every solution of a system of ODEs with irregular singular points is summable in this sense, but nevertheless it was shown in the next years that solutions of these systems are *multisummable*, that is, choosing a direction in the complex plane that avoids a finite number of directions, the formal series solutions can be uniquely decomposed as a sum of formal series, and a process of k-summability (with different values of k) can be applied to each of these summands in order to obtain a true holomorphic solution of the system.

Different essays were done in order to generalize this notion to several variables. Asymptotics and summability with parameters were used by different authors, from Wasow, Hukuhara, and others, but this was not a true complete notion of summability in several variables, that could be used, for instance, to study systems of partial differential equations. The first notion that clearly generalized that of Poincaré was due to H. Majima, who in 1983 presented what he called *strong asymptotic expansions*, and applied in 1984 to the study of integrable connections.

In this section, we shall recall the notion of strong asymptotic expansions. In the Gevrey case, his work was generalized by Haraoka [15]. Further developments of this notion were established by Zurro [16], Hernández [17], Galindo and Sanz [18] and the second author, among others.

Given $\mathbf{a} < \mathbf{b}$, and $\mathbf{R} > 0$, denote V the polysector $V(\mathbf{a}, \mathbf{b}; \mathbf{R})$.

Definition 1.1 A total family of coefficients in V is a family of holomorphic functions

$$\mathcal{F} = \{f_{\alpha_J}(\mathbf{z}_{J^c}) \in \mathcal{O}(V_{J^c}); \quad \emptyset \neq J \subseteq \{1, 2, \ldots, n\}, \ \alpha_J \in \mathbb{N}^J\}.$$

Given such a family, let us define, for $\mathbf{N} \in \mathbb{N}^n$, the $\mathbf{N}$-approximant of $\mathcal{F}$ as

$$\mathrm{App}_{\mathbf{N}}(\mathcal{F}) = \sum_{\emptyset \neq J \subseteq \{1,2,\ldots,n\}} (-1)^{\#J+1} \cdot \sum_{\substack{\alpha_J \in \mathbb{N}^J \\ \alpha_J < \mathbf{N}_J}} f_{\alpha_J}(\mathbf{z}_{J^c}) \cdot \mathbf{z}_J^{\alpha_J} \in \mathcal{O}(V),$$

where $\mathbb{N}^J$ denotes $\prod_{j \in J} \mathbb{N}$, $J^c = \{1, 2, \ldots, n\} \setminus J$.

Definition 1.2 Let V be a polysector in $\mathbb{C}^n$, $f \in \mathcal{O}(V)$, and $\mathcal{F}$ a total family of coefficients in V. We will say that f admits $\mathcal{F}$ as a strong asymptotic expansion in V if for every $\mathbf{N} \in \mathbb{N}^n$, and every strict subpolysector W of V (see Remark 1.1), there are constants $C(W, \mathbf{N}) > 0$ such that, if $\mathbf{z} \in W$,

$$|f(\mathbf{z}) - \mathrm{App}_{\mathbf{N}}(\mathcal{F})(\mathbf{z})| \leq C(W, \mathbf{N}) \cdot |\mathbf{z}^{\mathbf{N}}|.$$

If $\mathbf{s} = (s_1, s_2, \ldots, s_n) \in [0, \infty]^n$, the asymptotic expansion is called of $\mathbf{s}$-Gevrey type if there exists constants $C(W) > 0$ and $\mathbf{A}(W) > \mathbf{0}$, depending on the subpolysector W, such that $C(W, \mathbf{N})$ can be chosen as

$$C(W, \mathbf{N}) = C(W) \cdot A(W)^{|\mathbf{N}|} \cdot \mathbf{N}!^{\mathbf{s}}.$$

Remark 1.1 In this remark, and throughout the text, we will say that W is a strict subpolysector of V if it is bounded, and moreover,

$$\overline{W} \subseteq (V_1 \cup \{0\}) \times \cdots \times (V_n \cup \{0\}).$$

We will denote this situation as $W \prec V$.

Let us denote $\mathcal{A}(V)$ (resp. $\mathcal{A}_s(V)$) the set of functions on V admitting a strong asymptotic expansion (resp. of $\mathbf{s}$-Gevrey type). They are differential $\mathbb{C}$-algebras.

In particular, if $\beta \in \mathbb{N}^n$ and f admits a family $\mathcal{F}$ as a strong asymptotic expansion, the derivative $D^\beta f$ admits $\mathcal{G}$, where

$$\mathcal{G} = \left\{ g_{\alpha_J}(\mathbf{z}_{J^c}) \in \mathcal{O}(V_{J^c}); \quad g_{\alpha_J}(\mathbf{z}_{J^c}) = \frac{(\alpha_J + \beta_J)!}{\alpha_J!} \cdot D^{\beta_{J^c}} (f_{\alpha_J + \beta_J}(\mathbf{z}_{J^c})) \right\},$$

as strong asymptotic expansion. The unicity of the asymptotic expansion can be deduced from the following fact: take, for instance, $\mathbf{N} = (1, 0, \ldots, 0), J = \{1\}$. Then, we have that

$$|f(\mathbf{z}) - f_0(z_2, \ldots, z_n)| \le C(W, \mathbf{N}) \cdot |\mathbf{z_1}|.$$

in a proper subpolysector W. Taking limits when z_1 tends to 0, we have

$$f_0(z_2, \ldots, z_n) = \lim_{\substack{z_1 \to 0 \\ z_1 \in W}} f(z).$$

Stability under derivation allows us to conclude. Due to this unicity, the total family of coefficients of strong asymptotic expansion for a function f is denoted as $TA(f)$. Denoting $f_\alpha = f_{\alpha_{\{1,2,\ldots,n\}}}$, the formal power series

$$\sum_\alpha f_\alpha \mathbf{z}^\alpha$$

is the formal power series of asymptotic expansion of f, denoted as $FA(f)$. In the particular case, this series, when expanded with respect to any variable, has coefficients holomorphic in a common disk around the origin in the other variables, then it determines the family $TA(f)$, and we will say that f has the series $FA(f)$ as strong asymptotic expansion at the origin.

Definition 1.3 A total family of coefficients $\mathcal{F} = \{f_{\alpha_J}(\mathbf{z}_{J^c})\}$ in V is called *consistent* if each $f_{\alpha_J}(z_{J^c}) \in \mathcal{A}(V_{J^c})$ and moreover the family

$$\mathcal{F}_{\alpha_J} = \{f_{\alpha_J, \beta_{J'}}(\mathbf{z}_{(J \cup J')^c}); \quad \emptyset \ne J' \subseteq J^c\}$$

equals $TA(f_{\alpha_J}(\mathbf{z}_{J^c}))$. If $f \in \mathcal{A}(V)$, the family $TA(f)$ is a consistent one.

The following characterization turns out to be very useful in order to work with strong asymptotic expansions in polysectors:

Theorem 1.1 Let V be a polysector in $\mathbb{C}^n$ and $f \in \mathcal{O}(V)$. The following conditions are equivalent:

1. $f \in \mathcal{A}(V)$.
2. If $W \prec V$, every derivative $D^\beta f$ is bounded in W.
3. If $W \prec V$, the restriction $f|_W$ can be extended to the whole space $\mathbb{C}^n$ as a C^∞-function (considering $\mathbb{C}^n \cong \mathbb{R}^{2n}$).

Proof: (1) $\Rightarrow$ (2) is evident, as strongly asymptotically developable functions are bounded in subpolysectors and the $\mathbb{C}$-algebra $\mathcal{A}(V)$ is stable by derivation.

(2) $\Rightarrow$ (3). The subpolysector W is 1-regular in the sense of H. Whitney: for every $z_0 \in \overline{W}$, there exists a neighborhood U of z_0, and a constant $C > 0$ such that, if $\mathbf{x}, \mathbf{y} \in U \cap W$, a rectifiable curve γ exists in W joining $\mathbf{x}$ and $\mathbf{y}$ and such that its length $l(\gamma)$ satisfies a bound

$$l(\gamma) \le C \cdot |x - y|.$$

Given $z_0 \in \partial W$, let us take a sequence $\{z_n\}_{n=1}^{\infty}$ in W converging to z_0. If $r, s \in \mathbb{N}$, a curve γ_{rs} exists joining z_r and z_s, and such that $l(\gamma_{rs}) \le C \cdot |z_r - z_s|$. Then

$$|f(\mathbf{z}_r) - f(\mathbf{z}_s)| = \left| \int_a^b \left[\sum_{i=1}^{n} D^i f(\gamma_{rs}(t)) \cdot \gamma'_{rs,i}(t) \right] dt \right|$$

$$\le C_1 n l(\gamma_{rs}) \le C_1 n C |z_r - z_s|,$$

where C_1 is a bound for the first derivatives of f. The sequence $\{f(z_n)\}_{n=1}^{\infty}$ is a Cauchy sequence, so f can be extended to $\mathbf{z}_0$. The same argument allows us to extend to $\overline{W}$ all the derivatives of f.

As W is 1-regular, these extensions define a C^{∞}-function in the sense of Whitney, and the result follows.

(3) $\Rightarrow$ (1). Consider a C^{∞} extension F_W of $f|_W$, and define

$$f_{\alpha_J}(\mathbf{z}_{J^c}) := \frac{1}{\alpha_J!} \cdot D^{\alpha_J}(F_W(\mathbf{0}_J, \mathbf{z}_{J^c}))$$

on W_{J^c}, functions that patch together giving a holomorphic function in V_{J^c}. These functions define a total family of coefficients $\mathcal{F}$. Taylor integral formula allows us to show that, on W,

$$|f(\mathbf{z}) - \mathrm{App}_N(\mathcal{F})(\mathbf{z})| \le \frac{C_N}{N!} \cdot |\mathbf{z}^N|,$$

where C_N is a bound of $D^N f(\mathbf{z})$ on W. $\qquad\qquad\square$

In classical asymptotics in one variable, Borel–Ritt theorem is of great utility: It says that every formal power series is the asymptotic expansion of some function in an arbitrarily chosen sector. There is an analogue for strong asymptotic expansions, as follows:

Theorem 1.2 (Borel-Ritt) Given a consistent family $\mathcal{F}$ of coefficients on a polysector V, there exists a function $f \in \mathcal{O}(V)$ such that $TA(f) = \mathcal{F}$.

Sketch of proof: In [8], Majima proves a weaker result. More precisely, he shows that given a formal power series $\hat{f}(\mathbf{z}) \in \mathbb{C}[[\mathbf{z}]]$, there exists a function $f \in \mathcal{A}(V)$

such that $FA(f) = \hat{f}$. He follows the same idea as in the classical proof in one variable: from the expansion

$$\hat{f}(\mathbf{z}) = \sum_{\alpha \in \mathbb{N}^n} f_\alpha \mathbf{z}^\alpha,$$

he constructs a function

$$f(\mathbf{z}) = \sum_{\alpha \in \mathbb{N}^n} f_\alpha \cdot \prod_{j=1}^n \left(1 - \exp\left(\frac{d_{\alpha,j}}{z_j^{p_j}} \right) \right) \cdot \mathbf{z}^\alpha$$

defining $d_{\alpha,j}$ and p_j appropriately in order to guarantee that the previous expression defines a holomorphic function in the polysector, and the bounds of the definition of the strong asymptotic expansion are verified. A modification of this proof is presented in [9] to show the general case stated here. He employs induction on n, assuming in each step that part of the functions $f_{\alpha_j}(\mathbf{z}_{J^c})$ are zero.

Let us mention that another proof may be done as follows: a consistent family $\mathcal{F}$ verifies regularity conditions (in fact, it is formally holomorphic), and so, defines a C^∞-function in the sense of Whitney (see [19] for precise definitions and details). So, there exists $F(\mathbf{z})$, a C^∞ function, defined in a neighborhood of $\mathbf{0}$ in $\mathbb{C}^n$, such that the restriction of its derivatives coincide with the functions f_{α_j}. Truncated Laplace transform of $F(\mathbf{z})$ defines an element of $\mathcal{A}(V)$ and this allows us to conclude. $\qquad\square$

Strong asymptotic expansions may be defined from one variable asymptotic expansions using functional analysis techniques. For, let us consider the space $\mathcal{A}(V)$. For every $W \prec V$ and $\beta \in \mathbb{N}^n$, define

$$p_{W,\beta}(f) := \sup\{ |D^\beta f(\mathbf{z})|;\ \mathbf{z} \in W \}.$$

This number exists, by Theorem 1.1, and defines a family of seminorms $\{p_{W,\beta}\}_{W,\beta}$, that provides $\mathcal{A}(V)$ with a Frechet space structure. If E is a Frechet space and V a polysector, the space $\mathcal{A}(V;E)$ of strongly asymptotically developable functions with values in E may be defined. It turns out that there are natural isomorphisms $\mathcal{A}(V_1 \times V_2) \cong \mathcal{A}(V_1; \mathcal{A}(V_2))$ [17], and this allows us to make recurrence on the number of variables.

Returning to Theorem 1.1, let us denote $\mathcal{R}(W) = C^\infty(\overline{W}) \cap \mathcal{O}(W)$, that is, the space of holomorphic functions in W that are C^∞ in the sense of Whitney in the compact set $\overline{W}$. Due to the regularity of W, this implies that they can be extended as a C^∞-function to the whole space $\mathbb{C}^n$. So, we have that $\mathcal{A}(V) = \varprojlim_{W \prec V} \mathcal{R}(W)$. The space $\mathcal{R}(W)$, as a subspace of $C^\infty(W)$, is a nuclear space [20, 21], and hence, $\mathcal{A}(V)$ is also nuclear. As $\mathcal{A}(V)$ is dense in $C^\infty(W)$, it can be shown that in fact, $\mathcal{A}(V_1 \times V_2) \cong \mathcal{A}(V_1) \hat{\otimes} \mathcal{A}(V_2)$, where $\hat{\otimes}$ denotes the topological tensor product, as defined by A. Grothendieck. Precise details may be found in [22].

Let us comment briefly further properties of strong asymptotic expansions.

1. Consider a multidirection $\mathbf{d} = (d_1, d_2, \ldots, d_n)$ on a polysector $V(\mathbf{a}, \mathbf{b}; \mathbf{r})$; $d_j = e^{i\theta_j}$, where $a_j < \theta_j < b_j$. Assume that $f \in \mathcal{O}(V)$ is a holomorphic and bounded function having a strong asymptotic expansion on $\mathbf{d}$: the bounds are verified when restricting to this multidirection (which in fact defines a n-dimensional real space). Then $f \in \mathcal{A}(V)$, that is, the asymptotic expansion exists in the whole polysector. This result is shown in [23], and generalizes a result for the one variable case stated by Fruchard and Zhang [24].

2. From the sheaf of Whitney C^∞-functions, Honda and Prelli construct in [25] the sheaf of strongly asymptotically developable functions by applying a functor of specialization. This functional setting appears to be rather interesting for future applications, and it deserves further development.

Most of the main results presented in this section have been stated in the context of the so-called Poincaré asymptotics. In the Gevrey case, they are still valid, with more or less straightforward modifications. In the literature on the subject you can find complete statements and proofs. Let us, nevertheless, mention some interesting issues concerning the Gevrey case.

For, recall that in one variable, Watson's lemma says that if $f \in \mathcal{O}(V)$ has a series $\hat{f}$ as s-Gevrey asymptotic expansion, and the opening of the sector V is strictly greater than $s\pi$, then $f(z)$ is unique, and therefore is called the *sum* of $\hat{f}(z)$ in V.

In the context of strong asymptotic expansions, a similar result is the following one:

Theorem 1.3 Let V be a polysector, $f \in \mathcal{A}(V)$ having a total family of coefficients $\mathcal{F}$ as $\mathbf{s}$-Gevrey strong asymptotic expansion. Then, if for some j, the opening of V_j is greater than $s_j \pi$, f is unique.

In order to give a proof of this result it is enough to use the fact that $\mathcal{A}(V) \cong \mathcal{A}(V_j; \mathcal{A}(V_{j^c}))$, and techniques of topological vector spaces. In [26], the reader can find precise details in a rather more general setting.

1.3 Monomial Asymptotic Expansions

Several existing examples in the literature may lead us to a different notion of asymptotic expansions in several complex variables, more precisely, asymptotics expansions with respect to a monomial. Let us state here, as an example, one of these situations concerning resonant holomorphic foliations.

Martinet and Ramis [27] studied the analytic classification of resonant holomorphic foliations in dimension two, that is, foliations generated by a 1-form

$$\omega = x\,\mathrm{d}y + \xi y a(x, y)\mathrm{d}x,$$

where $a(0,0) = 1, \xi = p/q \in \mathbb{Q}_{>0}$. Such a 1-form turns out to be formally equivalent to a certain formal normal form. In order to study the analytical equivalence, they introduce the space of formal power series

$$\hat{B} = \left(\bigcup_{r>0} \mathcal{O}(D_r) \right) [[y]] \cap \left(\bigcup_{r>0} \mathcal{O}(D_r) \right) [[x]].$$

Elements $\hat{f} \in \hat{B}$ define a formally holomorphic map, in the sense of Łojasiewicz. Considering the map $\rho : \mathbb{C}^2 \to \mathbb{C}^3$ defined by $\rho(x,y) = (x^p y^q, x, y)$, define

$$\hat{B}_{p,q} = \bigcup_{r>0} \rho^*(\mathcal{O}(D_r \times D_r)[[u]]).$$

In fact, $\hat{B}_{p,q} = \hat{B}$ is a formal space. But, if considering s-Gevrey series in u, they turn out to be different. Moreover, from s-Gevrey asymptotic expansions in u, with coefficients in the variables x and y, functions in two variables may be defined as representing a kind of asymptotic expansions "in $x^p y^q$" Martinet and Ramis define also a notion of summability with respect to this monomial, but with the warning that "La notion de resommabilité à plusieurs variables reste peu claire [· · ·]" [27, Section 4.3]. The notion of monomial summability we will present here will clarify these definitions.

This example, and another ones that can be found in the literature (see, for instance, Chapter X in [28]) lead us to state a notion of asymptotic expansion and summability that depends on a monomial. This is what we will do briefly in the sequel.

Definition 1.4 Consider a monomial $x^p y^q$, $p, q \in \mathbb{N}^*$. A *monomial sector* in $x^p y^q$ is a set

$$\Pi_{p,q}(a,b;r) = \{(x,y) \in \mathbb{C}^2; \quad 0 < |x|, |y| < r, \ a < \arg(x^p y^q) < b\}.$$

By analogy, the number $b - a$ is called the *opening* of the monomial sector.

Given a monomial $x^p y^q$, denote $\mathcal{E}_R$ the set of analytic functions in a neighborhood of the origin $D(\mathbf{0}; R)$, such that its power series expansion at the origin is

$$\sum_{n,m} a_{nm} x^n y^m,$$

where $a_{nm} = 0$ if $n \geq p$ and $m \geq q$. $\mathcal{E}$ will denote the set $\bigcup_{R>0} \mathcal{E}_R$, with obvious identifications.

Definition 1.5 Let f be a holomorphic function defined over a monomial sector $\Pi = \Pi_{p,q}(a,b;R)$, and $\hat{f} \in \mathbb{C}[[x,y]]$. We will say that f has $\hat{f}$ as monomial asymptotic expansion at the origin, or $x^p y^q$-asymptotic expansion, if the following conditions are satisfied:

1. There exists $\tilde{R} > 0$ such that

$$\hat{f}(x, y) = \sum_{n=0}^{\infty} f_n(x, y) \cdot (x^p y^q)^n,$$

where $f_n(x, y) \in \mathcal{E}_{\tilde{R}}$.

2. For every proper subsector $\tilde{\Pi} \subseteq \Pi$ (i.e., $\tilde{\Pi} = \Pi_{p,q}(a', b'; r')$, with $a < a' < b' < b$, $r' < r$), and every $N \in \mathbb{N}$, there exists a constant $C(\tilde{\Pi}, N) > 0$ such that

$$\left| f(x, y) - \sum_{n=0}^{N-1} f_n(x, y)(x^p y^q)^n \right| \le C(\tilde{\Pi}, N) \cdot |x^p y^q|^N,$$

over $\tilde{\Pi}$.

The asymptotic expansion is called of s-Gevrey type if, moreover:

1. $\hat{f}(x, y)$ is of s-Gevrey type in $x^p y^q$, that is, there exists $R' < \tilde{R}$ such that

$$||f_n(x, y)||_{D(0;R')} \le K A^n n!^s,$$

for some constants K and $A > 0$ (here, $||\cdot||_D$ denotes the supremum norm on D).

2. $C(\tilde{\Pi}, N)$ can be chosen as $C(\tilde{\Pi}, N) = C B^N N!^s$, where C and B depend only on $\tilde{\Pi}$, but they do not depend on N.

The definition of monomial asymptotic expansion is equivalent to the following one: There exists a family of holomorphic functions defined in a fixed neighborhood of $\mathbf{0}$, $f_n(x, y) \in \mathcal{O}(\overline{D}(0; R))$, such that, given a monomial subsector $\tilde{\Pi} < \Pi$ and $N \in \mathbb{N}$,

$$|f(x, y) - f_N(x, y)| \le C(\tilde{\Pi}, N) \cdot |x^p y^q|^N,$$

on $\tilde{\Pi}$, for appropriate constants $C(\tilde{\Pi}, N)$.

Let us observe that if a formal series $\hat{f}(x, y)$ is the s-Gevrey monomial asymptotic expansion of a function $f(x, y)$, writing

$$\hat{f}(x, y) = \sum_{n,m} f_{nm} x^n y^m,$$

the coefficients of $\hat{f}(x, y)$ satisfy bounds

$$|f_{nm}| \le C A^{n+m} \cdot \min\{n!^{s/p}, m!^{s/q}\}.$$

Equivalently, $\hat{f}(x, y)$ is a $\frac{s}{p}$-Gevrey series in x (with coefficients holomorphic in a common neighborhood of 0 in y), and $\frac{s}{q}$-Gevrey series in y (with coefficients holomorphic in a common neighbourhood of 0 in x).

The main tools to study monomial asymptotics and summability are the following operators, T and $\hat{T}$, which allow us to reduce monomial asymptotic expansions to asymptotic expansions in one variable. Let us define first the operator $\hat{T}$, acting on the space of formal power series.

For, given a formal power series $\hat{f}(x, y) \in \mathbb{C}[[x, y]]$, consider the filtration by the powers of $x^p y^q$, and let us write uniquely

$$\hat{f}(x, y) = \sum_{n=0}^{\infty} \hat{f}_n(x, y)(x^p y^q)^n,$$

where $\hat{f}_n(x, y) \in \Delta(x^p y^q)$, the set of series such that $x^p y^q$ does not divide any of its terms. Define

$$\hat{T} : \mathbb{C}[[x, y]] \longrightarrow \Delta(x^p y^q)[[t]],$$

$$\hat{f}(x, y) \longmapsto \sum_{n=0}^{\infty} \hat{f}_n(x, y)t^n.$$

Note that $(\hat{T}\hat{f})(x^p y^q, x, y) = \hat{f}(x, y)$. A similar operator can be constructed in the analytic setting: consider $f(x, y) \in \mathcal{O}(\Pi)$, $\Pi = \Pi_{p,q}(a, b; r)$ being a $x^p y^q$-sector. Suppose first that $(p, q) = (1, 1)$. The function $f\left(\frac{t}{y}, y\right)$ is defined if $a < \arg t < b$, $\frac{|t|}{r} < |y| < r$, and so, it admits a Laurent expansion

$$f\left(\frac{t}{y}, y\right) = \sum_{n \in \mathbb{Z}} f_n(t)y^n.$$

If $n < 0$, the term $f_n(t)y^n$ can be rewritten, taking into account that $t = xy$, as

$$f_n(t)t^n \cdot \frac{1}{x^n}.$$

Define $g_n(t) = \frac{1}{t^n}f_{-n}(t)$, and consider

$$Tf(t, x, y) = \sum_{n>0} g_n(t)x^n + \sum_{n=0}^{\infty} f_n(t)y^n.$$

It is defined on $V(a, b; R) \times D(0; R)$, for appropriate $R > 0$ small enough, and it verifies $Tf(xy, x, y) = f(x, y)$.

For a general monomial $x^p y^q$, $f(x, y) \in \mathcal{O}(\Pi)$ can be decomposed as

$$f(x, y) = \sum_{\substack{0 \le i < p \\ 0 \le j < q}} x^i y^j f_{ij}(x^p, y^q),$$

where $f_{ij}(\zeta, \eta) \in \mathcal{O}(\tilde{\Pi})$, $\tilde{\Pi} = \Pi_{(1,1)}(a, b; \tilde{r})$. This decomposition is obtained explicitly solving a Vandermonde type linear system. Applying the previous operator T, we obtain a holomorphic function $Tf : V(a, b; R) \times D(\mathbf{0}; R) \to \mathbb{C}$ such that:

1. $Tf(x^p y^q, x, y) = f(x, y)$.
2. Fixing $t \in V(a, b; R)$, the Taylor expansion at the origin is an element of $\Delta(x^p y^q)$.

This function Tf is uniquely determined. It allows us to transfer asymptotic properties of f to asymptotic properties of Tf, in the new variable t. More precisely:

Theorem 1.4 Let $\Pi = \Pi_{p,q}(a, b; R)$ be a $x^p y^q$-sector, $\hat{f} \in \mathbb{C}[[x, y]]$. Assume that $\widehat{Tf}(t, x, y) = \sum_{n=0}^{\infty} f_n(x, y) t^n$, where $f_n(x, y) \in \Delta(x^p y^q)$, and all the functions $f_n(x, y)$ have a common disk of convergence. Then, the following conditions are satisfied:

1. f has $\hat{f}$ as $x^p y^q$-asymptotic expansion over Π.
2. Tf has $\widehat{Tf}$ as t-asymptotic expansion.

Moreover, if we assume that the asymptotic expansions are of s-Gevrey type, then the previous equivalences are still valid.

This transfer between monomial and classical summability allows us to define properly summability with respect to a monomial.

Definition 1.6 Let $s > 0$, $k = \frac{1}{s}$, and $\hat{f}(x, y) \in \mathbb{C}[[x, y]]$ be given. Let $\Pi = \Pi_{p,q}(a, b; R)$ be a $x^p y^q$-sector. We will say that $\hat{f}$ is $x^p y^q - k-$summable in Π if $b - a > s\pi$, and there exists $f(x, y) \in \mathcal{O}(\Pi)$ having $\hat{f}(x, y)$ as s-Gevrey asymptotic expansion in Π. An adaptation of Watson's lemma allows us to prove the unicity of such f. In fact, functions having the null series as s-Gevrey monomial asymptotic expansion with respect to $x^p y^q$ are exactly the functions that are exponentially small, that is, in each proper monomial subsector, they satisfy bounds

$$|f(x, y)| \leq C \exp\left(-\frac{A}{|x^p y^q|^{1/s}}\right).$$

$\hat{f}$ is called $x^p y^q - k-$summable in a direction $\theta \in \mathbb{R}$ if there exists a $x^p y^q$-sector Π, bisected by θ, of opening greater than $s\pi$, such that $\hat{f}$ in $x^p y^q - k-$summable in Π.

We will say shortly that $\hat{f}$ is $x^p y^q - k$-summable if it is $x^p y^q - k$-summable in every direction with finitely many exceptions modulo 2π. The space of $x^p y^q - k$-summable series will be denoted by $R_{1/k}^{(p,q)}$.

In one variable, it is crucial to establish the result about the "incompatibility" of summability in different levels (i.e., with respect to different values of k). More precisely, J.-P. Ramis shows that if a formal power series $\hat{f}(x) \in \mathbb{C}[[x]]$

is both k_1- and k_2-summable, with $k_1 \neq k_2$, then it is convergent (this result is known as Ramis' Tauberian theorem). In order to make a precise statement of a similar result for monomial summability, we must first notice that, if $M > 0$, there is a natural equality $R_{1/k}^{(p,q)} = R_{M/k}^{(Mp,Mq)}$. So, in order to compare two algebras $R_{1/k}^{(p,q)}$ and $R_{1/k'}^{(p',q')}$, we must take into account this fact and assume that there does not exist $M > 0$ such that $(Mp, Mq, M/k) = (p', q', 1/k')$. A particular case is the following:

Proposition 1.1   If $0 < k < k'$, then $R_{1/k}^{(p,q)} \cap R_{1/k'}^{(p,q)} = \mathbb{C}\{x, y\}$.

This is a consequence of the properties of the operator T: If $\hat{f} \in R_{1/k}^{(p,q)} \cap R_{1/k'}^{(p,q)}$, then $\hat{T}\hat{f}$ is both k'- and k-summable as a series in t with coefficients in $\Delta(x^p y^q)$. Applying Ramis' Tauberian theorem, $\hat{T}\hat{f}$ is convergent, so it is $\hat{f}$.

Consider now the general case:

Theorem 1.5   Consider the differential algebras $R_{1/k}^{(p,q)}$ and $R_{1/k'}^{(p',q')}$. Then, if the three numbers p/p', q/q', and k/k' are not equal, $R_{1/k}^{(p,q)} \cap R_{1/k'}^{(p',q')} = \mathbb{C}\{x, y\}$.

Proof: In [12], two different proofs of this interesting result are shown. We will follow schematically the first of them.

If $p/p' = q/q' \neq k/k'$, it is the previous proposition. Assume $p/p' \neq q/q'$.

If $k'/k > \max\{p/p', q/q'\}$, it is enough to observe that $\hat{T}\hat{f}$ is a k-summable series in t, and moreover, it is $\max\{p/p', q/q'\}/k'$- Gevrey, so it is convergent. Analogously, if $k'/k < \min\{p/p', q/q'\}$.

Suppose now that $p/p' \leq k'/k \leq q/q'$. Positive numbers s_1, s_2, s_1', s_2' exist such that

$$s_1 + s_2 = 1; \quad s_1/p = s_1'/p';$$
$$s_1' + s_2' = 1; \quad s_2/q = s_2'/q'.$$

Fixing (x_0, y_0), the formal series in z, $\hat{f}(z^{s_1/p} x_0, z^{s_2/q} y_0) = \hat{f}(z^{s_1'/p'} x_0, z^{s_2'/q'} y_0)$ is k- and k'-summable, so, if $k \neq k'$, it is convergent. From this, it can be deduced that $\hat{f}$ converges, using complex analysis standard arguments. If $k = k'$, consider a, b with $p/p' < a/b < q/q'$ and observe that $R_{1/k}^{(p,q)} = R_{b/k}^{(bp,bq)}$, $R_{1/k'}^{(p',q')} = R_{a/k'}^{(ap',aq')}$. □

Remark 1.2   As has been said before, another proof using properties of blow-ups is given in [12].

Let us finish this section linking monomial asymptotic expansions with strong asymptotic expansions. For, assume $p = q = 1$, and let us consider $f \in \mathcal{O}(\Pi)$, $\Pi = \Pi_{1,1}(a, b; R)$ a function admitting a monomial asymptotic expansion with respect to xy. Let V_1, V_2 be two sectors in $\mathbb{C}$, such that

$V_1 \times V_2 \subseteq \Pi$. If $W_1 < V_1$, $W_2 < V_2$, there exists $\tilde{\Pi} = \Pi_{1,1}(\tilde{a}, \tilde{b}; \tilde{R})$ a xy-sector strictly contained in Π, and such that $\overline{W}_1 \times \overline{W}_2 \subseteq \tilde{\Pi}$.

If $\hat{f}$ is the monomial asymptotic expansion of f in Π, $Tf \sim \hat{T}\hat{f}$ as asymptotic expansion in the variable t in $D \times V$, V being the sector $V(a, b; R)$. Restrict Tf to $\tilde{D} \times W$, $W = V(\tilde{a}, \tilde{b}; \tilde{R})$, $\overline{\tilde{D}} \subseteq D$; it can be extended as a C^∞-function to the whole space $\mathbb{C}^2 \times \mathbb{C}$. As $f(x, y) = Tf(x, y, xy)$, the restriction of f to $W_1 \times W_2$ admits a C^∞ extension to $\mathbb{C}^2$. As a consequence, f has a strong asymptotic expansion in $V_1 \times V_2$. We have shown:

Theorem 1.6 Monomial asymptotic expansion implies strong asymptotic expansion. More precisely, if a holomorphic function defined over a monomial sector Π has a monomial asymptotic expansion, then it has a strong asymptotic expansion when restricted to every polysector included in Π.

1.4 Monomial Summability for Singularly Perturbed Differential Equations

Let us apply the notion of monomial summability to the study of the so-called doubly singular systems of ODEs, summarizing here the main results of [10, 29]. For, let us consider a differential system as follows:

$$x^{p+1}\varepsilon^q \frac{\partial \mathbf{z}}{\partial x} = f(x, \varepsilon, \mathbf{z}), \tag{1.1}$$

where f is a holomorphic function $f : D \to \mathbb{C}^l$ defined in a neighborhood of $(0, 0, \mathbf{0}) \in \mathbb{C} \times \mathbb{C} \times \mathbb{C}^l$, $f(0, 0, \mathbf{0}) = \mathbf{0}$. Assume that this equation has invertible linear part, that is, $\frac{\partial f}{\partial \mathbf{z}}(0, 0, \mathbf{0})$ is an invertible matrix. Such a system has a unique formal solution

$$\hat{\mathbf{z}}(x, \varepsilon) = \sum_{n \geq 0} \mathbf{z}_{n*}(\varepsilon) x^n.$$

Implicit function theorem allows us to assume that $\mathbf{z}_{0*}(\varepsilon) = \mathbf{0}$, that is, $f_{0,0}(\varepsilon) = \mathbf{0}$. Assume, for simplicity, that $p = q = 1$, and write

$$f(x, \varepsilon, \mathbf{z}) = \sum_{i,j} f_{i,j}(\varepsilon) x^j \mathbf{z}^i.$$

Plugging in the series $\hat{\mathbf{z}}(x, \varepsilon)$ in the equation leads to a recurrence equation as

$$n\varepsilon \mathbf{z}_{n*}(\varepsilon) = f_{0,n+1}(\varepsilon) + f_{1,0}(\varepsilon)\mathbf{z}_{n+1,*}(\varepsilon) + \cdots,$$

and the $\mathbf{z}_{n*}(\varepsilon)$ turn out to be unique, holomorphic functions in a neighborhood of the origin. A majorant series argument shows that $\mathbf{z}(x, \varepsilon)$ is 1-Gevrey in x.

A similar argument, using Nagumo norms, shows that $\hat{\mathbf{z}}(x, \varepsilon)$ is also 1-Gevrey as a formal power series in ε. Precise details of these computations may be found in [29] and [10]. See also [30] for details concerning Nagumo norms and for a precise and detailed study of systems analogous to (1.1) but with $p = -1$.

Consider now the linear case, that is,

$$x^{p+1}\varepsilon^q \frac{\partial \mathbf{z}}{\partial x} = A(x,\varepsilon)\mathbf{z} + \mathbf{b}(x,\varepsilon). \tag{1.2}$$

This system, from the point of view of summability with respect to x and ε separately, has been studied in [29]. Again, assume $p = q = 1$, and apply Borel transform $\mathcal{B}$ to (1.2), with respect to x. If $\tilde{\mathbf{z}} = \hat{B}(\mathbf{z})$, we obtain an equation

$$\varepsilon\zeta\tilde{\mathbf{z}} - A_{0*}(\varepsilon)\tilde{\mathbf{z}} = \tilde{A}(\zeta,\varepsilon) * \tilde{\mathbf{z}} + \tilde{\mathbf{b}}(\zeta,\varepsilon),$$

where $\tilde{A}(\zeta,\varepsilon)$ denotes the Borel transform of $A(x,\varepsilon) - A_{0*}(\varepsilon)$, and $\tilde{\mathbf{b}}(\zeta,\varepsilon)$ the Borel transform of $\mathbf{b}$. As $A_{0*}(0)$ is invertible, this equation may be solved for ε small enough, and the solution extends in every direction that avoids the eigenvalues of $A_{0*}(0)$. If $d_1, d_2, \ldots, d_l$ are the directions of these eigenvalues, then we obtain the following theorem.

Theorem 1.7 $\hat{\mathbf{z}}(x,\varepsilon)$ is 1-summable in x in direction d, for every d such that

$$d_j < \arg\varepsilon + d < d_{j+1}, \quad \varepsilon \text{ small enough.}$$

Using the fact that summability implies unicity if the opening of the sectors are wide enough, it is also shown in the aforementioned paper that $\hat{\mathbf{z}}(x,\varepsilon)$ is 1-summable in ε, in every direction d such that

$$d_j < \arg x + d < d_{j+1}, \quad z \text{ small enough.}$$

Moreover we have the following lemma:

Lemma 1.1 If a formal power series in two variables,

$$\sum_{i,j=0}^{\infty} a_{ij}x^i y^j,$$

is 1-Gevrey on x, and 1-Gevrey on y, and $0 < s < 1$, then it is also a $(s, 1-s)$-Gevrey series in both variables, that is, there exists positive constants C and A, such that

$$|a_{ij}| \leq CA^{i+j}i!^s j!^{1-s}.$$

This result follows easily from the inequality

$$\min\{n!, m!\} \leq n!^s \cdot m!^{1-s}.$$

So, the problem of the summability in both variables x and ε can be stated. In this context, in [29] it is shown that $\hat{\mathbf{z}}(x,\varepsilon)$ is $(s, 1-s)$-summable. The sum is here defined via a weighted Borel–Laplace transform: from the formal solution

$$\hat{\mathbf{z}}(x,\varepsilon) = \sum_{i,j} z_{ij}x^i \varepsilon^j,$$

construct

$$\tilde{\mathbf{z}}_s(x, \varepsilon) = \hat{B}_{(s,1-s)}(\hat{z})(x, \epsilon) = \sum_{i,j} z_{ij} \frac{x^i \varepsilon^j}{\Gamma(1 + si + (1-s)j)}.$$

This turns out to be convergent, and a weighted-Laplace transform may be applied, in the form

$$\int_0^{\infty \cdot e^{id}} \tilde{\mathbf{z}}_s(xt^s, \varepsilon t^{1-s})e^{-t}dt,$$

from which the sum is obtained, for more details, see [29].

The nonlinear case is studied in [10], using monomial summability. More precisely, it is shown the following result:

Theorem 1.8 Consider the system of differential equations

$$x^{p+1}\varepsilon^q \frac{\partial \mathbf{z}}{\partial x} = f(x, \varepsilon, \mathbf{z}),$$

with $f : D \subseteq \mathbb{C} \times \mathbb{C} \times \mathbb{C}^l \to \mathbb{C}^l$, $\frac{\partial f}{\partial \mathbf{z}}(0, 0, \mathbf{0})$ an invertible matrix. Then, the unique formal power series solution is $x^p \varepsilon^q$-1-summable.

The proof is rather technical, and uses Banach fixed point theorem in an appropriate space obtained after applying the operator T to the system of differential equations.

Monomial summability may be related with Borel–Laplace transforms, with weighted variables, in order to construct the sum. Using this, it can be shown that Theorem 1.8 implies easily that the only formal power series solution in summable with respect to either the variable x, or with respect to the variable ε, obtaining similar results to those previously obtained in the linear case. In fact, in [31], the following result is shown:

Theorem 1.9 Let $\hat{f}$ be a $1/k$-Gevrey series in $x^p y^q$ (Definition 1.5). Then $\hat{f}$ is $x^p y^q - k$-summable in direction d if and only if for some $s \in (0, 1)$, $\hat{f}$ is $k - (s, 1 - s)$-summable in direction d.

So, both approaches to asymptotic expansions in monomial sectors coincide. We will not explore this in these pages, precise details are given in [12, 31].

1.5 Pfaffian Systems

One of the main applications for asymptotic expansions in several variables concerns doubly singular Pfaffian systems, with normal crossings, that is, systems of differential equations of the form

$$x^{p_1+1}y^{q_1} \frac{\partial \mathbf{z}}{\partial x} = f_1(x, y, \mathbf{z}), \tag{1.3}$$

$$x^{p_2} y^{q_2+1} \frac{\partial \mathbf{z}}{\partial y} = f_2(x, y, \mathbf{z}), \tag{1.4}$$

where $\mathbf{z} \in \mathbb{C}^l$, f_1, f_2 holomorphic functions defined in a neighborhood of the origin in $\mathbb{C} \times \mathbb{C} \times \mathbb{C}^l$, and p_1, q_1, p_2, $q_2 \in \mathbb{N}^*$. The system is said to satisfy the complete integrability condition, or, shortly, to be completely integrable, if the equality

$$x^{p_2} y^{q_2} \left[y \frac{\partial f_1}{\partial y}(x, y, \mathbf{z}) - q_1 f_1 \right] - x^{p_1} y^{q_1} \left[x \frac{\partial f_2}{\partial x}(x, y, \mathbf{z}) - p_2 f_2 \right] + \frac{\partial f_1}{\partial \mathbf{z}} f_2 - \frac{\partial f_2}{\partial \mathbf{z}} f_1 = 0$$

holds, for every $(x, y, \mathbf{z}) \in \mathbb{C} \times \mathbb{C} \times \mathbb{C}^l$. These systems have been studied by Gérard and Sibuya [7] and by Majima [9]. Let us summarize here some of the main results obtained by these authors, concerning mainly the asymptotic behavior of the solutions.

For instance, Gérard and Sibuya studied Pfaffian systems of the simplified form (with respect to the previous one):

$$x^{p+1} \frac{\partial z}{\partial x} = f_1(x, y, \mathbf{z}), \tag{1.5}$$

$$y^{q+1} \frac{\partial z}{\partial y} = f_2(x, y, \mathbf{z}), \tag{1.6}$$

with the hypothesis of the invertibility of the linear parts of both parts of the system, that is, of the matrices

$$\frac{\partial f_1}{\partial \mathbf{z}}(0, 0, \mathbf{0}) \quad \text{and} \quad \frac{\partial f_2}{\partial \mathbf{z}}(0, 0, \mathbf{0}),$$

and assuming complete integrability. It can be shown by indeterminate coefficients that there is a unique formal power series $\hat{f}(x, y)$ solution of (1.5) and (1.6). Write this solution as

$$\hat{f}(x, y) = \sum_{m \geq 0} f_{*m}(x) y^m = \sum_{n \geq 0} f_{n*}(y) x^n = \sum_{n, m \geq 0} f_{nm} x^n y^m.$$

This solution, by classical results on holomorphic ordinary differential equations (ODEs), turns out to be p-summable in x, and q-summable in y, considering summability with respect to one variable, parameterized by the other one. This second condition means that there exists a finite number of directions

$$\{d_1, d_2, \dots, d_n\},$$

such that if $d \notin \{d_1, d_2, \dots, d_n\}$, there exists $f_d(x, y) \in \mathcal{O}(D_1 \times V_d)$, where D_1 is a disk, $V_d = V\left(d - \frac{\pi}{2q} - \delta, d + \frac{\pi}{2q} + \delta; r\right)$, $\delta > 0$, with

$$\left| f_d(x, y) - \sum_{m < M} f_{*m}(x) y^m \right| \leq C A^M M!^{1/q} |y|^M.$$

Write

$$f_d(x, y) = \sum_{k=1}^{\infty} g_{d,k}(y) x^k,$$

with $g_{d,k}(y) \in \mathcal{O}(V_d)$. We have bounds

$$\left| g_k(y) - \sum_{m<M} f_{km} y^m \right| = \left| \frac{1}{2\pi i} \int_\gamma \frac{f_d(\omega, y) - \sum_{m<M} f_{*m}(\omega) y^m}{\omega^{k+1}} d\omega \right|$$

$$\leq \frac{1}{2\pi} 2\pi r \cdot \frac{C A^M M!^{1/q} |y|^M}{r^{k+1}} = \frac{C}{r^k} A^M M!^{1/q} |y|^M,$$

where γ denotes a circle of radius r around the origin. So, each function $g_k(y)$ turns out to be q-summable, with sum $\sum_m f_{km} y^m = f_{k*}(y)$. As it is also a convergent series, then $g_{d,k}(y)$ in fact converges in a neighborhood of $\mathbf{0}$. So, the family of functions $g_{d,k}(y)$ glue together defining a function that coincides with $g_k(y)$ around the origin, so $f_d(x, y)$ glue in $f(x, y)$, holomorphic in x, y, and solution of the equation. Then $\hat{f}(x, y)$ converges. This is an adaptation of the proof given by Sibuya [32].

In the framework of strong asymptotic expansions, Pfaffian systems have been studied by Majima [9]. He considers systems (1.3) and (1.4) assuming that at least one of the matrices $A = \frac{\partial f_1}{\partial z}(0, 0, \mathbf{0})$, $B = \frac{\partial f_2}{\partial z}(0, 0, \mathbf{0})$ is invertible, and proving several results. Under this assumption, and under complete integrability condition, he manages to show that the only formal solution of this system has a strong asymptotic expansion. In fact, this can be deduced from Majima's results about the existence of strongly asymptotically developable function solutions of systems of ODEs, and the hypothesis of complete integrability. More precisely, the following result is shown:

Proposition 1.2   Let us consider the Pfaffian system

$$x^{p_1+1} y^{q_1} \frac{\partial z}{\partial x} = f_1(x, y, \mathbf{z}), \tag{1.7}$$

$$x^{p_2} y^{q_2+1} \frac{\partial z}{\partial y} = f_2(x, y, \mathbf{z}). \tag{1.8}$$

Suppose that the system is completely integrable. Assume that $\mathcal{F}$ is a consistent total family of coefficients of strong asymptotic expansion, defining formally a solution of the Pfaffian system. Let $V_1 \times V_2$ be a strictly proper sector with respect to (A, B) (see Remark 1.3). Then, there exists $f \in \mathcal{A}(V)$ solution of the system, having the family $\mathcal{F}$ as the strong asymptotic expansion.

Remark 1.3   The condition of V being a strictly proper sector with respect to (A, B) means the following: V is contained in $\Pi_1 \cap \Pi_2$, where

$$\Pi_1 = \Pi_{p_1,q_1}(a_1, b_1; r_1); \quad \Pi_2 = \Pi_{p_2,q_2}(a_2, b_2; r_2),$$

such that the interval $[a_1, b_1]$ (considered in $\mathbb{S}^1$) does not contain completely a negative interval for A, that is, an interval where $\exp(-\frac{p_1 \lambda}{x^{p_1} y^{q_1}})$ is exponentially flat, λ being an eigenvalue of A, and the same with $[a_2, b_2]$, $\exp(-\frac{q_2 \mu}{x^{p_2} y^{q_2}})$, μ an eigenvalue of B.

The family $\mathcal{F}$ that appears in the statement can be obtained from a formal power series $\hat{f}(x, y)$ solution of the system, so, this would mean that such a formal power series is summable in appropriate polysectors. The existence of $\hat{f}(x, y)$ follows from the invertibility of A, B, and from the complete integrability of the system. These are very technical results, that we will not be developed further here, whose complete proofs are given in [9]. They can be generalized, assuming that at least one of the matrices A and B is invertible.

The conditions on the polysectors (they are strictly proper with respect to (A, B)) allows us to think that monomial asymptotics could be used. Moreover, if $(p_1, q_1) \neq (p_2, q_2)$, two different levels of summability seem to appear, which would imply convergence. But complete integrability condition implies very serious restrictions on the Pfaffian system (1.7), (1.8). More precisely, we have the following proposition.

Proposition 1.3 [33]

Let us consider the Pfaffian system (1.7), (1.8), and denote $A = \frac{\partial f_1}{\partial \mathbf{z}}(0, 0, \mathbf{0})$, $B = \frac{\partial f_2}{\partial \mathbf{z}}(0, 0, \mathbf{0})$. Then:

(1) If $p_2 < p_1$ or $q_2 < q_1$, then A is nilpotent.
(2) If $p_1 < p_2$ or $q_1 < q_2$, then B is nilpotent.
(3) If $p_1 = p_2 = p$, $q_1 = q_2 = q$. Then, for every eigenvalue μ of B, there exists an eigenvalue λ of A such that $q\lambda = p\mu$.

So, we must be careful when imposing integrability condition in these singularly perturbed Pfaffian systems. It may happen that this integrability condition forbids the system to have invertible matrices of the linear part at the origin. At least, this happens for a great number of values (p_1, p_2, q_1, q_2). Nevertheless, in the absence of complete integrability condition, even if the existence of a formal solution cannot be guaranteed, if it exists, we can provide useful information about it regarding summability. More precisely we have:

Theorem 1.10 Consider the Pfaffian system (1.7) and (1.8), with previous notations.

1. Suppose that a formal solution exists. If A and B are invertible, and $(p_1, q_1) \neq (p_2, q_2)$, then the solution is convergent.
2. If the system is completely integrable and A is invertible, there is a unique formal solution, $x^{p_1} y^{q_1}$-1-summable.

3. If the system is completely integrable and B is invertible, there is a unique formal solution, $x^{p_2} y^{q_2}$-1-summable.

References

1 Wasow, W. (1965) *Asymptotic Expansions for Ordinary Differential Equations*, John Wiley & Sons, Inc.
2 Ramis, J.-P. (1980) *Les séries k-sommables et leurs applications* Complex analysis, microlocal calculus and relativistic quantum theory (Proceedings of International Colloquium, Centre of Physics, Les Houches, 1979), Lecture Notes in Physics, vol. **126**, Springer-Verlag, Berlin, New York, pp. 178–199.
3 Écalle, J. (1981–1985) *Les fonctions résurgentes. Tome I, II, III*, Publications Mathématiques d'Orsay. Université de Paris-Sud, Département de Mathématique, Orsay.
4 Balser, W. (1994) *From Divergent Power Series to Analytic Functions*, Lecture Notes in Mathematics, No. 1582, Springer-Verlag, Berlin.
5 Braaksma, B.L.J. (1992) Multisummability of formal power series solutions of nonlinear meromorphic differential equations. *Ann. Inst. Fourier Grenoble*, **42**, 517–540.
6 Balser, W., Braaksma, B.L.J., Ramis, J.-P., and Sibuya, Y. (1991) Multisummability of formal power series solutions of linear ordinary differential equations. *Asympt. Anal.*, **5**, 27–45.
7 Gérard, R. and Sibuya, Y. (1979) *Étude de certains systèmes de Pfaff avec singularités*, Lecture Notes in Mathematics, vol. **172**, Springer-Verlag, Berlin, pp. 131–288.
8 Majima, H. (1983) Analogues of Cartan's decomposition theorem in asymptotic analysis. *Funk. Ekvac.*, **26**, 131–154.
9 Majima, H. (1984) *Asymptotic Analysis for Integrable Connections with Irregular Singular Points*, Lecture Notes in Mathematics vol. **1075**, Springer-Verlag, Berlin.
10 Canalis-Durand, M., Mozo-Fernández, J., and Schäfke, R. (2007) Monomial summability and doubly singular differential equations. *J. Differ. Equ.*, **233** (2), 485–511.
11 Fruchard, A. and Schäfke, R. (2013) *Composite Asymptotic Expansions*, Lecture Notes in Mathematics, vol. **2066**, Springer-Verlag.
12 Carrillo, S. (2016) Monomial summability through Borel-Laplace transforms. Applications to sngularly perturbed differential equations and Pfaffian systems. PhD thesis. University of Valladolid.
13 Ramis, J.-P. (1993) Séries divergentes et théories asymptotiques. *Bull. Soc. Math. France*, **121**, coll. Panoramas et Syntheses.
14 Ramis, J.-P. (2006) Les séries divergentes. *Pour la Sci.*, **350**, 132–139.

15 Haraoka, Y. (1989) Theorems of Sibuya-Malgrange type for Gevrey functions of several variables. *Funk. Ekvac.*, **32**, 365–388.

16 Zurro, M.A. (1997) A new Taylor type formula and C^∞ extensions for asymptotically developable functions. *Stud. Math.*, **123** (2), 151–163.

17 Hernández, J.A. (1994) Desarrollos asintóticos en polisectores. Problemas de existencia y unicidad (Asymptotic expansions in polysectors. Existence and uniqueness problems). PhD dissertation. Universidad de Valladolid, Spain.

18 Galindo, F. and Sanz, J. (1999) On strongly asymptotically developable functions and the Borel-Ritt theorem. *Stud. Math.*, **133**, 231–248.

19 Malgrange, B. (1967) *Ideals of Differentiable Functions*, Tata Institute of Fundamental Research Studies in Mathematics, No. 3, Tata Institute of Fundamental Research, Bombay; Oxford University Press, London, vii+106 pp.

20 Grothendieck, A. (1955) Produits tensoriels topologiques et espaces nucléaires. *Mem. Am. Math. Soc.*, **16**.

21 Douady, R. (1974) Produits tensoriels topologiques et espaces nucléaires. *Astérisque*, **16**, I.01I.25.

22 Mozo-Fernández, J. (1999) Topological tensor products and asymptotic developments. *Ann. Fac. Sci. Toulouse 6 Srie t.*, **8** (2), 281–295.

23 Lastra, A., Mozo-Fernández, J., and Sanz, J. (2012) Strong asymptotic expansions in a multidirection. *Funk. Ekvac.*, **55** (2), 317–345.

24 Fruchard, A. and Zhang, C. (1999) Remarques sur les développements asymptotiques. *Ann. Fac. Sci. Toulouse Math. (6)*, **8** (1), 91–115.

25 Honda, N. and Prelli, L. (2013) Multi-specialization and multi-asymptotic expansions. *Adv. Math.*, **232**, 432–498.

26 Sanz, J. (2003) Linear continuous extension operators for Gevrey classes on polysectors. *Glasgow Math. J.*, **45**, 199–216.

27 Martinet, J. and Ramis, J.-P. (1983) Classification analytique des équations différentielles non linéaires résonnantes du premier ordre. *Ann. Sci. É.N.S. 4e. sér., t*, **16** (4), 571–621.

28 Wasow, W. (1985) *Linear Turning Point Theory*, Applied Mathematical Sciences, vol. **54**, Springer-Verlag, New York.

29 Balser, W. and Mozo-Fernández, J. (2002) Multisummability of formal solutions of singular perturbation problems. *J. Differ. Equ.*, **183** (2), 526–545.

30 Canalis-Durand, M., Ramis, J.-P., Schäfke, R., and Sibuya, Y. (2000) Gevrey solutions of singularly perturbed differential equations. *J. Reine Angew. Math.*, **518**, 95–129.

31 Carrillo, S.A. and Mozo-Fernández, J. (2018) An extension of Borel-Laplace methods and monomial summability. *J. Math. Anal. Appl.*, **457**, 461–477.

32 Sibuya, Y. (1994) Convergence of formal solutions of meromorphic differential equations containing parameters. *Funk. Ekvac.*, **37** (2), 395–400.

33 Carrillo, S.A. and Mozo-Fernández, J. (2016) Tauberian properties for monomial summability with applications to Pfaffian systems. *J. Differ. Equ.*, **261** (12), 7237–7255.

2

Duality for Gaussian Processes from Random Signed Measures

Palle E.T. Jorgensen[1] and Feng Tian[2]

[1] *Department of Mathematics, The University of Iowa, Iowa City, IA 52242-1419, USA*
[2] *Department of Mathematics, Hampton University, Hampton, VA 23668, USA*

2000 Primary 47L60, 46N30, 65R10, 58J65, 81S25

2.1 Introduction

While reproducing kernel Hilbert spaces (RKHSs) play an important role in the analysis of Gaussian processes, there is a mismatch between the measurable setting of stochastic analysis on the one hand, and the more traditional setting of RKHSs on the other. To see this, recall that a RKHS is a Hilbert space $\mathcal{H}$ of functions on some prescribed set, say X, such that the values $f(x)$ of functions f from $\mathcal{H}$ are reproduced via the inner product in $\mathcal{H}$. By contrast, in stochastic analysis, one studies system of random variables, which is of course formulated in a measurable category, details are given below. Hence, the better framework for these applications is that of measure spaces $(X, \mathcal{B})$ and suitable Hilbert spaces of measurable functions defined on it. But, since measurable functions are not defined in the pointwise sense, the more traditional and geometric approach of Aronszajn et al. to RKHS theory is not amenable to the measurable category. Going back to the pioneering work of Aronszajn [1, 2], the motivation for the study of RKHSs was derived primarily from Greens function analysis in the study of partial differential equations (PDE) and from the study of Hilbert spaces of analytic functions Bergman et al., and for this, the pointwise setting appears natural and fruitful. We note that, historically, the applications of RKHS theory to probability theory came later, and so it is not surprising that, by necessity, the use of Hilbert space geometry is different there.

The purpose of the present chapter is to offer a duality theory for Gaussian processes, on the one hand, and to develop a remedy to the above mentioned problem with the reproducing property for Hilbert spaces of measurable functions on prescribed measure spaces $(X, \mathcal{B})$, on the other.

Discussion of the literature. The theory of RKHS and their applications is vast, and below we only make a selection. Readers will be able to find more

Mathematical Analysis and Applications: Selected Topics, First Edition.
Edited by Michael Ruzhansky, Hemen Dutta, and Ravi P. Agarwal.

cited there. As for the general theory of RKHS in the pointwise category, we find useful [3–7]. The applications include fractals [1, 8, 9]; probability theory [10–16]; and application to learning theory [17–20].

Organization of the chapter. This chapter is organized as follows: the sections in the first half of the chapter cover new research advances; and in order to make the whole chapter useful to a wider readership, including nonexperts, the later sections will therefore cover such background material as is needed, with an emphasis on the theory of Gaussian processes, and RKHSs, and especially on their interconnections.

2.2 Reproducing Kernel Hilbert Spaces (RKHSs) in the Measurable Category

Motivated by applications to stochastic processes, we propose here the following generalization of RKHSs in the usual sense of Aronszajn [1, 2] (also see [17–19]), but now to a measurable framework. In the traditional approach, the starting point is a positive definite (p.d.) function $K : X \times X \to \mathbb{C}$, which is defined pointwise, and X is a given set. The "p.d." requirement is that, for all $p \in \mathbb{N}$, and all $x_i \in X$, $1 \leq i \leq p$ the $p \times p$ matrix $(K(x_i, x_j))_{i,j=1}^p$ is positive semidefinite. The RKHS $\mathscr{H}$ is then defined as the Hilbert-completion of the space of functions (on X),

$$\sum_i \xi_i K(\cdot, x_i), \ \xi_i \in \mathbb{C} \tag{2.1}$$

with respect to $\|\cdot\|_{\mathscr{H}}$ defined as follows:

$$\left\| \sum_i \xi_i K(\cdot, x_i) \right\|_{\mathscr{H}}^2 = \sum_i \sum_j \overline{\xi_i} \xi_j K(x_i, x_j). \tag{2.2}$$

For recent applications, we refer to [20–22].

The measurable setting
Since stochastic processes are formulated in the language of probability, it is important that the elements in $\mathscr{H}$ be suitably determined measurable functions; and that the new inner product then be redefined in this measurable setting.

Definition 2.1 Let X be a locally compact Hausdorff space, and let $\mathscr{B}$ be the corresponding σ-algebra of Borel measurable subsets of X. Let λ be a positive measure on $(X, \mathscr{B})$, that is, λ is defined to be σ-additive on $\mathscr{B}$.

We shall consider Hilbert spaces $\mathscr{H} \subset L^2_{\text{loc}}(\lambda)$. (The subscript loc for "local" is important, that is, the requirement is that

$$\int_B |f|^2 \mathrm{d}\lambda < \infty \tag{2.3}$$

for all $B \in \mathscr{B}$ s.t. $\lambda(B) < \infty$, and all $f \in \mathscr{H}$.)

We shall denote by

$$\mathscr{B}_{\text{fin}} := \{B \in \mathscr{B} \mid \lambda(B) < \infty\}. \tag{2.4}$$

Definition 2.2 A pair $(\mathscr{H}, \lambda)$ as above is said to be *p.d.* if and only if, for all $B \in \mathscr{B}_{\text{fin}}$,

$$\sup_{f \in \mathscr{H},\ \|f\|_{\mathscr{H}} \leq 1} \left| \int_B f \, d\lambda \right| < \infty. \tag{2.5}$$

Remark 2.1 At the outset our setting is that of general σ-finite measure spaces $(X, \mathscr{B})$; but for some considerations it will be useful to assume that X be a locally compact Hausdorff space, and in that case the prescribed σ-algebra $\mathscr{B}$ will be the Borel subsets of X. We shall then be working with $C_c(X) =$ the continuous functions of compact support.

Remark 2.2 (The reproducing property) The RKHSs are of great use in part, because of what is called the *reproducing property*.

In the case when K is given as in (2.1) and (2.2) above, then the corresponding RKHS $\mathscr{H}(K)$ is a Hilbert space of functions f on X, and we have

$$f(x) = \langle f, K(\cdot, x) \rangle_{\mathscr{H}(X)}, \quad \forall x \in X, \ \forall f \in \mathscr{H}(K). \tag{2.6}$$

Now in the measurable category, (2.6) does not make sense since functions are not defined pointwise.

Remark 2.3 In the proofs below, we shall mostly restrict attention to real valued functions, but all arguments easily extend to the complex case.

Lemma 2.1 Suppose $(\mathscr{H}, \lambda)$ is as above.

(i) Then, for $\forall B \in \mathscr{B}_{\text{fin}}$, $\exists! K^B \in \mathscr{H}$ such that

$$\int_B f \, d\lambda = \langle f, K^B \rangle_{\mathscr{H}}, \quad \forall f \in \mathscr{H}. \tag{2.7}$$

(ii) Setting $K(\cdot, B) = K^B$, $B \in \mathscr{B}_{\text{fin}}$, note that $x \longmapsto K(x, \cdot)$ is then a measurable field of random signed measures on $(X, \mathscr{B})$, and that

$$\rho_{\mathscr{H}}(A, B) = \langle K^A, K^B \rangle_{\mathscr{H}}$$
$$= \int_A K(x, B) \, d\lambda(x) = \overline{\int_B K(y, A) \, d\lambda(y)}. \tag{2.8}$$

Then $\rho_{\mathscr{H}}$ is p.d. on $\mathscr{B}_{\text{fin}} \times \mathscr{B}_{\text{fin}}$.

Proof: (i) By the assumption (2.5), the mapping $\mathscr{H} \ni f \longrightarrow \int_B f \mathrm{d}\lambda$ is a bounded linear functional on $\mathscr{H}$, so by Riesz theorem, $\exists! K(\cdot, B) \in \mathscr{H}$ s.t. (2.7) holds.

For part (ii), note that (2.8) follows from (2.7) with $f = K^A$; and it is immediate that $\rho_{\mathscr{H}}$ is p.d. on $\mathscr{B}_{\mathrm{fin}} \times \mathscr{B}_{\mathrm{fin}}$. Indeed, for $\forall p \in \mathbb{N}$, $\forall (A_i)_{i=1}^p$ in $\mathscr{B}_{\mathrm{fin}}$, and $\forall (\xi_i)_{i=1}^p$ in $\mathbb{R}$, we have

$$\sum_i \sum_j \xi_i \xi_j \rho_{\mathscr{H}}(A_i, A_j) = \sum_i \sum_j \xi_i \xi_j \langle K^{A_i}, K^{A_j} \rangle_{\mathscr{H}}$$

$$= \left\| \sum_i \xi_i K^{A_i} \right\|_{\mathscr{H}}^2 \geq 0. \qquad \square$$

Remark 2.4 Let $(\mathscr{H}, \lambda)$ be as specified in Definitions 2.1 and 2.2. In this case, the corresponding reproducing property takes any one of the following equivalent forms (see the setting $(X, \mathscr{B}, \mathscr{H}, \lambda, K)$ as in Definitions 2.1 and 2.2):

(1) For all $B \in \mathscr{B}_{\mathrm{fin}}$, we have $\mathscr{H} \subset L^2_{\mathrm{loc}}(\lambda)$, and

$$\int_B f \mathrm{d}\lambda = \langle f(\cdot), K(\cdot, B) \rangle_{\mathscr{H}}, \quad \forall B \in \mathscr{B}_{\mathrm{fin}}, \forall f \in \mathscr{H}. \tag{2.9}$$

(2) Setting $K^\varphi(\cdot) = \int \varphi(y) K(\cdot, \mathrm{d}y)(\in \mathscr{H})$, then we have

$$\int f \varphi \mathrm{d}\lambda = \langle f, K^\varphi \rangle_{\mathscr{H}}, \quad \forall \varphi \in C_c(X), \forall f \in \mathscr{H}. \tag{2.10}$$

When $(\mathscr{H}, \lambda)$ is given as above, and $f \in \mathscr{H}$, we set μ_f to be the signed measure

$$\mu_f(B) := \langle f(\cdot), K(\cdot, B) \rangle_{\mathscr{H}}, \quad B \in \mathscr{B}_{\mathrm{fin}}. \tag{2.11}$$

We shall show below that $\mathrm{d}\mu_f = f \mathrm{d}\lambda$ locally (see Lemma 2.2).

Definition 2.3 Let (X, λ) be as above. Let $\rho_{\mathscr{H}}$ be the positive definition function on $\mathscr{B}_{\mathrm{fin}} \times \mathscr{B}_{\mathrm{fin}}$ as in (2.8). We denote $\mathscr{H}^{(Aronsz)}(\rho_{\mathscr{H}})$ the corresponding RKHS in the sense of Aronszajn's.

Lemma 2.2 Let $(\mathscr{H}, \lambda)$ be as above, that is, p.d. For $f \in \mathscr{H}$, we set μ_f to be the signed measure (2.11) on $\mathscr{B}_{\mathrm{fin}}$, that is, $\mu_f(B) := \langle f(\cdot), K(\cdot, B) \rangle_{\mathscr{H}}$, $B \in \mathscr{B}_{\mathrm{fin}}$; then $\mu_f \ll \lambda$, and $\mathrm{d}\mu_f/\mathrm{d}\lambda = f$ in $L^1_{\mathrm{loc}}(\lambda)$.

Proof: We now show that if $\lambda(A) = 0$, then $\mu_f(A) = 0$. But for $\forall B \in \mathscr{B}_{\mathrm{fin}}$, we have

$$0 = \int_A K(x, B) \mathrm{d}\lambda(x) = \int_B K(y, A) \mathrm{d}\lambda(y); \tag{2.12}$$

and so by approximation

$$\int \varphi(y) K(y, A) \mathrm{d}\lambda(y) = 0, \tag{2.13}$$

for $\forall \varphi \in C_c(X)$, and

$$\langle K^\varphi, K^A \rangle_{\mathscr{H}} = 0. \tag{2.14}$$

Since $\{K^\varphi \mid \varphi \in C_c(X)\}$ is dense in $\mathscr{H}$, we get $\langle f(\cdot), K(\cdot, A) \rangle_{\mathscr{H}} = \mu_f(A) = 0$, which is the desired conclusion.

But, by (2.13) and (2.14), we have

$$\int_X \varphi f \, d\lambda = \langle f, K^\varphi \rangle_{\mathscr{H}} = \int_X \varphi(y) \langle f, K(\cdot, dy) \rangle_{\mathscr{H}}$$

$$= \int_X \varphi(y) \, d\mu_f(y), \ \forall \varphi \in C_c(X),$$

and so the conclusion $d\mu_f / d\lambda = f$ in $L^1_{\text{loc}}(\lambda)$ follows. $\qquad\square$

Remark 2.5 (The Radon–Nikodym-derivative) Fix $(X, \mathscr{B})$, assume $(\mathscr{H}, \lambda)$ is p.d. Thus, λ is a positive measure on $(X, \mathscr{B})$ s.t. (2.5) holds; set $\mu_f(A) = \langle f(\ldots), K(\cdot, A) \rangle_{\mathscr{H}}$, $\forall A \in \mathscr{B}_{\text{fin}}, \forall f \in \mathscr{H}$.

Then, for a.e. x w.r.t λ, there exists $A_i \in \mathscr{B}_{\text{fin}}$, $i = 1, 2, \cdots$, s.t. $\lambda(A_i) \to 0$, $x \in \cap A_i$, so that (see [23] and Figure 2.1)

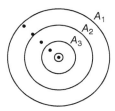

$$\lim_{i \to \infty} \frac{\mu_f(A_i)}{\lambda(A_i)} = f(x).$$

Figure 2.1 The sequence $(A_i)_{i=1}^\infty$ in $\mathscr{B}_{\text{fin}}$ s.t. $\lambda(A_i) \to 0$, as $i \to \infty$.

This is a way to make precise the local Radon–Nikodym derivative, $d\mu_f = f d\lambda$.

Corollary 2.1 Let λ be a positive regular measure on a locally compact Hausdorff space X with Borel σ-algebra $\mathscr{B}$; and let $\mathscr{H}$ be a Hilbert space s.t. (2.5) holds; that is $(\mathscr{H}, \lambda)$ is p.d. Then the functions $f \in \mathscr{H}$ are in $L^2_{\text{loc}}(\lambda)$.

Proof: Given $f \in \mathscr{H}$, we proved that the signed measures $\mu_f(B) = \langle f, K(\cdot, B) \rangle_{\mathscr{H}}$, $B \in \mathscr{B}_{\text{fin}}$ are well defined, and locally $\mu_f \ll \lambda$, with $d\mu_f / d\lambda = f$ (locally), that is, for $\forall A \in \mathscr{B}_{\text{fin}}$, we have

$$\mu_f(A) = \int_A f \, d\lambda. \qquad\square$$

Corollary 2.2 Let $(\mathscr{H}, \lambda)$ be as above; and let $\{f_i\}_{i \in \mathbb{N}}$ be a fixed ONB in $\mathscr{H}$, or a Parseval Frame, then, for all $B \in \mathscr{B}_{\text{fin}}$, we have:

$$K^B(x) = \sum_{i=1}^\infty \left(\int_B f_i \, d\lambda \right) f_i(x); \tag{2.15}$$

with convergence in the $\mathscr{H}$-norm, and

$$\|K^B\|_{\mathscr{H}}^2 = \sum_{i=1}^{\infty} \left| \int_B f_i \, d\lambda \right|^2. \tag{2.16}$$

Proof: Note that

$$\|K^B\|_{\mathscr{H}}^2 = \sum_{i=1}^{\infty} \langle K^B, f_i \rangle_{\mathscr{H}} \langle f_i, K^B \rangle_{\mathscr{H}} \quad \text{(Parseval)}$$

$$= \sum_{i=1}^{\infty} |\mu_{f_i}(B)|^2 = \sum_{i=1}^{\infty} \left| \int_B f_i \, d\lambda \right|^2. \qquad \square$$

Definition 2.4 Let $(X, \mathscr{B})$ be a measure space as above, and let λ be a positive measure on $(X, \mathscr{B})$. Let $X \ni x \longmapsto K(x, A)$, $A \in \mathscr{B}$, be a *random signed measure*. We say that (λ, K) is *p.d.* iff, for $\forall k$, $\forall \{A_i\}_{i=1}^k$, $A_i \in \mathscr{B}_{\text{fin}}$, $\{\xi_i\}_{i=1}^k$, $\xi_i \in \mathbb{C}$, we have:

$$\sum_i \sum_j \overline{\xi}_i \xi_j \int_{A_i} K(x, A_j) \, d\lambda(x) \geq 0. \tag{2.17}$$

When (2.17) is satisfied for a fixed pair (λ, K), let $\mathscr{H}$ be the corresponding Hilbert space completion. In this case, of course, $\lambda \in \mathscr{L}_+(\mathscr{H})$ (see Definition 2.5).

Theorem 2.1 Let $(\mathscr{H}, \lambda)$ be as above, that is, p.d.

(i) For all $f \in \mathscr{H}$, set $\mu_f(B) := \langle f(\cdot), K(\cdot, B) \rangle_{\mathscr{H}}$, $\forall B \in \mathscr{B}_{\text{fin}}$, that is, μ_f is a signed measure for $\forall f \in \mathscr{H}$. Then $\mu_f \ll \lambda$, $f \in L^2_{\text{loc}}(\lambda)$, and

$$\frac{d\mu_f}{d\lambda} = f \tag{2.18}$$

defined in $L^1_{\text{loc}}(\lambda)$, as a Raodon–Nikodym-derivative.

(ii) Let $\rho_{\mathscr{H}}$ be as in (2.8), that is, p.d. on $\mathscr{B}_{\text{fin}} \times \mathscr{B}_{\text{fin}}$, and let $\mathscr{H}^{(\text{Aronsz})}(\rho_{\mathscr{H}})$ be the corresponding RKHS. Then, for all $f \in \mathscr{H}$, we have $\mu_f \in \mathscr{H}^{(\text{Aronsz})}(\rho_{\mathscr{H}})$, with

$$\|\mu_f\|_{\mathscr{H}^{(\text{Aronsz})}(\rho_{\mathscr{H}})} = \|f\|_{\mathscr{H}}, \quad \forall f \in \mathscr{H}. \tag{2.19}$$

More precisely, the functions in $\mathscr{H}^{(\text{Aronsz})}(\rho_{\mathscr{H}})$ are signed measures on $(X, \mathscr{B})$; and the map

$$\mathscr{H} \ni f \xrightarrow[\cong]{T} \mu_f = f \, d\lambda \in \mathscr{H}^{(\text{Aronsz})}(\rho_{\mathscr{H}}), \tag{2.20}$$

is an isometric isomorphism from $\mathscr{H}$ onto $\mathscr{H}^{(\text{Aronsz})}(\rho_{\mathscr{H}})$.

Proof: For part (i), see the proof of Lemma 2.2.

To prove part (ii), we will need the following: □

Lemma 2.3 Set $Tf = \mu_f$, for all $f \in \mathscr{H}$, where μ_f is the signed measure on $(X, \mathscr{B})$, given by $\mu_f(A) = \langle f, K(\cdot, A)\rangle_{\mathscr{H}} \; \forall A \in \mathscr{B}_{\text{fin}}$; then

$$\mu_f \in \mathscr{H}^{(\text{Aronsz})}(\rho_{\mathscr{H}}), \quad \forall f \in \mathscr{H}. \tag{2.21}$$

Proof: Given $f \in \mathscr{H}$, we must prove that: $\forall p \in \mathbb{N}, \forall \xi_i \in \mathbb{R}$, and $\forall A_i \in \mathscr{B}_{\text{fin}}, 1 \leq i \leq p$, there is a constant $C_f < \infty$ s.t.

$$\left| \sum_i \xi_i \mu_f(A_i) \right|^2 \leq C_f \sum_i \sum_j \xi_i \xi_j \rho_{\mathscr{H}}(A_i, A_j), \tag{2.22}$$

where $\rho_{\mathscr{H}}(A_i, A_j) = \int_{A_i} K(x, A_j) d\lambda(x), \forall i, j = 1, \dots, p$.

Proof of (2.22). Set $\psi(x) = \sum_{i=1}^p \xi_i \chi_{A_i}(x)$, then

$$
\begin{aligned}
\text{LHS}_{(2.22)} &= \left| \left\langle f(\cdot), \int \psi(y) K(\cdot, dy) \right\rangle_{\mathscr{H}} \right|^2 \\
&\leq \|f\|_{\mathscr{H}}^2 \left\| \int \psi(y) K(\cdot, dy) \right\|_{\mathscr{H}}^2 \quad \text{(by Cauchy-Schwarz)} \\
&= \|f\|_{\mathscr{H}}^2 \left| \sum_i \sum_j \xi_i \xi_j \int_{A_i} K(x, A_j) d\lambda(x) \right| \\
&= \|f\|_{\mathscr{H}}^2 \sum_i \sum_j \xi_i \xi_j \rho_{\mathscr{H}}(A_i, A_j) \\
&= \|f\|_{\mathscr{H}}^2 \|\psi\|_{\mathscr{H}^{(\text{Aronsz})}}^2 = \text{RHS}_{(2.22)}. \quad \square
\end{aligned}
$$

Lemma 2.4 Let $T : \mathscr{H} \to \mathscr{H}^{(\text{Aronsz})}(\rho_{\mathscr{H}})$ be as in (2.20). Then

$$T^*(\rho_{\mathscr{H}}(A, \cdot)) = K(\cdot, A), \quad \forall A \in \mathscr{B}_{\text{fin}}.$$

Proof: Let $f \in \mathscr{H}$, and $\rho_{\mathscr{H}}(A, \cdot) \in \mathscr{H}^{(\text{Aronsz})}(\rho_{\mathscr{H}}), A \in \mathscr{B}_{\text{fin}}$; then

$$\langle Tf, \rho_{\mathscr{H}}(A, \cdot)\rangle_{\mathscr{H}^{(\text{Aronsz})}} = \mu_f(A) = \langle f(\cdot), K(\cdot, A)\rangle_{\mathscr{H}}$$

so that the conclusion follows. □

Lemma 2.5 Let $(\mathscr{H}, \lambda)$ be as above, then $span\{K(\cdot, A)\}_{A \in \mathscr{B}_{\text{fin}}}$ is dense in $\mathscr{H}$.

Proof: To see this, let $g \in \mathscr{H}$ and assume that $\langle g(\cdot), K(\cdot, A)\rangle_{\mathscr{B}_{\text{fin}}} = 0, \forall A \in \mathscr{B}_{\text{fin}}$. Then $\langle g, K^{\varphi}\rangle_{\mathscr{H}} = 0, \forall \varphi \in C_c(X)$, where $K^{\varphi}(\cdot) = \int \varphi(y) K(\cdot, dy)$. It follows that

$$\int \varphi d\mu_g = \int \varphi g d\lambda = 0,$$

which implies $g d\lambda = 0$, so that $g = 0$ a.e.-λ; thus $g = 0$. □

Proof: It remains to prove that T is an isometry, that is, $\|\mu_f\|_{\mathscr{H}\,(\text{Aronsz})} = \|f\|_{\mathscr{H}}$. Note that every $f \in \mathscr{H}$ is $\|\cdot\|_{\mathscr{H}}$-limit of finite sums

$$\sum \xi_i K(\cdot, A_i), \quad \xi_i \in \mathbb{R}, \; A_i \in \mathscr{B}_{\text{fin}}.$$

From the definition, $Tf = \mu_f$ is the $\|\cdot\|_{\mathscr{H}\,(\text{Aronsz})}$-limit of finite sums

$$\sum \xi_i \rho_{\mathscr{H}}(\cdot, A_i), \quad \xi_i \in \mathbb{R}, \; A_i \in \mathscr{B}_{\text{fin}}.$$

Now we have

$$\left\| \sum \xi_i K(\cdot, A_i) \right\|_{\mathscr{H}}^2 = \sum_i \sum_j \xi_i \xi_j \langle K(\cdot, A_i), K(\cdot, A_j) \rangle$$

$$= \sum_i \sum_j \xi_i \xi_j \rho_{\mathscr{H}}(A_i, A_j)$$

$$= \left\| \sum \xi_i \rho_{\mathscr{H}}(\cdot, A_i) \right\|_{\mathscr{H}\,(\text{Aronsz})}^2,$$

which yields the desired conclusion. □

2.3 Applications to Gaussian Processes

Given $(\mathscr{H}, \lambda)$, p.d. (see Definition 2.2), there is a centered Gaussian process $\{W_A\}_{A \in \mathscr{B}_{\text{fin}}}$, realized on a probability space $L^2(\Omega, \mathbb{P})$. The Gaussian process is determined by

$$\mathbb{E}[W_A] = 0, \quad W_A \in L^2(\Omega, \mathbb{P}), \quad \text{and} \tag{2.23}$$

$$\mathbb{E}[W_A W_B] = \rho_{\mathscr{H}}(A, B) = \langle K^A, K^B \rangle_{\mathscr{H}}, \tag{2.24}$$

$\forall A, B \in \mathscr{B}_{\text{fin}}$. Note that

$$\mathbb{E}[\exp(i W_A)] = \exp\left(-\frac{1}{2} \rho_{\mathscr{H}}(A, A) \right), \quad A \in \mathscr{B}_{\text{fin}}, \tag{2.25}$$

determines the process.

For all measurable functions φ on $(X, \mathscr{B})$, we may define the corresponding Ito-integral $\int \varphi \, dW$; details below.

Theorem 2.2 If $\{f_i\}_{i \in \mathbb{N}}$ is an ONB in $\mathscr{H}$, and $Z_i(\cdot) \sim N(0, 1)$, i.i.d., then

$$W_A = \sum_{i=1}^{\infty} \left(\int_A f_i \, d\lambda \right) Z_i, \quad \text{and} \tag{2.26}$$

$$\int \varphi \, dW = \sum_{i=1}^{\infty} \left(\int \varphi f_i \, d\lambda \right) Z_i. \tag{2.27}$$

Proof: This is an immediate application of the new RKHS $\mathscr{H} = \mathscr{H}^{\text{(Aronsz)}}$. Apply Kolmogorov's extension principle to $\rho_{\mathscr{H}}$, we get a Gaussian process $\{W_A\}_{A \in \mathscr{B}_{\text{fin}}}$ satisfying conditions (2.23) and (2.24), so that

$$\mathscr{H} \ni K^A = K(\cdot, A) \longmapsto W_A \in L^2(\Omega, \mathbb{P}) \tag{2.28}$$

is isometric from $\mathscr{H}$ into $L^2(\Omega, \mathbb{P})$, where $\|K^A\|_{\mathscr{H}}^2 = \mathbb{E}[|W_A|^2] = \rho_{\mathscr{H}}(A, A)$, $\forall A \in \mathscr{B}_{\text{fin}}$.

We introduce

$$K^{\varphi} = \int \varphi(y) K(\cdot, dy) (\in \mathscr{H})$$

as a dense space of vectors in $\mathscr{H}$; and then the new Ito isometry may be realized as

$$K^{\varphi} \longmapsto \int \varphi dW, \quad \|K^{\varphi}\|_{\mathscr{H}}^2 = \mathbb{E}\left[\left|\int \varphi dW\right|^2\right]. \tag{2.29}$$

Since $L^2(\Omega, \mathbb{P})$ is complete, we conclude that (2.29) extends to $\mathscr{H} \to L^2(\Omega, \mathbb{P})$ as an isometry. Hence, $\mathscr{H}$ is realized as a closed subspace in $L^2(\Omega, \mathbb{P})$.

Now (2.26)–(2.27) follow from Corollary 2.2.

Details: Given $\epsilon > 0$, pick a *measurable partition* of X, that is, $\{A_i\} \subset \mathscr{B}_{\text{fin}}$, $A_i \cap A_j = \emptyset$ if $i \neq j$, and $\cup A_i = X$, so that

$$\left\|K^{\varphi} - \sum_i \varphi(y_i) K(\cdot, A_i)\right\|_{\mathscr{H}} < \epsilon, \quad y_i \in A_i.$$

With a net α of partitions, $\max \lambda(A_i) \xrightarrow{\alpha} 0$, we get as an application of our $(\mathscr{H}, \lambda)$ analysis:

$$\left\|\sum_i \varphi(x_i) K(\cdot, A_i) - \sum_j \varphi(y_j) K(\cdot, B_j)\right\|_{\mathscr{H}} < \epsilon,$$

for all $\alpha = (A_i)$, $\beta = (B_j)$. By (2.28), and Kolmogorov, we then get

$$\mathbb{E}\left[\left|\sum_i \varphi(x_i) W_{A_i} - \sum_j \varphi(y_j) W_{B_j}\right|^2\right] < \epsilon.$$

Hence, the induced net $N_{\alpha}^{\varphi} = \sum_i \varphi(x_i) W_{A_i}$ is convergent in $L^2(\Omega, \mathbb{P})$, and the limit

$$\int \varphi dW = \lim_{\alpha} \sum_i \varphi(x_i) W_{A_i}$$

is the generalized Ito-integral; with this approach it is clear that (2.29) then yields

$$\mathbb{E}\left[\left|\int \varphi \mathrm{d} W\right|^2\right] = \|K^\varphi\|_{\mathscr{H}}^2 = \int \varphi(x) \int \varphi(y) K(x, \mathrm{d} y) \mathrm{d}\lambda(x)$$

$$= \int \varphi(x) \int \varphi(y) \mathbb{E}[W_x \mathrm{d} W_y].$$

□

Corollary 2.3 Let $(X, \mathscr{B}, \mathscr{H}, \lambda)$ be as above. Let $\{W_A\}_{A \in \mathscr{B}_{\mathrm{fin}}}$ be the Gaussian process with $\mathbb{E}[W_A] = 0$, $\mathbb{E}[W_A W_B] = \langle K^A, K^B \rangle_{\mathscr{H}} = \rho_{\mathscr{H}}(A, B)$. Then we get an Ito-integral $\int \varphi \mathrm{d} W$ s.t.

$$\mathbb{E}\left[\left|\int \varphi \mathrm{d} W\right|^2\right] = \|K^\varphi\|_{\mathscr{H}}^2 = \varphi K_\lambda \varphi. \tag{2.30}$$

More precisely,

$$\varphi K_\lambda \varphi = \int \varphi(x) \int \varphi(y) K(x, \mathrm{d} y) \mathrm{d}\lambda(x)$$

$$= \int \varphi(x) \int \varphi(y) \mathbb{E}[W_x \mathrm{d} W_y]. \tag{2.31}$$

Theorem 2.3 Let $(X, \mathscr{B}, \mathscr{H}, \lambda)$ be as above, and let $\{W_A\}_{A \in \mathscr{B}_{\mathrm{fin}}}$ be the corresponding Gaussian process. Denote by $\int \varphi \mathrm{d} W$ the Ito-integral constructed from Theorem 2.2. Then

$$\mathbb{E}\left[e^{i \int \varphi \mathrm{d} W}\right] = e^{-\frac{1}{2}\varphi K_\lambda \varphi} \tag{2.32}$$

where φ is a measurable function on $(X, \mathscr{B})$ s.t. $\varphi K_\lambda \varphi < \infty$ (2.30).

Proof: Let $A \in \mathscr{B}_{\mathrm{fin}}$, then

$$\mathbb{E}[W_A^{2n+1}] = 0,$$

$$\mathbb{E}[W_A^{2n}] = (2n - 1)!! \left(\int_A K(x, A) \mathrm{d}\lambda(x)\right)^n,$$

where

$$(2n - 1)!! = \frac{(2n)!}{2^n n!} = (2n - 1)(2n - 3) \cdots 5 \cdot 3 \cdot 1.$$

It follows that

$$\mathbb{E}\left[\left|\int \varphi \mathrm{d} W\right|^{2n}\right] = (2n - 1)!!(\varphi K_\lambda \varphi)^n.$$

Now

$$\mathbb{E}\left[e^{i\int \varphi dW}\right] = \sum_{n=0}^{\infty} \frac{(-1)^n}{(2n)!} \mathbb{E}\left[\left|\int \varphi dW\right|^{2n}\right]$$

$$= \sum_{n=0}^{\infty} \frac{(-1)^n}{n! 2^n} (\varphi K_\lambda \varphi)^n = e^{-\frac{1}{2}\varphi K_\lambda \varphi},$$

which is the desired conclusion. □

Remark 2.6 We get the results, even when there is *not* an obvious Gelfand triple in the framework of (2.32).

Theorem 2.4 Let $(X, \mathcal{B}, \mathcal{H}, \lambda)$ be as usual and let $\{W_A\}_{A \in \mathcal{B}_{\text{fin}}}$ be the corresponding Gaussian process realized on $L^2(\Omega, \mathbb{P})$. Set $K^\varphi(\cdot) = \int \varphi(y) K(\cdot, dy)$. Let $\{f_n\}_{n \in \mathbb{N}}$ be an ONB in $\mathcal{H}$. Then TFAE:

(i) $\int \varphi dW \in L^2(\Omega, \mathbb{P})$;
(ii) $K^\varphi \in \mathcal{H}$;
(iii) $\sum_{n=1}^{\infty} \left|\int \varphi f_n d\lambda\right|^2 < \infty$, and then

$$\mathbb{E}\left[e^{i\int \varphi dW}\right] = e^{-\frac{1}{2}\|K^\varphi\|_{\mathcal{H}}^2}. \tag{2.33}$$

If any one of the conditions (i)–(iii) holds, then

$$\mathbb{E}\left[\left|\int \varphi dW\right|^2\right] = \|K^\varphi\|_{\mathcal{H}}^2 = \varphi K_\lambda \varphi;$$

and if $(Z_n(\cdot))$ is a system of i.i.d $N(0, 1)$, we then have

$$\int \varphi dW = \sum_{n=1}^{\infty} \left(\int \varphi f_n d\lambda\right) Z_n. \tag{2.34}$$

Proof: See Theorems 2.2, 2.3, and Corollary 2.3 for details. □

Example 2.1 (i) Let λ be a positive measure on $(X, \mathcal{B})$, and consider the function ρ_λ on $\mathcal{B}_{\text{fin}}$ (see (2.4) and Section 2.2) defined by

$$\rho_\lambda(A, B) = \lambda(A \cap B), \quad \forall A, B \in \mathcal{B}_{\text{fin}}. \tag{2.35}$$

It is immediate that ρ is positive definite.

We denote by $\mathcal{H}^{(\text{Aronsz})}(\rho_\lambda)$ the associated RKHS in the sense of Aronszajn. The corresponding pair $(\mathcal{H}, \lambda)$ in the measurable category is $\mathcal{H} = L^2(X, \lambda)$, and the isometric isomorphism $\mathcal{H} \to \mathcal{H}^{(\text{Aronsz})}(\rho_\lambda)$, $f \to \mu_f$, with $\|f\|_{L^2(\lambda)} = \|\mu_f\|_{\mathcal{H}^{(\text{Aronsz})}(\rho_\lambda)}$ is simply $f \to f d\lambda$.

(ii) In the special case where $X = [0, 1] = $ the unit interval, and $\lambda = dx = $ the restriction of Lebesgue measure to $[0, 1]$, then the RKHS $\mathcal{H}^{(\text{Aronsz})}(\rho_\lambda)$ is the usual Cameron–Martin Hilbert space [20, 24, 25]

$$\{f \text{ measurable} \mid f' \in L^2(0, 1), f(0) = 0\},$$

and

$$\|f\|^2_{\mathscr{H}^{(\mathrm{Aronsz})}(\rho_\lambda)} = \int_0^1 |f'(x)|^2 \mathrm{d}x.$$

Note that this $\mathscr{H}^{(\mathrm{Aronsz})}(\rho_\lambda)$ is the RKHS of Brownian motion [3–5, 8].

Corollary 2.4 Let $(\Omega, \mathscr{F}, \mathbb{P})$ be a probability space, and let $(X, \mathscr{B})$ be a measure space. Let $\{W_A\}_{A \in \mathscr{B}_{\mathrm{fin}}}$ be a Gaussian process such that $W_A \in L^2(\Omega, \mathscr{F}, \mathbb{P})$ for $\forall A \in \mathscr{B}_{\mathrm{fin}}$, and $\mathbb{E}[W_A] = 0$. For measurable functions φ on $(X, \mathscr{B})$, let $\int_X \varphi \mathrm{d}W$ be the corresponding Ito-integral, defined for measurable functions φ such that $\mathbb{E}\left[\left|\int \varphi \mathrm{d}W\right|^2\right] < \infty$.

Let λ be a positive measure on $(X, \mathscr{B})$. Then there is a Hilbert space $\mathscr{H} \subset L^2_{\mathrm{loc}}(\lambda)$ such that $(\mathscr{H}, \lambda)$ is p.d. (see Definition 2.2), and

$$\mathbb{E}[W_A W_B] = \rho_{\mathscr{H}}(A, B) = \langle K^A, K^B \rangle_{\mathscr{H}}, \quad \forall A, B \in \mathscr{B}_{\mathrm{fin}}, \tag{2.36}$$

if and only if

$$\mathbb{E}\left[W_A \int \varphi \mathrm{d}W\right] = \int_A \varphi \mathrm{d}\lambda, \ \forall A \in \mathscr{B}_{\mathrm{fin}}, \quad \forall \varphi \in C_c(X). \tag{2.37}$$

Corollary 2.5 Let $(\mathscr{H}, \lambda)$, W, and $(\Omega, \mathscr{F}, \mathbb{P})$ be as above, that is, $\{W_A\}_{A \in \mathscr{B}_{\mathrm{fin}}}$ is a Gaussian process on $L^2(\Omega, \mathbb{P})$, indexed by $\mathscr{B}_{\mathrm{fin}}$, such that $\mathbb{E}\left[W_A \int \varphi \mathrm{d}W\right] = \int_A \varphi \mathrm{d}\lambda$ holds for all $\varphi \in C_c(X)$, where $\int \varphi \mathrm{d}W$ denotes the Ito-integral.

A function F in $L^2(\Omega, \mathbb{P})$ has the form $F = \int \varphi \mathrm{d}W$, for some $\varphi \in L^2_{\mathrm{loc}}(\lambda)$ if and only if there is a finite constant $C = C_F$ such that, for $\forall p$, $\forall \{\xi_i\}_1^p$, $\xi_i \in \mathbb{R}$, $\forall \{A_i\}_1^p$, $A_i \in \mathscr{B}_{\mathrm{fin}}$,

$$\left|\sum_{i=1}^p \xi_i \mathbb{E}[W_{A_i} F]\right|^2 \leq C \sum_i \sum_j \xi_i \xi_j \rho_{\mathscr{H}}(A_i, A_j). \tag{2.38}$$

2.4 Choice of Probability Space

In Section 2.3, we make reference to an as of yet unspecified probability space $(\Omega, \mathscr{F}, \mathbb{P})$, and a Gaussian process $\{W_B\}_{B \in \mathscr{B}_{\mathrm{fin}}}$, indexed by a fixed measure space $(X, \mathscr{B})$. In the present section, we shall keep the above assumptions placed on $(X, \mathscr{B})$. The notation $\mathscr{B}_{\mathrm{fin}}$ refers to a fixed positive measure λ on $(X, \mathscr{B})$, so $\mathscr{B}_{\mathrm{fin}} := \{B \in \mathscr{B} \mid \lambda(B) < \infty\}$.

We showed in Corollary 2.4, that when a Hilbert space $\mathscr{H}$ is given such that for all $B \in \mathscr{B}_{\mathrm{fin}}$,

$$\sup_{f \in \mathscr{H}, \|f\|_{\mathscr{H}} \leq 1} \left|\int_B f \mathrm{d}\lambda\right| < \infty, \tag{2.39}$$

that is, with a finite constant in (2.39) depending on $\mathscr{B}$, we then get a RKHS in the measurable category, and $\mathscr{H} \subset L^2_{\text{loc}}(\lambda)$.

As an application of Kolmogorov's extension principle, we showed that a Gaussian process $W = W^{\mathscr{H}}$, indexed by $\mathscr{B}_{\text{fin}}$, exists, subject to the conditions from (2.23) to (2.24). However, in general, there are many choices of probability spaces $(\Omega, \mathscr{F}, \mathbb{P})$, which realize this Gaussian process $W^{\mathscr{H}}$. To be more precise, the Gaussian process $W^{\mathscr{H}}$ satisfies the following conditions: Each random variable $W^{\mathscr{H}}_A$, $A \in \mathscr{B}_{\text{fin}}$ is Gaussian, with distribution $N(0, \|K^A\|^2_{\mathscr{H}})$ where $K^A \in \mathscr{H}$ is as specified in Lemma 2.1, so $W^{\mathscr{H}}_A$ is Gaussian with mean 0, and variance $\|K^A\|^2_{\mathscr{H}}$. But it also follows that, for $\forall k \in \mathbb{N}$, $\forall \{A_i\}^k_{i=1}$ from $\mathscr{B}_{\text{fin}}$, the system

$$(W^{\mathscr{H}}_{A_1}, W^{\mathscr{H}}_{A_2}, \cdots, W^{\mathscr{H}}_{A_k}) \tag{2.40}$$

has a *joint* Gaussian distribution on $\mathbb{R}^k$ with mean 0, and with $k \times k$ covariance matrix, given by the entries

$$\mathbb{E}[W^{\mathscr{H}}_{A_i} W^{\mathscr{H}}_{A_j}] = \langle K_{A_i}, K_{A_j} \rangle_{\mathscr{H}} = \rho_{\mathscr{H}}(A_i, A_j); \tag{2.41}$$

see Theorem 2.2 and Corollary 2.4.

Below, we identify two choices of probability space $(\Omega, \mathscr{F}, \mathbb{P})$, which serve to realize the process outlined in (2.40), and in Section 2.3.

Choice 1.

$$\Omega = \prod_{\mathbb{N}} \mathbb{R} = \text{the infinite Cartesian product;} \tag{2.42}$$

$$\mathscr{F} = \text{the cylinder } \sigma\text{-algebra of subsets of } \Omega; \tag{2.43}$$

$$\mathbb{P} = \mathsf{X}_{\mathbb{N}} N(0, 1), \tag{2.44}$$

that is, $\mathbb{P}$ is the infinite product measure on Ω constructed from the fixed standard Gaussian

$$g(x) = \frac{1}{\sqrt{2\pi}} e^{-\frac{1}{2}x^2} \text{ on } \mathbb{R}, \tag{2.45}$$

and each coordinate a copy of $\mathbb{R}$.

In this realization, we must use the representation $\{W^{\mathscr{H}}_B\}_{B \in \mathscr{B}_{\text{fin}}}$ from Theorem 2.2. Indeed, it follows from (2.26) to (2.27), and the i.i.d. condition for $Z_1, Z_2, \cdots$, that $\{W^{\mathscr{H}}_B\}_{B \in \mathscr{B}_{\text{fin}}}$ will then satisfy the conditions outlined in (2.40) and (2.41). Note that we may use the same $\mathbb{P}$ (2.44) for all the Gaussian processes $W^{\mathscr{H}}$, so for all possible Hilbert spaces $\mathscr{H}$ subject to (2.39).

Choice 2. This is the opposite extreme (of choice for the sample space Ω); hence, the probability measure $\mathbb{P}$ will depend on the choice of $\mathscr{H}$.

Details:

$$\Omega_{\mathfrak{M}} = \mathfrak{M}(X, \mathscr{B}) = \text{all signed measures on } (X, \mathscr{B}); \tag{2.46}$$

$$\mathscr{F}_{\mathfrak{M}} = \text{the cylinder } \sigma\text{-algebra of subsets of } \mathfrak{M}(X, \mathscr{B}); \tag{2.47}$$

$$\mathbb{P}^{(\mathscr{H})} = \text{a probability measure in } (\Omega_{\mathfrak{M}}, \mathscr{F}_{\mathfrak{M}}) \text{ (see (2.53) below).} \tag{2.48}$$

We now proceed to describe $\mathbb{P}^{(\mathscr{H})}$. The construction of $\mathbb{P}^{(\mathscr{H})}$ will have the property that the process $\{W_A^{\mathscr{H}}\}_{A \in \mathscr{B}_{\text{fin}}}$ takes the form

$$W_A^{\mathscr{H}}(\mu) = \mu(A), \tag{2.49}$$

for $\forall A \in \mathscr{B}_{\text{fin}}$, and $\forall \mu \in \mathfrak{M}(X, \mathscr{B})$.

Now fix $(\mathscr{H}, \lambda)$, subject to condition (2.39). Then for each finite system

$$\alpha := (A_1, \cdots, A_k), \qquad A_i \in \mathscr{B}_{\text{fin}}, \tag{2.50}$$

we consider $\pi_\alpha : \mathfrak{M}(X, \mathscr{B}) \longrightarrow \mathbb{R}^k$,

$$\mathfrak{M}(X, \mathscr{B}) \ni \mu \overset{\pi_\alpha}{\longmapsto} (\mu(A_1), \mu(A_1), \cdots, \mu(A_k)) \in \mathbb{R}^k. \tag{2.51}$$

Equip $\mathbb{R}^k$ with the joint Gaussian g_α with mean 0, and covariance matrix $[\rho_{\mathscr{H}}(A_i, A_j)]_{i,j=1}^k$, and we set $\mathbb{P}_\alpha$ the unique probability measure on $(\Omega_{\mathfrak{M}}, \mathscr{F}_{\mathfrak{M}})$ determined by

$$\mathbb{P}_\alpha \circ \pi_\alpha^{-1} = g_\alpha. \tag{2.52}$$

Using standard Kolmogorov extension theory, we note that the system of measures $\{P_\alpha\}$ is *consistent*, and that therefore, there is a unique probability measure $\mathbb{P}^{(\mathscr{H})}$ on $(\Omega_{\mathfrak{M}}, \mathscr{F}_{\mathfrak{M}})$ such that

$$\mathbb{P}^{(\mathscr{H})} \circ \pi_\alpha^{-1} = g_\alpha \tag{2.53}$$

for all α as in (2.50).

With this, it is then clear that (2.49) defines the desired Gaussian process realized in $(\Omega_{\mathfrak{M}}, \mathscr{F}_{\mathfrak{M}}, \mathbb{P}^{(\mathscr{H})})$ as per construction (see also (2.46)–(2.48)).

Equivalence of $\mathbb{P}^{(\mathscr{H}_1)}$ and $\mathbb{P}^{(\mathscr{H}_2)}$. Let $(X, \mathscr{B})$ be a fixed measure space specified as above; and let λ be a positive measure on $(X, \mathscr{B})$. Let $\mathscr{H}_i$, $i = 1, 2$, be two Hilbert spaces, $\mathscr{H}_i \subset L^2_{\text{loc}}(\lambda)$ such that both pairs $(\mathscr{H}_i, \lambda)$, $i = 1, 2$, are p.d.; in particular, $\mathscr{H}_i \in \mathscr{K}(\lambda)$, $i = 1, 2$.

Let $\mathbb{P}^{(\mathscr{H}_i)}$, $i = 1, 2$, be the corresponding probability measures on $(\Omega, \mathscr{F}) = (\Omega_{\mathfrak{M}}, \mathscr{F}_{\mathfrak{M}})$, as specified in (2.48). Recall, setting

$$W_B(\mu) := \mu(B), \qquad B \in \mathscr{B}_{\text{fin}}, \quad \mu \in \mathfrak{M}(X, \mathscr{B}), \tag{2.54}$$

we get that $\{W_B\}_{B \in \mathscr{B}_{\text{fin}}}$ is a Gaussian process; but it has different distributions relative to the two probability spaces $(\Omega_{\mathfrak{M}}, \mathscr{F}_{\mathfrak{M}}, \mathbb{P}^{(\mathscr{H}_i)})$, $i = 1, 2$; in particular,

$$\mathbb{E}_{\mathbb{P}^{(\mathscr{H}_i)}}[W_A W_B] = \rho_{\mathscr{H}_i}(A, B), \qquad i = 1, 2; \tag{2.55}$$

see (2.36) in Corollary 2.4.

The following result follows from our present Section 2.3, and a result of Jørsboe [13].

Corollary 2.6 Let $X, \mathscr{B}, \lambda, \mathscr{H}_i, \mathbb{P}^{(\mathscr{H}_i)}$, $i = 1, 2$, be as specified above. Then the two probability measures $\mathbb{P}^{(\mathscr{H}_i)}$ are equivalent if and only if there is an ONB

$\{f_n\}_{n\in\mathbb{N}}$ for $\mathcal{H}_2$; see Corollary 2.2, and scalars $c_n > 0$ with $\sum_{n=1}^{\infty}(1-c_n)^2 < \infty$, and

$$K^{(\mathcal{H}_1)}(x,A) = \sum_{n=1}^{\infty} c_n \left(\int_A f_n \mathrm{d}\lambda \right) f_n(x), \qquad (2.56)$$

for a.a. x (w.r.t λ), $A \in \mathcal{B}_{\mathrm{fin}}$.

2.5 A Duality

Fix a measurable space $(X, \mathcal{B})$ as in Definition 2.1, where X is locally compact and Hausdorff, and $\mathcal{B}$ is a Borel σ-algebra of subsets of X. We shall consider the following duality:

Definition 2.5 Fix a positive measure λ on $(X, \mathcal{B})$, and set

$$\mathcal{K}(\lambda) := \{\text{Hilbert space } \mathcal{H} \text{ on } (X, \mathcal{B}) \mid \text{For } \forall \varphi \in C_c(X)$$
$$\sup \left\{ \left| \int \varphi f \mathrm{d}\lambda \right|, \ \|f\|_{\mathcal{H}} \le 1 \right\} < \infty \right\}. \qquad (2.57)$$

If $\mathcal{H}$ is a Hilbert space of measurable functions on $(X, \mathcal{B})$, set

$$\mathcal{L}(\mathcal{H}) := \left\{ \text{signed measures } \lambda \mid \sup \left\{ \left| \int \varphi f \mathrm{d}\lambda \right|, \ \|f\|_{\mathcal{H}} \le 1 \right\} < \infty \right\}, \qquad (2.58)$$

and

$$\mathcal{L}_+(\mathcal{H}) = \{\lambda \in \mathcal{L}(\mathcal{H}) \mid \lambda \text{ is positive}\}. \qquad (2.59)$$

Lemma 2.6 The following are equivalent:

$$\sup \left\{ \left| \int_A f \mathrm{d}\lambda \right|, \ f \in \mathcal{H}, \ \|f\|_{\mathcal{H}} \le 1 \right\} < \infty, \qquad \forall A \in \mathcal{B}_{\mathrm{fin}}, \qquad (2.60)$$

$$\Updownarrow$$

$$\sup \left\{ \left| \int \varphi f \mathrm{d}\lambda \right|, \ f \in \mathcal{H}, \ \|f\|_{\mathcal{H}} \le 1 \right\} < \infty, \qquad \forall \varphi \in C_c(X). \qquad (2.61)$$

Proof: By Riesz' theorem, $(2.60) \iff \forall A \in \mathcal{B}_{\mathrm{fin}} \ \exists! \ K^A \in \mathcal{H}$ s.t.

$$\int_A f \mathrm{d}\lambda = \langle f, K^A \rangle_{\mathcal{H}}; \qquad (2.62)$$

and $(2.61) \iff \forall \varphi \in C_c(X) \ \exists! K^\varphi$ s.t.

$$\int \varphi f \mathrm{d}\lambda = \langle f, K^\varphi \rangle_{\mathcal{H}}. \qquad (2.63)$$

To show that (2.60) $\Longrightarrow$ (2.61): The RHS of (2.62) is σ-additive, so it is a signed measure on $(X, \mathscr{B})$, $\mu_f(A) := \langle f, K^A \rangle_{\mathscr{H}}$. We see that $\mu_f \ll \lambda, f \in L^1_{\text{loc}}(\lambda)$, $\mathrm{d}\mu_f = f\mathrm{d}\lambda$, and from (2.62),

$$
\begin{aligned}
\int f\varphi \mathrm{d}\lambda &= \int \varphi \mathrm{d}\mu_f \\
&= \int \varphi(y) \langle f(\cdot), K(\cdot, \mathrm{d}y) \rangle_{\mathscr{H}} \\
&= \left\langle f(\cdot), \int \varphi(y) K(\cdot, \mathrm{d}y) \right\rangle_{\mathscr{H}}, \quad \forall \varphi \in C_c(X);
\end{aligned}
$$

so that $K^\varphi(\cdot) = \int \varphi(y) K(\cdot, \mathrm{d}y) \in \mathscr{H}$, we see that (2.63) holds. It is clear that K^φ in (2.63) is unique.

The converse direction (2.61) $\Longrightarrow$ (2.60) is clear. □

Definition 2.6 (Multipliers) Let $(X, \mathscr{B})$ be as above. Fix $(\mathscr{H}, \lambda)$ such that $\lambda \in \mathscr{L}(\mathscr{H})$. We then have $\mathscr{H} \subset L^2_{\text{loc}}(\lambda)$. A measurable function g on $(X, \mathscr{B})$ is said to be a *multiplier* if the operator $M_g f = gf, f \in \mathscr{H}$, is bounded $\mathscr{H} \to \mathscr{H}$. The corresponding adjoint operator will be denoted M_g^*. The set of all multipliers will be denoted $Multp(\mathscr{H})$.

Theorem 2.5 Let $(X, \mathscr{B}, \mathscr{H}, \lambda)$ be as above, and let g be a measurable function on $(X, \mathscr{B})$, then TFAE:

(i) $g \in Multp(\mathscr{H})$, and
(ii) $g \, \mathrm{d}\lambda \in \mathscr{L}(\mathscr{H})$ (see Definition 2.5).

Proof: If $g \in Multp(\mathscr{H})$, then for all $A \in \mathscr{B}_{\text{fin}}$, and all $f \in \mathscr{H}$, we have

$$
\begin{aligned}
\int_A fg \mathrm{d}\lambda &= \langle gf, K(\cdot, A) \rangle_{\mathscr{H}} \\
&= \langle f, M_g^*(K(\cdot, A)) \rangle_{\mathscr{H}}.
\end{aligned}
\tag{2.64}
$$

But $M_g^*(K(\cdot, A)) \in \mathscr{H}$ by assumption, and so the condition for $g\mathrm{d}\lambda \in \mathscr{L}(\mathscr{H})$ is satisfied.

Using (2.66), the reader will note that the argument works in reverse, so we also have (ii) $\Longrightarrow$ (i). □

Corollary 2.7 Let $(X, \mathscr{B}, \mathscr{H}, \lambda)$ be as above, and suppose $g \in Multp(\mathscr{H})$, then

$$
K_{g\mathrm{d}\lambda}(\cdot, A) = M_g^*(K_\lambda(\cdot, A)), \quad \forall A \in \mathscr{B}_{\text{fin}}.
$$

Given $(X, \mathscr{B}, \lambda)$, let $\mathscr{H} \in \mathscr{K}(\lambda)$. Then we may define a random signed measure $K(\cdot, B)$, for all $B \in \mathscr{B}_{\text{fin}}$ (see Lemma 2.1).

There are two associated RKHSs:

- $\mathscr{H}$: Hilbert space of measurable functions on $(X, \mathscr{B})$;
- $\mathscr{H}^{(\text{Aronsz})}$: The RKHS determined by the positive definite kernel $\rho_{\mathscr{H}}$ on $\mathscr{B}_{\text{fin}}$ (2.8), where

$$\rho_{\mathscr{H}}(A, B) = \langle K^A, K^B \rangle_{\mathscr{H}}, \quad \forall A, B \in \mathscr{B}_{\text{fin}}.$$

Note that $\mathscr{H}^{(\text{Aronsz})}$ consists of functions on $\mathscr{B}_{\text{fin}}$, and it depends on both $\mathscr{H}$ and λ.

There are many solutions to $\mathscr{H} \in \mathscr{K}(\lambda)$, that is, to the Hilbert factorization problem (2.57), but they are all isomorphic to an associated Aronszajn RKHS. Thus, the two RKHSs are *isometrically isomorphic* via

$$\mathscr{H} \xrightarrow{\simeq} \mathscr{H}^{(\text{Aronsz})}, \quad f \longmapsto \mu_f. \tag{2.65}$$

See Figure 2.2.

The converse problem:

Given $(X, \mathscr{B})$ and a p.d. function $\rho : \mathscr{B} \times \mathscr{B} \to \mathbb{R}(\text{or } \mathbb{C})$, that is,

$$\sum_i \sum_j \xi_i \xi_j \rho(A_i, A_j) \geq 0 \tag{2.66}$$

holds for $\forall p \in \mathbb{N}, \forall \xi_i \in \mathbb{R}, \forall A_i \in \mathscr{B}, 1 \leq i \leq p$, then we get an Aronszajn RKHS $\mathscr{H}(\rho)$, consisting of functions on $\mathscr{B}$.

Not every p.d. function on $\mathscr{B}$ corresponds to a pair $(\mathscr{H}, \lambda)$. Given $\rho : \mathscr{B} \times \mathscr{B} \longrightarrow \mathbb{R}$ p.d., we place two additional conditions on it:

(i) $\forall A \in \mathscr{B}$, $\rho(\cdot, A)$ is a signed measure on $(X, \mathscr{B})$.
(ii) There exists a positive measure λ on $(X, \mathscr{B})$ s.t. $\rho(\cdot, A) \ll \lambda$ for all $A \in \mathscr{B}$, and denote the Radon–Nikodym derivative

$$K(\cdot, A) = \frac{d\rho(\cdot, A)}{d\lambda} \tag{2.67}$$

or equivalently,

$$d\rho(x, A) = K(x, A)d\lambda(x). \tag{2.68}$$

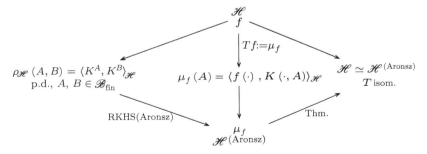

Figure 2.2 The RKHSs $\mathscr{H} \simeq \mathscr{H}^{(\text{Aronsz})}$, $\mathscr{H} \ni f \longmapsto \mu_f \in \mathscr{H}^{(\text{Aronsz})}$.

Then for $A, B \in \mathscr{B}_{\text{fin}}$ we have

$$\rho_{\mathscr{H}}(A, B) = \int_A K(y, B) \mathrm{d}\lambda(y) = \int_B K(x, A) \mathrm{d}\lambda(x).$$

This is a p.d. function on $\mathscr{B}_{\text{fin}} \times \mathscr{B}_{\text{fin}}$.

We get $\mathscr{H}$ from (K, λ) as defined in (2.57); and the isomorphism $\mathscr{H} \longrightarrow \mathscr{H}^{(\text{Aronsz})}(\rho_{\mathscr{H}}), f \longmapsto \mu_f$.

Theorem 2.6 $\lambda \in \mathscr{L}(\mathscr{K}(\lambda))$, and $\mathscr{H} \in \mathscr{K}(\mathscr{L}(\mathscr{H}))$.

Proof: See See Section 2.2, especially Lemmas 2.1, 2.2, and Theorem 2.1. □

2.A Stochastic Processes

… from its shady beginnings devising gambling strategies and counting corpses in medieval London, probability theory and statistical inference now emerge as better foundations for scientific models, especially those of the process of thinking and as essential ingredients of theoretical mathematics, …

— David Mumford. From: "The Dawning of the Age of Stochasticity." [26]

Early Roots

The interest in positive definite (p.d.) functions has at least three roots:

(1) Fourier analysis, and harmonic analysis more generally, including the noncommutative variant where we study unitary representations of groups [27–43].
(2) Optimization and approximation problems, involving for example, spline approximations as envisioned by Schöenberg [18, 44–49].
(3) Stochastic processes [24, 50–55].

Below, we sketch a few details regarding (3). A *stochastic process* is an indexed family of random variables based on a fixed probability space. In our present analysis, the processes will be indexed by some group G or by a subset of G. For example, $G = \mathbb{R}$, or $G = \mathbb{Z}$, corresponds to processes indexed by real time, or discrete time, respectively. A main tool in the analysis of stochastic processes is an associated *covariance function* (2.A.14).

A process $\{X_g \mid g \in G\}$ is called *Gaussian* if each random variable X_g is Gaussian, that is, its distribution is Gaussian. For Gaussian processes, we only need two moments. Therefore, if we normalize, setting the mean equal to 0, then the process is determined by its covariance function. In general the covariance function is a function on $G \times G$, or on a subset, but if the process is *stationary*, the covariance function will in fact be a p.d. function defined on G, or a subset of G.

We will be using three stochastic processes, Brownian motion, Brownian Bridge, and the Ornstein–Uhlenbeck process, all Gaussian, and Ito integrals.

We outline a brief sketch of these facts below.

A *probability space* is a triple $(\Omega, \mathscr{F}, \mathbb{P})$ where Ω is a set (sample space), $\mathscr{F}$ is a (fixed) σ-algebra of subsets of Ω, and $\mathbb{P}$ is a (σ-additive) probability measure defined on $\mathscr{F}$. (Elements E in $\mathscr{F}$ are "events," and $\mathbb{P}(E)$ represents the probability of the event E.)

A real valued *random variable* is a function $X : \Omega \to \mathbb{R}$ such that, for every Borel subset $A \subset \mathbb{R}$, we have that $X^{-1}(A) = \{\omega \in \Omega \mid X(\omega) \in A\}$ is in $\mathscr{F}$. Then

$$\mu_X(A) = \mathbb{P}(X^{-1}(A)), \quad A \in \mathscr{B} \tag{2.A.1}$$

defines a positive measure on $\mathbb{R}$; here $\mathscr{B}$ denotes the Borel σ-algebra of subsets of $\mathbb{R}$. This measure is called the *distribution* of X.

The following notation for the $\mathbb{P}$ integral of random variables $X(\cdot)$ will be used:

$$\mathbb{E}[X] := \int_\Omega X(\omega)\mathrm{d}\mathbb{P}(\omega),$$

denoted *expectation*. If μ_X is the distribution of X, and $\psi : \mathbb{R} \to \mathbb{R}$ is a Borel function, then

$$\int_\mathbb{R} \psi \, \mathrm{d}\mu_X = \mathbb{E}[\psi \circ X].$$

An example of a probability space is as follows:

$$\Omega = \prod_\mathbb{N} \{\pm 1\} = \text{infinite Cartesian product}$$

$$= \{\{\omega_i\}_{i \in \mathbb{N}} \mid \omega_i \in \{\pm 1\}, \, \forall i \in \mathbb{N}\}, \text{ and} \tag{2.A.2}$$

$\mathscr{F}$: subsets of Ω specified by a finite number of outcomes (called "cylinder sets".)

$\mathbb{P}$: the infinite-product measure corresponding to a fair coin $\left(\frac{1}{2}, \frac{1}{2}\right)$ measure for each outcome ω_i.

The transform

$$\widehat{\mu_X}(\lambda) = \int_\mathbb{R} e^{i\lambda x} \mathrm{d}\mu_X(x) \tag{2.A.3}$$

is called the *Fourier transform*, or the *generating function*.

Let a be fixed, $0 < a < 1$. A *random a-power series* is the function

$$X_a(\omega) = \sum_{i=1}^\infty \omega_i \, a^i, \quad \omega = (\omega_i) \in \Omega. \tag{2.A.4}$$

One checks that the generating function for X_a is as follows:

$$\widehat{d\mu_{X_a}}(\lambda) = \prod_{k=1}^{\infty} \cos(a^k \lambda), \qquad \lambda \in \mathbb{R} \tag{2.A.5}$$

where the R.H.S. in (2.A.5) is an infinite product. Note that it is easy to check independently that the R.H.S. in (2.A.5), $F_a(\lambda) = \prod_{k=1}^{\infty} \cos(a^k \lambda)$ is *p.d.* and continuous on $\mathbb{R}$, and so it determines a measure.

An indexed family of random variables is called a *stochastic process*.

Example 2.A.1 Brownian motion

Ω: all continuous real valued function on $\mathbb{R}$;
$\mathscr{F}$: subsets of Ω specified by a finite number of sample-points;
$\mathbb{P}$: Wiener-measure on $(\Omega, \mathscr{F})$.

For $\omega \in \Omega$, $t \in \mathbb{R}$, set $X_t(\omega) = \omega(t)$; then it is well known that $\{X_t\}_{t \in \mathbb{R}_+}$ is a Gaussian-random variable with the property that:

$$d\mu_{X_t}(x) = \frac{1}{\sqrt{2\pi t}} e^{-x^2/2t} dx, \qquad x \in \mathbb{R}, \ t > 0,$$

$$X_0 = 0, \text{ and} \tag{2.A.6}$$

whenever $0 \le t_1 < t_2 < \cdots < t_n$, then the random variables

$$X_{t_1}, X_{t_2} - X_{t_1}, \cdots, X_{t_n} - X_{t_{n-1}} \tag{2.A.7}$$

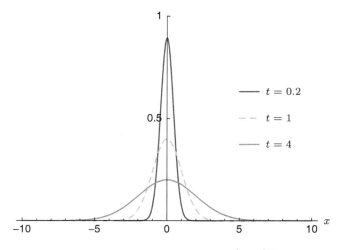

Figure 2.A.1 Gaussian distribution $d\mu_{X_t}(x) = \frac{1}{\sqrt{2\pi t}} e^{-x^2/2t} dx$, $t > 0$ (variance).

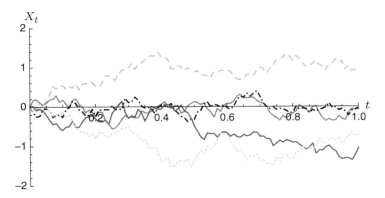

Figure 2.A.2 Monte-Carlo simulation of Brownian motion starting at $x = 0$, with five sample paths. ("Monte-Carlo simulation" refers to the use of computer-generated random numbers.)

are independent. (The R.H.S. in (2.A.6) is Gaussian distribution with mean 0 and variance $t > 0$, see Figure 2.A.1.)

In more detail, X_t satisfies:

(1) $\mathbb{E}[X_t] = 0$, for all t; mean zero;
(2) $\mathbb{E}[X_t^2] = t$, variance $= t$;
(3) $\mathbb{E}[X_s X_t] = s \wedge t$, the covariance function; and
(4) $\mathbb{E}[(X_{b_1} - X_{a_1})(X_{b_2} - X_{a_2})] = |[a_1, b_1] \cap [a_2, b_2]|$, for any pair of intervals.

This stochastic process is called *Brownian motion* (see Figure 2.A.2).

Lemma 2.A.1 (**The Ito integral**) Let $\{X_t\}_{t \in \mathbb{R}_+}$ be Brownian motion, and let $f \in L^2(\mathbb{R}_+)$. For partitions of $\mathbb{R}_+$, $\pi : \{t_i\}$, $t_i \le t_{i+1}$, consider the sums

$$S(\pi) := \sum_i f(t_i)(X_{t_i} - X_{t_{i-1}}) \in L^2(\Omega, \mathbb{P}). \tag{2.A.8}$$

Then the limit (in $L^2(\Omega, \mathbb{P})$) of the terms (2.A.8) exists, taking limit on the net of all partitions s.t. $\max_i (t_{i+1} - t_i) \to 0$. The limit is denoted

$$\int_0^\infty f(t) dX_t \in L^2(\Omega, \mathbb{P}), \tag{2.A.9}$$

and it is called the Ito-integral. The following isometric property holds:

$$\mathbb{E}\left[\left|\int_0^\infty f(t) dX_t\right|^2\right] = \int_0^\infty |f(t)|^2 dt. \tag{2.A.10}$$

Equation (2.A.10) is called the Ito-isometry.

Proof: We refer to [24] for an elegant presentation, but the key step in the proof involves the above-mentioned properties of Brownian motion. The first step is the verification of

$$\mathbb{E}[|S(\pi)|^2] = \sum_i |f(t_i)|^2 (t_i - t_{i-1}),$$

which is based on (2.A.7). $\qquad\square$

An application of Lemma 2.A.1: A p.d. function on an infinite-dimensional vector space.

Let S denote the real valued Schwartz functions. For $\varphi \in S$, set $X(\varphi) = \int_0^\infty \varphi(t)\mathrm{d}X_t$, the Ito integral from (2.A.9). Then we get the following:

$$\mathbb{E}[e^{iX(\varphi)}] = e^{-\frac{1}{2}\|\varphi\|_{L^2}^2}, \tag{2.A.11}$$

where $\mathbb{E}$ is the expectation w.r.t. Wiener-measure.

It is immediate that

$$F(\varphi) := e^{-\frac{1}{2}\|\varphi\|_{L^2}^2}, \tag{2.A.12}$$

that is, the R.H.S. in (2.A.11), is a p.d. function on S. To get from this, an associated probability measure (the Wiener measure $\mathbb{P}$) is nontrivial [24, 56, 57]: The dual of S, the tempered distributions S', turns into a measure space, $(S', \mathscr{F}, \mathbb{P})$ with the σ-algebra $\mathscr{F}$ generated by the cylinder sets in S'. With this we get an equivalent realization of Wiener measure (see the cited papers); now with the l.h.s. in (2.A.11) as $\mathbb{E}(\cdots) = \int_{S'} \cdots \mathrm{d}\mathbb{P}(\cdot)$. But the p.d. function F in (2.A.12) *cannot* be realized by a σ-additive measure on L^2, one must pass to a "bigger" infinite-dimensional vector space, hence S'. The system

$$S \hookrightarrow L_{\mathbb{R}}^2(\mathbb{R}) \hookrightarrow S' \tag{2.A.13}$$

is called a *Gelfand-triple*. The second right hand side inclusion $L^2 \hookrightarrow S'$ in (2.A.13) is obtained by dualizing $S \hookrightarrow L^2$, where S is given its Frchet topology [58].

Let G be a locally compact group, and let $(\Omega, \mathscr{F}, \mathbb{P})$ be a probability space, $\mathscr{F}$ a σ-algebra, and $\mathbb{P}$ a probability measure defined on $\mathscr{F}$. A stochastic L^2-process is a system of random variables $\{X_g\}_{g \in G}$, $X_g \in L^2(\Omega, \mathscr{F}, \mathbb{P})$. The *covariance function* $c_X : G \times G \to \mathbb{C}$ of the process is given by

$$c_X(g_1, g_2) = \mathbb{E}[\overline{X}_{g_1} X_{g_2}], \quad \forall(g_1, g_2) \in G \times G. \tag{2.A.14}$$

To simplify, we will assume that the mean $\mathbb{E}[X_g] = \int_\Omega X_g \mathrm{d}\mathbb{P}(\omega) = 0$ for all $g \in G$.

We say that (X_g) is stationary iff

$$c_X(hg_1, hg_2) = c_X(g_1, g_2), \quad \forall h \in G. \tag{2.A.15}$$

In this case, c_X is a function of $g_1^{-1} g_2$, that is,

$$\mathbb{E}[X_{g_1}, X_{g_2}] = c_X(g_1^{-1} g_2), \quad \forall g_1, g_2 \in G; \tag{2.A.16}$$

(setting $h = g_1^{-1}$ in (2.A.15).)

The covariance function of Brownian motion $\mathbb{E}[X_s X_t]$ is computed in Example 2.B.1 below.

2.B Overview of Applications of RKHSs

In a general setup, RKHSs were pioneered by Aronszajn in the 1950s [1, 2]; and subsequently, they have been used in a host of applications.

The key idea of Aronszajn is that a RKHS is a Hilbert space $\mathcal{H}_K$ of functions f on a set such that the values $f(x)$ are "reproduced" from f and a vector K_x in $\mathcal{H}_K$, in such a way that the inner product $\langle K_x, K_y \rangle =: K(x, y)$ is a p.d. kernel.

Since this setting is too general for many applications, it is useful to restrict the very general framework for RKHSs to concrete cases in the study of particular spectral theoretic problems; p.d. functions on groups is a case in point. Such specific issues arise in physics [59, 60] where one is faced with extending p.d. functions F, which are only defined on a subset of a given group.

Connections to Gaussian Processes

By a theorem of Kolmogorov, every Hilbert space may be realized as a (Gaussian) RKHS, see for example, [53, 61, 62], and Theorem 2.B.1.

Definition 2.B.1   A function c defined on a subset of a group G is said to be *positive definite* iff

$$\sum_{i=1}^{n} \sum_{j=1}^{n} \overline{\lambda_i} \lambda_j c(g_i^{-1} g_j) \geq 0 \tag{2.B.1}$$

for all $n \in \mathbb{N}$, and all $\{\lambda_i\}_{i=1}^{n} \subset \mathbb{C}$, $\{g_i\}_{i=1}^{n} \subset G$ with $g_i^{-1} g_j$ in the domain of c.

From (2.B.1), it follows that $F(g^{-1}) = \overline{F(g)}$, and $|F(g)| \leq F(e)$, for all g in the domain of F, where e is the neutral element in G.

We recall the following theorem of Kolmogorov. One direction is easy, and the other is the deep part:

Theorem 2.B.1   Kolmogorov   A function $c : G \to \mathbb{C}$ is positive definite if and only if there is a stationary Gaussian process $(\Omega, \mathcal{F}, \mathbb{P}, X)$ with mean zero, such that $c = c_X$, that is, $c(g_1, g_2) = \mathbb{E}[\overline{X}_{g_1} X_{g_2}]$ (2.A.14).

Proof: We refer to [53] for the nontrivial direction. To stress the idea, we include a proof of the easy part of the theorem: Assume $c = c_X$. Let $\{\lambda_i\}_{i=1}^{n} \subset \mathbb{C}$

and $\{g_i\}_{i=1}^n \subset G$, then we have

$$\sum_i \sum_j \overline{\lambda_i} \lambda_j c(g_i^{-1} g_j) = \mathbb{E}\left[\left|\sum \lambda_i X_{g_i}\right|^2\right] \geq 0,$$

that is, c is p.d. $\qquad\qquad\square$

Example 2.B.1   Let $\Omega = [0, 1]$, the closed unit interval, and let $\mathscr{H} :=$ the space of continuous functions ξ on Ω such that $\xi(0) = 0$, and $\xi' \in L^2(0, 1)$, where $\xi' = \frac{d}{dx}\xi$ is the weak derivative of ξ, that is, the derivative in the Schwartz-distribution sense. For $x, y \in \Omega$, set

$$K(x, y) = x \wedge y = \min(x, y); \text{ and}$$
$$K_x(y) = K(x, y). \tag{2.B.2}$$

Then in the sense of distribution, we have

$$(K_x)' = \chi_{[0,x]}; \tag{2.B.3}$$

that is, the indicator function of the interval $[0, x]$ (see Figure 2.B.1).
    For $\xi_1, \xi_2 \in \mathscr{H}$, set

$$\langle \xi_1, \xi_2 \rangle_{\mathscr{H}} := \int_0^1 \overline{\xi_1'(x)} \xi_2'(x) dx.$$

Since $L^2(0, 1) \subset L^1(0, 1)$, and $\xi(0) = 0$ for $\xi \in \mathscr{H}$, we see that

$$\xi(x) = \int_0^x \xi'(y) dy, \quad \xi' \in L^2(0, 1), \tag{2.B.4}$$

and $\mathscr{H}$ consists of continuous functions on Ω.

Claim 2.B.1   The Hilbert space $\mathscr{H}$ is a RKHS with $\{K_x\}_{x \in \Omega}$ as its kernel (G3).

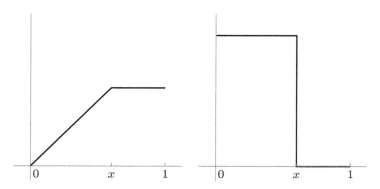

Figure 2.B.1 K_x and its distributional derivative.

Proof: Let $\xi \in \mathscr{H}$, then by (2.B.4), we have:

$$
\xi(x) = \int_0^1 \chi_{[0,x]}(y)\xi'(y)dy = \int_0^1 K_x'(y)\xi'(y)dy
$$
$$
= \langle K_x, \xi \rangle_{\mathscr{H}}, \quad \forall x \in \Omega. \qquad \square
$$

Remark 2.B.1 In the case of Example 2.B.1, the Gaussian process resulting from the p.d. kernel (2.B.2) is standard *Brownian motion*.

Reproducing Kernel Hilbert Spaces Induced by Graphs

By an electrical network (Figure 2.B.2), we mean a graph G of vertices and edges satisfying suitable conditions that allow for computation of voltage distribution from a network of prescribed resistors assigned to the edges in G. The mathematical axioms are prescribed in a way that facilitates the use of the laws of Kirchhoff and Ohm in computing voltage distributions and resistance distances in G. It will be more convenient to work with prescribed conductance functions c on G. Indeed with a choice of conductance function c specified we define two crucial tools for our analysis, a graph Laplacian $\Delta(= \Delta_c,)$ a discrete version of more classical notions of Laplacians, and an energy Hilbert space $\mathscr{H}_E$.

Large networks arise in both pure and applied mathematics, and more recently, they have become a current and fast developing research area [63–68]. Applications include a host of problems coming from graph theory, internet search engine algorithm, and social networks [69–72]. Hence, of the recent applications, there is a change in outlook from finite to infinite.

In traditional graph theoretical problems, the whole graph is given exactly, and we are then looking for relationships between its parameters, variables and functions; or for efficient computation algorithms. By contrast, for very large networks (like the Internet), variables are typically not given completely, and in most cases they may not even be well-defined. In such applications, related

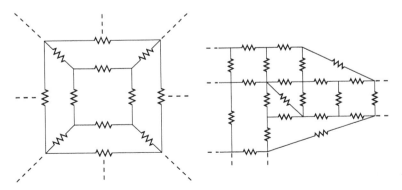

Figure 2.B.2 Examples of configuration of resistors in a network.

data can only be extracted by indirect means; hence, random variables and local sampling must be used as opposed to global processes.

The motivation derives from learning, where "learning" is understood broadly to include *machine learning* of suitable probability distribution or samples of training data. Machine learning is a field that has evolved from the study of pattern recognition and computational learning theory in artificial intelligence [17, 19, 48, 73, 74]. It explores the construction and study of algorithms that can learn from and make predictions on limited data. Such algorithms operate by building models from "training data" inputs in order to make data-driven predictions or decisions [75–77].

Recent development includes an analysis of a class of large (infinite) weighted graphs, via kernel theory, and an harmonic analysis of associated Perron-Frobenius transfer operators [20, 78]. One reason for this approach is that statistical features in such an analysis are best predicted by consideration of probability spaces corresponding to measures on infinite sample spaces. Moreover, the latter are best designed from consideration of infinite weighted graphs, as opposed to their finite counterparts. An example of statistical features that are relevant even for finite samples is long-range order, that is, the study of correlations between distant sites (vertices), and related phase-transitions, for example, sign-flips at distant sites. In designing efficient learning models, it is important to understand the possible occurrence of unexpected long-range correlations; for example, correlations between distant sites in a finite sample. Other applications of weighted graphs include statistical mechanics, such as infinite spin models, and large digital networks. It is natural to ask then how one best approaches analysis on "large" systems.

A second reason for the use of infinite sample-spaces, and associated stochastic models, is their applications in designing efficient sampling procedures. The interesting solutions will often occur first as vectors in an infinite-dimensional RKHS. Indeed, such RKHSs serve as powerful tools in the solution of a kernel-optimization problems with penalty terms. Once an optimal solution is obtained in infinite dimensions, one may then proceed to study its restrictions to suitably chosen finite subgraphs.

Graph Laplacians on Infinite Networks

For an infinite network $G = (V, E)$, where V is the vertex-set, E denotes edges, an assignment of weights is a positive symmetric function c on E (see Figures 2.B.3–2.B.5). In electrical network models, the function $c : E \longrightarrow \mathbb{R}_+$ represents conductance, and its reciprocal resistance. Hence fixing a conductance function is equivalent to an assignment of resistors on the edges E. Functions on V typically represent voltage distributions, and the harmonic functions are of special importance. There is a naturally defined reversible Markov process, and a corresponding graph Laplacian Δ_c [79–82] acting on an energy Hilbert space $\mathscr{H}_c$, computed from the conductance.

Figure 2.B.3 Covariance between vertices.

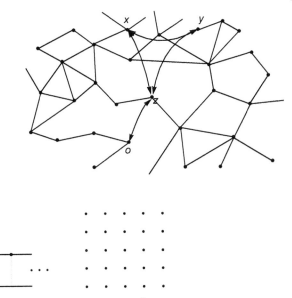

$V = \text{Band}$ $V = \mathbb{Z}^2$

Figure 2.B.4 Nonlinear system of vertices.

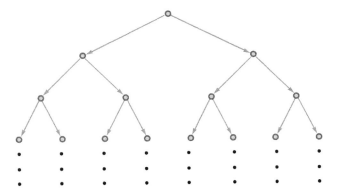

Figure 2.B.5 A binary tree model.

A framework for analyzing the spectral theoretic properties of graph Laplacians on infinite networks is established in [83, 84]. Especially, this is applied to networks with *a random time-varying* conductance function, where $c = c(t)$ is a second order stochastic process.

Key tools in this analysis are Hilbert space models adapted to the problems at hand. Since the emphasis is statistical models and their harmonic analysis, the setting will be infinite-dimensional Hilbert space, and careful choices of orthonormal bases (ONBs) [83–84]. The appropriate linear operators will be given by unbounded quadratic forms, but nonetheless have $\infty \times \infty$ matrix representations. In this analysis, choices of selfadjoint extensions play an important role. Reason: Only the selfadjoint extensions have spectral resolutions.

Discrete RKHSs

RKHSs have been studied extensively since the pioneering papers by Aronszajn. They further play an important role in the theory of partial differential operators (PDO); for example as Green's functions of second order elliptic PDOs. However, the literature so far has focused on the theory of kernel functions defined on continuous domains, either domains in Euclidean space, or complex domains in one or more variables. For these cases, the Dirac distributions δ_x do not have finite $\mathcal{H}$-norm. For RKHSs over discrete point distributions, it is reasonable to expect that δ_x will in fact have finite $\mathcal{H}$-norm.

An illustration from *neural networks*: An extreme learning machine (ELM) is a neural network configuration in which a hidden layer of weights are randomly sampled, and the object is then to determine analytically resulting output layer weights. Hence ELM may be thought of as an approximation to a network with infinite number of hidden units.

In general when reproducing kernels and their Hilbert spaces are used, one ends up with functions on a suitable set, and so far we believe that the dichotomy, discrete versus continuous, has not yet received sufficient attention. After all, a choice of sampling points in relevant optimization models based on kernel theory suggests the need for a better understanding of point masses as they are accounted for in the RKHS at hand.

We are concerned with a characterization of those RKHSs $\mathcal{H}$, which contain the Dirac masses δ_x for all points $x \in V$. Of the examples and applications where this question plays an important role, we emphasize three:

(1) Discrete Brownian motion-Hilbert spaces, that is, discrete versions of the Cameron–Martin Hilbert space;
(2) Energy-Hilbert spaces corresponding to graph-Laplacians; and finally
(3) RKHSs generated by binomial coefficients.

Acknowledgments

The coauthors thank the following colleagues for helpful and enlightening discussions: Professors Daniel Alpay, Sergii Bezuglyi, Ilwoo Cho, Paul Muhly, Myung-Sin Song, Wayne Polyzou, and members in the Math Physics seminar at The University of Iowa.

References

1 Aronszajn, N. (1943) La théorie des noyaux reproduisants et ses applications. I. *Proc. Cambridge Philos. Soc.*, **39**, 133–153. MR 0008639.

2 Aronszajn, N. (1950) Theory of reproducing kernels. *Trans. Am. Math. Soc.*, **68**, 337–404. MR 0051437.

3 Alpay, D., Bolotnikov, V., Dijksma, A., and de Snoo, H. (1993) On some operator colligations and associated reproducing kernel Hilbert spaces, in *Operator Extensions, Interpolation of Functions and Related Topics*, Operator Theory: Advances and Applications, vol. 61, Birkhäuser, Basel, pp. 1–27. MR 1246577 (94i:47018).

4 Alpay, D. and Dym, H. (1992) On reproducing kernel spaces, the Schur algorithm, and interpolation in a general class of domains, in *Operator Theory and Complex Analysis (Sapporo, 1991)*, Operator Theory: Advances and Applications, vol. 59, Birkhäuser, Basel, pp. 30–77. MR 1246809 (94j:46034).

5 Alpay, D. and Dym, H. (1993) On a new class of structured reproducing kernel spaces. *J. Funct. Anal.*, **111** (1), 1–28. MR 1200633 (94g:46035).

6 Lata, S., Mittal, M., and Paulsen, V.I. (2009) An operator algebraic proof of Agler's factorization theorem. *Proc. Am. Math. Soc.*, **137** (11), 3741–3748. MR 2529882.

7 Paulsen, V.I. and Raghupathi, M. (2016) *An Introduction to the Theory of Reproducing Kernel Hilbert Spaces*, Cambridge Studies in Advanced Mathematics, vol. 152, Cambridge University Press, Cambridge,. MR 3526117.

8 Alpay, D., Jorgensen, P., Seager, R., and Volok, D. (2013) On discrete analytic functions: products, rational functions and reproducing kernels. *J. Appl. Math. Comput.*, **41** (1–2), 393–426. MR 3017129.

9 Bezuglyi, S. and Handelman, D. (2014) Measures on Cantor sets: the good, the ugly, the bad. *Trans. Am. Math. Soc.*, **366** (12), 6247–6311. MR 3267010.

10 Saitoh, S. (2016) A reproducing kernel theory with some general applications, in *Mathematical Analysis, Probability and Applications—Plenary Lectures. ISAAC 2015*, Springer Proceedings in Mathematics & Statistics, vol. **177** (eds T. Qian and L. Rodino) Springer, Cham, pp. 151–182. MR 3571702.

11 Muandet, K., Sriperumbudur, B., Fukumizu, K., Gretton, A., and Schölkopf, B. (2016) Kernel mean shrinkage estimators. *J. Mach. Learn. Res.*, **17**, Paper No. 48, 41. MR 3504608.

12 Hsing, T. and Eubank, R. (2015) *Theoretical Foundations of Functional Data Analysis, With an Introduction to Linear Operators*, Wiley Series in Probability and Statistics, John Wiley & Sons, Ltd., Chichester. MR 3379106.

13 Jørsboe, O.G. (1968) *Equivalence or Singularity of Gaussian Measures on Function Spaces*, Various Publications Series, No. 4, Matematisk Institut, Aarhus Universitet, Aarhus. MR 0277027.

14 Machkouri, M.E., Es-Sebaiy, K., and Ouassou, I. (2017) On local linear regression for strongly mixing random fields. *J. Multivariate Anal.*, **156**, 103–115. MR 3624688.

15 Parussini, L., Venturi, D., Perdikaris, P., and Karniadakis, G.E. (2017) Multi-fidelity Gaussian process regression for prediction of random fields. *J. Comput. Phys.*, **336**, 36–50. MR 3622604.

16 Cortes, R.X., Martins, T.G., Prates, M.O., and Silva, B.A. (2017) Inference on dynamic models for non-Gaussian random fields using INLA. *Braz. J. Probab. Stat.*, **31** (1), 1–23. MR 3601658.

17 Smale, S. and Zhou, D.-X. (2004) Shannon sampling and function reconstruction from point values. *Bull. Am. Math. Soc. (N.S.)*, **41** (3), 279–305. MR 2058288.

18 Smale, S. and Zhou, D.-X. (2009) Geometry on probability spaces. *Constr. Approx.*, **30** (3), 311–323. MR 2558684.

19 Smale, S. and Zhou, D.-X. (2009) Online learning with Markov sampling. *Anal. Appl. (Singapore)*, **7** (1), 87–113. MR 2488871.

20 Jorgensen, P. and Tian, F. (2015) Discrete reproducing kernel Hilbert spaces: sampling and distribution of Dirac-masses. *J. Mach. Learn. Res.*, **16**, 3079–3114. MR 3450534.

21 Jorgensen, P.E.T., Pedersen, S., and Tian, F. (2015) *Harmonic Analysis of a Class of Reproducing Kernel Hilbert Spaces Arising from Groups*, Trends in Harmonic Analysis and its Applications, Contemporary Mathematics, vol. 650, American Mathematical Society, Providence, RI, pp. 157–197. MR 3441738.

22 Jorgensen, P., Pedersen, S., and Tian, F. (2016) *Extensions of Positive Definite Functions: Applications and Their Harmonic Analysis*, Lecture Notes in Mathematics, vol. 2160, Springer, Cham. MR 3559001.

23 Walter, R. (1987) *Real and Complex Analysis*, 3rd edn, McGraw-Hill Book Co., New York. MR 924157.

24 Hida, T. (1980) *Brownian Motion*, Applications of Mathematics, vol. 11, Springer-Verlag, New York. Translated from the Japanese by the author and T. P. Speed. MR 562914 (81a:60089).

25 Alpay, D., Jorgensen, P., and Volok, D. (2014) Relative reproducing kernel Hilbert spaces. *Proc. Am. Math. Soc.*, **142** (11), 3889–3895. MR 3251728.

26 Mumford, D. (2000) *The Dawning of the Age of Stochasticity*, Mathematics: Frontiers and Perspectives, American Mathematical Society, Providence, RI, pp. 197–218. MR 1754778 (2001e:01004).

27 Devinatz, A. (1959) On the extensions of positive definite functions. *Acta Math.*, **102**, 109–134. MR 0109992 (22 #875).

28 Nussbaum, A.E. (1975) Extension of positive definite functions and representation of functions in terms of spherical functions in symmetric spaces of noncompact type of rank 1. *Math. Ann.*, **215**, 97–116. MR 0385473 (52 #6334).

29 Walter, R. (1963) The extension problem for positive-definite functions. *Ill. J. Math.*, **7**, 532–539. MR 0151796 (27 #1779).

30 Walter, R. (1970) An extension theorem for positive-definite functions, *Duke Math. J.*, **37** (1), 49–53. MR 0254514 (40 #7722).

31 Jorgensen, P.E.T. (1986) Analytic continuation of local representations of Lie groups. *Pac. J. Math.*, **125** (2), 397–408. MR 863534 (88m:22030).

32 Jorgensen, P.E.T. (1989) Positive definite functions on the Heisenberg group. *Math. Z.*, **201** (4), 455–476. MR 1004167 (90m:22024).

33 Jorgensen, P.E.T. (1990) Extensions of positive definite integral kernels on the Heisenberg group. *J. Funct. Anal.*, **92** (2), 474–508. MR 1069255 (91m:22013).

34 Jorgensen, P.E.T. (1991) Integral representations for locally defined positive definite functions on Lie groups. *Int. J. Math.*, **2** (3), 257–286. MR 1104120 (92h:43017).

35 Kaniuth, E. and Lau, A.T. (2007) Extension and separation properties of positive definite functions on locally compact groups. *Trans. Am. Math. Soc.*, **359** (1), 447–463 (electronic). MR 2247899 (2007k:43008).

36 Krein, M.G. and Langer, H. (2014) Continuation of hermitian positive definite functions and related questions. *Integr. Equ. Oper. Theory*, **78** (1), 1–69. MR 3147401.

37 Klein, A. (1973/74) An explicitly solvable model in "Euclidean" field theory: the fixed source. *Z. Wahrscheinlichkeitstheorie Verw. Geb.*, **28**, 323–334. MR 0389089 (52 #9920).

38 Malamud, M.M. and Schmüdgen, K. (2012) Spectral theory of Schrödinger operators with infinitely many point interactions and radial positive definite functions. *J. Funct. Anal.*, **263** (10), 3144–3194. MR 2973337.

39 Ørsted, B. (1979) Induced representations and a new proof of the imprimitivity theorem. *J. Funct. Anal.*, **31** (3), 355–359. MR 531137 (80d:22007).

40 Osterwalder, K. and Schrader, R. (1973) Axioms for Euclidean Green's functions. *Commun. Math. Phys.*, **31**, 83–112. MR 0329492 (48 #7834).

41 Schmüdgen, K. (1986) A note on commuting unbounded selfadjoint operators affiliated to properly infinite von Neumann algebras. II. *Bull. London Math. Soc.*, **18** (3), 287–292. MR 829589 (87g:47079).

42 Schmüdgen, K. (1986) On commuting unbounded selfadjoint operators. IV. *Math. Nachr.*, **125**, 83–102. MR 847352 (88j:47026).

43 Schmüdgen, K. and Friedrich, J. (1984) On commuting unbounded selfadjoint operators. II. *Integr. Equ. Oper. Theory*, **7** (6), 815–867. MR 774726 (86i:47032).

44 Pólya, G. (1949) Remarks on Characteristic Functions. *Proceedings of the 1st Berkeley Conference on Mathematical Statistics and Probability*, pp. 115–123.

45 Schoenberg, I.J. (1938) Metric spaces and completely monotone functions. *Ann. Math. 2*, **39** (4), 811–841. MR 1503439.

46 Schoenberg, I.J. (1938) Metric spaces and positive definite functions. *Trans. Am. Math. Soc.*, **44** (3), 522–536. MR 1501980.

47 Schoenberg, I.J. (1964) Spline interpolation and the higher derivatives. *Proc. Natl. Acad. Sci. U.S.A.*, **51**, 24–28. MR 0160064 (28 #3278).

48 Smale, S. and Zhou, D.-X. (2007) Learning theory estimates via integral operators and their approximations. *Constr. Approx.*, **26** (2), 153–172. MR 2327597.

49 Poggio, T. and Smale, S. (2003) The mathematics of learning: dealing with data. *Not. Am. Math. Soc.*, **50** (5), 537–544.

50 Bochner, S. (1946) Finitely additive set functions and stochastic processes. *Proc. Natl. Acad. Sci. U.S.A.*, **32**, 259–261. MR 0017897 (8,215a).

51 Bernstein, S. (1946) Sur le théorème limite de la théorie des probabilités. *Bull. [Izvestiya] Math. Mech. Inst. Univ. Tomsk*, **3**, 174–190. MR 0019854 (8,471b).

52 Itô, K. (2004) *Stochastic Processes*, Springer-Verlag, Berlin. Lectures given at Aarhus University, Reprint of the 1969 original, Edited and with a foreword by Ole E. Barndorff-Nielsen and Ken-iti Sato. MR 2053326 (2005e:60002).

53 Parthasarathy, K.R. and Schmidt, K. (1975) Stable positive definite functions. *Trans. Am. Math. Soc.*, **203**, 161–174. MR 0370681 (51 #6907).

54 Applebaum, D. (2009) *Lévy Processes and Stochastic Calculus*, Cambridge Studies in Advanced Mathematics, vol. 116, 2nd edn, Cambridge University Press, Cambridge. MR 2512800 (2010m:60002).

55 Greville, T.N.E., Schoenberg, I.J., and Sharma, A. (1983) The Behavior of the Exponential Euler Spline $S_n(x; t)$ as $n \to \infty$ for Negative Values of the Base t. *2nd Edmonton Conference on Approximation Theory (Edmonton, Alta, 1982)*, CMS Conference Proceedings, vol. 3, American Mathematical Society, Providence, RI, pp. 185–198. MR 729330 (85c:41017).

56 Alpay, D. and Jorgensen, P.E.T. (2012) Stochastic processes induced by singular operators. *Numer. Funct. Anal. Optim.*, **33** (79), 708–735. MR 2966130.

57 Alpay, D., Jorgensen, P., and Levanony, D. (2011) A class of Gaussian processes with fractional spectral measures. *J. Funct. Anal.*, **261** (2), 507–541. MR 2793121 (2012e:60101).

58 Trèves, F. (2006) *Basic Linear Partial Differential Equations*, Dover Publications Inc., Mineola, NY. Reprint of the 1975 original. MR 2301309 (2007k:35004).

59 Falkowski, B.J. (1974) Factorizable and infinitely divisible PUA representations of locally compact groups. *J. Math. Phys.*, **15**, 1060–1066. MR 0372115 (51 #8332).

60 Jorgensen, P.E.T. (2007) The measure of a measurement. *J. Math. Phys.*, **48** (10), 103506, MR 2362879 (2008i:81011).

61 Itô, K. and McKean, H.P. Jr. (1965) *Diffusion Processes and Their Sample Paths*, Die Grundlehren der Mathematischen Wissenschaften, Band **125**, Academic Press Inc., Publishers, New York. MR 0199891 (33 #8031).

62 Sz-Nagy, B., Foias, C., Bercovici, H., and Kérchy, L. (2010) *Harmonic Analysis of Operators on Hilbert Space*, enlarged edn, Universitext, Springer, New York. MR 2760647 (2012b:47001).

63 Barrière, L., Comellas, F., and Dalfó, C. (2006) Fractality and the small-world effect in Sierpinski graphs. *J. Phys. A*, **39** (38), 11739–11753. MR 2275879 (2008i:28008).

64 Rodgers, G.J., Austin, K., Kahng, B., and Kim, D. (2005) Eigenvalue spectra of complex networks. *J. Phys. A*, **38** (43), 9431–9437. MR 2187996 (2006j:05186).

65 Konno, N., Masuda, N., Roy, R., and Sarkar, A. (2005) Rigorous results on the threshold network model. *J. Phys. A*, **38** (28), 6277–6291. MR 2166622 (2006j:82054).

66 Burioni, R. and Cassi, D. (2005) Random walks on graphs: ideas, techniques and results. *J. Phys. A*, **38** (8), R45–R78. MR 2119174 (2006b:82059).

67 Aristoff, D. and Radin, C. (2013) Emergent structures in large networks. *J. Appl. Probab.*, **50** (3), 883–888. MR 3102521.

68 Lovász, L. (2012) *Large Networks and Graph Limits*, American Mathematical Society Colloquium Publications, vol. 60, American Mathematical Society, Providence, RI. MR 3012035.

69 Pozrikidis, C. (2015) Application of a generalized Sherman-Morrison formula to the computation of network Green's functions and the construction of spanning trees. *J. Stat. Mech. Theory Exp.*, **2015** (5), P05007. MR 3360295.

70 Li, C. and Qu, Z. (2013) Distributed estimation of algebraic connectivity of directed networks. *Systems Control Lett.*, **62** (6), 517–524. MR 3054084.

71 Colin de Verdière, Y. (1998) *Spectres de graphes*, Cours Spécialisés [Specialized Courses], vol. 4, Société Mathématique de France, Paris,. MR 1652692 (99k:05108).

72 Trentelman, H.L., Takaba, K., and Monshizadeh, N. (2013) Robust synchronization of uncertain linear multi-agent systems. *IEEE Trans. Automat. Control*, **58** (6), 15111523. MR 3065133.

73 Niyogi, P., Smale, S., and Weinberger, S. (2011) A topological view of unsupervised learning from noisy data. *SIAM J. Comput.*, **40** (3), 646–663. MR 2810909 (2012h:62015).

74 Cucker, F. and Smale, S. (2002) On the mathematical foundations of learning. *Bull. Am. Math. Soc. (N.S.)*, **39** (1), 1–49 (electronic). MR 1864085.

75 Julien, M. (2015) Incremental majorization-minimization optimization with application to large-scale machine learning. *SIAM J. Optim.*, **25** (2), 829–855. MR 3335503.

76 Le Thi, H.A., Le, H.M., and Dinh, T.P. (2015) Feature selection in machine learning: an exact penalty approach using a difference of Convex function algorithm. *Mach. Learn.*, **101** (1–3), 163186. MR 3399224.

77 Bleich, J. (2015) Extensions and applications of ensemble-of-trees methods in machine learning. ProQuest LLC, Ann Arbor, MI, Thesis (PhD)–University of Pennsylvania. MR 3407331.

78 Jorgensen, P. and Pearse, E.P.J. (2013) A discrete Gauss-Green identity for unbounded Laplace operators, and the transience of random walks. *Israel J. Math.*, **196** (1), 113–160. MR 3096586.

79 Valerio, F. (2011) Improved biological network reconstruction using graph Laplacian regularization. *J. Comput. Biol.*, **18** (8), 987–996. MR 2823543 (2012d:92015).

80 Chen, Y., Pan, R., and Zhang, X. (2010) The Laplacian spectra of graphs and complex networks. *J. Univ. Sci. Technol. China*, **40** (12), 1236–1244. MR 2827605 (2012d:05222).

81 Tzeng, W.J. and Wu, F.Y. (2006) Theory of impedance networks: the two-point impedance and *LC* resonances. *J. Phys. A*, **39** (27), 8579–8591. MR 2241793 (2007d:94003).

82 Jorgensen, P.E.T. and Song, M.-S. (2013) Compactification of infinite graphs and sampling. *Sampl. Theory Signal Image Process.*, **12** (2-3), 139158. MR 3285408.

83 Jorgensen, P. and Tian, F. (2015) Infinite networks and variation of conductance functions in discrete Laplacians. *J. Math. Phys.*, **56** (4), 043506.

84 Jorgensen, P. and Tian, F. (2015) Frames and factorization of graph Laplacians. *Opusc. Math.*, **35** (3), 293.

85 Jorgensen, P.E.T. and Song, M.-S. (2007) Entropy encoding, Hilbert space, and KarhunenLoève transforms. *J. Math. Phys.*, **48** (10), 103503.

86 Dutkay, D.E., Picioroaga, G., and Song, M.-S. (2014) Orthonormal bases generated by Cuntz algebras. *J. Math. Anal. Appl.*, **409** (2), 1128–1139. MR 3103223.

87 Song, M.-S. (2006) *Wavelet Image Compression*, Operator Theory, Operator Algebras, and Applications, Contemporary Mathematics, vol. 414, American Mathematical Society, Providence, RI, pp. 41–73.

3

Many-Body Wave Scattering Problems for Small Scatterers and Creating Materials with a Desired Refraction Coefficient

Alexander G. Ramm

Department of Mathematics, Kansas State University, Manhattan, KS 66506-2602, USA

MSC: 35Q60; 74E99; 78A40; 78A45; 78A48 **PACS:** *02.30.Rz; 02.30.Mv; 41.20.Jb*

3.1 Introduction

In this chapter, we discuss a method for creating materials with a desired refraction coefficients. This method is proposed and developed by the author and is based on a series of his papers and on his monograph [1]. The author thinks that these results may be new for materials science people although the results were published in mathematical and mathematical physics Journals. This is the basic reason for including this chapter in this book. This chapter should be useful to materials science researchers, physicists and engineers.

Parts of this chapter are taken verbatim from the paper by the author [2]. The author thanks Springer for permission to use verbatim parts of the author's paper, see also monograph [3].

There is a large literature on wave scattering by small bodies, starting from Rayleigh's work (1871), [4–6]. For the problem of wave scattering by one body, an analytical solution was found only for the bodies of special shapes, for example, for balls and ellipsoids. If the scatterer is small, then the scattered field can be calculated analytically for bodies of arbitrary shapes, see [2, 7], and [1] where this theory is presented.

The many-body wave scattering problem was discussed in the literature, mostly numerically, in the cases when the number of scatterers is small or the influence on a particular particle of the waves scattered by other particles is negligible. This corresponds to the case when the distance d between neighboring particles is much larger than the wavelength λ, and the characteristic size a of a small body (particle) is much smaller than λ, that is, $d \gg \lambda$ and $a \ll \lambda$. By $k = \frac{2\pi}{\lambda}$, the wave number is denoted.

In this chapter, the much more difficult case is considered, when $a \ll d \ll \lambda$. In this case, the influence of the scattered field on a particular particle is essential, that is, *multiple scattering effects are essential.*

Mathematical Analysis and Applications: Selected Topics, First Edition.
Edited by Michael Ruzhansky, Hemen Dutta, and Ravi P. Agarwal.
© 2018 John Wiley & Sons, Inc. Published 2018 by John Wiley & Sons, Inc.

The derivations of the results, presented in this chapter, are rigorous. They are taken from the earlier papers of the author, cited in the list of references. Many formulas and arguments are taken from these papers, especially from the paper by the author [2]. Large parts of this chapter are taken verbatim, and monograph [1] is also used essentially. In this chapter, we do not discuss electromagnetic wave scattering by small bodies (particles). A detailed discussion of electromagnetic wave scattering by small perfectly conducting and impedance particles of an *arbitrary shape* is given in [1, 8], and also see [7].

A *physically novel* point in our theory is the following one:

While in the classical theory of wave scattering by small body of characteristic size a (e.g., in Rayleigh's theory) the scattering amplitude is $O(a^3)$ as $a \to 0$, in our theory for a small impedance particle the scattering amplitude is *much larger*: it is of the order $O(a^{2-\kappa})$, where $a \to 0$ and $\kappa \in [0, 1)$ are the parameters (see the text below formula (3.22) in this chapter).

Can this result be used in technology?

The practical applications of the theory, presented in this chapter, are immediate provided that the important practical problem of preparing small particles with the prescribed boundary impedance is solved.

The author thinks that an impedance boundary condition (BC) (condition (3.7)) must be physically (experimentally) realizable if this condition guarantees the uniqueness of the solution to the corresponding boundary problem. The impedance BC (3.7) guarantees the uniqueness of the solution to the scattering boundary problem (3.1)–(3.4) provided that $\text{Im}\zeta_1 \leq 0$.

Therefore, there should exist a practical (experimental) method for producing small particles with any boundary impedance ζ_1 satisfying the inequality $Im\zeta_1 \leq 0$.

The author asks the materials science specialists to contact him if they are aware of a method for practical (experimental) preparing (producing) small particles with the prescribed boundary impedance

The materials science researchers are not familiar with the author's papers on creating materials with a desired refraction coefficient because the author's theory was presented in the journals, which are not popular among materials science researchers.

Although the author's results were presented in many of the author's earlier publications, cited in references, the author hopes that *they will be not only new but practically useful for materials science researchers.*

The basic results of this section consist of:

(i) Derivation of analytic formulas for the scattering amplitude for the wave scattering problem by one small ($ka \ll 1$) impedance body *of an arbitrary shape*;

(ii) Solution to *many-body wave scattering problem* by small particles, embedded in an inhomogeneous medium, under the assumptions $a \ll d \ll \lambda$, where d is the minimal distance between neighboring particles;

(iii) Derivation of the equations for the limiting effective (self-consistent) field in an inhomogeneous medium in which many small particles are embedded, when $a \to 0$ and the number $M = M(a)$ of the small particles tends to infinity at an appropriate rate;

(iv) Derivation of linear algebraic system (LAS) for solving many-body wave scattering problems. These systems are not obtained in the standard way from boundary integral equations; they have physical meaning and give an efficient numerical method for solving many-body wave scattering problems in the case of small scatterers. In [8] for the first time, the many-body wave scattering problems were solved for billions of particles. This was not feasible earlier;

(v) Application of our results to creating materials with a desired refraction coefficient.

The order of the error estimates as $a \to 0$ is obtained. Our presentation follows very closely that in [2], but it is essentially self-contained. Our methods give powerful numerical methods for solving many-body wave scattering problems in the case when the scatterers are small but multiple scattering effects are essential [9–11]. In [9], the scattering problem is solved numerically for 10^{10} particles apparently for the first time.

In Sections 3.1–3.4 wave scattering by small impedance bodies is developed.

Let us formulate the wave scattering problems we deal with. First, let us consider a one-body scattering problem. Let D_1 be a bounded domain in $\mathbb{R}^3$ with a sufficiently smooth boundary S_1. The scattering problem consists of finding the solution to the problem:

$$(\nabla^2 + k^2)u = 0 \text{ in } D_1' := \mathbb{R}^3 \backslash D_1, \tag{3.1}$$

$$\Gamma u = 0 \text{ on } S_1, \tag{3.2}$$

$$u = u_0 + v, \tag{3.3}$$

where

$$u_0 = e^{ik\alpha \cdot x}, \quad \alpha \in S^2, \tag{3.4}$$

S^2 is the unit sphere in $\mathbb{R}^3$, u_0 is the incident field, v is the scattered field satisfying the radiation condition

$$v_r - ikv = o\left(\frac{1}{r}\right), \quad r := |x| \to \infty, v_r := \frac{\partial v}{\partial r}, \tag{3.5}$$

Γu is the BC of one of the following types

$$\Gamma u = \Gamma_1 u = u \quad \text{(Dirichlet BC)}, \tag{3.6}$$

$$\Gamma u = \Gamma_2 u = u_N - \zeta_1 u, \quad \text{Im}\zeta_1 \leq 0 \text{ (impedance BC)}, \tag{3.7}$$

where ζ_1 is a constant, N is the unit normal to S_1, pointing out of D_1, and

$$\Gamma u = \Gamma_3 u = u_N \quad \text{(Neumann BC)}. \tag{3.8}$$

It is well known [12, 13] that problem (3.1)–(3.3) has a unique solution. We now assume that

$$a := 0.5 \, \text{diam} D_1, \quad ka \ll 1, \tag{3.9}$$

which is the "smallness assumption" equivalent to $a \ll \lambda$, where λ is the wave length. We look for the solution to problem (3.1)–(3.3) of the form

$$u(x) = u_0(x) + \int_{S_1} g(x, t)\sigma_1(t)dt, \quad g(x, y) := \frac{e^{ik|x-y|}}{4\pi|x-y|}, \tag{3.10}$$

where dt is the element of the surface area of S_1. One can prove that the unique solution to the scattering problem (3.1)–(3.3) with any of the BCs (3.6)–(3.8) can be found in the form (3.10), and the function σ_1 in (3.10) is uniquely defined from the BC (3.2). The scattering amplitude $A(\beta, \alpha) = A(\beta, \alpha, k)$ is defined by the formula

$$v = \frac{e^{ikr}}{r} A(\beta, \alpha, k) + o\left(\frac{1}{r}\right), \quad r \to \infty, \quad \beta := \frac{x}{r}. \tag{3.11}$$

The equations for finding σ_1 are:

$$\int_{S_1} g(s, t)\sigma_1(t)dt = -u_0(s), \tag{3.12}$$

$$u_{0N} - \zeta_1 u_0 + \frac{A\sigma_1 - \sigma_1}{2} - \zeta_1 \int_{S_1} g(s, t)\sigma_1(t)dt = 0, \tag{3.13}$$

$$u_{0N} + \frac{A\sigma_1 - \sigma_1}{2} = 0, \tag{3.14}$$

respectively, for conditions (3.6)–(3.8). The operator A is defined as follows:

$$A\sigma := 2\int_{S_1} \frac{\partial}{\partial N_s} g(s, t)\sigma_1(t)dt. \tag{3.15}$$

Equations (3.12)–(3.14) are uniquely solvable, but there are no analytic formulas for their solutions for bodies of arbitrary shapes. However, if the body D_1 is small, $ka \ll 1$, one can rewrite (3.10) as

$$u(x) = u_0(x) + g(x, 0)Q_1 + \int_{S_1} [g(x, t) - g(x, 0)]\sigma_1(t)dt, \tag{3.16}$$

where

$$Q_1 := \int_{S_1} \sigma_1(t)dt, \tag{3.17}$$

and $0 \in D_1$ is the origin.

If $ka \ll 1$, then we prove that

$$|g(x, 0)Q_1| \gg \left|\int_{S_1} [g(x, t) - g(x, 0)]\sigma_1(t)dt\right|, \quad |x| > a. \tag{3.18}$$

Therefore, *the scattered field is determined outside D_1 by a single number Q_1.*

This number can be obtained analytically without solving (3.12) and (3.13). The case (3.14) requires a special approach by the reason discussed in detail later.

Let us give the results for (3.12) and (3.13) first. For (3.12), one has

$$Q_1 = \int_{S_1} \sigma_1(t)dt = -Cu_0(0)[1 + o(1)], \quad a \to 0, \tag{3.19}$$

where C *is the electric capacitance of a perfect conductor with the shape* D_1. For (3.13), one has

$$Q_1 = -\zeta_1|S_1|u_0(0)[1 + o(1)], \quad a \to 0, \tag{3.20}$$

where $|S_1|$ is the surface area of S_1. The scattering amplitude for problem (3.1)–(3.3) with $\Gamma = \Gamma_1$ (acoustically soft particle) is

$$A_1(\beta, \alpha) = -\frac{C}{4\pi}[1 + o(1)], \tag{3.21}$$

since

$$u_0(0) = e^{ika \cdot x}|_{x=0} = 1.$$

Therefore, in this case, the scattering is isotropic and of the order $O(a)$, *because the capacitance* $C = O(a)$.

The scattering amplitude for problem (3.1)–(3.3) with $\Gamma = \Gamma_2$ (*small impedance particles*) is:

$$A_2(\beta, \alpha) = -\frac{\zeta_1|S_1|}{4\pi}[1 + o(1)], \tag{3.22}$$

since $u_0(0) = 1$.

In this case, the scattering is also isotropic, and of the order $O(\zeta|S_1|)$.

If $\zeta_1 = O(1)$, then $A_2 = O(a^2)$, because $|S_1| = O(a^2)$. If $\zeta_1 = O\left(\frac{1}{a^\kappa}\right)$, $\kappa \in (0, 1)$, then $A_2 = O(a^{2-\kappa})$. The case $\kappa = 1$ was considered in [14].

The scattering amplitude for problem (3.1)–(3.3) with $\Gamma = \Gamma_3$ (acoustically hard particles) is

$$A_3(\beta, \alpha) = -\frac{k^2|D_1|}{4\pi}(1 + \beta_{pq}\beta_p\alpha_q), \quad \text{if } u_0 = e^{ika \cdot x}. \tag{3.23}$$

Here and below summation is understood over the repeated indices, $\alpha_q = \alpha \cdot e_q$, $\alpha \cdot e_q$ denotes the dot product of two vectors in $\mathbb{R}^3$, $p, q = 1, 2, 3$, $\{e_p\}$ is an orthonormal Cartesian basis of $\mathbb{R}^3$, $|D_1|$ is the volume of D_1, β_{pq} is the magnetic polarizability tensor defined as follows [7, p. 62]:

$$\beta_{pq} := \frac{1}{|D_1|}\int_{S_1} t_p\sigma_{1q}(t)dt, \tag{3.24}$$

σ_{1q} is the solution to the equation

$$\sigma_{1q}(s) = A_0\sigma_{1q} - 2N_q(s), \tag{3.25}$$

$N_q(s) = N(s) \cdot e_q$, $N = N(s)$ is the unit outer normal to S_1 at the point s, that is, the normal pointing out of D_1, and A_0 is the operator A at $k = 0$. For small bodies, $\|A - A_0\| = o(ka)$.

If $u_0(x)$ is an arbitrary field satisfying (3.1), not necessarily the plane wave $e^{ika \cdot x}$, then

$$A_3(\beta, \alpha) = \frac{|D_1|}{4\pi} \left(ik\beta_{pq} \frac{\partial u_0}{\partial x_q} \beta_p + \Delta u_0 \right). \tag{3.26}$$

The above formulas are derived in Section 3.2. In Section 3.3 we develop a theory for many-body wave scattering problem and derive the equations for effective field in the medium, in which many small particles are embedded, as $a \to 0$.

The results, presented in this chapter, are based on the earlier works of the author [1, 2, 7, 9, 12–34]. These results and methods of their derivation differ much from those published by other authors.

Our approach to homogenization-type theory is also different from the approaches of other authors [35, 36]. The differences are:

(i) no periodic structure in the problems is assumed,
(ii) the operators in our problems are non-selfadjoint and have continuous spectrum,
(iii) the limiting medium is not homogeneous and its parameters are not periodic,
(iv) the technique for passing to the limit is different from the one used in homogenization theory.

Let us summarize the results for one-body wave scattering.

Theorem 3.1   The scattering amplitude for the problem (3.1)–(3.4) for small body of an arbitrary shape is given by formulas (3.25)–(3.27), for the BCs Γ_1–Γ_3, respectively.

3.2   Derivation of the Formulas for One-Body Wave Scattering Problems

Let us recall the known result [12]

$$\frac{\partial}{\partial N_s^-} \int_{S_1} g(x, t)\sigma_1(t)\mathrm{d}t = \frac{A\sigma_1 - \sigma_1}{2} \tag{3.27}$$

concerning the limiting value of the normal derivative of single-layer potential from outside. Let $x_m \in D_m$, $t \in S_m$, S_m is the surface of D_m, $a = 0.5$ diamD_m.

In this section $m = 1$, and $x_m = 0$ is the origin.

We assume that $ka \ll 1$, $ad^{-1} \ll 1$, so $|x - x_m| = d \gg a$. Then

$$\frac{e^{ik|x-t|}}{4\pi|x - t|} = \frac{e^{ik|x-x_m|}}{4\pi|x - x_m|}e^{-ik(x-x_m)^o \cdot (t-x_m)}\left(1 + O\left(ka + \frac{a}{d}\right)\right), \qquad (3.28)$$

$$k|x - t| = k|x - x_m| - k(x - x_m)^o \cdot (t - x_m) + O\left(\frac{ka^2}{d}\right), \qquad (3.29)$$

where

$$d = |x - x_m|, \quad (x - x_m)^o := \frac{x - x_m}{|x - x_m|},$$

and

$$\frac{|x - t|}{|x - x_m|} = 1 + O\left(\frac{a}{d}\right). \qquad (3.30)$$

Let us derive estimate (3.19). Since $|t| \le a$ on S_1, one has

$$g(s, t) = g_0(s, t)(1 + O(ka)),$$

where $g_0(s, t) = \frac{1}{4\pi|s-t|}$. Since $u_0(s)$ is a smooth function, one has $|u_0(s) - u_0(0)| = O(a)$. Consequently, (3.12) can be considered as an equation for electrostatic charge distribution $\sigma_1(t)$ on the surface S_1 of a perfect conductor D_1, charged to the constant potential $-u_0(0)$ (up to a small term of the order $O(ka)$). It is known that the total charge $Q_1 = \int_{S_1} \sigma_1(t)dt$ of this conductor is equal to

$$Q_1 = -Cu_0(0)(1 + O(ka)), \qquad (3.31)$$

where C is the electric capacitance of the perfect conductor with the shape D_1.

Analytic formulas for electric capacitance C of a perfect conductor of an arbitrary shape, which allow to calculate C with a desired accuracy, are derived in [7]. For example, the zeroth approximation formula is:

$$C^{(0)} = \frac{4\pi|S_1|^2}{\int_{S_1}\int_{S_1}\frac{dsdt}{r_{st}}}, \quad r_{st} = |t - s|, \qquad (3.32)$$

and we assume in (3.32) that $\epsilon_0 = 1$, where ϵ_0 is the dielectric constant of the homogeneous medium in which the perfect conductor is placed. Formula (3.31) is formula (3.19). If $u_0(x) = e^{ika \cdot x}$, then $u_0(0) = 1$, and $Q_1 = -C(1 + O(ka))$. In this case,

$$A_1(\beta, \alpha) = \frac{Q_1}{4\pi} = -\frac{C}{4\pi}[1 + O(ka)],$$

which is formula (3.21).

Consider now wave scattering by an impedance particle.

Let us derive formula (3.20). Integrate (3.13) over S_1, use the divergence formula

$$\int_{S_1} u_{0N}ds = \int_{D_1} \nabla^2 u_0 dx = -k^2\int_{D_1} u_0 dx = k^2|D_1|u_0(0)[1 + o(1)], \qquad (3.33)$$

where $|D_1| = O(a^3)$, and the formula

$$-\zeta_1 \int_{S_1} u_0 ds = -\zeta_1 |S_1| u_0(0)[1 + o(1)], \qquad (3.34)$$

which is valid because the body D_1 is small: in this case, $u_0(s) \approx u_0(0)$. Furthermore $|\int_{S_1} g(s,t)ds| = O(a)$, so

$$\zeta_1 \int_{S_1} ds \int_{S_1} g(s,t)\sigma_1(t)dt = O(aQ_1). \qquad (3.35)$$

Therefore, the term (3.35) is negligible compared with Q_1 as $a \to 0$. Finally, if $ka \ll 1$, then $g(s,t) = g_0(s,t)(1 + ik|s - t| + \cdots)$, and

$$\frac{\partial}{\partial N_s} g(s,t) = \frac{\partial}{\partial N_s} g_0(s,t)[1 + O(ka)]. \qquad (3.36)$$

Denote by A_0 the operator

$$A_0 \sigma = 2 \int_{S_1} \frac{\partial g_0(s,t)}{\partial N_s} \sigma_1(t)dt. \qquad (3.37)$$

It is known from the potential theory [1] that

$$\int_{S_1} A_0 \sigma_1 ds = -\int_{S_1} \sigma_1(t)dt, \quad 2 \int_{S_1} \frac{\partial g_0(s,t)}{\partial N_s} ds = -1, \quad t \in S_1. \qquad (3.38)$$

Therefore,

$$\int_{S_1} ds \frac{A\sigma_1 - \sigma_1}{2} = -Q_1[1 + O(ka)]. \qquad (3.39)$$

Consequently, from formulas (3.33)–(3.39), one gets formula (3.22).

One can see that *the wave scattering by an impedance particle is isotropic, and the scattered field is of the order* $O(\zeta_1 |S_1|)$. *Since* $|S_1| = O(a^2)$, *one has* $O(\zeta_1 |S_1|) = O(a^{2-\kappa})$ *if* $\zeta_1 = O\left(\frac{1}{a^\kappa}\right)$, $\kappa \in [0, 1)$.

Consider now wave scattering by an acoustically hard small particle, that is, the problem with the Neumann BC.

In this case, we will prove that:

(i) The scattering is anisotropic,
(ii) It is defined not by a single number, as in the previous two cases, but by a tensor, and
(iii) The order of the scattered field is $O(a^3)$ as $a \to 0$, for a fixed $k > 0$, that is, the scattered field is much smaller than in the previous two cases.

Integrating over S_1 (3.14), one gets

$$Q_1 = \int_{D_1} \nabla^2 u_0 dx = \nabla^2 u_0(0)|D_1|[1 + o(1)], \quad a \to 0. \qquad (3.40)$$

Thus, $Q_1 = O(a^3)$. Therefore, the contribution of the term $e^{-ikx^o \cdot t}$ in formula (3.28) with $x_m = 0$ will be also of the order $O(a^3)$ and should be taken into account, *in contrast to the previous two cases.* Namely,

$$u(x) = u_0(x) + g(x, 0) \int_{S_1} e^{-ik\beta \cdot t} \sigma_1(t) dt, \quad \beta := \frac{x}{|x|} = x^o. \tag{3.41}$$

One has

$$\int_{S_1} e^{-ik\beta \cdot t} \sigma_1(t) dt = Q_1 - ik\beta_p \int_{S_1} t_p \sigma_1(t) dt, \tag{3.42}$$

where the terms of higher order of smallness are neglected and summation over index p is understood. The function σ_1 solves (3.14):

$$\sigma_1 = A\sigma_1 + 2u_{0N} = A\sigma_1 + 2ik\alpha_q N_q u_0(s), \quad s \in S_1 \tag{3.43}$$

if $u_0(x) = e^{ik\alpha \cdot x}$.

Comparing (3.43) with (3.25), using (3.24), and taking into account that $ka \ll 1$, one gets

$$-ik\beta_p \int_{S_1} t_p \sigma_1(t) dt = -ik\beta_p |D_1| \beta_{pq}(-ik\alpha_q) u_0(0)[1 + O(ka)]$$
$$= -k^2 |D_1| \beta_{pq} \beta_p \alpha_q u_0(0)[1 + O(ka)]. \tag{3.44}$$

From (3.40), (3.42), and (3.44), one gets formula (3.23), because $\nabla^2 u_0 = -k^2 u_0$.

If $u_0(x)$ is an arbitrary function, satisfying (3.1), then $ik\alpha_q$ in (3.43) is replaced by $\frac{\partial u_0}{\partial x_q}$, and $-k^2 u_0 = \Delta u_0$, which yields formula (3.26).

This completes the derivation of the formulas for the solution of scalar wave scattering problem by one small body on the boundary, of which the Dirichlet, or the impedance, or the Neumann boundary condition is imposed.

3.3 Many-Body Scattering Problem

In this section we assume that there are $M = M(a)$ small bodies (particles) D_m, $1 \leq m \leq M$, $a = 0.5\text{max diam}D_m$, $ka \ll 1$. The distance $d = d(a)$ between neighboring bodies is much larger than a, $d \gg a$, but *we do not assume that $d \gg \lambda$, so there may be many small particles on the distances of the order of the wavelength λ.*

This means that our medium with the embedded particles is not necessarily diluted.

We assume that the small bodies are embedded in an arbitrary large but finite domain D, $D \subset \mathbb{R}^3$, so $D_m \subset D$. Denote $D' := \mathbb{R}^3 \backslash D$ and $\Omega := \cup_{m=1}^{M} D_m$, $S_m := \partial D_m$, $\partial \Omega = \cup_{m=1}^{M} S_m$. By N, we denote a unit normal to $\partial \Omega$, pointing out of Ω; and by $|D_m|$ the volume of the body D_m is denoted.

The scattering problem consists of finding the solution to the following problem

$$(\nabla^2 + k^2) u = 0 \text{ in } \mathbb{R}^3 \backslash \Omega, \tag{3.45}$$

$$\Gamma u = 0 \text{ on } \partial\Omega, \tag{3.46}$$

$$u = u_0 + v, \tag{3.47}$$

where u_0 is the incident field, satisfying (3.45) in $\mathbb{R}^3$, for example, $u_0 = e^{ik\alpha \cdot x}$, $\alpha \in S^2$, and v is the scattered field, satisfying the radiation condition (3.5). The BC (3.46) can be of the types (3.6)–(3.8).

In the case of impedance BC (3.7), we assume that

$$u_N = \zeta_m u \text{ on } S_m, \quad 1 \leq m \leq M, \tag{3.48}$$

so the impedance may vary from one particle to another. We assume that

$$\zeta_m = \frac{h(x_m)}{a^\kappa}, \quad \kappa \in (0, 1), \tag{3.49}$$

where $x_m \in D_m$ is a point in D_m, and $h(x)$, $x \in D$, is a given function, which we can choose as we wish, subject to the condition $\text{Im} h(x) \leq 0$. For simplicity, we assume that $h(x)$ is a continuous function.

Let us make the following assumption about the distribution of small particles:

If $\Delta \subset D$ is an arbitrary open subset of D, then the number $\mathcal{N}(\Delta)$ of small particles in Δ, assuming the impedance BC, is:

$$\mathcal{N}_\zeta(\Delta) = \frac{1}{a^{2-\kappa}} \int_\Delta N(x) dx [1 + o(1)], \quad a \to 0, \tag{3.50}$$

where $N(x) \geq 0$ is a given function.

If the Dirichlet BC is assumed, then

$$\mathcal{N}_D(\Delta) = \frac{1}{a} \int_\Delta N(x) dx [1 + o(1)], \quad a \to 0. \tag{3.51}$$

The case of the Neumann BC will not be considered in this chapter, see [2].

We look for the solution to problem (3.45)–(3.47) with the Dirichlet BC of the form

$$u = u_0 + \sum_{m=1}^{M} \int_{S_m} g(x, t) \sigma_m(t) dt, \tag{3.52}$$

where $\sigma_m(t)$ are some functions to be determined from the boundary condition (3.46). It is proved in [14] that problem (3.45)–(3.47) has a unique solution of the form (3.52). For any $\sigma_m(t)$, function (3.52) solves (3.45) and satisfies condition (3.47). The BC (3.46) determines σ_m uniquely. However, if $M \gg 1$, then numerical solution of the system of integral equations for σ_m, where $1 \leq m \leq M$, which one gets from the BC (3.52), is practically not feasible.

To avoid this principal difficulty, we prove that the solution to scattering problem (3.45)–(3.47) is determined by M numbers

$$Q_m := \int_{S_m} \sigma_m(t) dt, \tag{3.53}$$

rather than M functions $\sigma_m(t)$. This allows one to drastically reduce the complexity of the numerical solution of the many-body scattering problems in the case of small particles.

This is possible to prove that if the particles D_m are small. We derive analytical formulas for Q_m as $a \to 0$.

Let us define the effective (self-consistent) field $u_e(x) = u_e^{(j)}(x)$, acting on the j-th particle, by the formula

$$u_e(x) := u(x) - \int_{S_j} g(x, t)\sigma_j(t)dt, \quad |x - x_j| \sim a. \tag{3.54}$$

Physically, this field acts on the j-th particle and is a sum of the incident field and the fields acting from all other particles:

$$u_e(x) = u_e^{(j)}(x) := u_0(x) + \sum_{m \neq j} \int_{S_m} g(x, t)\sigma_m(t)dt. \tag{3.55}$$

Let us rewrite (3.55) as follows:

$$u_e(x) = u_0(x) + \sum_{m \neq j}^{M} g(x, x_m)Q_m + \sum_{m \neq j}^{M} \int_{S_m} [g(x, t) - g(x, x_m)]\sigma_m(t)dt. \tag{3.56}$$

We want to prove that the last sum is negligible compared with the first one as $a \to 0$.

To prove this, let us give some estimates. One has $|t - x_m| \leq a$, $d = |x - x_m|$,

$$|g(x, t) - g(x, x_m)| = \max \left\{ O\left(\frac{a}{d^2}\right), O\left(\frac{ka}{d}\right) \right\}, \quad |g(x, x_m)| = O(1/d). \tag{3.57}$$

Therefore, if $|x - x_j| = O(a)$, then

$$\frac{\left| \int_{S_m} [g(x, t) - g(x, x_m)]\sigma_m(t)dt \right|}{|g(x, x_m)Q_m|} \leq O(ad^{-1} + ka). \tag{3.58}$$

One can also prove that

$$J_1/J_2 = O(ka + ad^{-1}), \tag{3.59}$$

where J_1 is the first sum in (3.56) and J_2 is the second sum in (3.56). Therefore, at any point $x \in \Omega' = \mathbb{R}^3 \backslash \Omega$, one has

$$u_e(x) = u_0(x) + \sum_{m=1}^{M} g(x, x_m)Q_m, \quad x \in \Omega', \tag{3.60}$$

where the terms of higher order of smallness are omitted.

3.3.1 The Case of Acoustically Soft Particles

If (3.46) is the Dirichlet condition, then, as we have proved in Section 3.2 (see formula (3.31)), one has

$$Q_m = -C_m u_e(x_m).$$ (3.61)

Thus,

$$u_e(x) = u_0(x) - \sum_{m=1}^{M} g(x, x_m) C_m u_e(x_m), \quad x \in \Omega'.$$ (3.62)

One has

$$u(x) = u_e(x) + o(1), \quad a \to 0,$$ (3.63)

so the full field and effective field are practically the same.

Let us write a LAS for finding unknown quantities $u_e(x_m)$:

$$u_e(x_j) = u_0(x_j) - \sum_{m \neq j}^{M} g(x_j, x_m) C_m u_e(x_m).$$ (3.64)

If M is not very large, say $M = O(10^3)$, then LAS (3.64) can be solved numerically, and formula (3.62) can be used for calculation of $u_e(x)$.

Consider the limiting case, when $a \to 0$. One can rewrite (3.64) as follows:

$$u_e(\xi_q) = u_0(\xi_q) - \sum_{p \neq q}^{P} g(\xi_q, \xi_p) u_e(\xi_p) \sum_{x_m \in \Delta_p} C_m,$$ (3.65)

where $\{\Delta_p\}_{p=1}^{P}$ is a union of cubes which forms a covering of D,

$$\max_p diam\Delta_p := b = b(a) \gg a,$$

$$\lim_{a \to 0} b(a) = 0.$$ (3.66)

By $|\Delta_p|$ we denote the volume (measure) of Δ_p, and ξ_p is the center of Δ_p, or a point x_p in an arbitrary small body D_p, located in Δ_p. Let us assume that there exists the limit

$$\lim_{a \to 0} \frac{\sum_{x_m \in \Delta_p} C_m}{|\Delta_p|} = C(\xi_p), \quad \xi_p \in \Delta_p.$$ (3.67)

For example, one may have

$$C_m = c(\xi_p)a$$ (3.68)

for all m such that $x_m \in \Delta_p$, where $c(x)$ is some function in D. If all D_m are balls of radius a, then $c(x) = 4\pi$. We have

$$\sum_{x_m \in \Delta_p} C_m = C_p a \mathcal{N}(\Delta_p) = C_p N(\xi_p) |\Delta_p| [1 + o(1)], \quad a \to 0,$$ (3.69)

so limit (3.67) exists, and

$$C(\xi_p) = c(\xi_p)N(\xi_p). \tag{3.70}$$

From (3.65), (3.68)–(3.70), one gets

$$u_e(\xi_q) = u_0(\xi_q) - \sum_{p \ne q} g(\xi_q, \xi_p)c(\xi_p)N(\xi_p)u_e(\xi_p)|\Delta_p|, \quad 1 \le p \le P. \tag{3.71}$$

LAS (3.71) can be considered as the *collocation method for solving integral equation*

$$u(x) = u_0(x) - \int_D g(x, y)c(y)N(y)u(y)\mathrm{d}y. \tag{3.72}$$

It is proved in [30] that

System (3.71) is uniquely solvable for all sufficiently small b(a), and the function

$$u_P(x) := \sum_{p=1}^{P} \chi_p(x)u_e(\xi_p) \tag{3.73}$$

converges in $L^\infty(D)$ to the unique solution of equation (3.72).

The function $\chi_p(x)$ in (3.73) is the characteristic function of the cube Δ_p: it is equal to 1 in Δ_p and vanishes outside Δ_p. Thus, if $a \to 0$, the solution to the many-body wave scattering problem in the case of the Dirichlet BC is well approximated by the unique solution of the integral equation (3.72).

Applying the operator $L_0 := \nabla^2 + k^2$ to (3.72), and using the formula $L_0g(x, y) = -\delta(x - y)$, where $\delta(x)$ is the delta-function, one gets

$$\nabla^2 u + k^2 u - q(x)u = 0 \text{ in } \mathbb{R}^3, \quad q(x) := c(x)N(x). \tag{3.74}$$

The physical conclusion is:

If one embeds $M(a) = O(1/a)$ small acoustically soft particles, which are distributed as in (3.51), then one creates, as $a \to 0$, a limiting medium, which is inhomogeneous and has a refraction coefficient $n^2(x) = 1 - k^{-2}q(x)$.

It is interesting from the physical point of view to note that

The limit, as $a \to 0$, of the total volume of the embedded particles is zero.

Indeed, the volume of one particle is $O(a^3)$, the total number M of the embedded particles is $O(a^3M) = O(a^2)$, and $\lim_{a \to 0} O(a^2) = 0$.

The second observation is: if (3.51) holds, then on a unit length straight line, there are $O\left(\frac{1}{a^{1/3}}\right)$ particles, so the distance between neighboring particles is $d = O(a^{1/3})$. If $d = O(a^\gamma)$ with $\gamma > \frac{1}{3}$, then the number of the embedded particles in a subdomain Δ_p is $O\left(\frac{1}{d^3}\right) = O(a^{-3\gamma})$. In this case, for $3\gamma > 1$, the limit in (3.69) is $C(\xi_p) = \lim_{a \to 0} c_p a O(a^{-3\gamma}) = \infty$. Therefore, the product of this limit by u remains finite only if $u = 0$ in D. Physically, this means that if the distances between neighboring perfectly soft particles are smaller than $O(a^{1/3})$, namely, they are $O(a^\gamma)$ with any $\gamma > \frac{1}{3}$, then $u = 0$ in D.

On the other hand, if $\gamma < \frac{1}{3}$, then the limit $C(\xi_p) = 0$, and $u = u_0$ in D, so that the embedded particles do not change, in the limit $a \to 0$, the properties of the medium.

This concludes our discussion of the scattering problem for many acoustically soft particles.

3.3.2 Wave Scattering by Many Impedance Particles

We assume now that (3.49) and (3.50) hold, use the exact BC (3.46) with $\Gamma = \Gamma_2$, that is,

$$u_{eN} - \zeta_m u_e + \frac{A_m \sigma_m - \sigma_m}{2} - \zeta_m \int_{S_m} g(s,t)\sigma_m(t)dt = 0, \tag{3.75}$$

and integrate (3.75) over S_m in order to derive an analytical asymptotic formula for $Q_m = \int_{S_m} \sigma_m(t)dt$.

We have

$$\int_{S_m} u_{eN}ds = \int_{D_m} \nabla^2 u_e dx = O(a^3), \tag{3.76}$$

$$\int_{S_m} \zeta_m u_e(s)ds = h(x_m)a^{-\kappa}|S_m|u_e(x_m)[1 + o(1)], \quad a \to 0, \tag{3.77}$$

$$\int_{S_m} \frac{A_m \sigma_m - \sigma_m}{2}ds = -Q_m[1 + o(1)], \quad a \to 0, \tag{3.78}$$

and

$$\zeta_m \int_{S_m} \int_{S_m} g(s,t)\sigma_m(t)dt = h(x_m)a^{1-\kappa}Q_m = o(Q_m), \quad 0 < \kappa < 1. \tag{3.79}$$

From (3.75) to (3.79), one finds

$$Q_m = -h(x_m)a^{2-\kappa}|S_m|a^{-2}u_e(x_m)[1 + o(1)]. \tag{3.80}$$

This yields the formula for the approximate solution to the wave scattering problem for many impedance particles:

$$u(x) = u_0(x) - a^{2-\kappa}\sum_{m=1}^{M} g(x, x_m)b_m h(x_m)u_e(x_m)[1 + o(1)], \tag{3.81}$$

where

$$b_m := |S_m|a^{-2}$$

are some positive numbers which depend on the geometry of S_m and are independent of a. For example, if all D_m are balls of radius a, then $b_m = 4\pi$.

A *LAS* for $u_e(x_m)$, analogous to (3.64), is

$$u_e(x_j) = u_0(x_j) - a^{2-\kappa}\sum_{m=1,m\neq j}^{M} g(x_j, x_m)b_m h(x_m)u_e(x_m). \tag{3.82}$$

The integral equation for the limiting effective field in the medium with embedded small particles, as $a \to 0$, is

$$u(x) = u_0(x) - b \int_D g(x, y)N(y)h(y)u(y)\mathrm{d}y, \tag{3.83}$$

where

$$u(x) = \lim_{a \to 0} u_e(x), \tag{3.84}$$

and we have assumed in (3.83) for simplicity that $b_m = b$ for all m, that is, all small particles are of the same size.

Applying operator $L_0 = \nabla^2 + k^2$ to equation (3.83), one finds the differential equation for the limiting effective field $u(x)$:

$$(\nabla^2 + k^2 - bN(x)h(x))u = 0 \text{ in } \mathbb{R}^3, \tag{3.85}$$

and u satisfies condition (3.47).

The conclusion is: the limiting medium is inhomogeneous, and its properties are described by the function

$$q(x) := bN(x)h(x). \tag{3.86}$$

This concludes our discussion of the wave scattering problem with many small impedance particles.

3.4 Creating Materials with a Desired Refraction Coefficient

Since the choice of the functions $N(x) \geq 0$ and $h(x)$, $\mathrm{Im}h(x) \leq 0$, is at our disposal, we can create the medium with a desired refraction coefficient by embedding many small impedance particles, with suitable impedances, according to the distribution law (3.50) with a suitable $N(x)$. The function

$$n_0^2(x) - k^{-2}q(x) = n^2(x) \tag{3.87}$$

is the refraction coefficient of the limiting medium, where $n_0^2(x)$ is the refraction coefficient of the original medium (see also Section 3.5). In (3.85), it is assumed that $n_0^2(x) = 1$. If $n_0^2(x) \neq 1$, then the operator $L_0 = \nabla^2 + k^2 n_0^2(x)$.

A recipe for creating material with a desired refraction coefficient can now be formulated.

Given a desired refraction coefficient $n^2(x)$, $\mathrm{Im}n^2(x) \geq 0$, one can find $N(x)$ and $h(x)$ so that (3.87) holds, where $q(x)$ is defined in (3.86), that is, one can create a material with a desired refraction coefficient by embedding into a given material many small particles with suitable boundary impedances and suitable distribution law.

3.5 Scattering by Small Particles Embedded in an Inhomogeneous Medium

Suppose that the operator $\nabla^2 + k^2$ in (3.1) and in (3.45) is replaced by the operator $L_0 = \nabla^2 + k^2 n_0^2(x)$, where $n_0^2(x)$ is a known function,

$$\text{Im } n_0^2(x) \geq 0. \tag{3.88}$$

The function $n_0^2(x)$ is the refraction coefficient of an inhomogeneous medium in which many small particles are embedded. The results, presented in Sections 3.1–3.3 remain valid if one replaces function $g(x, y)$ by the Green's function $G(x, y)$,

$$[\nabla^2 + k^2 n_0^2(x)]G(x, y) = -\delta(x - y), \tag{3.89}$$

satisfying the radiation condition. We assume that

$$n_0^2(x) = 1 \text{ in } D' := \mathbb{R}^3 \backslash D. \tag{3.90}$$

The function $G(x, y)$ is uniquely defined [14]. The derivations of the results remain essentially the same because

$$G(x, y) = g_0(x, y)[1 + O(|x - y|)], \quad |x - y| \to 0, \tag{3.91}$$

where $g_0(x, y) = \frac{1}{4\pi|x-y|}$. Estimates of $G(x, y)$ as $|x - y| \to 0$ and as $|x - y| \to \infty$ are obtained in [14]. Smallness of particles in an inhomogeneous medium with refraction coefficient $n_0^2(x)$ is described by the relation $k n_0 a \ll 1$, where $n_0 := \max_{x \in D} |n_0(x)|$, and $a = \max_{1 \leq m \leq M} \text{diam} D_m$.

3.6 Conclusions

Analytic formulas for the scattering amplitudes for wave scattering by a single small particle are derived for small acoustically soft, or hard, or impedance particles.

The equation for the effective field in the medium, in which many small particles are embedded, is derived in the limit $a \to 0$. *The physical assumptions $a \ll d \ll \lambda$ are such that the multiple scattering effects are essential.* The derivations are rigorous.

On the basis of the developed theory, efficient numerical methods are proposed for solving many-body wave scattering problems in the case of small scatterers. These methods allow one to solve the problems, which earlier were not possible to solve.

A method for creating materials with a desired refraction coefficient is given and rigorously justified. Its practical implementation requires development of a method for preparing small particles with prescribed boundary impedances.

The physically novel point, compared with the known results for wave scattering by small bodies, is the dependence on the size a of the small scatterer, which is much larger than $O(a^3)$, the Rayleigh-type dependence, see, for example, formula (3.22), where the dependence on a is $O(\zeta|S_1|) = O(a^{2-\kappa})$. The formulas for the wave scattering by small particles of an arbitrary shape for various types of the boundary conditions are new. The equations for the effective field in the medium, in which many small particles with various BCs are embedded, are new.

In this chapter, we did not discuss the EM (electromagnetic waves) scattering and the related problems of creating materials with a desired refraction coefficient [1, 8, 37].

References

1 Ramm, A.G. (2013) Scattering of acoustic and electromagnetic waves by small bodies of arbitrary shapes, in *Applications to Creating New Engineered Materials*, Momentum Press, New York.

2 Ramm, A.G. (2013) Many-body wave scattering problems in the case of small scatterers. *J. Appl. Math Comput.*, **41** (1–2), 473–500.

3 Ramm, A.G. (2017) Creating materials with a desired refraction coefficient, IOP Concise Physics, Morgan and Claypool Publishers, San Rafael, CA, USA.

4 Rayleigh, J. (1992) *Scientific Papers*, Cambridge University Press, Cambridge.

5 Landau, L. and Lifschitz, L. (1984) *Electrodynamics of Continuous Media*, Pergamon Press, Oxford.

6 van de Hulst, H.C. (1961) *Light Scattering by Small Particles*, Dover Publications, New York.

7 Ramm, A.G. (2005) *Wave Scattering by Small Bodies of Arbitrary Shapes*, World Science Publishers, Singapore.

8 Ramm, A.G. (2015) Scattering of EM waves by many small perfectly conducting or impedance bodies. *J. Math. Phys.*, **56** (N9), 091901.

9 Ramm, A.G. and Tran, N. (2015) A fast algorithm for solving scalar wave scattering problem by billions of particles. *J. Algorithms Optim.*, **3** (1), 1–13.

10 Andriychuk, M. and Ramm, A.G. (2011) Numerical solution of many-body wave scattering problem for small particles and creating materials with desired refraction coefficient, Chapter in the book *Numerical Simulations of Physical and Engineering Processes*, (edited by J. Awrejcewicz), InTech, Vienna, pp. 1–28. ISBN: 978-953-307-620-1.

11 Andriychuk, M. and Ramm, A.G. (2012) Scattering of electromagnetic waves by many thin cylinders: theory and computational modeling. *Opt. Commun.*, **285** (20), 4019–4026.

12 Ramm, A.G. (1986) *Scattering by Obstacles*, D. Reidel, Dordrecht.

13 Ramm, A.G. (2017) Scattering by obstacles and potentials, World Sci. Publishers, Singapore.

14 Ramm, A.G. (2007) Many-body wave scattering by small bodies and applications. *J. Math. Phys.*, **48**, 103511.

15 Ramm, A.G. (2007) Scattering by many small bodies and applications to condensed matter physics. *Eur. Phys. Lett.*, **80**, 44001.

16 Ramm, A.G. (2007) Wave scattering by small particles in a medium. *Phys. Lett. A*, **367**, 156–161.

17 Ramm, A.G. (2007) Wave scattering by small impedance particles in a medium. *Phys. Lett. A*, **368**, 164–172.

18 Ramm, A.G. (2007) Distribution of particles which produces a "smart" material. *J. Stat. Phys.*, **127**, 915–934.

19 Ramm, A.G. (2007) Distribution of particles which produces a desired radiation pattern. *Physica B*, **394**, 253–255.

20 Ramm, A.G. (2008) Creating wave-focusing materials. *LAJSS (Lat.-Am. J. Solids Struct.)*, **5**, 119–127.

21 Ramm, A.G. (2008) Electromagnetic wave scattering by small bodies. *Phys. Lett. A*, **372**, 4298–4306.

22 Ramm, A.G. (2008) Wave scattering by many small particles embedded in a medium. *Phys. Lett. A*, **372**, 3064–3070.

23 Ramm, A.G. (2009) Preparing materials with a desired refraction coefficient and applications, in *the book "Topics in Chaotic Systems: Selected Papers from Chaos 2008 International Conference"* (eds C. Skiadas and I. Dimotikalis), World Science Publishing, pp. 265–273.

24 Ramm, A.G. (2009) Preparing materials with a desired refraction coefficient. *Nonlinear Anal. Theory Methods Appl.*, **70**, e186–e190.

25 Ramm, A.G. (2009) Creating desired potentials by embedding small inhomogeneities. *J. Math. Phys.*, **50**, 123525.

26 Ramm, A.G. (2010) A method for creating materials with a desired refraction coefficient. *Int. J. Mod. Phys. B*, **24**, 5261–5268.

27 Ramm, A.G. (2010) Materials with a desired refraction coefficient can be created by embedding small particles into the given material. *Int. J. Struct. Changes Solids (IJSCS)*, **2**, 17–23.

28 Ramm, A.G. (2011) Wave scattering by many small bodies and creating materials with a desired refraction coefficient. *Afr. Mat.*, **22**, 33–55.

29 Ramm, A.G. (2011) Scattering by many small inhomogeneities and applications, in *the book "Topics in Chaotic Systems: Selected Papers from Chaos 2010 International Conference"* (eds C. Skiadas and I. Dimotikalis), World Science Publishing, pp. 41–52.

30 Ramm, A.G. (2010) Collocation method for solving some integral equations of estimation theory. *Int. J. Pure Appl. Math.*, **62**, 57–65.

31 Ramm, A.G. (2011) Scattering of scalar waves by many small particles. *AIP Adv.*, **1**, 022135.

32 Ramm, A.G. (2011) Scattering of electromagnetic waves by many thin cylinders. *Results Phys.*, **1** (1), 13–16.

33 Ramm, A.G. (2012) Electromagnetic wave scattering by many small perfectly conducting particles of an arbitrary shape. *Opt. Commun.*, **285** (18), 3679–3683.

34 Ramm, A.G. (2013) Wave scattering by many small bodies: transmission boundary conditions. *Rep. Math. Phys.*, **71** (3), 279–290.

35 Jikov, V., Kozlov, S., and Oleinik, O. (1994) *Homogenization of Differential Operators and Integral Functionals*, Springer-verlag, Berlin.

36 Marchenko, V. and Khruslov, E. (2006) *Homogenization of Partial Differential Equations*, Birkhäuser, Boston, MA.

37 Ramm, A.G. (2013) Scattering of electromagnetic waves by many nano-wires. *Mathematics*, **1**, 89–99.

4

Generalized Convex Functions and their Applications

Adem Kiliçman and Wedad Saleh

Department of Mathematics, Faculty of Science, Putra University of Malaysia, 43400 Serdang, Selangor, Malaysia

2010 AMS Subject Classification 26A51, 26D07, 26D15

4.1 Brief Introduction

Let $M \subseteq \mathbb{R}$ be an interval. A function $\varphi : M \subseteq \mathbb{R} \to \mathbb{R}$ is called a convex if for any $y_1, y_2 \in M$ and $\eta \in [0, 1]$,

$$\varphi(\eta y_1 + (1 - \eta)y_2) \leq \eta\varphi(y_1) + (1 - \eta)\varphi(y_2). \tag{4.1}$$

If the inequality (4.1) is the strict inequality, then φ is called a strict convex function.

From a geometrical point of view, a function φ is convex provided that the line segment connecting any two points of its graph lies on or above the graph. The function φ is strictly convex provided that the line segment connecting any two points of its graph lies above the graph. If $-\varphi$ is convex (resp. strictly convex), then φ is called concave (resp. strictly concave).

The convexity of functions have been widely used in many branches of mathematics, for example, in mathematical analysis, function theory, functional analysis, optimization theory, and so on. For a production function $x = \varphi(L)$, concacity of φ is expressed economically by saying that φ exhibits diminishing returns. While if φ is convex, then it exhibits increasing returns. Due to its applications and significant importance, the concept of convexity has been extended and generalized in several directions, see [1, 3].

Recently, the fractal theory has received significantly remarkable attention from scientists and engineers. In the sense of Mandelbrot, a fractal set is the one whose Hausdorff dimension strictly exceeds the topological dimension [4, 5]. Many researchers studied the properties of functions on fractal space and constructed many kinds of fractional calculus by using different approaches [1, 6, 7]. Particularly, in [8], Yang stated the analysis of local fractional functions

Mathematical Analysis and Applications: Selected Topics, First Edition.
Edited by Michael Ruzhansky, Hemen Dutta, and Ravi P. Agarwal.
© 2018 John Wiley & Sons, Inc. Published 2018 by John Wiley & Sons, Inc.

on fractal space systematically, which includes local fractional calculus and the monotonicity of function.

Throughout this chapter $\mathbb{R}^\alpha$ will be denoted as a real linear fractal set.

Definition 4.1 (**Mo et al. [15]**) A function $\varphi: M \subset \mathbb{R} \to \mathbb{R}^\alpha$ is called generalized convex if

$$\varphi(\eta y_1 + (1 - \eta)y_2) \leq \eta^\alpha \varphi(y_1) + (1 - \eta)^\alpha \varphi(y_2) \tag{4.2}$$

for all $y_1, y_2 \in M$, $\eta \in [0, 1]$ and $\alpha \in (0, 1]$.

It is called strictly generalized convex if the inequality (4.2) holds strictly whenever y_1 and y_2 are distinct points and $\eta \in (0, 1)$. If $-\varphi$ is generalized convex (resp., strictly generalized convex), then φ is generalized concave (respectively, strictly generalized concave).

In $\alpha = 1$, we have a convex function, that is, (4.1) is obtained.

Let $f \in {}_{a_1}I_{a_2}^{(\alpha)}$ be a generalized convex function (gECF) on $[a_1, a_2]$ with $a_1 < a_2$. Then,

$$f\left(\frac{a_1 + a_2}{2}\right) \leq \frac{\Gamma(1 + \alpha)}{(a_2 - a_1)^\alpha} {}_{a_1}I_{a_2}^{(\alpha)} f(x) \leq \frac{f(a_1) + f(a_2)}{2^\alpha} \tag{4.3}$$

is known as generalized Hermite–Hadmard's inequality [10]. Many authers paid attention to the study of generalized Hermite–Hadmard's inequality and generalized convex function, see [11, 12]. If $\alpha = 1$ in (4.3), then [13]

$$f\left(\frac{a_1 + a_2}{2}\right) \leq \frac{1}{a_2 - a_1} \int_{a_1}^{a_2} f(x)\mathrm{d}x \leq \frac{f(a_1) + f(a_2)}{2}, \tag{4.4}$$

which is known as classical Hermite–Hadamard's inequality, for more properties about this inequality we refer the interested readers to [14, 15].

4.2 Generalized E-Convex Functions

In 1999, Youness [16] introduced E-convexity of sets and functions, which have some important applications in various branches of mathematical sciences [17, 18]. However, Yang [19] showed that some results given by Youness [16] seem to be incorrect. Chen [20] extended E-convexity to a semi-E-convexity and discussed some of its properties. For more results on E-convex function or semi-E-convex function see [21, 25].

Definition 4.2 (**Youness [30]**)

(i) A set $B \subseteq \mathbb{R}^n$ is called a E-convex iff there exists $E: \mathbb{R}^n \to \mathbb{R}^n$ such that

$$\eta E(r_1) + (1 - \eta)E(r_2) \in B, \quad \forall r_1, \ r_2 \in B, \ \eta \in [0, 1].$$

(ii) A function $g: \mathbb{R}^n \to \mathbb{R}$ is called E-convex (ECF) on a set $B \subseteq \mathbb{R}^n$ iff there exists $E: \mathbb{R}^n \to \mathbb{R}^n$ and

$$g(\eta E(r_1) + (1-\eta)E(r_2)) \leq \eta g(E(r_1)) + (1-\eta)g(E(r_2)), \quad \forall r_1, r_2 \in B, \eta \in [0,1].$$

The following propositions were proved in [16].

Proposition 4.1

(i) Suppose that a set $B \subseteq \mathbb{R}^n$ is E-convex, then $E(B) \subseteq B$.
(ii) Assume that $E(B)$ is convex and $E(B) \subseteq B$, then B is E-convex.

Definition 4.3 A function $g : \mathbb{R}^n \to \mathbb{R}^\alpha$ is called gECF on a set $B \subseteq \mathbb{R}^n$ iff there exists a map $E : \mathbb{R}^n \to \mathbb{R}^n$ such that B is an E-convex set and

$$g(\eta E(r_1) + (1-\eta)E(r_2)) \leq \eta^\alpha g(E(r_1)) + (1-\eta)^\alpha g(E(r_2)), \tag{4.5}$$

$\forall r_1, r_2 \in B, \eta \in (0,1)$ and $\alpha \in (0,1]$. On the other hand, if

$$g(\eta E(x_1) + (1-\eta)E(x_2)) \geq \eta^\alpha g(E(x_1)) + (1-\eta)^\alpha g(E(x_1)),$$

$\forall x_1, x_2 \in B, \eta \in (0,1)$ and $\alpha \in (0,1]$, then g is called generalized E-concave on B. If the inequality sings in the previous two inequality are strict, then g is called generalized strictly E-convex and generalized strictly E-concave, respectively.

Proposition 4.2

(i) Every ECF on a convex set B is gECF, where $E = I$.
(ii) If $\alpha = 1$ in (4.5), then g is called ECF on a set B.
(iii) If $\alpha = 1$ and $E = I$ in (4.5), then g is called a convex function

The following two examples show that generalized E-convex function, which are not necessarily generalized convex.

Example 4.1 Assume that $B \subseteq \mathbb{R}^2$ is given as

$$B = \left\{ (x_1, x_2) \in \mathbb{R}^2 : \mu_1(0,0) + \mu_2(0,3) + \mu_3(2,1) \right\},$$

with $\mu_i > 0, \sum_{i=1}^{3} \mu_i = 1$ and define a map $E : \mathbb{R}^2 \to \mathbb{R}^2$ such as $E(x_1, x_2) = (0, x_2)$. The function $g : \mathbb{R}^2 \to \mathbb{R}^\alpha$ defined by

$$g(x_1, x_2) = \begin{cases} x_1^{3\alpha}, & x_2 < 1, \\ x_1^\alpha x_2^{3\alpha}, & x_2 \geq 1. \end{cases}$$

The function g is gECF on B, but is not generalized convex.

Remark 4.1 If $\alpha \to 0$ in the above example, then g goes to generalized convex function.

Example 4.2   Assume that $g : \mathbb{R} \to \mathbb{R}^\alpha$ is defined as

$$g(r) = \begin{cases} 1^\alpha, & r > 0, \\ (-r)^\alpha, & r \leq 0 \end{cases}$$

and assume that $E : \mathbb{R} \to \mathbb{R}$ is defined as $E(r) = -r^2$. Hence, $\mathbb{R}$ is an E-convex set and g is gECF, but is not generalized convex.

Theorem 4.1   Assume that $B \subseteq \mathbb{R}^n$ is an E-convex set and $g_1 : B \to \mathbb{R}$ is an ECF. If $g_2 : U \to \mathbb{R}^\alpha$ is nondecreasing generalized convex function such that the rang $g_1 \subset U$, then $g_2 o g_1$ is a gECF on B.

Proof: Since g_1 is ECF, then

$$g_1(\eta E(r_1) + (1 - \eta)E(r_2)) \leq \eta g_1(E(r_1)) + (1 - \eta)g_1(E(r_2)),$$

$\forall r_1, r_2 \in B$ and $\eta \in [0, 1]$. Also, since g_2 is non-decreasing generalized convex function, then

$$\begin{aligned} g_2 o g_1(\eta E(r_1) + (1 - \eta)E(r_2)) &\leq g_2[\eta g_1(E(r_1)) + (1 - \eta)g_1(E(r_2))] \\ &\leq \eta^\alpha g_2(g_1(E(r_1))) + (1 - \eta)^\alpha g_2(g_1(E(r_2))) \\ &= \eta^\alpha g_2 o g_1(E(r_1)) + (1 - \eta)^\alpha g_2 o g_1(E(r_2)), \end{aligned}$$

which implies that $g_2 o g_1$ is a gECF on B. Similarly, $g_2 o g_1$ is a strictly gECF if g_2 is a strictly non-decreasing generalized convex function. □

Theorem 4.2   Assume that $B \subseteq \mathbb{R}^n$ is an E-convex set, and $g_i : B \to \mathbb{R}^\alpha$, $i = 1, 2, \dots, l$ are generalized E-convex function. Then,

$$g = \sum_{i=1}^{l} k_i^\alpha g_i$$

is a gECF on B for all $k_i^\alpha \in \mathbb{R}^\alpha$.

Proof: Since $g_i, i = 1, 2, \dots, l$ are gECF, then

$$g_i(\eta E(r_1) + (1 - \eta)E(r_2)) \leq \eta^\alpha g_i(E(r_1)) + (1 - \eta)^\alpha g_i(E(r_2)),$$

$\forall r_1, r_2 \in B, \eta \in [0, 1]$, and $\alpha \in (0, 1]$. Then,

$$\begin{aligned} \sum_{i=1}^{l} k_i^\alpha g_i(\eta E(r_1) + (1 - \eta)E(r_2)) & \\ \leq \eta^\alpha \sum_{i=1}^{l} k_i^\alpha g_i(E(r_1)) + (1 - \eta)^\alpha \sum_{i=1}^{l} k_i^\alpha g_i(E(r_2)) & \\ = \eta^\alpha g(E(r_1)) + (1 - \eta)^\alpha g(E(r_2)). & \end{aligned}$$

Thus, g is a gECF. □

Definition 4.4   Assume that $B \subseteq \mathbb{R}^n$ is a convex set. A function $g : B \to \mathbb{R}^\alpha$ is called generalized quasi convex if

$$g(\eta r_1 + (1 - \eta)r_2) \leq \max \{g(r_1), g(r_2)\},$$

$\forall r_1, r_2 \in B$ and $\eta \in [0, 1]$.

Definition 4.5   Assume that $B \subseteq \mathbb{R}^n$ is an E-convex set. A function $g : B \to \mathbb{R}^\alpha$ is called

(i) Generalized E-quasiconvex function iff

$$g(\eta E(r_1) + (1 - \eta)E(r_2) \leq \max \{g(E(r_1)), E(g(r_2))\},$$

$\forall r_1, r_2 \in B$ and $\eta \in [0, 1]$.

(ii) Strictly generalized E-quasiconcave function iff

$$g(\eta E(r_1) + (1 - \eta)E(r_2) > \min \{g(E(r_1)), E(g(r_2))\},$$

$\forall r_1, r_2 \in B$ and $\eta \in [0, 1]$.

Theorem 4.3   Assume that $B \subseteq \mathbb{R}^n$ is an E-convex set, and $g_i : B \to \mathbb{R}^\alpha$, $i = 1, 2, \ldots, l$ are gECF. Then,

(i) The function $g : B \to \mathbb{R}^\alpha$ which is defined by $g(r) = \sup_{i \in I} g_i(r)$, $r \in B$ is a gECF on B.

(ii) If g_i, $i = 1, 2, \ldots, l$ are generalized E-quasiconvex functions on B, then the function g is a generalized E-quasiconvex function on B.

Proof:

(i) Due to g_i, $i \in I$ be gECF on B, then

$$
\begin{aligned}
g(\eta E(r_1) &+ (1 - \eta)E(r_2)) \\
&= \sup_{i \in I} g_i(\eta E(r_1) + (1 - \eta)E(r_2)) \\
&\leq \eta^\alpha \sup_{i \in I} g_i(E(r_1)) + (1 - \eta)^\alpha \sup_{i \in I} g_i(E(r_2)) \\
&= \eta^\alpha g(E(r_1)) + (1 - \eta)^\alpha g(E(r_2)).
\end{aligned}
$$

Hence, g is a gECF on B.

(ii) Since g_i, $i \in I$ are generalized E-quasiconvex functions on B, then

$$
\begin{aligned}
g(\eta E(x_1) + (1 - \eta)E(x_2)) &= \sup_{i \in I} g_i(\eta E(x_1) + (1 - \eta)E(x_2)) \\
&\leq \sup_{i \in I} \max\{g_i(E(x_1)), g_i(E(x_2))\} \\
&= \max\{\sup_{i \in I} g_i(E(x_1)), \sup_{i \in I} g_i(E(x_2))\} \\
&= \max\{g(E(x_1)), g(E(x_2))\}.
\end{aligned}
$$

Hence, g is a generalized E-quasiconvex function on B.

□

Considering $B \subseteq \mathbb{R}^n$ is a nonempty E-convex set. From Propostion 4.1(i), we get $E(B) \subseteq B$. Hence, for any $g : B \to \mathbb{R}^\alpha$, the restriction $\tilde{g} : E(B) \to \mathbb{R}^\alpha$ of g to E(B) defined by

$$\tilde{g}(\tilde{x}) = g(\tilde{x}), \quad \forall \tilde{x} \in E(B)$$

is well defined.

Theorem 4.4 Assume that $B \subseteq \mathbb{R}^n$, and $g : B \to \mathbb{R}^\alpha$ is a generalized E-quasiconvex function on B. Then, the restriction $\tilde{g} : U \to \mathbb{R}^\alpha$ of g to any nonempty convex subset U of E(B) is a generalized quasiconvex on U.

Proof: Assume that $x_1, x_2 \in U \subseteq E(B)$, then there exist $x_1^*, x_2^* \in B$ such that $x_1 = E(x_1^*)$ and $x_2 = E(x_2^*)$. Since U is a convex set, we have

$$\eta x_1 + (1 - \eta)x_2 = \eta E(x_1^*) + (1 - \eta)E(x_2^*) \in U, \quad \forall \eta \in [0, 1].$$

Therefore, we have

$$
\begin{aligned}
\tilde{g}(\eta x_1 + (1 - \eta)x_2) &= \tilde{g}(\eta E(x_1^*) + (1 - \eta)E(x_2^*)) \\
&\leq \max\{g(E(x_1^*)), g(E(x_2^*))\} \\
&= \max\{g(x_1), g(x_2)\} \\
&= \max\{\tilde{g}(x_1), \tilde{g}(x_2)\}.
\end{aligned}
$$

$\square$

Theorem 4.5 Assume that $B \subseteq \mathbb{R}^n$ is an E-convex set, and $E(B)$ is a convex set. Then, $g : B \to \mathbb{R}^\alpha$ is a generalized E-quasiconvex on B iff its restriction $\tilde{g} = g_{|_{E(B)}}$ is a generalized quasiconvex function on $E(B)$.

Proof: Due to Theorem 4.4, the if condition is true. Conversely, suppose that $x_1, x_2 \in B$, then $E(x_1), E(x_2) \in E(B)$ and $\eta E(x_1) + (1 - \eta)E(x_2) \in E(B) \subseteq B, \forall \eta \in [0, 1]$. Since $E(B) \subseteq B$, then

$$
\begin{aligned}
g(\eta E(x_1) + (1 - \eta)E(x_2)) &= \tilde{g}(\eta E(x_1) + (1 - \eta)E(x_2)) \\
&\leq \max\{\tilde{g}(E(x_1)), \tilde{g}(E(x_2))\} \\
&= \max\{g(E(x_1)), g(E(x_2))\}.
\end{aligned}
$$

$\square$

An analogous result to Theorem 4.4 for the generalized E-convex case is as follows:

Theorem 4.6 Assume that $B \subseteq \mathbb{R}^n$ is an E-convex set, and $g : B \to \mathbb{R}^\alpha$ is a gECF on B. Then, the restriction $\tilde{g} : U \to \mathbb{R}^\alpha$ of g to any nonempty convex subset U of E(B) is a gCF.

An analogous result to Theorem 4.5 for the generalized E-convex case is as follows:

Theorem 4.7   Assume that $B \subseteq \mathbb{R}^n$ is an E-convex set, and $E(B)$ is a convex set. Then, $g : B \to \mathbb{R}^\alpha$ is a gECF on B iff its restriction $\tilde{g} = g_{|_{E(B)}}$ is a gCF on $E(B)$.

The lower level set of $goE : B \to \mathbb{R}^\alpha$ is defined as

$$L_{r^\alpha}(goE) = \{x \in B : (goE)(x) = g(E(x)) \leq r^\alpha, r^\alpha \in \mathbb{R}^\alpha\}.$$

The lower level set of $\tilde{g} : E(B) \to \mathbb{R}^\alpha$ is defined as

$$L_{r^\alpha}(\tilde{g}) = \{\tilde{x} \in E(B) : \tilde{g}(\tilde{x}) = g(\tilde{x}) \leq r^\alpha, r^\alpha \in \mathbb{R}^\alpha\}.$$

Theorem 4.8   Suppose that $E(B)$ be a convex set. A function $g : B \to \mathbb{R}^\alpha$ is a generalized E-quasiconvex iff $L_{r^\alpha}(\tilde{g})$ of its restriction $\tilde{g} : E(B) \to \mathbb{R}^\alpha$ is a convex set for each $r^\alpha \in \mathbb{R}^\alpha$.

Proof: Due to $E(B)$ be a convex set, then for each $E(x_1), E(x_2) \in E(B)$, we have $\eta E(x_1) + (1 - \eta)E(x_2) \in E(B) \subseteq B$. Let $\tilde{x}_1 = E(x_1)$ and $\tilde{x}_2 = E(x_2)$. If $\tilde{x}_1, \tilde{x}_2 \in L_{r^\alpha}(\tilde{g})$, then $g(\tilde{x}_1) \leq r^\alpha$ and $g(\tilde{x}_2) \leq r^\alpha$. Thus,

$$\begin{aligned}
\tilde{g}(\eta \tilde{x}_1 + (1 - \eta)\tilde{x}_2) &= g(\eta \tilde{x}_1 + (1 - \eta)\tilde{x}_2) \\
&= g(\eta E(x_1) + (1 - \eta)E(x_2)) \\
&\leq \max\{g(E(x_1)), g(E(x_2))\} \\
&= \max\{g(\tilde{x}_1), g(\tilde{x}_2)\} \\
&= \max\{\tilde{g}(\tilde{x}_1), \tilde{g}(\tilde{x}_2)\} \\
&\leq r^\alpha,
\end{aligned}$$

which show that $\eta \tilde{x}_1 + (1 - \eta)\tilde{x}_2 \in L_{r^\alpha}(\tilde{g})$. Hence, $L_{r^\alpha}(\tilde{g})$ is a convex set.

Conversely, let $L_{r^\alpha}(\tilde{g})$ be a convex set for each $r^\alpha \in \mathbb{R}^\alpha$, that is, $\eta \tilde{x}_1 + (1 - \eta)\tilde{x}_2 \in L_{r^\alpha}(\tilde{g}), \forall \tilde{x}_1, \tilde{x}_2 \in L_{r^\alpha}(\tilde{g})$ and $r^\alpha = \max\{g(\tilde{x}_1), g(\tilde{x}_2)\}$. Thus,

$$\begin{aligned}
g(\eta E(x_1) + (1 - \eta)E(x_2)) &= \tilde{g}(\eta E(x_1) + (1 - \eta)E(x_2)) \\
&= \tilde{g}(\eta \tilde{x}_1 + (1 - \eta)\tilde{x}_2) \\
&\leq r^\alpha \\
&= \max\{g(\tilde{x}_1), g(\tilde{x}_2)\} \\
&= \max\{g(E(x_1)), g(E(x_2))\}.
\end{aligned}$$

Hence, g is a generalized E-quasiconvex.   □

Theorem 4.9   Let $B \subseteq \mathbb{R}^n$ be a nonempty E-convex set and let $g_1 : B \to \mathbb{R}^\alpha$ be a generalized E-quasiconvex on B. Suppose that $g_2 : \mathbb{R}^\alpha \to \mathbb{R}^\alpha$ is a nondecreasing function. Then, $g_2 o g_1$ is a generalized E-quasiconvex.

Proof: Since $g_1 : B \to \mathbb{R}^\alpha$ is generalized E-quasiconvex on B and $g_2 : \mathbb{R}^\alpha \to \mathbb{R}^\alpha$ is a nondecreasing function, then

$$(g_2 \circ g_1)(\eta E(x_1) + (1 - \eta)E(x_2)) = g_2(g_1(\eta E(x_1) + (1 - \eta)E(x_2)))$$
$$\leq g_2(\max\{g_1(E(x_1)), g_1(E(x_2))\})$$
$$= \max\{(g_2 \circ g_1)(E(x_1)), (g_2 \circ g_1)(E(x_2))\},$$

which shows that $g_2 \circ g_1$ is a generalized E-quasiconvex on B. □

Theorem 4.10   If the function g is a gECF on $B \subseteq \mathbb{R}^n$, then g is a generalized E-quasiconvex on B.

Proof: Assume that g is a gECF on B. Then,

$$g(\eta E(r_1) + (1 - \eta)E(r_2)) \leq \eta^\alpha g(E(r_1)) + (1 - \eta)^\alpha g(E(r_2))$$
$$\leq \eta^\alpha \max\{g(E(r_1)), g(E(r_2))\}$$
$$+ (1 - \eta)^\alpha \max\{g(E(r_1)), g(E(r_2))\}$$
$$= \max\{g(E(r_1)), g(E(r_2))\}.$$

□

4.3 E^α-Epigraph

Definition 4.6   Assume that $B \subseteq \mathbb{R}^n \times \mathbb{R}^\alpha$ and $E : \mathbb{R}^n \to \mathbb{R}^n$, then the set B is called E^α-convex set iff

$$(\eta E(x_1) + (1 - \eta)E(x_2), \eta^\alpha r_1^\alpha + (1 - \eta)^\alpha r_2^\alpha) \in B$$

$\forall (x_1, r_1^\alpha), (x_2, r_2^\alpha) \in B, \eta \in [0, 1]$, and $\alpha \in (0, 1]$.

Now, the E^α-epigraph of g is given by

$$epi_{E^\alpha}(g) = \{(E(x), r^\alpha) : x \in B, r^\alpha \in \mathbb{R}^\alpha, g(E(x)) \leq r^\alpha\}.$$

A sufficient condition for g to be a gECF is given by the following theorem:

Theorem 4.11   Let $E : \mathbb{R}^n \to \mathbb{R}^n$ be an idempoted map. Assume that $B \subseteq \mathbb{R}^n$ is an E-convex set and $epi_{E^\alpha}(g)$ is an E^α-convex set where $g : B \to \mathbb{R}^\alpha$, then g is a gECF on B.

Proof: Assume that $r_1, r_2 \in B$ and $(E(r_1), g(E(r_1))), (E(r_2), g(E(r_2))) \in epi_{E^\alpha}(g)$. Since $epi_{E^\alpha}(g)$ is E^α-convex set, we have

$$(\eta E(E(r_1)) + (1 - \eta)E(E(r_2)), \eta^\alpha g(E(r_1)) + (1 - \eta)^\alpha g(E(r_2))) \in epi_{E^\alpha}(g),$$

then

$$g(E(\eta E(r_1)) + (1 - \eta)E(E(r_2))) \leq \eta^\alpha g(E(r_1)) + (1 - \eta)^\alpha g(E(r_2)).$$

Due to $E : \mathbb{R}^n \to \mathbb{R}^n$ be an idempotent map, then

$$g(\eta E(r_1) + (1-\eta)E(r_2)) \le \eta^{\alpha} g(E(r_1)) + (1-\eta)^{\alpha} g(E(r_2)).$$

Hence, g is a gECF. □

Theorem 4.12 Assume that $\{B_i\}_{i \in I}$ is a family of E^{α}-convex sets. Then, their intersection $\cap_{i \in I} B_i$ is an E^{α}-convex set.

Proof: Considering $(x_1, r_1^{\alpha}), (x_2, r_2^{\alpha}) \in \cap_{i \in I} B_i$, then $(x_1, r_1^{\alpha}), (x_2, r_2^{\alpha}) \in B_i, \forall i \in I$. By E^{α}-convexity of $B_i, \forall i \in I$, then we have

$$(\eta E(x_1) + (1-\eta)E(x_2), \eta^{\alpha} r_1^{\alpha} + (1-\eta)^{\alpha} r_2^{\alpha}) \in B_i,$$

$\forall \eta \in [0,1]$ and $\alpha \in (0,1]$. Hence,

$$(\eta E(x_1) + (1-\eta)E(x_2), \eta^{\alpha} r_1^{\alpha} + (1-\eta)^{\alpha} r_2^{\alpha}) \in \cap_{i \in I} B_i.$$

□

The following theorem is a special case of Theorem 4.3(i) where $E : \mathbb{R}^n \to \mathbb{R}^n$ is an idempotent map.

Theorem 4.13 Assume that $E : \mathbb{R}^n \to \mathbb{R}^n$ is an idempotent map, and $B \subseteq \mathbb{R}^n$ is an E-convex set. Let $\{g_i\}_{i \in I}$ be a family function that is bounded from above. If $epi_{E^{\alpha}}(g_i)$ are E^{α}-convex sets, then the function g which defined by $g(x) = \sup_{i \in I} g_i(x), x \in B$ is a gECF on B.

Proof: Since

$$epi_{E^{\alpha}}(g_i) = \{(E(x), r^{\alpha}) : x \in B, r^{\alpha} \in \mathbb{R}^{\alpha}, g_i(E(x)) \le r^{\alpha}, i \in I\}$$

are E^{α}-convex set in $B \times \mathbb{R}^{\alpha}$, then

$$\cap_{i \in I} epi_{E^{\alpha}}(g_i) = \{(E(x), r^{\alpha}) : x \in B, r^{\alpha} \in \mathbb{R}^{\alpha}, g_i(E(x)) \le r^{\alpha}, i \in I\}$$
$$= \{(E(x), r^{\alpha}) : x \in B, r^{\alpha} \in \mathbb{R}^{\alpha}, g(E(x)) \le r^{\alpha}\}, \qquad (4.6)$$

where $g(E(x)) = \sup_{i \in I} g_i(E(x))$, also is E^{α}-convex set. Hence, $\cap_{i \in I} epi_{E^{\alpha}}(g_i)$ is an E^{α}-epigraph, then by Theorem 4.12, g is a generalized E-convex function on B. □

4.4 Generalized s-Convex Functions

There are many researchers who studied the properties of functions on fractal space and constructed many kinds of fractional calculus by using different approaches, see [26, 28].

In [10], two kinds of generalized *s*-convex functions on fractal sets are introduced as follows:

Definition 4.7

(i) A function $\varphi : \mathbb{R}_+ \to \mathbb{R}^\alpha$ is called a generalized s-convex $(0 < s < 1)$ in the first sense if

$$\varphi(\eta_1 y_1 + \eta_2 y_2) \leq \eta_1^{s\alpha} \varphi(y_1) + \eta_2^{s\alpha} \varphi(y_2) \tag{4.7}$$

for all $y_1, y_2 \in \mathbb{R}_+$ and all $\eta_1, \eta_2 \geq 0$ with $\eta_1^s + \eta_2^s = 1$, this class of functions is denoted by GK_s^1.

(ii) A function $\varphi : \mathbb{R}_+ \to \mathbb{R}^\alpha$ is called a generalized s-convex $(0 < s < 1)$ in the second sense if (4.7) holds for all $y_1, y_2 \in \mathbb{R}_+$ and all $\eta_1, \eta_2 \geq 0$ with $\eta_1 + \eta_2 = 1$, this class of functions is denoted by GK_s^2.

In the same paper [10], Mo and Sui proved that all functions that are generalized s-convex in the second sense, for $s \in (0, 1)$, are nonnegative.

If $\alpha = 1$ in Definition 4.7, then we have the classical s-convex functions in the first sense (second sense) see [13].

Also, Dragomir and Fitzatrick [13] demonstrated a variation of Hadamard's inequality, which holds for s-convex functions in the second sense.

Theorem 4.14 Assume that $\varphi : \mathbb{R}_+ \to \mathbb{R}_+$ is a s-convex function in the second sense, $0 < s < 1$ and $y_1, y_2 \in \mathbb{R}_+, y_1 < y_2$. If $\varphi \in L^1([y_1, y_2])$, then

$$2^{s-1} \varphi \left(\frac{y_1 + y_2}{2} \right) \leq \frac{1}{y_2 - y_1} \int_{y_1}^{y_2} \varphi(z) dz \leq \frac{\varphi(y_1) + \varphi(y_2)}{s + 1}. \tag{4.8}$$

If we set $k = \frac{1}{s+1}$, then it is the best possible in the second inequality in (4.8).

A variation of generalized Hadamard's inequality that holds for generalized s-convex functions in the second sense [1].

Theorem 4.15 Assume that $\varphi : \mathbb{R}_+ \to \mathbb{R}_+^\alpha$ is a generalized s-convex function in the second sense where $0 < s < 1$ and $y_1, y_2 \in \mathbb{R}_+$ with $y_1 < y_2$. If $\varphi \in L^1([y_1, y_2])$, then

$$2^{\alpha(s-1)} \varphi \left(\frac{y_1 + y_2}{2} \right) \leq \frac{\Gamma(1 + \alpha)}{(y_2 - y_1)^\alpha} \, _{y_1} I_{y_2}^{(\alpha)} \varphi(x)$$

$$\leq \frac{\Gamma(1 + s\alpha)\Gamma(1 + \alpha)}{\Gamma(1 + (s + 1)\alpha)} (\varphi(y_1) + \varphi(y_2)). \tag{4.9}$$

Proof: We know that φ is generalized s-convex in the second sense, which lead to

$$\varphi(\eta y_1 + (1 - \eta) y_2) \leq \eta^{\alpha s} \varphi(y_1) + (1 - \eta)^{\alpha s} \varphi(y_2), \forall \eta \in [0, 1].$$

Then, the following inequality can be written:

$$\Gamma(1+\alpha)\ {}_{0}I_{1}^{(\alpha)}\varphi(\eta y_1 + (1-\eta)y_2) \le \varphi(y_1)\Gamma(1+\alpha){}_{0}I_{1}^{(\alpha)}\eta^{as}$$
$$+ \varphi(y_2)\Gamma(1+\alpha){}_{0}I_{1}^{(\alpha)}(1-\eta)^{as}$$
$$= \frac{\Gamma(1+s\alpha)\Gamma(1+\alpha)}{\Gamma(1+(s+1)\alpha)}(\varphi(y_2)+y_2).$$

By considering $z = \eta y_1 + (1-\eta)y_2$. Then

$$\Gamma(1+\alpha)\ {}_{0}I_{1}^{(\alpha)}\varphi(\eta y_1 + (1-\eta)y_2) = \frac{\Gamma(1+\alpha)}{(y_1-y_2)^{\alpha}}\ {}_{y_2}I_{y_1}^{(\alpha)}\varphi(x)$$
$$= \frac{\Gamma(1+\alpha)}{(y_2-y_1)^{\alpha}}\ {}_{y_1}I_{y_2}^{(\alpha)}\varphi(z).$$

Here

$$\frac{\Gamma(1+\alpha)}{(y_2-y_1)^{\alpha}}\ {}_{y_1}I_{y_2}^{(\alpha)}\varphi(x) \le \frac{\Gamma(1+s\alpha)\Gamma(1+\alpha)}{\Gamma(1+(s+1)\alpha)}(\varphi(y_1)+\varphi(y_2)).$$

Then, the second inequality in (4.9) is given. Now

$$\varphi\left(\frac{z_1+z_2}{2}\right) \le \frac{\varphi(z_1)+\varphi(z_2)}{2^{as}}, \quad \forall z_1, z_2 \in I. \tag{4.10}$$

Let $z_1 = \eta y_1 + (1-\eta)y_2$ and $z_2 = (1-\eta)y_1 + \eta y_2$ with $\eta \in [0,1]$. Hence, by applying (4.10), the next inequalty holds

$$\varphi\left(\frac{y_1+y_2}{2}\right) \le \frac{\varphi(\eta y_1 + (1-\eta)y_2) + \varphi((1-\eta)y_1 + \eta y_2)}{2^{as}}, \quad \forall \eta \in [0,1].$$

So

$$\frac{1}{\Gamma(1+\alpha)}\int_0^1 \varphi\left(\frac{y_1+y_2}{2}\right)(d\eta)^{\alpha} \le \frac{1}{2^{\alpha(s-1)}(y_2-y_1)^{\alpha}}\ {}_{y_1}I_{y_2}^{(\alpha)}\varphi(z).$$

Then,

$$2^{\alpha(s-1)}\varphi\left(\frac{y_1+y_2}{2}\right) \le \frac{\Gamma(1+\alpha)}{(y_2-y_1)^{\alpha}}\ {}_{y_1}I_{y_2}^{(\alpha)}\varphi(z).$$

$\square$

Lemma 4.1 Assume that $\varphi : [y_1, y_2] \subset \mathbb{R} \to \mathbb{R}^{\alpha}$ is a local fractional derivative of order α ($\varphi \in D_{\alpha}$) on (y_1, y_2) with $y_1 < y_2$. If $\varphi^{(2\alpha)} \in C_{\alpha}[y_1, y_2]$, then the following equality holds:

$$\frac{\Gamma(1+2\alpha)[\Gamma(1+\alpha)]^2}{2^{\alpha}(y_2-y_1)}\ {}_{y_1}I_{y_2}^{(\alpha)}\varphi(x) - \frac{\Gamma(1+2\alpha)}{2^{\alpha}}\varphi\left(\frac{y_1+y_2}{2}\right)$$
$$= \frac{(y_2-y_1)^{2\alpha}}{16^{\alpha}}\left[{}_{0}I_{1}^{(\alpha)}\gamma^{2\alpha}\varphi^{(2\alpha)}\left(\gamma\frac{y_1+y_2}{2} + (1-\gamma)y_1\right)\right.$$
$$\left. + {}_{0}I_{1}^{(\alpha)}(\gamma-1)^{2\alpha}\varphi^{(2\alpha)}\left(\gamma y_2 + (1-\gamma)\frac{y_1+y_2}{2}\right)\right].$$

Proof: From the local fractional integration by parts, we get

$$
\begin{aligned}
B_1 &= \frac{1}{\Gamma(1+\alpha)} \int_0^1 \gamma^{2\alpha} \varphi^{(2\alpha)} \left(\gamma \frac{y_1+y_2}{2} + (1-\gamma)y_1 \right) (d\gamma)^\alpha \\
&= \left(\frac{2}{y_2-y_1} \right)^\alpha \varphi^{(\alpha)} \left(\frac{y_1+y_2}{2} \right) \\
&\quad - \Gamma(1+2\alpha) \left(\frac{2}{y_2-y_1} \right)^{2\alpha} \gamma^\alpha \varphi \left(\gamma \frac{y_1+y_2}{2} + (1-\gamma)y_1 \right) \Big|_0^1 \\
&\quad + \Gamma(1+2\alpha)\Gamma(1+\alpha) \left(\frac{2}{b-a} \right)^{2\alpha} \int_0^1 \varphi \left(\gamma \frac{y_1+y_2}{2} + (1-\gamma)y_1 \right) (d\gamma)^\alpha \\
&= \left(\frac{2}{y_2-y_1} \right)^\alpha \varphi^{(\alpha)} \left(\frac{y_1+y_2}{2} \right) - \Gamma(1+2\alpha) \left(\frac{2}{y_2-y_1} \right)^{2\alpha} \varphi \left(\frac{y_1+y_2}{2} \right) \\
&\quad + \Gamma(1+2\alpha)\Gamma(1+\alpha) \left(\frac{2}{y_2-y_1} \right)^{2\alpha} \int_0^1 \varphi \left(\gamma \frac{y_1+y_2}{2} + (1-\gamma)y_1 \right) (d\gamma)^\alpha.
\end{aligned}
$$

Setting $x = \gamma \frac{y_1+y_2}{2} + (1-\gamma)y_1$, for $\gamma \in [0,1]$ and multiply both sides in the last equation by $\frac{(y_2-y_1)^{2\alpha}}{16^\alpha}$, we get

$$
\begin{aligned}
B_1 &= \frac{(y_2-y_1)^{2\alpha}}{16^\alpha} {}_0 I_1^{(\alpha)} \gamma^{2\alpha} \varphi^{(2\alpha)} \left(\gamma \frac{y_1+y_2}{2} + (1-\gamma)y_1 \right) \\
&= \frac{(y_2-y_1)^\alpha}{8^\alpha} \varphi^{(\alpha)} \left(\frac{y_1+y_2}{2} \right) - \frac{\Gamma(1+2\alpha)}{4^\alpha} \varphi \left(\frac{y_1+y_2}{2} \right) \\
&\quad + \frac{\Gamma(1+2\alpha)\Gamma(1+\alpha)}{2^\alpha(y_2-y_1)^\alpha} \int_{y_1}^{y_1+y_2/2} \varphi(x)(dx)^\alpha.
\end{aligned}
$$

By the similar way, also we have

$$
\begin{aligned}
B_2 &= \frac{(y_2-y_1)^{2\alpha}}{16^\alpha} {}_0 I_1^{(\alpha)} (\gamma-1)^{2\alpha} \varphi^{(2\alpha)} \left(\gamma y_2 + (1-\gamma)\frac{y_1+y_2}{2} \right) \\
&= -\frac{(y_2-y_1)^\alpha}{8^\alpha} \varphi^{(\alpha)} \left(\frac{y_1+y_2}{2} \right) - \frac{\Gamma(1+2\alpha)}{4^\alpha} \varphi \left(\frac{y_1+y_2}{2} \right) \\
&\quad + \frac{\Gamma(1+2\alpha)\Gamma(1+\alpha)}{2^\alpha(y_2-y_1)^\alpha} \int_{\frac{y_1+y_2}{2}}^{y_2} \varphi(x)(dx)^\alpha.
\end{aligned}
$$

Thus, adding B_1 and B_2, we get the desired result. $\qquad\square$

Theorem 4.16 Assume that $\varphi : U \subset [0,\infty) \to \mathbb{R}^\alpha$ such that $\varphi \in D_\alpha$ on $Int(U)$ ($Int(U)$ is the interior of U) and $\varphi^{(2\alpha)} \in C_\alpha[y_1,y_2]$, where $y_1, y_2 \in U$ with $y_1 < y_1$. If $|\varphi|$ is generalized s-convex on $[y_1,y_2]$, for some fixed $0 < s \leq 1$, then the following inequality holds:

$$\left| \frac{\Gamma(1+2\alpha)}{2^\alpha} \varphi\left(\frac{y_1+y_2}{2}\right) - \frac{\Gamma(1+2\alpha)[\Gamma(1+\alpha)]^2}{2^\alpha(y_2-y_1)^\alpha} {}_{y_1}I^{(\alpha)}_{y_2}\varphi(x) \right|$$

$$\leq \frac{(y_2-y_1)^{2\alpha}}{16^\alpha} \left\{ \frac{2^\alpha\Gamma(1+(s+2)\alpha)}{\Gamma(1+(s+3)\alpha)} \left| \varphi^{(2\alpha)}\left(\frac{y_1+y_2}{2}\right) \right| + \left[\frac{\Gamma(1+s\alpha)}{\Gamma(1+(s+1)\alpha)} \right. \right.$$

$$\left. \left. - 2^\alpha \frac{\Gamma(1+(s+1)\alpha)}{\Gamma(1+(s+2)\alpha)} + \frac{\Gamma(1+(s+2)\alpha)}{\Gamma(1+(s+3)\alpha)} \right] \left[|\varphi^{(2\alpha)}(y_1)| + |\varphi^{(2\alpha)}(y_2)| \right] \right\} \quad (4.11)$$

$$\leq \frac{(y_2-y_1)^{2\alpha}}{16^\alpha} \left\{ \frac{2^{\alpha(2-s)}\Gamma(1+(s+2)\alpha)}{\Gamma(1+(s+3)\alpha)} \frac{\Gamma(1+s\alpha)\Gamma(1+\alpha)}{\Gamma(1+(s+1)\alpha)} + \frac{\Gamma(1+s\alpha)}{\Gamma(1+(s+1)\alpha)} \right.$$

$$\left. - \frac{2^\alpha\Gamma(1+(s+1)\alpha)}{\Gamma(1+(s+2)\alpha)} + \frac{\Gamma(1+(s+2)\alpha)}{\Gamma(1+(s+3)\alpha)} \right\} \left[|\varphi^{(2\alpha)}(y_1)| + |\varphi^{(2\alpha)}(y_2)| \right]. \quad (4.12)$$

Proof: From Lemma 4.1, we have

$$\left| \frac{\Gamma(1+2\alpha)}{2^\alpha} \varphi\left(\frac{y_1+y_2}{2}\right) - \frac{\Gamma(1+2\alpha)[\Gamma(1+\alpha)]^2}{2^\alpha(y_2-y_1)^\alpha} {}_{y_1}I^{(\alpha)}_{y_2}\varphi(x) \right|$$

$$\leq \frac{(y_2-y_1)^{2\alpha}}{16^\alpha} \left[{}_0I^{(\alpha)}_1 \gamma^{2\alpha} \left| \varphi^{(2\alpha)}\left(\gamma\frac{y_1+y_2}{2} + (1-\gamma)y_1\right) \right| \right.$$

$$\left. + {}_0I^{(\alpha)}_1(\gamma-1)^{2\alpha} \left| \varphi^{(2\alpha)}\left(\gamma y_2 + (1-\gamma)\frac{y_1+y_2}{2}\right) \right| \right]$$

$$\leq \frac{(y_2-y_1)^{2\alpha}}{16^\alpha} {}_0I^{(\alpha)}_1 \gamma^{2\alpha} \left[\gamma^{\alpha s} \left| \varphi^{(2\alpha)}\left(\frac{y_1+y_2}{2}\right) \right| + (1-\gamma)^{\alpha s}|\varphi^{(2\alpha)}(y_1)| \right]$$

$$+ \frac{(y_2-y_1)^{2\alpha}}{16^\alpha} {}_0I^{(\alpha)}_1(\gamma-1)^{2\alpha} \left[\gamma^{\alpha s}|\varphi^{(2\alpha)}(y_2)| + (1-\gamma)^{\alpha s} \left| \varphi^{(2\alpha)}\left(\frac{y_1+y_2}{2}\right) \right| \right]$$

$$= \frac{(y_2-y_1)^{2\alpha}}{16^\alpha} \left\{ \frac{\Gamma(1+(s+2)\alpha)}{\Gamma(1+(s+3)\alpha)} \left| \varphi^{(2\alpha)}\left(\frac{y_1+y_2}{2}\right) \right| \right.$$

$$\left. + \left[\frac{\Gamma(1+\alpha s)}{\Gamma(1+(s+1)\alpha)} - 2^\alpha\frac{\Gamma(1+(s+1)\alpha)}{\Gamma(1+(s+2)\alpha)} + \frac{\Gamma(1+(s+2)\alpha)}{\Gamma(1+(s+3)\alpha)} \right] |\varphi^{(2\alpha)}(y_1)| \right\}$$

$$+ \frac{(y_2-y_1)^{2\alpha}}{16^\alpha} \left\{ \frac{\Gamma(1+(s+2)\alpha)}{\Gamma(1+(s+3)\alpha)} \left| \varphi^{(2\alpha)}\left(\frac{y_1+y_2}{2}\right) \right| \right.$$

$$\left. + \left[\frac{\Gamma(1+\alpha s)}{\Gamma(1+(s+1)\alpha)} - 2^\alpha\frac{\Gamma(1+(s+1)\alpha)}{\Gamma(1+(s+2)\alpha)} + \frac{\Gamma(1+(s+2)\alpha)}{\Gamma(1+(s+3)\alpha)} \right] |\varphi^{(2\alpha)}(a_2)| \right.$$

$$\left. + \frac{\Gamma(1+(s+2)\alpha)}{\Gamma(1+(s+3)\alpha)} \left| \varphi^{(2\alpha)}\left(\frac{y_1+y_2}{2}\right) \right| \right\}$$

$$= \frac{(y_2 - y_1)^{2\alpha}}{16^\alpha} \left\{ \frac{2^\alpha \Gamma(1 + (s+2)\alpha)}{\Gamma(1 + (s+3)\alpha)} \left| \varphi^{(2\alpha)} \left(\frac{y_1 + y_2}{2} \right) \right| + \left[\frac{\Gamma(1 + s\alpha)}{\Gamma(1 + (s+1)\alpha)} \right. \right.$$

$$\left. \left. -2^\alpha \frac{\Gamma(1 + (s+1)\alpha)}{\Gamma(1 + (s+2)\alpha)} + \frac{\Gamma(1 + (s+1)\alpha)}{\Gamma(1 + (s+3)\alpha)} \right] \left[|\varphi^{(2\alpha)}(y_1)| + |\varphi^{(2\alpha)}(y_2)| \right] \right\}.$$

This proves inequality (4.11). Since

$$2^{\alpha(s-1)} \varphi^{(2\alpha)} \left(\frac{y_1 + y_2}{2} \right) \le \frac{\Gamma(1 + s\alpha)\Gamma(1 + \alpha)}{\Gamma(1 + (s+1)\alpha)} (\varphi^{(2\alpha)}(y_1) + \varphi^{(2\alpha)}(y_2)),$$

then

$$\left| \frac{\Gamma(1 + 2\alpha)}{2^\alpha} \varphi \left(\frac{y_1 + y_2}{2} \right) - \frac{\Gamma(1 + 2\alpha)[\Gamma(1 + \alpha)]^2}{2^\alpha (y_2 - y_1)^\alpha} \, {}_{y_1} I_{y_2}^{(\alpha)} \varphi(x) \right|$$

$$\le \frac{(y_2 - y_1)^{2\alpha}}{16^\alpha} \left\{ \frac{2^\alpha \Gamma(1 + (s+2)\alpha)}{\Gamma(1 + (s+3)\alpha)} \frac{2^{-\alpha(s-1)} \Gamma(1 + s\alpha)\Gamma(1 + \alpha)}{\Gamma(1 + (s+1)\alpha)} \right.$$

$$\times \left[|\varphi^{(2\alpha)}(y_1)| + |\varphi^{(2\alpha)}(y_2)| \right]$$

$$+ \left[\frac{\Gamma(1 + s\alpha)}{\Gamma(1 + (s+1)\alpha)} - \frac{2^\alpha \Gamma(1 + (s+1)\alpha)}{\Gamma(1 + (s+2)\alpha)} + \frac{\Gamma(1 + (s+2)\alpha)}{\Gamma(1 + (s+3)\alpha)} \right]$$

$$\left. \times \left[|\varphi^{(2\alpha)}(y_1)| + |\varphi^{(2\alpha)}(y_2)| \right] \right\}$$

$$= \frac{(y_2 - y_1)^{2\alpha}}{16^\alpha} \left\{ \frac{2^{\alpha(2-s)} \Gamma(1 + (s+2)\alpha)}{\Gamma(1 + (s+3)\alpha)} \frac{\Gamma(1 + s\alpha)\Gamma(1 + \alpha)}{\Gamma(1 + (s+1)\alpha)} + \frac{\Gamma(1 + s\alpha)}{\Gamma(1 + (s+1)\alpha)} \right.$$

$$\left. - \frac{2^\alpha \Gamma(1 + (s+1)\alpha)}{\Gamma(1 + (s+2)\alpha)} + \frac{\Gamma(1 + (s+2)\alpha)}{\Gamma(1 + (s+3)\alpha)} \left[|\varphi^{(2\alpha)}(y_1)| + |\varphi^{(2\alpha)}(y_2)| \right] \right\}$$

Thus, we get the inequality (4.12) and the proof is complete. $\qquad\square$

Remark 4.2

1. When $\alpha = 1$, Theorem 4.16 reduce to Theorem 2 in [29].
2. If $s = 1$ in Theorem 4.16, then

$$\left| \frac{\Gamma(1 + 2\alpha)}{2^\alpha} \varphi \left(\frac{y_1 + y_2}{2} \right) - \frac{\Gamma(1 + 2\alpha)[\Gamma(1 + \alpha)]^2}{2^\alpha (y_2 - y_1)^\alpha} \, {}_{y_1} I_{y_2}^{(\alpha)} \varphi(x) \right|$$

$$\le \frac{(y_2 - y_1)^{2\alpha}}{16^\alpha} \left\{ \frac{2^\alpha \Gamma(1 + 3\alpha)}{\Gamma(1 + 4\alpha)} \left| \varphi^{(2\alpha)} \left(\frac{y_1 + y_2}{2} \right) \right| + \left[\frac{\Gamma(1 + \alpha)}{\Gamma(1 + 2\alpha)} \right. \right.$$

$$\left. \left. - 2^\alpha \frac{\Gamma(1 + 2\alpha)}{\Gamma(1 + 3\alpha)} + \frac{\Gamma(1 + 3\alpha)}{\Gamma(1 + 4\alpha)} \right] \left[|\varphi^{(2\alpha)}(y_1)| + |\varphi^{(2\alpha)}(y_2)| \right] \right\}$$

$$\le \frac{(y_2 - y_1)^{2\alpha}}{16^\alpha} \left\{ \frac{2^\alpha \Gamma(1 + 3\alpha)}{\Gamma(1 + 4\alpha)} \frac{[\Gamma(1 + \alpha)]^2}{\Gamma(1 + 2\alpha)} + \frac{\Gamma(1 + \alpha)}{\Gamma(1 + 2\alpha)} \right.$$

$$\left. - \frac{2^\alpha \Gamma(1 + 2\alpha)}{\Gamma(1 + 3\alpha)} + \frac{\Gamma(1 + 3\alpha)}{\Gamma(1 + 4\alpha)} \right\} \left[|\varphi^{(2\alpha)}(y_1)| + |\varphi^{(2\alpha)}(y_2)| \right]. \qquad (4.13)$$

3. If $s = 1$ and $\alpha = 1$ in Theorem 4.16, then

$$\left| \varphi\left(\frac{y_1 + y_2}{2}\right) - \frac{1}{y_2 - y_1} \int_{y_1}^{y_2} \varphi(x)dx \right|$$

$$\leq \frac{(y_2 - y_1)^2}{192} \left\{ 6\left|\varphi''\left(\frac{y_1 + y_2}{2}\right)\right| + |\varphi''(y_1)| + |\varphi''(y_2)| \right\}$$

$$\leq \frac{(y_2 - y_1)^2}{48} \{|\varphi''(y_1)| + |\varphi''(y_2)|\}.$$

We give a new upper bound of the left generalized Hadamard's inequality for generalized s-convex functions in the following theorem:

Theorem 4.17 Assume that $\varphi : U \subset [0, \infty) \to \mathbb{R}^\alpha$ such that $\varphi \in D_\alpha$ on $Int(U)$ and $\varphi^{(2\alpha)} \in C_\alpha[y_1, y_2]$, where $y_1, y_2 \in U$ with $y_1 < y_2$. If $|\varphi^{(2\alpha)}|^{p_2}$ is generalized s-convex on $[y_1, y_2]$, for some fixed $0 < s \leq 1$ and $p_2 > 1$ with $\frac{1}{p_1} + \frac{1}{p_2} = 1$, then the following inequality holds:

$$\left| \frac{\Gamma(1 + 2\alpha)}{2^\alpha} \varphi\left(\frac{y_1 + y_2}{2}\right) - \frac{\Gamma(1 + 2\alpha)[\Gamma(1 + \alpha)]^2}{2^\alpha(y_2 - y_1)^\alpha} {}_{y_1}I_{y_2}^{(\alpha)} \varphi(x) \right|$$

$$\leq \frac{(y_2 - y_1)^{2\alpha}}{16^\alpha} \left[\frac{\Gamma(1 + s\alpha)}{\Gamma(1 + (s+1)\alpha)} \right]^{1/p_2} \left[\frac{\Gamma(1 + 2p_1\alpha)}{\Gamma(1 + (2p_1 + 1)\alpha)} \right]^{1/p_1}$$

$$\times \left[\left(\left|\varphi^{(2\alpha)}\left(\frac{y_1 + y_2}{2}\right)\right|^{p_2} + |\varphi^{(2\alpha)}(y_1)|^{p_2} \right)^{1/p_2} \right.$$

$$\left. + \left(\left|\varphi^{(2\alpha)}\left(\frac{y_1 + y_2}{2}\right)\right|^{p_2} + |\varphi^{(2\alpha)}(y_2)|^{p_2} \right)^{1/p_2} \right]. \tag{4.14}$$

Proof: Let $p_1 > 1$, then from Lemma 4.1 and using generalized Hölder's inequality [8], we obtain

$$\left| \frac{\Gamma(1 + 2\alpha)}{2^\alpha} \varphi\left(\frac{y_1 + y_2}{2}\right) - \frac{\Gamma(1 + 2\alpha)[\Gamma(1 + \alpha)]^2}{2^\alpha(y_2 - y_1)^\alpha} {}_{y_1}I_{y_2}^{(\alpha)} \varphi(x) \right|$$

$$\leq \frac{(y_2 - y_1)^{2\alpha}}{16^\alpha} \left\{ {}_0I_1^{(\alpha)} \gamma^{2\alpha} \left| \varphi^{(2\alpha)}\left(\gamma\frac{y_1 + y_2}{2} + (1 - \gamma)y_1 \right) \right| \right.$$

$$\left. + {}_0I_1^{(\alpha)}(\gamma - 1)^{2\alpha} \left| \varphi^{(2\alpha)}\left(\gamma y_2 + (1 - \gamma)\frac{y_1 + y_2}{2} \right) \right| \right\}$$

$$\leq \frac{(y_2 - y_1)^{2\alpha}}{16^\alpha} ({}_0I_1^{(\alpha)} \gamma^{2p_1\alpha})^{1/p_1}$$

$$\times \left({}_0I_1^{(\alpha)} \left| \varphi^{(2\alpha)} \left(\gamma \frac{y_1 + y_2}{2} + (1 - \gamma)y_1 \right) \right|^{p_2} \right)^{1/p_2}$$

$$+ \frac{(y_2 - y_1)^{2\alpha}}{16^{\alpha}} ({}_0I_1^{(\alpha)}(1 - \gamma)^{2p_1\alpha})^{1/p_1}$$

$$\times \left({}_0I_1^{(\alpha)} \left| \varphi^{(2\alpha)} \left(\gamma y_2 + (1 - \gamma)\frac{y_1 + y_2}{2} \right) \right|^{p_2} \right)^{1/p_2}.$$

Since $|\varphi^{(2\alpha)}|^{p_2}$ is generalized s-convex, then

$$
{}_0I_1^{(\alpha)} \left| \varphi^{(2\alpha)} \left(\gamma \frac{y_1 + y_2}{2} + (1 - \gamma)y_1 \right) \right|^{p_2}
$$
$$
\leq \frac{\Gamma(1 + s\alpha)}{\Gamma(1 + (s + 1)\alpha)} \left| \varphi^{(2\alpha)} \left(\frac{y_1 + y_2}{2} \right) \right|^{p_2} + \frac{\Gamma(1 + s\alpha)}{\Gamma(1 + (s + 1)\alpha)} |\varphi^{(2\alpha)}(y_1)|^{p_2},
$$

which means

$$
{}_0I_1^{(\alpha)} \left| \varphi^{(2\alpha)} \left(\gamma y_2 + (1 - \gamma)\frac{y_1 + y_2}{2} \right) \right|^{p_2}
$$
$$
\leq \frac{\Gamma(1 + s\alpha)}{\Gamma(1 + (s + 1)\alpha)} |\varphi^{(2\alpha)}(y_2)|^{p_2}
$$
$$
+ \frac{\Gamma(1 + s\alpha)}{\Gamma(1 + (s + 1)\alpha)} \left| \varphi^{(2\alpha)} \left(\frac{y_1 + y_2}{2} \right) \right|^{p_2}.
$$

Hence

$$
\left| \frac{\Gamma(1 + 2\alpha)}{2^{\alpha}} \varphi \left(\frac{y_1 + y_2}{2} \right) - \frac{\Gamma(1 + 2\alpha)[\Gamma(1 + \alpha)]^2}{2^{\alpha}(y_2 - y_1)^{\alpha}} \; {}_{y_1}I_{y_2}^{(\alpha)} \varphi(x) \right|
$$
$$
\leq \frac{(y_2 - y_1)^{2\alpha}}{16^{\alpha}} \left[\frac{\Gamma(1 + s\alpha)}{\Gamma(1 + (s + 1)\alpha)} \right]^{1/p_2} \left[\frac{\Gamma(1 + 2p_1\alpha)}{\Gamma(1 + (2p_1 + 1)\alpha)} \right]^{1/p_1}
$$
$$
\times \left\{ \left[\left| \varphi^{(2\alpha)} \left(\frac{y_1 + y_2}{2} \right) \right|^{p_2} + |\varphi^{(2\alpha)}(y_1)|^{p_2} \right]^{1/p_2} \right.
$$
$$
\left. + \left[\left| \varphi^{(2\alpha)} \left(\frac{y_1 + y_2}{2} \right) \right|^{p_2} + |\varphi^{(2\alpha)}(y_2)|^{p_2} \right]^{1/p_2} \right\}.
$$

The proof is complete. $\qquad\qquad\qquad\qquad\qquad\qquad\qquad\qquad\qquad\qquad \square$

Remark 4.3 If $s = 1$ in Theorem 4.17, then

$$\left| \frac{\Gamma(1+2\alpha)}{2^\alpha} \varphi \left(\frac{y_1 + y_2}{2} \right) - \frac{\Gamma(1+2\alpha)[\Gamma(1+\alpha)]^2}{2^\alpha (y_2 - y_1)^\alpha} \, {}_{y_1}I_{y_2}^{(\alpha)} \varphi(x) \right|$$

$$\leq \frac{(y_2 - y_1)^{2\alpha}}{16^\alpha} \left[\frac{\Gamma(1+\alpha)}{\Gamma(1+2\alpha)} \right]^{1/p_2} \left[\frac{\Gamma(1+2p_1\alpha)}{\Gamma(1+(2p_1+1)\alpha)} \right]^{1/p_1}$$

$$\times \left\{ \left[\left| \varphi^{(2\alpha)} \left(\frac{y_1 + y_2}{2} \right) \right|^{p_2} + |\varphi^{(2\alpha)}(y_1)|^{p_2} \right]^{1/p_2} \right.$$

$$\left. + \left[\left| \varphi^{(2\alpha)} \left(\frac{y_1 + y_2}{2} \right) \right|^{p_2} + |\varphi^{(2\alpha)}(y_2)|^{p_2} \right]^{1/p_2} \right\}. \tag{4.15}$$

Corollary 4.1 Assume that $\varphi : U \subset [0, \infty) \to \mathbb{R}^\alpha$ such that $\varphi \in D_\alpha$ on Int(U) and $\varphi^{(2\alpha)} \in C_\alpha[y_1, y_2]$, where $y_1, y_2 \in U$ with $y_1 < y_1$. If $|\varphi^{(2\alpha)}|^{p_2}$ is generalized s-convex on $[y_1, y_2]$, for some fixed $0 < s \leq 1$ and $p_2 > 1$ with $\frac{1}{p_1} + \frac{1}{p_2} = 1$, then the following inequality holds:

$$\left| \frac{\Gamma(1+2\alpha)}{2^\alpha} \varphi \left(\frac{y_1 + y_2}{2} \right) - \frac{\Gamma(1+2\alpha)[\Gamma(1+\alpha)]^2}{2^\alpha (y_2 - y_1)^\alpha} \, {}_{y_1}I_{y_2}^{(\alpha)} \varphi(x) \right|$$

$$\leq \frac{(y_2 - y_1)^{2\alpha}}{16^\alpha} \frac{[\Gamma(1+s\alpha)]^{1/p_2}}{[\Gamma(1+(s+1)\alpha)]^{2/p_2}} \left[\frac{\Gamma(1+2p_1\alpha)}{\Gamma(1+(2p_1+1)\alpha)} \right]^{1/p_1}$$

$$\times \{ [(2^{\alpha(1-s)}\Gamma(1+s\alpha)\Gamma(1+\alpha) + \Gamma(1+(s+1)\alpha))^{1/p_2}$$

$$+ 2^{\alpha(1-s)/p_2}[\Gamma(1+\alpha)]^{1/p_2}[\Gamma(1+\alpha)]^{1/p_2}][|\varphi^{(2\alpha)}(y_1)| + |\varphi^{(2\alpha)}(y_2)|] \}.$$

Proof: Since $|\varphi^{(2\alpha)}|^{p_2}$ is generalized s-convex, then

$$2^{\alpha(s-1)} \varphi^{(2\alpha)} \left(\frac{y_1 + y_2}{2} \right) \leq \frac{\Gamma(1+s\alpha)\Gamma(1+\alpha)}{\Gamma(1+(s+1)\alpha)} (\varphi^{(2\alpha)}(y_1) + \varphi^{(2\alpha)}(y_2)).$$

Hence, using (4.14), we get

$$\left| \frac{\Gamma(1+2\alpha)}{2^\alpha} \varphi \left(\frac{y_1 + y_2}{2} \right) - \frac{\Gamma(1+2\alpha)[\Gamma(1+\alpha)]^2}{2^\alpha (y_2 - y_1)^\alpha} \, {}_{y_1}I_{y_2}^{(\alpha)} \varphi(x) \right|$$

$$\leq \frac{(y_2 - y_1)^{2\alpha}}{16^\alpha} \frac{[\Gamma(1+s\alpha)]^{1/p_2}}{[\Gamma(1+(s+1)\alpha)]^{2/p_2}} \left[\frac{\Gamma(1+2p_1\alpha)}{\Gamma(1+(2p_1+1)\alpha)} \right]^{1/p_1}$$

$$\times \{ [(2^{\alpha(1-s)}\Gamma(1+s\alpha)\Gamma(1+\alpha) + \Gamma(1+(s+1)\alpha))|\varphi^{(2\alpha)}(y_1)|^{p_2}$$

$$+ 2^{\alpha(1-s)}\Gamma(1+s\alpha)\Gamma(1+\alpha)|\varphi^{(2\alpha)}(y_2)|^{p_2}]^{1/p_2}$$

$$+ [2^{\alpha(1-s)}\Gamma(1+s\alpha)\Gamma(1+\alpha)|\varphi^{(2\alpha)}(y_1)|^{p_2}$$

$$+ (2^{\alpha(1-s)}\Gamma(1+s\alpha)\Gamma(1+\alpha) + \Gamma(1+(s+1)\alpha))|\varphi^{(2\alpha)}(y_2)|^{p_2}]^{1/q} \}$$

and since $\displaystyle\sum_{i=1}^{k} (x_i + z_i)^{\alpha n} \le \sum_{i=1}^{k} x_i^{\alpha n} + \sum_{i=1}^{k} z_i^{\alpha n}$ for $0 < n < 1, x_i, z_i \ge 0; \forall 1 \le i \le k$,

then we have

$$
\left| \frac{\Gamma(1+2\alpha)}{2^\alpha} \varphi\left(\frac{y_1+y_2}{2}\right) - \frac{\Gamma(1+2\alpha)[\Gamma(1+\alpha)]^2}{2^\alpha(y_2-y_1)^\alpha} \,_{y_1}I_{y_2}^{(\alpha)} \varphi(x) \right|
$$

$$
\le \frac{(y_2-y_1)^{2\alpha}}{16^\alpha} \frac{[\Gamma(1+s\alpha)]^{1/p_2}}{[\Gamma(1+(s+1)\alpha)]^{2/p_2}} \left[\frac{\Gamma(1+2p_1\alpha)}{\Gamma(1+(2p_1+1)\alpha)} \right]^{1/p_1}
$$

$$
\times \{ [(2^{\alpha(1-s)}\Gamma(1+s\alpha)\Gamma(1+\alpha) + \Gamma(1+(s+1)\alpha))^{1/p_2}|\varphi^{(2\alpha)}(y_1)|
$$

$$
+ 2^{\alpha(1-s)/p_2}[\Gamma(1+s\alpha)]^{1/p_2}[\Gamma(1+\alpha)]^{1/p_2}|\varphi^{(2\alpha)}(y_2)|]
$$

$$
+ [2^{\alpha(1-s)/p_2}[\Gamma(1+s\alpha)]^{1/p_2}[\Gamma(1+\alpha)]^{1/p_2}|\varphi^{(2\alpha)}(y_1)|
$$

$$
+ (2^{\alpha(1-s)}\Gamma(1+s\alpha)\Gamma(1+\alpha) + \Gamma(1+(s+1)\alpha))^{1/p_2}|\varphi^{(2\alpha)}(y_2)|]\}.
$$

where $0 < \frac{1}{p_2} < 1$ for $p_2 > 1$. By a simple calculation, we obtain the required result. □

Now, a new upper bound of the left generalized Hadamard's inequality for generalized s-concave functions is given in the following theorem:

Theorem 4.18 Assume that $\varphi : U \subset [0, \infty) \to \mathbb{R}^\alpha$ such that $\varphi \in D_\alpha$ on $Int(U)$ and $\varphi^{(2\alpha)} \in C_\alpha[y_1, y_2]$, where $y_1, y_2 \in U$ with $y_1 < y_2$. If $|\varphi^{(2\alpha)}|^{p_2}$ is generalized s-convex on $[y_1, y_2]$, for some fixed $0 < s \le 1$ and $p_2 > 1$ with $\frac{1}{p_1} + \frac{1}{p_2} = 1$, then the following inequality holds:

$$
\left| \frac{\Gamma(1+2\alpha)}{2^\alpha} \varphi\left(\frac{y_1+y_2}{2}\right) - \frac{\Gamma(1+2\alpha)[\Gamma(1+\alpha)]^2}{2^\alpha(y_2-y_1)^\alpha} \,_{y_1}I_{y_2}^{(\alpha)} \varphi(x) \right|
$$

$$
\le \frac{2^{\alpha(s-1)/p_2}(y_2-y_1)^{2\alpha}}{16^\alpha(\Gamma(1+\alpha))^{1/p_2}} \left[\frac{\Gamma(1+2p_1\alpha)}{\Gamma(1+(2p_1+1)\alpha)} \right]^{1/p_1}
$$

$$
\times \left[\left| \varphi^{(2\alpha)}\left(\frac{3y_1+y_2}{4}\right) \right| + \left| \varphi^{(2\alpha)}\left(\frac{y_1+3y_2}{4}\right) \right| \right].
$$

Proof: From Lemma 4.1 and using the generalized Hölder inequality for $p_2 > 1$ and $\frac{1}{p_1} + \frac{1}{p_2} = 1$, we get

$$
\left| \frac{\Gamma(1+2\alpha)}{2^\alpha} \varphi\left(\frac{y_1+y_2}{2}\right) - \frac{\Gamma(1+2\alpha)[\Gamma(1+\alpha)]^2}{2^\alpha(a_2-a_1)^\alpha} \,_{y_1}I_{y_2}^{(\alpha)} \varphi(x) \right|
$$

$$
\le \frac{(y_2-y_1)^{2\alpha}}{16^\alpha} \left[{}_0I_1^{(\alpha)}\gamma^{2\alpha} \left| \varphi^{(2\alpha)}\left(\gamma\frac{y_1+y_2}{2} + (1-\gamma)y_1\right) \right| \right.
$$

$$+ {_0}I_1^{(\alpha)} (\gamma - 1)^{2\alpha} \left| \varphi^{(2\alpha)} \left(\gamma y_2 + (1 - \gamma) \frac{y_1 + y_2}{2} \right) \right| \Bigg]$$

$$\leq \frac{(y_2 - y_1)^{2\alpha}}{16^\alpha} ({_0}I_1^{(\alpha)} \gamma^{2p_1\alpha})^{1/p_1} \left({_0}I_1^{(\alpha)} \left| \varphi^{(2\alpha)} \left(\gamma \frac{y_1 + y_2}{2} + (1 - \gamma) y_1 \right) \right|^{p_2} \right)^{1/p_2}$$

$$+ \frac{(y_2 - y_1)^{2\alpha}}{16^\alpha} ({_0}I_1^{(\alpha)} (\gamma - 1)^{2p_1\alpha})^{1/p_1} \left({_0}I_1^{(\alpha)} \left| \varphi^{(2\alpha)} \left(\gamma y_2 + (1 - \gamma) \frac{y_1 + y_2}{2} \right) \right|^{p_2} \right)^{1/p_2}.$$

Since $|\varphi^{(2\alpha)}|^{p_2}$ is generalized s-concave, then

$$_0I_1^{(\alpha)} \left| \varphi^{(2\alpha)} \left(\gamma \frac{y_1 + y_2}{2} + (1 - \gamma) y_1 \right) \right|^{p_2} \leq \frac{2^{\alpha(s-1)}}{\Gamma(1 + \alpha)} \left| \varphi^{(2\alpha)} \left(\frac{3y_1 + y_2}{4} \right) \right|^{p_2} \tag{4.16}$$

also

$$_0I_1^{(\alpha)} \left| \varphi^{(2\alpha)} \left(\gamma y_2 + (1 - \gamma) \frac{y_1 + y_2}{2} \right) \right|^{p_2} \leq \frac{2^{\alpha(s-1)}}{\Gamma(1 + \alpha)} \left| \varphi^{(2\alpha)} \left(\frac{y_1 + 3y_2}{4} \right) \right|^{p_2}. \tag{4.17}$$

From (4.16) and (4.17), we observe that

$$\left| \frac{\Gamma(1 + 2\alpha)}{2^\alpha} \varphi \left(\frac{y_1 + y_2}{2} \right) - \frac{\Gamma(1 + 2\alpha)[\Gamma(1 + \alpha)]^2}{2^\alpha (y_2 - y_1)^\alpha} {_{y_1}}I_{y_2}^{(\alpha)} \varphi(x) \right|$$

$$\leq \frac{(y_2 - y_1)^{2\alpha}}{16^\alpha} \left[\frac{\Gamma(1 + 2p_1\alpha)}{\Gamma(1 + (2p_1 + 1)\alpha)} \right]^{1/p_1} \frac{2^{\alpha(s-1)/p_2}}{(\Gamma(1 + \alpha))^{1/p_2}} \left| \varphi^{(2\alpha)} \left(\frac{3y_1 + y_2}{4} \right) \right|$$

$$+ \frac{(y_2 - y_1)^{2\alpha}}{16^\alpha} \left[\frac{\Gamma(1 + 2p_1\alpha)}{\Gamma(1 + (2p_1 + 1)\alpha)} \right]^{1/p_1} \frac{2^{\alpha(s-1)/p_2}}{(\Gamma(1 + \alpha))^{1/p_2}} \left| \varphi^{(2\alpha)} \left(\frac{y_1 + 3y_2}{4} \right) \right|$$

$$= \frac{2^{\alpha(s-1)/p_2} (y_2 - y_1)^{2\alpha}}{16^\alpha (\Gamma(1 + \alpha))^{1/p_2}} \left[\frac{\Gamma(1 + 2p_1\alpha)}{\Gamma(1 + (2p_1 + 1)\alpha)} \right]^{1/p_1}$$

$$\times \left[\left| \varphi^{(2\alpha)} \left(\frac{3y_1 + y_2}{4} \right) \right| + \left| \varphi^{(2\alpha)} \left(\frac{y_1 + 3y_2}{4} \right) \right| \right]$$

the proof is complete. □

Remark 4.4

1. If $\alpha = 1$ in Theorem 4.18, then

$$\left| \varphi \left(\frac{y_1 + y_2}{2} \right) - \frac{1}{y_2 - y_1} \int_{y_1}^{y_2} \varphi(x) dx \right|$$

$$\leq \frac{2^{s-1/q} (y_2 - y_1)^2}{16} \left[\frac{1}{\Gamma(2p_1 + 1)} \right]^{1/p_1} \left[\left| \varphi'' \left(\frac{3y_1 + y_2}{4} \right) \right| + \left| \varphi'' \left(\frac{y_1 + 3y_2}{4} \right) \right| \right]$$

2. If $s = 1$ and $\frac{1}{3} < \left[\frac{\Gamma(1+2p_1\alpha)}{\Gamma(1+(2p_1+1)\alpha)} \right]^{1/p_1} < 1, p_1 > 1$ in Theorem 4.18, then

$$\left| \frac{\Gamma(1+2\alpha)}{2^\alpha} \varphi\left(\frac{y_1+y_2}{2} \right) - \frac{\Gamma(1+2\alpha)[\Gamma(1+\alpha)]^2}{2^\alpha(y_2-y_1)^\alpha} \, {}_{y_1}I_{y_2}^{(\alpha)} \varphi(x) \right|$$

$$\leq \frac{(y_2-y_1)^{2\alpha}}{16^\alpha(\Gamma(1+\alpha))^{1/p_2}} \left[\left| \varphi^{(2\alpha)}\left(\frac{3y_1+y_2}{4} \right) \right| + \left| \varphi^{(2\alpha)}\left(\frac{y_1+3y_2}{4} \right) \right| \right].$$

4.5 Applications to Special Means

As in [30], some generalized means are considered such as: $A(y_1,y_2) = \frac{y_1^\alpha+y_2^\alpha}{2^\alpha}$, $y_1, y_2 \geq 0$; $L_n(y_1,y_2) = \left[\frac{\Gamma(1+n\alpha)}{\Gamma(1+(n+1)\alpha)} \left(y_2^{(n+1)\alpha} - y_1^{(n+1)\alpha} \right) \right]^{1/n}$, $n \in \mathbb{Z}\{-1,0\}$, $y_1, y_2 \in \mathbb{R}$, $y_1 \neq y_2$. In [10], the following example was given: let $0 < s < 1$ and $y_1^\alpha, y_2^\alpha, y_3^\alpha \in \mathbb{R}^\alpha$. Defining for $x \in \mathbb{R}_+$,

$$\varphi(b) = \begin{cases} y_1^\alpha, & b = 0, \\ y_2^\alpha b^{s\alpha} + y_3^\alpha, & b > 0. \end{cases}$$

If $y_2^\alpha \geq 0^\alpha$ and $0^\alpha \leq y_3^\alpha \leq y_1^\alpha$, then $\varphi \in GK_s^2$.

Proposition 4.3 Let $0 < y_1 < y_2$ and $s \in (0,1)$. Then

$$\left| \frac{\Gamma(1+2\alpha)}{2^\alpha} A^s(y_1,y_2) - \frac{\Gamma(1+2\alpha)[\Gamma(1+\alpha)]^2}{2^\alpha(y_2-y_1)^\alpha} L_s^s(y_1,y_2) \right|$$

$$\leq \frac{(y_2-y_1)^{2\alpha}}{16^\alpha} \left| \frac{\Gamma(1+s\alpha)}{\Gamma(1+(s-2)\alpha)} \right| \left\{ \frac{2^\alpha \Gamma(1+3\alpha)[\Gamma(1+\alpha)]^2}{\Gamma(1+4\alpha)\Gamma(1+2\alpha)} \right.$$

$$+ \frac{\Gamma(1+\alpha)}{\Gamma(1+2\alpha)} - \frac{2^\alpha \Gamma(1+2\alpha)}{\Gamma(1+3\alpha)} + \frac{\Gamma(1+3\alpha)}{\Gamma(1+4\alpha)} \right\} \left[|y_1|^{(s-2)\alpha} + |y_2|^{(s-2)\alpha} \right].$$

Proof: The result follows from Remark 4.13(2) with $\varphi : [0,1] \to [0^\alpha, 1^\alpha]$, $\varphi(x) = x^{s\alpha}$ and when $\alpha = 1$, we have the following inequalitly:

$$\left| A^s(y_1,y_2) - \frac{1}{y_2-y_1} L_s^s(y_1,y_2) \right| \leq \frac{(y_2-y_1)^2 |s(s-1)|}{48} \{ |y_1|^{s-2} + |y_2|^{s-2} \}.$$

$$(4.18)$$

$\square$

Proposition 4.4 Let $0 < y_1 < y_2$ and $s \in (0,1)$. Then

$$\left| \frac{\Gamma(1+2\alpha)}{2^\alpha} A^s(y_1, y_2) - \frac{\Gamma(1+2\alpha)[\Gamma(1+\alpha)]^2}{2^\alpha (y_2 - y_1)^\alpha} L_s^s(y_1, y_2) \right|$$

$$\leq \frac{(y_2 - y_1)^{2\alpha}}{16^\alpha} \left[\frac{\Gamma(1+\alpha)}{\Gamma(1+2\alpha)} \right]^{1/p_2} \left| \frac{\Gamma(1+s\alpha)}{\Gamma(1+(s-2)\alpha)} \right| \left[\frac{\Gamma(1+2p_1\alpha)}{\Gamma(1+(2p_1+1)\alpha)} \right]^{1/p_1}$$

$$\times \left[\left(\left| \frac{y_1 + y_2}{2} \right|^{(s-2)p_2\alpha} + |y_1|^{(s-2)p_2\alpha} \right)^{1/p_2} + \left(\left| \frac{y_1 + y_2}{2} \right|^{(s-2)p_2\alpha} + |y_2|^{(s-2)p_2\alpha} \right)^{1/p_2} \right],$$

where $p_2 > 1$ and $\frac{1}{p_1} + \frac{1}{p_2} = 1$.

Proof: The result follows (4.15) with $\varphi : [0,1] \to [0^\alpha, 1^\alpha]$, $\varphi(x) = x^{s\alpha}$ and when $\alpha = 1$, we have the following inequalitly:

$$\left| A^s(y_1, y_2) - \frac{1}{y_2 - y_1} L_s^s(y_1, y_2) \right|$$

$$\leq \frac{(y_2 - y_1)^2 |s(s-1)|}{2^{1/p_2} 16(2p_1 + 1)^{1/p_1}} \left\{ \left(\left| \frac{y_1 + y_2}{2} \right|^{(s-2)p_2} + |y_1|^{(s-2)p_2} \right)^{1/p_2} \right.$$

$$\left. + \left(\left| \frac{y_1 + y_2}{2} \right|^{(s-2)p_2} + |y_2|^{(s-2)p_2} \right)^{1/p_2} \right\}, \tag{4.19}$$

□

where $A(y_1, y_2)$ and $L_n(y_1, y_2)$ in (4.18) and (4.19) are known as

1. Arithmetic mean: $A(y_1, y_2) = \frac{y_1 + y_2}{2}$, $y_1, y_2 \in \mathbb{R}^+$;
2. Logarithmic mean: $L(y_1, y_2) = \frac{y_1 - y_2}{\ln |y_1| - \ln |y_2|}$, $|y_1| \neq y_2, y_1, a_2 \neq 0, y_1, y_2 \in \mathbb{R}^+$;
3. Generalized Log-mean: $L_n(y_1, y_2) = \left[\frac{y_2^{n+1} - y_1^{n+1}}{(n+1)(y_2 - y_1)} \right]^{1/n}$, $n \in \mathbb{Z} \setminus \{-1, 0\}$, $y_1, y_2 \in \mathbb{R}^+$.

Now, we give application to wave equation on Cantor sets: the wave equation on Cantor sets (local fractional wave equation) was given by [8]

$$\frac{\partial^{2\alpha} f(x,t)}{\partial t^{2\alpha}} = A^{2\alpha} \frac{\partial^{2\alpha} f(x,t)}{\partial x^{2\alpha}}. \tag{4.20}$$

Following (4.20), a wave equation on Cantor sets was proposed as follows [31]:

$$\frac{\partial^{2\alpha} f(x,t)}{\partial t^{2\alpha}} = \frac{x^{2\alpha}}{\Gamma(1+2\alpha)} \frac{\partial^{2\alpha} f(x,t)}{\partial x^{2\alpha}}, \quad 0 \leq \alpha \leq 1, \tag{4.21}$$

where $f(x,t)$ is a fractal wave function and the initial value is given by $f(x,0) = \frac{x^\alpha}{\Gamma(1+\alpha)}$. The solution of (4.21) is given as $f(x,t) = \frac{x^\alpha}{\Gamma(1+\alpha)} + \frac{t^{2\alpha}}{\Gamma(1+2\alpha)}$. By using (4.1), we have

$$
\frac{\Gamma(1+2\alpha)\Gamma(1+\alpha)}{2^\alpha(y_2-y_1)^\alpha} \int_{y_1}^{y_2} f(x,t)(dt)^\alpha - \frac{\Gamma(1+2\alpha)}{2^\alpha} f\left(x, \frac{y_1+y_2}{2}\right)
$$

$$
= \frac{(y_2-y_1)^\alpha}{8^\alpha\Gamma(1+2\alpha)} \left[\left(\frac{2}{y_2-y_1}\right)^{2\alpha} {}_{y_1}I_{\frac{y_2+y_1}{2}}^{(\alpha)}(t-y_1)^{2\alpha}x^{2\alpha}\frac{\partial^{2\alpha}f(x,t)}{\partial x^\alpha} \right.
$$

$$
\left. + {}_{y_1}I_{\frac{y_2+y_1}{2}}^{(\alpha)}\left(\frac{2(t-y_1)}{y_2-y_1}-1\right)^{2\alpha}x^{2\alpha}\frac{\partial^{2\alpha}f(x,t)}{\partial x^\alpha} \right].
$$

References

1 Kılıçman, A. and Saleh, W. (2015) Some inequalities for generalized s-convex functions. *JP J. Geom. Topol.*, **17**, 63–82.

2 Grinalatt, M. and Linnainmaa, J.T. (2011) Jensen's inquality, parameter uncertainty, and multiperiod investment. *Rev. Asset Pric. Stud.*, **1** (1), 1–34.

3 Ruel, J.J. and Ayres, M.P. (1999) Jensen's inequality predicts effects of environmental variation. *Trends Ecol. Evol.*, **14** (9), 361–366.

4 Edgar, G.A. (1998) *Integral, Probability, and Fractal Measures*, Springer, New York.

5 Kolwankar, K.M. and Gangal, A.D. (1999) Local fractional calculus: a calculus for fractal space-time, in *Fractals: Theory and Applications in Engineering*, Springer, London, pp. 171–181.

6 Carpinteri, A., Chiaia, B., and Cornetti, P. (2001) Static-kinematic duality and the principle of virtual work in the mechanics of fractal media. *Comput. Methods Appl. Mech. Eng.*, **191** (12), 3–19.

7 Zhao, Y., Cheng, D.F., and Yang, X.J. (2013) Approximation solutions for local fractional Schrödinger equation in the on-dimensional Cantorian system. *Adv. Math. Phys.*, **2013**, 5.

8 Yang, X.J. (2012) *Advanced Local Fractional Calculus and its Applications*, World Science Publisher, New York.

9 Mo, H., Sui, X., and Yu, D. (2014) Generalized convex functions on fractal sets and two related inequalities. *Abstr. Appl. Anal.*, Article ID 636751, 7 p.

10 Mo, H. and Sui, X. (2014) Generalized s-convex functions on fractal sets. *Abstr. Appl. Anal.*, Article ID 254731, 8 p.

11 Kılıçman, A. and Saleh, W. (2015) Notions of generalized s-convex functions on fractal sets. *J. Inequ. Appl.*, **2015** (1), 312.

12 Mo, H. and Sui, X. (2015) Hermite-Hadamard type inequalities for generalized s-convex functions on real linear fractal set $\mathbb{R}^\alpha$ ($0 <? < 1$) (arXiv:1506.07391).

13 Dragomir, S.S. and Fitzpatrick, S. (1999) The Hadamards inequality for s-convex functions in the second sense. *Demonstratio Math.*, **32** (4), 687–696.

14 Dragomir, S.S. (2001) On Hadamards inequality for convex functions on the co-ordinates in a rectangle from the plane. *Taiwan. J. Math.*, **5**, 775–788.

15 Hua, J., Xi, B., and Qi, F. (2014) Inequalities of Hermite-Hadamard type involving an s-convex function with applications. *Appl. Math. Comput.*, **246**, 752–760.

16 Youness, E.A. (1999) E-convex set, E-convex functions and E-convex pro-gramming. *J. Optim. Theory Appl.*, **102**, 439–450.

17 Abou-Tair, I. and Sulaiman, W.T. (1999) Inequalities via convex functions. *J. Math. Math. Sci.*, **22** (3), 543–546.

18 Noor, M.A. (1994) Fuzzy preinvex functions. *Fuzzy Sets Syst.*, **64**, 95–104.

19 Yang, X.M. (2001) On E-convex set, E-convex functions and E-convex programming. *J. Optim. Theory Appl.*, **109**, 699–703.

20 Chen, X. (2002) Some properties of semi-E-convex functions. *J. Math. Anal. Appl.*, **275**, 251–262.

21 Kılıçman, A. and Saleh, W. (2015) On geodesic strongly E-convex sets and geodesic strongly E-convex functions. *J. Inequa. Appl.*, **2015** (1), 1–10.

22 Fulga, C. and Preda, V. (2009) Nonlinear programming with E-preinvex and local E-preinvex functions. *Eur. J. Oper. Res.*, **192** (3), 737–743.

23 Iqbal, A., Ahmad, I., and Ali, S. (2011) Some properties of geodesic semi-E-convex functions. *Nonlinear Anal. Theory Method Appl.*, **74** (17), 6805–6813.

24 Iqbal, A., Ali, S., and Ahmad, I. (2012) On geodesic E-convex sets, geodesic E-convex functions and E-epigraphs. *J. Optim Theory Appl.*, **55** (1), 239–251.

25 Syau, Y.R. and Lee, E.S. (2005) Some properties of E-convex functions. *Appl. Math. Lett.*, **18**, 1074–1080.

26 Baleanu, D., Srivastava, H.M., and Yang, X.J. (2015) Local fractional varia-tional iteration algorithms for the parabolic Fokker-Planck equation defined on Cantor sets. *Prog. Fract. Differ. Appl.*, **1** (1), 1–11.

27 Sarikaya, M.Z., Tunc, T., and Budak, Hu. (2016) On generalized some inte-gral inequalities for local fractional integrals. *Appl. Math. Comput.*, **276**, 316–323.

28 Yang, X.J., Baleanu, D., and Srivastava, H.M. (2015) Local fractional similar-ity solution for the diffusion equation defined on Cantor sets. *Appl. Math. Lett.*, **47**, 54–60.

29 Özdemir, M.E., Yıldız, Ç., Akdemir, A.O., and Set, E. (2013) On some inequalities for s-convex functions and applications. *J. Inequ. Appl.*, **2013** (1), 1–11.

30 Pearce, C.E.M. and Pecaric, J. (2000) Inequalities for differentiable mappings with application to specialmeans and quadrature formulae. *Appl. Math. Lett.*, **13** (2), 51–55.

31 Yang, A.M., Yang, X.-J., and Li, Z.B. (2013) Local fractional series expansion method for solving wave and diffusion equations on Cantor sets. *Abstract Appl. Anal.*, Article ID 351057, 5 p.

5

Some Properties and Generalizations of the Catalan, Fuss, and Fuss–Catalan Numbers

Feng Qi[1,2,3] *and Bai-Ni Guo*[4]

[1] *Institute of Mathematics, Henan Polytechnic University, Jiaozuo, Henan, 454010, China*
[2] *College of Mathematics, Inner Mongolia University for Nationalities, Tongliao, Inner Mongolia, 028043, China*
[3] *Department of Mathematics, College of Science, Tianjin Polytechnic University, Tianjin, 300387, China*
[4] *School of Mathematics and Informatics, Henan Polytechnic University, Jiaozuo, Henan, 454010, China*

2010 AMS Subject Classification *05A15; 05A19; 05A20; 11B75; 11B83; 11C08; 11R33; 26A06; 26A48; 26A51; 26D15; 30E20; 33B10; 33B15; 33C05; 33C20; 34E05; 41A60; 44A10.*

5.1   The Catalan Numbers

5.1.1   A Definition of the Catalan Numbers

The Catalan numbers C_n for $n \geq 0$ form a sequence of natural numbers (positive integers) that occur in tree enumeration problems in combinatorics. The Catalan number C_{n-2} is the number of ways that a regular n-gon is divided into $n - 2$ triangles if different orientations are counted separately. The first 11 Catalan numbers C_n for $0 \leq n \leq 10$ are

$$1, \quad 1, \quad 2, \quad 5, \quad 14, \quad 42, \quad 132, \quad 429, \quad 1430, \quad 4862, \quad 16796.$$

For more information, please refer to the monographs [1, 2], the websites [3, 4], and the closely related references therein.

5.1.2   The History of the Catalan Numbers

The Catalan numbers C_n had been used in Qing Dynasty of China by the Mongolian mathematician Ming An Tu by 1730 (see [5, 6]). The sequence of Catalan numbers C_n was described in the eighteenth century by Leonhard Euler. The sequence is named after the Belgian mathematician Eugène Charles Catalan. For detailed information, please refer to the monographs [1, 2], the websites [3, 4], and the closely related references therein.

Mathematical Analysis and Applications: Selected Topics, First Edition.
Edited by Michael Ruzhansky, Hemen Dutta, and Ravi P. Agarwal.
© 2018 John Wiley & Sons, Inc. Published 2018 by John Wiley & Sons, Inc.

5.1.3 A Generating Function of the Catalan Numbers

The Catalan numbers C_n can be generated by the elementary function

$$\frac{2}{1 + \sqrt{1 - 4x}} = \frac{1 - \sqrt{1 - 4x}}{2x} = \sum_{n=0}^{\infty} C_n x^n = 1 + x + 2x^2 + 5x^3 + \cdots.$$

(5.1)

For more information, please refer to the monographs [1, 2], the websites [3, 4], and the closely related references therein.

5.1.4 Some Expressions of the Catalan Numbers

Explicit formulas of C_n for $n \geq 0$ include

$$C_n = \frac{1}{n+1} \binom{2n}{n} = \frac{(2n)!}{n!(n+1)!} = \frac{1}{n} \binom{2n}{n-1} = \frac{2^n(2n-1)!!}{(n+1)!}$$

$$= \prod_{k=2}^{n} \frac{n+k}{k} = \frac{1}{n+1} \sum_{k=0}^{n} \binom{n}{k}^2 = \binom{2n}{n} - \binom{2n}{n+1}$$

(5.2)

$$= (-1)^n 2^{2n+1} \binom{1/2}{n+1} = \frac{1}{(n+1)!} \prod_{k=1}^{n} (4k - 2)$$

$$= {}_2F_1(1 - n, -n; 2; 1) = \frac{4^n \Gamma(n + 1/2)}{\sqrt{\pi}\, \Gamma(n+2)},$$

where

$$\Gamma(z) = \int_0^{\infty} t^{z-1} e^{-t} \mathrm{d}t, \quad \Re(z) > 0$$

is the classical Euler gamma function and

$$_pF_q(a_1, \ldots, a_p; b_1, \ldots, b_q; z) = \sum_{n=0}^{\infty} \frac{(a_1)_n \cdots (a_p)_n}{(b_1)_n \cdots (b_q)_n} \frac{z^n}{n!}$$

is the generalized hypergeometric series, which are defined for complex numbers $a_i \in \mathbb{C}$ and $b_i \in \mathbb{C}\backslash\{0, -1, -2, \ldots\}$, for positive integers $p, q \in \mathbb{N}$, and in terms of the rising factorials

$$(x)_n = \begin{cases} x(x+1)(x+2)\cdots(x+n-1), & n \geq 1, \\ 1, & n = 0. \end{cases}$$

(5.3)

These expressions are collected in the monographs [1, 2], on the websites [3, 4], and in the closely related references therein.

Recently, with the help of the Faá di Bruno formula (5.17) and the special values of the Bell polynomials of the second kind (5.18) in Section 5.1.9, the following two alternative expressions were obtained in [7, 8] by directly computing the nth derivative of the generating function $\frac{2}{1+\sqrt{1-4x}}$ in (5.1).

Theorem 5.1 **(7 and [8, theorem 3])** For $n \geq 0$, the Catalan numbers C_n can be expressed by

$$C_n = (-1)^n \frac{2^n}{n!} \sum_{k=0}^{n} \frac{1}{2^k} \sum_{\ell=0}^{k} (-1)^{\ell} \binom{k}{\ell} \prod_{m=0}^{n-1} (\ell - 2m) \tag{5.4}$$

$$= \frac{2^n}{n!} \sum_{k=0}^{n} \frac{k!}{2^k} \binom{2n - k - 1}{2(n - k)} [2(n - k) - 1]!!. \tag{5.5}$$

We note that the Catalan numbers C_n can also be represented as

$$C_n = \frac{4^n}{(n+1)!} \left(\frac{1}{2} \right)_n,$$

where $(x)_n$ is the rising factorial defined by (5.3).

5.1.5 Integral Representations of the Catalan Numbers

In [9, p. 413, proposition 2.1], [10, p. 10], and [1, 2, 4, 11], the integral representation

$$C_n = \frac{1}{2\pi} \int_0^4 \sqrt{\frac{4 - x}{x}} \, x^n dx \tag{5.6}$$

was listed. In [10, p. 10], another integral representation

$$C_n = \frac{2^{2n+5}}{\pi} \int_0^1 \frac{x^2 (1 - x^2)^{2n}}{(1 + x^2)^{2n+3}} dx$$

was given. Recently, by applying the Cauchy integral formula in the theory of complex functions to the generating function $\frac{2}{1+\sqrt{1-4x}}$ in (5.1), the following alternative integral representations of the Catalan numbers C_n was established.

Theorem 5.2 **([12, theorem 1.4])** For $x \in \left(-\infty, \frac{1}{4} \right]$, we have

$$\frac{1}{1 + \sqrt{1 - 4x}} = \frac{1}{2\pi} \int_0^{\infty} \frac{\sqrt{t}}{(t + 1/4)(t - x + 1/4)} dt.$$

Consequently, the Catalan numbers C_n for $n \geq 0$ can be represented by

$$C_n = \frac{1}{\pi} \int_0^{\infty} \frac{\sqrt{t}}{(t + 1/4)^{n+2}} dt = \frac{2}{\pi} \int_0^{\infty} \frac{t^2}{(t^2 + 1/4)^{n+2}} dt. \tag{5.7}$$

By employing the last expression in (5.2) and an integral representation for the logarithm of the gamma function $\Gamma(x)$, an integral representation of the Catalan function

$$C_x \triangleq \frac{4^x \Gamma \left(x + \frac{1}{2} \right)}{\sqrt{\pi} \, \Gamma(x + 2)}$$

was presented in [13] as follows.

Theorem 5.3 ([13, theorem 1]) For $x \geq 0$, we have

$$C_x = \frac{e^{3/2} 4^x (x + 1/2)^x}{\sqrt{\pi}\ (x + 2)^{x+3/2}}$$

$$\times \exp\left[\int_0^\infty \left(\frac{1}{e^t - 1} - \frac{1}{t} + \frac{1}{2}\right) \frac{e^{-t/2} - e^{-2t}}{t} e^{-xt} dt\right]. \tag{5.8}$$

The integral representations in (5.6), (5.7), and (5.8) have been further generalized in Theorems 5.22 and 5.23 below.

5.1.6 Asymptotic Expansions of the Catalan Function

It is known that the Catalan function C_x has an asymptotic expansion [3, 14, 15] and [1, pp. 110–111]

$$C_x \sim \frac{4^x}{\sqrt{\pi}} \left(\frac{1}{x^{3/2}} - \frac{9}{8} \frac{1}{x^{5/2}} + \frac{145}{128} \frac{1}{x^{7/2}} + \cdots\right), \quad x \to \infty. \tag{5.9}$$

What is the general term of the asymptotic expansion (5.9)? This question was answered in [16, 17] as follows.

Theorem 5.4 ([16, theorem 2] and [17, theorem 2.1]) The Catalan function C_x has the asymptotic expansion

$$C_x \sim \frac{4^x}{\sqrt{\pi}} \sum_{k=0}^\infty (-1)^k \frac{B_k^{(-1/2)}(1/2)}{k!} \frac{\Gamma(k+3/2)}{\Gamma(3/2)} \frac{1}{x^{k+3/2}}, \quad x \to \infty, \tag{5.10}$$

where $B_k^{(\sigma)}(x)$ denote the generalized Bernoulli polynomials generated by

$$e^{xz} \left(\frac{z}{e^z - 1}\right)^\sigma = \sum_{k=0}^\infty \frac{B_k^{(\sigma)}(x)}{k!} z^k, \quad \sigma \in \mathbb{C}, \quad |z| < 2\pi. \tag{5.11}$$

The Catalan function C_x and the Catalan numbers C_n have also the following expansions.

Theorem 5.5 ([18, theorem 1.2]) The Catalan function C_x has the exponential expansion

$$C_x = \frac{e^{3/2} 4^x \left(x + \frac{1}{2}\right)^x}{\sqrt{\pi}\ (x + 2)^{x+3/2}}$$

$$\times \exp\left\{\sum_{j=1}^\infty \frac{B_{2j}}{2j(2j-1)} \left[\frac{1}{\left(x + \frac{1}{2}\right)^{2j-1}} - \frac{1}{(x+2)^{2j-1}}\right]\right\}, \tag{5.12}$$

where B_{2j} are the Bernoulli numbers which can be generated by

$$\frac{x}{e^x - 1} = \sum_{i=0}^{\infty} B_i \frac{x^i}{i!} = 1 - \frac{x}{2} + \sum_{j=1}^{\infty} B_{2j} \frac{x^{2j}}{(2j)!}, \quad |x| < 2\pi. \tag{5.13}$$

Theorem 5.6 ([18, theorem 1.1])   For $n \geq 0$, the Catalan numbers C_n have the expansion

$$C_n = \frac{4^{n+1}}{\pi} \left[\frac{1}{2n + 1} - \sum_{k=1}^{\infty} \frac{(2k - 3)!!}{2^k k!} \frac{1}{2n + 2k + 1} \right]. \tag{5.14}$$

In the survey paper [19], there are plenty of closely related references on asymptotic expansions of the Catalan numbers C_n.

The expansions (5.10), (5.12), and (5.14) have been generalized by (5.28), (5.29), and (5.31) below.

5.1.7   Complete Monotonicity of the Catalan Numbers

From Refs. [20, pp. 372–373] and [21, p. 108, definition 4], we recall that a sequence $\{\mu_n\}_{0 \leq n \leq \infty}$ is said to be completely monotonic if its elements are non-negative and its successive differences are alternatively non-negative, that is,

$$(-1)^k \Delta^k \mu_n \geq 0$$

for $n, k \geq 0$, where

$$\Delta^k \mu_n = \sum_{m=0}^{k} (-1)^m \binom{k}{m} \mu_{n+k-m}.$$

From Ref. [21, p. 163, definition 14a], we recall that a completely monotonic sequence $\{a_n\}_{n \geq 0}$ is minimal if it ceases to be completely monotonic when a_0 is decreased.

Theorem 5.7 ([18, theorem 1.1] and [12, theorem 1.4])   For any $N \in \{0\} \cup \mathbb{N}$, the sequences $\left\{ \frac{C_n}{4^n} \right\}_{n \geq 0}$ and

$$\left\{ \frac{4}{\pi} \left[\frac{1}{2n + 1} - \sum_{k=1}^{N} \frac{(2k - 3)!!}{2^k k!} \frac{1}{2n + 2k + 1} \right] - \frac{C_n}{4^n} \right\}_{n \geq 0} \tag{5.15}$$

are completely monotonic and minimal, where an empty sum is understood to be 0.

Theorem 5.8 ([18, theorem 1.2]) The sequences

$$\left\{ \ln\left[\frac{\sqrt{\pi}\,(n+2)^{n+3/2}}{e^{3/2}4^n(n+1/2)^n}C_n \right] \right\}_{n\geq 0} \quad \text{and} \quad \left\{ \frac{(n+2)^{n+3/2}}{4^n(n+1/2)^n}C_n \right\}_{n\geq 0}$$

(5.16)

are completely monotonic and minimal.

Corollary 5.1 ([18, corollary 1.1]) The sequence $\left\{ \frac{C_n}{4^n} \right\}_{n\geq 0}$ and the one in (5.15) are convex and the second sequence in (5.16) is logarithmically convex.

5.1.8 Inequalities of the Catalan Numbers and Function

By virtue of the above asymptotic expansions, integral representations, and complete monotonicity of the Catalan numbers C_n and the Catalan function C_x, one can derive the following inequalities, including determinantal inequalities and product inequalities of the Catalan numbers C_n and the Catalan function C_x.

Theorem 5.9 ([18, theorem 1.1]) For all $n, m \geq 0$, we have

$$C_n < \frac{4^{n+1}}{\pi}\left[\frac{1}{2n+1} - \sum_{k=1}^{m} \frac{(2k-3)!!}{2^k k!} \frac{1}{2n+2k+1} \right].$$

Theorem 5.10 ([18, theorem 1.2]) The double inequality

$$\frac{e^{3/2}4^x\left(x+\frac{1}{2}\right)^x}{\sqrt{\pi}\,(x+2)^{x+3/2}} \exp\left\{ \sum_{j=1}^{2m} \frac{B_{2j}}{2j(2j-1)}\left[\frac{1}{\left(x+\frac{1}{2}\right)^{2j-1}} - \frac{1}{(x+2)^{2j-1}} \right] \right\}$$

$$< C_x < \frac{e^{3/2}4^x\left(x+\frac{1}{2}\right)^x}{\sqrt{\pi}\,(x+2)^{x+3/2}}$$

$$\times \exp\left\{ \sum_{j=1}^{2m-1} \frac{B_{2j}}{2j(2j-1)}\left[\frac{1}{\left(x+\frac{1}{2}\right)^{2j-1}} - \frac{1}{(x+2)^{2j-1}} \right] \right\}$$

is valid for $m \in \mathbb{N}$, where B_{2j} are the Bernoulli numbers defined by (5.13).

Theorem 5.11 ([12, theorem 1.5]) Let $m, n \in \mathbb{N}$ and a_k for $1 \leq k \leq m$ be non-negative integers. Denote $\mathcal{C}_n = n!C_n$. Then

$$\left| (-1)^{a_i+a_j}\mathcal{C}_{n+a_i+a_j} \right|_m \geq 0 \quad \text{and} \quad \left| \mathcal{C}_{n+a_i+a_j} \right|_m \geq 0,$$

where $|e_{kj}|_m$ denotes a determinant of order m with elements e_{kj}.

Theorem 5.12 ([12, **theorem 1.4**]) If $m \geq 1$ and $a_0, a_1, \ldots, a_m$ be non-negative integers, then

$$\left(\frac{C_{a_0}}{4^{a_0}} \right)^{m-1} \frac{C_{\sum_{k=0}^m a_k}}{4^{\sum_{k=0}^m a_k}} \geq \prod_{k=1}^m \frac{C_{a_0+a_k}}{4^{a_0+a_k}} \quad \text{and} \quad \left| \frac{C_{a_i+a_j}}{4^{a_i+a_j}} \right|_m \geq 0.$$

Theorem 5.13 ([18, **theorem 1.1**]) For $m \in \mathbb{N}$ and any non-negative integers $a_0, a_1, \ldots, a_m$, we have

$$\left(\frac{4}{\pi} \frac{1}{2a_0 + 1} - \frac{C_{a_0}}{4^{a_0}} \right)^{m-1} \left(\frac{4}{\pi} \frac{1}{1 + 2\sum_{k=0}^m a_k} - \frac{C_{\sum_{k=0}^m a_k}}{4^{\sum_{k=0}^m a_k}} \right)$$

$$\geq \prod_{k=1}^m \left[\frac{4}{\pi} \frac{1}{2(a_0 + a_k) + 1} - \frac{C_{a_0+a_k}}{4^{a_0+a_k}} \right]$$

and

$$\left| \frac{4}{\pi} \frac{1}{2(a_i + a_j) + 1} - \frac{C_{a_i+a_j}}{4^{a_i+a_j}} \right|_m \geq 0.$$

Theorem 5.14 ([18, **theorem 1.2**]) For $m \in \mathbb{N}$ and any non-negative integers $a_0, a_1, \ldots, a_m$, we have

$$\left\{ \ln \left[\frac{\sqrt{\pi} \, (a_0 + 2)^{a_0 + 3/2}}{e^{3/2} 4^{a_0} \left(a_0 + \frac{1}{2} \right)^n} C_{a_0} \right] \right\}^{m-1}$$

$$\times \ln \left[\frac{\sqrt{\pi} \left(2 + \sum_{k=0}^m a_k \right)^{\frac{3}{2} + \sum_{k=0}^m a_k}}{e^{3/2} 4^{\sum_{k=0}^m a_k} \left(\frac{1}{2} + \sum_{k=0}^m a_k \right)^{\sum_{k=0}^m a_k}} C_{\sum_{k=0}^m a_k} \right]$$

$$\geq \prod_{k=1}^m \ln \left[\frac{\sqrt{\pi} \, (a_0 + a_k + 2)^{a_0 + a_k + 3/2}}{e^{3/2} 4^{a_0+a_k} \left(a_0 + a_k + \frac{1}{2} \right)^n} C_{a_0+a_k} \right],$$

$$\left[\frac{(a_0 + 2)^{a_0 + 3/2}}{4^{a_0} (a_0 + 1/2)^n} C_{a_0} \right]^{m-1} \left[\frac{\left(2 + \sum_{k=0}^m a_k \right)^{\frac{3}{2} + \sum_{k=0}^m a_k}}{4^{\sum_{k=0}^m a_k} \left(\frac{1}{2} + \sum_{k=0}^m a_k \right)^{\sum_{k=0}^m a_k}} C_{\sum_{k=0}^m a_k} \right]$$

$$\geq \prod_{k=1}^m \left[\frac{(a_0 + a_k + 2)^{a_0 + a_k + 3/2}}{4^{a_0+a_k} (a_0 + a_k + 1/2)^n} C_{a_0+a_k} \right],$$

$$\left| \ln \left[\frac{\sqrt{\pi} \, (a_i + a_j + 2)^{a_i + a_j + 3/2}}{e^{3/2} 4^{a_i+a_j} (a_i + a_j + 1/2)^{a_i+a_j}} C_{a_i+a_j} \right] \right|_m \geq 0,$$

and

$$\left| \frac{(a_i + a_j + 2)^{a_i + a_j + 3/2}}{4^{a_i + a_j}(a_i + a_j + 1/2)^{a_i + a_j}} C_{a_i + a_j} \right|_m \geq 0.$$

Theorem 5.15 ([12, theorem 1.7]) If $\ell \geq 0, n \geq k \geq m, k \geq n - k$, and $m \geq n - m$, then

$$\frac{C_{\ell + k} C_{\ell + n - k}}{C_{\ell + m} C_{\ell + n - m}} \geq \frac{(\ell + m)!(\ell + n - m)!}{(\ell + k)!(\ell + n - k)!}.$$

For $n, m \in \mathbb{N}$ and $\ell \geq 0$, let

$$\begin{aligned}
\mathcal{G}_{n,m,\ell} &= C_{\ell + n + 2m}(C_\ell)^2 - C_{\ell + n + m}C_{\ell + m}C_\ell \\
&\quad - C_{\ell + n}C_{\ell + 2m}C_\ell + C_{\ell + n}(C_{\ell + m})^2, \\
\mathcal{H}_{n,m,\ell} &= C_{\ell + n + 2m}(C_\ell)^2 - 2C_{\ell + n + m}C_{\ell + m}C_\ell + C_{\ell + n}(C_{\ell + m})^2, \\
\mathcal{I}_{n,m,\ell} &= C_{\ell + n + 2m}(C_\ell)^2 - 2C_{\ell + n}C_{\ell + 2m}C_\ell + C_{\ell + n}(C_{\ell + m})^2.
\end{aligned}$$

Then we have

$$\mathcal{G}_{n,m,\ell} \geq 0, \quad \mathcal{H}_{n,m,\ell} \geq 0,$$
$$\mathcal{H}_{n,m,\ell} \gtreqless \mathcal{G}_{n,m,\ell} \quad \text{when } m \lesseqgtr n,$$

and

$$\mathcal{I}_{n,m,\ell} \geq \mathcal{G}_{n,m,\ell} \geq 0 \quad \text{when } n \geq m.$$

Theorem 5.16 ([12, theorem 1.6]) Suppose that $\lambda = (\lambda_1, \lambda_2, \ldots, \lambda_n) \in \mathbb{R}^n$ and $\mu = (\mu_1, \mu_2, \ldots, \mu_n) \in \mathbb{R}^n$. A sequence λ is said to be majorized by μ (in symbols $\lambda \preceq \mu$) if

$$\sum_{\ell=1}^k \lambda_{[\ell]} \leq \sum_{\ell=1}^k \mu_{[\ell]}$$

for $k = 1, 2, \ldots, n - 1$ and

$$\sum_{\ell=1}^n \lambda_\ell = \sum_{\ell=1}^n \mu_\ell,$$

where $\lambda_{[1]} \geq \lambda_{[2]} \geq \cdots \geq \lambda_{[n]}$ and $\mu_{[1]} \geq \mu_{[2]} \geq \cdots \geq \mu_{[n]}$ are, respectively, the components of λ and μ in decreasing order. A sequence λ is said to be strictly majorized by μ (in symbols $\lambda \prec \mu$) if λ is not a permutation of μ.

Let $m \in \mathbb{N}$ and let λ and μ be two m-tuples of non-negative integers such that $\lambda \preceq \mu$. Then

$$\left| \prod_{i=1}^m C_{n + \lambda_i} \right| \leq \left| \prod_{i=1}^m C_{n + \mu_i} \right|.$$

Consequently,

(1) the infinite sequence $\{C_n\}_{n\geq 0}$ is logarithmically convex,
(2) the inequality

$$C^n_{\ell+k} \leq C^k_{\ell+n} C^{n-k}_{\ell}$$

is valid for $\ell \geq 0$ and $n > k > 0$.

5.1.9 The Bell Polynomials of the Second Kind and the Bessel Polynomials

In combinatorial analysis, the Bell polynomials of the second kind, also known as the partial Bell polynomials, denoted by $B_{n,k}(x_1, x_2, \ldots, x_{n-k+1})$, can be defined by

$$B_{n,k}(x_1, x_2, \ldots, x_{n-k+1}) = \sum_{\substack{1 \leq i \leq n, \ell_i \in \{0\} \cup \mathbb{N} \\ \sum_{i=1}^{n} i\ell_i = n \\ \sum_{i=1}^{n} \ell_i = k}} \frac{n!}{\prod_{i=1}^{n-k+1} \ell_i!} \prod_{i=1}^{n-k+1} \left(\frac{x_i}{i!}\right)^{\ell_i}$$

for $n \geq k \geq 0$. The Faà di Bruno formula can be described in [22, p. 139, theorem C] in terms of the Bell polynomials of the second kind $B_{n,k}(x_1, x_2, \ldots, x_{n-k+1})$ by

$$\frac{d^n}{dx^n} f \circ g(x) = \sum_{k=0}^{n} f^{(k)}(g(x)) B_{n,k}(g'(x), g''](x), \ldots, g^{(n-k+1)}(x)). \tag{5.17}$$

The Bessel polynomials y_n were defined [23] by

$$y_n(x) = \sum_{k=0}^{n} b_{n,k} x^k = \sum_{k=0}^{n} \frac{(n+k)!}{(n-k)!k!} \left(\frac{x}{2}\right)^k.$$

The first 5 Bessel polynomials $y_n(x)$ for $0 \leq n \leq 4$ are

$$y_0(x) = 1, \quad y_1(x) = x + 1, \quad y_2(x) = 3x^2 + 3x + 1,$$
$$y_3(x) = 15x^3 + 15x^2 + 6x + 1,$$
$$y_4(x) = 105x^4 + 105x^3 + 45x^2 + 10x + 1.$$

For more information on the Bessel polynomials $y_n(x)$, please refer to the websites [24–26] and the closely related references therein.

Recently, two explicit formulas for special values

$$B_{n,k}((-1)!!, 1!!, 3!!, \ldots, [2(n-k)-1]!!) \tag{5.18}$$

were presented and the quantities (5.18) were connected with coefficients $b_{n,k}$ of the Bessel polynomials $y_n(x)$.

Theorem 5.17   (7 and [8, theorems 1 and 2])   For $n \geq k \geq 0$, we have

$$B_{n,k}((-1)!!, 1!!, 3!!, \ldots, [2(n-k)-1]!!)$$

$$= \frac{(-1)^n}{k!} \sum_{\ell=0}^{k} (-1)^\ell \binom{k}{\ell} \prod_{m=0}^{n-1} (\ell - 2m),$$

$$B_{n,k}((-1)!!, 1!!, 3!!, \ldots, [2(n-k)-1]!!)$$

$$= \binom{2n-k-1}{2(n-k)} [2(n-k)-1]!!,$$

and

$$b_{n,k} = B_{n+1,n-k+1}((-1)!!, 1!!, 3!!, \ldots, (2k-1)!!).$$

The above explicit formulas for special values (5.18) can be applied to find the expressions (5.4) and (5.5) and to compute the nth derivatives of a kind of elementary functions.

Theorem 5.18   (7 and [8, theorem 4])   Let $g(x) = \sqrt{a+bx}$ for $a, b \in \mathbb{R}$ and $b \neq 0$ and let $n \in \mathbb{N}$. Then the Bell polynomials of the second kind $B_{n,k}$ satisfy

$$B_{n,k}(g'(x), g''](x), \ldots, g^{(n-k+1)}(x))$$

$$= (-1)^{n+k} [2(n-k)-1]!! \left(\frac{b}{2}\right)^n \binom{2n-k-1}{2(n-k)} \frac{1}{(a+bx)^{n-k/2}}.$$

Consequently, for $n \geq 0$, we have

$$\frac{d^n \sin \sqrt{x}}{dx^n} = \frac{(-1)^n}{(2x)^n} \sum_{k=0}^{n} (-1)^k [2(n-k)-1]!!$$

$$\times \binom{2n-k-1}{2(n-k)} x^{k/2} \sin\left(\sqrt{x} + \frac{k\pi}{2}\right),$$

$$\frac{d^n \cos \sqrt{x}}{dx^n} = \frac{(-1)^n}{(2x)^n} \sum_{k=0}^{n} (-1)^k [2(n-k)-1]!!$$

$$\times \binom{2n-k-1}{2(n-k)} x^{k/2} \cos\left(\sqrt{x} + \frac{k\pi}{2}\right),$$

$$\frac{d^n e^{\sqrt{x}}}{dx^n} = \frac{(-1)^n}{(2x)^n} e^{\sqrt{x}} \sum_{k=0}^{n} (-1)^k [2(n-k)-1]!! \binom{2n-k-1}{2(n-k)} x^{k/2},$$

and

$$\frac{d^n \ln(1 \pm \sqrt{x})}{dx^n} = \frac{(-1)^{n+1}}{(2x)^n} \sum_{k=0}^{n} (\pm 1)^k (k-1)! [2(n-k)-1]!!$$

$$\times \binom{2n-k-1}{2(n-k)} \left(\frac{\sqrt{x}}{1 \pm \sqrt{x}}\right)^k.$$

For $n \geq 1$, we have

$$\frac{d^n \arcsin x}{dx^n} = -\frac{d^n \arccos x}{dx^n} = \frac{d^{n-1}}{dx^{n-1}}\left(\frac{1}{\sqrt{1-x^2}}\right) = \frac{1}{(2x)^n}$$

$$\times \sum_{k=0}^{n-1} 2^{k+1}(2k-1)!!(n-k-1)! \binom{n-1}{k}\binom{k}{n-k-1}\left(\frac{x^2}{1-x^2}\right)^{k+1/2}.$$

Consequently, for $n \geq 1$, we have

$$\frac{d^n \arcsin \sqrt{x}}{dx^n} = -\frac{d^n \arccos \sqrt{x}}{dx^n}$$

$$= \frac{(-1)^n}{(2x)^n} \sum_{k=1}^{n} \frac{(-1)^k}{2^k}[2(n-k)-1]!! \binom{2n-k-1}{2(n-k)}\frac{1}{(\sqrt{x})^k}$$

$$\times \sum_{\ell=0}^{k-1} 2^{\ell+1}(2\ell-1)!!(k-\ell-1)! \binom{k-1}{\ell}\binom{\ell}{k-\ell-1}\left(\frac{x^2}{1-x^2}\right)^{\ell+1/2}.$$

In recent years, several explicit formulas of special values for the Bell polynomials of the second kind $B_{n,k}(x_1, x_2, \ldots, x_{n-k+1})$ were recovered, discovered, and applied in [27–33] and references cited therein.

5.2 The Catalan–Qi Function

The Catalan numbers C_n can be generalized alternatively.

5.2.1 The Fuss Numbers

A generalization of the Catalan numbers C_n was defined in [34–36] by

$$_p d_n = \frac{1}{n}\binom{pn}{n-1} = \frac{1}{(p-1)n+1}\binom{pn}{n}, \quad n, p \geq 1.$$

It is obvious that $_2 d_n = C_n$.

In [1, pp. 375–376], the generalization $_{p+1} d_n$ of the Catalan numbers C_n is denoted by $C(n, p)$ for $p \geq 0$ and is called the generalized Catalan numbers.

In [1, pp. 377–378], the Fuss numbers

$$F(m, n) = \frac{1}{mn+1}\binom{mn+1}{n}$$

were given and discussed. It is apparent that $F(2, n) = C_n$.

5.2.2 A Definition of the Catalan–Qi Function

For $\Re(a), \Re(b) > 0$ and $\Re(z) \geq 0$, let

$$C(a, b; z) = \frac{\Gamma(b)}{\Gamma(a)}\left(\frac{b}{a}\right)^z \frac{\Gamma(z+a)}{\Gamma(z+b)}. \tag{5.19}$$

Specially, we have

$$C(a, b; n) = \left(\frac{b}{a}\right)^n \frac{(a)_n}{(b)_n}.$$

It is not difficult to verify that $C\left(\frac{1}{2}, 2; n\right) = C_n$ and

$$\left(\frac{n+1}{2}\right)^{(m-1)n} C(n+1, 2; (m-1)n) = {_m}d_n = C(n, m-1)$$

for $m, n \geq 1$. Therefore, the quantity $C(a, b; z)$ defined by (5.19) is an alternative and analytical generalization of the Catalan numbers C_n, the Catalan function C_x, and the Fuss numbers $F(m, n)$.

For uniqueness and convenience of referring to the quantity $C(a, b; z)$, we call $C(a, b; z)$ the Catalan–Qi function and, when taking $z = n \geq 0$, call $C(a, b; n)$ the Catalan–Qi numbers.

The quantity $C(a, b; z)$ was first introduced and called the Catalan–Qi function in [37, Remark 1] which has been formally published in [38, p. 939].

5.2.3 Some Identities of the Catalan–Qi Function

Some identities of the Catalan numbers C_n can be generalized to ones of the Catalan–Qi function $C(a, b; z)$.

In [3] and related references therein, the identities

$$C_{n+1} = \frac{2(2n+1)}{n+2} C_n, \quad C_n = \frac{1}{(n+1)!} \prod_{k=1}^{n} (4k-2),$$

$$\sum_{n=1}^{\infty} \frac{C_n}{4^n} = 1, \quad \sum_{n=0}^{\infty} C_n \frac{x^{2n}}{(2n)!} = \frac{I_1(2x)}{x},$$

and

$$e^{2x}[I_0(2x) - I_1(2x)] = \sum_{n=0}^{\infty} C_n \frac{x^n}{n!}$$

of the Catalan numbers C_n were listed, where

$$I_\nu(z) = \sum_{k=0}^{\infty} \frac{1}{k! \Gamma(\nu + k + 1)} \left(\frac{z}{2}\right)^{2k+\nu}, \quad \nu \in \mathbb{R}, z \in \mathbb{C}$$

denotes the modified Bessel function of the first kind [39, p. 375, 9.6.10]. Corresponding to these identities, the following identities of the Catalan–Qi function $C(a, b; z)$ were obtained.

Theorem 5.19 ([38, theorem 1.5]) For $n \geq 0$ and $\Re(z) \geq 0$, we have

$$C(a, b; z+1) = \frac{bz+a}{az+b} C(a, b; z), \quad C(a, b; n) = \left(\frac{b}{a}\right)^n \prod_{k=0}^{n-1} \frac{a+k}{b+k},$$

$$\sum_{n=1}^{\infty} \left(\frac{a}{b}\right)^n C(a,b;n) = \frac{a}{b-a-1}, \quad b > a+1 > 1,$$

$$\sum_{n=0}^{\infty} C(a,b;n)\frac{x^{2n}}{(2n)!} = {}_1F_2\left(a; \frac{1}{2}, b; \frac{b}{4a}x^2\right),$$

$$\sum_{n=0}^{\infty} C(a,b;n)\frac{x^n}{n!} = {}_1F_1\left(a; b; \frac{b}{a}x\right).$$

By Theorem 5.19, we can regard the functions

$${}_1F_2\left(a; \frac{1}{2}, b; \frac{b}{4a}x^2\right) \quad \text{and} \quad {}_1F_1\left(a; b; \frac{b}{a}x\right) \tag{5.20}$$

as generating functions of the Catalan–Qi numbers $C(a,b;n)$.

In 1928, J. Touchard derived an identity

$$C_{n+1} = \sum_{k=0}^{\lfloor \frac{n}{2} \rfloor} \binom{n}{2k} 2^{n-2k} C_k, \tag{5.21}$$

where $\lfloor x \rfloor$ denotes the floor function whose value is the largest integer less than or equal to x, see [83, p. 472; 23, p. 319]. For the proof of the formula (5.21) by virtue of the generating function (5.1), see [1, pp. 319–320].

In 1987, when attending a summer program at Hope College, Holland, Michigan, United States, D. Jonah presented that

$$\binom{n+1}{m} = \sum_{k=0}^{m} \binom{n-2k}{m-k} C_k, \quad n \geq 2m, \quad n \in \mathbb{N}. \tag{5.22}$$

See [41, p. 214] and [1, pp. 324–326].

In 1990, Hilton and Pedersen generalized the identity (5.22) for an arbitrary real number n and any integer $m \geq 0$ [41, p. 214], [1, p. 327].

In 2009, J. Koshy provided in [1, p. 322] another recursive formula

$$C_n = \sum_{k=1}^{\lfloor \frac{n+1}{2} \rfloor} (-1)^{k-1} \binom{n-k+1}{k} C_{n-k}. \tag{5.23}$$

We observe that the identity (5.23) can be rearranged as

$$\sum_{k=\lceil \frac{n-1}{2} \rceil}^{n} (-1)^k \binom{k+1}{n-k} C_k = 0,$$

where $\lceil x \rceil$ stands for the ceiling function which gives the smallest integer not less than x.

The identities (5.21), (5.22), and (5.23) for the Catalan numbers C_n can be generalized to ones for the Catalan–Qi numbers $C(a,b;n)$, respectively.

Theorem 5.20 ([42, theorem 1]) For $a, b > 0$, $n \in \mathbb{N}$, and $n \geq 2m \geq 0$, the Catalan–Qi numbers $C(a, b; n)$ satisfy

$$_3F_2\left(a, \frac{1-n}{2}, -\frac{n}{2}; b, \frac{1}{2}; 1\right) = \sum_{k=0}^{\lfloor \frac{n}{2} \rfloor} \binom{n}{2k}\left(\frac{a}{b}\right)^k C(a, b; k),$$

$$_4F_3\left(1, a, -m, m-n; b, \frac{1-n}{2}, -\frac{n}{2}; \frac{b}{4a}\right) = \frac{1}{\binom{n}{m}} \sum_{k=0}^{m} \binom{n-2k}{m-k} C(a, b; k),$$

and

$$_3F_2\left(1-b-n, -\frac{n+1}{2}, -\frac{n}{2}; -n-1, 1-a-n; \frac{4a}{b}\right)$$

$$= \frac{1}{C(a, b; n)} \sum_{k=\lceil \frac{n-1}{2} \rceil}^{n} (-1)^{n-k} \binom{k+1}{n-k} C(a, b; k).$$

5.2.4 Integral Representations of the Catalan–Qi Function

Some integral representations of the Catalan numbers C_n in Section 5.1.5 can be generalized as the following forms.

Theorem 5.21 ([16, p. 3, eq. (10)] and [17, p. 3, eq. (1.9)]) For $b > a > 0$ and $x \geq 0$, we have

$$C(a, b; x) = \frac{1}{B(a, b-a)}\left(\frac{b}{a}\right)^x \int_0^\infty (1 - e^{-u})^{b-a-1} e^{-(x+a)u} du,$$

where $B(z, w)$ denotes the classical beta function

$$B(z, w) = \int_0^1 t^{z-1}(1-t)^{w-1} dt = \int_0^\infty \frac{t^{z-1}}{(1+t)^{z+w}} dt \tag{5.24}$$

for $\Re(z), \Re(w) > 0$.

Theorem 5.22 ([16, theorem 4] and [17, theorem 3.1]) For $b > a > 0$ and $x \geq 0$, the Catalan–Qi function $C(a, b; x)$ has integral representations

$$C(a, b; x) = \left(\frac{a}{b}\right)^{b-1} \frac{1}{B(a, b-a)} \int_0^{b/a} \left(\frac{b}{a} - t\right)^{b-a-1} t^{x+a-1} dt \tag{5.25}$$

and

$$C(a, b; x) = \left(\frac{a}{b}\right)^a \frac{1}{B(a, b-a)} \int_0^\infty \frac{t^{b-a-1}}{(t+a/b)^{x+b}} dt. \tag{5.26}$$

Theorem 5.23 (**[38, theorem 1.4]**)   For $\Re(a), \Re(b) > 0$ and $\Re(z) \geq 0$, we have

$$C(a, b; z) = \frac{\Gamma(b)}{\Gamma(a)} \left(\frac{b}{a} \right)^z \frac{(z+a)^z}{(z+b)^{z+b-a}}$$

$$\times \exp \left[b - a + \int_0^\infty \frac{1}{t} \left(\frac{1}{1 - e^{-t}} - \frac{1}{t} - a \right) (e^{-at} - e^{-bt}) e^{-zt} dt \right]. \qquad (5.27)$$

In Ref. [4], it was said that the integral representation (5.6) means that the Catalan numbers C_n are a solution of the Hausdorff moment problem on the interval $[0, 4]$ instead of $[0, 1]$. Analogously, we guess that the integral representation (5.25) probably means that the Catalan–Qi numbers $C(a, b; n)$ are a solution of the Hausdorff moment problem on the interval $\left[0, \frac{b}{a}\right]$ instead of $[0, 1]$ and $[0, 4]$.

5.2.5   Asymptotic Expansions of the Catalan–Qi Function

The asymptotic expansions in Section 5.1.6 can be generalized as follows.

Theorem 5.24 (**[16, theorem 2]** and **[17, theorem 2.1]**)   For $b > a > 0$, the Catalan–Qi function $C(a, b; x)$ has the asymptotic expansion

$$C(a, b; x) \sim \frac{1}{B(a, b-a)} \left(\frac{b}{a} \right)^x$$

$$\times \sum_{k=0}^{\infty} (-1)^k B_k^{(a-b+1)}(a) \frac{\Gamma(b-a+k)}{k!} \frac{1}{x^{k+b-a}} \qquad (5.28)$$

as $x \to \infty$, where $B(x, y)$ is the classical beta function defined by (5.24) and $B_k^{(\sigma)}(x)$ is the generalized Bernoulli polynomials defined by (5.11).

Theorem 5.25 (**[16, theorem 3]** and **[17, theorem 2.2]**)   The Catalan–Qi function $C(a, b; x)$ has the exponential expansion

$$C(a, b; x) = \frac{\Gamma(b)}{\Gamma(a)} \left(\frac{b}{a} \right)^x \sqrt{\frac{x+b}{x+a}} \left[I(x+a, x+b) \right]^{a-b}$$

$$\times \exp \left[\sum_{j=1}^{\infty} \frac{B_{2j}}{2j(2j-1)} \left(\frac{1}{(x+a)^{2j-1}} - \frac{1}{(x+b)^{2j-1}} \right) \right], \qquad (5.29)$$

where $I(\alpha, \beta)$ denotes the exponential mean defined by

$$I(\alpha, \beta) = \frac{1}{e} \left(\frac{\beta^\beta}{\alpha^\alpha} \right)^{1/(\beta-\alpha)} \qquad (5.30)$$

for $\alpha, \beta > 0$ with $\alpha \neq \beta$ and B_{2j} are the Bernoulli numbers defined by (5.13).

Theorem 5.26 **([16, theorem 5] and [17, theorem 3.2])** For $b > a > 0$, we have

$$C(a, b; x) = \frac{1}{B(a, b - a)} \left(\frac{b}{a}\right)^x \sum_{k=0}^{\infty} (-1)^k \frac{\langle b - a - 1\rangle_k}{k!} \frac{1}{x + a + k}, \quad (5.31)$$

where

$$\langle x \rangle_n = \prod_{k=0}^{n-1} (x - k) = \begin{cases} x(x - 1) \cdots (x - n + 1), & n \geq 1 \\ 1, & n = 0 \end{cases}$$

is the falling factorial.

5.2.6 Complete Monotonicity of the Catalan–Qi Function

From Refs. [20, chapter XIII], [43, chapter 1], and [21, chapter IV], we recall that an infinitely differentiable function f is said to be completely monotonic on an interval I if it satisfies $0 \leq (-1)^k f^{(k)}(x) < \infty$ on I for all $k \geq 0$

$$0 \leq (-1)^k f^{(k)}(x) < \infty$$

on I for all $k \geq 0$.

From Refs. [43–46], we recall that an infinitely differentiable and positive function f is said to be logarithmically completely monotonic on an interval I if $0 \leq (-1)^k [\ln f(x)]^{(k)} < \infty$ holds on I for all $k \in \mathbb{N}$. For more information on logarithmically completely monotonic functions, please refer to [47–50] and plenty of references therein.

It is known [45, 46, 48] that a logarithmically completely monotonic function must be completely monotonic.

Theorem 5.27 **([51, theorem 1.2] and 52)** Let $a, b > 0$ and $x \geq 0$. Then the function $[C(a, b; x)]^{\pm 1}$ is logarithmically completely monotonic

(1) with respect to $a > 0$ if and only if $x \gtrless 1$,
(2) with respect to $b > 0$ if and only if $x \lessgtr 1$.

Theorem 5.28 **([16, theorem 7] and [17, theorem 4.2])** Let $a, b > 0$ and $x \geq 0$. Then

(1) the unique zero x_0 of the equation

$$\frac{\psi(x + b) - \psi(x + a)}{\ln b - \ln a} = 1$$

satisfies $x_0 \in \left(0, \frac{1}{2}\right)$, where ψ is the logarithmic derivative of the gamma function Γ;

(2) when $b > a$, the function $C(a, b; x)$ is decreasing in $x \in [0, x_0)$, increasing in $x \in (x_0, \infty)$, and logarithmically convex in $x \in [0, \infty)$;

(3) when $b < a$, the function $C(a, b; x)$ is increasing in $x \in [0, x_0)$, decreasing in $x \in (x_0, \infty)$, and logarithmically concave in $x \in [0, \infty)$.

Theorem 5.29 **([16, theorems 8 and 9] and [17, theorems 4.3 and 4.4])** For $b > a > 0$, the functions

$$\left(\frac{a}{b}\right)^x C(a, b; x) \quad \text{and} \quad \left(\frac{a}{b}\right)^x \frac{(x+b)^{x+b-a}}{(x+a)^x} C(a, b; x)$$

are logarithmically completely monotonic on $[0, \infty)$.

Theorem 5.30 **([16, theorem 5] and [17, theorem 3.2])** The function

$$(-1)^{\lfloor b-a \rfloor} \left[\left(\frac{a}{b}\right)^x C(a, b; x) - \frac{1}{B(a, b-a)} \sum_{k=0}^{N} \frac{(-1)^k}{k!} \frac{\langle b - a - 1 \rangle_k}{x + a + k} \right]$$

for $N \in \{0\} \cup \mathbb{N}$ and $b > a > 0$ is completely monotonic in $x \in [0, \infty)$, where $\lfloor x \rfloor$ denotes the floor function whose value is the largest integer less than or equal to x.

Theorem 5.31 **([16, theorem 6] and [17, theorem 4.1])** The function

$$C^{\pm 1}(a, b; x) = \begin{cases} 1, & x = 0 \\ [C(a, b; x)]^{\pm 1/x}, & x > 0 \end{cases}$$

is logarithmically completely monotonic with respect to $x \in [0, \infty)$ if and only if $a \gtrless b$.

Now consider the function

$$\mathscr{C}_{a,b;x}(t) = C(a + t, b + t; x), \quad t, x \geq 0, \quad a, b > 0.$$

Theorem 5.32 **([53, theorem 1.1])** For $x \geq 0$ and $a, b > 0$,

(1) the function $\mathscr{C}_{a,b;x}(t)$ is logarithmically completely monotonic on $[0, \infty)$ if and only if either $0 \leq x \leq 1$ and $a \leq b$ or $x \geq 1$ and $a \geq b$,
(2) the function $\dfrac{1}{\mathscr{C}_{a,b;x}(t)}$ is logarithmically completely monotonic on $[0, \infty)$ if and only if either $0 \leq x \leq 1$ and $a \geq b$ or $x \geq 1$ and $a \leq b$.

Is the function

$$\mathscr{C}_{a,b;x;\alpha,\beta}(t) = C(a + \alpha t, b + \beta t; x), \quad x \geq 0, \quad a, b > 0$$

of logarithmically complete monotonicity in $t \in [0, \infty)$?

Theorem 5.33 **(54)** If and only if $\alpha = 0$ and $\beta > 0$, or $\alpha > 0$ and $\beta = 0$, or $\alpha = \beta > 0$, the function $\mathscr{C}_{a,b;x;\alpha,\beta}(t)$ is of the following logarithmically complete monotonicity:

(1) the function $[C(a, b; x)]^{\pm 1}$ is logarithmically completely monotonic
   (a) with respect to $a > 0$ if and only if $x \gtrless 1$,
   (b) with respect to $b > 0$ if and only if $x \lessgtr 1$,

(2) the function $\mathscr{C}_{a,b;x}(t)$ is logarithmically completely monotonic on $[0, \infty)$ if and only if either $0 \le x \le 1$ and $a \le b$ or $x \ge 1$ and $a \ge b$,

(3) the function $\frac{1}{\mathscr{C}_{a,b;x}(t)}$ is logarithmically completely monotonic on $[0, \infty)$ if and only if either $0 \le x \le 1$ and $a \ge b$ or $x \ge 1$ and $a \le b$.

5.2.7 Schur-Convexity of the Catalan–Qi Function

Recall from Refs. [55, p. 80] and [56, pp. 75–76] that a function f with n arguments defined on I^n is called Schur-convex if $f(x) \le f(y)$ for each two n-tuples $x = (x_1, \dots, x_n)$ and $y = (y_1, \dots, y_n)$ on I^n such that $x \prec y$ holds, where I is an interval with nonempty interior. A function f is Schur-concave if and only if $-f$ is Schur-convex.

Theorem 5.34 ([51, theorem 1.1] and 52) For $a, b > 0$ and $x \ge 0$, let

$$F_x(a, b) = |\ln C(a, b; x)|.$$

Then the function $F_x(a, b)$ is Schur-convex in $(a, b) \in (0, \infty) \times (0, \infty)$ for all $x \ge 0$. In other words, if and only if $(a_1, b_1) \preceq (a_2, b_2)$, the inequality

$$|\ln C(a_1, b_1; x)| \le |\ln C(a_2, b_2; x)|$$

is valid for all $x \ge 0$.

5.2.8 Generating Functions of the Catalan–Qi Numbers

We discovered that the function $_2F_1\left(a, 1; b; \frac{bt}{a}\right)$ is a generating function of the Catalan–Qi numbers $C(a, b; n)$.

Theorem 5.35 ([16, theorem 10] and [17, theorem 5.1]) For $a, b > 0$ and $n \ge 0$, the Catalan–Qi numbers $C(a, b; n)$ can be generated by

$$_2F_1\left(a, 1; b; \frac{bt}{a}\right) = \sum_{n=0}^{\infty} C(a, b; n) t^n$$

and, conversely, satisfy

$$C(a, b; n) = (-1)^n \sum_{k=0}^{n} (-1)^k \binom{n}{k} {}_2F_1\left(a, -k; b; -\frac{b}{a}\right).$$

The last two formulas in Theorem 5.19 show that the functions in (5.20) can also be regarded as the generating functions of the Catalan–Qi numbers $C(a, b; n)$.

5.2.9 A Double Inequality of the Catalan–Qi Function

From (5.29), one can deduce the following double inequality for the Catalan–Qi function $C(a, b; x)$.

Theorem 5.36 ([16, theorem 11] and [17, theorem 6.1]) The Catalan–Qi function $C(a, b; x)$ satisfies the double inequality

$$\exp\left[\sum_{j=1}^{2m} \frac{B_{2j}}{2j(2j-1)} \left(\frac{1}{(x+a)^{2j-1}} - \frac{1}{(x+b)^{2j-1}} \right) \right]$$

$$< \frac{\Gamma(a)}{\Gamma(b)} \left(\frac{a}{b} \right)^x \sqrt{\frac{x+a}{x+b}} \frac{C(a, b; x)}{[I(x+a, x+b)]^{a-b}}$$

$$< \exp\left[\sum_{j=1}^{2m-1} \frac{B_{2j}}{2j(2j-1)} \left(\frac{1}{(x+a)^{2j-1}} - \frac{1}{(x+b)^{2j-1}} \right) \right],$$

where B_i for $i \in \mathbb{N}$ are the Bernoulli numbers defined by (5.13) and I is the exponential mean defined by (5.30).

5.2.10 The q-Catalan–Qi Numbers and Properties

Mainly motivated by the paper [16], Zou introduced [57, 58] the q-analogue of the Catanlan–Qi numbers, called the q-Catalan–Qi numbers and, among other things, obtained supercongruences for the Catalan–Qi numbers $C(a, b; k)$.

5.2.11 The Catalan Numbers and the k-Gamma and k-Beta Functions

Motivated by the papers [9, 59, 60] and the closely related references therein, Qi and his coauthors, Akkurt and Yildirim, established more explicit formulas and integral representations of the Catalan numbers C_k and presented a class of parametric integrals in terms of the k-gamma and k-beta functions in [61].

5.2.12 Series Identities Involving the Catalan Numbers

In [62], Yin and Qi discovered several series identities involving the Catalan numbers C_k, the Catalan function C_x, the Riemanian zeta function $\zeta(z)$, and the alternative Hurwitz zeta function. For more information on related topics, please refer to [63, 64] and the closely related references therein.

5.3 The Fuss–Catalan Numbers

5.3.1 A Definition of the Fuss–Catalan Numbers

In combinatorial mathematics and statistics, the Fuss–Catalan numbers $A_n(p, r)$ were defined [65, 66] as numbers of the form

$$A_n(p, r) = \frac{r}{np+r} \binom{np+r}{n} = \frac{r\Gamma(np+r)}{\Gamma(n+1)\Gamma(n(p-1)+r+1)}.$$

It is easy to see that

$$A_n(p, 1) = F(p, n), \quad A_n(2, 1) = C_n, \quad A_{n-1}(p, p) = {}_pd_n = C(n, p).$$

This means that the Fuss–Catalan numbers $A_n(p, r)$ is a unified generalization of the Catalan numbers C_n, the Fuss numbers $F(m, n)$, and the generalized Catalan numbers $C(m, n - 1) = {}_n d_m$.

There have been some literature, such as [66–78], on the investigation of the Fuss–Catalan numbers $A_n(p, r)$.

5.3.2 A Product-Ratio Expression of the Fuss–Catalan Numbers

The Fuss–Catalan numbers $A_n(p, r)$ can be represented as a product-ratio expression and other forms in terms of the Catalan–Qi numbers $C(a, b; n)$.

Theorem 5.37 ([79, theorem 1.1]) For $n, r \geq 0$ and $p > 1$, we have

$$A_n(p, r) = r^n \frac{\prod_{k=1}^{p} C\left(\frac{k+r-1}{p}, 1; n\right)}{\prod_{k=1}^{p-1} C\left(\frac{k+r}{p-1}, 1; n\right)}. \tag{5.32}$$

For $n, p \in \mathbb{N}$ and $r \geq 0$, we have

$$A_n(p, r) = \frac{r}{nB(n(p-1) + 1, n)} \frac{(np)^{r-1}}{[n(p-1) + 1]^r} C(np, n(p-1) + 1; r).$$

For $r + 1 > n > 0$ and $p \geq 0$, we have

$$A_n(p, r) = \frac{1}{nB(n, r-n+1)} \left(\frac{r}{r-n+1}\right)^{np} C(r, r-n+1; np).$$

When $r + 1 > n \geq 1$ and $p \geq 0$, we have

$$A_n(p, r) = \frac{1}{n} \frac{[B(r+1-n, n)]^{p-1}}{[B(r, n)]^p} \prod_{k=0}^{n-1} C\left(\frac{r+k}{n}, \frac{r-n+k+1}{n}; p\right).$$

For $n \geq 2$, $r + 1 > n$, and $p \in \mathbb{N}$, we have

$$A_n(p, r) = rp^{1/2} B(n-1, 2) \frac{[B(r+1-n, n-1)]^{n-1}}{[B(r+1-n+p, n-1)]^n}$$

$$\times \prod_{k=0}^{p-1} C\left(\frac{r+k}{p}, \frac{r+k+1-n}{p}; n\right).$$

5.3.3 Complete Monotonicity of the Fuss–Catalan Numbers

Applying the product-ratio expression (5.32) and some properties of the Catalan–Qi function $C(a, b; x)$, one can find several properties of the Fuss–Catalan numbers $A_n(p, r)$, including monotonicity, logarithmic convexity, complete monotonicity, and minimality.

Theorem 5.38   ([79, theorem 1.2])   When $p \geq r > 0$,

(1) the sequence $\{A_n(p,r)\}_{n \geq 0}$,

$$A_n(p,r) = \begin{cases} 1, & n = 0, \\ \dfrac{1}{\sqrt[n]{A_n(p,r)}}, & n \in \mathbb{N}, \end{cases}$$

is decreasing, logarithmic convex, completely monotonic, and minimal;

(2) the sequence of the Fuss–Catalan numbers $\{A_n(p,r)\}_{n \geq 0}$ is increasing and logarithmic convex.

5.3.4   A Double Inequality for the Fuss–Catalan Numbers

By applying a double inequality of the beta function $B(x,y)$ from the papers [80] and [81, pp. 78–81, section 3], a double inequality for the Fuss–Catalan numbers $A_n(p,r)$ was obtained.

Theorem 5.39   ([79, theorem 1.3])   For $n \geq 2$ and $p, r \in \mathbb{N}$, we have

$$A_n(p,r) \geq \frac{r[n(p-1)+r+1]}{n}. \tag{5.33}$$

When $m \triangleq \min\{D(n-1)D(n(p-1)+r+1), b_A(n-1)[n(p-1)+r+1]\} < 1$ for $n \geq 2$ and $p, r \in \mathbb{N}$, we have

$$A_n(p,r) \leq \frac{r[n(p-1)+r+1]}{n(1-m)},$$

where

$$D(x) = \frac{x-1}{\sqrt{2x-1}} \quad \text{and} \quad b_A = \max_{x \geq 1}\left[\frac{1}{x^2} - \frac{\Gamma^2(x)}{\Gamma(2x)}\right] = 0.08731\ldots.$$

for $x \geq 1$.

By some inequalities for the beta function $B(x,y)$ surveyed in the paper [82], one can derive more inequalities for the Fuss–Catalan numbers $A_n(p,r)$ and others.

5.4   The Fuss–Catalan–Qi Function

5.4.1   A Definition of the Fuss–Catalan–Qi Function

In the early morning of September 15, 2015, a unified generalization of the Catalan numbers C_k, the generalized Catalan numbers $C(n,m)$, the Fuss numbers $F(m, n)$, the Fuss–Catalan numbers $A_n(p, r)$, and the Catalan–Qi function $C(a, b; z)$ was framed out eventually and successfully by Qi in [83, section 2]. This generalization can be described by a five-variable function

$$Q(a, b; p, q; z) = \frac{\Gamma(b)}{\Gamma(a)} \left(\frac{b}{a}\right)^{(q-p+1)z} [\Gamma(z+1)]^{q-p} \frac{\Gamma(pz+a)}{\Gamma(qz+b)}, \tag{5.34}$$

where $\Re(a), \Re(b) > 0$, $\Re(p), \Re(q) > 0$, and $\Re(z) \geq 0$.

It is easy to see that

$$Q\left(\frac{1}{2}, 2; 1, 1; n\right) = Q(1, 2; 2, 1; n) = C_n,$$

$$Q(p, p+1; p, p-1; n-1) = {}_p d_n = C(n, p-1),$$

$$Q(r, r+1; p, p-1; n) = A_n(p, r),$$

$$Q(a, b; 1, 1; z) = C(a, b; z).$$

Accordingly, the function $Q(a, b; p, q; z)$ is a unified generalization of the Catalan numbers C_n, the generalized Catalan numbers $C(n, m)$, the Fuss numbers $F(m, n)$, the Fuss–Catalan numbers $A_n(p, r)$, and the Catalan–Qi function $C(a, b; z)$.

For uniqueness and convenience of referring to the quantity $Q(a, b; p, q; z)$, we call $Q(a, b; p, q; z)$ the Fuss–Catalan–Qi function and, when taking $z = n \geq 0$, call $Q(a, b; p, q; n)$ the Fuss–Catalan–Qi numbers.

5.4.2 A Product-Ratio Expression of the Fuss–Catalan–Qi Function

Similar to the product-ratio expression (5.32) for the Fuss–Catalan numbers $A_n(p, r)$, a product-ratio expression for the Fuss–Catalan–Qi function $Q(a, b; p, q; z)$ can also be established.

Theorem 5.40 ([83, theorem 3.1]) For $\Re(a), \Re(b) > 0$ and $\Re(z) \geq 0$, when $p, q \in \mathbb{N}$, we have

$$Q(a, b; p, q; z) = \left[\left(\frac{b}{a}\right)^{q-p+1} \frac{\Gamma(b)\Gamma(p+a)}{\Gamma(a)\Gamma(q+b)}\right]^z \frac{\prod_{k=0}^{p-1} C\left(\frac{k+a}{p}, 1; z\right)}{\prod_{k=0}^{q-1} C\left(\frac{k+b}{q}, 1; z\right)}. \tag{5.35}$$

We observe that

$$\left(\frac{b}{a}\right)^{q-p+1} \frac{\Gamma(b)\Gamma(p+a)}{\Gamma(a)\Gamma(q+b)} = Q(a, b; p, q; 1), \quad \Re(a), \Re(b), \Re(p), \Re(q) > 0.$$

Hence, the product-ratio expression (5.35) can be rewritten as

$$\frac{Q(a, b; p, q; z)}{[Q(a, b; p, q; 1)]^z} = \frac{\prod_{k=0}^{p-1} C\left(\frac{k+a}{p}, 1; z\right)}{\prod_{k=0}^{q-1} C\left(\frac{k+b}{q}, 1; z\right)}.$$

5.4.3 Integral Representations of the Fuss–Catalan–Qi Function

Making use of the integral representation (5.27), one can derive integral representations of the Fuss–Catalan–Qi function $Q(a, b; p, q; z)$.

Theorem 5.41 ([83, theorem 4.1]) For $\mathfrak{R}(a), \mathfrak{R}(b) > 0$ and $\mathfrak{R}(z) \geq 0$, when $p, q \in \mathbb{N}$, we have

$$Q(a, b; p, q; z) = (2\pi)^{(q-p)/2} \frac{\Gamma(b)}{\Gamma(a)} \frac{p^{a-1/2}}{q^{b-1/2}} (z+1)^{(q-p)(z+1/2)+(a-b)}$$

$$\times \left[\left(\frac{b}{a} \right)^{q-p+1} \frac{\prod_{k=0}^{p-1}(a + pz + k)}{\prod_{k=0}^{q-1}(b + qz + k)} \right]^{z} \exp \left\{ \frac{p-q}{2} + (b-a) \right.$$

$$+ \int_{0}^{\infty} \frac{e^{-zt}}{t} \left[\left(\frac{1}{1-e^{-t}} - \frac{1}{t} - \frac{a}{p} \right) \left(\frac{1-e^{-t}}{1-e^{-t/p}} e^{-at/p} - pe^{-t} \right) \right.$$

$$- \left(\frac{1}{1-e^{-t}} - \frac{1}{t} - \frac{b}{q} \right) \left(\frac{1-e^{-t}}{1-e^{-t/q}} e^{-bt/q} - qe^{-t} \right)$$

$$- \left(\frac{e^{-t/p}}{p(1-e^{-t/p})} - \frac{e^{-t}}{1-e^{-t}} \right) \frac{1-e^{-t}}{1-e^{-t/p}} e^{-at/p}$$

$$+ \left. \left(\frac{e^{-t/q}}{q(1-e^{-t/q})} - \frac{e^{-t}}{1-e^{-t}} \right) \frac{1-e^{-t}}{1-e^{-t/q}} e^{-bt/q} + \frac{p-q}{2} e^{-t} \right] dt \right\}.$$

Theorem 5.42 ([83, theorem 4.2]) For $a, b > 0$ and $x \geq 0$, when $p, q \in \mathbb{N}$, we have

$$Q(a, b; p, q; x) = (2\pi)^{(q-p)/2} e^{(p-q)/2+b-a} \frac{\Gamma(b)}{\Gamma(a)} \left(\frac{b}{a} \right)^{(q-p+1)x}$$

$$\times (x+1)^{(q-p)(x+1/2)} \frac{p^{px+a-1/2}}{q^{qx+b-1/2}} \frac{\prod_{k=0}^{p-1} \left(x + \frac{a+k}{p} \right)^{x+(a+k)/p-1/2}}{\prod_{k=0}^{q-1} \left(x + \frac{b+k}{q} \right)^{x+(b+k)/q-1/2}}$$

$$\times \exp \left\{ \int_{0}^{\infty} \left[\frac{(q-p)e^{-t}}{1-e^{-t}} + \frac{e^{-at/p}}{1-e^{-t/p}} - \frac{e^{-bt/q}}{1-e^{-t/q}} \right] \beta(t)(1-e^{-t})e^{-xt} dt \right\},$$

where

$$\beta(t) = \frac{1}{t} \left(\frac{1}{e^t - 1} - \frac{1}{t} + \frac{1}{2} \right). \tag{5.36}$$

Theorem 5.43 ([83, theorem 4.3]) For $a, b, p, q > 0$ and $x \geq 0$, we have

$$Q(a, b; p, q; x) = \frac{\Gamma(b)}{\Gamma(a)} (2\pi)^{(q-p)/2} e^{p-q+b-a} \left(\frac{b}{a} \right)^{(q-p+1)x}$$

$$\times (x+1)^{(q-p)(x+1/2)} \frac{(px+a)^{px+a-1/2}}{(qx+b)^{qx+b-1/2}}$$

$$\times \exp\left\{\int_0^\infty \beta(t)[(q-p)e^{-(x+1)t} - e^{-(qx+b)t} + e^{-(px+a)t}]\,\mathrm{d}t\right\},$$

where $\beta(t)$ is defined by (5.36).

5.4.4 Complete Monotonicity of the Fuss–Catalan–Qi Function

From Ref. [44], we recall that, if $f^{(k)}(x)$ for some nonnegative integer k is completely monotonic on an interval I, but $f^{(k-1)}(x)$ is not completely monotonic on I, then $f(x)$ is called a completely monotonic function of the kth order on I.

Stimulated by the above definition and those main results in the papers [44–46, 84], we introduced the concept of logarithmically completely monotonic functions of the kth order.

Definition 5.1 ([83, definition 5.1]) For a positive function $f(x)$ on an interval I, if $[\ln f(x)]^{(k)}$ for some nonnegative integer k is completely monotonic on an interval I, but $[\ln f(x)]^{(k-1)}$ is not completely monotonic on I, then we call $f(x)$ a logarithmically completely monotonic function of the k-th order on I.

Theorem 5.44 ([83, theorem 5.1]) The function

$$Q(a, b; q, q; x) = \frac{\Gamma(b)}{\Gamma(a)}\left(\frac{b}{a}\right)^x \frac{\Gamma(qx+a)}{\Gamma(qx+b)}, \qquad a, b, p > 0, \quad x \geq 0$$

satisfies the following conclusions:

(1) if $a < b$ and $q \leq \frac{\ln b - \ln a}{\psi(b) - \psi(a)}$, the function $Q(a, b; q, q; x)$ is increasing on $[0, \infty)$;

(2) if $a > b$ and $q \leq \frac{\ln b - \ln a}{\psi(b) - \psi(a)}$, the function $Q(a, b; q, q; x)$ is decreasing on $[0, \infty)$;

(3) if $a < b$ and $q > \frac{\ln b - \ln a}{\psi(b) - \psi(a)}$, the function $Q(a, b; q, q; x)$ has a unique minimum on $(0, \infty)$;

(4) if $a > b$ and $q > \frac{\ln b - \ln a}{\psi(b) - \psi(a)}$, the function $Q(a, b; q, q; x)$ has a unique maximum on $(0, \infty)$;

(5) if and only if $a \lessgtr b$, the function $[Q(a, b; q, q; x)]^{\pm}$ is logarithmically completely monotonic of the second order on $[0, \infty)$; in particular, if and only if $a \lessgtr b$, the function $[Q(a, b; q, q; x)]^{\pm 1}$ is logarithmically convex on $[0, \infty)$.

5.5 Some Properties for Ratios of Two Gamma Functions

By studying the Catalan numbers C_n, the Catalan–Qi function $C(a, b; z)$, the Fuss–Catalan numbers $A_n(p, r)$, and the Fuss–Catalan–Qi function $Q(a, b; p, q, z)$, we acquired, as by-products, some properties for ratios of two gamma functions. About known results for ratios of two gamma functions, please refer to the expository and survey articles [48, 83–85]] and the closely related references therein.

5.5.1 An Integral Representation and Complete Monotonicity

Step by step, we obtained [12, 89] an integral representation and complete monotonicity of a function involving a ratio of two gamma functions.

Theorem 5.45 ([89, theorem 2] and [12, theorem 3]) Let

$$F_{a,b}(x) = \frac{\Gamma(x + a)}{(x + a)^x} \frac{(x + b)^{x+b-a}}{\Gamma(x + b)}, \quad a, b \in \mathbb{R}, a \neq b, x > -\min\{a, b\}.$$

For $a, b > 0$, the following conclusions are valid:

(1) the function $F_{a,b}(x)$ has the exponential representation

$$F_{a,b}(x) = \exp\left[b - a + \int_0^\infty \frac{1}{t}\left(a + \frac{1}{t} - \frac{1}{1 - e^{-t}}\right)(e^{-bt} - e^{-at})e^{-xt}dt\right]$$

on $[0, \infty)$;

(2) the function $[F_{a,b}(x)]^{\pm 1}$ is logarithmically completely monotonic on $[0, \infty)$ if and only if

$$(a, b) \in D_{\pm}(a, b) = \{(a, b) : a \gtreqqless b, a \geq 1\} \cup \left\{(a, b) : a \lesseqqgtr b, a \leq \frac{1}{2}\right\}.$$

5.5.2 An Exponential Expansion for the Ratio of Two Gamma Functions

As a by-product of investigating the Catalan–Qi function, we discovered [16, 17] an exponential expansion for the ratio $\frac{\Gamma(a)}{\Gamma(b)}$ of two gamma functions.

Theorem 5.46 ([16, theorem 3] and [17, theorem 2.2]) For $a, b > 0$, we have

$$\frac{\Gamma(a)}{\Gamma(b)} = \frac{a^{a-1/2}}{b^{b-1/2}} \exp\left[\sum_{j=1}^\infty \frac{B_{2j}}{2j(2j-1)}\left(\frac{1}{a^{2j-1}} - \frac{1}{b^{2j-1}}\right)\right]. \tag{5.37}$$

5.5.3 A Double Inequality for the Ratio of Two Gamma Functions

From Ref. (5.37), we can readily derive a double inequality for the ratio $\frac{\Gamma(a)}{\Gamma(b)}$ of two gamma functions.

Theorem 5.47 ([16, theorem 11] and [17, theorem 6.1]) For $a, b > 0$, we have

$$\sqrt{\frac{b}{a}}\ [I(a, b)]^{a-b} \exp\left[\sum_{j=1}^{2m} \frac{B_{2j}}{2j(2j-1)}\left(\frac{1}{a^{2j-1}} - \frac{1}{b^{2j-1}}\right)\right] < \frac{\Gamma(a)}{\Gamma(b)}$$

$$< \sqrt{\frac{b}{a}} \, [I(a,b)]^{a-b} \exp\left[\sum_{j=1}^{2m-1} \frac{B_{2j}}{2j(2j-1)} \left(\frac{1}{a^{2j-1}} - \frac{1}{b^{2j-1}} \right) \right],$$

where B_i for $i \in \mathbb{N}$ are the Bernoulli numbers defined by (5.13) and I is the exponential mean defined by (5.30).

5.6 Some New Results on the Catalan Numbers

Recently, some new results on the Catalan numbers C_n were obtained in the papers [90, 91] and the closely related references therein.

5.7 Open Problems

Till now, to the best of our knowledge, the following problems posed in [92] still keep open.

(1) What is the combinatorial interpretation of the Catalan–Qi function?
(2) What is the combinatorial interpretation of the Fuss–Catalan–Qi function?
(3) How can one obtain more properties of the Fuss–Catalan–Qi function?
(4) As mentioned on page 16, does the integral representation (5.25) mean that the Catalan–Qi numbers $C(a, b; n)$ are a solution of the Hausdorff moment problem on the interval $\left[0, \frac{b}{a}\right]$ instead of $[0, 1]$ and $[0, 4]$?
(5) Can one similarly interpret the integral representation (5.26)?
(6) The two-parameter sequence of non-negative integers

$$S_{m,n} = \frac{(2m)!(2n)!}{(m+n)!m!n!}$$

is a generalization of the Catalan numbers C_n. These numbers $S_{m,n}$ are named super Catalan numbers by Ira Gessel. It is clear that $S_{1,n} = 2C_n$. For $m = n$, the numbers $S_{m,m}$ have an easy combinatorial description. However, other combinatorial descriptions are only known for $m = 2, 3$, and it is an open problem to find a general combinatorial interpretation (see the website [4]). Is the super Catalan numbers $S_{m,n}$ a special case of the Fuss–Catalan–Qi function $Q(a, b; p, q; z)$?
(7) In [57, 58], Zou considered the q-Catalan–Qi numbers and the q-analogous of the Catalan–Qi numbers. Similarly, can one consider the q-analogous of the Fuss–Catalan–Qi function $Q(a, b; p, q; z)$ defined by (5.34) and investigate their properties and applications?

Remark 5.1 This chapter is a corrected and revised version of the preprint [92].

Acknowledgments

The original version of this chapter was ever reported on November 21, 2015 as an invited talk of the 29th Conference of the Jangjeon Mathematical Society held November 20–21, 2015 at Pukyong National University and reported on November 25, 2015 at Kwangwoon University in South Korea by the first author. The first author appreciates Professor Daekyun Kim for his warming invitation and financial support for attending the conference and thanks Professors Dmitry V. Dolgy, Dae San Kim, Seog-Hoon Rim, Jong Jin Seo, Dr. Hyuck-In Kwon, and other friends in South Korea for their sincere hospitality.

References

1 Koshy, T. (2009) *Catalan Numbers with Applications*, Oxford University Press, Oxford.

2 Stanley, R.P. (2015) *Catalan Numbers*, Cambridge University Press, New York. doi: 10.1017/CBO9781139871495.

3 Stanley, R. and Weisstein, E.W. *Catalan number*, From MathWorld–A Wolfram Web Resource, http://mathworld.wolfram.com/CatalanNumber.html (accessed 16 October 2017).

4 WikiPedia *Catalan number*, From the Free Encyclopedia, https://en.wikipedia.org/wiki/Catalan_number (accessed 16 October 2017).

5 Larcombe, P.J. (1999/2000) The 18th century Chinese discovery of the Catalan numbers. *Math. Spectr.*, **32** (1), 5–7.

6 Luo, J.-J. (1988) Ming Antu, the first inventor of Catalan numbers in the world. *J. Inner Mongolia Univ. (Acta Scientiarum Naturalium Universitatis NeiMongol)*, **19** (2), 239–245.

7 Qi, F., Shi, X.-T., and Liu, F.-F. (2015) Several Formulas for Special Values of the Bell Polynomials of the Second Kind and Applications. ResearchGate Technical Report. doi: 10.13140/RG.2.1.3230.1927.

8 Qi, F., Shi, X.-T., Liu, F.-F., and Kruchinin, D.V. (2017) Several formulas for special values of the Bell polynomials of the second kind and applications. *J. Appl. Anal. Comput.*, 7 (3), 857–871. doi: 10.11948/2017054.

9 Dana-Picard, T. (2005) Parametric integrals and Catalan numbers. *Int. J. Math. Edu. Sci. Technol.*, **36** (4), 410–414. doi: 10.1080/00207390412331321603.

10 Nkwanta, A. and Tefera, A. (2013) Curious relations and identities involving the Catalan generating function and numbers. *J. Integer Seq.*, **16** (9), Article 13.9.5, 15 pp.

11 Penson, K.A. and Sixdeniers, J.-M. (2001) Integral representations of Catalan and related numbers. *J. Integer Seq.*, **4** (2), Article 01.2.5, 6 pp.

12 Qi, F., Shi, X.-T., and Liu, F.-F. (2015) An Integral Representation, Complete Monotonicity, and Inequalities of the Catalan Numbers. ResearchGate Technical Report. doi: 10.13140/RG.2.1.3754.4806.

13 Shi, X.-T., Liu, F.-F., and Qi, F. (2015) An integral representation of the Catalan numbers. *Glob. J. Math. Anal.*, **3** (3), 130–133. 10.14419/gjma.v3i3.5055.

14 Graham, R.L., Knuth, D.E., and Patashnik, O. (1994) *Concrete Mathematics—A Foundation for Computer Science*, 2nd edn, Addison-Wesley Publishing Company, Reading, MA.

15 Vardi, I. (1991) *Computational Recreations in Mathematica*, Addison-Wesley, Redwood City, CA.

16 Qi, F., Mahmoud, M., Shi, X.-T., and Liu, F.-F. (2016) Some properties of the Catalan–Qi function related to the Catalan numbers. *SpringerPlus*, **5**, 1126, 20 p. doi: 10.1186/s40064-016-2793-1.

17 Qi, F., Mahmoud, M., Shi, X.-T., and Liu, F.-F. (2015) Some Properties of the Catalan–Qi Function Related to the Catalan Numbers. ResearchGate Technical Report. doi: 10.13140/RG.2.1.3810.7369.

18 Qi, F. (2015) Asymptotic Expansions, Complete Monotonicity, and Inequalities of the Catalan Numbers. ResearchGate Technical Report. doi: 10.13140/RG.2.1.4371.6321.

19 Elezović, N. (2015) Asymptotic expansions of gamma and related functions, binomial coefficients, inequalities and means. *J. Math. Inequal.*, **9** (4), 1001–1054. doi: 10.7153/jmi-09-81.

20 Mitrinović, D.S., Pečarić, J.E., and Fink, A.M. (1993) *Classical and New Inequalities in Analysis*, Kluwer Academic Publishers, Dordrecht, Boston, MA, London. doi: 10.1007/978-94-017-1043-5.

21 Widder, D.V. (1941) *The Laplace Transform*, Princeton Mathematical Series, vol. **6**, Princeton University Press, Princeton, NJ.

22 Comtet, L. (1974) *Advanced Combinatorics: The Art of Finite and Infinite Expansions*, Revised and Enlarged Edition, D. Reidel Publishing Co., Dordrecht and Boston, MA.

23 Krall, H.L. and Frink, O. (1949) A new class of orthogonal polynomials: the Bessel polynomials. *Trans. Am. Math. Soc.*, **65**, 100–115.

24 Sloane, N.J.A. Triangle of coefficients of Bessel polynomials, From The On-Line Encyclopedia of Integer Sequences, http://oeis.org/A001497 (accessed 16 October 2017).

25 Stanley, R. and Weisstein, E.W. Bessel polynomial, From MathWorld–A Wolfram Web Resource, http://mathworld.wolfram.com/BesselPolynomial .html (accessed 16 October 2017).

26 WikiPedia, Bessel polynomials, From the Free Encyclopedia, https://en.wikipedia.org/wiki/Bessel_polynomials (accessed 16 Ocotber 2017).

27 Guo, B.-N. and Qi, F. (2015) An explicit formula for Bernoulli numbers in terms of Stirling numbers of the second kind. *J. Anal. Number Theory*, **3** (1), 27–30. doi: 10.12785/jant/030105.

28 Qi, F. and Guo, B.-N. (2017) Explicit formulas for special values of the Bell polynomials of the second kind and for the Euler numbers and polynomials. *Mediterr. J. Math.*, **14** (3), Article 140, 14 pp. doi: 10.1007/s00009-017-0939-1.

29 Qi, F. (2015) Derivatives of tangent function and tangent numbers. *Appl. Math. Comput.*, **268**, 844–858. doi: 10.1016/j.amc.2015.06.123.

30 Qi, F. and Chapman, R.J. (2016) Two closed forms for the Bernoulli polynomials. *J. Number Theory*, **159**, 89–100. doi: 10.1016/j.jnt.2015.07.021.

31 Qi, F. and Guo, B.-N. (2016) Viewing some ordinary differential equations from the angle of derivative polynomials, Preprints, 12 p. 10.20944/preprints201610.0043.v1.

32 Qi, F. and Zheng, M.-M. (2015) Explicit expressions for a family of the Bell polynomials and applications. *Appl. Math. Comput.*, **258**, 597–607. doi: 10.1016/j.amc.2015.02.027.

33 Wei, C.-F. and Qi, F. (2015) Several closed expressions for the Euler numbers. *J. Inequal. Appl.*, **219**, 8 p. doi: 10.1186/s13660-015-0738-9.

34 Hilton, P. and Pedersen, J. (1991) Catalan numbers, their generalization, and their uses. *Math. Intelligencer*, **13** (2), 64–75. doi: 10.1007/BF03024089.

35 Klarner, D.A. (1970) Correspondences between plane trees and binary sequences. *J. Comb. Theory*, **9**, 401–411.

36 McCarthy, J. (1992) Catalan numbers. Letter to the editor: "Catalan numbers, their generalization, and their uses" [Math. Intelligencer 13 (1991), no. 2, 64–75] by P. Hilton and J. Pedersen. *Math. Intelligencer*, **14** (2), 5.

37 Qi, F., Shi, X.-T., and Liu, F.-F. (2015) An exponential representation for a function involving the gamma function and originating from the Catalan numbers, ResearchGate Research. doi: 10.13140/RG.2.1.1086.4486.

38 Qi, F., Shi, X.-T., Mahmoud, M., and Liu, F.-F. (2017) The Catalan numbers: a generalization, an exponential representation, and some properties. *J. Comput. Anal. Appl.*, **23** (5), 937–944.

39 Abramowitz, M. and Stegun, I.A. (eds) (1972) *Handbook of Mathematical Functions with Formulas, Graphs, and Mathematical Tables*, National Bureau of Standards, Applied Mathematics Series, vol. **55**, 10th printing, Washington, DC.

40 Touchard, J. (1928) Sur certaines équations fonctionnelles. Proceedings of the International Mathematical Congress, Toronto, 1924, Vol. 1, pp. 465–472.

41 Hilton, P. and Pedersen, J. (1990) The ballot problem and Catalan numbers. *Nieuw Arch. Voor Wiskunde*, **7–8**, 209–216.

42 Mahmoud, M. and Qi, F. (2016) Three identities of the Catalan–Qi numbers. *Mathematics*, **4** (2), Article 35, 7. doi: 10.3390/math4020035.

43 Schilling, R.L., Song, R., and Vondraček, Z. (2012) *Bernstein Functions—Theory and Applications*, de Gruyter Studies in Mathematics, vol. **37**, 2nd edn, Walter de Gruyter, Berlin. doi: 10.1515/9783110269338.

44 Atanassov, R.D. and Tsoukrovski, U.V. (1988) Some properties of a class of logarithmically completely monotonic functions.*C.R. Acad. Bulgare Sci.*, **41** (2), 21–23.

45 Qi, F. and Chen, C.-P. (2004) A complete monotonicity property of the gamma function. *J. Math. Anal. Appl.*, **296**, 603–607. doi: 10.1016/j.jmaa.2004.04.026.

46 Qi, F. and Guo, B.-N. (2004) Complete monotonicities of functions involving the gamma and digamma functions. *RGMIA Res. Rep. Coll.*, **7** (1), Art. 8, 63–72. http://rgmia.org/v7n1.php.

47 Qi, F., Guo, S., and Guo, B.-N. (2010) Complete monotonicity of some functions involving polygamma functions. *J. Comput. Appl. Math.*, **233** (9), 2149–2160. doi: 10.1016/j.cam.2009.09.044.

48 Qi, F. and Li, W.-H. (2015) A logarithmically completely monotonic function involving the ratio of gamma functions. *J. Appl. Anal. Comput.*, **5** (4), 626–634. doi: 10.11948/2015049.

49 Qi, F., Luo, Q.-M., and Guo, B.-N. (2013) Complete monotonicity of a function involving the divided difference of digamma functions. *Sci. China Math.*, **56** (11), 2315–2325. doi: 10.1007/s11425-012-4562-0.

50 Qi, F., Wei, C.-F., and Guo, B.-N. (2012) Complete monotonicity of a function involving the ratio of gamma functions and applications. *Banach J. Math. Anal.*, **6** (1), 35–44. doi: 10.15352/bjma/1337014663.

51 Qi, F., Shi, X.-T., Mahmoud, M., and Liu, F.-F. (2016) Schur-convexity of the Catalan–Qi function related to the Catalan numbers. *Tbilisi Math. J.*, **9** (2), 141–150. doi: 10.1515/tmj-2016-0026.

52 Qi, F., Shi, X.-T., Mahmoud, M., and Liu, F.-F. (2015) Schur-convexity of the Catalan–Qi function related to the Catalan numbers. ResearchGate Technical Report. doi: 10.13140/RG.2.1.2434.4802.

53 Qi, F. and Guo, B.-N. (2016) Logarithmically complete monotonicity of Catalan–Qi function related to Catalan numbers. *Cogent Math.*, **3**, 1179379, 6 p. doi: 10.1080/23311835.2016.1179379.

54 Qi, F. and Guo, B.-N. (2016) Logarithmically complete monotonicity of a function related to the Catalan–Qi function. *Acta Univ. Sapientiae Math.*, **8** (1), 93–102. doi: 10.1515/ausm-2016-0006.

55 Marshall, A.W., Olkin, I., and Arnold, B.C. (2011) *Inequalities: Theory of Majorization and its Applications*, 2nd edn, Springer-Verlag, New York, Dordrecht, Heidelberg, London. doi: 10.1007/978-0-387-68276-1.

56 Pečarić, J.E., Proschan, F., and Tong, Y.L. (1992) *Convex Functions, Partial Orderings, and Statistical Applications*, Mathematics in Science and Engineering, vol. **187**, Academic Press.

57 Zou, Q. (2016) Analogues of several identities and supercongruences for the Catalan–Qi numbers. *J. Inequal. Spec. Funct.*, **7** (4), 235–241.

58 Zou, Q. (2017) The q-binomial inverse formula and a recurrence relation for the q-Catalan–Qi numbers. *J. Math. Anal.*, **8** (1), 176–182.

59 Qi, F. (2016) An improper integral with a square root, Preprints, 8 p. doi: 10.20944/preprints201610.0089.v1.

60 Qi, F. (2017) Parametric integrals, the Catalan numbers, and the beta function. *Elem. Math.*, **72** (3), 103–110. doi: 10.4171/EM/332.

61 Qi, F., Akkurt, A., and Yildirim, H. (2018) Catalan numbers, k-gamma and k-beta functions, and parametric integrals. *J. Comput. Anal. Appl.*, **25** (6), 1036–1042.

62 Yin, L. and Qi, F. (2017) Several series identities involving the Catalan numbers, Preprints, 11 p. doi: 10.20944/preprints201703.0029.v1.

63 Abel, U. (2016) Reciprocal Catalan sums: solution to problem 11765. *Am. Math. Mon.*, **123** (4), 405–406. doi: 10.4169/amer.math.monthly.123.4.399.

64 Amdeberhan, T., Guan, X., Jiu, L., Moll, V.H., and Vignat, C. (2016) A series involving Catalan numbers: proofs and demonstrations. *Elem. Math.*, **71** (3), 109–121. doi: 10.4171/EM/306.

65 Fuss, N.I. (1791) Solutio quaestionis, quot modis polygonum n laterum in polygona m laterum, per diagonales resolvi queat. *Nova Acta Acad. Sci. Petropolitanae*, **9**, 243–251.

66 WikiPedia, *Fuss–Catalan number*, From the Free Encyclopedia, https://en .wikipedia.org/wiki/Fuss--Catalan_number (accessed 16 October 2017).

67 Alexeev, N., Götze, F., and Tikhomirov, A. (2010) Asymptotic distribution of singular values of powers of random matrices. *Lith. Math. J.*, **50** (2), 121–132. doi: 10.1007/s10986-010-9074-4.

68 Aval, J.-C. (2008) Multivariate Fuss–Catalan numbers. *Discrete Math.*, **308** (20), 4660–4669. doi: 10.1016/j.disc.2007.08.100.

69 Bisch, D. and Jones, V. (1997) Algebras associated to intermediate subfactors. *Invent. Math.*, **128** (1), 89–157. doi: 10.1007/s002220050137.

70 Gordon, I.G. and Griffeth, S. (2012) Catalan numbers for complex reflection groups. *Am. J. Math.*, **134** (6), 1491–1502. doi: 10.1353/ajm.2012.0047.

71 Lin, C.-H. (2011) Some combinatorial interpretations and applications of Fuss–Catalan numbers. *ISRN Discrete Math.*, **2011**, Article ID 534628, 8 p. doi: 10.5402/2011/534628.

72 Liu, D.-Z., Song, C.-W., and Wang, Z.-D. (2011) On explicit probability densities associated with Fuss–Catalan numbers. *Proc. Am. Math. Soc.*, **139** (10), 3735–3738. doi: 10.1090/S0002-9939-2011-11015-3.

73 Młotkowski, W. (2010) Fuss–Catalan numbers in noncommutative probability. *Doc. Math.*, **15**, 939–955.

74 Młotkowski, W., Penson, K.A., and .Zyczkowski, K. (2013) Densities of the Raney distributions. *Doc. Math.*, **18**, 1573–1596.

75 Pak, I. Catalan Numbers Page, http://www.math.ucla.edu/ pak/lectures/Cat/pakcat.htm (accessed 16 October 2017).

76 Przytycki, J.H. and Sikora, A.S. (2000) Polygon dissections and Euler, Fuss, Kirkman, and Cayley numbers. *J. Comb. Theory Ser. A*, **92** (1), 68–76. doi: 10.1006/jcta.1999.3042.

77 Stump, C. (2008) q, t-Fuß-Catalan Numbers for Complex Reflection Groups. 20th Annual International Conference on Formal Power Series and Algebraic Combinatorics (FPSAC 2008), Discrete Mathematics & Theoretical Computer Science Proc., AJ, Assoc., Nancy, pp. 295–306.

78 Stump, C. (2010) q, t-Fuß-Catalan numbers for finite reflection groups. *J. Algebraic Comb.*, **32** (1), 67–97. doi: 10.1007/s10801-009-0205-0.

79 Qi F. and Cerone, P. (2017) Some properties of the Fuss–Catalan numbers. *Preprints*, 14 pp. doi: 10.20944/preprints201708.0056.v1.

80 Alzer, H. (2001) Sharp inequalities for the beta function. *Indag. Math. (N.S.)*, **12** (1), 15–21. doi: 10.1016/S0019-3577(01)80002-1.

81 Cerone, P. (2007) Special functions: approximations and bounds. *Appl. Anal. Discrete Math.*, **1** (1), 72–91. doi: 10.2298/AADM0701072C.

82 Grenié, L. and Molteni, G. (2015) Inequalities for the beta function. *Math. Inequal. Appl.*, **18** (4), 1427–1442. doi: 10.7153/mia-18-111.

83 Qi, F., Shi, X.-T., and Cerone, P. (2015) A unified generalization of the Catalan, Fuss, Fuss–Catalan numbers and Catalan–Qi function, ResearchGate Working Paper. doi: 10.13140/RG.2.1.3198.6000.

84 Guo, B.-N. and Qi, F. (2010) A property of logarithmically absolutely monotonic functions and the logarithmically complete monotonicity of a power-exponential function. *Politehn. Univ. Bucharest Sci. Bull. Ser. A Appl. Math. Phys.*, **72** (2), 21–30.

85 Qi, F. (2010) Bounds for the ratio of two gamma functions. *J. Inequal. Appl.*, **2010**, 84. Article ID 493058. doi: 10.1155/2010/493058.

86 Qi, F. (2014) Bounds for the ratio of two gamma functions: from Gautschi's and Kershaw's inequalities to complete monotonicity. *Turkish J. Anal. Number Theory*, **2** (5), 152–164. doi: 10.12691/tjant-2-5-1.

87 Qi, F. and Luo, Q.-M. (2012) Bounds for the ratio of two gamma functions—From Wendel's and related inequalities to logarithmically completely monotonic functions. *Banach J. Math. Anal.*, **6** (2), 132–158. doi: 10.15352/bjma/1342210165.

88 Qi, F. and Luo, Q.-M. (2013) Bounds for the ratio of two gamma functions: from Wendel's asymptotic relation to Elezović-Giordano-Pečarić's theorem. *J. Inequal. Appl.*, **2013**, 542, 20 p. doi: 10.1186/1029-242X-2013-542.

89 Liu, F.-F., Shi, X.-T., and Qi, F. (2015) A logarithmically completely monotonic function involving the gamma function and originating from the Catalan numbers and function. *Glob. J. Math. Anal.*, **3** (4), 140–144. doi: 10.14419/gjma.v3i4.5187.

90 Qi, F., Zou, Q., and Guo, B.-N. (2017) Some identities and a matrix inverse related to the Chebyshev polynomials of the second kind and the Catalan numbers. *Preprints*, 25 pp. doi: 10.20944/preprints201703.0209.v2.

91 Qi, F. and Guo, B.-N. (2017) Integral representations of the Catalan numbers and their applications, *Mathematics* **5** (3), Article 40, 31 pp. doi: 10.3390/math5030040.

92 Qi, F. (2015) Some properties and generalizations of the Catalan, Fuss, and Fuss–Catalan numbers, ResearchGate Research. doi: 10.13140/RG.2.1.1778.3128.

6

Trace Inequalities of Jensen Type for Self-adjoint Operators in Hilbert Spaces: A Survey of Recent Results

Silvestru Sever Dragomir[1,2]

[1] *Mathematics Department, College of Engineering & Science, Victoria University, Melbourne 8001, Australia*
[2] *DST-NRF Centre of Excellence in the Mathematical and Statistical Sciences, School of Computer Science and Applied Mathematics, University of the Witwatersrand, Johannesburg 2000, South Africa*

2010 AMS Subject Classification 47A63; 47A99

6.1 Introduction

6.1.1 Jensen's Inequality

Let A be a self-adjoint operator on the complex Hilbert space $(H, \langle ., . \rangle)$ with the spectrum $\mathrm{Sp}(A)$ included in the interval $[m, M]$ for some real numbers $m < M$ and let $\{E_\lambda\}_\lambda$ be its *spectral family*. Then for any continuous function $f : [m, M] \to \mathbb{C}$, it is well known that we have the following *spectral representation in terms of the Riemann–Stieltjes integral* (see, for instance, [1, p. 257]):

$$\langle f(A)x, y \rangle = \int_{m-0}^{M} f(\lambda) \mathrm{d}(\langle E_\lambda x, y \rangle) \tag{6.1}$$

and

$$\| f(A)x \|^2 = \int_{m-0}^{M} |f(\lambda)|^2 \mathrm{d} \| E_\lambda x \|^2 \tag{6.2}$$

for any $x, y \in H$.

The function $g_{x,y}(\lambda) := \langle E_\lambda x, y \rangle$ is of *bounded variation* on the interval $[m, M]$ and

$$g_{x,y}(m - 0) = 0 \quad \text{while} \quad g_{x,y}(M) = \langle x, y \rangle$$

for any $x, y \in H$. It is also well known that $g_x(\lambda) := \langle E_\lambda x, x \rangle$ is *monotonic nondecreasing* and *right continuous* on $[m, M]$ for any $x \in H$.

Mathematical Analysis and Applications: Selected Topics, First Edition.
Edited by Michael Ruzhansky, Hemen Dutta, and Ravi P. Agarwal.
© 2018 John Wiley & Sons, Inc. Published 2018 by John Wiley & Sons, Inc.

The following result that provides an operator version for the Jensen inequality may be found, for instance, in Mond and Pečarić [2] (see also [3, p. 5]):

Theorem 6.1 Let A be a self-adjoint operator on the Hilbert space H and assume that $\mathrm{Sp}(A) \subseteq [m, M]$ for some scalars m, M with $m < M$. If f is a convex function on $[m, M]$, then

$$f(\langle Ax, x \rangle) \leq \langle f(A)x, x \rangle \tag{MP}$$

for each $x \in H$ with $\|x\| = 1$.

As a special case of Theorem 6.1 we have the following Hölder–McCarthy inequality:

Theorem 6.2 ([4]) Let A be a self-adjoint positive operator on a Hilbert space H. Then for all $x \in H$ with $\|x\| = 1$:

(i) $\langle A^r x, x \rangle \geq \langle Ax, x \rangle^r$ for all $r > 1$;
(ii) $\langle A^r x, x \rangle \leq \langle Ax, x \rangle^r$ for all $0 < r < 1$;
(iii) If A is invertible, then $\langle A^r x, x \rangle \geq \langle Ax, x \rangle^r$ for all $r < 0$.

The following reverse for (MP) that generalizes the scalar Lah–Ribarić inequality for convex functions is well known, see, for instance, [3, p. 57]:

Theorem 6.3 Let A be a self-adjoint operator on the Hilbert space H and assume that $\mathrm{Sp}(A) \subseteq [m, M]$ for some scalars m, M with $m < M$. If f is a convex function on $[m, M]$, then

$$\langle f(A)x, x \rangle \leq \frac{M - \langle Ax, x \rangle}{M - m} f(m) + \frac{\langle Ax, x \rangle - m}{M - m} f(M) \tag{LR}$$

for each $x \in H$ with $\|x\| = 1$.

The following result that provides a reverse of the Jensen inequality has been obtained in [5]:

Theorem 6.4 ([5]) Let I be an interval and $f : I \to \mathbb{R}$ be a convex and differentiable function on $\mathring{I}$ (the interior of I) whose derivative f' is continuous on $\mathring{I}$. If A is a self-adjoint operators on the Hilbert space H with $\mathrm{Sp}(A) \subseteq [m, M] \subset \mathring{I}$, then

$$(0 \leq) \langle f(A)x, x \rangle - f(\langle Ax, x \rangle) \leq \langle f'(A)Ax, x \rangle - \langle Ax, x \rangle \langle f'(A)x, x \rangle \tag{6.3}$$

for any $x \in H$ with $\|x\| = 1$.

Perhaps more convenient reverses of (MP) are the following inequalities that have been obtained in the same paper [5]:

Theorem 6.5 ([5]) Let I be an interval and $f : I \to \mathbb{R}$ be a convex and differentiable function on $\mathring{I}$ (the interior of I) whose derivative f' is continuous on $\mathring{I}$. If A is a self-adjoint operators on the Hilbert space H with $\mathrm{Sp}(A) \subseteq [m, M] \subset \mathring{I}$, then

$$(0 \le)\langle f(A)x, x \rangle - f(\langle Ax, x \rangle) \tag{6.4}$$

$$\le \begin{cases} \dfrac{1}{2}(M - m)\big[\| f'(A)x\|^2 - \langle f'(A)x, x\rangle^2\big]^{1/2} \\[2mm] \dfrac{1}{2}(f'(M) - f'(m))\big[\|Ax\|^2 - \langle Ax, x\rangle^2\big]^{1/2} \end{cases}$$

$$\le \dfrac{1}{4}(M - m)(f'(M) - f'(m))$$

for any $x \in H$ with $\|x\| = 1$.

We also have the inequality

$$(0 \le)\langle f(A)x, x \rangle - f(\langle Ax, x \rangle) \tag{6.5}$$

$$\le \dfrac{1}{4}(M - m)(f'(M) - f'(m))$$

$$- \begin{cases} [\langle Mx - Ax, Ax - mx\rangle\langle f'(M)x - f'(A)x, f'(A)x - f'(m)x\rangle]^{1/2}, \\[2mm] \left| \langle Ax, x\rangle - \dfrac{M + m}{2} \right| \left| \langle f'(A)x, x\rangle - \dfrac{f'(M) + f'(m)}{2} \right| \end{cases}$$

$$\le \dfrac{1}{4}(M - m)(f'(M) - f'(m))$$

for any $x \in H$ with $\|x\| = 1$.

Moreover, if $m > 0$ and $f'(m) > 0$, then we also have

$$(0 \le)\langle f(A)x, x \rangle - f(\langle Ax, x \rangle) \tag{6.6}$$

$$\le \begin{cases} \dfrac{1}{4}\dfrac{(M - m)(f'(M) - f'(m))}{\sqrt{Mmf'(M)f'(m)}}\langle Ax, x\rangle\langle f'(A)x, x\rangle, \\[2mm] \left(\sqrt{M} - \sqrt{m}\right)\left(\sqrt{f'(M)} - \sqrt{f'(m)}\right)[\langle Ax, x\rangle\langle f'(A)x, x\rangle]^{1/2}, \end{cases}$$

for any $x \in H$ with $\|x\| = 1$.

In [6], we obtained the following operator version for Slater's inequality as well as a reverse of it:

Theorem 6.6 ([6]) Let I be an interval and $f : I \to \mathbb{R}$ be a convex and differentiable function on $\mathring{I}$ (the interior of I) whose derivative f' is continuous on $\mathring{I}$. If A is a self-adjoint operator on the Hilbert space H with $\mathrm{Sp}(A) \subseteq [m, M] \subset \mathring{I}$

and $f'(A)$ is a positive invertible operator on H then

$$0 \leq f\left(\frac{\langle Af'(A)x, x\rangle}{\langle f'(A)x, x\rangle}\right) - \langle f(A)x, x\rangle \tag{6.7}$$

$$\leq f'\left(\frac{\langle Af'(A)x, x\rangle}{\langle f'(A)x, x\rangle}\right)\left[\frac{\langle Af'(A)x, x\rangle - \langle Ax, x\rangle\langle f'(A)x, x\rangle}{\langle f'(A)x, x\rangle}\right]$$

for any $x \in H$ with $\|x\| = 1$.

For other similar results, see [6].

For some inequalities for convex functions, see [7–14]. For inequalities for functions of self-adjoint operators, see [2, 5, 6, 15–26] and the books [3, 27, 28].

In order to state our results concerning some trace inequalities for convex functions of self-adjoint operators on Hilbert space $(H, \langle ., .\rangle)$, we need some preparations as follows.

6.1.2 Traces for Operators in Hilbert Spaces

Let $(H, \langle \cdot, \cdot \rangle)$ be a complex Hilbert space and $\{e_i\}_{i \in I}$ an *orthonormal basis* of H. We say that $A \in B(H)$ is a *Hilbert–Schmidt operator* if

$$\sum_{i \in I} \|Ae_i\|^2 < \infty. \tag{6.8}$$

It is well known that, if $\{e_i\}_{i \in I}$ and $\{f_j\}_{j \in J}$ are orthonormal bases for H and $A \in B(H)$ then

$$\sum_{i \in I} \|Ae_i\|^2 = \sum_{j \in I} \|Af_j\|^2 = \sum_{j \in I} \|A^*f_j\|^2 \tag{6.9}$$

showing that the definition (6.8) is independent of the orthonormal basis and A is a Hilbert–Schmidt operator iff A^* is a Hilbert–Schmidt operator.

Let $B_2(H)$ the set of Hilbert–Schmidt operators in $B(H)$. For $A \in B_2(H)$, we define

$$\|A\|_2 := \left(\sum_{i \in I} \|Ae_i\|^2\right)^{1/2} \tag{6.10}$$

for $\{e_i\}_{i \in I}$ an orthonormal basis of H. This definition does not depend on the choice of the orthonormal basis.

Using the triangle inequality in $l^2(I)$, one checks that $B_2(H)$ is a *vector space* and that $\|\cdot\|_2$ is a norm on $B_2(H)$, which is usually called in the literature as the *Hilbert–Schmidt norm*.

Denote *the modulus* of an operator $A \in B(H)$ by $|A| := (A^*A)^{1/2}$. Because $\||A|x\| = \|Ax\|$ for all $x \in H$, A is Hilbert–Schmidt iff $|A|$ is Hilbert–Schmidt and $\|A\|_2 = \||A|\|_2$. From (6.9) we have that if $A \in B_2(H)$, then $A^* \in B_2(H)$ and $\|A\|_2 = \|A^*\|_2$.

The following theorem collects some of the most important properties of Hilbert–Schmidt operators:

Theorem 6.7 We have the following properties:

(i) $(B_2(H), \|\cdot\|_2)$ is a Hilbert space with inner product

$$\langle A, B \rangle_2 := \sum_{i \in I} \langle Ae_i, Be_i \rangle = \sum_{i \in I} \langle B^*Ae_i, e_i \rangle \tag{6.11}$$

and the definition does not depend on the choice of the orthonormal basis $\{e_i\}_{i \in I}$.

(ii) We have the inequalities

$$\|A\| \leq \|A\|_2 \tag{6.12}$$

for any $A \in B_2(H)$ and

$$\|AT\|_2, \|TA\|_2 \leq \|T\| \|A\|_2 \tag{6.13}$$

for any $A \in B_2(H)$ and $T \in B(H)$.

(iii) $B_2(H)$ is an operator ideal in $B(H)$, that is,

$$B(H)B_2(H)B(H) \subseteq B_2(H).$$

(iv) $B_{fin}(H)$, the space of operators of finite rank, is a dense subspace of $B_2(H)$.

(v) $B_2(H) \subseteq K(H)$, where $K(H)$ denotes the algebra of compact operators on H.

If $\{e_i\}_{i \in I}$ an orthonormal basis of H, we say that $A \in B(H)$ is *trace class* if

$$\|A\|_1 := \sum_{i \in I} \langle |A|e_i, e_i \rangle < \infty. \tag{6.14}$$

The definition of $\|A\|_1$ does not depend on the choice of the orthonormal basis $\{e_i\}_{i \in I}$. We denote by $B_1(H)$ the set of trace class operators in $B(H)$.

The following proposition holds:

Proposition 6.1 If $A \in B(H)$, then the following are equivalent:

(i) $A \in B_1(H)$;
(ii) $|A|^{1/2} \in B_2(H)$;
(iii) A (or $|A|$) is the product of two elements of $B_2(H)$.

The following properties are also well known:

Theorem 6.8 With the above notations:

(i) We have

$$\|A\|_1 = \|A^*\|_1 \text{ and } \|A\|_2 \leq \|A\|_1 \tag{6.15}$$

for any $A \in B_1(H)$.

(ii) $B_1(H)$ is an operator ideal in $B(H)$, that is,

$$B(H)B_1(H)B(H) \subseteq B_1(H).$$

(iii) We have

$$B_2(H)B_2(H) = B_1(H).$$

(iv) We have

$$\|A\|_1 = \sup\{|\langle A, B\rangle_2| \, |B \in B_2(H), \, \|B\| \le 1\}.$$

(v) $(B_1(H), \|\cdot\|_1)$ is a Banach space.

(vi) We have the following isometric isomorphisms

$$B_1(H) \cong K(H)^* \quad \text{and} \quad B_1(H)^* \cong B(H),$$

where $K(H)^*$ is the dual space of $K(H)$ and $B_1(H)^*$ is the dual space of $B_1(H)$.

We define the *trace* of a trace class operator $A \in B_1(H)$ to be

$$\text{tr}(A) := \sum_{i \in I} \langle Ae_i, e_i\rangle, \tag{6.16}$$

where $\{e_i\}_{i \in I}$ an orthonormal basis of H. Note that this coincides with the usual definition of the trace if H is finite dimensional. We observe that the series (6.16) converges absolutely and it is independent from the choice of basis.

The following result collects some properties of the trace:

Theorem 6.9 We have the following properties:

(i) If $A \in B_1(H)$ then $A^* \in B_1(H)$ and

$$\text{tr}(A^*) = \overline{\text{tr}(A)}. \tag{6.17}$$

(ii) If $A \in B_1(H)$ and $T \in B(H)$, then $AT, TA \in B_1(H)$ and

$$\text{tr}(AT) = \text{tr}(TA) \quad \text{and} \quad |\text{tr}(AT)| \le \|A\|_1 \|T\|. \tag{6.18}$$

(iii) $\text{tr}(\cdot)$ is a bounded linear functional on $B_1(H)$ with $\|\text{tr}\| = 1$.

(iv) If $A, B \in B_2(H)$ then $AB, BA \in B_1(H)$ and $\text{tr}(AB) = \text{tr}(BA)$.

(v) $B_{fin}(H)$ is a dense subspace of $B_1(H)$.

Utilizing the trace notation, we obviously have that

$$\langle A, B\rangle_2 = \text{tr}(B^*A) = \text{tr}(AB^*) \quad \text{and} \quad \|A\|_2^2 = \text{tr}(A^*A) = \text{tr}(|A|^2)$$

for any $A, B \in B_2(H)$.

For the theory of trace functionals and their applications the reader is referred to [29]. For some classical trace inequalities see [30–33], which are continuations of the work of Bellman [34]. For related works the reader can refer to [30, 35–42].

6.2 Jensen's Type Trace Inequalities

6.2.1 Some Trace Inequalities for Convex Functions

Consider the orthonormal basis $\mathcal{E} := \{e_i\}_{i \in I}$ in the complex Hilbert space $(H, \langle \cdot, \cdot \rangle)$ and for a nonzero operator $B \in B_2(H)$ let introduce the subset of indices from I defined by

$$I_{\mathcal{E},B} := \{i \in I : Be_i \neq 0\}.$$

We observe that $I_{\mathcal{E},B}$ is nonempty for any nonzero operator B and if $\ker(B) = 0$, that is, B is injective, then $I_{\mathcal{E},B} = I$. We also have for $B \in B_2(H)$ that

$$\text{tr}(|B|^2) = \text{tr}(B^*B) = \sum_{i \in I}\langle B^*Be_i, e_i \rangle = \sum_{i \in I} \|Be_i\|^2 = \sum_{i \in I_{\mathcal{E},B}} \|Be_i\|^2.$$

Theorem 6.10 ([43]) Let A be a self-adjoint operator on the Hilbert space H and assume that $\text{Sp}(A) \subseteq [m, M]$ for some scalars m, M with $m < M$. If f is a continuous convex function on $[m, M]$, $\mathcal{E} := \{e_i\}_{i \in I}$ is an orthonormal basis in H and $B \in B_2(H) \setminus \{0\}$, then $\frac{\text{tr}(|B|^2A)}{\text{tr}(|B|^2)} \in [m, M]$ and

$$f\left(\frac{\text{tr}(|B|^2A)}{\text{tr}(|B|^2)}\right) \text{tr}(|B|^2) \tag{6.19}$$

$$\leq J_{\varepsilon}(f; A, B) \leq \text{tr}(|B|^2f(A))$$

$$\leq \frac{1}{M-m}\left(f(m)\text{tr}\left[|B|^2(M1_H - A)\right] + f(M)\text{tr}\left[|B|^2(A - m1_H)\right]\right),$$

where

$$J_{\varepsilon}(f; A, B) := \sum_{i \in I_{\mathcal{E},B}} f\left(\frac{\langle B^*ABe_i, e_i \rangle}{\|Be_i\|^2}\right)\|Be_i\|^2. \tag{6.20}$$

Proof: Since $\text{Sp}(A) \subseteq [m, M]$, then $m\|y\|^2 \leq \langle Ay, y \rangle \leq M\|y\|^2$ for any $y \in H$. Therefore

$$m\|Be_i\|^2 \leq \langle ABe_i, Be_i \rangle \leq M\|Be_i\|^2$$

for any $i \in I$, which implies that

$$m\sum_{i \in I} \|Be_i\|^2 \leq \sum_{i \in I}\langle ABe_i, Be_i \rangle \leq M\sum_{i \in I} \|Be_i\|^2$$

and we conclude that $\frac{\text{tr}(|B|^2A)}{\text{tr}(|B|^2)} \in [m, M]$.

By Jensen's inequality (MP) we have

$$f\left(\frac{\langle Ay, y \rangle}{\|y\|^2}\right) \leq \frac{\langle f(A)y, y \rangle}{\|y\|^2} \tag{6.21}$$

for any $y \in H \setminus \{0\}$.

Let F be a finite part of $I_{\mathcal{E},B}$. Then for any $i \in F$ we have from (6.21) that

$$f\left(\frac{\langle ABe_i, Be_i\rangle}{\|Be_i\|^2}\right) \le \frac{\langle f(A)Be_i, Be_i\rangle}{\|Be_i\|^2},$$

which is equivalent to

$$f\left(\frac{\langle B^*ABe_i, e_i\rangle}{\|Be_i\|^2}\right)\|Be_i\|^2 \le \langle B^*f(A)Be_i, e_i\rangle. \tag{6.22}$$

Summing over $i \in F$, we get

$$\sum_{i\in F} f\left(\frac{\langle B^*ABe_i, e_i\rangle}{\|Be_i\|^2}\right)\|Be_i\|^2 \le \sum_{i\in F}\langle B^*f(A)Be_i, e_i\rangle. \tag{6.23}$$

Using Jensen's discrete inequality for finite sums and for the positive weights w_i

$$f\left(\frac{\sum_{i\in F}w_i u_i}{\sum_{i\in F}w_i}\right) \le \frac{\sum_{i\in F}w_i f(u_i)}{\sum_{i\in F}w_i},$$

we have

$$f\left(\frac{\sum_{i\in F}\frac{\langle B^*ABe_i, e_i\rangle}{\|Be_i\|^2}\|Be_i\|^2}{\sum_{i\in F}\|Be_i\|^2}\right) \le \frac{\sum_{i\in F}f\left(\frac{\langle B^*ABe_i, e_i\rangle}{\|Be_i\|^2}\right)\|Be_i\|^2}{\sum_{i\in F}\|Be_i\|^2},$$

which is equivalent to

$$f\left(\frac{\sum_{i\in F}\langle B^*ABe_i, e_i\rangle}{\sum_{i\in F}\|Be_i\|^2}\right)\sum_{i\in F}\|Be_i\|^2 \le \sum_{i\in F}f\left(\frac{\langle B^*ABe_i, e_i\rangle}{\|Be_i\|^2}\right)\|Be_i\|^2. \tag{6.24}$$

Therefore, for any F a finite part of $I_{\mathcal{E},B}$ we have from (6.23) that

$$f\left(\frac{\sum_{i\in F}\langle B^*ABe_i, e_i\rangle}{\sum_{i\in F}\|Be_i\|^2}\right)\sum_{i\in F}\|Be_i\|^2 \le \sum_{i\in F}f\left(\frac{\langle B^*ABe_i, e_i\rangle}{\|Be_i\|^2}\right)\|Be_i\|^2 \tag{6.25}$$

$$\le \sum_{i\in F}\langle B^*f(A)Be_i, e_i\rangle.$$

By the continuity of f, we then have from (6.25) that

$$f\left(\frac{\sum_{i\in I_{\mathcal{E},B}}\langle B^*ABe_i, e_i\rangle}{\sum_{i\in I_{\mathcal{E},B}}\|Be_i\|^2}\right)\sum_{i\in I_{\mathcal{E},B}}\|Be_i\|^2 \tag{6.26}$$

$$\le \sum_{i\in I_{\mathcal{E},B}}f\left(\frac{\langle B^*ABe_i, e_i\rangle}{\|Be_i\|^2}\right)\|Be_i\|^2 \le \sum_{i\in I_{\mathcal{E},B}}\langle B^*f(A)Be_i, e_i\rangle$$

and since $B \in \mathcal{B}_2(H) \setminus \{0\}$, then also

$$\sum_{i \in I_{\varepsilon,B}} \|Be_i\|^2 = \sum_{i \in I} \|Be_i\|^2 = \mathrm{tr}(|B|^2),$$

$$\sum_{i \in I_{\varepsilon,B}} \langle B^*ABe_i, e_i \rangle = \sum_{i \in I} \langle B^*ABe_i, e_i \rangle = \mathrm{tr}(|B|^2A)$$

and

$$\sum_{i \in I_{\varepsilon,B}} \langle B^*f(A)Be_i, e_i \rangle = \sum_{i \in I} \langle B^*f(A)Be_i, e_i \rangle = \mathrm{tr}(|B|^2f(A)).$$

From (6.26), we then get the first and the second inequalities in (6.19).

From (LR), we also have

$$\langle f(A)y, y \rangle \le \frac{1}{M - m}[\langle (M1_H - A)y, y \rangle f(m) + \langle (A - m1_H)y, y \rangle f(M)] \tag{6.27}$$

for any $y \in H$.

This implies that

$$\langle f(A)Be_i, Be_i \rangle \tag{6.28}$$
$$\le \frac{1}{M - m}[\langle (M1_H - A)Be_i, Be_i \rangle f(m) + \langle (A - m1_H)Be_i, Be_i \rangle f(M)]$$

for any $i \in I$.

By summation, we have

$$\sum_{i \in I} \langle f(A)Be_i, Be_i \rangle$$

$$\le \frac{1}{M - m}\left[f(m) \sum_{i \in I} \langle (M1_H - A)Be_i, Be_i \rangle + f(M) \sum_{i \in I} \langle (A - m1_H)Be_i, Be_i \rangle \right]$$

and the last part of (6.19) is proved. □

Remark 6.1 We observe that the quantities

$$J_s(f; A, B) = \sup_{\varepsilon} J_\varepsilon(f; A, B) \quad \text{and} \quad J_i(f; A, B) = \inf_{\varepsilon} J_\varepsilon(f; A, B)$$

are finite and satisfy the bounds

$$f\left(\frac{\mathrm{tr}(|B|^2A)}{\mathrm{tr}(|B|^2)} \right) \mathrm{tr}(|B|^2) \le J_i(f; A, B) \tag{6.29}$$

$$\le J_s(f; A, B) \le \mathrm{tr}(|B|^2f(A)).$$

We have the following version for nonnegative operators $P \ge 0$, that is, P satisfies the condition $\langle Px, x \rangle \ge 0$ for any $x \in H$.

Corollary 6.1 ([43]) Let A be a self-adjoint operator on the Hilbert space H and assume that $\mathrm{Sp}(A) \subseteq [m, M]$ for some scalars m, M with $m < M$. If f is a continuous convex function on $[m, M]$, $\mathcal{E} := \{e_i\}_{i \in I}$ is an orthonormal basis in H and $P \in B_1(H) \setminus \{0\}, P \geq 0$ then $\frac{\mathrm{tr}(PA)}{\mathrm{tr}(P)} \in [m, M]$ and

$$f\left(\frac{\mathrm{tr}(PA)}{\mathrm{tr}(P)}\right)\mathrm{tr}(P) \tag{6.30}$$

$$\leq K_\varepsilon(f; A, P) \leq \mathrm{tr}(Pf(A))$$

$$\leq \frac{1}{M - m}(f(m)\mathrm{tr}[P(M1_H - A)] + f(M)\mathrm{tr}[P(A - m1_H)]),$$

where

$$K_\varepsilon(f; A, P) := \sum_{i \in I_{\varepsilon,P}} f\left(\frac{\langle P^{1/2}AP^{1/2}e_i, e_i\rangle}{\langle Pe_i, e_i\rangle}\right)\langle Pe_i, e_i\rangle$$

and

$$I_{\mathcal{E},P} := \{i \in I : P^{1/2}e_i \neq 0\}$$

Moreover, the quantities

$$K_i(f; A, P) := \inf_\varepsilon K_\varepsilon(f; A, P) \text{ and } K_s(f; A, P) := \sup_\varepsilon K_\varepsilon(f; A, P)$$

are finite and satisfy the bounds

$$f\left(\frac{\mathrm{tr}(PA)}{\mathrm{tr}(P)}\right)\mathrm{tr}(P) \leq K_i(f; A, P) \leq K_s(f; A, P) \leq \mathrm{tr}(Pf(A)). \tag{6.31}$$

The finite-dimensional case is of interest.

Let $\mathcal{M}_n(\mathbb{C})$ be the space of all square matrices of order n with complex elements.

Corollary 6.2 ([43]) Let $A \in \mathcal{M}_n(\mathbb{C})$ be a Hermitian matrix and assume that $\mathrm{Sp}(A) \subseteq [m, M]$ for some scalars m, M with $m < M$. If f is a continuous convex function on $[m, M]$, $\mathcal{E} := \{e_i\}_{i \in \{1,\ldots,n\}}$ is an orthonormal basis in $\mathbb{C}^n$, then $\frac{1}{n}\mathrm{tr}(A) \in [m, M]$ and

$$nf\left(\frac{\mathrm{tr}(A)}{n}\right) \leq J_\varepsilon(f; A) \leq \mathrm{tr}(f(A)) \tag{6.32}$$

$$\leq \frac{1}{M - m}[f(m)\mathrm{tr}(MI_n - A) + f(M)\mathrm{tr}(A - mI_n)],$$

where

$$J_\varepsilon(f; A) := \sum_{i=1}^n f(\langle Ae_i, e_i\rangle)$$

and I_n is the identity matrix in $\mathcal{M}_n(\mathbb{C})$.

Remark 6.2 The second inequality in (6.32), namely

$$\sum_{i=1}^{n} f(\langle Ae_i, e_i \rangle) \leq \text{tr}(f(A))$$

for any $\{e_i\}_{i \in \{1,\dots,n\}}$ an orthonormal basis in $\mathbb{C}^n$ is known in literature as Peierls Inequality. For a different proof and some applications, see, for instance, [44].

6.2.2 Some Functional Properties

If we denote by $\mathcal{B}_1^+(H)$ the convex cone of nonnegative operators from $\mathcal{B}_1(H)$ we can consider the functional $\sigma_{f,A} : \mathcal{B}_1^+(H) \backslash \{0\} \to [0, \infty)$ defined by

$$\sigma_{f,A}(P) := \text{tr}(Pf(A)) - \text{tr}(P)f\left(\frac{\text{tr}(PA)}{\text{tr}(P)}\right) \geq 0, \tag{6.33}$$

where A is a self-adjoint operator on the Hilbert space H with $\text{Sp}(A) \subseteq [m, M]$ for some scalars $m, M (m < M)$ and f is a continuous convex function on $[m, M]$.

One can easily observe that, if f is a continuous strictly convex function on $[m, M]$, then the inequality is strict in (6.33).

Theorem 6.11 ([43]) Let A be a self-adjoint operator on the Hilbert space H with $\text{Sp}(A) \subseteq [m, M]$ for some scalars m, M with $m < M$ and f is a continuous convex function on $[m, M]$.

(i) For any $P, Q \in \mathcal{B}_1^+(H) \backslash \{0\}$ we have

$$\sigma_{f,A}(P + Q) \geq \sigma_{f,A}(P) + \sigma_{f,A}(Q) (\geq 0), \tag{6.34}$$

that is, $\sigma_{f,A}(\cdot)$ is a superadditive functional on $\mathcal{B}_1^+(H) \backslash \{0\}$.

(ii) For any $P, Q \in \mathcal{B}_1^+(H) \backslash \{0\}$ with $P \geq Q$ we have

$$\sigma_{f,A}(P) \geq \sigma_{f,A}(Q) (\geq 0), \tag{6.35}$$

that is, $\sigma_{f,A}(\cdot)$ is a monotonic nondecreasing functional on $\mathcal{B}_1^+(H) \backslash \{0\}$.

(iii) If there exists the real numbers $\gamma, \Gamma > 0$ such that $\Gamma Q \geq P \geq \gamma Q$ with $P, Q \in \mathcal{B}_1^+(H) \backslash \{0\}$, then

$$\Gamma \sigma_{f,A}(Q) \geq \sigma_{f,A}(P) \geq \gamma \sigma_{f,A}(Q) (\geq 0). \tag{6.36}$$

Proof:

(i) Let $P, Q \in \mathcal{B}_1^+(H) \backslash \{0\}$. Then, we have

$$\sigma_{f,A}(P + Q) = \text{tr}((P + Q)f(A)) - \text{tr}(P + Q)f\left(\frac{\text{tr}((P + Q)A)}{\text{tr}(P + Q)}\right) \tag{6.37}$$

$$= \text{tr}(Pf(A)) + \text{tr}(Pf(A))$$

$$- [\text{tr}(P) + \text{tr}(Q)]f\left(\frac{\text{tr}(PA) + \text{tr}(QA)}{\text{tr}(P) + \text{tr}(Q)}\right).$$

By the convexity of f we have

$$f\left(\frac{\operatorname{tr}(PA) + \operatorname{tr}(QA)}{\operatorname{tr}(P) + \operatorname{tr}(Q)}\right) = f\left(\frac{\operatorname{tr}(P)\frac{\operatorname{tr}(PA)}{\operatorname{tr}(P)} + \operatorname{tr}(Q)\frac{\operatorname{tr}(QA)}{\operatorname{tr}(Q)}}{\operatorname{tr}(P) + \operatorname{tr}(Q)}\right) \qquad (6.38)$$

$$\leq \frac{\operatorname{tr}(P)f\left(\frac{\operatorname{tr}(PA)}{\operatorname{tr}(P)}\right) + \operatorname{tr}(Q)f\left(\frac{\operatorname{tr}(QA)}{\operatorname{tr}(Q)}\right)}{\operatorname{tr}(P) + \operatorname{tr}(Q)}.$$

Making use of (6.37) and (6.38) we have

$$\sigma_{f,A}(P + Q) \geq \operatorname{tr}(Pf(A)) + \operatorname{tr}(Pf(A))$$

$$- [\operatorname{tr}(P) + \operatorname{tr}(Q)]\frac{\operatorname{tr}(P)f\left(\frac{\operatorname{tr}(PA)}{\operatorname{tr}(P)}\right) + \operatorname{tr}(Q)f\left(\frac{\operatorname{tr}(QA)}{\operatorname{tr}(Q)}\right)}{\operatorname{tr}(P) + \operatorname{tr}(Q)}$$

$$= \operatorname{tr}(Pf(A)) + \operatorname{tr}(Pf(A))$$

$$- \operatorname{tr}(P)f\left(\frac{\operatorname{tr}(PA)}{\operatorname{tr}(P)}\right) - \operatorname{tr}(Q)f\left(\frac{\operatorname{tr}(QA)}{\operatorname{tr}(Q)}\right)$$

$$= \sigma_{f,A}(P) + \sigma_{f,A}(Q)$$

and the inequality (6.34) is proved.

(ii) Let $P, Q \in B_1^+(H)\setminus\{0\}$ with $P \geq Q$. Then on applying the superadditivity property of $\sigma_{f,A}$ for $P - Q \geq 0$ and $Q \geq 0$ we have

$$\sigma_{f,A}(P) = \sigma_{f,A}(P - Q + Q) \geq \sigma_{f,A}(P - Q) + \sigma_{f,A}(Q) \geq \sigma_{f,A}(Q)$$

and the inequality (6.35) is proved.

(iii) If $P \geq \gamma Q$, then by the monotonicity property of $\sigma_{f,A}$ we have

$$\sigma_{f,A}(P) \geq \sigma_{f,A}(\gamma Q) = \gamma \sigma_{f,A}(Q)$$

and a similar inequality for Γ. $\qquad\square$

We have the following particular case of interest:

Corollary 6.3 Let $A \in \mathcal{M}_n(\mathbb{C})$ be a Hermitian matrix and assume that $\operatorname{Sp}(A) \subseteq [m, M]$ for some scalars m, M with $m < M$. If f is a continuous convex function on $[m, M]$, there exists the real numbers $\gamma, \Gamma > 0$ such that $\Gamma I_n \geq P \geq \gamma I_n$ with P positive definite, where I_n is the identity matrix, then

$$\Gamma\left[\operatorname{tr}(f(A)) - nf\left(\frac{\operatorname{tr}(A)}{n}\right)\right] \geq \operatorname{tr}(Pf(A)) - \operatorname{tr}(P)f\left(\frac{\operatorname{tr}(PA)}{\operatorname{tr}(P)}\right) \qquad (6.39)$$

$$\geq \gamma\left[\operatorname{tr}(f(A)) - nf\left(\frac{\operatorname{tr}(A)}{n}\right)\right] (\geq 0).$$

The following theorem also holds:

Theorem 6.12 ([43]) Let A be a self-adjoint operator on the Hilbert space H with $\mathrm{Sp}(A) \subseteq [m, M]$ for some scalars m, M with $m < M$ and f is a continuous convex function on $[m, M]$. For $p \geq 1$, the functional $\psi_{p,f,A} : \mathcal{B}_1^+(H) \setminus \{0\} \to [0, \infty)$ defined by

$$\psi_{p,f,A}(P) := [\mathrm{tr}(P)]^{1-\frac{1}{p}} \sigma_{f,A}(P)$$

is superadditive on $\mathcal{B}_1^+(H) \setminus \{0\}$.

Proof: First, we observe that the following elementary inequality holds:

$$(\alpha + \beta)^p \geq (\leq) \alpha^p + \beta^p \tag{6.40}$$

for any $\alpha, \beta \geq 0$ and $p \geq 1$ $(0 < p < 1)$.

Indeed, if we consider the function $f_p : [0, \infty) \to \mathbb{R}$, $f_p(t) = (t + 1)^p - t^p$ we have $f_p'(t) = p[(t + 1)^{p-1} - t^{p-1}]$. Observe that for $p > 1$ and $t > 0$ we have that $f_p'(t) > 0$ showing that f_p is strictly increasing on the interval $[0, \infty)$. Now for $t = \frac{\alpha}{\beta} (\beta > 0, \alpha \geq 0)$ we have $f_p(t) > f_p(0)$ giving that $\left(\frac{\alpha}{\beta} + 1\right)^p - \left(\frac{\alpha}{\beta}\right)^p > 1$, that is, the desired inequality (6.40).

For $p \in (0, 1)$ we have that f_p is strictly decreasing on $[0, \infty)$, which proves the second case in (6.40).

Now, since $\sigma_{f,A}(\cdot)$ is superadditive on $\mathcal{B}_1^+(H) \setminus \{0\}$ and $p \geq 1$ then by (6.40) we have

$$\sigma_{f,A}^p(P + Q) \geq [\sigma_{f,A}(P) + \sigma_{f,A}(Q)]^p \geq \sigma_{f,A}^p(P) + \sigma_{f,A}^p(Q) \tag{6.41}$$

for any $P, Q \in \mathcal{B}_1^+(H) \setminus \{0\}$.

Utilizing (6.41) and the additivity property of $\mathrm{tr}(\cdot)$ on $\mathcal{B}_1^+(H) \setminus \{0\}$ we have

$$\frac{\sigma_{f,A}^p(P + Q)}{\mathrm{tr}(P + Q)} \geq \frac{\sigma_{f,A}^p(P) + \sigma_{f,A}^p(Q)}{\mathrm{tr}(P) + \mathrm{tr}(Q)} \tag{6.42}$$

$$= \frac{\mathrm{tr}(P)\frac{\sigma_{f,A}^p(P)}{\mathrm{tr}(P)} + \mathrm{tr}(Q)\frac{\sigma_{f,A}^p(Q)}{\mathrm{tr}(Q)}}{\mathrm{tr}(P) + \mathrm{tr}(Q)}$$

$$= \frac{\mathrm{tr}(P)\left(\frac{\sigma_{f,A}(P)}{\mathrm{tr}^{1/p}(P)}\right)^p + \mathrm{tr}(Q)\left(\frac{\sigma_{f,A}(Q)}{\mathrm{tr}^{1/q}(Q)}\right)^p}{\mathrm{tr}(P) + \mathrm{tr}(Q)} =: I$$

for any $P, Q \in \mathcal{B}_1^+(H) \setminus \{0\}$.

Since for $p \geq 1$ the power function $g(t) = t^p$ is convex, then

$$I \geq \left(\frac{\mathrm{tr}(P)\frac{\sigma_{f,A}(P)}{\mathrm{tr}^{1/p}(P)} + \mathrm{tr}(Q)\frac{\sigma_{f,A}(Q)}{\mathrm{tr}^{1/q}(Q)}}{\mathrm{tr}(P) + \mathrm{tr}(Q)}\right)^p \tag{6.43}$$

$$= \left(\frac{\mathrm{tr}^{1-1/p}(P)\sigma_{f,A}(P) + \mathrm{tr}^{1-1/q}(Q)\sigma_{f,A}(Q)}{\mathrm{tr}(P + Q)}\right)^p$$

for any $P, Q \in \mathcal{B}_1^+(H) \setminus \{0\}$.

By combining (6.42) with (6.43), we get

$$\frac{\sigma^p_{f,A}(P+Q)}{\operatorname{tr}(P+Q)} \geq \left(\frac{\operatorname{tr}^{1-1/p}(P)\sigma_{f,A}(P) + \operatorname{tr}^{1-1/q}(Q)\sigma_{f,A}(Q)}{\operatorname{tr}(P+Q)}\right)^p,$$

which is equivalent to

$$\frac{\sigma_{f,A}(P+Q)}{\operatorname{tr}^{1/p}(P+Q)} \geq \frac{\operatorname{tr}^{1-1/p}(P)\sigma_{f,A}(P) + \operatorname{tr}^{1-1/q}(Q)\sigma_{f,A}(Q)}{\operatorname{tr}(P+Q)} \tag{6.44}$$

for any $P, Q \in \mathcal{B}^+_1(H)\backslash\{0\}$.

Finally, if we multiply (6.44) by $\operatorname{tr}(P+Q) > 0$ we get

$$\psi_{p,f,A}(P+Q) \geq \psi_{p,f,A}(P) + \psi_{p,f,A}(Q)$$

for any $P, Q \in \mathcal{B}^+_1(H)\backslash\{0\}$ and the proof is complete. $\square$

Corollary 6.4 With the assumptions of Theorem 6.12, the two parameters $P, Q \geq 1$ functional $\psi_{p,q,f,A} : \mathcal{B}^+_1(H)\backslash\{0\} \to [0, \infty)$ defined by

$$\psi_{p,q,f,A}(P) := [\operatorname{tr}(P)]^{q(1-\frac{1}{p})}\sigma^q_{f,A}(P)$$

is superadditive on $\mathcal{B}^+_1(H)\backslash\{0\}$.

Proof: Observe that $\psi_{p,q,f,A}(P) = [\psi_{p,f,A}(P)]^q$ for $P \in \mathcal{B}^+_1(H)\backslash\{0\}$. Therefore, by Theorem 6.12 and the inequality (6.40) for $q \geq 1$ we have that

$$\psi_{p,q,f,A}(P+Q) = [\psi_{p,f,A}(P+Q)]^q$$

$$\geq [\psi_{p,f,A}(P) + \psi_{p,f,A}(Q)]^q$$

$$\geq [\psi_{p,f,A}(P)]^q + [\psi_{p,f,A}(Q)]^q = \psi_{p,q,f,A}(P) + \psi_{p,q,f,A}(Q)$$

for any $P, Q \in \mathcal{B}^+_1(H)\backslash\{0\}$ and the statement is proved. $\square$

Remark 6.3 If we consider the functional

$$\tilde{\psi}_{p,f,A}(P) := [\operatorname{tr}(P)]^{p-1}\sigma^p_{f,A}(P)$$

then, for $p \geq 1$, $\tilde{\psi}_{p,f,A}(\cdot)$ is superadditive on $\mathcal{B}^+_1(H)\backslash\{0\}$.

Corollary 6.5 With the assumptions of Theorem 6.12 and for parameter $p \geq 1$, if there exists the real numbers $\gamma, \Gamma > 0$ such that $\Gamma Q \geq P \geq \gamma Q$ with $P, Q \in \mathcal{B}^+_1(H)\backslash\{0\}$, then

$$\Gamma^{2-\frac{1}{p}}[\operatorname{tr}(Q)]^{1-\frac{1}{p}}\sigma_{f,A}(Q) \geq [\operatorname{tr}(P)]^{1-\frac{1}{p}}\sigma_{f,A}(P) \tag{6.45}$$

$$\geq \gamma^{2-\frac{1}{p}}[\operatorname{tr}(Q)]^{1-\frac{1}{p}}\sigma_{f,A}(Q)(\geq 0).$$

The case of finite-dimensional spaces is as follows:

Corollary 6.6 Let $A \in M_n(\mathbb{C})$ be a Hermitian matrix and assume that $Sp(A) \subseteq [m, M]$ for some scalars m, M with $m < M$. If f is a continuous convex function on $[m, M]$, there exists the real numbers $\gamma, \Gamma > 0$ such that $\Gamma I_n \geq P \geq \gamma I_n$ with P positive definite, then

$$\Gamma^{2-\frac{1}{p}} n^{1-\frac{1}{p}} \left[\text{tr}(f(A)) - nf\left(\frac{\text{tr}(A)}{n}\right) \right] \tag{6.46}$$

$$\geq [\text{tr}(P)]^{1-\frac{1}{p}} \left[\text{tr}(Pf(A)) - \text{tr}(P)f\left(\frac{\text{tr}(PA)}{\text{tr}(P)}\right) \right]$$

$$\geq \gamma^{2-\frac{1}{p}} n^{1-\frac{1}{p}} \left[\text{tr}(f(A)) - nf\left(\frac{\text{tr}(A)}{n}\right) \right] (\geq 0)$$

for any $p \geq 1$.

The following result also holds:

Theorem 6.13 ([43]) Let A be a self-adjoint operator on the Hilbert space H with $Sp(A) \subseteq [m, M]$ for some scalars m, M with $m < M$ and f is a continuous strictly convex function on $[m, M]$. For $p \in (0, 1)$, the functional $\chi_{p,f,A}$: $B_1^+(H)\backslash\{0\} \to [0, \infty)$ defined by

$$\chi_{p,f,A}(P) := \frac{[\text{tr}(P)]^{1-(1/p)}}{\sigma_{f,A}(P)}$$

is subadditive on $B_1^+(H)\backslash\{0\}$.

Proof: Let $s := -p \in (-1, 0)$. For $s < 0$, we have the following inequality:

$$(\alpha + \beta)^s \leq \alpha^s + \beta^s \tag{6.47}$$

for any $\alpha, \beta > 0$.

Indeed, by the convexity of the function $f_s(t) = t^s$ on $(0, \infty)$ with $s < 0$ we have that

$$(\alpha + \beta)^s \leq 2^{s-1}(\alpha^s + \beta^s)$$

for any $\alpha, \beta > 0$ and since, obviously, $2^{s-1}(\alpha^s + \beta^s) \leq \alpha^s + \beta^s$, then (6.47) holds true.

Taking into account that $\sigma_{f,A}(\cdot)$ is superadditive and $s \in (-1, 0)$ we have

$$\sigma_{f,A}^s(P + Q) \leq [\sigma_{f,A}(P) + \sigma_{f,A}(Q)]^s \leq \sigma_{f,A}^s(P) + \sigma_{f,A}^s(Q) \tag{6.48}$$

for any $P, Q \in B_1^+(H)\backslash\{0\}$.

Since $\text{tr}(\cdot)$ is additive on $\mathcal{B}_1^+(H)\backslash\{0\}$, then by (6.49) we have

$$\frac{\sigma_{f,A}^s(P+Q)}{\text{tr}(P+Q)} \leq \frac{\sigma_{f,A}^s(P)+\sigma_{f,A}^s(Q)}{\text{tr}(P)+\text{tr}(Q)} \tag{6.49}$$

$$= \frac{\text{tr}(P)\left(\frac{\sigma_{f,A}(P)}{\text{tr}^{1/s}(P)}\right)^s + \text{tr}(Q)\left(\frac{\sigma_{f,A}(Q)}{\text{tr}^{1/s}(Q)}\right)^s}{\text{tr}(P)+\text{tr}(Q)}$$

$$= \frac{\text{tr}(P)\left(\frac{\text{tr}^{1/s}(P)}{\sigma_{f,A}(P)}\right)^{-s} + \text{tr}(Q)\left(\frac{\text{tr}^{1/s}(Q)}{\sigma_{f,A}(Q)}\right)^{-s}}{\text{tr}(P)+\text{tr}(Q)} =: J$$

for any $P, Q \in \mathcal{B}_1^+(H)\backslash\{0\}$.

By the concavity of the function $g(t) = t^{-s}$ with $s \in (-1,0)$, we also have

$$J \leq \left[\frac{\text{tr}(P)\frac{\text{tr}^{1/s}(P)}{\sigma_{f,A}(P)} + \text{tr}(Q)\frac{\text{tr}^{1/s}(Q)}{\sigma_{f,A}(Q)}}{\text{tr}(P)+\text{tr}(Q)}\right]^{-s} \tag{6.50}$$

for any $P, Q \in \mathcal{B}_1^+(H)\backslash\{0\}$.

Making use of (6.49) and (6.50), we get

$$\frac{\sigma_{f,A}^s(P+Q)}{\text{tr}(P+Q)} \leq \left[\frac{\text{tr}(P)\frac{\text{tr}^{1/s}(P)}{\sigma_{f,A}(P)} + \text{tr}(Q)\frac{\text{tr}^{1/s}(Q)}{\sigma_{f,A}(Q)}}{\text{tr}(P)+\text{tr}(Q)}\right]^{-s}$$

for any $P, Q \in \mathcal{B}_1^+(H)\backslash\{0\}$, and by taking the power $-1/s > 0$ we get

$$\frac{\sigma_{f,A}^{-1}(P+Q)}{\text{tr}^{-1/s}(P+Q)} \leq \frac{\frac{\text{tr}^{1+1/s}(P)}{\sigma_{f,A}(P)} + \frac{\text{tr}^{1+1/s}(Q)}{\sigma_{f,A}(Q)}}{\text{tr}(P)+\text{tr}(Q)},$$

which is equivalent to

$$\frac{\text{tr}^{1+1/s}(P+Q)}{\sigma_{f,A}(P+Q)} \leq \frac{\text{tr}^{1+1/s}(P)}{\sigma_{f,A}(P)} + \frac{\text{tr}^{1+1/s}(Q)}{\sigma_{f,A}(Q)}$$

for any $P, Q \in \mathcal{B}_1^+(H)\backslash\{0\}$.

This completes the proof. $\square$

The following result may be stated as well:

Corollary 6.7 With the assumptions of Theorem 6.13, the two parameters $0 < p, q < 1$ functional $\chi_{p,q,f,A} : \mathcal{B}_1^+(H)\backslash\{0\} \to [0,\infty)$ defined by

$$\chi_{p,q,f,A}(P) = \frac{\text{tr}^{q\left(1-\frac{1}{p}\right)}(P)}{\sigma_{f,A}^q(P)}$$

is subadditive on $\mathcal{B}_1^+(H)\backslash\{0\}$.

Remark 6.4 If we consider the functional $\tilde{\chi}_{p,f,A}(P) = \frac{\mathrm{tr}^{p-1}(P)}{\sigma^p_{f,A}(P)}$ for $0 < p < 1$, then $\tilde{\chi}_{p,f,A}(\cdot)$ is also subadditive on $B_1^+(H) \setminus \{0\}$.

6.2.3 Some Examples

We consider the power function $f : (0, \infty) \to (0, \infty), f(t) = t^r$ with $t \in \mathbb{R} \setminus \{0\}$. For $r \in (-\infty, 0) \cup [1, \infty), f$ is convex while for $r \in (0, 1), f$ is concave.

Let $r \geq 1$ and A be a self-adjoint operator on the Hilbert space H and assume that $\mathrm{Sp}(A) \subseteq [m, M]$ for some scalars m, M with $0 \leq m < M$. If $\mathcal{E} := \{e_i\}_{i \in I}$ is an orthonormal basis in H and $P \in B_1^+(H) \setminus \{0\}$ then

$$[\mathrm{tr}(PA)]^r [\mathrm{tr}(P)]^{1-r} \tag{6.51}$$

$$\leq K_\varepsilon(r; A, P) \leq \mathrm{tr}(PA^r)$$

$$\leq \frac{1}{M - m}(m^r \, \mathrm{tr}[P(M1_H - A)] + M^r \mathrm{tr}[P(A - m1_H)]),$$

where

$$K_\varepsilon(r; A, P) := \sum_{i \in I_{\mathcal{E},P}} \langle P^{1/2} A P^{1/2} e_i, e_i \rangle^r \langle P e_i, e_i \rangle^{1-r}.$$

Moreover, the quantities

$$K_i(r; A, P) := \inf_\varepsilon K_\varepsilon(r; A, P) \quad \text{and} \quad K_s(r; A, P) := \sup_\varepsilon K_\varepsilon(r; A, P)$$

are finite and satisfy the bounds

$$[\mathrm{tr}(PA)]^r [\mathrm{tr}(P)]^{1-r} \leq K_i(r; A, P) \leq K_s(r; A, P) \leq \mathrm{tr}(PA^r). \tag{6.52}$$

Now, if we take $A = P, P \in B_1^+(H) \setminus \{0\}$, then by (6.51) we have

$$[\mathrm{tr}(P^2)]^r [\mathrm{tr}(P)]^{1-r} \leq K_\varepsilon(r; P) \leq \mathrm{tr}(P^{r+1}) \tag{6.53}$$

where

$$K_\varepsilon(r; P) := \sum_{i \in I_{\mathcal{E},P}} \langle P^2 e_i, e_i \rangle^r \langle P e_i, e_i \rangle^{1-r}.$$

If we consider the functional $\sigma_{r,A} : B_1^+(H) \setminus \{0\} \to [0, \infty)$ defined by

$$\sigma_{r,A}(P) := \mathrm{tr}(PA^r) - [\mathrm{tr}(PA)]^r [\mathrm{tr}(P)]^{1-r} \geq 0, \tag{6.54}$$

where A is a self-adjoint operator on the Hilbert space H with $\mathrm{Sp}(A) \subseteq [m, M] \subset [0, \infty)$, then $\sigma_{r,A}(\cdot)$ is superadditive, monotonic nondecreasing and if there exists the real numbers $\gamma, \Gamma > 0$ such that $\Gamma Q \geq P \geq \gamma Q$ with $P, Q \in B_1^+(H) \setminus \{0\}$, then

$$\Gamma \sigma_{r,A}(Q) \geq \sigma_{r,A}(P) \geq \gamma \sigma_{r,A}(Q)(\geq 0). \tag{6.55}$$

Consider the convex function $f : (0, \infty) \to (0, \infty), f(t) = -\ln t$ and let A be a self-adjoint operator on the Hilbert space H and assume that $\mathrm{Sp}(A) \subseteq [m, M]$

for some scalars m, M with $0 < m < M$. If $\mathcal{E} := \{e_i\}_{i \in I}$ is an orthonormal basis in H and $P \in \mathcal{B}_1^+(H) \setminus \{0\}$ then

$$\left(\frac{\mathrm{tr}(PA)}{\mathrm{tr}(P)}\right)^{\mathrm{tr}(P)} \geq L_{\mathcal{E}}(A, P) \geq \exp[\mathrm{tr}(P \ln A)] \tag{6.56}$$

$$\geq m^{\frac{\mathrm{tr}[P(M1_H - A)]}{M - m}} M^{\frac{\mathrm{tr}[P(A - m1_H)]}{M - m}},$$

where

$$L_{\mathcal{E}}(A, P) := \prod_{i \in I_{\mathcal{E}, P}} \left(\frac{\langle P^{1/2} A P^{1/2} e_i, e_i \rangle}{\langle P e_i, e_i \rangle}\right)^{\langle P e_i, e_i \rangle}.$$

Moreover, the quantities

$$L_i(A, P) := \inf_{\mathcal{E}} L_{\mathcal{E}}(A, P) \text{ and } L_s(A, P) := \sup_{\mathcal{E}} L_{\mathcal{E}}(A, P)$$

are finite and satisfy the bounds

$$\left(\frac{\mathrm{tr}(PA)}{\mathrm{tr}(P)}\right)^{\mathrm{tr}(P)} \geq L_s(A, P) \geq L_i(A, P) \geq \exp[\mathrm{tr}(P \ln A)]. \tag{6.57}$$

Now, if we take $A = P, P \in \mathcal{B}_1^+(H) \setminus \{0\}$, then by (6.56) we get

$$\left(\frac{\mathrm{tr}(P^2)}{\mathrm{tr}(P)}\right)^{\mathrm{tr}(P)} \geq L_{\mathcal{E}}(P) \geq \exp[\mathrm{tr}(P \ln P)] \tag{6.58}$$

where

$$L_{\mathcal{E}}(P) := \prod_{i \in I_{\mathcal{E}, P}} \left(\frac{\langle P^2 e_i, e_i \rangle}{\langle P e_i, e_i \rangle}\right)^{\langle P e_i, e_i \rangle}.$$

Consider the functional $\delta_A : \mathcal{B}_1^+(H) \setminus \{0\} \to (0, \infty)$ defined by

$$\delta_A(P) := \frac{\left(\frac{\mathrm{tr}(PA)}{\mathrm{tr}(P)}\right)^{\mathrm{tr}(P)}}{\exp(\mathrm{tr}(P \ln A))} \geq 1,$$

where A is a self-adjoint operator on the Hilbert space H and such that $\mathrm{Sp}(A) \subseteq [m, M]$ for some scalars m, M with $0 < m < M$.

Observe that

$$\sigma_{-\ln, A}(P) := \ln\left(\frac{\mathrm{tr}(PA)}{\mathrm{tr}(P)}\right)^{\mathrm{tr}(P)} - \ln[\exp(\mathrm{tr}(P \ln A))] = \ln[\delta_A(P)]$$

for $P \in \mathcal{B}_1^+(H) \setminus \{0\}$.

Utilizing the properties of $\sigma_{-\ln, A}(\cdot)$, we conclude that $\delta_A(\cdot)$ is supermultiplicative, that is,

$$\delta_A(P + Q) \geq \delta_A(P)\delta_A(Q) \geq 1$$

for any $P, Q \in \mathcal{B}_1^+(H) \setminus \{0\}$. The functional $\delta_A(\cdot)$ is also monotonic nondecreasing on $\mathcal{B}_1(H) \setminus \{0\}$.

Consider the convex function $f(t) = t \ln t$ and let A be a Self-adjoint operator on the Hilbert space H and assume that $\mathrm{Sp}(A) \subseteq [m, M]$ for some scalars m, M with $0 < m < M$. If $\mathcal{E} := \{e_i\}_{i \in I}$ is an orthonormal basis in H and $P \in \mathcal{B}_1^+(H) \setminus \{0\}$ then

$$\left(\frac{\mathrm{tr}(PA)}{\mathrm{tr}(P)} \right)^{\mathrm{tr}(PA)} \leq I_{\mathcal{E}}(A, P) \leq \exp[\mathrm{tr}(PA \ln A)] \tag{6.59}$$

$$\leq m^{\frac{m \mathrm{tr}[P(M1_H - A)]}{M - m}} M^{\frac{M \mathrm{tr}[P(A - m1_H)]}{M - m}},$$

where

$$I_{\mathcal{E}}(A, P) := \prod_{i \in I_{\mathcal{E}, P}} \left(\frac{\langle P^{1/2} A P^{1/2} e_i, e_i \rangle}{\langle P e_i, e_i \rangle} \right)^{\langle P^{1/2} A P^{1/2} e_i, e_i \rangle}.$$

Moreover, the quantities

$$I_i(A, P) := \inf_{\mathcal{E}} I_{\mathcal{E}}(A, P) \quad \text{and} \quad I_s(A, P) := \sup_{\mathcal{E}} I_{\mathcal{E}}(A, P)$$

are finite and satisfy the bounds

$$\left(\frac{\mathrm{tr}(PA)}{\mathrm{tr}(P)} \right)^{\mathrm{tr}(PA)} \leq I_i(A, P) \leq I_s(A, P) \leq \exp[\mathrm{tr}(PA \ln A)]. \tag{6.60}$$

Now, if we take $A = P, P \in \mathcal{B}_1^+(H) \setminus \{0\}$, then by (6.59) we get

$$\left(\frac{\mathrm{tr}(P^2)}{\mathrm{tr}(P)} \right)^{\mathrm{tr}(P^2)} \leq I_{\mathcal{E}}(P) \leq \exp[\mathrm{tr}(P^2 \ln P)], \tag{6.61}$$

where

$$I_{\mathcal{E}}(P) := \prod_{i \in I_{\mathcal{E}, P}} \left(\frac{\langle P^2 e_i, e_i \rangle}{\langle P e_i, e_i \rangle} \right)^{\langle P^2 e_i, e_i \rangle}.$$

Observe that for $f(t) = t \ln t$ we have

$$\sigma_{(\cdot) \ln(\cdot), A}(P) = \mathrm{tr}(PA \ln A) - \mathrm{tr}(PA) \ln \left(\frac{\mathrm{tr}(PA)}{\mathrm{tr}(P)} \right)$$

$$= \ln \left[\frac{\exp[\mathrm{tr}(PA \ln A)]}{\left(\frac{\mathrm{tr}(PA)}{\mathrm{tr}(P)} \right)^{\mathrm{tr}(PA)}} \right]$$

for any $P \in \mathcal{B}_1^+(H) \setminus \{0\}$.

Consider the functional $\lambda_A : \mathcal{B}_1^+(H) \setminus \{0\} \to (0, \infty)$ defined by

$$\lambda_A(P) := \frac{\exp[\mathrm{tr}(PA \ln A)]}{\left(\frac{\mathrm{tr}(PA)}{\mathrm{tr}(P)} \right)^{\mathrm{tr}(PA)}} \geq 1.$$

Utilizing the properties of $\sigma_{(\cdot)\ln(\cdot),A}(\cdot)$ we can conclude that $\lambda_A(\cdot)$ is supermultiplicative and monotonic nondecreasing on $\mathcal{B}_1^+(H)\setminus\{0\}$.

6.2.4 More Inequalities for Convex Functions

We recall the *gradient inequality* for the convex function $f:[m,M]\to\mathbb{R}$, namely

$$f(\varsigma)-f(\tau)\geq\delta_f(\tau)(\varsigma-\tau) \tag{6.62}$$

for any $\varsigma,\tau\in[m,M]$ where $\delta_f(\tau)\in[f'_-(\tau),f'_+(\tau)]$, (for $\tau=m$ we take $\delta_f(\tau)=f'_+(m)$ and for $\tau=M$ we take $\delta_f(\tau)=f'_-(M)$). Here $f'_+(m)$ and $f'_-(M)$ are the lateral derivatives of the convex function f.

The following result holds:

Theorem 6.14 ([45]) Let A be a self-adjoint operator on the Hilbert space H and assume that $\mathrm{Sp}(A)\subseteq[m,M]$ for some scalars m,M with $m<M$. If f is a continuous convex function on $[m,M]$ and $B\in\mathcal{B}_2(H)\setminus\{0\}$, then we have $\frac{\mathrm{tr}(|B|^2A)}{\mathrm{tr}(|B|^2)}\in[m,M]$,

$$\delta_f\left(\frac{\mathrm{tr}(|B|^2A)}{\mathrm{tr}(|B|^2)}\right)\frac{\mathrm{tr}(|B^*|^2A)-\mathrm{tr}(|B|^2A)}{\mathrm{tr}(|B|^2)} \tag{6.63}$$

$$\leq\frac{\mathrm{tr}(|B^*|^2f(A))}{\mathrm{tr}(|B|^2)}-f\left(\frac{\mathrm{tr}(|B|^2A)}{\mathrm{tr}(|B|^2)}\right),$$

where

$$\delta_f\left(\frac{\mathrm{tr}(|B|^2A)}{\mathrm{tr}(|B|^2)}\right)\in\left[f'_-\left(\frac{\mathrm{tr}(|B|^2A)}{\mathrm{tr}(|B|^2)}\right),f'_+\left(\frac{\mathrm{tr}(|B|^2A)}{\mathrm{tr}(|B|^2)}\right)\right]$$

and the Jensen's inequality

$$f\left(\frac{\mathrm{tr}(|B|^2A)}{\mathrm{tr}(|B|^2)}\right)\leq\frac{\mathrm{tr}(|B|^2f(A))}{\mathrm{tr}(|B|^2)}. \tag{6.64}$$

Proof: Let $\mathcal{E}:=\{e_i\}_{i\in I}$ be an orthonormal basis in H. Utilizing the gradient inequality (6.62) we get

$$f(\varsigma)-f\left(\frac{\mathrm{tr}(|B|^2A)}{\mathrm{tr}(|B|^2)}\right)\geq\delta_f\left(\frac{\mathrm{tr}(|B|^2A)}{\mathrm{tr}(|B|^2)}\right)\left(\varsigma-\frac{\mathrm{tr}(|B|^2A)}{\mathrm{tr}(|B|^2)}\right) \tag{6.65}$$

for any $\varsigma\in[m,M]$, since obviously, by $\mathrm{Sp}(A)\subseteq[m,M]$ we have

$$m\|Be_i\|^2\leq\langle ABe_i,Be_i\rangle\leq M\|Be_i\|^2,$$

for $i\in I$, which, by summation shows that

$$\frac{\mathrm{tr}(|B|^2A)}{\mathrm{tr}(|B|^2)}\in[m,M].$$

The inequality (6.65) implies in the operator order of $\mathcal{B}(H)$ that

$$f(A) - f\left(\frac{\operatorname{tr}(|B|^2 A)}{\operatorname{tr}(|B|^2)}\right) 1_H \geq \delta_f\left(\frac{\operatorname{tr}(|B|^2 A)}{\operatorname{tr}(|B|^2)}\right)\left(A - \frac{\operatorname{tr}(|B|^2 A)}{\operatorname{tr}(|B|^2)} 1_H\right),$$
(6.66)

which can be written as

$$\langle f(A)y, y\rangle - f\left(\frac{\operatorname{tr}(|B|^2 A)}{\operatorname{tr}(|B|^2)}\right) \langle y, y\rangle \tag{6.67}$$

$$\geq \delta_f\left(\frac{\operatorname{tr}(|B|^2 A)}{\operatorname{tr}(|B|^2)}\right)\left(\langle Ay, y\rangle - \frac{\operatorname{tr}(|B|^2 A)}{\operatorname{tr}(|B|^2)}\langle y, y\rangle\right)$$

for any $y \in H$. This inequality is also of interest in itself.

Taking in (6.67) $y = Be_i$, we get

$$\langle f(A)Be_i, Be_i\rangle - f\left(\frac{\operatorname{tr}(|B|^2 A)}{\operatorname{tr}(|B|^2)}\right) \langle Be_i, Be_i\rangle$$

$$\geq \delta_f\left(\frac{\operatorname{tr}(|B|^2 A)}{\operatorname{tr}(|B|^2)}\right)\left(\langle ABe_i, Be_i\rangle - \frac{\operatorname{tr}(|B|^2 A)}{\operatorname{tr}(|B|^2)}\langle Be_i, Be_i\rangle\right),$$

which is equivalent to

$$\langle B^* f(A)Be_i, e_i\rangle - f\left(\frac{\operatorname{tr}(|B|^2 A)}{\operatorname{tr}(|B|^2)}\right) \langle |B|^2 e_i, e_i\rangle \tag{6.68}$$

$$\geq \delta_f\left(\frac{\operatorname{tr}(|B|^2 A)}{\operatorname{tr}(|B|^2)}\right)\left(\langle B^* ABe_i, e_i\rangle - \frac{\operatorname{tr}(|B|^2 A)}{\operatorname{tr}(|B|^2)}\langle |B|^2 e_i, e_i\rangle\right)$$

for any $i \in I$.

Summing in (6.68), we get

$$\sum_{i \in I}\langle B^* f(A)Be_i, e_i\rangle - f\left(\frac{\operatorname{tr}(|B|^2 A)}{\operatorname{tr}(|B|^2)}\right) \sum_{i \in I}\langle |B|^2 e_i, e_i\rangle \tag{6.69}$$

$$\geq \delta_f\left(\frac{\operatorname{tr}(|B|^2 A)}{\operatorname{tr}(|B|^2)}\right)\left(\sum_{i \in I}\langle B^* ABe_i, e_i\rangle - \frac{\operatorname{tr}(|B|^2 A)}{\operatorname{tr}(|B|^2)}\sum_{i \in I}\langle |B|^2 e_i, e_i\rangle\right).$$

However,

$$\sum_{i \in I}\langle B^* f(A)Be_i, e_i\rangle = \sum_{i \in I}\langle BB^* f(A)e_i, e_i\rangle$$

$$= \sum_{i \in I}\langle |B^*|^2 f(A)e_i, e_i\rangle = \operatorname{tr}(|B^*|^2 f(A))$$

and

$$\sum_{i \in I}\langle B^* ABe_i, e_i\rangle = \sum_{i \in I}\langle BB^* Ae_i, e_i\rangle = \operatorname{tr}(|B^*|^2 A).$$

By (6.69), we get

$$
\operatorname{tr}(|B^*|^2 f(A)) - f\left(\frac{\operatorname{tr}(|B|^2 A)}{\operatorname{tr}(|B|^2)}\right) \operatorname{tr}(|B|^2) \tag{6.70}
$$
$$
\geq \delta_f\left(\frac{\operatorname{tr}(|B|^2 A)}{\operatorname{tr}(|B|^2)}\right)(\operatorname{tr}(|B^*|^2 A) - \operatorname{tr}(|B|^2 A)),
$$

and the inequality (6.63) is thus proved.

Taking in (6.67) $y = B^* e_i$, we also get

$$
\langle f(A)B^* e_i, B^* e_i \rangle - f\left(\frac{\operatorname{tr}(|B|^2 A)}{\operatorname{tr}(|B|^2)}\right) \langle B^* e_i, B^* e_i \rangle
$$
$$
\geq \delta_f\left(\frac{\operatorname{tr}(|B|^2 A)}{\operatorname{tr}(|B|^2)}\right)\left(\langle AB^* e_i, B^* e_i \rangle - \frac{\operatorname{tr}(|B|^2 A)}{\operatorname{tr}(|B|^2)}\langle B^* e_i, B^* e_i \rangle\right),
$$

which is equivalent to

$$
\langle Bf(A)B^* e_i, e_i \rangle - f\left(\frac{\operatorname{tr}(|B|^2 A)}{\operatorname{tr}(|B|^2)}\right) \langle BB^* e_i, e_i \rangle \tag{6.71}
$$
$$
\geq \delta_f\left(\frac{\operatorname{tr}(|B|^2 A)}{\operatorname{tr}(|B|^2)}\right)\left(\langle BAB^* e_i, e_i \rangle - \frac{\operatorname{tr}(|B|^2 A)}{\operatorname{tr}(|B|^2)}\langle BB^* e_i, e_i \rangle\right)
$$

for any $i \in I$.

Summing in (6.71), we get

$$
\sum_{i \in I}\langle Bf(A)B^* e_i, e_i \rangle - f\left(\frac{\operatorname{tr}(|B|^2 A)}{\operatorname{tr}(|B|^2)}\right) \sum_{i \in I}\langle BB^* e_i, e_i \rangle \tag{6.72}
$$
$$
\geq \delta_f\left(\frac{\operatorname{tr}(|B|^2 A)}{\operatorname{tr}(|B|^2)}\right)\left(\sum_{i \in I}\langle BAB^* e_i, e_i \rangle - \frac{\operatorname{tr}(|B|^2 A)}{\operatorname{tr}(|B|^2)}\sum_{i \in I}\langle BB^* e_i, e_i \rangle\right).
$$

Since

$$
\sum_{i \in I}\langle Bf(A)B^* e_i, e_i \rangle = \operatorname{tr}(Bf(A)B^*) = \operatorname{tr}(B^* Bf(A)) = \operatorname{tr}(|B|^2 f(A)),
$$

$$
\sum_{i \in I}\langle BB^* e_i, e_i \rangle = \operatorname{tr}(BB^*) = \operatorname{tr}(B^* B) = \operatorname{tr}(|B|^2)
$$

and

$$
\sum_{i \in I}\langle BAB^* e_i, e_i \rangle = \operatorname{tr}(BAB^*) = \operatorname{tr}(B^* BA) = \operatorname{tr}(|B|^2 A),
$$

then by (6.72), we get

$$
\operatorname{tr}(|B|^2 f(A)) - f\left(\frac{\operatorname{tr}(|B|^2 A)}{\operatorname{tr}(|B|^2)}\right) \operatorname{tr}(|B|^2) \geq 0
$$

and the inequality (6.64) is obtained. □

Remark 6.5 The inequality (6.64) is obviously not as good as the first part of (6.19). However, it is the natural alternative of Jensen's inequality for trace and provides simple and nice examples for various convex functions of interest. The proof here is also simpler than the one from [43] and has some natural reverses as follows.

Corollary 6.8 Let A be a self-adjoint operator on the Hilbert space H and assume that $\mathrm{Sp}(A) \subseteq [m, M]$ for some scalars m, M with $m < M$. If f is a continuous convex function on $[m, M]$ and $P \in \mathcal{B}_1(H) \setminus \{0\}, P \geq 0$ then $\frac{\mathrm{tr}(PA)}{\mathrm{tr}(P)} \in [m, M]$ and

$$f\left(\frac{\mathrm{tr}(PA)}{\mathrm{tr}(P)}\right) \leq \frac{\mathrm{tr}(Pf(A))}{\mathrm{tr}\,(P)}. \tag{6.73}$$

The proof follows by either (6.63) or (6.64) on choosing $B = P^{1/2}$, $P \in \mathcal{B}_1(H) \setminus \{0\}, P \geq 0$.

6.3 Reverses of Jensen's Trace Inequality

6.3.1 A Reverse of Jensen's Inequality

The following lemma is of interest in itself:

Lemma 6.1 ([45]) Let S be a self-adjoint operator such that $\gamma 1_H \leq S \leq \Gamma 1_H$ for some real constants $\Gamma \geq \gamma$. Then for any $B \in \mathcal{B}_2(H) \setminus \{0\}$, we have

$$0 \leq \frac{\mathrm{tr}(|B|^2 S^2)}{\mathrm{tr}(|B|^2)} - \left(\frac{\mathrm{tr}(|B|^2 S)}{\mathrm{tr}(|B|^2)}\right)^2 \tag{6.74}$$

$$\leq \frac{1}{2}(\Gamma - \gamma)\frac{1}{\mathrm{tr}(|B|^2)}\mathrm{tr}\left(|B|^2\left|S - \frac{\mathrm{tr}(|B|^2 S)}{\mathrm{tr}(|B|^2)}1_H\right|\right)$$

$$\leq \frac{1}{2}(\Gamma - \gamma)\left[\frac{\mathrm{tr}(|B|^2 S^2)}{\mathrm{tr}(|B|^2)} - \left(\frac{\mathrm{tr}(|B|^2 S)}{\mathrm{tr}(|B|^2)}\right)^2\right]^{1/2} \leq \frac{1}{4}(\Gamma - \gamma)^2.$$

Proof: The first inequality follows by Jensen's inequality (6.64) for the convex function $f(t) = t^2$.

Now, observe that

$$\frac{1}{\mathrm{tr}(|B|^2)}\mathrm{tr}\left(|B|^2\left(S - \frac{\Gamma + \gamma}{2}1_H\right)\left(S - \frac{\mathrm{tr}(|B|^2 S)}{\mathrm{tr}(|B|^2)}1_H\right)\right) \tag{6.75}$$

$$= \frac{1}{\mathrm{tr}(|B|^2)}\mathrm{tr}\left(|B|^2 S\left(S - \frac{\mathrm{tr}(|B|^2 S)}{\mathrm{tr}(|B|^2)}1_H\right)\right)$$

$$-\frac{\Gamma+\gamma}{2}\frac{1}{\text{tr}(|B|^2)}\text{tr}\left(|B|^2\left(S-\frac{\text{tr}(|B|^2S)}{\text{tr}(|B|^2)}1_H\right)\right)$$

$$=\frac{\text{tr}(|B|^2S^2)}{\text{tr}(|B|^2)}-\left(\frac{\text{tr}(|B|^2S)}{\text{tr}(|B|^2)}\right)^2$$

since, obviously

$$\text{tr}\left(|B|^2\left(S-\frac{\text{tr}(|B|^2S)}{\text{tr}(|B|^2)}1_H\right)\right)=0.$$

Now, since $\gamma 1_H \leq S \leq \Gamma 1_H$ then

$$\left|S-\frac{\Gamma+\gamma}{2}1_H\right|\leq\frac{1}{2}(\Gamma-\gamma).$$

Taking the modulus in (6.75) and using the properties of trace, we have

$$\frac{\text{tr}(|B|^2S^2)}{\text{tr}(|B|^2)}-\left(\frac{\text{tr}(|B|^2S)}{\text{tr}(|B|^2)}\right)^2 \qquad (6.76)$$

$$=\frac{1}{\text{tr}(|B|^2)}\left|\text{tr}\left(|B|^2\left(S-\frac{\Gamma+\gamma}{2}1_H\right)\left(S-\frac{\text{tr}(|B|^2S)}{\text{tr}(|B|^2)}1_H\right)\right)\right|$$

$$\leq\frac{1}{\text{tr}(|B|^2)}\text{tr}\left(|B|^2\left|\left(S-\frac{\Gamma+\gamma}{2}1_H\right)\left(S-\frac{\text{tr}(|B|^2S)}{\text{tr}(|B|^2)}1_H\right)\right|\right)$$

$$\leq\frac{1}{2}(\Gamma-\gamma)\frac{1}{\text{tr}(|B|^2)}\text{tr}\left(|B|^2\left|S-\frac{\text{tr}(|B|^2S)}{\text{tr}(|B|^2)}1_H\right|\right),$$

which proves the first part of (6.74).

By Schwarz inequality for trace, we also have

$$\frac{1}{\text{tr}(|B|^2)}\text{tr}\left(|B|^2\left|S-\frac{\text{tr}(|B|^2S)}{\text{tr}(|B|^2)}1_H\right|\right) \qquad (6.77)$$

$$\leq\left[\frac{1}{\text{tr}(|B|^2)}\text{tr}\left(|B|^2\left(S-\frac{\text{tr}(|B|^2S)}{\text{tr}(|B|^2)}1_H\right)^2\right)\right]^{1/2}$$

$$=\left[\frac{\text{tr}(|B|^2S^2)}{\text{tr}(|B|^2)}-\left(\frac{\text{tr}(|B|^2S)}{\text{tr}(|B|^2)}\right)^2\right]^{1/2}.$$

From (6.76) and (6.77), we get

$$\frac{\text{tr}(|B|^2S^2)}{\text{tr}(|B|^2)}-\left(\frac{\text{tr}(|B|^2S)}{\text{tr}(|B|^2)}\right)^2$$

$$\leq\frac{1}{2}(\Gamma-\gamma)\left[\frac{\text{tr}(|B|^2S^2)}{\text{tr}(|B|^2)}-\left(\frac{\text{tr}(|B|^2S)}{\text{tr}(|B|^2)}\right)^2\right]^{1/2},$$

which implies that

$$\left[\frac{\mathrm{tr}(|B|^2 S^2)}{\mathrm{tr}(|B|^2)} - \left(\frac{\mathrm{tr}(|B|^2 S)}{\mathrm{tr}(|B|^2)} \right)^2 \right]^{1/2} \le \frac{1}{2}(\Gamma - \gamma).$$

By (6.77), we then obtain

$$\frac{1}{\mathrm{tr}(|B|^2)} \mathrm{tr} \left(|B|^2 \left| S - \frac{\mathrm{tr}(|B|^2 S)}{\mathrm{tr}(|B|^2)} 1_H \right| \right)$$

$$\le \left[\frac{\mathrm{tr}(|B|^2 S^2)}{\mathrm{tr}(|B|^2)} - \left(\frac{\mathrm{tr}(|B|^2 S)}{\mathrm{tr}(|B|^2)} \right)^2 \right]^{1/2} \le \frac{1}{2}(\Gamma - \gamma)$$

that proves the last part of (6.74).   □

Remark 6.6   Let S be a self-adjoint operator such that $\gamma 1_H \le S \le \Gamma 1_H$ for some real constants $\Gamma \ge \gamma$. Then for any $P \in B_1(H) \setminus \{0\}, P \ge 0$ we have

$$0 \le \frac{\mathrm{tr}(PS^2)}{\mathrm{tr}(P)} - \left(\frac{\mathrm{tr}(PS)}{\mathrm{tr}(P)} \right)^2 \tag{6.78}$$

$$\le \frac{1}{2}(\Gamma - \gamma) \frac{1}{\mathrm{tr}(P)} \mathrm{tr} \left(P \left| S - \frac{\mathrm{tr}(PS)}{\mathrm{tr}(P)} 1_H \right| \right)$$

$$\le \frac{1}{2}(\Gamma - \gamma) \left[\frac{\mathrm{tr}(PS^2)}{\mathrm{tr}(P)} - \left(\frac{\mathrm{tr}\,(PS)}{\mathrm{tr}(P)} \right)^2 \right]^{1/2} \le \frac{1}{4}(\Gamma - \gamma)^2.$$

The following result provides reverses for the inequalities (6.63) and (6.64):

Theorem 6.15   ([45])   Let A be a self-adjoint operator on the Hilbert space H and assume that $\mathrm{Sp}(A) \subseteq [m, M]$ for some scalars m, M with $m < M$. If f is a continuously differentiable convex function on $[m, M]$ and $B \in B_2(H) \setminus \{0\}$, then we have

$$\frac{\mathrm{tr}(|B^*|^2 f(A))}{\mathrm{tr}(|B|^2)} - f \left(\frac{\mathrm{tr}(|B|^2 A)}{\mathrm{tr}(|B|^2)} \right) \tag{6.79}$$

$$\le \frac{\mathrm{tr}(|B^*|^2 f'(A)A)}{\mathrm{tr}(|B|^2)} - \frac{\mathrm{tr}(|B|^2 A)}{\mathrm{tr}(|B|^2)} \frac{\mathrm{tr}(|B^*|^2 f'(A))}{\mathrm{tr}(|B|^2)}$$

and

$$0 \le \frac{\mathrm{tr}(|B|^2 f(A))}{\mathrm{tr}(|B|^2)} - f \left(\frac{\mathrm{tr}(|B|^2 A)}{\mathrm{tr}(|B|^2)} \right) \tag{6.80}$$

$$\le \frac{\mathrm{tr}(|B|^2 f'(A)A)}{\mathrm{tr}(|B|^2)} - \frac{\mathrm{tr}(|B|^2 A)}{\mathrm{tr}(|B|^2)} \frac{\mathrm{tr}(|B|^2 f'(A))}{\mathrm{tr}(|B|^2)} =: \mathcal{K}(f', B, A).$$

Moreover, we have

$$\mathcal{K}(f', B, A) \tag{6.81}$$

$$\leq \begin{cases} \frac{1}{2}[f'(M) - f'(m)]\dfrac{\text{tr}\left(|B|^2 \left|A - \frac{\text{tr}(|B|^2 A)}{\text{tr}(|B|^2)} 1_H\right|\right)}{\text{tr}(|B|^2)} \\[3em] \frac{1}{2}(M - m)\dfrac{\text{tr}\left(|B|^2 \left|f'(A) - \frac{\text{tr}(|B|^2 f'(A))}{\text{tr}(|B|^2)} 1_H\right|\right)}{\text{tr}(|B|^2)} \end{cases}$$

$$\leq \begin{cases} \frac{1}{2}[f'(M) - f'(m)]\left[\dfrac{\text{tr}(|B|^2 A^2)}{\text{tr}(|B|^2)} - \left(\dfrac{\text{tr}(|B|^2 A)}{\text{tr}(|B|^2)}\right)^2\right]^{1/2} \\[3em] \frac{1}{2}(M - m)\left[\dfrac{\text{tr}(|B|^2 [f'(A)]^2)}{\text{tr}(|B|^2)} - \left(\dfrac{\text{tr}(|B|^2 f'(A))}{\text{tr}(|B|^2)}\right)^2\right]^{1/2} \end{cases}$$

$$\leq \frac{1}{4}[f'(M) - f'(m)](M - m).$$

Proof: By the gradient inequality, we have

$$f(\tau) - f(\varsigma) \leq f'(\tau)(\tau - \varsigma) \tag{6.82}$$

for any $\tau, \varsigma \in [m, M]$.

This inequality implies in the operator order

$$f(A) - f\left(\frac{\text{tr}(|B|^2 A)}{\text{tr}(|B|^2)}\right) 1_H \leq f'(A)\left(A - \frac{\text{tr}(|B|^2 A)}{\text{tr}(|B|^2)} 1_H\right)$$

that is equivalent to

$$\langle f(A)y, y \rangle - f\left(\frac{\text{tr}(|B|^2 A)}{\text{tr}(|B|^2)}\right)\langle y, y \rangle \tag{6.83}$$

$$\leq \langle f'(A)Ay, y \rangle - \frac{\text{tr}(|B|^2 A)}{\text{tr}(|B|^2)}\langle f'(A)y, y \rangle$$

for any $y \in H$, which is of interest in itself as well.

Let $\mathcal{E} := \{e_i\}_{i \in I}$ be an orthonormal basis in H. If we take in (6.83) $y = Be_i$ and sum, then we get

$$\sum_{i \in I} \langle f(A)Be_i, Be_i \rangle - f\left(\frac{\text{tr}(|B|^2 A)}{\text{tr}(|B|^2)}\right) \sum_{i \in I} \langle Be_i, Be_i \rangle$$

$$\leq \sum_{i \in I} \langle f'(A)ABe_i, Be_i \rangle - \frac{\text{tr}(|B|^2 A)}{\text{tr}(|B|^2)} \sum_{i \in I} \langle f'(A)Be_i, Be_i \rangle,$$

which is equivalent to

$$\sum_{i\in I}\langle B^*f(A)Be_i,e_i\rangle - f\left(\frac{\mathrm{tr}(|B|^2A)}{\mathrm{tr}(|B|^2)}\right)\sum_{i\in I}\langle B^*Be_i,e_i\rangle$$

$$\leq \sum_{i\in I}\langle B^*f'(A)ABe_i,e_i\rangle - \frac{\mathrm{tr}(|B|^2A)}{\mathrm{tr}(|B|^2)}\sum_{i\in I}\langle B^*f'(A)Be_i,e_i\rangle$$

and the inequality (6.79) is obtained.

If we take in (6.83) $y = B^*e_i$ and sum, then we get

$$\sum_{i\in I}\langle f(A)B^*e_i,B^*e_i\rangle - f\left(\frac{\mathrm{tr}(|B|^2A)}{\mathrm{tr}(|B|^2)}\right)\sum_{i\in I}\langle B^*e_i,B^*e_i\rangle$$

$$\leq \sum_{i\in I}\langle f'(A)AB^*e_i,B^*e_i\rangle - \frac{\mathrm{tr}(|B|^2A)}{\mathrm{tr}(|B|^2)}\sum_{i\in I}\langle f'(A)B^*e_i,B^*e_i\rangle$$

that is equivalent to

$$\sum_{i\in I}\langle Bf(A)B^*e_i,e_i\rangle - f\left(\frac{\mathrm{tr}(|B|^2A)}{\mathrm{tr}(|B|^2)}\right)\sum_{i\in I}\langle BB^*e_i,e_i\rangle \qquad (6.84)$$

$$\leq \sum_{i\in I}\langle Bf'(A)AB^*e_i,e_i\rangle - \frac{\mathrm{tr}(|B|^2A)}{\mathrm{tr}(|B|^2)}\sum_{i\in I}\langle Bf'(A)B^*e_i,e_i\rangle$$

and the inequality (6.80) is obtained.

Now, since f is continuously convex on $[m,M]$, then f' is monotonic non-decreasing on $[m,M]$ and $f'(m) \leq f'(t) \leq f'(M)$ for any $t \in [m,M]$. We also observe that

$$\frac{1}{\mathrm{tr}(|B|^2)}\mathrm{tr}\left(|B|^2\left[f'(A)-\frac{f'(m)+f'(M)}{2}1_H\right]\left[A-\frac{\mathrm{tr}(|B|^2A)}{\mathrm{tr}(|B|^2)}1_H\right]\right) \qquad (6.85)$$

$$= \frac{1}{\mathrm{tr}(|B|^2)}\mathrm{tr}\left(|B|^2f'(A)\left[A-\frac{\mathrm{tr}(|B|^2A)}{\mathrm{tr}(|B|^2)}1_H\right]\right)$$

$$- \frac{f'(m)+f'(M)}{2}\frac{1}{\mathrm{tr}(|B|^2)}\mathrm{tr}\left(|B|^2\left[A-\frac{\mathrm{tr}(|B|^2A)}{\mathrm{tr}(|B|^2)}1_H\right]\right)$$

$$= \mathcal{K}(f',B,A).$$

Since

$$\left|f'(A)-\frac{f'(m)+f'(M)}{2}1_H\right| \leq \frac{1}{2}[f'(M)-f'(m)]1_H,$$

then by taking the modulus in (6.85) and utilizing the properties of trace we have

$$0 \leq \mathcal{K}(f',B,A) \qquad (6.86)$$

$$\leq \frac{1}{\mathrm{tr}(|B|^2)}$$

$$\times \operatorname{tr}\left(|B|^2 \left[f'(A) - \frac{f'(m) + f'(M)}{2}1_H\right]\left[A - \frac{\operatorname{tr}(|B|^2 A)}{\operatorname{tr}(|B|^2)}1_H\right]\right)$$

$$\leq \frac{1}{2}[f'(M) - f'(m)]\frac{1}{\operatorname{tr}(|B|^2)}\operatorname{tr}\left(|B|^2\left|A - \frac{\operatorname{tr}(|B|^2 A)}{\operatorname{tr}(|B|^2)}1_H\right|\right),$$

and the first inequality in the first branch of (6.81) is proved.

We have $m1_H \leq A \leq M1_H$ and by applying Lemma 6.1 we can state that

$$\frac{1}{\operatorname{tr}(|B|^2)}\operatorname{tr}\left(|B|^2\left|A - \frac{\operatorname{tr}(|B|^2 A)}{\operatorname{tr}(|B|^2)}1_H\right|\right) \qquad (6.87)$$

$$\leq \left[\frac{\operatorname{tr}(|B|^2 A^2)}{\operatorname{tr}(|B|^2)} - \left(\frac{\operatorname{tr}(|B|^2 A)}{\operatorname{tr}(|B|^2)}\right)^2\right]^{1/2} \leq \frac{1}{2}(M - m).$$

Making use of (6.86) and (6.87), we deduce the second and the third inequalities in the first branch of (6.81).

We observe that $\mathcal{K}(f', B, A)$ can also be represented as

$$\mathcal{K}(f', B, A)$$
$$= \frac{1}{\operatorname{tr}(|B|^2)}\operatorname{tr}\left(|B|^2\left[f'(A) - \frac{\operatorname{tr}(|B|^2 f'(A))}{\operatorname{tr}(|B|^2)}1_H\right]\left(A - \frac{m + M}{2}1_H\right)\right).$$

Applying a similar argument as above for this representation, we get the second branch of the inequality (6.81).

The proof is complete. □

Corollary 6.9 Let A be a self-adjoint operator on the Hilbert space H and assume that $\operatorname{Sp}(A) \subseteq [m, M]$ for some scalars m, M with $m < M$. If f is a continuously differentiable convex function on $[m, M]$ and $P \in \mathcal{B}_1(H)\setminus\{0\}, P \geq 0$, then we have

$$0 \leq \frac{\operatorname{tr}(Pf(A))}{\operatorname{tr}(P)} - f\left(\frac{\operatorname{tr}(PA)}{\operatorname{tr}(P)}\right) \qquad (6.88)$$

$$\leq \frac{\operatorname{tr}(Pf'(A)A)}{\operatorname{tr}(P)} - \frac{\operatorname{tr}(PA)}{\operatorname{tr}(P)} \cdot \frac{\operatorname{tr}(Pf'(A))}{\operatorname{tr}(P)}$$

$$\leq \begin{cases} \dfrac{1}{2}[f'(M) - f'(m)]\dfrac{\operatorname{tr}\left(P\left|A - \frac{\operatorname{tr}(PA)}{\operatorname{tr}(P)}1_H\right|\right)}{\operatorname{tr}(P)} \\[3ex] \dfrac{1}{2}(M - m)\dfrac{\operatorname{tr}\left(P\left|f'(A) - \frac{\operatorname{tr}(Pf'(A))}{\operatorname{tr}(P)}1_H\right|\right)}{\operatorname{tr}(P)} \end{cases}$$

$$
\leq
\begin{cases}
\dfrac{1}{2}[f'(M) - f'(m)] \left[\dfrac{\operatorname{tr}(PA^2)}{\operatorname{tr}(P)} - \left(\dfrac{\operatorname{tr}(PA)}{\operatorname{tr}(P)} \right)^2 \right]^{1/2} \\[4mm]
\dfrac{1}{2}(M - m) \left[\dfrac{\operatorname{tr}(P[f'(A)]^2)}{\operatorname{tr}(P)} - \left(\dfrac{\operatorname{tr}(Pf'(A))}{\operatorname{tr}(P)} \right)^2 \right]^{1/2}
\end{cases}
$$

$$
\leq \frac{1}{4}[f'(M) - f'(m)](M - m).
$$

Remark 6.7 Let $\mathcal{M}_n(\mathbb{C})$ be the space of all square matrices of order n with complex elements and $A \in \mathcal{M}_n(\mathbb{C})$ be a Hermitian matrix such that $\operatorname{Sp}(A) \subseteq [m, M]$ for some scalars m, M with $m < M$. If f is a continuously differentiable convex function on $[m, M]$, then by taking $P = I_n$, the identity matrix, in (6.88) we get

$$
0 \leq \frac{\operatorname{tr}(f(A))}{n} - f\left(\frac{\operatorname{tr}(A)}{n} \right) \tag{6.89}
$$

$$
\leq \frac{\operatorname{tr}(f'(A)A)}{n} - \frac{\operatorname{tr}(A)}{n} \frac{\operatorname{tr}(f'(A))}{n}
$$

$$
\leq
\begin{cases}
\dfrac{1}{2}[f'(M) - f'(m)] \dfrac{\operatorname{tr}\left(\left| A - \frac{\operatorname{tr}(A)}{n} I_n \right| \right)}{n} \\[4mm]
\dfrac{1}{2}(M - m) \dfrac{\operatorname{tr}\left(\left| f'(A) - \frac{\operatorname{tr}(f'(A))}{n} I_n \right| \right)}{n}
\end{cases}
$$

$$
\leq
\begin{cases}
\dfrac{1}{2}[f'(M) - f'(m)] \left[\dfrac{\operatorname{tr}(A^2)}{n} - \left(\dfrac{\operatorname{tr}(A)}{n} \right)^2 \right]^{1/2} \\[4mm]
\dfrac{1}{2}(M - m) \left[\dfrac{\operatorname{tr}([f'(A)]^2)}{n} - \left(\dfrac{\operatorname{tr}(f'(A))}{n} \right)^2 \right]^{1/2}
\end{cases}
$$

$$
\leq \frac{1}{4}[f'(M) - f'(m)](M - m).
$$

6.3.2 Some Examples

We consider the power function $f : (0, \infty) \to (0, \infty)$, $f(t) = t^r$ with $t \in \mathbb{R} \setminus \{0\}$. For $r \in (-\infty, 0) \cup [1, \infty)$, f is convex while for $r \in (0, 1)$, f is concave. Denote $B_1^+(H) := \{P \text{ with } P \in B_1(H) \text{ and } P \geq 0\}$.

Let $r \geq 1$ and A be a self-adjoint operator on the Hilbert space H and assume that $\operatorname{Sp}(A) \subseteq [m, M]$ for some scalars m, M with $0 \leq m < M$. If $P \in B_1^+(H) \setminus \{0\}$,

then

$$0 \le \frac{\mathrm{tr}(PA^r)}{\mathrm{tr}(P)} - \left(\frac{\mathrm{tr}(PA)}{\mathrm{tr}(P)} \right)^r \tag{6.90}$$

$$\le r \left[\frac{\mathrm{tr}(PA^r)}{\mathrm{tr}(P)} - \frac{\mathrm{tr}(PA)}{\mathrm{tr}(P)} \frac{\mathrm{tr}(PA^{r-1})}{\mathrm{tr}(P)} \right]$$

$$\le \begin{cases} \frac{1}{2} r(M^{r-1} - m^{r-1}) \dfrac{\mathrm{tr}\left(P \left| A - \frac{\mathrm{tr}(PA)}{\mathrm{tr}(P)} 1_H \right| \right)}{\mathrm{tr}(P)} \\[4mm] \frac{1}{2} r(M - m) \dfrac{\mathrm{tr}\left(P \left| A^{r-1} - \frac{\mathrm{tr}(PA^{r-1})}{\mathrm{tr}(P)} 1_H \right| \right)}{\mathrm{tr}(P)} \end{cases}$$

$$\le \begin{cases} \frac{1}{2} r(M^{r-1} - m^{r-1}) \left[\dfrac{\mathrm{tr}(PA^2)}{\mathrm{tr}(P)} - \left(\dfrac{\mathrm{tr}(PA)}{\mathrm{tr}(P)} \right)^2 \right]^{1/2} \\[4mm] \frac{1}{2} r(M - m) \left[\dfrac{\mathrm{tr}(PA^{2(r-1)})}{\mathrm{tr}(P)} - \left(\dfrac{\mathrm{tr}(PA^{r-1})}{\mathrm{tr}(P)} \right)^2 \right]^{1/2} \end{cases}$$

$$\le \frac{1}{4} r(M^{r-1} - m^{r-1})(M - m).$$

Consider the convex function $f : (0, \infty) \to (0, \infty), f(t) = -\ln t$ and let A be a self-adjoint operator on the Hilbert space H and assume that $\mathrm{Sp}(A) \subseteq [m, M]$ for some scalars m, M with $0 < m < M$. If $P \in B_1^+(H) \setminus \{0\}$, then

$$0 \le \ln \left(\frac{\mathrm{tr}(PA)}{\mathrm{tr}(P)} \right) - \frac{\mathrm{tr}(P \ln A)}{\mathrm{tr}(P)} \tag{6.91}$$

$$\le \frac{\mathrm{tr}(PA)}{\mathrm{tr}(P)} \frac{\mathrm{tr}(PA^{-1})}{\mathrm{tr}(P)} - 1$$

$$\le \begin{cases} \dfrac{M - m}{2mM} \dfrac{\mathrm{tr}\left(P \left| A - \frac{\mathrm{tr}(PA)}{\mathrm{tr}(P)} 1_H \right| \right)}{\mathrm{tr}(P)} \\[4mm] \frac{1}{2}(M - m) \dfrac{\mathrm{tr}\left(P \left| A^{-1} - \frac{\mathrm{tr}(PA^{-1})}{\mathrm{tr}(P)} 1_H \right| \right)}{\mathrm{tr}(P)} \end{cases}$$

$$\le \begin{cases} \dfrac{M - m}{2mM} \left[\dfrac{\mathrm{tr}(PA^2)}{\mathrm{tr}(P)} - \left(\dfrac{\mathrm{tr}(PA)}{\mathrm{tr}(P)} \right)^2 \right]^{1/2} \\[4mm] \frac{1}{2}(M - m) \left[\dfrac{\mathrm{tr}(PA^{-2})}{\mathrm{tr}(P)} - \left(\dfrac{\mathrm{tr}(PA^{-1})}{\mathrm{tr}(P)} \right)^2 \right]^{1/2} \end{cases}$$

$$\le \frac{(M - m)^2}{4mM}.$$

Consider the convex function $f(t) = t \ln t$ and let A be a Self-adjoint operator on the Hilbert space H and assume that $\mathrm{Sp}(A) \subseteq [m, M]$ for some scalars m, M with $0 < m < M$. If $P \in B_1^+(H) \backslash \{0\}$, then

$$0 \le \frac{\mathrm{tr}(PA \ln A)}{\mathrm{tr}(P)} - \frac{\mathrm{tr}(PA)}{\mathrm{tr}(P)} \ln\left(\frac{\mathrm{tr}(PA)}{\mathrm{tr}(P)}\right) \tag{6.92}$$

$$\le \frac{\mathrm{tr}(PA \ln(eA))}{\mathrm{tr}(P)} - \frac{\mathrm{tr}(PA)}{\mathrm{tr}(P)} \frac{\mathrm{tr}(P \ln(eA))}{\mathrm{tr}(P)}$$

$$\le \begin{cases} \frac{1}{2} \ln\left(\frac{M}{m}\right) \dfrac{\mathrm{tr}\left(P\left|A - \frac{\mathrm{tr}(PA)}{\mathrm{tr}(P)} 1_H\right|\right)}{\mathrm{tr}(P)} \\[2ex] \frac{1}{2}(M - m) \dfrac{\mathrm{tr}\left(P\left|\ln(eA) - \frac{\mathrm{tr}(P \ln(eA))}{\mathrm{tr}(P)} 1_H\right|\right)}{\mathrm{tr}(P)} \end{cases}$$

$$\le \begin{cases} \frac{1}{2} \ln\left(\frac{M}{m}\right) \left[\dfrac{\mathrm{tr}(PA^2)}{\mathrm{tr}(P)} - \left(\dfrac{\mathrm{tr}(PA)}{\mathrm{tr}(P)}\right)^2\right]^{1/2} \\[2ex] \frac{1}{2}(M - m) \left[\dfrac{\mathrm{tr}(P[\ln(eA)]^2)}{\mathrm{tr}(P)} - \left(\dfrac{\mathrm{tr}(P \ln(eA))}{\mathrm{tr}(P)}\right)^2\right]^{1/2} \end{cases}$$

$$\le \frac{1}{4}(M - m) \ln\left(\frac{M}{m}\right).$$

6.3.3 Further Reverse Inequalities for Convex Functions

The following reverses of Jensen's trace inequality also hold:

Theorem 6.16 ([46]) Let A be a self-adjoint operator on the Hilbert space H and assume that $\mathrm{Sp}(A) \subseteq [m, M]$ for some scalars m, M with $m < M$. If f is a continuos convex function on $[m, M]$ and $P \in B_1(H) \backslash \{0\}, P \ge 0$ is such that $\frac{\mathrm{tr}(PA)}{\mathrm{tr}(P)} \in (m, M)$ then we have

$$0 \le \frac{\mathrm{tr}(Pf(A))}{\mathrm{tr}(P)} - f\left(\frac{\mathrm{tr}(PA)}{\mathrm{tr}(P)}\right) \tag{6.93}$$

$$\le \frac{\left(M - \frac{\mathrm{tr}(PA)}{\mathrm{tr}(P)}\right)\left(\frac{\mathrm{tr}(PA)}{\mathrm{tr}(P)} - m\right)}{M - m} \Psi_f\left(\frac{\mathrm{tr}(PA)}{\mathrm{tr}(P)}; m, M\right)$$

$$\le \frac{\left(M - \frac{\mathrm{tr}(PA)}{\mathrm{tr}(P)}\right)\left(\frac{\mathrm{tr}(PA)}{\mathrm{tr}(P)} - m\right)}{M - m} \sup_{t \in (m,M)} \Psi_f(t; m, M)$$

$$\le \left(M - \frac{\mathrm{tr}(PA)}{\mathrm{tr}(P)}\right)\left(\frac{\mathrm{tr}(PA)}{\mathrm{tr}(P)} - m\right) \frac{f'_-(M) - f'_+(m)}{M - m}$$

$$\le \frac{1}{4}(M - m)[f'_-(M) - f'_+(m)],$$

where $\Psi_f(\cdot; m, M) : (m, M) \to \mathbb{R}$ is defined by

$$\Psi_f(t; m, M) = \frac{f(M) - f(t)}{M - t} - \frac{f(t) - f(m)}{t - m}.$$

We also have

$$0 \leq \frac{\mathrm{tr}(Pf(A))}{\mathrm{tr}(P)} - f\left(\frac{\mathrm{tr}(PA)}{\mathrm{tr}(P)}\right) \tag{6.94}$$

$$\leq \frac{\left(M - \frac{\mathrm{tr}(PA)}{\mathrm{tr}(P)}\right)\left(\frac{\mathrm{tr}(PA)}{\mathrm{tr}(P)} - m\right)}{M - m} \Psi_f\left(\frac{\mathrm{tr}(PA)}{\mathrm{tr}(P)}; m, M\right)$$

$$\leq \frac{1}{4}(M - m)\Psi_f\left(\frac{\mathrm{tr}(PA)}{\mathrm{tr}(P)}; m, M\right)$$

$$\leq \frac{1}{4}(M - m) \sup_{t \in (m, M)} \Psi_f(t; m, M)$$

$$\leq \frac{1}{4}(M - m)[f'_-(M) - f'_+(m)]$$

for any $P \in \mathcal{B}_1(H) \setminus \{0\}, P \geq 0$ such that $\frac{\mathrm{tr}(PA)}{\mathrm{tr}(P)} \in (m, M)$.

Proof: Since f is convex, then we have

$$f(t) = f\left(\frac{m(M - t) + M(t - m)}{M - m}\right) \leq \frac{(M - t)f(m) + (t - m)f(M)}{M - m}$$

for any $t \in [m, M]$.

This scalar inequality implies, by utilizing the spectral representation of continuous functions of self-adjoint operators, the following inequality

$$f(A) \leq \frac{f(m)(M1_M - A) + f(M)(A - m1_H)}{M - m} \tag{6.95}$$

in the operator order of $\mathcal{B}(H)$.

Utilizing the properties of the trace and the inequality (6.95), we have

$$\frac{\mathrm{tr}(Pf(A))}{\mathrm{tr}(P)} - f\left(\frac{\mathrm{tr}(PA)}{\mathrm{tr}(P)}\right) \tag{6.96}$$

$$= \frac{\mathrm{tr}(Pf(A))}{\mathrm{tr}(P)} - f\left(\frac{\mathrm{tr}\left(P\frac{m(M1_H - A) + M(A - 1_H m)}{M - m}\right)}{\mathrm{tr}(P)}\right)$$

$$\leq \frac{\mathrm{tr}\left(P\frac{f(m)(M1_M - A) + f(M)(A - m1_H)}{M - m}\right)}{\mathrm{tr}(P)}$$

$$- f\left(\frac{\mathrm{tr}\left(P\frac{m(M1_H - A) + M(A - 1_H m)}{M - m}\right)}{\mathrm{tr}(P)}\right)$$

$$= \frac{\left(M - \frac{\text{tr}(PA)}{\text{tr}(P)}\right)f(m) + \left(\frac{\text{tr}(PA)}{\text{tr}(P)} - m\right)f(M)}{M - m}$$

$$-f\left(\frac{\left(M - \frac{\text{tr}(PA)}{\text{tr}(P)}\right)m + \left(\frac{\text{tr}(PA)}{\text{tr}(P)} - m\right)M}{M - m}\right)$$

$$=: B(f, P, A, m, M)$$

for any $P \in \mathcal{B}_1(H)\setminus\{0\}$, $P \geq 0$.

By denoting

$$\Delta_f(t; m, M) := \frac{(t - m)f(M) + (M - t)f(m)}{M - m} - f(t), \quad t \in [m, M],$$

we have

$$\Delta_f(t; m, M) = \frac{(t - m)f(M) + (M - t)f(m) - (M - m)f(t)}{M - m} \tag{6.97}$$

$$= \frac{(t - m)f(M) + (M - t)f(m) - (M - t + t - m)f(t)}{M - m}$$

$$= \frac{(t - m)[f(M) - f(t)] - (M - t)[f(t) - f(m)]}{M - m}$$

$$= \frac{(M - t)(t - m)}{M - m}\Psi_f(t; m, M)$$

for any $t \in (m, M)$.

Therefore,

$$B(f, P, A, m, M) = \frac{\left(M - \frac{\text{tr}(PA)}{\text{tr}(P)}\right)\left(\frac{\text{tr}(PA)}{\text{tr}(P)} - m\right)}{M - m}\Psi_f\left(\frac{\text{tr}(PA)}{\text{tr}(P)}; m, M\right), \tag{6.98}$$

provided that $\frac{\text{tr}(PA)}{\text{tr}(P)} \in (m, M)$.

If $\frac{\text{tr}(PA)}{\text{tr}(P)} \in (m, M)$, then

$$\Psi_f\left(\frac{\text{tr}(PA)}{\text{tr}(P)}; m, M\right) \tag{6.99}$$

$$\leq \sup_{t \in (m, M)} \Psi_f(t; m, M)$$

$$= \sup_{t \in (m, M)} \left[\frac{f(M) - f(t)}{M - t} - \frac{f(t) - f(m)}{t - m}\right]$$

$$\leq \sup_{t \in (m, M)} \left[\frac{f(M) - f(t)}{M - t}\right] + \sup_{t \in (m, M)} \left[-\frac{f(t) - f(m)}{t - m}\right]$$

$$
= \sup_{t \in (m,M)} \left[\frac{f(M) - f(t)}{M - t} \right] - \inf_{t \in (m,M)} \left[\frac{f(t) - f(m)}{t - m} \right]
$$

$$
= f'_-(M) - f'_+(m),
$$

which by (6.96) and (6.98) produces the second, third, and fourth inequalities in (6.93).

Since, obviously

$$
\frac{1}{M - m} \left(M - \frac{\mathrm{tr}(PA)}{\mathrm{tr}(P)} \right) \left(\frac{\mathrm{tr}(PA)}{\mathrm{tr}(P)} - m \right) \le \frac{1}{4}(M - m),
$$

then the last part of (6.93) also holds.

The second part of the theorem is clear and the details are omitted. $\quad\square$

The following result also holds:

Theorem 6.17 ([46]) Let A be a self-adjoint operator on the Hilbert space H and assume that $\mathrm{Sp}(A) \subseteq [m, M]$ for some scalars m, M with $m < M$. If f is a continuos convex function on $[m, M]$ then for all $P \in B_1(H) \setminus \{0\}, P \ge 0$ we have that $\frac{\mathrm{tr}(PA)}{\mathrm{tr}(P)} \in [m, M]$ and

$$
0 \le \frac{\mathrm{tr}(Pf(A))}{\mathrm{tr}(P)} - f\left(\frac{\mathrm{tr}(PA)}{\mathrm{tr}(P)} \right) \tag{6.100}
$$

$$
\le 2\max \left\{ \frac{M - \frac{\mathrm{tr}(PA)}{\mathrm{tr}(P)}}{M - m}, \frac{\frac{\mathrm{tr}(PA)}{\mathrm{tr}(P)} - m}{M - m} \right\} \left[\frac{f(m) + f(M)}{2} - f\left(\frac{m + M}{2} \right) \right]
$$

$$
\le 2 \left[\frac{f(m) + f(M)}{2} - f\left(\frac{m + M}{2} \right) \right].
$$

Proof: Since $m1_H \le A \le M1_H$, it follows that $m\,\mathrm{tr}(P) \le \mathrm{tr}(PA) \le M\,\mathrm{tr}(P)$ for any $P \in B_1(H) \setminus \{0\}, P \ge 0$, which shows that $\frac{\mathrm{tr}(PA)}{\mathrm{tr}(P)} \in [m, M]$.

Further on, we recall the following result (see, for instance, [11]) that provides a refinement and a reverse for the weighted Jensen's discrete inequality:

$$
n \min_{i \in \{1,\dots,n\}} \{p_i\} \left[\frac{1}{n} \sum_{i=1}^{n} f(x_i) - f\left(\frac{1}{n} \sum_{i=1}^{n} x_i \right) \right] \tag{6.101}
$$

$$
\le \frac{1}{P_n} \sum_{i=1}^{n} p_i f(x_i) - f\left(\frac{1}{P_n} \sum_{i=1}^{n} p_i x_i \right)
$$

$$
\le n \max_{i \in \{1,\dots,n\}} \{p_i\} \left[\frac{1}{n} \sum_{i=1}^{n} f(x_i) - f\left(\frac{1}{n} \sum_{i=1}^{n} x_i \right) \right],
$$

where $f : C \to \mathbb{R}$ is a convex function defined on the convex subset C of the linear space X, $\{x_i\}_{i \in \{1,\dots,n\}} \subset C$ are vectors and $\{p_i\}_{i \in \{1,\dots,n\}}$ are nonnegative numbers with $P_n := \sum_{i=1}^{n} p_i > 0$.

For $n = 2$, we deduce from (6.101) that

$$2 \min\{t, 1 - t\} \left[\frac{f(x) + f(y)}{2} - f\left(\frac{x+y}{2}\right) \right] \tag{6.102}$$

$$\leq tf(x) + (1 - t)f(y) - f(tx + (1 - t)y)$$

$$\leq 2 \max\{t, 1 - t\} \left[\frac{f(x) + f(y)}{2} - f\left(\frac{x+y}{2}\right) \right]$$

for any $x, y \in C$ and $t \in [0, 1]$.

If we use the second inequality in (6.102) for the convex function $f : I \to \mathbb{R}$ where $m, M \in \mathbb{R}$, $m < M$ with $[m, M] = I$, we have for $x = m, y = M$ and $t = \frac{M - \frac{\text{tr}(PA)}{\text{tr}(P)}}{M - m}$ that

$$B(f, P, A, m, M) = \frac{\left(M - \frac{\text{tr}(PA)}{\text{tr}(P)}\right)f(m) + \left(\frac{\text{tr}(PA)}{\text{tr}(P)} - m\right)f(M)}{M - m}$$

$$- f\left(\frac{m\left(M - \frac{\text{tr}(PA)}{\text{tr}(P)}\right) + M\left(\frac{\text{tr}(PA)}{\text{tr}(P)} - m\right)}{M - m} \right)$$

$$\leq 2 \max \left\{ \frac{M - \frac{\text{tr}(PA)}{\text{tr }(P)}}{M - m}, \frac{\frac{\text{tr}(PA)}{\text{tr}(P)} - m}{M - m} \right\}$$

$$\times \left[\frac{f(m) + f(M)}{2} - f\left(\frac{m+M}{2}\right) \right].$$

Making use of (6.96), we deduce the first inequality in (6.100).

Since

$$\max \left\{ \frac{M - \frac{\text{tr}(PA)}{\text{tr}(P)}}{M - m}, \frac{\frac{\text{tr}(PA)}{\text{tr}(P)} - m}{M - m} \right\} \leq 1,$$

the last part of (6.100) is also proved. $\qquad\square$

6.3.4 Some Examples

For $p > 1$ and $0 < m < M < \infty$, consider the convex function $f(t) = t^p$ defined on $[m, M]$. Then $\Psi_p(\cdot; m, M) : (m, M) \to \mathbb{R}$ is defined by

$$\Psi_p(t; m, M) = \frac{M^p - t^p}{M - t} - \frac{t^p - m^p}{t - m}$$

$$= \frac{t(M^p - m^p) - t^p(M - m) - mM(M^{p-1} - m^{p-1})}{(M - t)(t - m)}.$$

Let A be a nonnegative self-adjoint operator on the Hilbert space H and assume that $\text{Sp}(A) \subseteq [m, M]$ for some scalars m, M with $0 \leq m < M$. If

$P \in \mathcal{B}_1(H) \setminus \{0\}, P \geq 0$ such that $\frac{\operatorname{tr}(PA)}{\operatorname{tr}(P)} \in (m, M)$, then we have from (6.93) that

$$0 \leq \frac{\operatorname{tr}(PA^p)}{\operatorname{tr}(P)} - \left(\frac{\operatorname{tr}(PA)}{\operatorname{tr}(P)} \right)^p \tag{6.103}$$

$$\leq \frac{\left(M - \frac{\operatorname{tr}(PA)}{\operatorname{tr}(P)} \right) \left(\frac{\operatorname{tr}(PA)}{\operatorname{tr}(P)} - m \right)}{M - m} \Psi_p \left(\frac{\operatorname{tr}(PA)}{\operatorname{tr}(P)}; m, M \right)$$

$$\leq \frac{\left(M - \frac{\operatorname{tr}(PA)}{\operatorname{tr}(P)} \right) \left(\frac{\operatorname{tr}(PA)}{\operatorname{tr}(P)} - m \right)}{M - m} \sup_{t \in (m,M)} \Psi_p(t; m, M)$$

$$\leq p \left(M - \frac{\operatorname{tr}(PA)}{\operatorname{tr}(P)} \right) \left(\frac{\operatorname{tr}(PA)}{\operatorname{tr}(P)} - m \right) \frac{M^{p-1} - m^{p-1}}{M - m}$$

$$\leq \frac{1}{4} p (M - m)(M^{p-1} - m^{p-1})$$

and from (6.94) that

$$0 \leq \frac{\operatorname{tr}(PA^p)}{\operatorname{tr}(P)} - \left(\frac{\operatorname{tr}(PA)}{\operatorname{tr}(P)} \right)^p \tag{6.104}$$

$$\leq \frac{\left(M - \frac{\operatorname{tr}(PA)}{\operatorname{tr}(P)} \right) \left(\frac{\operatorname{tr}(PA)}{\operatorname{tr}(P)} - m \right)}{M - m} \Psi_p \left(\frac{\operatorname{tr}(PA)}{\operatorname{tr}(P)}; m, M \right)$$

$$\leq \frac{1}{4}(M - m) \Psi_p \left(\frac{\operatorname{tr}(PA)}{\operatorname{tr}(P)}; m, M \right)$$

$$\leq \frac{1}{4}(M - m) \sup_{t \in (m,M)} \Psi_p(t; m, M)$$

$$\leq \frac{1}{4} p (M - m)(M^{p-1} - m^{p-1}).$$

For $p = 2$, we have

$$\Psi_2(t; m, M) = \frac{M^2 - t^2}{M - t} - \frac{t^2 - m^2}{t - m} = M - m$$

and by (6.103) we get

$$0 \leq \frac{\operatorname{tr}(PA^2)}{\operatorname{tr}(P)} - \left(\frac{\operatorname{tr}(PA)}{\operatorname{tr}(P)} \right)^2 \leq \left(M - \frac{\operatorname{tr}(PA)}{\operatorname{tr}(P)} \right) \left(\frac{\operatorname{tr}(PA)}{\operatorname{tr}(P)} - m \right) \tag{6.105}$$

$$\leq \frac{1}{4}(M - m)^2$$

for any $P \in \mathcal{B}_1(H) \setminus \{0\}, P \geq 0$.

Making use of the inequality (6.100) we have

$$0 \le \frac{\operatorname{tr}(PA^p)}{\operatorname{tr}(P)} - \left(\frac{\operatorname{tr}(PA)}{\operatorname{tr}(P)} \right)^p \tag{6.106}$$

$$\le 2\max\left\{ \frac{M - \frac{\operatorname{tr}(PA)}{\operatorname{tr}(P)}}{M - m}, \frac{\frac{\operatorname{tr}(PA)}{\operatorname{tr}(P)} - m}{M - m} \right\} \left[\frac{m^p + M^p}{2} - \left(\frac{m + M}{2} \right)^p \right]$$

$$\le 2 \left[\frac{m^p + M^p}{2} - \left(\frac{m + M}{2} \right)^p \right]$$

for any positive operator A with $\operatorname{Sp}(A) \subseteq [m, M]$ and for any $P \in \mathcal{B}_1(H) \backslash \{0\}$, $P \ge 0$.

In particular, for $p = 2$ we get

$$0 \le \frac{\operatorname{tr}(PA^2)}{\operatorname{tr}(P)} - \left(\frac{\operatorname{tr}(PA)}{\operatorname{tr}(P)} \right)^2 \tag{6.107}$$

$$\le \frac{1}{2}(M - m)\max\left\{ M - \frac{\operatorname{tr}(PA)}{\operatorname{tr}(P)}, \frac{\operatorname{tr}(PA)}{\operatorname{tr}(P)} - m \right\}$$

$$\le \frac{1}{2}(M - m)^2.$$

Since

$$\max\left\{ M - \frac{\operatorname{tr}(PA)}{\operatorname{tr}(P)}, \frac{\operatorname{tr}(PA)}{\operatorname{tr}(P)} - m \right\}$$

$$= \frac{1}{2}(M - m) + \left| \frac{\operatorname{tr}(PA)}{\operatorname{tr}(P)} - \frac{1}{2}(m + M) \right|,$$

then the second inequality in (6.107) is not as good as the second inequality in (6.105).

For $p = -1$ and $0 < m < M < \infty$ consider the convex function $f(t) = t^{-1}$ defined on $[m, M]$. Then $\Psi_{-1}(\cdot; m, M) : (m, M) \to \mathbb{R}$ is defined by

$$\Psi_{-1}(t; m, M) = \frac{M^{-1} - t^{-1}}{M - t} - \frac{t^{-1} - m^{-1}}{t - m} = \frac{M - m}{mMt}.$$

The definition of $\Psi_{-1}(\cdot; m, M)$ can be extended to the closed interval $[m, M]$. We also have that

$$\sup_{t \in (m, M)} \Psi_{-1}(t; m, M) = \frac{M - m}{m^2 M}.$$

From the inequality (6.93), we get

$$0 \le \frac{\operatorname{tr}(PA^{-1})}{\operatorname{tr}(P)} - \frac{\operatorname{tr}(P)}{\operatorname{tr}(PA)} \tag{6.108}$$

$$\le \frac{\left(M - \frac{\operatorname{tr}(PA)}{\operatorname{tr}(P)} \right)\left(\frac{\operatorname{tr}(PA)}{\operatorname{tr}(P)} - m \right)}{mM} \frac{\operatorname{tr}(P)}{\operatorname{tr}(PA)}.$$

$$\leq \frac{1}{m^2 M} \left(M - \frac{\text{tr}(PA)}{\text{tr}(P)} \right) \left(\frac{\text{tr}(PA)}{\text{tr}(P)} - m \right)$$

$$\leq \frac{1}{4} \frac{(M - m)^2 (M + m)}{m^2 M^2},$$

while from (6.94) we get

$$0 \leq \frac{\text{tr}(PA^{-1})}{\text{tr}(P)} - \frac{\text{tr}(P)}{\text{tr}(PA)} \tag{6.109}$$

$$\leq \frac{\left(M - \frac{\text{tr}(PA)}{\text{tr}(P)} \right) \left(\frac{\text{tr}(PA)}{\text{tr}(P)} - m \right)}{mM} \frac{\text{tr}(P)}{\text{tr}(PA)}$$

$$\leq \frac{1}{4} \frac{(M - m)^2}{mM} \frac{\text{tr}(P)}{\text{tr}(PA)} \leq \frac{1}{4} \frac{(M - m)^2}{m^2 M}$$

for any positive definite operator A with $\text{Sp}(A) \subseteq [m, M]$ and $P \in \mathcal{B}_1(H) \backslash \{0\}$, $P \geq 0$. Since $m > 0$, then $\text{tr}(PA) \geq m\text{tr}(P) > 0$.

From the inequality (6.100), we have

$$0 \leq \frac{\text{tr}(PA^{-1})}{\text{tr}(P)} - \frac{\text{tr}(P)}{\text{tr}(PA)} \tag{6.110}$$

$$\leq \frac{(M - m)^2}{mM(m + M)} \max \left\{ \frac{M - \frac{\text{tr}(PA)}{\text{tr}(P)}}{M - m}, \frac{\frac{\text{tr}(PA)}{\text{tr}(P)} - m}{M - m} \right\}$$

$$\leq \frac{(M - m)^2}{mM(m + M)}$$

for any positive definite operator A with $\text{Sp}(A) \subseteq [m, M]$ and any $P \in \mathcal{B}_1(H) \backslash \{0\}, P \geq 0$.

In order to compare the upper bounds provided by (6.109) and (6.110) consider the difference

$$\Delta(m, M) := \frac{1}{4} \frac{(M - m)^2}{m^2 M} - \frac{(M - m)^2}{mM(m + M)}$$

$$= \frac{(M - m)^2}{mM} \left(\frac{1}{4m} - \frac{1}{m + M} \right) = \frac{(M - m)^2 (M - 3m)}{4m^2 M(m + M)},$$

where $0 < m < M$.

We observe that if $M < 3m$, then the upper bound provided by (6.109) is better than the bound provided by (6.110). The conclusion is the other way around if $M \geq 3m$.

If we consider the convex function $f(t) = -\ln t$ defined on $[m, M] \subset (0, \infty)$, then $\Psi_{-\ln}(\cdot; m, M) : (m, M) \to \mathbb{R}$ is defined by

$$
\begin{aligned}
\Psi_{-\ln}(t; m, M) &= \frac{-\ln M + \ln t}{M - t} - \frac{-\ln t + \ln m}{t - m} \\
&= \frac{(M - m) \ln t - (M - t) \ln m - (t - m) \ln M}{(M - t)(t - m)} \\
&= \ln \left(\frac{t^{M-m}}{m^{M-t} M^{t-m}} \right)^{\frac{1}{(M-t)(t-m)}}.
\end{aligned}
$$

Utilizing the inequality (6.93), we have

$$
0 \le \ln \left(\frac{\mathrm{tr}(PA)}{\mathrm{tr}(P)} \right) - \frac{\mathrm{tr}(P \ln A)}{\mathrm{tr}(P)} \tag{6.111}
$$

$$
\le \frac{1}{M - m} \ln \left(\frac{\left(\frac{\mathrm{tr}(PA)}{\mathrm{tr}(P)} \right)^{M-m}}{m^{M - \frac{\mathrm{tr}(PA)}{\mathrm{tr}(P)}} M^{\frac{\mathrm{tr}(PA)}{\mathrm{tr}(P)} - m}} \right)
$$

$$
\le \frac{\left(M - \frac{\mathrm{tr}(PA)}{\mathrm{tr}(P)} \right) \left(\frac{\mathrm{tr}(PA)}{\mathrm{tr}(P)} - m \right)}{M - m} \sup_{t \in (m, M)} \Psi_{-\ln}(t; m, M)
$$

$$
\le \frac{1}{Mm} \left(M - \frac{\mathrm{tr}(PA)}{\mathrm{tr}(P)} \right) \left(\frac{\mathrm{tr}(PA)}{\mathrm{tr}(P)} - m \right) \le \frac{(M - m)^2}{4mM}
$$

for any positive definite operator A with $\mathrm{Sp}(A) \subseteq [m, M]$ and $P \in \mathcal{B}_1(H) \setminus \{0\}$, $P \ge 0$.

From (6.94), we have

$$
0 \le \ln \left(\frac{\mathrm{tr}(PA)}{\mathrm{tr}(P)} \right) - \frac{\mathrm{tr}(P \ln A)}{\mathrm{tr}(P)} \tag{6.112}
$$

$$
\le \frac{1}{M - m} \ln \left(\frac{\left(\frac{\mathrm{tr}(PA)}{\mathrm{tr}(P)} \right)^{M-m}}{m^{M - \frac{\mathrm{tr}(PA)}{\mathrm{tr}(P)}} M^{\frac{\mathrm{tr}(PA)}{\mathrm{tr}(P)} - m}} \right)
$$

$$
\le \frac{1}{4} \frac{(M - m)}{\left(M - \frac{\mathrm{tr}(PA)}{\mathrm{tr}(P)} \right) \left(\frac{\mathrm{tr}(PA)}{\mathrm{tr}(P)} - m \right)} \ln \left(\frac{\left(\frac{\mathrm{tr}(PA)}{\mathrm{tr}(P)} \right)^{M-m}}{m^{M - \frac{\mathrm{tr}(PA)}{\mathrm{tr}(P)}} M^{\frac{\mathrm{tr}(PA)}{\mathrm{tr}(P)} - m}} \right)
$$

$$
\le \frac{1}{4} (M - m) \sup_{t \in (m, M)} \Psi_{-\ln}(t; m, M)
$$

$$
\le \frac{(M - m)^2}{4mM}
$$

for any positive definite operator A with $\mathrm{Sp}(A) \subseteq [m, M]$ and $P \in \mathcal{B}_1(H) \setminus \{0\}$, $P \ge 0$.

From the inequality (6.100), we get

$$0 \leq \ln\left(\frac{\text{tr}(PA)}{\text{tr}(P)}\right) - \frac{\text{tr}(P\ln A)}{\text{tr}(P)} \tag{6.113}$$

$$\leq \max\left\{\frac{M - \frac{\text{tr}(PA)}{\text{tr}(P)}}{M - m}, \frac{\frac{\text{tr}(PA)}{\text{tr}(P)} - m}{M - m}\right\} \ln\left(\frac{\left(\frac{m+M}{2}\right)^2}{mM}\right)$$

$$\leq \ln\left(\frac{\left(\frac{m+M}{2}\right)^2}{mM}\right)$$

for any positive definite operator A with $\text{Sp}(A) \subseteq [m, M]$ and $P \in B_1(H)\setminus\{0\}$, $P \geq 0$.

We observe that, since $\ln x \leq x - 1$ for any $x > 0$, then

$$\ln\left(\frac{\left(\frac{m+M}{2}\right)^2}{mM}\right) \leq \frac{\left(\frac{m+M}{2}\right)^2}{mM} - 1 = \frac{(M - m)^2}{4mM},$$

which shows that the absolute upper bound for

$$\ln\left(\frac{\text{tr}(PA)}{\text{tr}(P)}\right) - \frac{\text{tr}(P\ln A)}{\text{tr}(P)}$$

provided by the inequality (6.113) is better than the one provided by (6.112).

6.3.5 Reverses of Hölder's Inequality

We have the following result:

Theorem 6.18 ([46]) Assume that $p, q > 1$ with $\frac{1}{p} + \frac{1}{q} = 1$. Let S be a positive operator that commutes with Q, a positive invertible operator and such that there exists the constants $k, K > 0$ with

$$k1_H \leq SQ^{1-q} \leq K1_H. \tag{6.114}$$

If $S^p, Q^q \in B_1(H)$, then we have

$$0 \leq [\text{tr}(S^p)]^{1/p}[\text{tr}(Q^q)]^{1/q} - \text{tr}(SQ) \leq B_p(k, K)\text{tr}(Q^q), \tag{6.115}$$

where

$$B_p(k, K) = \begin{cases} \dfrac{1}{4^{1/p}} p^{1/p}(K - k)^{1/p}(K^{p-1} - k^{p-1})^{1/p}, \\ \\ 2^{1/p}\left[\dfrac{k^p + K^p}{2} - \left(\dfrac{k + K}{2}\right)^p\right]^{1/p}. \end{cases} \tag{6.116}$$

Proof: If we write the inequality

$$0 \le \frac{\text{tr}(PA^p)}{\text{tr}(P)} - \left(\frac{\text{tr}(PA)}{\text{tr}(P)}\right)^p \le \frac{1}{4}p(M-m)(M^{p-1}-m^{p-1}) \qquad (6.117)$$

for the operators $P = Q^q$ and $A = SQ^{1-q}$ then we get

$$0 \le \frac{\text{tr}(Q^q(SQ^{1-q})^p)}{\text{tr}(Q^q)} - \left(\frac{\text{tr}(Q^qSQ^{1-q})}{\text{tr}(Q^q)}\right)^p \qquad (6.118)$$

$$\le \frac{1}{4}p(K-k)(K^{p-1}-k^{p-1}).$$

Observe that, by the properties of trace we have

$$\text{tr}(Q^qSQ^{1-q}) = \text{tr}(SQ^{1-q}Q^q) = \text{tr}(SQ).$$

It is known, see, for instance, [47, p. 356–358], that if A and B are two *commuting bounded self-adjoint operators* on the complex Hilbert space H, then there exists a bounded self-adjoint operator T on H and two bounded functions φ and ψ such that $A = \varphi(T)$ and $B = \psi(T)$. Moreover, if $\{E_\lambda\}$ is the spectral family over the closed interval $[0, 1]$ for the self-adjoint operator T, then $T = \int_{0-}^{1} \lambda dE_\lambda$, where the integral is taken in the Riemann–Stieltjes sense, the functions φ and ψ are summable with respect with $\{E_\lambda\}$ on $[0, 1]$ and

$$A = \varphi(T) = \int_{0-}^{1} \varphi(\lambda)dE_\lambda \quad \text{and} \quad B = \psi(T) = \int_{0-}^{1} \psi(\lambda)dE_\lambda.$$

Now, if A and B are as above with $\text{Sp}(A)$, $\text{Sp}(B) \subseteq J$ an interval of real numbers, then for any continuous functions $f, g : J \to \mathbb{C}$ we have the representations

$$f(A) = \int_{0-}^{1} (f\circ\varphi)(\lambda)dE_\lambda \quad \text{and} \quad g(B) = \int_{0-}^{1} (g\circ\psi)(\lambda)dE_\lambda.$$

If we apply the above property to the commuting self-adjoint operators S and Q, then we have two positive functions φ and ψ such that $S = \varphi(T)$ and $Q = \psi(T)$. Moreover, using the integral representation for functions of self-adjoint operators, we have

$$Q^q(SQ^{1-q})^p = [\psi(T)]^q(\varphi(T)[\psi(T)]^{1-q})^p$$

$$= \int_{0-}^{1} [\psi(\lambda)]^q(\varphi(\lambda)[\psi(\lambda)]^{1-q})^p dE_\lambda$$

$$= \int_{0-}^{1} [\psi(\lambda)]^q[\varphi(\lambda)]^p[\psi(\lambda)]^{(1-q)p} dE_\lambda$$

$$= \int_{0-}^{1} [\varphi(\lambda)]^p[\psi(\lambda)]^{q+p-qp} dE_\lambda = \int_{0-}^{1} [\varphi(\lambda)]^p dE_\lambda = S^p.$$

Therefore, the inequality (6.118) is equivalent to

$$0 \le \frac{\mathrm{tr}(S^p)}{\mathrm{tr}(Q^q)} - \left(\frac{\mathrm{tr}(SQ)}{\mathrm{tr}(Q^q)}\right)^p \le \frac{1}{4}p(K-k)(K^{p-1}-k^{p-1}), \qquad (6.119)$$

which is of interest in itself.

From this inequality, we have

$$\mathrm{tr}(S^p)[\mathrm{tr}(Q^q)]^{p-1} \le (\mathrm{tr}(SQ))^p + \frac{1}{4}p(K-k)(K^{p-1}-k^{p-1})[\mathrm{tr}(Q^q)]^p.$$

Taking the power $1/p \in (0,1)$ and using the property that

$$(\alpha + \beta)^r \le \alpha^r + \beta^r, \quad \text{where } \alpha, \beta \ge 0 \text{ and } r \in (0,1),$$

we get

$$[\mathrm{tr}(S^p)]^{1/p}[\mathrm{tr}(Q^q)]^{(p-1)/p}$$
$$\le \left[(\mathrm{tr}(SQ))^p + \frac{1}{4}p(K-k)(K^{p-1}-k^{p-1})[\mathrm{tr}(Q^q)]^p\right]^{1/p}$$
$$\le \mathrm{tr}(SQ) + \frac{1}{4^{1/p}}p^{1/p}(K-k)^{1/p}(K^{p-1}-k^{p-1})^{1/p}[\mathrm{tr}(Q^q)],$$

that is,

$$[\mathrm{tr}(S^p)]^{1/p}[\mathrm{tr}(Q^q)]^{1/q} - \mathrm{tr}(SQ)$$
$$\le \frac{1}{4^{1/p}}p^{1/p}(K-k)^{1/p}(K^{p-1}-k^{p-1})^{1/p}[\mathrm{tr}(Q^q)]$$

The second part follows from the inequality

$$0 \le \frac{\mathrm{tr}(PA^p)}{\mathrm{tr}(P)} - \left(\frac{\mathrm{tr}(PA)}{\mathrm{tr}(P)}\right)^p \le 2\left[\frac{m^p + M^p}{2} - \left(\frac{m+M}{2}\right)^p\right]$$

and the details are omitted. □

Remark 6.8 We observe that under the previous assumptions, from any upper bound for the difference

$$0 \le \frac{\mathrm{tr}(PA^p)}{\mathrm{tr}(P)} - \left(\frac{\mathrm{tr}(PA)}{\mathrm{tr}(P)}\right)^p,$$

we can deduce in a similar way an upper bound for the Hölder's difference

$$0 \le [\mathrm{tr}(S^p)]^{1/p}[\mathrm{tr}\,(Q^q)]^{1/q} - \mathrm{tr}(SQ).$$

Also, if the commutativity property of the operators S and Q is dropped, then we can prove that

$$0 \le [\mathrm{tr}(Q^q(SQ^{1-q})^p)]^{1/p}[\mathrm{tr}(Q^q)]^{1/q} - \mathrm{tr}(SQ) \le B_p(k,K)\mathrm{tr}(Q^q) \qquad (6.120)$$

with the same $B_p(k,K)$. However, the noncommutative case of the second inequality in (6.115) is an open question for the author.

The following reverse of Schwarz inequality holds:

Corollary 6.10 Let S be a positive operator that commutes with Q, a positive invertible operator and such that there exists the constants $k, K > 0$ with

$$k1_H \le SQ^{-1} \le K1_H. \tag{6.121}$$

If $S^2, Q^2 \in B_1(H)$, then we have

$$0 \le [\text{tr}(S^2)]^{1/2}[\text{tr }(Q^2)]^{1/2} - \text{tr}(SQ) \le \frac{\sqrt{2}}{2}(K - k)\text{tr}(Q^2). \tag{6.122}$$

Remark 6.9 If we take $p = q = 2$ in (6.120) and drop the commutativity assumption, then we get

$$0 \le [\text{tr}(QSQ^{-1}S)]^{1/2}[\text{tr}(Q^2)]^{1/2} - \text{tr}(SQ) \le \frac{\sqrt{2}}{2}(K - k)\text{tr}(Q^2),$$

provided that (6.121) holds true.

Also, if we use the inequality (6.105), then we have

$$0 \le \text{tr}(QSQ^{-1}S)\text{tr}(Q^2) - [\text{tr}(SQ)]^2 \tag{6.123}$$

$$\le (K\text{tr}(Q^2) - \text{tr}(SQ))\,(\text{tr}(SQ) - k\,\text{tr}(Q^2)) \le \frac{1}{4}(K - k)^2[\text{tr }(Q^2)]^2$$

provided that (6.115) holds true.

6.4 Slater's Type Trace Inequalities

6.4.1 Slater's Type Inequalities

Suppose that I is an interval of real numbers with interior $\mathring{I}$ and $f : I \to \mathbb{R}$ is a convex function on I. Then f is continuous on $\mathring{I}$ and has finite left and right derivatives at each point of $\mathring{I}$. Moreover, if $x, y \in \mathring{I}$ and $x < y$, then $f'_-(x) \le f'_+(x) \le f'_-(y) \le f'_+(y)$, which shows that both f'_- and f'_+ are nondecreasing function on $\mathring{I}$. It is also known that a convex function must be differentiable except for at most countably many points.

For a convex function $f : I \to \mathbb{R}$, the subdifferential of f denoted by ∂f is the set of all functions $\varphi : I \to [-\infty, \infty]$ such that $\varphi(\mathring{I}) \subset \mathbb{R}$ and

$$f(x) \ge f(a) + (x - a)\varphi(a) \quad \text{for any } x, a \in I.$$

It is also well known that if f is convex on I, then ∂f is nonempty, $f'_-, f'_+ \in \partial f$ and if $\varphi \in \partial f$, then

$$f'_-(x) \le \varphi(x) \le f'_+(x) \quad \text{for any } x \in \mathring{I}.$$

In particular, φ is a nondecreasing function. If f is differentiable and convex on $\mathring{I}$, then $\partial f = \{f'\}$.

The following result is well known in the literature as *Slater inequality*:

Theorem 6.19 ([48]) If $f : I \to \mathbb{R}$ is a nonincreasing (nondecreasing) convex function, $x_i \in I, p_i \geq 0$ with $P_n := \sum_{i=1}^{n} p_i > 0$ and $\sum_{i=1}^{n} p_i \varphi(x_i) \neq 0$, where $\varphi \in \partial f$, then

$$\frac{1}{P_n} \sum_{i=1}^{n} p_i f(x_i) \leq f \left(\frac{\sum_{i=1}^{n} p_i x_i \varphi(x_i)}{\sum_{i=1}^{n} p_i \varphi(x_i)} \right). \tag{6.124}$$

As pointed out in [10, p. 208], the monotonicity assumption for the derivative φ can be replaced with the condition

$$\frac{\sum_{i=1}^{n} p_i x_i \varphi(x_i)}{\sum_{i=1}^{n} p_i \varphi(x_i)} \in I, \tag{6.125}$$

which is more general and can hold for suitable points in I and for not necessarily monotonic functions.

The following result holds:

Theorem 6.20 ([49]) Let I be an interval and $f : I \to \mathbb{R}$ be a convex and differentiable function on $\mathring{I}$ (the interior of I) whose derivative f' is continuous on $\mathring{I}$. If A is a self-adjoint operator on the Hilbert space H with $\mathrm{Sp}(A) \subseteq [m, M] \subset \mathring{I}$ and $f'(A)$ is a positive invertible operator on H, then

$$0 \leq f \left(\frac{\mathrm{tr}[PAf'(A)]}{\mathrm{tr}[Pf'(A)]} \right) - \frac{\mathrm{tr}[Pf(A)]}{\mathrm{tr}(P)} \tag{6.126}$$

$$\leq f' \left(\frac{\mathrm{tr}[PAf'(A)]}{\mathrm{tr}[Pf'(A)]} \right) \left(\frac{\mathrm{tr}[PAf'(A)]}{\mathrm{tr}[Pf'(A)]} - \frac{\mathrm{tr}(PA)}{\mathrm{tr}(P)} \right)$$

for any $P \in \mathcal{B}_1^+(H) \backslash \{0\}$.

Proof: Since f is convex and differentiable on $\mathring{I}$, then we have

$$f'(s)(t - s) \leq f(t) - f(s) \leq f'(t)(t - s) \tag{6.127}$$

for any $t, s \in [m, M]$.

Now, if we fix $t \in [m, M]$ and use the continuous functional calculus for the operator A, then we have

$$tf'(A) - Af'(A) \leq f(t) \cdot 1_H - f(A) \leq f'(t)t \cdot 1_H - f'(t)A \tag{6.128}$$

for any $t \in [m, M]$.

If we apply the property of the trace to the inequality (6.128), then we have

$$t \, \mathrm{tr}[Pf'(A)] - \mathrm{tr}[PAf'(A)] \leq f(t) \mathrm{tr}(P) - \mathrm{tr}[Pf(A)] \tag{6.129}$$

$$\leq f'(t)t \, \mathrm{tr}(P) - f'(t) \mathrm{tr}(PA)$$

for any $P \in \mathcal{B}_1^+(H) \backslash \{0\}$.

Now, since A is self-adjoint with $m1_H \leq A \leq M1_H$ and $f'(A)$ is positive, then

$$mf'(A) \leq Af'(A) \leq Mf'(A).$$

If we apply again the property of the trace, then we get

$$m\,\mathrm{tr}[Pf'(A)] \leq \mathrm{tr}[PAf'(A)] \leq M\,\mathrm{tr}[Pf'(A)],$$

which shows that

$$t_0 := \frac{\mathrm{tr}[PAf'(A)]}{\mathrm{tr}[Pf'(A)]} \in [m, M].$$

Observe that since $f'(A)$ is a positive invertible operator on H, then $\mathrm{tr}[Pf'(A)] > 0$ for any $P \in \mathcal{B}_1^+(H)\backslash\{0\}$.

Finally, if we put $t = t_0$ in (6.129), then we get

$$\frac{\mathrm{tr}[PAf'(A)]}{\mathrm{tr}\,[Pf'(A)]}\mathrm{tr}[Pf'(A)] - \mathrm{tr}[PAf'(A)] \tag{6.130}$$

$$\leq f\left(\frac{\mathrm{tr}[PAf'(A)]}{\mathrm{tr}[Pf'(A)]}\right)\mathrm{tr}\,(P) - \mathrm{tr}[Pf(A)]$$

$$\leq f'\left(\frac{\mathrm{tr}[PAf'(A)]}{\mathrm{tr}[Pf'(A)]}\right)\frac{\mathrm{tr}[PAf'(A)]}{\mathrm{tr}[Pf'(A)]}\mathrm{tr}(P)$$

$$- f'\left(\frac{\mathrm{tr}[PAf'(A)]}{\mathrm{tr}[Pf'(A)]}\right)\mathrm{tr}(PA),$$

which is equivalent to the desired result (6.126). $\qquad\square$

Remark 6.10 It is important to observe that, the condition that $f'(A)$ is a positive invertible operator on H can be replaced with the more general assumption that

$$\frac{\mathrm{tr}[PAf'(A)]}{\mathrm{tr}[Pf'(A)]} \in \mathring{I} \quad \text{and} \quad \mathrm{tr}[Pf'(A)] \neq 0 \tag{6.131}$$

for any $P \in \mathcal{B}_1^+(H)\backslash\{0\}$, which may be easily verified for particular convex functions f in various examples as follows.

Also, as pointed out by the referee, if $\langle f'(A)x, x\rangle > 0$ for any $x \in H, x \neq 0$, then $\mathrm{tr}[Pf'(A)] > 0$ for any $P \in \mathcal{B}_1^+(H)\backslash\{0\}$ and the inequality (6.126) is valid as well.

Remark 6.11 Now, if the function is concave on $\mathring{I}$ and the condition (6.131) holds, then we have the inequalities

$$0 \leq \frac{\mathrm{tr}[Pf(A)]}{\mathrm{tr}(P)} - f\left(\frac{\mathrm{tr}[PAf'(A)]}{\mathrm{tr}[Pf'(A)]}\right) \tag{6.132}$$

$$\leq f'\left(\frac{\mathrm{tr}[PAf'(A)]}{\mathrm{tr}[Pf'(A)]}\right)\left(\frac{\mathrm{tr}(PA)}{\mathrm{tr}(P)} - \frac{\mathrm{tr}[PAf'(A)]}{\mathrm{tr}[Pf'(A)]}\right)$$

for any $P \in \mathcal{B}_1^+(H)\backslash\{0\}$.

Utilizing the inequality (6.132) for the concave function $f : (0, \infty) \to \mathbb{R}$, $f(t) = \ln t$, then we can state that

$$0 \le \frac{\text{tr}(P \ln A)}{\text{tr}(P)} - \ln\left(\frac{\text{tr}(P)}{\text{tr}(PA^{-1})}\right) \le \frac{\text{tr}(PA^{-1})}{\text{tr}(P)}\frac{\text{tr}(PA)}{\text{tr}(P)} - 1 \qquad (6.133)$$

for any positive invertible operator A and P with $P \in \mathcal{B}_1^+(H) \setminus \{0\}$.

Utilizing the inequality (6.126) for the convex function $f : (0, \infty) \to \mathbb{R}$, $f(t) = t^{-1}$, then we can state that

$$0 \le \frac{\text{tr}(PA^{-2})}{\text{tr}(PA^{-1})} - \frac{\text{tr}(PA^{-1})}{\text{tr}(P)} \le \frac{\text{tr}(PA)}{\text{tr}(P)}\frac{\text{tr}(PA^{-2})}{\text{tr}(PA^{-1})} - \frac{\text{tr}(PA^{-1})}{\text{tr}(PA^{-2})} \qquad (6.134)$$

for any positive invertible operator A and P with $P \in \mathcal{B}_1^+(H) \setminus \{0\}$.

If we take $B = A^{-1}$ in (6.134), then we get the equivalent inequality

$$0 \le \frac{\text{tr}(PB^2)}{\text{tr}(PB)} - \frac{\text{tr}(PB)}{\text{tr}(P)} \le \frac{\text{tr}(PB^2)}{\text{tr}(PB)}\frac{\text{tr}(PB^{-1})}{\text{tr}(P)} - \frac{\text{tr}(PB)}{\text{tr}(PB^2)} \qquad (6.135)$$

for any positive invertible operator B and P with $P \in \mathcal{B}_1(H) \setminus \{0\}$.

If we write the inequality (6.126) for the convex function $f(t) = \exp(\alpha t)$ with $\alpha \in \mathbb{R} \setminus \{0\}$, then we get

$$0 \le \exp\left(\alpha \frac{\text{tr}[PA \exp(\alpha A)]}{\text{tr}[P \exp(\alpha A)]}\right) - \frac{\text{tr}[P \exp(\alpha A)]}{\text{tr}(P)} \qquad (6.136)$$

$$\le \alpha \exp\left(\alpha \frac{\text{tr}[PA \exp(\alpha A)]}{\text{tr}[P \exp(\alpha A)]}\right)\left(\frac{\text{tr}[PA \exp(\alpha A)]}{\text{tr}[P \exp(\alpha A)]} - \frac{\text{tr}(PA)}{\text{tr}(P)}\right)$$

for any self-adjoint operator A and $P \in \mathcal{B}_1^+(H) \setminus \{0\}$.

6.4.2 Further Reverses

We use the following Grüss' type inequalities [50]:

Lemma 6.2 ([50]) Let S be a self-adjoint operator with $m1_H \le S \le M1_H$ and $f : [m, M] \to \mathbb{C}$ a continuous function of bounded variation on $[m, M]$. For any $C \in \mathcal{B}(H)$ and $P \in \mathcal{B}_1^+(H) \setminus \{0\}$ we have the inequality

$$\left|\frac{\text{tr}(Pf(S)C)}{\text{tr}(P)} - \frac{\text{tr}(Pf(S))}{\text{tr}(P)}\frac{\text{tr}(PC)}{\text{tr}(P)}\right| \qquad (6.137)$$

$$\le \frac{1}{2}\bigvee_m^M(f)\frac{1}{\text{tr}(P)}\text{tr}\left(\left|\left(C - \frac{\text{tr}(PC)}{\text{tr}(P)}1_H\right)P\right|\right)$$

$$\le \frac{1}{2}\bigvee_m^M(f)\left[\frac{\text{tr}(P|C|^2)}{\text{tr}(P)} - \left|\frac{\text{tr}(PC)}{\text{tr}(P)}\right|^2\right]^{1/2},$$

where $\bigvee_m^M(f)$ is the total variation of f on the interval.

If the function $f : [m, M] \to \mathbb{C}$ is Lipschitzian with the constant $L > 0$ on $[m, M]$, that is,

$$|f(t) - f(s)| \leq L|t - s|$$

for any $t, \ s \in [m, M]$, then

$$\left| \frac{\text{tr}(Pf(S)C)}{\text{tr}(P)} - \frac{\text{tr}(Pf(S))}{\text{tr}(P)} \frac{\text{tr}(PC)}{\text{tr}(P)} \right| \qquad (6.138)$$

$$\leq L \left\| S - \frac{\text{tr}(PS)}{\text{tr}(P)} 1_H \right\| \frac{1}{\text{tr}(P)} \text{tr} \left(\left| \left(C - \frac{\text{tr}(PC)}{\text{tr}(P)} 1_H \right) P \right| \right)$$

$$\leq L \left\| S - \frac{\text{tr}(PS)}{\text{tr}(P)} 1_H \right\| \left[\frac{\text{tr}(P|C|^2)}{\text{tr}(P)} - \left| \frac{\text{tr}(PC)}{\text{tr}(P)} \right|^2 \right]^{1/2}$$

for any $C \in \mathcal{B}(H)$ and $P \in \mathcal{B}_1^+(H) \setminus \{0\}$.

Proof: For the sake of completeness we give here a simple proof.

We observe that, for any $\lambda \in \mathbb{C}$ we have

$$\frac{1}{\text{tr}(P)} \text{tr} \left[P(A - \lambda 1_H) \left(C - \frac{\text{tr}(PC)}{\text{tr}(P)} 1_H \right) \right] \qquad (6.139)$$

$$= \frac{1}{\text{tr}(P)} \text{tr} \left[PA \left(C - \frac{\text{tr}(PC)}{\text{tr}(P)} 1_H \right) \right]$$

$$- \frac{\lambda}{\text{tr}(P)} \text{tr} \left[P \left(C - \frac{\text{tr}(PC)}{\text{tr}(P)} 1_H \right) \right]$$

$$= \frac{\text{tr}(PAC)}{\text{tr}(P)} - \frac{\text{tr}(PA)}{\text{tr}(P)} \frac{\text{tr}(PC)}{\text{tr}(P)}.$$

Taking the modulus in (6.139) and utilizing the properties of the trace, we have

$$\left| \frac{\text{tr}(PAC)}{\text{tr}(P)} - \frac{\text{tr}(PA)}{\text{tr}(P)} \frac{\text{tr}(PC)}{\text{tr}(P)} \right| \qquad (6.140)$$

$$= \frac{1}{\text{tr}(P)} \left| \text{tr} \left[P(A - \lambda 1_H) \left(C - \frac{\text{tr}(PC)}{\text{tr}(P)} 1_H \right) \right] \right|$$

$$= \frac{1}{\text{tr}(P)} \left| \text{tr} \left[(A - \lambda 1_H) \left(C - \frac{\text{tr}(PC)}{\text{tr}(P)} 1_H \right) P \right] \right|$$

$$\leq \|A - \lambda 1_H\| \frac{1}{\text{tr}(P)} \text{tr} \left(\left| \left(C - \frac{\text{tr}(PC)}{\text{tr}(P)} 1_H \right) P \right| \right)$$

for any $\lambda \in \mathbb{C}$.

From the inequality (6.140), we have

$$\left| \frac{\mathrm{tr}(Pf(S)C)}{\mathrm{tr}(P)} - \frac{\mathrm{tr}(Pf(S))}{\mathrm{tr}(P)} \frac{\mathrm{tr}(PC)}{\mathrm{tr}\,(P)} \right| \tag{6.141}$$

$$\leq \| f(S) - \lambda 1_H \| \frac{1}{\mathrm{tr}(P)} \mathrm{tr} \left(\left| \left(C - \frac{\mathrm{tr}(PC)}{\mathrm{tr}(P)} 1_H \right) P \right| \right)$$

for any $\lambda \in \mathbb{C}$.

From (6.141), we get

$$\left| \frac{\mathrm{tr}(Pf(S)C)}{\mathrm{tr}(P)} - \frac{\mathrm{tr}(Pf(S))}{\mathrm{tr}(P)} \frac{\mathrm{tr}(PC)}{\mathrm{tr}\,(P)} \right| \tag{6.142}$$

$$\leq \left\| f(S) - \frac{f(m) + f(M)}{2} 1_H \right\| \frac{1}{\mathrm{tr}(P)} \mathrm{tr} \left(\left| \left(C - \frac{\mathrm{tr}(PC)}{\mathrm{tr}(P)} 1_H \right) P \right| \right).$$

Since f is of bounded variation on $[m, M]$, then we have

$$\left| f(t) - \frac{f(m) + f(M)}{2} \right| = \left| \frac{f(t) - f(m) + f(t) - f(M)}{2} \right| \tag{6.143}$$

$$\leq \frac{1}{2} [|f(t) - f(m)| + |f(M) - f(t)|] \leq \frac{1}{2} \bigvee_{m}^{M} (f)$$

for any $t \in [m, M]$.

From (6.143), we get in the order $\mathcal{B}(H)$ that

$$\left| f(S) - \frac{f(m) + f(M)}{2} 1_H \right| \leq \frac{1}{2} \bigvee_{m}^{M} (f) 1_H,$$

which implies that

$$\left\| f(S) - \frac{f(m) + f(M)}{2} 1_H \right\| \leq \frac{1}{2} \bigvee_{m}^{M} (f). \tag{6.144}$$

Making use of (6.143) and (6.144), we get the first inequality in (6.137).

The second part is obvious by the Schwarz inequality for traces

$$\frac{\mathrm{tr} \left(\left| \left(C - \frac{\mathrm{tr}(PC)}{\mathrm{tr}(P)} 1_H \right) P \right| \right)}{\mathrm{tr}(P)} \leq \left(\frac{\mathrm{tr} \left(\left| \left(C - \frac{\mathrm{tr}(PC)}{\mathrm{tr}(P)} 1_H \right) P^{1/2} \right|^2 \right)}{\mathrm{tr}(P)} \right)^{1/2},$$

and by noticing that

$$\frac{\mathrm{tr} \left(\left| \left(C - \frac{\mathrm{tr}(PC)}{\mathrm{tr}(P)} 1_H \right) P^{1/2} \right|^2 \right)}{\mathrm{tr}(P)} = \frac{\mathrm{tr}(P|C|^2)}{\mathrm{tr}(P)} - \left| \frac{\mathrm{tr}(PC)}{\mathrm{tr}(P)} \right|^2 \tag{6.145}$$

for any $C \in \mathcal{B}(H)$ and $P \in \mathcal{B}_1^+(H) \backslash \{0\}$.

From (6.141), we also have

$$\left| \frac{\operatorname{tr}(Pf(S)C)}{\operatorname{tr}(P)} - \frac{\operatorname{tr}(Pf(S))}{\operatorname{tr}(P)} \frac{\operatorname{tr}(PC)}{\operatorname{tr}(P)} \right| \tag{6.146}$$

$$\leq \left\| f(S) - f\left(\frac{\operatorname{tr}(SP)}{\operatorname{tr}(P)} \right) 1_H \right\| \frac{1}{\operatorname{tr}(P)} \operatorname{tr}\left(\left| \left(C - \frac{\operatorname{tr}(PC)}{\operatorname{tr}(P)} 1_H \right) P \right| \right)$$

any $C \in \mathcal{B}(H)$ and $P \in \mathcal{B}_1^+(H) \setminus \{0\}$.

Since

$$|f(t) - f(s)| \leq L|t - s|$$

for any $t, s \in [m, M]$, then we have in the order $\mathcal{B}(H)$ that

$$|f(S) - f(s)1_H| \leq L|S - s1_H|$$

for any $s \in [m, M]$. In particular, we have

$$\left| f(S) - f\left(\frac{\operatorname{tr}(SP)}{\operatorname{tr}(P)} \right) 1_H \right| \leq L \left| S - \frac{\operatorname{tr}(SP)}{\operatorname{tr}(P)} 1_H \right|,$$

which implies that

$$\left\| f(S) - f\left(\frac{\operatorname{tr}(SP)}{\operatorname{tr}(P)} \right) 1_H \right\| \leq L \left\| S - \frac{\operatorname{tr}(SP)}{\operatorname{tr}(P)} 1_H \right\|$$

and by (6.146) we get the first inequality in (6.138).

The second part is obvious. □

We also have the following reverse of Schwarz inequality [50]:

Lemma 6.3 ([50]) If C is a self-adjoint operator with $k1_H \leq C \leq K1_H$ for some real numbers $k < K$, then

$$0 \leq \frac{\operatorname{tr}(PC^2)}{\operatorname{tr}(P)} - \left(\frac{\operatorname{tr}(PC)}{\operatorname{tr}(P)} \right)^2 \tag{6.147}$$

$$\leq \frac{1}{2}(K - k) \frac{1}{\operatorname{tr}(P)} \operatorname{tr}\left(\left| \left(C - \frac{\operatorname{tr}(PC)}{\operatorname{tr}(P)} 1_H \right) P \right| \right)$$

$$\leq \frac{1}{2}(K - k) \left[\frac{\operatorname{tr}(PC^2)}{\operatorname{tr}(P)} - \left(\frac{\operatorname{tr}(PC)}{\operatorname{tr}(P)} \right)^2 \right]^{1/2} \leq \frac{1}{4}(K - k)^2$$

for any $P \in \mathcal{B}_1^+(H) \setminus \{0\}$.

Proof: If we take in (6.137) $f(t) = t$ and $S = C$ we get

$$\left| \frac{\operatorname{tr}(PC^2)}{\operatorname{tr}(P)} - \left(\frac{\operatorname{tr}(PC)}{\operatorname{tr}(P)} \right)^2 \right| \tag{6.148}$$

$$\leq \frac{1}{2}(K-k)\frac{1}{\mathrm{tr}(P)}\mathrm{tr}\left(\left|\left(C-\frac{\mathrm{tr}(PC)}{\mathrm{tr}(P)}1_H\right)P\right|\right)$$

$$\leq \frac{1}{2}(K-k)\left[\frac{\mathrm{tr}(PC^2)}{\mathrm{tr}(P)}-\left(\frac{\mathrm{tr}(PC)}{\mathrm{tr}(P)}\right)^2\right]^{1/2}.$$

Since by (6.145), we have

$$\frac{\mathrm{tr}(PC^2)}{\mathrm{tr}(P)}-\left(\frac{\mathrm{tr}(PC)}{\mathrm{tr}(P)}\right)^2 \geq 0,$$

then by (6.148) we get

$$0 \leq \frac{\mathrm{tr}(PC^2)}{\mathrm{tr}(P)}-\left(\frac{\mathrm{tr}(PC)}{\mathrm{tr}(P)}\right)^2 \tag{6.149}$$

$$\leq \frac{1}{2}(K-k)\frac{1}{\mathrm{tr}(P)}\mathrm{tr}\left(\left|\left(C-\frac{\mathrm{tr}(PC)}{\mathrm{tr}(P)}1_H\right)P\right|\right)$$

$$\leq \frac{1}{2}(K-k)\left[\frac{\mathrm{tr}(PC^2)}{\mathrm{tr}(P)}-\left(\frac{\mathrm{tr}(PC)}{\mathrm{tr}(P)}\right)^2\right]^{1/2}.$$

Utilizing the inequality between the first and last term in (6.149), we also have

$$\left[\frac{\mathrm{tr}(PC^2)}{\mathrm{tr}(P)}-\left(\frac{\mathrm{tr}(PC)}{\mathrm{tr}(P)}\right)^2\right]^{1/2} \leq \frac{1}{2}(K-k),$$

which proves the last part of (6.147). □

Theorem 6.21 ([49]) Let I be an interval and $f:I \rightarrow \mathbb{R}$ be a convex and differentiable function on $\mathring{I}$ whose derivative f' is continuous on $\mathring{I}$. If A is a self-adjoint operator on the Hilbert space H with $\mathrm{Sp}(A) \subseteq [m,M] \subset \mathring{I}$ and $f'(A)$ is a positive invertible operator on H, or

$$\frac{\mathrm{tr}[PAf'(A)]}{\mathrm{tr}[Pf'(A)]} \in \mathring{I}, \quad \mathrm{tr}[Pf'(A)] \neq 0$$

for any $P \in \mathcal{B}_1^+(H)\setminus\{0\}$, then

$$0 \leq f\left(\frac{\mathrm{tr}[PAf'(A)]}{\mathrm{tr}[Pf'(A)]}\right)-\frac{\mathrm{tr}[Pf(A)]}{\mathrm{tr}(P)} \tag{6.150}$$

$$\leq \frac{\mathrm{tr}(P)}{\mathrm{tr}[Pf'(A)]}f'\left(\frac{\mathrm{tr}[PAf'(A)]}{\mathrm{tr}[Pf'(A)]}\right)L(P,A,f'(A))$$

for any $P \in \mathcal{B}_1^+(H)\setminus\{0\}$, where

$$L(P,A,f'(A)) := \frac{\mathrm{tr}[PAf'(A)]}{\mathrm{tr}(P)}-\frac{\mathrm{tr}(PA)}{\mathrm{tr}(P)}\frac{\mathrm{tr}\,[Pf'(A)]}{\mathrm{tr}(P)}$$

$$\leq \begin{cases} \frac{1}{2}(f'(M) - f'(m))\dfrac{1}{\operatorname{tr}(P)}\operatorname{tr}\left(\left|\left(A - \dfrac{\operatorname{tr}(PA)}{\operatorname{tr}(P)}1_H\right)P\right|\right) \\[4mm] \frac{1}{2}(M - m)\dfrac{1}{\operatorname{tr}(P)}\operatorname{tr}\left(\left|\left(f'(A) - \dfrac{\operatorname{tr}(Pf'(A))}{\operatorname{tr}(P)}1_H\right)P\right|\right) \end{cases}$$

$$\leq \begin{cases} \frac{1}{2}(f'(M) - f'(m))\left[\dfrac{\operatorname{tr}(PA^2)}{\operatorname{tr}(P)} - \left(\dfrac{\operatorname{tr}(PA)}{\operatorname{tr}(P)}\right)^2\right]^{1/2} \\[5mm] \frac{1}{2}(M - m)\left[\dfrac{\operatorname{tr}(P[f'(A)]^2)}{\operatorname{tr}(P)} - \left(\dfrac{\operatorname{tr}(Pf'(A))}{\operatorname{tr}(P)}\right)^2\right]^{1/2} \end{cases}$$

$$\leq \frac{1}{4}(f'(M) - f'(m))(M - m).$$

Proof: Utilizing Lemmas 6.2 and 6.3, we have

$$0 \leq \frac{\operatorname{tr}(Pf'(A)A)}{\operatorname{tr}(P)} - \frac{\operatorname{tr}(Pf'(A))}{\operatorname{tr}(P)}\frac{\operatorname{tr}(PA)}{\operatorname{tr}(P)} \tag{6.151}$$

$$\leq \frac{1}{2}(f'(M) - f'(m))\frac{1}{\operatorname{tr}(P)}\operatorname{tr}\left(\left|\left(A - \frac{\operatorname{tr}(PA)}{\operatorname{tr}(P)}1_H\right)P\right|\right)$$

$$\leq \frac{1}{2}(f'(M) - f'(m))\left[\frac{\operatorname{tr}(PA^2)}{\operatorname{tr}(P)} - \left(\frac{\operatorname{tr}(PA)}{\operatorname{tr}(P)}\right)^2\right]^{1/2}$$

$$\leq \frac{1}{4}(f'(M) - f'(m))(M - m)$$

and

$$0 \leq \frac{\operatorname{tr}(Pf'(A)A)}{\operatorname{tr}(P)} - \frac{\operatorname{tr}(Pf'(A))}{\operatorname{tr}(P)}\frac{\operatorname{tr}(PA)}{\operatorname{tr}(P)} \tag{6.152}$$

$$\leq \frac{1}{2}(M - m)\frac{1}{\operatorname{tr}(P)}\operatorname{tr}\left(\left|\left(f'(A) - \frac{\operatorname{tr}(Pf'(A))}{\operatorname{tr}(P)}1_H\right)P\right|\right)$$

$$\leq \frac{1}{2}(M - m)\left[\frac{\operatorname{tr}(P[f'(A)]^2)}{\operatorname{tr}(P)} - \left(\frac{\operatorname{tr}(Pf'(A))}{\operatorname{tr}(P)}\right)^2\right]^{1/2}$$

$$\leq \frac{1}{4}(f'(M) - f'(m))(M - m)$$

for any $P \in \mathcal{B}_1^+(H)\setminus\{0\}$.

The positivity of

$$\frac{\operatorname{tr}(Pf'(A)A)}{\operatorname{tr}(P)} - \frac{\operatorname{tr}(Pf'(A))}{\operatorname{tr}(P)}\frac{\operatorname{tr}(PA)}{\operatorname{tr}(P)}$$

follows by Čebyšev's trace inequality for synchronous functions of self-adjoint operators, see [51]. □

The case of convex and monotonic functions is as follows:

Corollary 6.11 Let I be an interval and $f : I \to \mathbb{R}$ be a convex and differentiable function on $\mathring{I}$ whose derivative f' is continuous on $\mathring{I}$. If A is a self-adjoint operator on the Hilbert space H with $\mathrm{Sp}(A) \subseteq [m, M] \subset \mathring{I}$ and $f'(m) > 0$, then

$$0 \leq f\left(\frac{\mathrm{tr}[PAf'(A)]}{\mathrm{tr}[Pf'(A)]}\right) - \frac{\mathrm{tr}[Pf(A)]}{\mathrm{tr}(P)} \leq \frac{f'(M)}{f'(m)} L(P, A, f'(A)) \tag{6.153}$$

for any $P \in \mathcal{B}_1^+(H) \backslash \{0\}$.

The proof follows by (6.150) observing that

$$0 \leq \frac{\mathrm{tr}(P)}{\mathrm{tr}[Pf'(A)]} f'\left(\frac{\mathrm{tr}[PAf'(A)]}{\mathrm{tr}[Pf'(A)]}\right) \leq \frac{f'(M)}{f'(m)}$$

for any $P \in \mathcal{B}_1^+(H) \backslash \{0\}$.

If we consider the monotonic nondecreasing convex function $f(t) = t^p$ with $p \geq 1$ and $t \geq 0$, then by (6.153) we have the sequence of inequalities

$$0 \leq \left(\frac{\mathrm{tr}(PA^p)}{\mathrm{tr}(PA^{p-1})}\right)^p - \frac{\mathrm{tr}(PA^p)}{\mathrm{tr}(P)} \tag{6.154}$$

$$\leq p\left(\frac{M}{m}\right)^{p-1}\left(\frac{\mathrm{tr}(PA^p)}{\mathrm{tr}(P)} - \frac{\mathrm{tr}(PA)}{\mathrm{tr}(P)}\frac{\mathrm{tr}(PA^{p-1})}{\mathrm{tr}(P)}\right)$$

$$\leq \frac{1}{2}p^2\left(\frac{M}{m}\right)^{p-1}$$

$$\times \begin{cases} (M^{p-1} - m^{p-1})\frac{1}{\mathrm{tr}(P)}\mathrm{tr}\left(\left|\left(A - \frac{\mathrm{tr}(PA)}{\mathrm{tr}(P)}1_H\right)P\right|\right) \\ (M - m)\frac{1}{\mathrm{tr}(P)}\mathrm{tr}\left(\left|\left(A^{p-1} - \frac{\mathrm{tr}(PA^{p-1})}{\mathrm{tr}(P)}1_H\right)P\right|\right) \end{cases}$$

$$\leq \frac{1}{2}p^2\left(\frac{M}{m}\right)^{p-1}$$

$$\times \begin{cases} (M^{p-1} - m^{p-1})\left[\frac{\mathrm{tr}(PA^2)}{\mathrm{tr}(P)} - \left(\frac{\mathrm{tr}(PA)}{\mathrm{tr}(P)}\right)^2\right]^{1/2} \\ (M - m)\left[\frac{\mathrm{tr}(PA^{2(p-1)})}{\mathrm{tr}(P)} - \left(\frac{\mathrm{tr}(PA^{p-1})}{\mathrm{tr}(P)}\right)^2\right]^{1/2} \end{cases}$$

$$\leq \frac{1}{4}p^2\left(\frac{M}{m}\right)^{p-1}(M^{p-1} - m^{p-1})(M - m)$$

for any $P \in \mathcal{B}_1^+(H) \backslash \{0\}$ and A with $\mathrm{Sp}(A) \subseteq [m, M] \subset (0, \infty)$.

Theorem 6.22 **([49])** Let I be an interval and $f : I \to \mathbb{R}$ be a convex and twice differentiable function on $\mathring{I}$ whose second derivative f'' is bounded on $\mathring{I}$, that is, there is a positive constant K such that $0 \leq f''(t) \leq K$ for any $t \in \mathring{I}$. If A is a self-adjoint operator on the Hilbert space H with $\mathrm{Sp}(A) \subseteq [m, M] \subset \mathring{I}$ and $f'(A)$ is a positive invertible operator on H, or

$$\frac{\mathrm{tr}[PAf'(A)]}{\mathrm{tr}[Pf'(A)]} \in \mathring{I}, \quad \mathrm{tr}[Pf'(A)] \neq 0$$

for any $P \in \mathcal{B}_1^+(H) \setminus \{0\}$, then

$$0 \leq f\left(\frac{\mathrm{tr}[PAf'(A)]}{\mathrm{tr}[Pf'(A)]}\right) - \frac{\mathrm{tr}[Pf(A)]}{\mathrm{tr}(P)} \tag{6.155}$$

$$\leq K \left\| A - \frac{\mathrm{tr}(PA)}{\mathrm{tr}(P)} 1_H \right\| \frac{1}{\mathrm{tr}(P)} \mathrm{tr}\left(\left| \left(A - \frac{\mathrm{tr}(PA)}{\mathrm{tr}(P)} 1_H \right) P \right| \right)$$

$$\times \frac{\mathrm{tr}(P)}{\mathrm{tr}[Pf'(A)]} f'\left(\frac{\mathrm{tr}[PAf'(A)]}{\mathrm{tr}[Pf'(A)]}\right)$$

$$\leq K \left\| A - \frac{\mathrm{tr}(PA)}{\mathrm{tr}(P)} 1_H \right\| \left[\frac{\mathrm{tr}(PA^2)}{\mathrm{tr}(P)} - \left(\frac{\mathrm{tr}(PA)}{\mathrm{tr}(P)}\right)^2 \right]^{1/2}$$

$$\times \frac{\mathrm{tr}(P)}{\mathrm{tr}[Pf'(A)]} f'\left(\frac{\mathrm{tr}[PAf'(A)]}{\mathrm{tr}[Pf'(A)]}\right)$$

$$\leq \frac{1}{2}(M - m)K \left\| A - \frac{\mathrm{tr}(PA)}{\mathrm{tr}(P)} 1_H \right\| \frac{\mathrm{tr}(P)}{\mathrm{tr}[Pf'(A)]} f'\left(\frac{\mathrm{tr}[PAf'(A)]}{\mathrm{tr}[Pf'(A)]}\right)$$

for any $P \in \mathcal{B}_1^+(H) \setminus \{0\}$.

Proof: From (6.150) we have

$$0 \leq f\left(\frac{\mathrm{tr}[PAf'(A)]}{\mathrm{tr}[Pf'(A)]}\right) - \frac{\mathrm{tr}[Pf(A)]}{\mathrm{tr}(P)} \tag{6.156}$$

$$\leq \frac{\mathrm{tr}(P)}{\mathrm{tr}[Pf'(A)]} f'\left(\frac{\mathrm{tr}[PAf'(A)]}{\mathrm{tr}[Pf'(A)]}\right) L(P, A, f'(A))$$

for any $P \in \mathcal{B}_1^+(H) \setminus \{0\}$.
From (6.138) we also have

$$(0 \leq) L(P, A, f'(A)) \tag{6.157}$$

$$\leq K \left\| A - \frac{\mathrm{tr}(PA)}{\mathrm{tr}(P)} 1_H \right\| \frac{1}{\mathrm{tr}(P)} \mathrm{tr}\left(\left| \left(A - \frac{\mathrm{tr}(PA)}{\mathrm{tr}(P)} 1_H \right) P \right| \right)$$

$$\leq K \left\| A - \frac{\mathrm{tr}(PA)}{\mathrm{tr}(P)} 1_H \right\| \left[\frac{\mathrm{tr}(PA^2)}{\mathrm{tr}(P)} - \left(\frac{\mathrm{tr}(PA)}{\mathrm{tr}(P)}\right)^2 \right]^{1/2}$$

for any $P \in \mathcal{B}_1^+(H) \setminus \{0\}$.

Therefore, by (6.156) and (6.157) we get

$$
0 \leq f\left(\frac{\mathrm{tr}[PAf'(A)]}{\mathrm{tr}[Pf'(A)]}\right) - \frac{\mathrm{tr}[Pf(A)]}{\mathrm{tr}(P)}
$$

$$
\leq K\left\|A - \frac{\mathrm{tr}(PA)}{\mathrm{tr}(P)}1_H\right\|\frac{1}{\mathrm{tr}(P)}\mathrm{tr}\left(\left|\left(A - \frac{\mathrm{tr}(PA)}{\mathrm{tr}(P)}1_H\right)P\right|\right)
$$

$$
\times \frac{\mathrm{tr}(P)}{\mathrm{tr}[Pf'(A)]}f'\left(\frac{\mathrm{tr}[PAf'(A)]}{\mathrm{tr}[Pf'(A)]}\right)
$$

$$
\leq K\left\|A - \frac{\mathrm{tr}(PA)}{\mathrm{tr}(P)}1_H\right\|\left[\frac{\mathrm{tr}(PA^2)}{\mathrm{tr}(P)} - \left(\frac{\mathrm{tr}(PA)}{\mathrm{tr}(P)}\right)^2\right]^{1/2}
$$

$$
\times \frac{\mathrm{tr}(P)}{\mathrm{tr}[Pf'(A)]}f'\left(\frac{\mathrm{tr}[PAf'(A)]}{\mathrm{tr}[Pf'(A)]}\right)
$$

that proves the second and third inequalities in (6.155).
The last part follows by Lemma 6.3. ☐

The inequality (6.155) can be also written for the convex function $f(t) = t^p$ with $p \geq 1$ and $t \geq 0$, however the details are not presented here.

References

1 Helmberg, G. (1969) *Introduction to Spectral Theory in Hilbert Space*, John Wiley & Sons, Inc., New York.

2 Mond, B. and Pečarić, J. (1993) Convex inequalities in Hilbert space. *Houston J. Math.*, **19**, 405–420.

3 Furuta, T., Mićić Hot, J., Pečarić, J., and Seo, Y. (2005) MondPečarić Method in Operator Inequalities, Inequalities for Bounded Selfadjoint Operators on a Hilbert Space, Element, Zagreb.

4 McCarthy, C.A. (1967) c_p, Israel. *J. Math.*, **5**, 249–271.

5 Dragomir, S.S. (2010) Some reverses of the Jensen inequality for functions of selfadjoint operators in Hilbert spaces. *J. Inequal. Appl.*, **2010**, Article ID 496821. Preprint RGMIA Res. Rep. Coll., 11 (e) (2008), Art. 15.

6 Dragomir, S.S. (2011) Some Slater's type inequalities for convex functions of selfadjoint operators in Hilbert spaces. *Rev. Un. Mat. Argentina*, **52** (1), 109–120. Preprint RGMIA Res. Rep. Coll., 11 (e) (2008), Art. 7.

7 Dragomir, S.S. (1999/2000) A converse result for Jensen's discrete inequality via Gruss' inequality and applications in information theory. *Ann. Univ. Oradea Fasc. Mat.*, 7, 178–189.

8 Dragomir, S.S. (2001) On a reverse of Jessen's inequality for isotonic linear functionals. *J. Inequal. Pure Appl. Math.*, **2** (3), Article 36.

9 Dragomir, S.S. (2003) A Grüss type inequality for isotonic linear functionals and applications. *Demonstratio Math.*, **36** (3), 551–562. Preprint RGMIA Res. Rep. Coll. 5 (2002), Suplement, Art. 12.

10 Dragomir, S.S. (2004) *Discrete Inequalities of the Cauchy-Bunyakovsky-Schwarz Type*, Nova Science Publishers, New York.

11 Dragomir, S.S. (2006) Bounds for the normalized Jensen functional. *Bull. Aust. Math. Soc.*, **74** (3), 471–476.

12 Dragomir, S.S. (2008) Bounds for the deviation of a function from the chord generated by its extremities. *Bull. Aust. Math. Soc.*, **78** (2), 225–248.

13 Dragomir, S.S. and Ionescu, N.M. (1994) Some converse of Jensen's inequality and applications. *Rev. Anal. Numér. Théor. Approx.*, **23** (1), 71–78. MR:1325895 (96c:26012).

14 Simić, S. (2008) On a global upper bound for Jensen's inequality. *J. Math. Anal. Appl.*, **343**, 414–419.

15 Dragomir, S.S. (2009) Some inequalities for convex functions of selfadjoint operators in Hilbert spaces. *Filomat*, **23** (3), 81–92. Preprint RGMIA Res. Rep. Coll., 11 (e) (2008), Art. 10.

16 Dragomir, S.S. (2009) Some Jensen's type inequalities for twice differentiable functions of selfadjoint operators in Hilbert spaces. *Filomat*, **23** (3), 211–222. Preprint RGMIA Res. Rep. Coll., 11 (e) (2008), Art. 13.

17 Dragomir, S.S. (2010) Some new Grüss' type inequalities for functions of selfadjoint operators in Hilbert spaces. *Sarajevo J. Math.* 6, **18** (1), 89–107. Preprint RGMIA Res. Rep. Coll., 11 (e) (2008), Art. 12.

18 Dragomir, S.S. (2010) New bounds for the Čebyšev functional of two functions of selfadjoint operators in Hilbert spaces. *Filomat*, **24** (2), 27–39.

19 Dragomir, S.S. (2011) Some Jensen's type inequalities for log-convex functions of selfadjoint operators in Hilbert spaces. *Bull. Malays. Math. Sci. Soc.*, **34**, 3. Preprint RGMIA Res. Rep. Coll., 13 (2010), Sup. Art. 2.

20 Dragomir, S.S. (2011) Hermite-Hadamard's type inequalities for operator convex functions. *Appl. Math. Comput.*, **218**, 766–772. Preprint RGMIA Res. Rep. Coll., 13(2010), No. 1, Art. 7.

21 Dragomir, S.S. (2012) Hermite-Hadamard's type inequalities for convex functions of selfadjoint operators in Hilbert spaces. *Linear Algebra Appl.*, **436** (5), 1503–1515. Preprint RGMIA Res. Rep. Coll., 13 (2010), Sup. Art. 1.

22 Dragomir, S.S. (2011) New Jensen's type inequalities for differentiable log-convex functions of selfadjoint operators in Hilbert spaces. *Sarajevo J. Math.*, **19** (1), 67–80. Preprint RGMIA Res. Rep. Coll., 13 (2010), Sup. Art. 2.

23 Matković, A., Pečarić, J., and Perić, I. (2006) A variant of Jensen's inequality of Mercer's type for operators with applications. *Linear Algebra Appl.*, **418** (2-3), 551–564.

24 Mićić, J., Seo, Y., Takahasi, S.-E., and Tominaga, M. (1999) Inequalities of Furuta and Mond-Pečarić. *Math. Inequal. Appl.*, **2**, 83–111.

25 Mond, B. and Pečarić, J. (1993) On some operator inequalities. *Indian J. Math.*, **35**, 221–232.

26 Mond, B. and Pečarić, J. (1994) Classical inequalities for matrix functions. *Utilitas Math.*, **46**, 155–166.

27 Dragomir, S.S. (2012) *Operator Inequalities of the Jensen, Čebyšev and Grüss Type*, Springer Briefs in Mathematics, Springer-Verlag, New York, xii+121. ISBN: 978-1-4614-1520-6.

28 Dragomir, S.S. (2012) *Operator Inequalities of Ostrowski and Trapezoidal Type*, Springer Briefs in Mathematics, Springer-Verlag, New York, x+112. ISBN: 978-1-4614-1778-1.

29 Simon, B. (1979) *Trace Ideals and Their Applications*, Cambridge University Press, Cambridge.

30 Chang, D. (1999) A matrix trace inequality for products of Hermitian matrices. *J. Math. Anal. Appl.*, **237**, 721–725.

31 Coop, I.D. (1994) On matrix trace inequalities and related topics for products of Hermitian matrix. *J. Math. Anal. Appl.*, **188**, 999–1001.

32 Neudecker, H. (1992) A matrix trace inequality. *J. Math. Anal. Appl.*, **166**, 302–303.

33 Yang, Y. (1988) A matrix trace inequality. *J. Math. Anal. Appl.*, **133**, 573–574.

34 Bellman, R. (1980) Some inequalities for positive definite matrices, in *General Inequalities 2, Proceedings of the 2nd International Conference on General Inequalities* (ed. E.F. Beckenbach), Birkhäuser, Basel, pp. 89–90.

35 Ando, T. (1995) Matrix Young inequalities. *Oper. Theory Adv. Appl.*, **75**, 33–38.

36 Belmega, E.V., Jungers, M., and Lasaulce, S. (2010) A generalization of a trace inequality for positive definite matrices. *Aust. J. Math. Anal. Appl.*, **7** (2), Art. 26, 5 p.

37 Furuichi, S. and Lin, M. (2010) Refinements of the trace inequality of Belmega, Lasaulce and Debbah. *Aust. J. Math. Anal. Appl.*, **7** (2), Art. 23, 4 p.

38 Lee, H.D. (2008) On some matrix inequalities. *Korean J. Math.*, **16** (4), 565–571.

39 Liu, L. (2007) A trace class operator inequality. *J. Math. Anal. Appl.*, **328**, 1484–1486.

40 Manjegani, S. (2007) Hölder and Young inequalities for the trace of operators. *Positivity*, **11**, 239–250.

41 Shebrawi, K. and Albadawi, H. (2008) Operator norm inequalities of Minkowski type. *J. Inequal. Pure Appl. Math.*, **9** (1), 1–10, Article 26.

42 Ulukök, Z. and Türkmen, R. (2010) On some matrix trace inequalities. *J. Inequal. Appl.*, 8. Art. ID 201486, pp.

43 Dragomir, S.S. (2016) Some trace inequalities for convex functions of selfadjoint operators in Hilbert spaces. *Korean J. Math.*, **24** (2), 273–296. Preprint RGMIA Res. Rep. Coll., 17 (2014), Art 115.

44 Carlen, E.A. (2010) Trace inequalities and quantum entropy: an introductory course, in *Entropy and the Quantum*, Contemporary Mathematics, vol. **529**, American Mathematical Society, Providence, RI, pp. 73–140.

45 Dragomir, S.S. (2016) Jensen's type trace inequalities for convex functions of selfadjoint operators in Hilbert spaces. *Facta Univ. Niš, Ser. Math. Inf.*, **31** (5), 981–998. Preprint RGMIA Res. Rep. Coll. 17 (2014), Art. 116.

46 Dragomir, S.S. (2016) Reverse Jensen's type trace inequalities for convex functions of selfadjoint operators in Hilbert spaces. *Ann. Math. Sil.*, **30**, 39–62. Preprint RGMIA Res. Rep. Coll. 17 (2014), Art. 118.

47 Riesz, F. and Sz-Nagy, B. (1990) *Functional Analysis*, Dover Publications, New York.

48 Slater, M.S. (1981) A companion inequality to Jensen's inequality. *J. Approx. Theory*, **32**, 160–166.

49 Dragomir, S.S. (2016) Some Slater's type trace inequalities for convex functions of selfadjoint operators in Hilbert spaces. *Toyama Math. J.*, **38**, 75–99. Preprint RGMIA Res. Rep. Coll. 17 Art. 117.

50 Dragomir, S.S. (2016) Some Grüss' type inequalities for trace of operators in Hilbert spaces. *Oper. Matrices*, **10** (4), 923–943. Preprint RGMIA Res. Rep. Coll., 17 (2014), Art 114.

51 Dragomir, S.S. (2016) Some trace inequalities of Čebyšev type for functions of operators in Hilbert spaces. *Linear Multilinear Algebra*, **64** (9), 1800–1813. Preprint RGMIA Res. Rep. Coll., 17 (2014), Art 111.

7

Spectral Synthesis and Its Applications

László Székelyhidi[1,2]

[1] *Institute of Mathematics, University of Debrecen, H-4010 Debrecen, Hungary*
[2] *Department of Mathematics, University of Botswana, Gaborone, Botswana*

Mathematics Subject Classification (2010), Primary 43A45, 43A90, 22D15; Secondary 43A60, 43A65, 20N20

7.1 Introduction

Spectral analysis and spectral synthesis deal with the description of translation invariant function spaces over locally compact Abelian groups. We consider the space $\mathscr{C}(G)$ of all complex-valued continuous functions on a locally compact Abelian group G, which is a locally convex topological linear space with respect to the pointwise linear operations (addition, multiplication with scalars) and to the topology of uniform convergence on compact sets. Continuous homomorphisms of G into the additive topological group of complex numbers and into the multiplicative topological group of nonzero complex numbers, respectively, are called *additive* and *exponential functions*. A function is a *polynomial* if it belongs to the algebra generated by the continuous additive functions. An *exponential monomial* is the product of a polynomial and an exponential.

It turns out that exponential functions, or more generally, exponential monomials can be considered as basic building blocks of varieties. A given variety may or may not contain any exponential function or exponential monomial. If every nonzero subvariety of a given variety contains an exponential function, then we say that *spectral analysis* holds for the variety. An exponential function in a variety can be considered as a kind of *spectral value* and the set of all exponential functions in a variety is called the *spectrum* of the variety. It follows that spectral analysis for a variety means that the spectrum of any nonzero subvariety is nonempty. On the other hand, the set of all exponential monomials contained in a variety is called the *spectral set* of the variety. It turns out that if a nonzero exponential monomial belongs to a variety, then the exponential function appearing in the representation of this exponential monomial belongs to the variety, too. Hence, if the spectral set of a variety is nonzero, then also the spectrum of the variety is nonempty and spectral analysis holds. There is,

Mathematical Analysis and Applications: Selected Topics, First Edition.
Edited by Michael Ruzhansky, Hemen Dutta, and Ravi P. Agarwal.
© 2018 John Wiley & Sons, Inc. Published 2018 by John Wiley & Sons, Inc.

however, an even stronger property of some varieties, namely, if the spectral set of the variety spans a dense subspace of the variety. In this case, we say that the variety is *synthesizable*. If every subvariety of a given variety is synthesizable, then we say that *spectral synthesis* holds for the variety. It follows that, for a given nonzero variety, spectral synthesis implies spectral analysis. If spectral analysis, respectively, spectral synthesis holds for every variety on an Abelian group, then we say that *spectral analysis*, respectively, *spectral synthesis* holds on the Abelian group. A famous and pioneering result of Schwartz [1] exhibits the situation by stating that if the group is the reals with the Euclidean topology, then spectral values do exist, that is, any nonzero variety contains an exponential function. In other words, in this case the spectrum is nonempty, spectral analysis holds. Furthermore, spectral synthesis also holds in this situation: there are sufficiently many exponential monomials in any variety in the sense that their linear hull is dense in the variety.

An interesting particular case is presented by discrete Abelian groups. Here the problem seems to be purely algebraic: all complex functions are continuous, and convergence is meant in the pointwise sense. The archetype is the additive group of integers: in this case, the closed translation invariant function spaces can be characterized by systems of homogeneous linear difference equations with constant coefficients. It is known that these function spaces are spanned by exponential monomials corresponding to the characteristic values of the equation, together with their multiplicities. In this sense, the classical theory of homogeneous linear difference equations with constant coefficients can be considered as spectral analysis and spectral synthesis on the additive group of integers.

The next simplest case is the case of systems of homogeneous linear difference equations with constant coefficients in several variables. This case is settled by the famous result of Lefranc [2] about spectral synthesis on $\mathbb{Z}^n$, the lattice group.

At this point, the reader may ask the natural question: what about general Abelian groups? In his 1965 paper [3], Elliot presented a theorem on spectral synthesis for arbitrary Abelian groups. However, in 1987, Gajda [4] called the present author's attention to the fact that the proof of Elliot's theorem had serious gaps. Since then several efforts have been made to solve the problem of discrete spectral analysis and spectral synthesis on Abelian groups. In the subsequent paragraphs, we exhibit the development of the theory of discrete spectral synthesis until the present status.

An equally important problem arises concerning the possible extension of the fundamental theorem of Schwartz to more general locally compact Abelian groups. The basic question is: is it possible to extend the theorem at least to $\mathbb{R}^n$? Unfortunately, counterexamples in [5] show that such a direct extension is impossible for $n \geq 2$. Nevertheless, some recent results of the present author show that still we can find a more sophisticated extension if "translation invariance" is replaced by "invariance under group action." A detailed discussion about these ideas will be presented below.

The subject of the above-mentioned investigations is spectral analysis and spectral synthesis on the whole group – nevertheless, the same problems can be studied on particular varieties, too. In fact, spectral analysis or spectral synthesis on the whole group, or what is the same, for all varieties on the group clearly depends on the structure of the given topological Abelian group: it holds or does not hold on topologically isomorphic groups simultaneously. However, some particular varieties may possess spectral analysis or synthesis without the whole group having this property. A classical result in this direction is due to Malgrange [6], which is spectral synthesis for the solution space of homogeneous linear partial differential equations with constant coefficients. A generalization of this theorem is the famous result of Ehrenpreis [7], also known as the Principal Ideal Theorem. This result has also been generalized by Elliott [8]. Nevertheless, general results about spectral analysis and spectral synthesis for particular varieties are still missing.

Based on some ideas in [7], the present author initiated a detailed study on spectral analysis and spectral synthesis in terms of the group algebra in the discrete case, or more generally, of the algebra of compactly supported Radon measures in the locally compact case. It turns out that general results about spectral synthesis properties of particular varieties can be obtained using this annihilator method [9]. The details of the annihilator method will be presented below.

Yet another possible way is to generalize the group structure. We summarize the formulation of the basic problems of spectral analysis and synthesis in the hypergroup-setting. The basics of this theory can be found in the monograph [10]. In addition, we present here some recent results on different types of hypergroups.

Finally, we illustrate some possible applications of spectral analysis and spectral synthesis in the theory of functional equations, difference equations, and differential equations. This can be found in the last section.

7.2 Basic Concepts and Function Classes

In this chapter $\mathbb{N}, \mathbb{Z}, \mathbb{R}, \mathbb{C}$ denotes the set of natural numbers (zero included), the set of integers, the set of real numbers, and the set of complex numbers, respectively. For a locally compact group G, we denote by $\mathscr{C}(G)$, the locally convex topological vector space of all continuous complex- valued functions defined on G, equipped with the pointwise operations and with the topology of uniform convergence on compact sets, also known as *topology of compact convergence*. For each function f in $\mathscr{C}(G)$, we define $\check{f}$ by $\check{f}(x) = f(x^{-1})$, whenever x is in G. For a subset H in $\mathscr{C}(G)$, we denote by $\check{H}$ the set of all functions $\check{f}$ with f in H. By a *ring*, we always mean a commutative ring with unit.

It is known that the dual of $\mathscr{C}(G)$ can be identified with the space $\mathscr{M}_c(G)$ of all compactly supported complex Borel measures on G which is equipped with the pointwise operations and with the weak*-topology. The pairing between $\mathscr{C}(G)$ and $\mathscr{M}_c(G)$ is given by the formula

$$\langle \mu, f \rangle = \int f \, d\mu.$$

The following theorem, describing the dual of $\mathscr{M}_c(G)$ is fundamental. The proof can be found in [11, 17.6, p. 155] (see also [9, theorem 3.43, p. 48]).

Theorem 7.1 Let G be a locally compact group. For every weak*-continuous linear functional $F: \mathscr{M}_c(G) \to \mathbb{C}$, there exists a continuous function $f: G \to \mathbb{C}$ such that $F(\mu) = \mu(f)$ for each μ in $\mathscr{M}_c(G)$.

In fact, the function f in this theorem is uniquely determined by F, as it is clear from the following result, which is a standard application of the Hahn–Banach Theorem in functional analysis.

Theorem 7.2 Let G be a locally compact group. The finitely supported complex measures form a weak*-closed subspace in $\mathscr{M}_c(G)$.

In spectral synthesis, the dual pair $\mathscr{C}(G)$ and $\mathscr{M}_c(G)$ will play the fundamental role. For each μ in $\mathscr{M}_c(G)$, we define $\check{\mu}$ by the equation $\check{\mu}(f) = \mu(\check{f})$ whenever f is in $\mathscr{C}(G)$. For every subset K in $\mathscr{M}_c(G)$, the symbol $\check{K}$ denotes the set of all measures of the form $\check{\mu}$ with μ in K. The *orthogonal complement* of the subset H in $\mathscr{C}(G)$ is the set of all measures μ in $\mathscr{M}_c(G)$ satisfying $\mu(f) = 0$ for each f in H, and it is denoted by $H^\perp$. The dual concept is the orthogonal complement of a set K in $\mathscr{M}_c(G)$ of all functions f in $\mathscr{C}(G)$ satisfying $\mu(f) = 0$ for every μ in K, and it is denoted by $K^\perp$. Obviously, $H^\perp$, respectively, $K^\perp$ is a closed subspace in $\mathscr{M}_c(G)$, respectively, in $\mathscr{C}(G)$.

Convolution on $\mathscr{M}_c(G)$ is defined by

$$\int f \, d(\mu * \nu) = \int f(xy) \, d\mu(x) \, d\nu(y)$$

for each μ, ν in $\mathscr{M}_c(G)$ and x in G. Convolution converts the linear space $\mathscr{M}_c(G)$ into a topological algebra with unit δ_e, e being the identity in G.

We also define convolution of measures in $\mathscr{M}_c(G)$ with arbitrary functions in $\mathscr{C}(G)$ by the similar formula

$$\mu * f(x) = \int f(xy^{-1}) \, d\mu(y)$$

for each μ in $\mathscr{M}_c(G)$, f in $\mathscr{C}(G)$, and x in G. The linear operators $f \mapsto \mu * f$ on $\mathscr{C}(G)$ are called *convolution operators*. It is easy to see that equipped with the action $f \mapsto f * \mu$ the space $\mathscr{C}(G)$ is a left topological module over $\mathcal{M}_c(G)$. For each subset K in $\mathcal{M}_c(G)$ and H in $\mathscr{C}(G)$, we use the notation

$$K * H = \{\mu * f : \mu \in K, f \in H\}.$$

We recall that for each subset H in $\mathscr{C}(G)$ the *annihilator* of H in $\mathcal{M}_c(G)$ is the set

$$\operatorname{Ann} H = \{\mu : \mu * f = 0 \quad \text{for each} \quad f \in H\}.$$

We also have the dual concept: for every subset K in $\mathcal{M}_c(G)$, the *annihilator* of K in $\mathscr{C}(G)$ is the set

$$\operatorname{Ann} K = \{f : \mu * f = 0 \quad \text{for each} \quad \mu \in K\}.$$

In the rest of this section, we suppose that G is *commutative* and we apply additive notation in G. The identity in G will be denoted by 0. *Translation* with the element y in G is the operator mapping the function f in $\mathscr{C}(G)$ onto its *translate* $\tau_y f$ defined by $\tau_y f(x) = f(x + y)$ for each x in G. Clearly, τ_y is a convolution operator, namely, it is the convolution with the measure δ_{-y}: we have $\tau_y f = \delta_{-y} * f$. A subset of $\mathscr{C}(G)$ is called translation invariant, if it contains all translates of its elements. A closed linear subspace of $\mathscr{C}(G)$ is called a *variety* on G, if it is translation invariant. Obviously, every closed submodule of $\mathscr{C}(G)$ is a variety, in fact, varieties are exactly the closed submodules of $\mathscr{C}(G)$. As it is easy to see, $\check{V}$ is a variety whenever V is a variety. For each function f, the smallest variety containing f is called the *variety generated by f*; or simply the variety of f, and it is denoted by $\tau(f)$, which is clearly the intersection of all varieties containing f.

The following simple relation between orthogonal complements and annihilators has been proved in [12]. It shows that we can use orthogonal complements and annihilators alternatively, for varieties, they are – in some sense – equivalent.

Theorem 7.3 For each variety V in $\mathscr{C}(G)$, its annihilator $\operatorname{Ann} V$ is a closed ideal in $\mathcal{M}_c(G)$, and $\operatorname{Ann} V = (\check{V})^\perp$. Similarly, for each ideal I in $\mathcal{M}_c(G)$, its annihilator $\operatorname{Ann} I$ is a variety in $\mathscr{C}(G)$, and $\operatorname{Ann} I = (\check{I})^\perp$.

The following theorem gives important information about the duality of varieties and annihilators.

Theorem 7.4 For each variety $V \subseteq W$ in $\mathscr{C}(G)$, we have $\operatorname{Ann} V \supseteq \operatorname{Ann} W$ and for each ideal $I \subseteq J$ in $\mathcal{M}_c(G)$, we have $\operatorname{Ann} I \supseteq \operatorname{Ann} J$. In addition, we have $\operatorname{Ann}(\operatorname{Ann} V) = V$ and $\operatorname{Ann}(\operatorname{Ann} I) \supseteq I$. In particular, $V \neq W$ implies $\operatorname{Ann} V \neq \operatorname{Ann} W$.

Proof: Let $V \subseteq W$ be varieties in $\mathscr{C}(G)$ and let $I \subseteq J$ be ideals in $\mathcal{M}_c(G)$. For every μ in $\operatorname{Ann} W$ and for each f in V, we have that f is in W, hence $\mu * f = 0$.

This proves that μ is in $\mathrm{Ann}\, V$, and $\mathrm{Ann}\, V \supseteq \mathrm{Ann}\, W$. Similarly, if f is in $\mathrm{Ann}\, J$ and μ is in I, then μ is in J, hence $\mu * f = 0$, which proves that f is in $\mathrm{Ann}\, I$, and $\mathrm{Ann}\, I \supseteq \mathrm{Ann}\, J$.

Assume that f is in V and μ is in $\mathrm{Ann}\, V$, then, by definition, $\mu * f = 0$, hence f is in $\mathrm{Ann}(\mathrm{Ann}\, V)$, which proves $\mathrm{Ann}(\mathrm{Ann}\, V) \supseteq V$. Similarly, we have $\mathrm{Ann}(\mathrm{Ann}\, I) \supseteq I$.

Suppose now that $\mathrm{Ann}(\mathrm{Ann}\, V) \subsetneqq V$. Consequently, there is a function f in $\mathrm{Ann}(\mathrm{Ann}\, V)$ such that f is not in V. By the Hahn–Banach Theorem, there is a λ in $\mathcal{M}_c(G)$ such that $\check{\lambda}(f) \neq 0$, and $\check{\lambda}$ vanishes on V. This means

$$(\varphi * \lambda)(0) = \int \varphi(-y)\, d\lambda(y) = \lambda(\check{\varphi}) = \check{\lambda}(\varphi) = 0,$$

whenever φ is in V. As V is a variety, this implies, by the previous theorem, that λ is in $\mathrm{Ann}\, V$, in particular, $\lambda * f = 0$, a contradiction. This proves $\mathrm{Ann}(\mathrm{Ann}\, V) = V$, which also implies $\mathrm{Ann}\, V \neq \mathrm{Ann}\, W$, whenever $V \neq W$. $\square$

We note that for ideals in $\mathcal{M}_c(G)$, the equality $\mathrm{Ann}(\mathrm{Ann}\, I) = I$ does not hold in general. For this, by Theorem 7.3, it is enough to show that $I^{\perp\perp} = I$ does not hold, in general. A counterexample can be found in [13]. Nevertheless, the following theorem holds true [13].

Theorem 7.5 Let G be a discrete Abelian group. Then $\mathrm{Ann}(\mathrm{Ann}\, I) = I$ holds for every ideal I in $\mathcal{M}_c(G)$.

For further properties of the variety–ideal correspondence, see [9, 12]. Here we recall the following result.

Theorem 7.6 Let G be a locally compact group.

1. For each family $(V_\gamma)_{\gamma \in \Gamma}$ of varieties in $\mathscr{C}(G)$, we have

$$\mathrm{Ann}\left(\sum_{\gamma \in \Gamma} V_\gamma\right) = \bigcap_{\gamma \in \Gamma} \mathrm{Ann}\, V_\gamma, \quad \left(\sum_{\gamma \in \Gamma} V_\gamma\right)^{\perp} = \bigcap_{\gamma \in \Gamma} V_\gamma^{\perp}.$$

2. For each family $(I_\gamma)_{\gamma \in \Gamma}$ of ideals in $\mathcal{M}_c(G)$, we have

$$\mathrm{Ann}\left(\sum_{\gamma \in \Gamma} I_\gamma\right) = \bigcap_{\gamma \in \Gamma} \mathrm{Ann}\, I_\gamma, \quad \left(\sum_{\gamma \in \Gamma} I_\gamma\right)^{\perp} = \bigcap_{\gamma \in \Gamma} I_\gamma^{\perp}.$$

We note that here $\sum_{\gamma \in \Gamma} V_\gamma$ denotes the *topological sum* of the family of varieties $(V_\gamma)_{\gamma \in \Gamma}$, that is, the closure of the union of the sums of finite subfamilies. However, $\sum_{\gamma \in \Gamma} I_\gamma$ denotes the *algebraic sum* of the family of ideals $(I_\gamma)_{\gamma \in \Gamma}$, that is, the ideal generated by the sums of finite subfamilies.

The basic building blocks of spectral analysis and spectral synthesis are the exponentials. Let G be a locally compact Abelian group and $m : G \to \mathbb{C}$

a continuous homomorphism of G into the multiplicative group of nonzero complex numbers. Then we call m and *exponential* on G. Clearly, every exponential is *normalized* in the sense that $m(0) = 1$. The following result is a simple characterization of exponentials.

Theorem 7.7   Let G be a locally compact Abelian group and $m : G \to \mathbb{C}$ a continuous function. Then the following statements are equivalent:

(i) m is an exponential.
(ii) m is nonzero and for each x, y in G we have

$$m(x + y) = m(x)m(y). \tag{7.1}$$

(iii) m is a normalized common eigenfunction of all translation operators.
(iv) m is a normalized common eigenfunction of all convolution operators.
(v) m is normalized and $\tau(m)$ is one dimensional.

Proof: The first two statements are obviously equivalent, by the definition. By (7.1), we have $\tau_y m = m(y) \cdot m$, hence (ii) implies (iii). If (iii) holds, then there exists a mapping $\lambda : G \to \mathbb{C}$ such that

$$m(x + y) = \delta_{-y} * m(x) = \lambda(y)m(x)$$

for each x, y in G. With $x = 0$, we have $\lambda = m$, hence (ii) holds. If m satisfies (iv), then it satisfies (iii). Conversely, if (iii) holds for m, then m is an exponential and we have

$$\mu * m(x) = \int_G m(x - y) \, d\mu(y) = m(x) \int_G m(-y) \, d\mu(y)$$

for each measure μ, hence we have (iv). It follows that all the conditions (i)–(iv) are equivalent. Finally, if m is normalized and $\tau(m)$ is one dimensional, then obviously, each translate of m is a constant multiple of m, hence we have (iii), and (ii) clearly implies (v). The theorem is proved.   □

Using translation, one introduces *modified differences* in the following manner: for each f and y in G, we let

$$\Delta_{f;y} = \delta_{-y} - f(y)\delta_0,$$

which is an element of $\mathcal{M}_c(G)$. Products of modified differences will be denoted in the following way: for each function f, for every natural number n, and for arbitrary $y_1, y_2, \ldots, y_{n+1}$ in G, we write

$$\Delta_{f;y_1,y_2,\ldots,y_{n+1}} = \Pi_{i=1}^{n+1} \left(\delta_{-y_i} - f(y_i)\delta_0 \right),$$

where Π denotes convolution. In the case $f \equiv 1$, we use the simplified notation Δ_y for $\Delta_{1;y}$ and we call it *difference*. Accordingly, we write $\Delta_{y_1,y_2,\ldots,y_{n+1}}$ for $\Delta_{1;y_1,y_2,\ldots,y_{n+1}}$. For a given function f, the closed ideal generated by all modified differences of the form $\Delta_{f;y}$ with y in G is denoted by M_f. The following theorem is of fundamental importance.

Theorem 7.8 Let G be a locally compact Abelian group and $f : G \to \mathbb{C}$ a continuous function. The ideal M_f is proper if and only if f is an exponential. In this case, the ideal $M_f = \operatorname{Ann}\tau(f)$ is maximal, and $\mathcal{M}_c(G)/M_f$ is topologically isomorphic to the complex field.

Proof: As M_f is closed, we have $\operatorname{Ann}(\operatorname{Ann}M_f) = M_f$ and $M_f^{\perp\perp} = M_f$ [12].

Suppose that f is an exponential. Then $f \neq 0$, and

$$\Delta_{f;y} * f(x) = f(x + y) - f(y)f(x) = 0$$

for each x, y in G, hence f is in $\operatorname{Ann}M_f$. As $\tau(f)$ consists of all constant multiples of f, we infer that $\tau(f) \subseteq \operatorname{Ann}M_f$. Moreover, if φ is in $\operatorname{Ann}M_f$, then we have

$$0 = \Delta_{f;y} * \varphi(x) = \varphi(x + y) - f(y)\varphi(x)$$

for each x, y in G. It follows $\varphi = \varphi(0) \cdot f$, hence φ is in $\tau(f)$. We conclude that $\tau(f) = \operatorname{Ann}M_f$, and $M_f = \operatorname{Ann}\tau(f)$.

We define the mapping $\Phi_f : \mathcal{M}_c(G) \to \mathbb{C}$ by

$$\Phi_f(\mu) = \mu(\check{f}) = \int f(-y) \, d\mu(y)$$

for each μ in $\mathcal{M}_c(G)$. Then Φ_f is a linear mapping, $\Phi_f(\delta_0) = 1$, and for each μ, v in $\mathcal{M}_c(G)$, we have

$$\Phi_f(\mu * v) = \int f(-x - y) d\mu(x) \, dv(y)$$

$$= \int f(-x) \, d\mu(x) \int f(-y) \, dv(y) = \Phi_f(\mu) \cdot \Phi_f(v),$$

hence Φ_f is an algebra homomorphism. Obviously, Φ_f maps $\mathcal{M}_c(G)$ onto $\mathbb{C}$, hence it is a multiplicative linear functional. We infer that $\operatorname{Ker} \Phi_f$ is a maximal ideal and $\mathcal{M}_c(G)/\operatorname{Ker} \Phi_f$ is isomorphic to the complex field $\mathbb{C}$. For each μ in $\operatorname{Ker} \Phi_f$, we have $\mu(\check{f}) = 0$; hence for each complex number c, we have

$$cf * \mu(x) = c \int f(x - y) \, d\mu(y) = cf(x)\mu(\check{f}) = 0,$$

consequently μ is in $\operatorname{Ann}\tau(f) = M_f$. It follows $\operatorname{Ker} \Phi_f \subseteq M_f$, which implies that M_f is a maximal ideal. We also have that $\operatorname{Ker} \Phi_f$ is closed, hence Φ_f is continuous. As Φ_f is also open, we have that $\mathcal{M}_c(G)/M_f$ is topologically isomorphic to the complex field.

Finally, if M_f is proper, then $\operatorname{Ann}M_f$ is nonzero. Let $\varphi \neq 0$ be a function in $\operatorname{Ann}M_f$, then we have

$$0 = \Delta_{f;y} * \varphi(x) = \varphi(x + y) - f(y)\varphi(x),$$

and in the same way as above we conclude that f is an exponential. The theorem is proved. ☐

Given a complex algebra A, we call a maximal ideal M in A an exponential maximal ideal, if the residue algebra A/M is isomorphic to the complex field. If A is a topological ring, then we require the isomorphism to be topological. From the above proof, it is clear that if G is a locally compact Abelian group, then each exponential maximal ideal in $\mathscr{M}_c(G)$ is of the form $M_m = \text{Ann}\,\tau(m)$ with some exponential m. In particular, every exponential maximal ideal is closed.

The following theorem presents another characterization of exponentials in terms of the annihilators of their varieties. The statements follow easily from the previous theorem and from the connection between annihilators and orthogonals [12].

Theorem 7.9 Let G be a locally compact Abelian group and $m : G \to \mathbb{C}$ a continuous function. Then the following statements are equivalent:

(i) m is an exponential.
(ii) m is normalized and $\text{Ann}\,\tau(m)$ is an exponential maximal ideal.
(iii) m is normalized and $\tau(m)^{\perp}$ is an exponential maximal ideal.
(iv) m is a normalized and $\mathscr{M}_c(G)/\text{Ann}\,\tau(m)$ is topologically isomorphic to the complex field.
(v) m is a normalized and $\mathscr{M}_c(G)/\tau(m)^{\perp}$ is topologically isomorphic to the complex field.

The most important function class in spectral analysis and spectral synthesis is the class of exponential monomials. There are different ways to define exponential monomials, the most common one is the following. Let G be a locally compact Abelian group. The continuous function $a : G \to \mathbb{C}$ is called *additive* if it is a homomorphism of G into the additive group of complex numbers. The continuous function $f : G \to \mathbb{C}$ is called an *exponential monomial* if it has the form

$$f(x) = P(a_1(x), a_2(x), \dots, a_k(x))m(x), \tag{7.2}$$

where k is a positive integer, P is an ordinary complex polynomial in k variables, $a_1, a_2, \dots, a_k$ are additive functions, and m is an exponential. Linear combinations of exponential monomials are called *exponential polynomials*. A fundamental result about exponential polynomials is the following [9]:

Theorem 7.10 Let G be a locally compact Abelian group. The continuous function f is an exponential polynomial if and only if $\tau(f)$ is finite dimensional.

It is easy to show that exponential monomials of the above form satisfy the functional equation

$$\Delta_{m;y_1,y_2,\dots,y_{n+1}} * f(x) = 0 \tag{7.3}$$

for each $x, y_1, y_2, \ldots, y_{n+1}$ in G with some natural number n. However, the converse is not necessarily true: not every solution of (7.3) generates a finite-dimensional variety. Continuous functions $f : G \to \mathbb{C}$ satisfying (7.3) are called *generalized exponential monomials*. They are exponential monomials exactly in the case if $\tau(f)$ is finite dimensional. The following characterization of generalized exponential monomials is of basic importance.

Theorem 7.11 Let G be a locally compact Abelian group. The continuous function f is an generalized exponential monomial if and only if there is an exponential maximal ideal M in $\mathscr{M}_c(G)$ and a natural number n such that

$$M^{n+1} \subseteq \operatorname{Ann} \tau(f). \tag{7.4}$$

Proof: If f satisfies (7.3), then $\tau(f)$ is annihilated by all measures of the form $\Delta_{m; y_1, y_2, \ldots, y_{n+1}}$. On the other hand, these measures generate a dense subset in M_m^{n+1}, hence (7.4) is satisfied with $M = M_m$, which is an exponential maximal ideal. Conversely, if (7.4) holds for some exponential maximal ideal M and natural number n, then $M = M_m$ for some exponential m, by Theorem 7.9, further we have (7.3), as $\Delta_{m; y_1, y_2, \ldots, y_{n+1}}$ is in M_m^{n+1} for each $y_1, y_2, \ldots, y_{n+1}$. $\quad\square$

We note that if f is nonzero, then the exponential in (7.3) is uniquely determined. Indeed, if f is nonzero, then $\operatorname{Ann} \tau(f)$ is a proper ideal, hence it is included in a maximal ideal M_0. It follows, by (7.4), that $M_m^{n+1} \subseteq M_0$. As M_0 is maximal, it is also prime, hence $M_m \subseteq M_0$, which implies $M_m = M_0$, by maximality. Thus a unique exponential m is assigned to every nonzero generalized exponential monomial f, and f is said to be *associated with m*, or we say that f is a *generalized m-exponential monomial*. The smallest n satisfying (7.3) is called the *degree* of f. We do not associate a degree to the generalized exponential monomial identically zero but it is considered as a generalized m-exponential monomial for every exponential m.

Linear combinations of generalized exponential monomials are called *generalized exponential polynomials* and linear combinations of exponential monomials are called *exponential polynomials*.

The following characterization theorems [14] are the fundamentals of the annihilator method on discrete Abelian groups – they characterize exponential monomials and exponential polynomials completely in terms of the residue algebra with respect to the annihilator, which plays the role of a coordinate ring.

Theorem 7.12 Let G be a discrete Abelian group. The function $f : G \to \mathbb{C}$ is an exponential monomial if and only if $\mathbb{C}G/\tau(f)^\perp$ is a local Noether ring with nilpotent exponential maximal ideal, or equivalently, it is a local Artin ring with exponential maximal ideal.

Theorem 7.13 Let G be an Abelian group. The function $f : G \to \mathbb{C}$ is an exponential polynomial if and only if $\mathbb{C}G/\tau(f)^\perp$ is an Artin ring with exponential maximal ideals. The function $f : G \to \mathbb{C}$ is a generalized exponential polynomial if and only if $\mathbb{C}G/\tau(f)^{\perp}$ is a semilocal ring with exponential maximal ideals and nilpotent Jacobson radical.

7.3   Discrete Spectral Synthesis

An important special case is the case of discrete Abelian groups. The use of the discrete topology makes things somewhat simpler: the function space consists of all complex-valued functions on G and the topology is that of the pointwise convergence. Compactly supported measures are identified with finitely supported functions, and every ideal in $\mathcal{M}_c(G)$ is closed with respect to the weak*-topology. As a consequence, we have the following result [13, 14].

Theorem 7.14   Let G be an Abelian group, V a variety on G and I an ideal in $\mathbb{C}G$. Then we have

$$V^{\perp\perp} = V, \quad I^{\perp\perp} = I .$$

Obviously, the following types of problems arise: the first is about a characterization of those varieties having spectral analysis or synthesis, or being synthesizable, and the second is about the characterization of those Abelian groups having spectral analysis or synthesis. As it is easy to prove that spectral analysis or synthesis holds or does not hold on isomorphic Abelian groups simultaneously, one may expect a completely group-theoretic solution for the second type of problems – the answer depends merely on the structure of the group. As we shall see, however, this is not the case concerning the first type of problems.

First, we have the following simple result.

Lemma 7.1   If spectral analysis, respectively, spectral synthesis holds on an Abelian group, then it holds on every subgroup of it, too.

Proof: Let G be an Abelian group on which spectral analysis holds, and let H be a subgroup. If V is a nonzero variety on H, and

$$W = \{ f : f \text{ is in } \mathscr{C}(G) \quad \text{and} \quad \tau_y|f_H \text{ is in } V \text{ for each } y \text{ in } G \} ,$$

then V is a variety on G. Let U be a subset of G containing exactly one element in each coset of H. If $f \neq 0$ is in V, further we define $g(u + x) = f(x)$ for u in U and x in H, then obviously g is in W, hence W is nonzero. By assumption, W contains an exponential m. Then $m|_H$ belongs to V, hence spectral analysis holds on H.

The statement about spectral synthesis can be proved in a similar way.   □

The following theorem shows that, in spectral analysis, exponentials can be replaced by nonzero exponential monomials.

Theorem 7.15 Let G be an Abelian group. A variety on G contains an exponential if and only if it contains a nonzero (generalized) exponential monomial.

Proof: As every exponential is a (generalized) exponential monomial, the necessity is obvious. Now suppose that the variety V contains the nonzero generalized m-exponential monomial φ of degree n. It follows that $\tau(\varphi) \subseteq V$, which implies $\operatorname{Ann} V \subseteq \operatorname{Ann} \tau(\varphi)$. By definition, we have $M_m^{n+1} \subseteq \operatorname{Ann} \tau(\varphi) \subseteq M_m$, and we infer $\operatorname{Ann} V \subseteq M_m$, consequently, we have $\tau(m) = \operatorname{Ann} M_m \subseteq V$ and the statement is proved. □

As a consequence, in the light of the results in section 12 in [9], in particular in Section 12.7, exponential polynomials are exactly those functions which generate finite-dimensional varieties. Consequently, spectral analysis, synthesizability and spectral synthesis for a variety can be reformulated in the following manner:

Theorem 7.16 Let G be an Abelian group. Spectral synthesis holds for a nonzero variety on G if and only if it has a nonzero finite-dimensional subvariety. A variety on G is synthesizable if and only if all finite-dimensional subvarieties span a dense subspace in it.

Using the characterization theorems in the previous section, we can reformulate the basic spectral analysis and synthesis problems in the language of annihilators.

Theorem 7.17 Let G be an Abelian group. Then spectral analysis holds for the variety V on G if and only if every maximal ideal of $\mathcal{M}_c(G)/\operatorname{Ann} V$ is exponential.

Proof: Clearly, the condition is equivalent to that every maximal ideal of $\mathcal{M}_c(G)$ containing $\operatorname{Ann} V$ is exponential. Suppose spectral analysis holds for V and let M be a maximal ideal with $\operatorname{Ann} V \subseteq M$. Then $\operatorname{Ann} M$ is a nonzero subvariety in V, hence there is an exponential m in $\operatorname{Ann} M$, and we infer $\operatorname{Ann} M_m = \tau(m) \subseteq \operatorname{Ann} M$, consequently $M \subseteq M_m$. By maximality, we deduce $M = M_m$, hence M is an exponential maximal ideal.

Conversely, suppose that every maximal ideal containing $\operatorname{Ann} V$ is exponential and let V_0 be a nonzero subvariety in V. We have $\operatorname{Ann} V \subseteq \operatorname{Ann} V_0$ and $\operatorname{Ann} V_0$ is a proper ideal. Let M be a maximal ideal in $\mathcal{M}_c(G)$ containing $\operatorname{Ann} V_0$, then, by assumption, M is exponential, hence $M = M_m$ for some exponential m. We conclude $\tau(m) = \operatorname{Ann} M_m \subseteq V_0$, thus m is in V_0 and the theorem is proved. □

The synthesizability of a variety can also be reformulated in terms of the annihilator of the variety, using the characterization in Theorem 7.12.

Theorem 7.18 Let G be an Abelian group. Then the variety V on G is synthesizable if and only if

$$\text{Ann}\, V = \bigcap_{\text{Ann}\, V \subseteq M} \bigcap_{n=0}^{\infty} (\text{Ann}\, V + M^{n+1}) \tag{7.5}$$

holds, where the outer intersection extends for all exponential maximal ideals M which contain $\text{Ann}\, V$ and $\mathcal{M}_c(G)/M^{n+1}$ is a local Noether ring.

Despite the simplification when restricting ourselves to discrete Abelian groups, for a relatively long time the only relevant result about spectral analysis and spectral synthesis was the one due to Lefranc [2] which says that spectral synthesis holds on $\mathbb{Z}^n$. The original proof of this sufficient condition for spectral analysis depends on the primary decomposition in Noether rings. Here we give another simple proof which uses the following generalization of Krull's Intersection Theorem [15]:

Theorem 7.19 Let R be a Noether ring. Then the intersection of all positive powers of all maximal ideals in R is zero.

Proof: We claim that

$$\bigcap_{M} \bigcap_{n=1}^{\infty} M^n = 0, \tag{7.6}$$

where the first intersection is extended to all maximal ideals M in $\mathbb{C}G$. We note that one classical version of Krull's Intersection claims that in a Noether ring R we have

$$\bigcap_{n=1}^{\infty} \left(\bigcap_{M} M \right)^n = 0, \tag{7.7}$$

and here the second intersection is extended to all maximal ideals M in R. We note that it is obvious that the left side of (7.6) is a subset of the left side of (7.7) in any ring.

We use the following form of Krull's Intersection Theorem [[16], theorem 2, p. 356]: Let R be a Noether ring and I and ideal in R. If x is in $\bigcap_{n=1}^{\infty} I^n$, then x is in xI. We apply this result to prove our statement in the following way: let M be a maximal ideal in R and suppose that x is in $\bigcap_{n=1}^{\infty} M^n$. Then, by the result quoted above, we have that x is in xM. Hence there exists an m in M such that $x = xm$, or $x(1 - m) = 0$. It follows that $1 - m$ annihilates x, hence the annihilator of x is not a subset of M. Consequently, if x is in $\bigcap_M \bigcap_{n=1}^{\infty} M^n$, then the annihilator of x is not contained in any maximal ideal, hence it is R, and $x = 0$. The theorem is proved. □

We need the following simple result.

Theorem 7.20 Let G be an Abelian group. Then $\mathbb{C}G$ is Noetherian if and only if G is finitely generated. If $\mathbb{C}G$ is Noetherian, then every maximal ideal of $\mathbb{C}G$ is exponential.

Proof: Suppose that X is a finite generating set for G. Then $X \cup X^{-1}$ generates $\mathbb{C}G$, whence [[17], corollary 7.10., p. 82] yields that $\mathbb{C}G$ is Noetherian and its maximal ideals are exponential.

For a subgroup H of G, we regard $\mathbb{C}H$ as a subalgebra of $\mathbb{C}G$. Letting S be a set of coset representatives of H in G, the set I_H of finite sums of terms $\delta_s * \mu$ with s in S and μ in the augmentation ideal of $\mathbb{C}H$ is an ideal of $\mathbb{C}G$ that does not depend on the specific choice of S. If K is a proper subgroup of H, then I_K is properly contained in I_H. Hence, for $\mathbb{C}G$ to be Noetherian, G must be "subgroup Noetherian," that is every strictly ascending chain of subgroups must be finite. In particular, G is finitely generated. □

Theorem 7.21 Let G be an Abelian group and V a variety on G such that $\mathcal{M}_c(G)/\mathrm{Ann}\, V$ is a Noether ring. Then spectral synthesis holds for V.

Proof: We shall use Theorem 7.12. Obviously, $\mathcal{M}_c(G)/M^{n+1}$ is always a local ring with nilpotent maximal ideal M', which is the image of M under the natural homomorphism of $\mathcal{M}_c(G)$ onto $\mathcal{M}_c(G)/M^{n+1}$. In our case, by Theorem 7.20, M and M' are exponential and $\mathcal{M}_c(G)/M^{n+1}$ is Noetherian. Synthesizability of V means, by Theorem 7.12, that

$$\mathrm{Ann}\, V = \bigcap_{\mathrm{Ann}\, V \in M} \bigcap_{n=0}^{\infty} (\mathrm{Ann}\, V + M^{n+1}), \tag{7.8}$$

where the sum extends to all maximal ideals M containing $\mathrm{Ann}\, V$. This means that in the residue ring we have to show that

$$0 = \bigcap_{M'} \bigcap_{n=0}^{\infty} M'^{n+1}, \tag{7.9}$$

where the first intersection extends to the set of all maximal ideals M' in the ring $R = \mathbb{C}G/\mathrm{Ann}\, V$. But this follows from Theorem 7.19. Hence V is synthesizable. Now we can repeat our argument for every subvariety of V, and the theorem is proved. □

Corollary 7.1 **(Lefranc)** Spectral synthesis holds on every finitely generated Abelian group.

Having Lefranc's result a reasonable question can be formulated: is there an Abelian group having spectral synthesis which is not finitely generated? This question remained open for almost a decade until the paper [3] of R. J. Elliott has been published in which the author presented a theorem claiming that spectral synthesis holds on any Abelian group. Unfortunately, the proof of Elliott's theorem had a gap which has not been filled after several unsuccessful attempts. Not to mention that without spectral synthesis even the spectral analysis problem for nonfinitely generated Abelian groups remained still open. The first relevant result in this respect has been published in [18] about spectral analysis on torsion groups, and then in [19] spectral synthesis on torsion groups has also been proved. As there are nonfinitely generated torsion groups, our above question has been answered in the positive.

The characterization problem of those discrete Abelian groups having spectral analysis has finally been settled in [20] where the authors proved the following result.

Theorem 7.22   Spectral analysis holds on an Abelian group if and only if the torsion-free rank of the group is less than the continuum.

We recall that the *torsion-free rank* of the Abelian group G is the cardinality of a maximal linearly independent system of elements. Independence is meant over the integers, that is, the set $x_1, x_2, \ldots, x_n$ of elements is called *linearly independent*, if an equation

$$k_1 x_1 + k_2 x_2 + \cdots + k_n x_n = 0$$

with some integers $k_1, k_2, \ldots, k_n$ implies $k_1 = k_2 = \cdots = k_n = 0$.

The proof of the above theorem given below depends on the characterization of spectral analysis in terms of the annihilators given in Theorem 7.17. In their proof in [20], the authors used the following lemma, which is a variant of [[21], theorem 5.2, p. 32] and is contained in problem 10(b) of [[22], chapter V. Section 3, p. 373].

Lemma 7.2   Suppose that K is a field, k is a subfield, and X is a subset of K such that $\max(|X|, \omega) < |k|$, and $K = k[X]$; that is, K, as a ring, is generated by k and X. Then K is an algebraic extension of k.

Proof: (of Theorem 7.22) First we prove the sufficiency. Suppose that G is an Abelian group and its torsion-free rank satisfies $r_0(G) < 2^\omega$. By Theorem 7.17, it is enough to show that every maximal ideal M in $\mathbb{C}G$ is exponential, or, what is the same, that the field $K = \mathbb{C}G/M$ is isomorphic to the complex field. We prove this in the subsequent paragraphs.

Let $F: \mathbb{C}G \to \mathbb{C}G/M$ denote the natural homomorphism. Then it is obvious that $k = \{c \cdot \delta_0 : c \in \mathbb{C}\}$ is a subfield in $\mathbb{C}G$, which is isomorphic to $\mathbb{C}$. Consequently, $c \cdot \delta_0 \mapsto F(c \cdot \delta_0)$ is an isomorphism from k into K, as $c \neq 0$ implies that $c \cdot \delta_0$ is not in M, hence $F(c \cdot \delta_0) \neq 0$ in K. It follows that $F(k)$ is a subfield in K isomorphic to $\mathbb{C}$, and if we identify it with $\mathbb{C}$, then K is an extension of $\mathbb{C}$. Let for each x in G

$$m(x) = F(\delta_x) \,,$$

then we have for each x, y in G

$$m(x + y) = F(\delta_{x+y}) = F(\delta_x * \delta_y) = F(\delta_x)F(\delta_y) = m(x)m(y) \ .$$

Obviously, K is the ring generated by $\mathbb{C}$ and by the elements $m(x)$ with x in G, in other words $K = \mathbb{C}[m(G)]$.

Let T denote the torsion subgroup of G, then G/T is torsion free. Let X be a subset in G containing exactly one element from each coset of T. We show that the cardinality $|X|$ of X satisfies $|X| < 2^\omega$. Supposing the contrary, by assumption, G/T contains a free Abelian group of rank $\geq 2^\omega$. Let Y be a set of generators of this free group with $|Y| \geq 2^\omega$. Let $\psi : G \to G/T$ denote the natural homomorphism, and select a point from each set $\psi^{-1}(y)$ with y in Y. Since the set Z of these points is linearly independent, it follows that the torsion-free rank of G is at least $|Z| \geq 2^\omega$, a contradiction. Consequently, $|X| < 2^\omega$.

Every element g in G can be written in the form $g = x + t$ with x in X and t in T. For t in T, there is a natural number n with $nt = 0$, hence $1 = m(0) = m(nt) = m(t)^n$, and we infer that $m(t)$ is an nth root of unity. In particular, $m(t)$ is a complex number for each t in T, that is, $m(T)$ is in $\mathbb{C}$. It follows that $K = \mathbb{C}[m(X)]$, and here $|m(X)| \leq |X| < 2^\omega = |\mathbb{C}|$. By Lemma 7.2, we have that K is an algebraic extension of $\mathbb{C}$. As $\mathbb{C}$ is algebraically closed, we conclude that $K = \mathbb{C}$.

For the necessity, we note that if the torsion-free rank of G is not less than 2^ω, then G contains a free subgroup of rank 2^ω. Hence, by Corollary 7.1, it is enough to show that spectral analysis does not hold on the free Abelian group of rank 2^ω. Although this is just a reformulation of problem 10(c) of [[22], chapter V. Section 3, p. 373], here is a direct proof by showing that if G is the free Abelian group of rank 2^ω, then there is a maximal ideal M in $\mathbb{C}G$ such that $\mathbb{C}G/M$ is not isomorphic to the complex field [20]. Indeed, let X be the set of generators of G, then $|X| = 2^\omega$. There is a surjective mapping $\phi : X \to \mathbb{C}(t)$, $\mathbb{C}(t)$ being the quotient field of the complex polynomial ring. Then there is a unique homomorphism $F : \mathbb{C}G \to \mathbb{C}(t)$ with $F(\delta_x) = \phi(x)$ for each x in X. As $\mathbb{C}(t)$ is a field, hence $\mathrm{Ker}\, F$ is a maximal ideal in $\mathbb{C}G$. However, $\mathbb{C}G/\mathrm{Ker}\, F \cong \mathbb{C}(t)$, hence it is not isomorphic to $\mathbb{C}$. This means that not every maximal ideal in $\mathbb{C}G$ is exponential, hence, by Theorem 7.17, spectral analysis fails to hold on G. Our proof is complete. □

Concerning spectral synthesis, in [23], the present author gave a counterexample for Elliott's result showing that not just the proof had a gap but, in fact, Elliott's theorem was not true. In the same paper, influenced by the previous result, the following conjecture has been formulated: spectral synthesis holds on an Abelian group if and only if the group has finite torsion-free rank. The necessity of this condition has been verified by the counterexample and the sufficiency remained an open problem for a couple of years. Now we settle the necessity part of the conjecture [24]. We need the following concept. Let G be an Abelian group. The variety V on G is called *decomposable* if it is the sum of two proper subvarieties. Here the "sum" is meant in the topological sense, that is, the algebraic sum of the two subvarieties is a dense subspace in V. We call V *indecomposable* if it is not decomposable. Clearly, by Theorem 7.6, V is indecomposable if and only if its annihilator is not the intersection of two ideals both being different from it. Using this concept, we can derive the following result.

Theorem 7.23 Let G be an Abelian group and V be a variety on G. If V is indecomposable, and synthesizable, then V is the variety of an exponential monomial.

Proof: If V is synthesizable, then it is the topological sum of all subvarieties generated by exponential monomials belonging to V, by definition. This means that we have, by Theorem 7.6,

$$V^{\perp} = \bigcap_{\varphi \in V} \tau(\varphi)^{\perp} , \tag{7.10}$$

where the intersection is extended to all exponential monomials φ in V. As V is indecomposable, hence, by definition, $V^{\perp}$ is equal to one of the factors of the intersection on the right side, that is $V^{\perp} = \tau(\varphi)^{\perp}$ for some exponential monomial φ in V. It follows $V = \tau(\varphi)$. $\qquad\square$

In particular, finite-dimensional indecomposable varieties are generated by exponential monomials. We note that the converse is also true: by theorem 12.27 on p. 176 in [9], it follows that exponential monomials are exactly those functions generating finite-dimensional indecomposable varieties.

From the above theorem, we conclude immediately that the variety of a generalized exponential monomial is synthesizable if and only if the generalized exponential monomial is an exponential monomial. As a consequence, we obtain the following theorem.

Theorem 7.24 Let G be an Abelian group and V a variety on G. If V contains a generalized exponential monomial which is not an exponential monomial, then spectral synthesis does not hold for V.

Let G denote the Abelian group of all finitely supported complex sequences equipped with pointwise addition. G is the direct sum of countably many copies of $\mathbb{Z}$. We define $f : G \to \mathbb{C}$ by

$$f(x) = \sum_{k=0}^{\infty} x_k^2$$

for $x = (x_0, x_1, \ldots, x_n \ldots)$ in G. We can write f in the form

$$f = \sum_{k=0}^{\infty} p_k^2,$$

where $p_k : G \to \mathbb{Z}$ is the projection to the kth copy for $k = 0, 1, \ldots$. Then it is an elementary calculation to check that the variety of f is generated by the functions $1, p_k$ and f, where $k = 0, 1, \ldots$, hence it is infinite dimensional, as the projections are linearly independent. On the other hand, it is easy to see that

$$\Delta_{1;y,z,w} * f = 0$$

holds for each y, z, w in G, where 1 denotes the exponential identically 1 on G. Hence f is a generalized exponential monomial, which is not an exponential monomial, consequently, its variety is not synthesizable, by Theorem 7.24. It follows that spectral synthesis does not hold on G. We note however, that, by Theorem 7.22, spectral analysis holds on G, as its torsion-free rank is countable. Hence we have an example for an Abelian group on which spectral analysis holds and spectral synthesis fails to hold. We also note that the construction of f works on any Abelian group with infinite torsion-free rank; hence, we have the following corollary:

Corollary 7.2 If spectral synthesis holds on an Abelian group, then the torsion-free rank of the group is finite.

As we noted above the converse of this statement has been formulated in [23] as a conjecture.

The above results on discrete spectral synthesis focus on the whole group. However, the problem about characterizing those varieties possessing spectral analysis or spectral synthesis is still open, although there are particular results. An obvious necessary condition for a variety to have spectral synthesis is that it does not contain a generalized exponential monomial which is not an exponential monomial but at this moment it is unclear if this condition is sufficient, too. Nevertheless, the existing examples suggest that only the existence of such functions violates spectral synthesis for a variety.

The first important result about spectral synthesis for special varieties is due to Malgrange [6] establishing synthesizability for the variety formed by the solution space of linear homogeneous partial differential equations with constant coefficients [25]. In [7], this result has been generalized by L. Ehrenpreis to the famous Principal Ideal Theorem for varieties in the space of infinitely differentiable functions whose annihilator is a principal ideal. Finally, in [8], a similar result has been proved for varieties on locally compact Abelian groups. The latter result uses the structure theory of locally compact Abelian groups. Here we give a simpler proof for a generalization of the discrete version of this theorem [26]. We recall that a continuous function $f : G \to \mathbb{C}$ on the locally compact Abelian group is an exponential monomial if and only if it has the form

$$f(x) = P(a_1(x), a_2(x), \ldots, a_n(x))m(x),$$

where $m : G \to \mathbb{C}$ is an exponential, $a_1, a_2, \ldots, a_n : G \to \mathbb{C}$ are additive functions and $P : \mathbb{C}^n \to \mathbb{C}$ is a polynomial in n variables. When dealing with exponential monomials and polynomials on Abelian groups, we may always suppose that all additive functions in their above representation are taken from a fixed basis of the linear space of all additive functions on G. In this case, the above representation is unique – assuming f is not identically zero, that is, the exponential m and the polynomial P is uniquely determined by f. In particular, the total degree of P is the degree of f. For more details, see [9].

Theorem 7.25 Let G be a discrete Abelian group. Then every variety that has a finitely generated annihilator ideal is synthesizable.

Proof: Suppose that the annihilator $I = \text{Ann}\, V$ of the variety V is generated by the measures $\mu_1, \mu_2, \ldots, \mu_n$. It is clear that the solution space of the system of functional equations

$$\mu_i * f = 0, \quad i = 1, 2, \ldots, n \tag{7.11}$$

coincides with the variety V. Indeed, $f : G \to \mathbb{C}$ is in $V = \text{Ann}\,\text{Ann}\, V = \text{Ann}\, I$ if and only if f annihilates the generators of I, which is equivalent to the system (7.11). Hence it is enough to show that in the solution space of this system of functional equations the exponential monomial solutions span a dense subspace. We will show this in the sequel.

Let F denote the (finitely generated) subgroup generated by the supports of the measures μ_i, $i = 1, 2, \ldots, n$. We suppose that a linearly independent set of additive functions on F is given and all exponential monomials on F are built up from this set. Given a finitely generated subgroup $H \supseteq F$ in G let E_H denote all finite linear combinations of those exponential monomials on H, which are solutions of the system (7.11) on H. In other words, E_H is the set of all *exponential polynomial* solutions of (7.11) on the group H. By Corollary 7.1, the set E_H is dense in the set of all solutions of (7.11) on H. It is obvious [27] that every exponential monomial on H can be extended to an exponential monomial on G: the set of all such possible extensions of the exponential polynomials in the sets E_H will be denoted by E, where H runs through all finitely generated subgroups of G containing F. We show that every function in E is a solution of (7.11) on G. In fact, it is enough prove the following statement: let $\varphi : G \to \mathbb{C}$ be an exponential monomial. Then φ is a solution of (7.11) on G if and only if the restriction of φ to F is a solution of (7.11) on the group F. The necessity of this condition is obvious. To prove the sufficiency suppose that φ has the following form:

$$\varphi(x) = P(a_1(x), a_2(x), \ldots, a_k(x)) m(x)$$

for each x in G, where $m : G \to \mathbb{C}$ is an exponential, the functions $a_i : G \to \mathbb{C}$ for $i = 1, 2, \ldots, k$ are additive and their restrictions to F are linearly independent, and $P : \mathbb{C}^k \to \mathbb{C}$ is an ordinary complex polynomial in k variables.

Suppose that φ is a solution of (7.11) on F, that is, we have for each x in F and for $j = 1, 2, \ldots, n$:

$$0 = \varphi * \mu_j(x) = \sum_{y \in G} \varphi(x - y)\mu_j(y) \tag{7.12}$$

$$= m(x) \sum_{y \in F} P(a_1(x - y), a_2(x - y), \ldots, a_k(x - y))m(-y)\mu_j(y).$$

For the sake of simplicity, we shall use standard multiindex notation and $a : G \to \mathbb{C}^k$ will denote the vector valued function $a = (a_1, a_2, \ldots, a_k)$, further

$$|\alpha| = \alpha_1 + \alpha_2 + \ldots + \alpha_k,$$
$$\partial = (\partial_1, \partial_2, \ldots, \partial_k), \quad \partial^\alpha = \partial_1^{\alpha_1} \partial_2^{\alpha_2} \cdots \partial_k^{\alpha_k},$$
$$a(x)^\alpha = a_1(x)^{\alpha_1} a_2(x)^{\alpha_2} \cdots a_k(x)^{\alpha_k}.$$

Using Taylor's Formula, we have

$$P(a(x - y)) = \sum_{|\alpha| \leq \deg P} \frac{1}{\alpha!} \partial^\alpha P(a(-y)) a(x)^\alpha.$$

We insert this into (7.12) to obtain

$$\sum_{|\alpha| \leq \deg P} \frac{1}{\alpha!} a(x)^\alpha \sum_{y \in F} \partial^\alpha P(a(-y))m(-y)\mu_j(y) = 0$$

for each x in G. As the additive functions $a_1, a_2, \ldots, a_k$ are linearly independent on F, hence so are the functions a^α for different multiindices α [28], lemma 2.7, p. 29. Consequently, the above equation implies that

$$\sum_{y \in F} \partial^\alpha P(a(-y))m(-y)\mu_j(y) = 0 \tag{7.13}$$

holds for $j = 1, 2, \ldots, n$. Multiplying these equations by $\frac{1}{\alpha!}m(x)a(x)^\alpha$ for any x in G and summing up for $|\alpha| \leq \deg P$, we obtain (7.12) for every x in G, which means that the exponential monomial φ is a solution of the system (7.11) on G.

We have proved that every function in E is an exponential polynomial solution of the system (7.12). Now let f be any solution of (7.12), further let $x_1, x_2, \ldots, x_k$ be arbitrary elements in G and $\varepsilon > 0$ a given number. If H denotes the subgroup generated by $x_1, x_2, \ldots, x_k$ and F, then H is finitely generated and the restriction of f is a solution of the system (7.12) on H. By assumption, there exists an exponential polynomial $\varphi_0 : H \to \mathbb{C}$ in E_H such that $|f(x_j) - \varphi_0(x_j)| < \varepsilon$ for $j = 1, 2, \ldots, k$. On the other hand, any extension φ of φ_0 to an exponential monomial on G belongs to E, and it satisfies the same inequalities, as $\varphi = \varphi_0$ on H. This proves that the exponential polynomial solutions of the system (7.12) form a dense subset in the set of all solutions of (7.12), and our theorem is proved. □

The purpose of the study of spectral synthesis for varieties on discrete Abelian groups is to find satisfactory necessary and sufficient conditions.

One step toward aim is presented in the recent paper [29], which is based on the following concept. A complex algebra A will be called *residually finite-dimensional* if, for each nonzero element a of A, there is an ideal of A of finite codimension not containing a. In other words, the algebra A is residually finite dimensional if and only if the intersection of the kernels of all algebra homomorphisms of A to finite-dimensional complex algebras is zero. Roughly speaking, A has *sufficiently many* finite-dimensional representations. Using this concept, we have the following characterization of synthesizable varieties [29].

Theorem 7.26   Let G be an Abelian group and let V be a variety on G. Then V is synthesizable if and only if $\mathbb{C}G/V^{\perp}$ is residually finite dimensional.

Proof: Let $I = V^{\perp}$. Suppose that V is synthesizable. Then V is the closure of the linear span of its finite-dimensional subvarieties. Thus if a is in $\mathbb{C}G\backslash I$, then there is a finite-dimensional subvariety U of V such that a is not in $U^{\perp}$. Letting $J = U^{\perp}$, we have $I \subseteq J$ and, obviously, U and $(\mathbb{C}G/U)^{*}$ are isomorphic as $\mathbb{C}G$-modules. In particular, $\mathbb{C}G/J$ has finite dimension. Since $\mathbb{C}G/J \cong (\mathbb{C}G/I)/(I/J)$ hence $\mathbb{C}G/I$ has turned out to be residually finite dimensional.

Conversely, assume that $\mathbb{C}G/I$ is residually finite-dimensional. Let W be the closure of the subspace spanned by the exponential monomials contained in V; note that W is a subvariety of V. Suppose that $W \neq V$. Then, by Theorem 7.14, there is an element a in $W^{\perp}\backslash I$. On the other hand, our assumption necessitates the existence of an ideal J of $\mathbb{C}G$ containing I such that $\mathbb{C}G/J$ has finite dimension and a is not in J. Let $U = J^{\perp}$. Evoking Theorem 7.14 and the module isomorphism of U and $(\mathbb{C}G/U)^{*}$, we see that U is a finite-dimensional subvariety of V such that a is not in $U^{\perp}$. However, a finite-dimensional $\mathbb{C}G$-module is spanned by its one-generator indecomposable submodules, whence $U \subseteq W \subseteq \{a\}^{\perp}$, a contradiction completing the proof.   □

As an application of this concept we obtain the result below.

Corollary 7.3   Let V, W be synthesizable varieties on the Abelian group G. Then the variety $V + W$ is synthesizable.

We note that here $V + W$ means the *topological sum* of V and W, that is, the closure of the algebraic sum.

Proof: We have to show that $\mathbb{C}G/(V + W)^{\perp}$ is residually finite dimensional. By Theorem 7.14, we have $(V + W)^{\perp} = V^{\perp} \cap W^{\perp}$. Let b be in $\mathbb{C}G$ but not in $V^{\perp} \cap W^{\perp}$. If b is not in $V^{\perp}$, then there is an ideal L of $\mathbb{C}G$ and of finite codimension such that $V^{\perp} \subseteq L$ and b is not in L. If b is in $V^{\perp}$, then b is not in $W^{\perp}$ and similarly, there is an ideal L of $\mathbb{C}G$ of finite codimension containing $W^{\perp}$ to which b does not belong. In both cases, there is an ideal containing $V^{\perp} \cap W^{\perp}$ of finite codimension to which b does not belong, hence $\mathbb{C}G/(V+W)^\perp$ is residually finite dimensional. □

Theorems 7.19–7.21 and Lefranc's theorem 7.1 suggest a deep relation between spectral synthesis and the Noetherian property. Nevertheless, simple examples show that the Noetherian property of $\mathbb{C}G$ is not necessary for spectral synthesis on G [[30], theorem 3.3]. To understand the situation better, we need the following concept: the ring R is called *locally Noetherian* if, for every maximal ideal M in R, the localization of R at M is Noetherian. Clearly, every Noetherian ring is locally Noetherian [[31], corollary 6.5, p. 15]. We illustrate the usefulness of these concepts by presenting a short proof of the result in [13] characterizing discrete Abelian groups with spectral synthesis. In fact, we prove the following equivalence result:

Theorem 7.27 Let G be an Abelian group. Then the following conditions are equivalent:

  (i) G has finite torsion-free rank.
 (ii) $\mathbb{C}G$ is locally Noetherian.
(iii) Every homomorphic image of $\mathbb{C}G$ is residually finite dimensional.
 (iv) Spectral synthesis holds on G.

Proof: Suppose that G has finite torsion-free rank. By theorem A (i) in [32], $\mathbb{C}G$ is locally Noetherian.

Now suppose that $\mathbb{C}G$ is locally Noetherian and let I be an ideal in $\mathbb{C}G$ and let b be an element of $\mathbb{C}G$ which is not in I. We need to show that $\mathbb{C}G/I$ is residually finite dimensional. Let $J = (I : b)$, the set of those x in $\mathbb{C}G$ for which bx is in I. Then J is a proper ideal of $\mathbb{C}G$ containing I. Let M be a maximal ideal of $\mathbb{C}G$ containing J. By Theorem 7.22 and [20], spectral analysis holds for G and Theorem 7.17 yields that M is exponential. Let $S = \mathbb{C}G\backslash M$. The unique maximal ideal of $\mathbb{C}G_M$, the localization of $\mathbb{C}G$ at M, is $S^{-1}M$. By Proposition 3.1 in [17],

$$S^{-1}(I + M^n) = S^{-1}I + S^{-1}M^n = S^{-1}I + (S^{-1}M)^n$$

for n in $\mathbb{N}$, and by assumption, the quotient $\mathbb{C}G_M/S^{-1}I$, which will be called R is a Noetherian local ring whose unique maximal ideal is $S^{-1}M/S^{-1}I$. Applying Krull's Intersection Theorem 7.19 to R, we obtain that $\bigcap_{n\in\mathbb{N}} S^{-1}(I + M^{n+1}) = S^{-1}I$.

Consider the image $\frac{b}{1}$ of b under the natural homomorphism of $\mathbb{C}G$ onto $\mathbb{C}G_M$. If $\frac{b}{1}$ is in $S^{-1}I$, then bs belongs to I for some s in S. However, this implies that s belongs to J, a contradiction because of $J \subseteq M$.

It follows that $\frac{b}{1}$ is not in $S^{-1}I + (S^{-1}M)^{n+1}$ for some positive integer n. Now letting $L = S^{-1}I + (S^{-1}M)^{n+1}$, the codimension of L in $\mathbb{C}G_M$ is finite because $\mathbb{C}G_M$ is Noetherian. Furthermore, $\mathbb{C}G_M/L$ becomes a $\mathbb{C}G$-module by the obvious definition $r(v + L) = \frac{r}{1}v + L$ whenever v is in $\mathbb{C}G_M$ and r is in $\mathbb{C}G$. Since I annihilates this module, while b does not, $\mathbb{C}G/I$ emerges as residually finite dimensional.

If every homomorphic image of $\mathbb{C}G$ is residually finite dimensional, then, by Theorem 7.26, every variety on G is synthesizable, hence spectral synthesis holds on G. Finally, if spectral synthesis holds on G, then, by Corollary 7.2, the torsion-free rank of G is finite. $\qquad \square$

As it has been shown in [[33], theorem 3], the torsion-free rank of an Abelian group is finite if and only if the ring of polynomial functions on the group is Noetherian. A side-result of the previous theorem is the following:

Theorem 7.28 The complex group algebra of an Abelian group is locally Noetherian if and only if the ring of polynomial functions on the group is Noetherian.

In fact, both of these properties are equivalent to spectral synthesis on the group. Another related result follows.

Theorem 7.29 Let G be an Abelian group possessing spectral analysis. Spectral synthesis holds on G if and only if for each maximal ideal M in $\mathbb{C}G$ the module M/M^2 is a finite-dimensional complex vector space.

Proof: First we recall that in any case M/M^2 is a complex vector space with the addition defined by

$$(\mu + M^2) + (\nu + M^2) = (\mu + \nu) + M^2$$

and with the multiplication with scalars defined by

$$c \cdot (\mu + M^2) = c\mu + M^2$$

for each μ, ν in M and c in $\mathbb{C}$. Suppose first that spectral synthesis holds on G and M is a maximal ideal in $\mathbb{C}G$. By spectral analysis, M is an exponential maximal ideal, and $\mathbb{C}G/M$ is isomorphic to the complex field. In particular, M is the annihilator of the variety of an exponential m: $M = \operatorname{Ann} \tau(m)$. Then we define the mapping $\Phi : \mathbb{C}G/M^2 \to \mathbb{C}G/M$ in the following manner: for each μ in $\mathbb{C}G$ we let

$$\Phi(\mu + M^2) = \mu + M.$$

If $\mu + M^2 = \mu' + M^2$, then $\mu - \mu'$ is in M^2, hence it is in M, which implies $\mu + M = \mu' + M$, that is, $\Phi(\mu + M^2) = \Phi(\mu' + M^2)$. It follows that the definition of Φ is correct and it is easy to see that Φ is a $\mathbb{C}G$-module homomorphism of $\mathbb{C}G/M^2$ onto $\mathbb{C}G/M$. Clearly, the kernel of Φ consists of those elements $\mu + M^2$ of $\mathbb{C}G/M^2$ for which

$$\Phi(\mu + M^2) = M$$

holds, that is Ker $\Phi = M/M^2$. By the Homomorphism Theorem, we have the isomorphism

$$\mathbb{C}G/M^2 M/M^2 \cong \mathbb{C}G/M. \tag{7.14}$$

In particular, this is an isomorphism of complex vector spaces. As spectral synthesis holds on G the torsion-free rank of G is finite. The torsion-free rank of G is equal to the dimension of Hom $(G, \mathbb{C})$, of the linear space of all complex additive functions on G. It follows that the linear space spanned by all exponential monomials of degree at most 1 corresponding to m is finite dimensional, too. This vector space is exactly Ann M^2. The algebraic dual space of the vector space Ann M^2 is isomorphic to $\mathbb{C}G/\text{Ann Ann} M^2 = \mathbb{C}G/M^2$. It follows that $\mathbb{C}G/M^2$ is a finite-dimensional vector space. As $\mathbb{C}G/M$ is isomorphic to the complex vector space, we infer from (7.14) that the vector space M/M^2 is finite dimensional.

To prove the converse statement first, we observe that in (7.14) the vector spaces M/M^2 and $\mathbb{C}G/M$ are finite dimensional; hence, $\mathbb{C}G/M^2$ has finite dimension, too. Repeating the above argument, we have that the dimensions of $\mathbb{C}G/M^2$ and Ann M^2 are equal, hence Hom $(G, \mathbb{C}) \subsetneq \text{Ann} M^2$ is finite dimensional. We infer that the torsion-free rank of G is finite, hence spectral synthesis holds on G. $\qquad\square$

An important result about synthesizability of a given variety is the following sufficient condition. For the proof, see [29].

Theorem 7.30 Let H be a subgroup of the Abelian group G and let V be a variety in $\mathscr{C}(G)$. Suppose that $V^\perp$ is generated by $V^\perp \cap \mathbb{C}H$ as an ideal in $\mathbb{C}G$ and that $\mathbb{C}H/(V^\perp \cap \mathbb{C}H)$ is residually finite dimensional. Then V is synthesizable.

We note that for a given variety V in $\mathscr{C}(G)$ and a for a subgroup H in G the set V_H of all the restrictions of the functions in V is a variety in H, as it is proved in [13], moreover $V_H^\perp = V^\perp \cap \mathbb{C}H$ in the sense that the functions in $V^\perp \cap \mathbb{C}H$ are considered as functions on H (being zero outside H). We may call the variety V_H the *restriction* of the variety V to H. Hence, by Theorem 7.26, this theorem can be formulated in the following way: if the annihilator of a variety on G is generated by the annihilator of its restriction to a subgroup H, and this restriction is synthesizable, then the variety is synthesizable.

An immediate corollary of the above results yields another sufficient condition for the synthesizability of a variety [29].

Corollary 7.4 Let G be an Abelian group and let V be a variety on G with the property that $V^\perp$ is generated by $V^\perp \cap \mathbb{C}H$ for some subgroup H of G of finite torsion-free rank. Then V is synthesizable.

We remark that using Corollary 7.4, we can give a short proof of Theorem 7.25 [[26], theorem 5] stating that a variety is synthesizable if its orthogonal complement is finitely generated. Let $V^\perp$ be the ideal of $\mathbb{C}G$ generated by the elements

$a_1, a_2, \ldots, a_n$. The subgroup H of G generated by the union of the supports of $a_1, a_2, \ldots, a_n$ is finitely generated and $V^\perp$ is generated by $V^\perp \cap \mathbb{C}H$.

7.4 Nondiscrete Spectral Synthesis

In the previous section, we presented the basic results about spectral analysis and synthesis on discrete Abelian groups. The situation on nondiscrete locally compact Abelian groups is more sophisticated. The fundamental result is due to Schwartz [1] stating spectral analysis and synthesis on the reals equipped with the Euclidean topology. A reasonable question is to ask whether this theorem can be extended to more general locally compact Abelian groups. We have seen that the basic tools are available on any locally compact Abelian group: exponentials and exponential monomials are well-defined in the obvious manner just we always impose the additional continuity property on these functions. For the successful application of the annihilator technique, however, a crucial point is the identity $\operatorname{Ann}\operatorname{Ann} I = I$ for the ideals in the measure algebra, which does not hold, in general (see, e.g., the counterexample in [13]). Hence the following natural question arises on an arbitrary locally compact Abelian group: what is the necessary and sufficient condition for the equality $\operatorname{Ann}\operatorname{Ann} I = I$, whenever I is an ideal in $\mathcal{M}_c(G)$? We obviously have the following lemmas.

Lemma 7.3 Let G be a locally compact Abelian group, and let I be an ideal in $\mathcal{M}_c(G)$. Then $\operatorname{Ann} I = \operatorname{Ann}\operatorname{Ann}\operatorname{Ann} I$.

Proof: Let $V = \operatorname{Ann} I$, then V is a variety on G, hence, by Theorem 7.4, we have $\operatorname{Ann}\operatorname{Ann} V = V$. It follows $\operatorname{Ann} I = V = \operatorname{Ann}\operatorname{Ann} V = \operatorname{Ann}\operatorname{Ann}\operatorname{Ann} I$. □

Lemma 7.4 Let G be a locally compact group, and let V be a variety on G. Then $\operatorname{Ann} V$ is weak*-closed in $\mathcal{M}_c(G)$.

Proof: As $\operatorname{Ann} V = (\check{V})^\perp$ hence it is enough to show that $V^\perp$ is weak*-closed in $\mathcal{M}_c(G)$ for each variety V. Let $(\mu_\gamma)_{\gamma \in \Gamma}$ be a generalized sequence in $V^\perp$ which converges to μ in the weak*-topology. This means that for each f in $\mathcal{C}(G)$ the generalized sequence $(\langle \mu_\gamma, f \rangle)_{\gamma \in \Gamma}$ converges to $\langle \mu, f \rangle$. If f is in V, then $\langle \mu_\gamma, f \rangle = 0$ for each γ in Γ, and we infer $\langle \mu, f \rangle = 0$, hence μ is in $V^\perp$. □

Now we can answer our previous question.

Theorem 7.31 Let G be a locally compact group, and let I be an ideal in $\mathcal{M}_c(G)$. Then we have $\operatorname{Ann}\operatorname{Ann} I = I$ if and only if I is weak*-closed.

Proof: By the previous lemma, the annihilator of each variety is weak*-closed, hence $\operatorname{Ann}\operatorname{Ann} I$, as the annihilator of the variety $\operatorname{Ann} I$, is weak*-closed, which proves the necessity of our condition.

As in the previous theorem, we switch from annihilators to orthogonal complements. Thus, we suppose that I is weak*-closed, and I is a proper subset of $I^{\perp\perp}$. Let μ be in $I^{\perp\perp}$ such that μ is not in I. As the space $\mathcal{M}_c(G)$ with the weak*-topology is locally convex, hence, by the Hahn–Banach Theorem, there is a linear functional ξ in $\mathcal{M}_c(G)^*$, which annihilates I and $\langle \xi, \mu \rangle \neq 0$. By Theorem 7.1, it follows that there is an f in $\mathscr{C}(G)$ with $\langle \xi, v \rangle = \langle v, f \rangle$ for each v in $\mathcal{M}_c(G)$. We infer $\langle \mu, f \rangle = \langle \xi, \mu \rangle \neq 0$, and μ is in $I^{\perp\perp}$, hence f is not in $I^{\perp\perp\perp}$. On the other hand, $\langle v, f \rangle = \langle \xi, v \rangle = 0$ for each v in I, as ξ annihilates I, which implies that f is in $I^{\perp}$, however, this contradicts to Lemma 7.3. Our theorem is proved. $\qquad\square$

The fact that the reflexivity property of orthogonal complements and annihilators holds for weak*-closed ideals only, makes things more complicated in nondiscrete locally compact Abelian groups. In particular, the annihilator method can be applied with restrictions only since not every ideal, especially not every maximal ideal in $\mathcal{M}_c(G)$ is the annihilator of some variety.

Theorem 7.32 Let G be an Abelian group. Then spectral analysis holds for the variety V on G if and only if every weak*-closed maximal ideal of $\mathcal{M}_c(G)/\mathrm{Ann}\, V$ is exponential.

Proof: Clearly, the condition is equivalent to that every weak*-closed maximal ideal of $\mathcal{M}_c(G)$ containing $\mathrm{Ann}\, V$ is exponential. Suppose spectral analysis holds for V and let M be a weak*-closed maximal ideal with $\mathrm{Ann}\, V \subseteq M$. Then $\mathrm{Ann}\, M$ is a nonzero subvariety in V, hence there is an exponential m in $\mathrm{Ann}\, M$, and we infer $\mathrm{Ann}\, M_m = \tau(m) \subseteq \mathrm{Ann}\, M$, consequently $M \subseteq M_m$. By maximality, we deduce that $M = M_m$, hence M is an exponential maximal ideal.

Conversely, suppose that every weak*-closed maximal ideal containing $\mathrm{Ann}\, V$ is exponential and let V_0 be a nonzero subvariety in V. We have $\mathrm{Ann}\, V \subseteq \mathrm{Ann}\, V_0$ and $\mathrm{Ann}\, V_0$ is a proper ideal. Let M be a weak*-closed maximal ideal in $\mathcal{M}_c(G)$ containing $\mathrm{Ann}\, V_0$, then, by assumption, M is exponential, hence $M = M_m$ for some exponential m. We conclude $\tau(m) = \mathrm{Ann}\, M_m \subseteq V_0$, thus m is in V_0 and the theorem is proved. $\qquad\square$

The first application of spectral synthesis ideas is due to B. Malgrange who proved the following theorem in [25] (see also [6]):

Theorem 7.33 For every nonzero linear partial differential operator with constant coefficients $P(D)$ in $\mathbb{R}^n$, the solution space of the partial differential equation $P(D)f = 0$ is synthesizable.

It follows that every solution of this partial differential equation is the uniform limit on compact sets of linear combinations of exponential monomial solutions of the form $x \mapsto p(x) \exp\langle \lambda, x \rangle$, where p is a complex polynomial in n variables, λ is a complex vector in $\mathbb{C}^n$ and $\langle \lambda, x \rangle = \sum_{k=1}^{n} \lambda_k \cdot x_k$. As such solutions can be computed easily by solving simple algebraic equations for the components of λ and the coefficients of p we obtain an effective method to describe a dense set in the solution space. The above theorem has been generalized by Ehrenpreis [7] to the following form, which is sometimes referred to as the "Principal Ideal Theorem":

Theorem 7.34 Every variety in $\mathscr{E}(\mathbb{C}^n)$ is synthesizable if its annihilator is a principal ideal.

Here $\mathscr{E}(\mathbb{C}^n)$ denotes the Schwartz's space of infinitely differentiable complex-valued functions on $\mathbb{C}^n$.

Although different generalizations under various additional conditions appeared on nondiscrete locally compact Abelian groups [3, 8], but despite several efforts an extension of Schwartz's spectral synthesis result from $\mathbb{R}$ to $\mathbb{R}^n$ was not available. Finally, in [5], Gurevich published the following result:

Theorem 7.35

1. For each natural number $n \geq 2$, there exist compactly supported measures μ, ν such that the exponential monomial solutions of the system of functional equations

$$\mu * f = 0, \quad \nu * f = 0$$

   do not span a dense subspace in the solution space of this system. In other words, spectral synthesis fails to hold in $\mathbb{R}^n$ for $n \geq 2$.
2. For each natural number $n \geq 2$, there exist compactly supported measures $\mu_1, \mu_2, \dots, \mu_6$ such that the system

$$\mu_k * f = 0, \quad k = 1, 2, \dots, 6$$

   has no exponential monomial solution. In other words, spectral analysis fails to hold in $\mathbb{R}^n$ for $n \geq 2$.

The counterexamples of Gurevich show that in the space of continuous functions in several variables the classical approach does not work. In the following section, we present another reasonable way to generalize Schwartz's result for higher dimensions.

7.5 Spherical Spectral Synthesis

Let G be a locally compact group. Then *convolution* in $\mathscr{M}_c(G)$ is defined by

$$\langle \mu * \nu, f \rangle = \int f(xy) \, d\mu(x) \, d\nu(y)$$

for each μ, v in $\mathcal{M}_c(G)$ and x in G. Convolution converts the linear space $\mathcal{M}_c(G)$ into a topological algebra with unit δ_e, e being the identity in G. In general, δ_x denotes the point mass with support $\{x\}$ for each x in G. In addition, $\check{\mu}$ is defined by

$$\langle \check{\mu}, f \rangle = \langle \mu, \check{f} \rangle$$

for each μ in $\mathcal{M}_c(G)$ and f in $\mathscr{C}(G)$.

Convolution of measures in $\mathcal{M}_c(G)$ with arbitrary functions in $\mathscr{C}(G)$ is defined by the similar formula

$$\mu * f(x) = \int f(y^{-1}x) \, d\mu(y)$$

for each μ in $\mathcal{M}_c(G)$, f in $\mathscr{C}(G)$ and x in G. With the action $f \mapsto \mu * f$, the space $\mathscr{C}(G)$ is a topological left module over $\mathcal{M}_c(G)$.

In what follows K will always denote a compact subgroup in G with normalized Haar measure ω. We recall that ω is left invariant, right invariant, and inversion invariant. The function f in $\mathscr{C}(G)$ is called K-*invariant*, if it satisfies

$$f(kxl) = f(x),$$

whenever x is in G and k, l are in K. All K-invariant functions form a closed subspace in the topological vector space $\mathscr{C}(G)$, which we denote by $\mathscr{C}(G//K)$. Clearly, $\check{f}$ is K-invariant if and only if f is K-invariant.

For each f in $\mathscr{C}(G)$, the function defined by

$$f^{\#}(x) = \int_K \int_K f(kxl) \, d\omega(k) \, d\omega(l)$$

for x in G is called the *projection* of f. Then the projection $f \mapsto f^{\#}$ is a continuous linear mapping of $\mathscr{C}(G)$ onto $\mathscr{C}(G//K)$ with $f^{\#\#} = f^{\#}$ and $(f^{\#})^{\vee} = (\check{f})^{\#}$ for each f in $\mathscr{C}(G)$, further f is K-invariant if and only if $f^{\#} = f$.

The *projection* $\mu^{\#}$ of the measure μ is defined by the equation

$$\langle \mu^{\#}, f \rangle = \langle \mu, f^{\#} \rangle = \int_G \int_K \int_K f(kxl) \, d\omega(k) \, d\omega(l) \, d\mu(x)$$

for each f in $\mathscr{C}(G)$. Clearly, μ^* is a measure. A measure μ is called K-*invariant*, if $\mu^* = \mu$. Obviously, the projection $\mu \mapsto \mu^{\#}$ is the adjoint of $f \mapsto f^{\#}$, hence it is a continuous linear mapping of $\mathcal{M}_c(G)$ onto the set $\mathcal{M}_c(G//K)$ of all K-invariant measures with the properties $\mu^{\#\#} = \mu^{\#}$, and $(\mu^{\#})^{\vee} = (\check{\mu})^{\#}$ for each μ in $\mathcal{M}_c(G)$. In particular, μ is K-invariant if and only if $\mu^{\#} = \mu$.

As a special case, the projection of the point mass δ_y defined by

$$\langle \delta_y^{\#}, f \rangle = f^{\#}(y) = \int_K f(kyl) \, d\omega(k) \, d\omega(l).$$

Using these measures, we introduce K-translation with y in G for each f in $\mathscr{C}(G)$ as the function $\tau_y f$ defined by the equation

$$\tau_y f(x) = \delta_{y^{-1}}^{\#} * f(x) = \int_K \int_K f(kylx) \, d\omega(k) \, d\omega(l)$$

for each x in G. In particular, for each K-invariant function f, we have

$$\tau_y f(x) = \int_K f(ykx)\, d\omega(k)$$

whenever x, y are in G. Similarly, we define

$$\tau_y \mu = \delta_{y^{-1}}^{\#} * \mu$$

for each μ in $\mathcal{M}_c(G//K)$.

A subset H in $\mathcal{C}(G//K)$ is called K-*invariant*, if for each f in H and y in G the function $\tau_y f$ is in H. A closed K-invariant linear subspace of $\mathcal{C}(G//K)$ is called a K-*variety*. Clearly, the intersection of any family of K-varieties is a K-variety. The intersection of all K-varieties including the K-invariant function f is called the K-variety *generated by* f and it is denoted by $\tau(f)$. This is the closure of the linear space spanned by all K-translates of f.

In the following theorem, we describe the space of all continuous linear functionals of the space $\mathcal{C}(G//K)$, that is, the topological dual $\mathcal{C}(G//K)^*$. This result will also justify the notation $\mathcal{M}_c(G//K)$ for the set of all K-invariant measures.

Theorem 7.36 The dual of $\mathcal{C}(G//K)$ is topologically isomorphic to the space of the restrictions of all K-invariant measures in $\mathcal{M}_c(G)$ to $\mathcal{C}(G//K)$.

Proof: Suppose that λ is in $\mathcal{C}(G//K)^*$ and we define

$$\langle \tilde{\lambda}, f \rangle = \langle \lambda, f^{\#} \rangle$$

for each f in $\mathcal{C}(G)$. Then obviously $\tilde{\lambda}$ is in $\mathcal{M}_c(G)$. We have

$$\langle (\tilde{\lambda})^{\#}, f \rangle = \langle \tilde{\lambda}, f^{\#} \rangle = \langle \lambda, f^{\#\#} \rangle = \langle \lambda, f^{\#} \rangle = \langle \tilde{\lambda}, f \rangle$$

for each f in $\mathcal{C}(G)$, hence $\tilde{\lambda}$ is K-invariant. If f is in $\mathcal{C}(G//K)$, then $f = f^{\#}$, and we infer

$$\langle \tilde{\lambda}, f \rangle = \langle \tilde{\lambda}, f^{\#} \rangle = \langle \lambda, f^{\#\#} \rangle = \langle \lambda, f^{\#} \rangle = \langle \lambda, f \rangle,$$

hence $\tilde{\lambda}$ coincides with λ on $\mathcal{C}(G//K)$. This shows that λ is the restriction of a K-invariant measure in $\mathcal{M}_c(G)$.

Conversely, it is clear that the restriction of every K-invariant measure in $\mathcal{M}_c(G)$ to $\mathcal{C}(G//K)$ is a linear functional on $\mathcal{C}(G//K)$, hence it is an element of $\mathcal{C}(G//K)^*$. Finally, it is obvious that the mapping $\lambda \mapsto \tilde{\lambda}$ is a topological isomorphism of $\mathcal{C}(G//K)^*$ onto the space of the restrictions of all K-invariant measures in $\mathcal{M}_c(G)$ to $\mathcal{C}(G//K)$. $\square$

It follows from this theorem that with the convolution of measures $\mathcal{M}_c(G//K)$ is a topological algebra with unit δ_e. It is easy to see that all finitely supported K-invariant measures form a dense subalgebra in $\mathcal{M}_c(G//K)$. Further, $\mathcal{C}(G//K)$ is a left module over $\mathcal{M}_c(G//K)$ with the ordinary convolution

$f \mapsto \mu * f$. Closed submodules of this module are exactly the K-varieties. Left, right and two-sided ideals in $\mathcal{M}_c(G//K)$ are called left K-ideal, right K-ideal and K-ideal, respectively. It is easy to see that all finitely supported K-invariant measures form a dense subalgebra in $\mathcal{M}_c(G//K)$.

As the role of orthogonal complements and annihilators is equally important for our purposes we introduce these concepts in this new setting, too. The following result has been proved in [34].

Theorem 7.37 Let I be a left K-ideal. Then

$$I^\perp = \{f : f \in \mathscr{C}(G//K), \ \langle \mu, f \rangle = 0 \quad \text{for each } \mu \in I\}$$

is a K-variety. Let V be a K-variety. Then

$$V^\perp = \{\mu : \mu \in \mathcal{M}_c(G//K), \langle \mu, f \rangle = 0 \quad \text{for each } f \in V\}$$

is a closed left K-ideal.

The K-variety $I^\perp$ is called the *orthogonal complement* of I and the closed left ideal $V^\perp$ is called the *orthogonal complement* of the K-variety V. The following duality theorem is of fundamental importance [34].

Theorem 7.38 For each K-variety V, we have $V^{\perp\perp} = V$. For each closed left K-ideal I, we have $I^{\perp\perp} = I$.

Proof: Obviously, we have $V^{\perp\perp} \supseteq V$ and $I^{\perp\perp} \supseteq I$.

Suppose now that $V^{\perp\perp} \supsetneq V$. Consequently, there is a function f in $V^{\perp\perp}$ such that f is not in V. By the Hahn–Banach Theorem [35], and by Theorem 7.36, there is a λ in $\mathcal{M}_c(G//K)$ such that $\langle \lambda, f \rangle \neq 0$, and λ vanishes on V. This means that λ is in $V^\perp$, and f is not in $V^{\perp\perp}$, a contradiction. Hence $V^{\perp\perp} = V$.

Similarly, suppose that $I^{\perp\perp} \supsetneq I$ and let μ be in $I^{\perp\perp}$ such that μ is not in I. As I is closed, by the Hahn–Banach Theorem, there is a linear functional Λ in the dual space $\mathcal{M}_c(G//K)^*$ such that Λ annihilates I but $\Lambda(\mu) \neq 0$. By Theorem 7.1, every weak*-continuous linear functional on a dual space arises from an element of the original space, that is, there is an f_Λ in $\mathscr{C}(G//K)$ with $\Lambda(\nu) = \langle \nu, f_\Lambda \rangle$ holds for each for each ν in $\mathcal{M}_c(G//K)$. As Λ annihilates I, we have

$$\Lambda(\nu) = \langle \nu, f_\Lambda \rangle = 0$$

for each ν in I, hence f_Λ is in $I^\perp$. On the other hand,

$$0 \neq \Lambda(\mu) = \langle \mu, f_\Lambda \rangle,$$

a contradiction, as μ is in $I^{\perp\perp}$. The proof is complete. □

As K-invariant measures have the property that, considering $\mathscr{C}(G)$ as a module over the algebra $\mathcal{M}_c(G)$, they leave $\mathscr{C}(G//K)$ invariant, hence $\mathscr{C}(G//K)$ itself is a topological module over the algebra $\mathcal{M}_c(G//K)$. Then the annihilators of subsets in $\mathscr{C}(G//K)$, respectively, in $\mathscr{M}_c(G//K)$ have the usual meaning from module theory [17]. Let V, respectively, I be a K-variety, respectively, a left K-ideal. Then the *annihilator* of V in $\mathscr{M}_c(G//K)$, respectively, of I in $\mathscr{C}(G//K)$ is defined as

$$\operatorname{Ann} V = \{\mu : \mu \in \mathscr{M}_c(G//K) \text{ and } \mu * f = 0 \quad \text{for each } f \text{ in } V\},$$

respectively,

$$\operatorname{Ann} I = \{f : f \in \mathscr{C}(G//K) \text{ and } \mu * f = 0 \quad \text{for each } \mu \text{ in } I\}.$$

Clearly, $\operatorname{Ann} V$, respectively, $\operatorname{Ann} I$ are closed subspaces in $\mathscr{M}_c(G//K)$, respectively, in $\mathscr{C}(G//K)$. Using the notation

$$\check{H} = \{\check{f} : f \in H\}, \quad \check{L} = \{\check{\mu} : \mu \in L\},$$

for each subset H in $\mathscr{C}(G)$ and L in $\mathscr{M}_c(G)$, we have the above simple relations between orthogonals and annihilators which can be verified by easy calculation [34].

Theorem 7.39 For each K-variety V and left K-ideal I, we have

$$\operatorname{Ann} V = (\check{V})^{\perp}, \quad \operatorname{Ann} I = (\check{I})^{\perp}.$$

By Theorem 7.38, we deduce the following corollary.

Corollary 7.5 For each K-variety V and closed K-ideal I, we have

$$\operatorname{Ann}(\operatorname{Ann} V) = V, \quad \operatorname{Ann}(\operatorname{Ann} I) = I.$$

Given the locally compact group G and its compact subgroup K we shall call (G, K) a *Gelfand pair* if the algebra $\mathscr{M}_c(G//K)$ is commutative. For basic knowledge on Gelfand pairs, the reader should consult with [36, 37]. We note that the commutativity of G is not necessary for (G, K) to be a Gelfand pair. A simple necessary and sufficient condition, as a consequence of the earlier results, can be given as follows.

Theorem 7.40 (G, K) is a Gelfand pair if and only if the measures $(\delta_y^{\#})_{y \in G}$ form a commuting family.

It is known [37] that if (G, K) is a Gelfand pair, then G is unimodular. We also have the following sufficient conditions of Gelfand [37]:

Theorem 7.41 Assume that there is a continuous involutive automorphism $\theta : G \to G$ such that $\theta(x)$ is in $Kx^{-1}K$ for each x in G. Then (G, K) is a Gelfand pair.

Corollary 7.6 Assume that there is a continuous involutive automorphism $\theta: G \to G$ such that the subgroup of all θ-fixed points K is compact, and every element x of G can be written in the form $x = ky$ with k in K and $\theta(y) = y^{-1}$. Then (G, K) is a Gelfand pair.

Proof: Let $x = ky$ with k in K and $\theta(y) = y^{-1}$, then

$$\theta(x) = \theta(k) \cdot \theta(y) = k \cdot y^{-1} = k \cdot x^{-1} \cdot k \in Kx^{-1}K,$$

hence the statement follows from the previous theorem. $\qquad\qquad\square$

An important special case is obtained when the group G is a semidirect product. Let N be a locally compact group and K a compact topological group of continuous automorphisms of N. Hence, as a group, K is a subgroup of the group Aut (N) of all continuous automorphisms of N, and K is equipped with a compact topology which is compatible with the group structure. We consider the semidirect product $G = K \ltimes N$, where the operation is defined by

$$(k, n) \cdot (l, m) = (k \cdot l, k(m) \cdot n) \tag{7.15}$$

for each $(k, n), (l, m)$ in G. Here $k(m)$ denotes the image of the element m in N under the automorphism k in K. The identity of this group is (id, e), where id is the identity mapping on N and e is the identity element of N, further, the inverse of (k, n) is

$$(k, n)^{-1} = (k^{-1}, k^{-1}(n^{-1})).$$

In general, K is topologically isomorphic to the compact subgroup given by $\{(k, e): k \in K\}$ of G, and N is topologically isomorphic to the normal subgroup given by $\{(id, n): n \in N\}$. We note that G is not necessarily commutative even if N and K is commutative.

Suppose now that N is commutative and we write the operation in N as addition with $e = o$. If we define $\theta: G \to G$ by

$$\theta(k, n) = (k, -n)$$

for each k in K and n in N, then we have that

$$\theta[(k, n)(l, m)] = \theta(kl, k(m) + n) = (kl, -k(m) - n)$$

$$= (k, -n)(l, -m) = \theta(k, n)\theta(l, m),$$

that is, θ is a continuous involutive automorphism of G. On the other hand, we have

$$\theta(k, n) = (k, -n) = (k, o)(k^{-1}, k^{-1}(-n))(k, o) = (k, o)(k, n)^{-1}(k, o),$$

that is, $\theta(k, n)$ is in $K \cdot (k, n)^{-1} \cdot K$ for each (k, n) in G. By Theorem 7.41, we have the following result.

Corollary 7.7 With the above notation $G = K \ltimes N$ we have that (G, K) is a Gelfand pair.

In this case, we can say – somewhat loosely – that (N, K) is a Gelfand pair. The continuous function $f : K \times N \to \mathbb{C}$ is K-invariant if and only if

$$f(k, n) = f((k', o)(k, n)(l', o)) = f(k'kl', k'(n))$$

for each k, k', l' in K and x in $\mathbb{N}$. With the choice $k' = l$ and $l' = k^{-1}l^{-1}$ it follows that $f(k, n) = f(id, l(n))$ for each k, l in K and n in N. This means that K-invariant functions depend on the second variable only, that is, they can be identified with continuous functions on N by restriction $f \mapsto f|_N$, and this restriction is invariant with respect to the action of K on N: $f|_N(k(n)) = f|_N(n)$ for each k in K and n in N. These functions on N are called K-*radial functions* and $\mathscr{C}(G//K)$ will be identified with the space $\mathscr{C}_K(N)$ of all K-radial functions. Hence f is in $\mathscr{C}_K(N)$ if and only if $f : N \to \mathbb{C}$ is a continuous function satisfying $f(k(n)) = f(n)$ for each k in K and n in N. The dual $\mathscr{M}_c(G//K)$ is the space of K-*radial measures* μ on G satisfying

$$\langle \mu, f \rangle = \int_G \int_K \int_K f(k'kl', k'(n)) \, d\omega(k') \, d\omega(l') \, d\mu(k, n)$$

for each continuous function $f : K \times N \to \mathbb{C}$. Clearly, K-radial measures can be identified with those measures μ on N, which satisfy

$$\int_N f(k(n)) \, d\mu(n) = \int_N f(n) \, d\mu(n)$$

for each continuous function $f : N \to \mathbb{C}$ and for every k in K. The space of K-radial measures will be denoted by $\mathscr{M}_K(N)$.

Given a continuous function $f : K \times N \to \mathbb{C}$ its K-projection is the function

$$f^\#(n) = \int_K f(kk'l', k(n)) \, d\omega(k) \, d\omega(l') = \int_K f(k, k(n)) \, d\omega(k).$$

For each (k, m) in $K \times N$, the K-radial measure $\delta^\#_{(k,m)}$ is independent of k: $\delta^\#_{(k,m)} = \delta^\#_m$, and for each K-radial function f, we have

$$\tau_m f(n) = \delta^\#_{-m} * f(n) = \int_K f(n + k(m)) \, d\omega(k)$$

Now suppose that X is a finite-dimensional linear space over $\mathbb{K} = \mathbb{R}$ or $\mathbb{C}$, and $GL(X)$ denotes the *general linear group* of all invertible linear operators on X. Clearly, we may identify $GL(X)$ with the topological group of all invertible $d \times d$ matrices with entries in $\mathbb{K}$, where d is the dimension of X. Suppose that K is a compact subgroup of $GL(X)$ and we consider the set of all *affine transformations* on X of the form

$$x \mapsto k \cdot x + v$$

for x in X, where k is from K and v is from X. Each transformation of this type can be uniquely described by the pair (k, v). All these affine transformations form a group for composition: the product of (k, v) and (l, w) in this order is given by

$$x \mapsto k \cdot (l \cdot x + w) + v = (k \cdot l) \cdot x + k \cdot w + v,$$

which is the affine transformation $(k \cdot l, k \cdot w + v)$. Consequently, the group of affine transformation can be identified with the semidirect product Aff $K = K \ltimes X$ which is called the *affine group of K on X*.

For instance, the *Poincaré group* Aff $O(1, 3)$ is the affine group of the *Lorentz group* $O(1, 3)$: $O(1, 3) \ltimes \mathbb{R}^{1,3}$, where $O(1, 3)$ is the *isometry group* of the real vector space $\mathbb{R}^{1,3} = \mathbb{R} \oplus \mathbb{R}^3$ equipped with the quadratic form, the *indefinite inner product*

$$\langle v, w \rangle = v_0 w_0 - \sum_{j=1}^{3} v_j w_j,$$

where $v = (v_0, v_1, v_2, v_3)$ and $w = (w_0, w_1, w_2, w_3)$.

One obtains further important special cases with the choice $N = \mathbb{R}^n$ and $K = O(n)$, or $K = SO(n)$, or $N = \mathbb{C}^n$ and $K = U(n)$, or $K = SU(n)$. In all these cases, $(\mathbb{R}^n, K)$ is a Gelfand pair, and the space $\mathscr{C}(G//K)$, respectively, $\mathscr{M}_c(G//K)$ will be identified with $\mathscr{C}_K(\mathbb{R}^n)$, respectively, $\mathscr{M}_K(\mathbb{R}^n)$. Convolution in $\mathscr{M}_K(\mathbb{R}^n)$, and between $\mathscr{M}_K(\mathbb{R}^n)$ and $\mathscr{C}_K(\mathbb{R}^n)$ is the same as in $\mathscr{M}_c(\mathbb{R}^n)$, and between $\mathscr{M}_c(\mathbb{R}^n)$ and $\mathscr{C}(\mathbb{R}^n)$.

From now on, we suppose that (G, K) is a Gelfand pair. For every f in $\mathscr{C}(G//K)$ and for each y in G, the K-invariant measure

$$D_{f;y} = \delta_{y^{-1}}^{\#} - f(y)\delta_e$$

is called the *modified K-spherical difference* corresponding to f and y [9, 12, 14]. Given f in $\mathscr{C}(G//K)$ the closed K-ideal generated by all modified K-spherical differences of the form $D_{f;y}$ with y in G will be denoted by M_f.

Theorem 7.42 Let f be in $\mathscr{C}(G//K)$. The K-ideal M_f is proper if and only if we have $f(e) = 1$, and f satisfies

$$\int_K f(xky) \, d\omega(k) = f(x)f(y) \tag{7.16}$$

for each x, y in G. In this case, M_f is a maximal ideal and $\mathscr{M}_c(G//K)/M_f \cong \mathbb{C}$.

Proof: Suppose that M_f is a proper ideal. Then $V = M_f^{\perp}$ is a nonzero K-variety, by Theorems 7.37 and 7.38. Let $\varphi \neq 0$ be in V, then we have for each x, y, z in G

$$0 = \langle D_{f;y}, \delta_z^{\#} * \varphi \rangle = \langle \delta_{y^{-1}}^{\#}, \delta_z^{\#} * \varphi \rangle - f(y) \, \delta_z^{\#} * \varphi(e)$$

$$= (\delta_z^{\#} * \varphi)^{\#}(y^{-1}) - f(y) \int_K \int_K \varphi(kz^{-1}l) \, d\omega(k) \, d\omega(l)$$

$$= \int_K \int_K \int_K \varphi(z^{-1}ll'y^{-1}k') \, d\omega(l) \, d\omega(l') \, d\omega(k') - f(y)$$

$$\int_K \int_K \varphi(kz^{-1}l) \, d\omega(k) \, d\omega(l)$$

$$= \int_K \varphi(z^{-1}ly^{-1}) \, d\omega(l) - f(y) \int_K \varphi(z^{-1}l) \, d\omega(l)$$

$$= \int_K \varphi(z^{-1}ly^{-1}) \, d\omega(l) - f(y)\varphi(z^{-1}).$$

The substitution $z = e$ gives $\varphi(y^{-1}) = f(y)\varphi(e)$ thus $\varphi(e) \neq 0$ and we obtain equation (7.16) and $f(e) = 1$.

We have proved that V consists of all constant multiples of $\check{f}$. In particular, V is a one-dimensional vector space, which implies, by Theorem 7.38, that M_f is a maximal ideal. We show that $\mathcal{M}_c(G//K)/M_f$, as an algebra, is isomorphic to the algebra of complex numbers.

For each μ in $\mathcal{M}_c(G//K)$, we define

$$\Phi(\mu) = \langle \check{\mu}, f \rangle.$$

Clearly, $\Phi : \mathcal{M}_c(G//K) \to \mathbb{C}$ is a surjective continuous linear functional. For μ, ν in $\mathcal{M}_c(G//K)$, we have

$$\Phi(\mu * \nu) = \langle \mu * \nu, \check{f} \rangle = \int_G \int_G \check{f}(uv) \, d\mu(u) \, d\nu(v) = \int_G \int_G \check{f}(uv) \, d\mu^\#(u) \, d\nu^\#(v)$$

$$= \int_G \int_G \int_K \int_K \int_K \int_K \check{f}(kull'vk') \, d\omega(k) \, d\omega(l) \, d\omega(l') \, d\omega(k') \, d\mu(u) \, d\nu(v)$$

$$= \int_G \int_G \int_K \check{f}(ulv) \, d\omega(l) \, d\mu(u) \, d\nu(v) = \int_G \int_G \int_K f(ulv) \, d\omega(l) \, d\check{\mu}(u) \, d\check{\nu}(v)$$

$$= \int_G \int_G f(u)f(v) \, d\check{\mu}(u) \, d\check{\nu}(v) = \Phi(\mu)\Phi(\nu),$$

that is, Φ is an algebra homomorphism. On the other hand,

$$\Phi(D_{f;y}) = \Phi(\delta_{y^{-1}}^\#) - f(y)$$

$$= \langle (\delta_{y^{-1}}^\#)^\vee, f \rangle - f(y) = \langle \delta_y^\#, f \rangle - f(y) = f^\#(y) - f(y) = 0,$$

as f is K-invariant. It follows that M_f is a subset of the kernel of Φ. As Φ is nonzero, its kernel is a maximal ideal, hence, in fact, M_f is the kernel of Φ. As Φ is surjective, we have $\mathcal{M}_c(G//K)/M_f \cong \mathbb{C}$.

Now we suppose that $f(e) = 1$, and $\check{f}$ satisfies (7.16). Then we have for each y in G:

$$\langle D_{f;y}, \check{f} \rangle = \int_G \check{f}(t) \, d\delta_{y^{-1}}^\#(t) - f(y)f(e)$$

$$= \int_G \int_K \int_K f(kt^{-1}l) \, d\omega(k) \, d\omega(l) \, d\delta_{y^{-1}}^\#(t) - f(y)$$

$$= \int_K \int_K f(kyl) \, d\omega(k) \, d\omega(l) - f(y) = 0,$$

which proves that $\check{f}$ is in $M_f^\perp$, hence M_f is proper, by Theorem 7.38. The theorem is proved. □

We call the nonzero K-invariant function f a *K-spherical function*, if it satisfies equation (7.16) for each x, y in G. In this case, $f(e) = 1$. By the previous theorem, there is a one-to-one correspondence between K-spherical functions on G and those maximal ideals of the algebra $\mathcal{M}_c(G//K)$ whose residue algebra is isomorphic to $\mathbb{C}$. Such maximal ideals are called – in accordance with the terminology in the commutative case – *exponential maximal ideals* [9]. In other words, a maximal ideal in $\mathcal{M}_c(G//K)$ is exponential, if it is the kernel of a continuous algebra homomorphism of $\mathcal{M}_c(G//K)$ onto $\mathbb{C}$. In particular, every exponential maximal ideal is closed. It follows immediately that, in the case of commutative G, the K-spherical functions are exactly the exponentials of the group G/K. For more about spherical functions, see [36–41].

We shall call the function f in $\mathscr{C}(G)$ *normalized* if $f(e) = 1$. Using this terminology, we have the following characterization of K-spherical functions.

Theorem 7.43   The f be a continuous K-invariant function. Then the following statements are equivalent:

(i) The function f is a K-spherical function.
(ii) The function f is nonzero and satisfies (7.16) for each x, y in G.
(iii) The function f is normalized and for each K-invariant measure μ there exists a complex number λ_μ such that $\mu * f = \lambda_\mu \cdot f$.
(iv) The function f is a common normalized eigenfunction of all translation operators τ_y with y in G.
(v) The function f is normalized and the ideal M_f is an exponential maximal ideal.
(vi) The function f is normalized and the mapping $\mu \mapsto \langle \mu, \check{f} \rangle$ is a nonzero multiplicative functional of the algebra $\mathcal{M}_c(G//K)$ with kernel M_f.

Proof: The first two statements are equivalent, by definition.

Suppose that f is nonzero and satisfies (7.16) for each x, y in G. Let μ be a K-invariant measure, then we have

$$\mu * f(x) = \int_G f(y^{-1}x)\,\mathrm{d}\mu(y) = \int_G \check{f}(x^{-1}y)\,\mathrm{d}\mu(y) = \int_G \check{f}(x^{-1}y)\,\mathrm{d}\mu^\#(y)$$

$$= \int_G \int_K \int_K \check{f}(x^{-1}kyl)\,\mathrm{d}\omega(k)\,\mathrm{d}\omega(l)\,\mathrm{d}\mu(y) = \int_G \int_K \check{f}(x^{-1}ky)\,\mathrm{d}\omega(k)\,\mathrm{d}\mu(y)$$

$$= \int_G \int_K f(y^{-1}kx)\,\mathrm{d}\omega(k)\,\mathrm{d}\mu(y) = \int_G f(y^{-1})\,\mathrm{d}\mu(y) \cdot f(x),$$

which proves (iii) with $\lambda_\mu = \int_G f(y^{-1})\,\mathrm{d}\mu(y)$.

As $\tau_y f = \delta_{y^{-1}} * f$, (iii) obviously implies (iv).

If f is a common normalized eigenfunction of all translation operators τ_y with y in G, then $\tau(f)$ is a one-dimensional variety. It is easy to see that $\tau(\check{f})$ is a one-dimensional variety, too. Hence $\tau(\check{f})^\perp$ is a maximal ideal in $\mathcal{M}_c(G//K)$, by Theorem 7.38. In the proof of Theorem 7.42 we have seen that $\tau(\check{f})^\perp = M_f$. The proof of the statement that M_f is exponential is included in the proof of Theorem 7.42, too.

Now we suppose that the ideal M_f is an exponential maximal ideal and f is normalized. Then we define $\Phi(\mu) = \langle \mu, \check{f} \rangle$ for each μ in $\mathcal{M}_c(G//K)$. We can perform the same calculation as in the proof of Theorem 7.42 to show that Φ is a multiplicative functional of the algebra $\mathcal{M}_c(G//K)$. As

$$\Phi(\delta_e^{\#}) = \langle \delta_e, \check{f} \rangle = \check{f}(e) = f(e) = 1,$$

hence Φ is nonzero. The statement about the kernel of Φ is proved in Theorem 7.42, too.

Finally, we suppose that the mapping $\mu \mapsto \langle \mu, \check{f} \rangle$ is a nonzero multiplicative functional of the algebra $\mathcal{M}_c(G//K)$ with kernel M_f and f is normalized. We have for each x in G

$$f(x) = \langle \delta_{x^{-1}}^{\#}, \check{f} \rangle = \Phi(\delta_{x^{-1}}^{\#}),$$

hence

$$\begin{aligned}
f(x)f(y) &= \Phi(\delta_{x^{-1}}^{\#})\Phi(\delta_{y^{-1}}^{\#}) = \Phi(\delta_{x^{-1}}^{\#} * \delta_{y^{-1}}^{\#}) = \langle \delta_{x^{-1}}^{\#} * \delta_{y^{-1}}^{\#}, \check{f} \rangle \\
&= \int_G \int_G \check{f}(uv)\, \mathrm{d}\delta_{x^{-1}}^{\#}(u)\, \mathrm{d}\delta_{y^{-1}}^{\#}(v) \\
&= \int_G \int_G \int_K \int_K \int_K \int_K \check{f}(kulk'vl')\, \mathrm{d}\omega(k)\ \mathrm{d}\omega(l)\ \mathrm{d}\omega(k') \\
&\qquad \mathrm{d}\omega(l')\, \mathrm{d}\delta_{x^{-1}}(u)\, \mathrm{d}\delta_{y^{-1}}(v) \\
&= \int_G \int_G \int_K \check{f}(ulv)\, \mathrm{d}\omega(l)\, \mathrm{d}\delta_{x^{-1}}(u)\, \mathrm{d}\delta_{y^{-1}}(v) \\
&= \int_K \check{f}(x^{-1}ly^{-1})\, \mathrm{d}\omega(l) = \int_K f(ylx)\, \mathrm{d}\omega(l),
\end{aligned}$$

and the theorem is proved. $\qquad\qquad\square$

Now we present the extension of Fourier transformation [9, 42, 43] to the spherical case. For each K-invariant measure μ and K-spherical function s, we define

$$\hat{\mu}(s) = \langle \mu, \check{s} \rangle. \tag{7.17}$$

The function $\hat{\mu}: S_K(G) \to \mathbb{C}$ is called the *K-spherical Fourier–Laplace transform* of μ. The mapping $\mu \mapsto \hat{\mu}$ is called *K-spherical Fourier–Laplace transformation*. The following lemma is obvious.

Lemma 7.5 All spherical Fourier–Laplace transforms form a commutative algebra with unit with respect to pointwise operations.

Proof: Clearly, we have $\hat{\mu} + \hat{\nu} = (\mu + \nu)\hat{}$, $(c \cdot \mu)\hat{} = c \cdot \hat{\mu}$, and $\hat{\mu} \cdot \hat{\nu} = (\mu * \nu)\hat{}$, for each μ, ν in $\mathcal{M}_c(G//K)$ and for any complex number c. Further, we have $\hat{\delta}_e = 1$.
□

The algebra of all K-spherical Fourier–Laplace transforms will be called the *K-spherical Fourier algebra* of G and it is denoted by $\mathcal{A}(G//K)$. The following statement is obvious, by the previous lemma.

Theorem 7.44 The K-spherical Fourier–Laplace transformation is an algebra homomorphism of $\mathcal{M}_c(G//K)$ onto $\mathcal{A}(G//K)$.

As a simple example we consider the case $G = K \ltimes \mathbb{R}$, where $K = O(1)$. By the above results, $(\mathbb{R}, K)$ is a Gelfand pair. Clearly, $K = O(1) = \{-1, 1\}$ with the multiplication. The continuous function $f : \mathbb{R} \to \mathbb{C}$ is K-radial if and only if $f(x) = f(|x|)$ holds for each x in $\mathbb{R}$, further the compactly supported measure μ on $\mathbb{R}$ is radial, if

$$\int_{\mathbb{R}} f(|x|)\, \mathrm{d}\mu(x) = \int_{\mathbb{R}} f(x)\, \mathrm{d}\mu(x)$$

holds for each continuous function $f : \mathbb{R} \to \mathbb{C}$.

Clearly, the K-projection of the continuous function $f : K \times \mathbb{R} \to \mathbb{C}$ is given by

$$f^{\#}(x) = \frac{1}{2}[f(-1, -x) + f(1, x)] \tag{7.18}$$

for each x in $\mathbb{R}$. We also have

$$\tau_y f(x) = \delta^{\#}_{-y} * f(x) = \int_K f(x + k(y))\, \mathrm{d}\omega(k) = \frac{1}{2}[f(x + y) + f(x - y)]$$

for every K-radial function f and for each x, y in $\mathbb{R}$. This is the K-translation with y in $\mathbb{R}$. The closed linear subspace V in $\mathcal{C}(G//K)$ is a K-variety if and only for each y in G it contains the function $x \mapsto \varphi(x + y) + \varphi(x - y)$ whenever φ is in V. In particular, V is a one-dimensional K-variety if and only if it consists of all constant multiples of a nonzero continuous even function $\varphi : \mathbb{R} \to \mathbb{C}$ satisfying

$$\varphi(x + y) + \varphi(x - y) = 2\psi(y)\varphi(x) \tag{7.19}$$

for each x, y in $\mathbb{R}$, where $\psi : \mathbb{R} \to \mathbb{C}$ is continuous and nonzero. Clearly, $\varphi(0) \neq 0$, and $\varphi(x) = \varphi(0)\psi(x)$. It follows that $\psi : \mathbb{R} \to \mathbb{C}$ is even, too, and it satisfies d'Alembert's functional equation

$$\psi(x + y) + \psi(x - y) = 2\psi(y)\psi(x) \tag{7.20}$$

for each x, y in $\mathbb{R}$. In particular, the nonzero continuous even function $\varphi : \mathbb{R} \to \mathbb{C}$ is a K-spherical function if and only if it is a solution of the functional equation (7.20) for each x, y in $\mathbb{R}$. We conclude that $\psi(x) = \cosh \lambda x$ with some complex number λ, and these are the generating functions of one-dimensional K-varieties. Clearly, different complex numbers λ correspond to different K-spherical functions, hence $S_K(\mathbb{R})$ can be identified with the set of complex numbers. The K-spherical Fourier–Laplace transform of the compactly supported K-radial measure μ is the function $\hat{\mu} : \mathbb{C} \to \mathbb{C}$ defined by

$$\hat{\mu}(\lambda) = \int_{\mathbb{R}} \cosh \lambda x \; d\mu(x) \tag{7.21}$$

for each λ in $\mathbb{C}$. We show that, in this case, the mapping $\mu \mapsto \hat{\mu}$ is an algebra isomorphism from $\mathcal{M}_c(G//K)$ to $\mathcal{A}(G//K)$. By Theorem 7.44, it is enough to prove injectivity. Suppose that $\hat{\mu}(\lambda) = 0$ for each λ in $\mathbb{C}$. This means

$$\int_{\mathbb{R}} \cosh \lambda x \; d\mu(x) = 0 \tag{7.22}$$

for each λ in $\mathbb{C}$. On the other hand, as μ is K-radial, we have, by (7.18), for each λ in $\mathbb{C}$

$$\int_{\mathbb{R}} \sinh \lambda x \; d\mu(x) = \int_{\mathbb{R}} \sinh \lambda x \; d\mu^{\#}(x) = \int_{\mathbb{R}} \sinh^{\#} \lambda x \; d\mu(x)$$

$$= \frac{1}{2} \int_{\mathbb{R}} [\sinh \lambda x + \sinh(-\lambda x)] \; d\mu(x) = 0. \tag{7.23}$$

This implies $\int_{\mathbb{R}} e^{\lambda x} \; d\mu(x) = 0$ for each λ in $\mathbb{C}$, that is, the ordinary Fourier–Laplace transform of μ vanishes. We conclude $\mu = 0$.

As another example, we let $N = \mathbb{R}^n$ and $K = SO(n)$, the *special orthogonal group* and we consider the semidirect product $G = \text{Aff } SO(n) = SO(n) \ltimes \mathbb{R}^n$ which is called the *group of Euclidean motions* [37]. The elements (k, a) in G can be thought as the product of a rotation k in $SO(n)$ and a translation by a in $\mathbb{R}^n$. Hence the pair $g = (k, a)$ operates on $\mathbb{R}^n$ by the rule

$$g \cdot x = k \cdot x + a$$

for each x in $\mathbb{R}^n$. By Corollary 7.7, we conclude that $(\mathbb{R}^n, SO(n))$ is a Gelfand pair. K-radial functions are those continuous functions $f : \mathbb{R}^n \to \mathbb{C}$ satisfying

$$f(x) = f(k \cdot x)$$

whenever x is in $\mathbb{R}^n$ and k is a real orthogonal $n \times n$ matrix with determinant $+1$. Similarly, the compactly supported measure μ is K-radial if and only if it satisfies

$$\int_{\mathbb{R}^n} f(x) \; d\mu(x) = \int_{\mathbb{R}^n} f(k \cdot x) \; d\mu(x)$$

for each continuous function $f : \mathbb{R}^n \to \mathbb{C}$ and for each real orthogonal $n \times n$ matrix k with determinant $+1$. The following theorem is fundamental [37, 44].

Theorem 7.45 Let $n \geq 2$. The K-radial function $\varphi : \mathbb{R}^n \to \mathbb{C}$ is a K-spherical function if and only if it is $\mathscr{C}^\infty$, it is an eigenfunction of the Laplacian, and $\varphi(0) = 1$.

For each μ in $\mathscr{M}_c(\mathbb{R})$ and f in $\mathscr{C}_K(\mathbb{R}^n)$ we define μ_K by the equation

$$\langle \mu_K, f \rangle = \langle \mu, f_0 \rangle,$$

where the function $f_0 : \mathbb{R} \to \mathbb{C}$ is given by

$$f_0(r) = f(r, 0, 0, \ldots, 0)$$

for each r in $\mathbb{R}$. As f is K-radial, and $K = SO(n)$ acts transitively on the unit sphere in $\mathbb{R}^n$ we have that $\|x\| = \|y\|$ implies $f(x) = f(y)$ for x, y in $\mathbb{R}^n$. In particular, $f(\pm\|x\|, 0, 0, \ldots, 0) = f(x)$, and

$$f_0(\pm\|x\|) = f(x)$$

holds for each x in $\mathbb{R}^n$.

The function φ in $\mathscr{C}(G//K)$ is called a *generalized K-spherical monomial*, or simply *generalized K-monomial* if there exists a K-spherical function s and a natural number n such that we have for each $x, y_1, y_2, \ldots, y_{n+1}$

$$D_{s;y_1} * D_{s;y_2} * \cdots * D_{s;y_{n+1}} * \varphi(x) = 0. \tag{7.24}$$

By the following result, if φ is nonzero, then the K-spherical function s in the previous equation is uniquely determined by φ. More exactly, we have:

Lemma 7.6 The nonzero function φ in $\mathscr{C}(G//K)$ is a generalized K-monomial if and only if there exists a unique exponential maximal ideal M in $\mathscr{M}_c(G//K)$ and a natural number n such that

$$M^{n+1} \subseteq \operatorname{Ann} \tau(\varphi).$$

Proof: Let $\varphi \neq 0$ be a generalized K-monomial. Then there exists a K-spherical function s and a natural number n such that (7.24) holds for each $x, y_1, y_2, \ldots, y_{n+1}$ in G, and, by the commutativity of $\mathscr{M}_c(G//K)$, we have

$$D_{s;y_1} * D_{s;y_2} * \cdots * D_{s;y_{n+1}} * \psi(x) = 0$$

whenever ψ is in $\tau(\varphi)$. As the measures $D_{s;y_1} * D_{s;y_2} * \cdots * D_{s;y_{n+1}}$ generate the ideal whose closure is M_s^{n+1}, we infer that $M_s^{n+1} \subseteq \operatorname{Ann} \tau(\varphi)$. As φ is nonzero, hence $\operatorname{Ann} \tau(\varphi)$ is proper and there exists a maximal ideal M in $\mathscr{M}_c(G//K)$ such that $\operatorname{Ann} \tau(\varphi) \subseteq M$, which implies

$$M_s^{n+1} \subseteq M.$$

Maximal ideals are prime, hence we conclude $M_s = M$, which is an exponential maximal ideal. If N is a maximal K-ideal with the property that

$$N^{k+1} \subseteq \operatorname{Ann} \tau(\varphi)$$

for some natural number n, then we have

$$N^{k+1} \subseteq M_s,$$

and M_s is prime, hence we conclude $N = M_s$.

The converse statement is obvious. □

If s is a K-spherical function and $M_s^{n+1} \subseteq \operatorname{Ann} \tau(\varphi) \subseteq M_s$ holds for some natural number n, then we say that the generalized K-monomial φ *corresponds to* s, or is *associated with* s. By the above lemma, s is unique. The smallest natural number n with this property is called the *degree* of φ. The zero function is a generalized K-monomial and we may assign the degree 0 to it. Then, for instance, constant multiples of K-spherical functions are exactly the K-monomials of degree 0. First degree generalized K-monomials associated with the K-spherical function s are the K-invariant continuous solutions of the functional equation

$$D_{s;y} * D_{s;z} * f(x) = 0,$$

more explicitly

$$\int_K \int_K f(zkylx) \, d\omega(k) \, d\omega(l) + s(y)s(z)f(x)$$
$$= s(y) \int_K f(zkx) \, d\omega(k) + s(z) \int_K f(ykx) \, d\omega(l) \tag{7.25}$$

for each x, y, z in G. We have the following characterization result of K-spherical monomials of degree at most one.

Theorem 7.46 Given the K-spherical function $s : G \to \mathbb{C}$ the K-invariant function $f : G \to \mathbb{C}$ is a K-monomial of degree at most one corresponding to s if and only if it satisfies

$$\int_K f(xky) \, d\omega(k) + s(x)s(y)f(e) = s(x)f(y) + s(y)f(x) \tag{7.26}$$

for each x, y in G.

Proof: If we put $x = e$ in (7.25), then we get (7.26). Conversely, suppose that f satisfies (7.26). Substituting ylz for y in (7.26) and then integrating with respect to l over K, we get

$$\int_K \int_K f(xkylz) \, d\omega(k) \, d\omega(l) + s(x)s(y)f(z) \tag{7.27}$$
$$= s(z)s(y)f(x) + s(z)s(x)f(y) + 2s(y)s(x)f(z) - 2s(x)s(y)s(z)f(e)$$

for each x, y, z. On the other hand, from (7.26), we have the equations

$$s(x) \int_K f(ykz) \, d\omega(k) = s(x)s(z)f(y) + s(x)s(y)f(z) - s(x)s(y)s(z)f(e), \tag{7.28}$$

and

$$s(y)\int_K f(xkz)\, d\omega(k) = s(y)s(z)f(x) + s(x)s(y)f(z) - s(x)s(y)s(z)f(e). \quad (7.29)$$

Adding (7.28) and (7.29), we get the statement. □

In the case $f(e) = 0$ above, we obtain the functional equation

$$\int_K f(xky)\, d\omega(k) = s(x)f(y) + s(y)f(x), \quad (7.30)$$

which is a "spherical" analog of the sine functional equation. The K-invariant solutions f of this equation are called *s-sine functions*.

We call a generalized K-spherical monomial simply a *K-spherical monomial*, or *K-monomial*, if it generates a finite-dimensional K-variety. Hence every generalized K-monomial of degree 0 is a K-monomial, as it generates a one-dimensional K-variety. In the case of commutative G, the K-monomials are exactly the exponential monomials on G/K.

By the definition, the set of all generalized K-spherical monomials of degree n corresponding to the K-spherical function s is $\operatorname{Ann} M_s^{n+1}$ and the set of all generalized K-spherical monomials corresponding to s is $\bigcup_{n\in\mathbb{N}} \operatorname{Ann} M_s^{n+1}$. Further, given the K-variety V the set of all generalized K-spherical monomials of degree n corresponding to the K-spherical function s in V is $V \cap \operatorname{Ann} M_s^{n+1}$, and the set of all generalized K-spherical monomials corresponding to s in V is $V \cap \bigcup_{n\in\mathbb{N}} \operatorname{Ann} M_s^{n+1}$.

To get a better understanding of K-monomials, we describe these functions in the case of the affine group of $K = SO(n)$ on $\mathbb{R}^n$. We shall use the following notation: if $n \geq 2$ then, for each complex number λ we denote by s_λ the K-spherical function which is the eigenfunction of the Laplacian corresponding to the eigenvalue λ, by Theorem 7.45. If $n = 1$, then we let $s_\lambda(x) = \exp(\lambda \cdot x)$. Clearly, for each x in $\mathbb{R}^n$ the function $\lambda \mapsto s_\lambda(x)$ is an entire function. Then for each natural number k we denote by $s_\lambda^{(k)}$ the k-th derivative of this entire function. In other words, $s_\lambda^{(k)}(x)$ denotes the k-th derivative of s_λ with respect to the parameter λ. In [44], we have proved the following statement.

Theorem 7.47 For each complex number λ and natural number k, any K-monomial associated with s_λ of degree at most k is a linear combination of the functions $s_\lambda^{(j)}$ with $j = 0, 1, \dots, k$.

It follows that the linear span of any K-monomial associated with s_λ of degree at most k is identical with the linear space spanned by the derivatives of s_λ with respect to the parameter λ up to the order k.

Now we are in the position to define K-spectral analysis and K-spectral synthesis for a K-variety.

Let V be a K-variety. We say that *K-spectral analysis* holds for V, if in every nonzero sub-K-variety of V, there is a K-spherical function. If G is commutative, then this is equivalent to spectral analysis for the variety V in $\mathscr{C}(G/K)$.

Theorem 7.48 Let V be a K-variety. K-spectral analysis holds for V if and only if every maximal ideal in $\mathcal{M}_c(G//K)$ containing Ann V is exponential. In other words, K-spectral analysis holds for V if and only if every maximal ideal in the residue algebra $\mathcal{M}_c(G//K)/\mathrm{Ann}\,V$ is exponential.

Proof: If K-spectral analysis holds for V and M is a maximal K-ideal such that $\mathrm{Ann}\,V \subseteq M$, then obviously $V = \mathrm{Ann}\,\mathrm{Ann}\,V \supseteq \mathrm{Ann}\,M$. Hence $\mathrm{Ann}\,M$ is a nonzero sub-K-variety of V, which includes a K-spherical function s. It follows that $\mathrm{Ann}\,\tau(s)$, which is a maximal K-ideal, is a superset of M, hence they are equal: $\mathrm{Ann}\,\tau(s) = M$. As $\mathrm{Ann}\,\tau(s)$ is exponential, the necessity part of the theorem is proved.

Suppose now that every maximal K-ideal containing $\mathrm{Ann}\,V$ is exponential and let W be a nonzero sub-K-variety in V. Then $\mathrm{Ann}\,W \supseteq \mathrm{Ann}\,V$, hence every maximal K-ideal containing $\mathrm{Ann}\,W$ also contains $\mathrm{Ann}\,V$, thus it is exponential. Let M be one of them; then we have $M = M_s$ for some K-spherical function s, further clearly $\mathrm{Ann}\,W \supseteq M_s = \mathrm{Ann}\,\tau(s)$. We conclude $\tau(s) \subseteq W$, and the theorem is proved. $\square$

This is the analog of the spectral analysis theorems 14.2 and 14.3 on p. 203 in [9].

We say that *K-spectral analysis holds on G*, or *on the Gelfand pair* (G, K) if K-spectral analysis holds for each K-variety on G. This means that K-spectral analysis holds for $\mathscr{C}(G//K)$. If G is commutative, then this is exactly the same as spectral analysis on the group G/K.

Corollary 7.8 K-spectral analysis holds on G if and only if every maximal ideal in $\mathcal{M}_c(G//K)$ is exponential.

For instance, if G is a discrete Abelian group and K is any finite subgroup, then $\mathcal{M}_c(G//K)$ is isomorphic to $\mathcal{M}_c(G/K)$. In this case, the condition of the theorem is satisfied if and only if the torsion-free rank of the group G/K is less than the continuum [20].

Let V be a K-variety. We say that V is *K-synthesizable*, if all K-spherical monomials span a dense subspace in V. We say that *K-spectral synthesis* holds for V, if every sub-K-variety of V is K-synthesizable. We say that *K-spectral synthesis* holds on G, or *on the Gelfand pair* (G, K) if every K-variety on G is K-synthesizable. It is obvious that K-spectral synthesis implies K-spectral analysis for a K-variety. Clearly, K-synthesizability and K-spectral synthesis reduce to synthesizability and spectral synthesis on G/K if G is commutative.

If, for instance, G is a discrete Abelian group, and K is a finite subgroup, then K-spectral synthesis holds on G if and only if the torsion-free rank of G/K is finite [13].

For synthesizability of varieties, we have the following result [9, 12, 14].

Theorem 7.49 The nonzero K-variety V is K-synthesizable if and only if

$$\operatorname{Ann} V = \bigcap_{M} \bigcap_{n \in \mathbb{N}} (\operatorname{Ann} V + M^{n+1}),$$

where the first intersection is taken for all exponential maximal K-ideals M containing $\operatorname{Ann} V$ and $\mathcal{M}_c(G//K)/M^{n+1}$ is finite dimensional.

Proof: By definition of synthesizability and by the remarks at the end of the previous section, the K-variety V is synthesizable if and only if

$$V = \sum_{M} \sum_{n \in \mathbb{N}} (V \cap \operatorname{Ann} M^{n+1}).$$

On the other hand, applying theorem 8 on p. 6 in [12], our statement follows. □

By this theorem, the K-synthesizability of a K-variety can be expressed in an equivalent way in terms of the annihilator of the K-variety. We can introduce the following definition: let R be a commutative complex topological algebra with unit. The proper closed ideal I in R is called *synthesizable* if

$$I = \bigcap_{M} \bigcap_{n \in \mathbb{N}} (I + M^{n+1}), \tag{7.31}$$

where the first intersection is taken for all exponential maximal ideals M containing I and R/M^{n+1} is finite dimensional. Accordingly, we say that spectral analysis holds on R, if every maximal ideal is exponential, and spectral synthesis holds on R, if every closed ideal I in R satisfies the above equation. In particular, K-spectral analysis holds on the Gelfand pair (G, K) if spectral analysis holds on $\mathcal{M}_c(G//K)$, and K-spectral synthesis holds on G if and only if this spectral synthesis holds on $\mathcal{M}_c(G//K)$. Roughly speaking, we have reformulated the problem of spectral analysis and spectral synthesis in terms of the measure algebra and "removed" the underlying group. The following theorem is a simple consequence of the above definitions [34, 44].

Theorem 7.50 Let R, Q be commutative complex topological algebras with unit. If spectral analysis, respectively, spectral synthesis holds on R, and there exists a continuous surjective homomorphism $\Phi : R \to Q$, then spectral analysis, respectively, spectral synthesis holds on Q.

Proof: Let M be a maximal ideal with Q, then $M = \Phi(N)$ with some ideal N in R such that $N = \Phi^{-1}(M)$. Let $\psi : Q \to Q/M$ denote the natural mapping, then ψ is continuous and open. We define

$$F(r) = \psi(\Phi(r))$$

for each r in R, then $F: R \to Q/M$ is a continuous homomorphism. Clearly, F is surjective. If $F(r) = 0$, then $\Phi(r)$ is in Ker $\psi = M$, that is, r is in N. It follows that $R/N \cong Q/M$, a field, hence N is a maximal ideal. If N is exponential, then M is exponential, too, which proves the statement about spectral analysis.

Let J be a proper closed ideal in Q and let $I = \Phi^{-1}(J)$. Then I is a proper closed ideal in R, hence it is synthesizable, by assumption. It follows that (7.31) holds. Then we have

$$J = \bigcap_{\Phi(M)} \bigcap_{n \in \mathbb{N}} (J + \Phi(M)^{n+1}), \tag{7.32}$$

and here the first intersection extends for all maximal ideals $\Phi(M)$ containing J. Indeed, the left hand side is clearly a subset of the right hand side. Suppose now that $q = \Phi(r)$ is not in J, then r is not in I. By (7.31), there exists a maximal ideal M with $I \subseteq M$, and a natural number n_0 such that r is not in $I + M^{n_0+1}$, hence $q = \Phi(r)$ is not in $J + \Phi(M)^{n_0+1}$. It follows that (7.32) holds.

What is left is to show that $Q/\Phi(M)^{n+1}$ is finite dimensional for every maximal ideal M with $I \subseteq M$ and for each natural number n. Now we shall define $F: R/M^{n+1} \to Q/\Phi(M)^{n+1}$ by

$$F(r + M^{n+1}) = \Phi(r) + \Phi(M)^{n+1}$$

for each r in R. We have to show that the value of F is independent of the choice of r in the coset $r + M^{n+1}$. Suppose that $r - r_1$ is in M^{n+1}, that is $r - r_1 = \sum x_1 x_2 \cdots x_{n+1}$, where the sum is finite, and $x_1, x_2, \dots, x_{n+1}$ is in M. Then

$$\Phi(r) = \Phi(r_1) + \sum \Phi(x_1) \Phi(x_2) \cdots \Phi(x_{n+1}),$$

hence $\Phi(r)$ and $\Phi(r_1)$ are in the same coset of $\Phi(M)^{n+1}$. As F is clearly a surjective homomorphism, we infer that $Q/\Phi(M)^{n+1}$ is finite dimensional and the proof is complete. $\qquad\square$

Using the above results, we can extend L. Schwartz's spectral synthesis result in the following form [44].

Theorem 7.51 Let n be a positive integer and let $K = SO(n)$. Then K-spectral synthesis holds for the Gelfand pair $(\text{Aff}SO(n), \mathbb{R}^n)$.

According to our convention, this means that K-spectral synthesis holds on $SO(n) \ltimes \mathbb{R}^n$. Clearly, for $n = 1$ we have $SO(n) = \{id\}$, hence $SO(n) \ltimes \mathbb{R}^n$ is identical with $\mathbb{R}$, further K-spectral synthesis coincides with ordinary spectral synthesis. We can also formulate the following corollary. Here, as above, for each complex λ, we denote by s_λ the normalized K-radial eigenfunction of the Laplacian corresponding to the eigenvalue λ.

Theorem 7.52 Let $f: \mathbb{R}^n \to \mathbb{C}$ be a K-radial function. Then f is the uniform limit on compact sets of a sequence of functions which are linear combinations of functions of the form $s_\lambda^{(k)}$ in the K-variety of f.

7.6   Spectral Synthesis on Hypergroups

The basic problems of spectral analysis and synthesis can be formulated in the hypergroup setting as well. Indeed, hypergroups present a generalization of groups where translation operators are available hence the fundamental concepts in spectral analysis and synthesis make sense. In this section, we present some results about spectral analysis and synthesis on hypergroups.

We start with some facts about hypergroups. For more details about hypergroups, see [10, 45]. The concept of DJS–hypergroup (according to the initials of C. F. Dunkl, R. I. Jewett, and R. Spector) is due to Lasser [46]. One begins with a locally compact Hausdorff space K, the space $\mathcal{M}(K)$ of all finite complex regular measures on K, the space $\mathcal{M}_c(K)$ of all compactly supported measures in $\mathcal{M}(K)$, the space $\mathcal{M}^1(K)$ of all probability measures in $\mathcal{M}(K)$, and the space $\mathcal{M}_c^1(K)$ of all compactly supported probability measures in $\mathcal{M}(K)$. The point mass concentrated at x is denoted by δ_x. Suppose that we have the following:

- (H^*) There is a continuous mapping $(x, y) \mapsto \delta_x * \delta_y$ from $K \times K$ into $\mathcal{M}_c^1(K)$, the latter being endowed with the weak*-topology with respect to the space of compactly supported complex-valued continuous functions on K. This mapping is called convolution.
- $(\check{H})$ There is an involutive homeomorphism $x \mapsto \check{x}$ from K to K. This mapping is called involution.
- (He) There is a fixed element e in K. This element is called identity.

Identifying x by δ_x the mapping in (H^*) has a unique extension to a continuous bilinear mapping from $\mathcal{M}(K) \times \mathcal{M}(K)$ to $\mathcal{M}(K)$. The involution on K extends to an involution on $\mathcal{M}(K)$. Then a DJS–hypergroup, or simply hypergroup is a quadruple $(K, *, \check{\,}, e)$ satisfying the following axioms: for any x, y, z in K, we have

1. (H1) $\delta_x * (\delta_y * \delta_z) = (\delta_x * \delta_y) * \delta_z$;
2. (H2) $(\delta_x * \delta_y)^{\check{\,}} = \delta_{\check{y}} * \delta_{\check{x}}$;
3. (H3) $\delta_x * \delta_e = \delta_e * \delta_x = \delta_x$;
4. (H4) e is in the support of $\delta_x * \delta_{\check{y}}$ if and only if $x = y$;
5. (H5) the mapping $(x, y) \mapsto \text{supp}\,(\delta_x * \delta_y)$ from $K \times K$ into the space of nonvoid compact subsets of K is continuous, the latter being endowed with the Michael–topology [45].

If the topology of K is discrete, then we call the hypergroup discrete. If $\delta_x * \delta_y = \delta_y * \delta_x$ holds for all x, y in K, then we call the hypergroup commutative. If $\check{x} = x$ holds for all x in K, then we call the hypergroup Hermitian. By (H2), any Hermitian hypergroup is commutative. For instance, if $K = G$ is a locally compact Hausdorff group, $\delta_x * \delta_y = \delta_{xy}$ for all x, y in K, $\check{x}$ is the inverse of x, and e is the identity of G, then we obviously have a hypergroup $(K, *, \check{\,}, e)$, which is commutative if and only if the group G is commutative. The same does not work if $K = S$ is a locally compact Hausdorff semigroup with identity, and involution is the identity mapping, as (H4) does not hold, in general. However, not every hypergroup originates in this way.

In any hypergroup K, we identify x by δ_x and we define the right translation operator T_y by the element y in K according to the formula:

$$T_y f(x) = \int_K f \; d(\delta_x * \delta_y),$$

for any f integrable with respect to $\delta_x * \delta_y$. In particular, T_y is defined for any continuous complex-valued function on K. Similarly, we can define left translation operators. Sometimes one uses the suggestive notation

$$f(x * y) = \int_K f(t) \; d(\delta_x * \delta_y)(t)$$

for each x, y in K. Obviously, in case of commutative hypergroups the simple term translation operator is used. The function $T_y f$ is the translate of f by y.

The presence of translation operators on commutative hypergroups leads to the concept of variety. Let K be a locally compact Hausdorff space and let $\mathscr{C}(K)$ denote the locally convex topological vector space of all continuous, complex-valued functions on K, equipped with the pointwise linear operations and with the topology of uniform convergence on compact sets. The dual of this space can be identified with $\mathscr{M}_c(K)$, the latter being endowed with the weak*-topology with respect to the space of complex-valued continuous functions on K. If, in addition, K is equipped with a commutative hypergroup structure, then a subset H of $\mathscr{C}(K)$ is called translation invariant, if for any f in H the function $T_y f$ belongs to H for all y in K. A closed, translation invariant subspace of $\mathscr{C}(K)$ is called a variety. For any f in $\mathscr{C}(K)$, the variety generated by f is the closed subspace generated by all translates of f, which is denoted by $\tau(f)$. Clearly, $\tau(f)$ is the intersection of all varieties containing f.

The space $\mathscr{M}_c(K)$ is a locally convex topological vector space, which bears a natural algebra structure, corresponding to the convolution of measures. It is easy to see [45], that for any continuous function f in $\mathscr{C}(K)$ the function $(x, y) \mapsto f(x * y)$ is continuous. For any measures μ, ν in $\mathscr{M}_c(K)$ and for each f in $\mathscr{C}(K)$, we let

$$(\mu * \nu)(f) = \int_K \int_K f(x * y) \; d\mu(x) \; d\nu(y).$$

Then $\mu * \nu$ is an element of $\mathscr{M}_c(K)$, which is called the convolution of μ and ν. The space $\mathscr{M}_c(K)$ equipped with the pointwise linear operations and with the convolution is a commutative algebra with unit δ_e.

For each μ in $\mathscr{M}_c(K)$ and f in $\mathscr{C}(K)$, we define

$$\mu * f(x) = \int_K f(x * \check{y}) \; d\mu(y),$$

the *convolution* of μ and f. It is easy to see that $\mu * f$ is continuous. With this action of $\mathscr{M}_c(K)$ on $\mathscr{C}(K)$, the latter is a topological left module. As finitely supported measures form a weak*-dense subspace in $\mathscr{M}_c(K)$ it follows easily that for commutative K, the varieties are exactly the closed submodules of this module. Similarly to the group case, the annihilator of a variety is a closed ideal in $\mathscr{M}_c(K)$ and the annihilator of an ideal in $\mathscr{M}_c(K)$ is a variety in $\mathscr{C}(K)$. By the Hahn–Banach–theorem, we have the following basic result:

Theorem 7.53 Let K be a commutative hypergroup. For every variety V in $\mathscr{C}(K)$ and for each closed ideal I in $\mathscr{M}_c(K)$, we have

$$\operatorname{Ann}\operatorname{Ann} V = V, \quad \operatorname{Ann}\operatorname{Ann} I = I.$$

From now on, we shall suppose that every hypergroup is *commutative* and the identity usually will be denoted by o.

For any closed linear subspace V in $\mathscr{C}(K)$, its orthogonal complement $V^\perp$ in $\mathscr{M}_c(K)$ is the set of all measures from $\mathscr{M}_c(K)$, which vanish on V. Clearly, it is a closed linear subspace of $\mathscr{M}_c(K)$. The dual correspondence is also true: the orthogonal complement $I^\perp$ of any closed linear subspace I of $\mathscr{M}_c(K)$, that is, the set of all elements of $\mathscr{C}(K)$, which belong to the kernel of all linear functionals in I, is a closed linear subspace of $\mathscr{C}(K)$. Using the notation

$$\check{H} = \{\check{f} : f \in H\}, \quad \text{and} \quad \check{J} = \{\check{\mu} : \ \mu \in J\}$$

whenever H is a subset in $\mathscr{C}(K)$ and J is a subset in $\mathscr{M}_c(K)$, we have

$$\operatorname{Ann} V = (\check{V})^\perp, \quad \text{and} \quad \operatorname{Ann} I = (\check{I})^\perp$$

for each variety V in $\mathscr{C}(K)$ and ideal I in $\mathscr{M}_c(K)$. By the above theorem, this implies the obvious relations $V = V^{\perp\perp}$ and $I = I^{\perp\perp}$ for any variety V in $\mathscr{C}(K)$ and for each ideal I of $\mathscr{M}_c(K)$.

Let V be a variety in $\mathscr{C}(K)$. We say that spectral analysis holds for V, if every nonzero subvariety of V contains a one-dimensional variety. If spectral analysis holds for every variety in $\mathscr{C}(K)$, then we say that spectral analysis holds for the hypergroup K. It turns out that one-dimensional varieties are related to certain basic functions exactly in the same way as in the group case. These basic functions are also called *exponentials*: they are those nonidentically zero continuous functions $m : K \to \mathbb{C}$ satisfying $m(x * y) = m(x)m(y)$. We emphasize that the meaning of $m(x * y)$ is the following integral:

$$m(x * y) = \int_K m(t)\,\mathrm{d}(\delta_x * \delta_y)(t)$$

for each x, y in K. Hence exponential functions are characterized by the integral equation

$$\int_K m(t)\,\mathrm{d}(\delta_x * \delta_y)(t) = m(x)m(y),$$

which holds for each x, y in K. It is important to mention that, in contrast to the case of groups, exponentials on hypergroups may take zero values. The following example may be interesting.

Example: Let K be a two element set: $K = \{0, 1\}$, where 0 and 1 are different. We describe all possible hypergroup structures on K. We choose 0 as the identity element, then we must have $0 * 0 = 0$, $0 * 1 = 1$, $1 * 0 = 1$, hence the only nontrivial convolution to be defined is $1 * 1$. In particular, every hypergroup structure on K is commutative. Clearly, $\check{0} = 0$ in any hypergroup. If $\check{1} = 0$, then $1 = (\check{1})^\vee = \check{0} = 0$, a contradiction. Hence the only involution on K is the identity. We have to define $1 * 1 = \delta_1 * \delta_1$ to be a probability measure, which must have the form

$$\delta_1 * \delta_1 = \theta \cdot \delta_0 + (1 - \theta) \cdot \delta_1,$$

where $0 \leq \theta \leq 1$. If $\theta = 0$, then we have $1 * 1 = 0$, which is the two element group. If $\theta = 1$, then $1 * 1 = 1$, but this contradicts the axiom (H4). Nevertheless, it is easy to check that for any choice of $0 \leq \theta < 1$ we have a hypergroup, which is a group if and only if $\theta = 0$. Now we determine all exponentials m on this hypergroup. On any hypergroup every exponential is 1 at the identity, hence we have $m(0) = 1$. On the other hand, we must have

$$m(1)^2 = m(1)m(1) = \theta m(0) + (1 - \theta)m(1) = \theta + (1 - \theta)m(1),$$

which is a quadratic equation for θ. Solving it we have that either $m(1) = 1$ which gives the exponential identically 1, or $m(\theta) = -\theta$. Hence exponentials on this hypergroup are in a one-to-one correspondence with the $[-1, 0]$ interval and the pointwise product of two exponentials is an exponential if and only if K is a group, or at least one of them is identically 1.

Similarly to the group case, exponentials can be characterized by the properties listed in the following theorem. Recall that a function $f : K \to \mathbb{C}$ is called normalized if $f(o) = 1$.

Theorem 7.54 Let K be a commutative hypergroup and $f : K \to \mathbb{C}$ a continuous function. Then the following statements are equivalent:

1. f is an exponential.
2. f is normalized and $\tau(f)$ is one dimensional.
3. f is normalized and $\mathcal{M}_c(K)/\mathrm{Ann}\,\tau(f) \cong \mathbb{C}$.

The proof is similar to that of Theorem 7.9. We note that the third property implies that $\mathrm{Ann}\,\tau(f)$ is a maximal ideal. Like in the group case, maximal ideals whose residue field is topologically isomorphic to the complex field are called *exponential* maximal ideals. Of course, we can use this concept in any commutative topological algebra with unit.

As soon as exponentials are available on hypergroups the Fourier–Laplace transformation can be successfully utilized, too. Given the commutative hypergroup K the *Fourier–Laplace transform* $\hat{\mu}$ of μ in $\mathcal{M}_c(K)$ is defined by

$$\hat{\mu}(m) = \int_K \check{m} \, d\mu$$

for each exponential m on K. Let $\mathscr{E}(K)$ denote the set of all exponentials on K, then $\hat{\mu} : \mathscr{E}(K) \to \mathbb{C}$ is a continuous function for each μ in $\mathscr{M}_c(K)$ if $\mathscr{E}(K)$ is equipped with the topology of uniform convergence on compact sets. Indeed, if μ in $\mathscr{M}_c(K)$ is given and $(m_i)_{i \in I}$ is a net in $\mathscr{E}(K)$ converging to the exponential m, then for each $\varepsilon > 0$ there is an i_0 in I such that $|\check{m}_i(x) - \check{m}(x)| < \varepsilon$ whenever $i \geq i_0$ and x is in supp μ. Then we have

$$|\hat{\mu}(m_i) - \hat{\mu}(m)| = \left| \int_K (\check{m}_i - \check{m}) \, \mathrm{d}\mu \right|$$

$$\leq \int_{\text{supp}\,\mu} |\check{m}_i(x) - \check{m}(x)| \, \mathrm{d}|\mu|(x) \leq \varepsilon \cdot |\mu(K)|.$$

The basic properties of the mapping $\mu \mapsto \hat{\mu}$ are formulated in the following theorem.

Theorem 7.55 The Fourier–Laplace transform $\mu \mapsto \hat{\mu}$ is an injective algebra homomorphism of $\mathscr{M}_c(K)$ onto an algebra of continuous functions on $\mathscr{E}(K)$.

Proof: For μ, ν in $\mathscr{M}_c(K)$, we have

$$(\mu * \nu)\hat{}(m) = \int_K \check{m}(x) \, \mathrm{d}(\mu * \nu)(x) = \int_K \int_K \check{m}(x * y) \, \mathrm{d}\mu(x) \, \mathrm{d}\nu(y)$$

$$= \int_K \check{m}(x) \, \mathrm{d}\mu(x) \cdot \int_K \check{m}(y) \, \mathrm{d}\nu(y) = \hat{\mu}(m) \cdot \hat{\nu}(m)$$

for each exponential m. Injectivity follows from the injectivity of the Fourier transform [[45], theorem 2.2.24]. All the other statements are obvious. $\square$

The range of the Fourier–Laplace transform is called the *Fourier–Laplace algebra* of K.

Using the previous theorem, we can formulate our first theorem on spectral analysis on commutative hypergroups which is the analog of Theorem 7.32.

Theorem 7.56 Let K be a commutative hypergroup. Then spectral analysis holds for the variety V on K if and only if every weak*-closed maximal ideal of $\mathscr{M}_c(K)/\text{Ann}\,V$ is exponential.

A reasonable question arises: does the existence of nonzero finite-dimensional varieties imply spectral analysis? In other words: does spectral analysis hold for nonzero finite-dimensional varieties? The answer is yes, as it was proved in [47]. Here we give another simple proof for this statement.

Theorem 7.57 On commutative hypergroups spectral analysis holds for every nonzero finite-dimensional variety.

Proof: Let V be a nonzero finite-dimensional variety on the commutative hypergroup K. All translation operators τ_y with y in K form a commuting family of linear operators on the finite-dimensional vector space V. It is well-known from linear algebra then in this case all operators in the family have a common eigenvector in V. In other words, there exist a nonzero f in V such that $\tau_y f = \lambda(y) \cdot f$ for each y in K with some function $\lambda : K \to \mathbb{C}$. We have

$$f(x * y) = \tau_y f(x) = \lambda(y) \cdot f(x)$$

for each x, y in K. Putting $x = o$ we infer $f(y) = f(o)\lambda(y)$, in particular, $f(o) \neq 0$. It follows $f(o) \cdot \lambda(x * y) = f(o)\lambda(x) \cdot \lambda(y)$, which implies that $\lambda \neq 0$, λ is an exponential and λ is in V. □

By this theorem, spectral analysis for a variety on a commutative hypergroup means that every nonzero subvariety has a nonzero *finite-dimensional* subvariety.

We illustrate spectral analysis on a particular class of commutative hypergroups: polynomial hypergroups in one variable. The formal definition follows. Let $(a_n)_{n\in\mathbb{N}}$, $(b_n)_{n\in\mathbb{N}}$ and $(c_n)_{n\in\mathbb{N}}$ be real sequences with the following properties: $c_n > 0$, $b_n \geq 0$, $a_{n+1} \geq 0$ for all n in $\mathbb{N}$, moreover $a_0 = 0$, and $a_n + b_n + c_n = 1$ for all n in $\mathbb{N}$. We define the sequence of polynomials $(P_n)_{n\in\mathbb{N}}$ by $P_0(x) = 1$, $P_1(x) = x$, and by the recursive formula

$$xP_n(x) = a_n P_{n-1}(x) + b_n P_n(x) + c_n P_{n+1}(x)$$

for all $n \geq 1$ and x in $\mathbb{R}$. The following theorem holds [45].

Theorem 7.58 If the sequence of polynomials $(P_n)_{n\in\mathbb{N}}$ satisfies the above conditions, then there exist constants $c(n, m, k)$ for all n, m, k in $\mathbb{N}$ such that

$$P_n P_m = \sum_{k=|n-m|}^{n+m} c(n, m, k) P_k$$

holds for all n, m in $\mathbb{N}$.

Proof: By the theorem of Favard [48, 49], the conditions on the sequence of polynomials $(P_n)_{n\in\mathbb{N}}$ imply that there exists a probability measure μ on $[-1, 1]$ such that $(P_n)_{n\in\mathbb{N}}$ forms an orthogonal system on $[-1, 1]$ with respect to μ. As P_n has degree n, we have

$$P_n P_m = \sum_{k=0}^{n+m} c(n, m, k) P_k$$

for all n, m in $\mathbb{N}$, where

$$c(n, m, k) = \frac{\int_{-1}^1 P_k P_n P_m \, d\mu}{\int_{-1}^1 P_k^2 \, d\mu}$$

holds for all n, m, k in $\mathbb{N}$. The orthogonality of $(P_n)_{n \in \mathbb{N}}$ with respect to μ implies $c(n, m, k) = 0$ for $k > n + m$ or $n > m + k$ or $m > n + k$. Hence our statement is proved. $\square$

The formula in the theorem is called linearization formula, and the coefficients $c(n, m, k)$ are called linearization coefficients. The recursive formula for the sequence $(P_n)_{n \in \mathbb{N}}$ implies $P_n(1) = 1$ for all n in $\mathbb{N}$, hence we have

$$\sum_{k=|n-m|}^{n+m} c(n, m, k) = 1$$

for all n in $\mathbb{N}$. If the linearization is nonnegative, that is, the linearization coefficients are nonnegative: $c(n, m, k) \geq 0$ for all n, m, k in $\mathbb{N}$, then we can define a hypergroup structure on $\mathbb{N}$ by the following rule:

$$\delta_n * \delta_m = \sum_{k=|n-m|}^{n+m} c(n, m, k) \delta_k$$

for all n, m in $\mathbb{N}$, with involution as the identity mapping and with o as 0. The resulting discrete Hermitian (hence commutative) hypergroup is called the polynomial hypergroup associated with the sequence $(P_n)_{n \in \mathbb{N}}$. We can denote it by $(\mathbb{N}, (P_n)_{n \in \mathbb{N}})$.

We mention here an easy consequence of the linearization formula in polynomial hypergroups. Namely, let $\varphi(n) = P_n^{(k)}(\lambda)$ for all n in $\mathbb{N}$ with some nonnegative integer k and complex number λ. Then we have

$$\varphi(n * 1) = \lambda P_n^{(k)}(\lambda) + k P_n^{(k-1)}(\lambda)$$

for all n in $\mathbb{N}$. Here $P_n^{(-1)}$ is meant to be 0.

Now we describe the general form of exponential functions on polynomial hypergroups [45, 50].

Theorem 7.59 Let K be the polynomial hypergroup associated with the sequence of polynomials $(P_n)_{n \in \mathbb{N}}$. The function $\varphi : \mathbb{N} \to \mathbb{C}$ is an exponential on K if and only if there exists a complex number λ such that

$$\varphi(n) = P_n(\lambda)$$

holds for all n in $\mathbb{N}$.

Proof: First of all we remark that if a sequence of polynomials $(P_n)_{n \in \mathbb{N}}$ satisfies a recursion of the form

$$P_n(x) P_m(x) = \sum_{k=0}^{n+m} c(n, m, k) P_k(x)$$

with some real or complex coefficients $c(n, m, k)$ for all real x, then the recursion holds for all complex x. Let λ be a complex number and let $\varphi(n) = P_n(\lambda)$ for any n in $\mathbb{N}$. Then by the definition of convolution we have for any m, n in $\mathbb{N}$

$$\varphi(\delta_n * \delta_m) = \sum_{k=|n-m|}^{n+m} c(n, m, k)\varphi(k)$$

$$= \sum_{k=|n-m|}^{n+m} c(n, m, k)P_k(\lambda) = P_n(\lambda)P_m(\lambda) = \varphi(n)\varphi(m),$$

hence φ is exponential.

Conversely, let φ be an exponential on K and we define $\lambda = \varphi(1)$. By the exponential property, we have for all positive integer n that

$$\lambda\varphi(n) = \varphi(1)\varphi(n) = \varphi(\delta_1 * \delta_n) = \sum_{k=n-1}^{n+1} c(n, 1, k)\varphi(k)$$

$$= c(n, 1, n - 1)\varphi(n - 1) + c(n, 1, n)\varphi(n) + c(n, 1, n + 1)\varphi(n + 1).$$

As the same recursion holds for $n \mapsto P_n(\lambda)$, further $\varphi(0) = 1 = P_0(\lambda)$ and $\varphi(1) = \lambda = P_1(\lambda)$, hence $\varphi(n) = P_n(\lambda)$ for all n in $\mathbb{N}$ and the theorem is proved. $\square$

Now we can prove that spectral analysis holds for any polynomial hypergroup.

Theorem 7.60 Spectral analysis holds on any polynomial hypergroup.

Proof: Let K be the hypergroup associated with the sequence of polynomials $(P_n)_{n\in\mathbb{N}}$ and let V be a nonzero variety in $\mathscr{C}(K)$. We remark that in this case, $\mathscr{C}(K)$ is the set of all complex-valued functions on $\mathbb{N}$, equipped with the pointwise linear operations and with the topology of pointwise convergence and $\mathscr{M}_c(K)$ is the set of all finitely supported complex measures on $\mathbb{N}$. By Theorem 7.14, the orthogonal complement $V^\perp$ of V is a proper closed ideal in $\mathscr{M}_c(K)$. By the convolution formula, the Fourier–Laplace transforms of the elements of $V^\perp$ form a proper ideal in the ring of the Fourier–Laplace transforms of all elements of $\mathscr{M}_c(K)$. By Theorem 7.59, the set of all exponentials of K can be identified by $\mathbb{C}$. For any μ in $\mathscr{M}_c(K)$ and for any λ in $\mathbb{C}$ we have

$$\hat{\mu}(\lambda) = \int_{\mathbb{N}} P_n(\lambda) \, d\mu(n).$$

As μ is finitely supported, hence $\hat{\mu}$ is a complex polynomial on $\mathbb{C}$. We can see easily, that any complex polynomial on $\mathbb{C}$ can be written in the form $\hat{\mu}$ with some μ in $\mathscr{M}_c(K)$. Indeed, if p is a complex polynomial on $\mathbb{C}$ of degree n, then it can be written in the form $p = \sum_{k=0}^{n} c_k P_k$ with some complex constants c_k ($k = 0, 1, \ldots, n$). Then we have

$$p = \left(\sum_{k=0}^{n} c_k \delta_k \right)^{\wedge}.$$

This means that the Fourier–Laplace transforms of the elements of $V^{\perp}$ form a proper ideal in the ring of all complex polynomials on $\mathbb{C}$. By Hilbert's Nullstellensatz [[51], theorem 7.17], there exists a complex λ_0 such that $\hat{\mu}(\lambda_0) = 0$ for all μ in $V^{\perp}$. By definition, this means that

$$\int_{\mathbb{N}} P_n(\lambda) \, d\mu(n) = 0$$

for each μ in $V^{\perp}$. In other words, the exponential $n \mapsto P_n(\lambda)$ is in $V^{\perp\perp} = V$ and the theorem is proved. $\square$

As Hilbert's Nullstellensatz works on the polynomial ring $\mathbb{C}[x_1, x_2, \ldots, x_n]$ in several variables, using the same idea we can prove spectral analysis on every polynomial hypergroup in several variables. For the details, see [10].

Let V be a variety in $\mathscr{C}(K)$. We say that V is *synthesizable*, if V is the topological sum of finite-dimensional varieties. This means, that there is a set $(V_\gamma)_{\gamma \in \Gamma}$ of finite-dimensional subvarieties of V such that all element f of the form

$$f = f_{\gamma_1} + f_{\gamma_2} + \cdots + f_{\gamma_n}$$

with some positive integer n, with some elements $\gamma_1, \gamma_2, \ldots, \gamma_n$ in Γ and with some functions f_{γ_i} in V_{γ_i} form a dense subset in V. In other words, all finite-dimensional subvarieties in V span a dense subspace. We say that *spectral synthesis holds for a variety*, if every subvariety of it is synthesizable. If spectral synthesis holds for every variety on K, then we say that *spectral synthesis holds on K*. If K is a locally compact Abelian group, then finite-dimensional varieties are spanned by exponential monomials, so it seems to be reasonable to define exponential monomials on hypergroups, too. For this purpose, we shall use the modified differences as we did in the group case. Given the commutative hypergroup K for each exponential $m : K \to \mathbb{C}$ and y in K we define

$$\Delta_{m;y} = \delta_{\check{y}} - m(y)\delta_o,$$

and

$$\Delta_{m;y_1, y_2, \ldots, y_{n+1}} = \Pi_{k=1}^{n+1} \Delta_{m;y_k}$$

where the product $\Pi_{k=1}^{n+1}$ refers to convolution. The closure of the ideal generated by all measures $\Delta_{m;y}$ with y in K will be denoted by M_m. Finally, it turns out that $M_m = \text{Ann}\,\tau(m)$, exactly what we have in the group case.

Modified differences can be used to define exponential monomials, by Theorem 7.11. We call the continuous function $f : K \to \mathbb{C}$ on the commutative hypergroup K a *generalized exponential monomial*, if there is an exponential m and a natural number n such that

$$M_m^{n+1} \subseteq \operatorname{Ann} \tau(f).$$

In other words, for each $y_1, y_2, \dots, y_{n+1}$ in K we have

$$\Delta_{m;y_1,y_2,\dots,y_{n+1}} * f = 0.$$

We can prove exactly in the same way as in the group case that if $f \neq 0$, then m is uniquely determined by f, and in that case we say that f is *associated with m*. The smallest n with the above property is called the *degree* of f. Linear combinations of generalized exponential monomials are called *generalized exponential polynomials* and a generalized exponential monomial, or polynomial is simply called an *exponential monomial, or polynomial*, if it generates a finite-dimensional variety. Clearly, these concepts coincide with the old ones in the group case. However, we underline that, in general, exponential monomials on hypergroups do not have a representation of the form (7.34).

Having the concept of exponential monomials we can reformulate synthesizability and spectral synthesis on hypergroups, following the pattern in the group case.

Theorem 7.61   Let K be a commutative hypergroup and V a variety on K. The variety V is synthesizable if and only if exponential monomials in V span a dense subspace. Spectral synthesis holds on K if and only if in every variety on K exponential monomials span a dense subspace.

Similarly to Theorem 7.57, spectral synthesis can be proved for finite-dimensional varieties [52].

Theorem 7.62   On commutative hypergroups spectral synthesis holds for every finite-dimensional variety.

Spectral synthesis on polynomial hypergroups has also been proved [50, 53]. We present here the result for polynomial hypergroups in one variable.

Theorem 7.63   Spectral synthesis holds on any polynomial hypergroup.

Proof: Let K be the polynomial hypergroup associated with the sequence of polynomials $(P_n)_{n \in \mathbb{N}}$. First we show that the variety generated by the function $n \mapsto P_n^{(k)}(\lambda)$ is finite dimensional for each nonnegative integer k and for any complex number λ. Let $\psi(n) = P_n^{(k)}(\lambda)$ for any n, k in $\mathbb{N}$ and λ in $\mathbb{C}$. Then, by the linearization formula, we have

$$\psi(n * m) = \sum_{j=0}^{k} \binom{k}{j} P_n^{(j)}(\lambda) P_m^{(k-j)}(\lambda)$$

for all m, n in $\mathbb{N}$, which yields the statement.

Now we know (see the proof of Theorem 7.60) that for each variety V in $\mathscr{C}(K)$ the Fourier–Laplace transforms of the elements of $V^{\perp}$ form a proper ideal in the polynomial ring over $\mathbb{C}$. We denote this ideal by J. It is known, by the Noether–Lasker Primary Decomposition Theorem [[51], theorem 7.21] that in this case there exist complex numbers $\lambda_1, \lambda_2, \dots, \lambda_k$ and nonnegative integers $m_1, m_2, \dots, m_k$ such that a polynomial p belongs to J if and only if $p^{(j)}(\lambda_i) = 0$ holds for $i = 1, 2, \dots, k$ and $j = 0, 1, \dots, m_i$. This means, that the measure μ in $\mathscr{M}_c(K)$ is in $V^{\perp}$ if and only if

$$\hat{\mu}^{(j)}(\lambda_i) = \int_K P_n^{(j)}(\lambda) \, d\mu(n) = 0$$

holds for $i = 1, 2, \dots, k$ and $j = 0, 1, \dots, m_i$. In other words, the family of functions $n \mapsto P_n^{(j)}(\lambda_i)$ spans a dense subspace in $V^{\perp\perp} = V$ for $i = 1, 2, \dots, k$ and $j = 0, 1, \dots, m_i$. As these functions generate finite-dimensional varieties, our statement is proved. □

7.7 Applications

Spectral analysis and synthesis is about the description of translation invariant function spaces. To find applications of these results, we have to find examples where translation invariant function spaces show up. One obvious example is presented by the solution spaces of convolution type functional equations. The first applications of spectral analysis and synthesis results on this area have been presented in the monographs [10, 28, 54]. Convolution type functional equations and systems of this type arise in the following way. Let G be a locally compact Abelian group and let Λ be a nonempty set of measures in $\mathscr{M}_c(G)$. The system of convolution type functional equations associated to Λ is the following:

$$\mu * f = 0, \quad \mu \in \Lambda, \tag{7.33}$$

where f in $\mathscr{C}(G)$ is the unknown function. The *solution space* of (7.33) is the set of all functions f satisfying (7.33). In fact, the solution space of (7.33) is the annihilator of the set Λ, that is, $\operatorname{Ann}\Lambda$. It is obvious, that every variety V arises in this way: by the identity $\operatorname{Ann}\operatorname{Ann} V = V$, which holds for every variety it follows that each variety is the solution space of some Λ, namely it is the solution space of $\operatorname{Ann} V$. In other words, spectral analysis and synthesis is the study of the description of solution spaces of systems of convolution type functional equations. This description is based on the exponential monomials in the solution space, that is, on the exponential monomial solutions. If spectral analysis holds for $\operatorname{Ann}\Lambda$, then we know that there are exponential monomial solutions, and if $\operatorname{Ann}\Lambda$ is synthesizable, then every solution of (7.33) is the uniform limit of some net formed by exponential polynomial solutions. Hence it is clear that methods for finding exponential monomial solutions of convolution type equation systems are of great importance for those areas where such systems arise. Typical examples are systems of homogeneous linear difference or differential equations with constant coefficients, systems of homogeneous linear partial difference or differential equations with constant coefficients, and so on. In the subsequent paragraphs, we show some methods how to apply spectral analysis and synthesis methods for the solution of such systems of equations, in particular, how to find their exponential monomial solutions.

Our first observation is that if the exponential monomial

$$f(x) = P(a_1(x), a_2(x), \dots, a_k(x))m(x),$$ (7.34)

is a solution of (7.33), then m is a solution, too. Hence the first step is to find all exponential solutions, the *spectrum* of (7.33). Suppose that m is a solution, then we have

$$0 = \mu * m(x) = \int_G m(x - y) \, d\mu(y) = m(x) \int \check{m} \, d\mu = \hat{\mu}(m),$$

that is, m is a zero of the Fourier–Laplace transforms of all measures in Λ. If $\hat{\Lambda}$ denotes the ideal formed by the Fourier–Laplace transforms of all measures in Λ, then our first step is to find the *zero set* $Z(\hat{\Lambda})$ of this ideal. Spectral analysis for $\mathrm{Ann}\,\Lambda$ means that this zero set is nonempty – roughly speaking, a kind of "Nullstellensatz" holds. We consider the following simple example: find all continuous solutions $f : G \to \mathbb{C}$ of d'Alembert's functional equation

$$f(x + y) + f(x - y) = 2f(x)f(y)$$ (7.35)

where G is a locally compact Abelian group and the equation holds for each x, y in G. This equation can be written as

$$\left(\frac{\delta_{-y} + \delta_y}{2} - f(y)\delta_0 \right) * f(x) = 0.$$

We should rather consider the following system of equations:

$$\left(\frac{\delta_{-y} + \delta_y}{2} - g(y)\delta_0 \right) * f(x) = 0,$$

where g is a continuous function on G. Here Λ is the set of all measures of the form $\frac{\delta_{-y} + \delta_y}{2} - g(y)\delta_0$ where y is in G. If we know that spectral analysis holds for $\mathrm{Ann}\,\Lambda$, then there is an exponential m in $\mathrm{Ann}\,\Lambda$, hence we have

$$m(x + y) + m(x - y) = 2m(x)g(y),$$

which implies $g(y) = \frac{m(y) + m(-y)}{2}$ for each y in G. Going back to (7.35), we conclude that

$$f(x) = \frac{m(x) + m(-x)}{2}$$

with some exponential, and simple substitution shows that indeed, this is a solution for any exponential. We recall that this argument works only if spectral analysis holds for the variety Ann Λ. If G is a discrete Abelian group than this is the case if, for instance, G is finitely generated – by Lefranc's theorem. Nevertheless, every G is a direct limit of finitely generated subgroups, and using a simple argument, we can save the above solution method: no matter if spectral analysis holds on G, the solution, as a direct limit of solutions of the above type, is the same.

After finding the spectrum of Ann Λ the next step is to determine the spectral set, that is, the exponential monomials corresponding to the different exponentials in the spectrum. An important observation is that if the function f in (7.34) is a solution of (7.33), then every function of the form

$$\varphi(x) = \partial_1^{\alpha_1} \partial_2^{\alpha_2} \cdots \partial_k^{\alpha_k} P(a_1(x), a_2(x), \dots, a_k(x)) m(x)$$

is a solution, too, with any nonnegative integers $\alpha_1, \alpha_2, \dots, \alpha_k$. Hence we can find step by step the solutions of the form $x \mapsto a(x)m(x)$, $x \mapsto a(x)b(x)m(x)$, and so on, where $a, b, \dots$ are linearly independent additive functions. For instance, in the case of the function $x \mapsto a(x)m(x)$, we have the equation

$$0 = \int_G a(x - y)m(x - y) \, d\mu(y) = a(x)m(x) \int_G \check{m} \, d\mu + m(x) \int_G \check{am} \, d\mu$$

for each x in G. As m and am are linearly independent, we infer that

$$\hat{\mu}(m) = \int_G \check{m} \, d\mu = 0, \quad \int_G (a\,m)^{\check{}} \, d\mu = 0.$$

Continuing in this manner, we have good chance to describe the spectral set of the equation.

Finally, if Ann Λ is synthesizable, then we need to find all functions which are uniform limits on compact sets of nets from the linear span of the spectral set. In some cases, the above idea with direct limits along finitely generated subgroups can be applied. An illustrative example is the following one [55]: we consider the functional equation:

$$\Delta_{y_1, y_2, \dots, y_{n+1}} * f(x) = 0 \tag{7.36}$$

for each $x, y_1, y_2, \dots, y_{n+1}$ in G, a discrete Abelian group. The functional equation (7.36) is called *Fréchet's equation*. Together with (7.36), we consider the functional equation

$$\Delta_y^{n+1} * f(x) = 0, \tag{7.37}$$

where n is a natural number and x, y are arbitrary in G. Clearly, every solution of (7.36) satisfies (7.37). The converse was proved by Székelyhidi [28, 54] and Ž. Djokovič [56]. This result has been used several times in functional equations; it can be considered as a basic theorem on polynomials. The proof of Djokovič's theorem depends on some identities on ordinary polynomials in several variables. Here we can give an alternative short proof depending on spectral synthesis.

Theorem 7.64 Let G be an Abelian group. If the function $f : G \to \mathbb{C}$ satisfies the functional equation (7.37) for each x, y in G, then f satisfies (7.36) for each x, y in G.

Proof: We note that, obviously, the set of all solutions of (7.36), and that of (7.37) is a variety. First we show that (7.37) implies (7.36) if $G = \mathbb{Z}^k$, where k is a positive integer. By Lefranc's theorem, it is enough to show that every exponential monomial satisfying (7.37) also satisfies (7.36). Supposing that for some nonzero p the exponential polynomial $\varphi = p \cdot m$ is in V it follows that m is in V, and we have

$$0 = \Delta_y^{n+1} * m(x) = m(x)(m(y) - 1)^{n+1}$$

for each x, y in G, hence $m = 1$, that is φ is a polynomial, that is, it is the restriction of a polynomial $P : \mathbb{R}^m \to \mathbb{C}$ to $\mathbb{Z}^m$, and we have

$$\sum_{k=0}^{n+1} (-1)^{n+1-k} \binom{n+1}{k} P(x + ky) = \Delta_y^{n+1} * P(x) = 0 \tag{7.38}$$

for each x, y in $\mathbb{Z}^m$. As P is a polynomial, this equation extends to $\mathbb{R}^m$, hence (7.38) holds for each x, y in $\mathbb{R}^m$. Introducing the notation $y = (y_1, y_2, \dots, y_m)$ and applying the differential operator $\partial_1^{\alpha_1} \partial_2^{\alpha_2} \cdots \partial_m^{\alpha_m}$ with respect to the components of y where the natural numbers $\alpha_1, \alpha_2, \dots, \alpha_m$ satisfy $\alpha_1 + \alpha_2 + \cdots + \alpha_m = n + 1$, we obtain by induction that

$$\partial_1^{\alpha_1} \partial_2^{\alpha_2} \cdots \partial_m^{\alpha_m} P(x) = 0$$

holds for each x in $\mathbb{Z}^m$. In other words, the $n + 1$-th differential of P vanishes. It follows that P is a polynomial of degree at most n, hence it obviously satisfies (7.36) for each $x, y_1, y_2, \dots, y_{n+1}$ in $\mathbb{Z}^m$, as it was to be proved.

Now we suppose that G is a finitely generated Abelian group. Then G is a homomorphic image of $\mathbb{Z}^m$ for some positive integer m. Let $\Phi : \mathbb{Z}^m \to G$ a surjective homomorphism and we define

$$F(u) = f(\Phi(u))$$

for each u in $\mathbb{Z}^m$. It is straightforward to check that

$$\Delta_v^{n+1} * F(u) = \Delta_{\Phi(v)}^{n+1} * f(\Phi(u)) = 0$$

for each u, v in $\mathbb{Z}^m$. By the above proof, we infer that

$$0 = \Delta_{v_1, v_2, \dots, v_{n+1}} * F(u) = \Delta_{\Phi(v_1), \Phi(v_2), \dots, \Phi(v_{n+1})} * f(\Phi(u)) = 0$$

holds for each $u, v_1, v_2, \dots, v_{n+1}$ in $\mathbb{Z}^m$. As Φ is surjective, we obtain that f satisfies (7.36).

Finally, we suppose that G is an arbitrary Abelian group and $f : G \to \mathbb{C}$ satisfies (7.37) for each x, y in G. Let $x, y_1, y_2, \dots, y_{n+1}$ be arbitrary in G and let H denote the subgroup of G generated by these elements. Obviously, the restriction of f to H satisfies (7.37) on H, and H is finitely generated, hence, by our proof above, f satisfies (7.36) on H. As $x, y_1, y_2, \ldots, y_{n+1}$ are arbitrary in G, we conclude that f satisfies (7.36) on G and the theorem is proved. $\quad\square$

Acknowledgments

Research was supported by OTKA Grant No. K111651 and by the University of Botswana CDU Research Project.

References

1 Schwartz, L. (1947) Théorie générale des fonctions moyenne-périodiques. *Ann. Math. 2*, **48**, 857–929.

2 Lefranc, M. (1958) L'analyse harmonique dans Z_n. *C.R. Acad. Sci. Paris*, **246**, 1951–1953.

3 Elliott, R.J. (1965) Two notes on spectral synthesis for discrete Abelian groups. *Proc. Cambridge Philos. Soc.*, **61**, 617–620.

4 Gajda, Z. (1987) Private communication. Hamburg–Rissen.

5 Gurevič, D.I. (1975) Counterexamples to a problem of L. Schwartz. *Funk. Anal. Priložen.*, **9** (2), 29–35.

6 Malgrange, B. (1954) Equations aux dérivées partielles à coefficients constants. II. Equations avec second membre. *C.R. Acad. Sci. Paris*, **238**, 196–198.

7 Ehrenpreis, L. (1955) Mean periodic functions. I. Varieties whose annihilator ideals are principal. *Am. J. Math.*, **77**, 293–328.

8 Elliott, R.J. (1965) Some results in spectral synthesis. *Proc. Cambridge Philos. Soc.*, **61**, 395–424.

9 Székelyhidi, L. (2014) *Harmonic and Spectral Analysis*, World Scientific Publishing Co. Pte. Ltd., Hackensack, NJ.

10 Székelyhidi, L. (2013) *Functional Equations on Hypergroups*, World Scientific Publishing Co. Pte. Ltd., Hackensack, NJ.

11 Kelley, J.L. and Namioka, I. (1976) *Linear Topological Spaces*, Springer-Verlag, New York.

12 Székelyhidi, L. (2016) Annihilator methods for spectral synthesis on locally compact Abelian groups. *Monatsh. Math.*, **180** (2), 357–371.

13 Laczkovich, M. and Székelyhidi, L. (2007) Spectral synthesis on discrete Abelian groups. *Math. Proc. Cambridge Philos. Soc.*, **143** (1), 103–120.

14 Székelyhidi, L. (2014) Annihilator methods in discrete spectral synthesis. *Acta Math. Hungar.*, **143** (2), 351–366.

15 Székelyhidi, L. and Wilkens, B. (2016) Spectral analysis and synthesis on varieties. *J. Math. Anal. Appl.*, **433** (2), 1329–1332.

16 Perdry, H. (2004) An elementary proof of Krull's intersection theorem. *Am. Math. Mon.*, **111** (4), 356–357.

17 Atiyah, M.F. and Macdonald, I.G. (1969) *Introduction to Commutative Algebra*, Addison-Wesley Publishing Co., Reading, MA, London, Don Mills, Ont.

18 Székelyhidi, L. (2001) A Wiener Tauberian theorem on discrete Abelian torsion group. *Ann. Acad. Paedagog. Crac., Stud. Math.*, **4**, 147–150.

19 Bereczky, A. and Székelyhidi, L. (2005) Spectral synthesis on torsion groups. *J. Math. Anal. Appl.*, **304** (2), 607–613.

20 Laczkovich, M. and Székelyhidi, G. (2005) Harmonic analysis on discrete abelian groups. *Proc. Am. Math. Soc.*, **133** (6), 1581–1586.

21 Matsumura, H. (1986) *Commutative Ring Theory*, Cambridge University Press, Cambridge.

22 Bourbaki, N. (1972) *Elements of Mathematics. Commutative Algebra*, Hermann, Paris.

23 Székelyhidi, L. (2004) The failure of spectral synthesis on some types of discrete Abelian groups. *J. Math. Anal. Appl.*, **291** (2), 757–763.

24 Horváth, G., Székelyhidi, L., and Wilkens, B. (2014) Non-synthesizable varieties. *J. Math. Anal. Appl.*, **417** (1), 394–399.

25 Malgrange, M. (1954) Sur quelques propriétés des équations de convolution. *C.R. Acad. Sci. Paris*, **238**, 2219–2221.

26 Székelyhidi, L. (2016) On the principal ideal theorem and spectral synthesis on discrete Abelian groups. *Acta Math. Hungar.*, **150** (1), 228–233.

27 Székelyhidi, L. (2000) On the extension of exponential polynomials. *Math. Bohem.*, **125** (3), 365–370.

28 Székelyhidi, L. (1991) *Convolution Type Functional Equations on Topological Abelian Groups*, World Scientific Publishing Co. Inc., Teaneck, NJ.

29 Székelyhidi, L. and Wilkens, B. (2017) Spectral synthesis and residually finite-dimensional algebras. *J. Algebra Appl.*, **16** (10), 1750200.

30 Székelyhidi, L. (2015) Spectral synthesis on special varieties. *Ann. Univ. Sci. Budapest. Sect. Comput.*, **44**, 29–36.

31 Nagata, M. (1975) *Local Rings*, Robert E. Krieger Publishing Co., Huntington, NY.

32 Brewer, J.W., Costa, D.L., and Lady, E.L. (1975) Prime ideals and localization in commutative group rings. *J. Algebra*, **34**, 300–308.

33 Székelyhidi, L. (2012) Noetherian rings of polynomial functions on Abelian groups. *Aequationes Math.*, **84** (1–2), 41–50.

34 Székelyhidi, L. On spectral synthesis in several variables, *arXiv:1607.07079* *[math.FA]*.

35 Rudin, W. (1991) *Functional Analysis*, International Series in Pure and Applied Mathematics, 2nd edn, McGraw-Hill Inc., New York.

36 Dieudonné, J. (1978) *Treatise on Analysis*, vol. VI, Academic Press Inc. [Harcourt Brace Jovanovich Publishers], New York.

37 van Dijk, G. (2009) *Introduction to Harmonic Analysis and Generalized Gelfand Pairs*, de Gruyter Studies in Mathematics, Walter de Gruyter & Co., Berlin.

38 Helgason, S. (1962) *Differential Geometry and Symmetric Spaces*, Academic Press, New York, London.

39 Takeuchi, M. (1994) *Modern Spherical Functions*, Translations of Mathematical Monographs, vol. **135**, American Mathematical Society, Providence, RI.

40 Helgason, S. (2000) *Groups and Geometric Analysis*, American Mathematical Society, Providence, RI.

41 Sakellaridis, Y. (2013) Spherical functions on spherical varieties. *Am. J. Math.*, **135** (5), 1291–1381.

42 Stein, E.M. and Weiss, G. (1971) *Introduction to Fourier Analysis on Euclidean Spaces*, Princeton University Press, Princeton, NJ.

43 Rudin, W. (1990) *Fourier Analysis on Groups*, Wiley Classics Library, John Wiley & Sons, Inc., New York.

44 Székelyhidi, L. (2017) Spherical spectral synthesis. *Acta Math. Hungar.*, **153** (1), 120–142.

45 Bloom, W.R. and Heyer, H. (1995) *Harmonic Analysis of Probability Measures on Hypergroups*, de Gruyter Studies in Mathematics, de Gruyter, Berlin, New York.

46 Ross, K.A. (1998) Hypergroups and signed hypergroups, in *Harmonic Analysis and Hypergroups (Delhi, 1995)*, Trends in Mathematics, Birkhäuser Boston, Boston, MA, pp. 77–91.

47 Székelyhidi, L. and Vajday, L. (2010) Spectral analysis on commutative hypergroups. *Aequationes Math.*, **80** (1-2), 223–226.

48 Favard, J. (1935) Sur les polynomes de Tchebycheff. *C.R. Acad. Sci. Paris*, **200**, 2052–2053.

49 Shohat, J. (1936) The relation of the classical orthogonal polynomials to the polynomials of Appel. *Am. J. Math.*, **58**, 453–464.

50 Székelyhidi, L. (2004) Spectral analysis and spectral synthesis on polynomial hypergroups. *Monatsh. Math.*, **141** (1), 33–43.

51 Jacobson, N. (1953) *Lectures in Abstract Algebra: Linear Algebra*, vol. II, D. Van Nostrand Co., Inc., Toronto, New York, London.

52 Székelyhidi, L. and Vajday, L. (2016) Spectral synthesis on commutative hypergroups. *Ann. Univ. Sci. Sect. Comput.*, **45**, 111–117.

53 Székelyhidi, L. (2008) Spectral synthesis on multivariate polynomial hypergroups. *Monatsh. Math.*, **153** (2), 145–152.

54 Székelyhidi, L. (2006) *Discrete Spectral Synthesis and its Applications*, Springer Monographs in Mathematics, Springer, Dordrecht.

55 Székelyhidi, L. (2014) On Fréchet's functional equation. *Monatsh. Math.*, **175** (4), 639–643.

56 Djoković, Dv.Z. (1969/1970) A representation theorem for $(X_1 - 1)$ $(X_2 - 1) \cdots (X_n - 1)$ and its applications. *Ann. Polon. Math.*, **22**, 189–198.

8

Various Ulam–Hyers Stabilities of Euler–Lagrange–Jensen General $(a, b; k = a + b)$-Sextic Functional Equations

John Michael Rassias[1] and Narasimman Pasupathi[2]

[1] *Pedagogical Department E.E., Section of Mathematics and Informatics, National and Capodistrian University of Athens, Athens 15342, Greece*
[2] *Department of Mathematics, Thiruvalluvar University College of Arts and Science, Tirupattur 635 901, Tamil Nadu, India*

2010 AMS Subject Classification 39B52, 32B72, 32B82

8.1 Brief Introduction

One of the most interesting questions in the theory of functional equations concerning the famous Ulam stability problem is: when is it true that a mapping satisfying a functional equation approximately, must be close to an exact solution of the given functional equation?

The first stability problem was raised by Ulam [1] during his talk at the University of Wisconsin in 1940. In fact we are given a group $(G_1, \cdot)$ and let $(G_2, *)$ be a metric group with the metric $d(\cdot, \cdot)$. Given $\epsilon > 0$, does there exist a $\delta > 0$, such that if a mapping $h : G_1 \to G_2$ satisfies the inequality $d(h(x \cdot y), h(x) * h(y)) < \delta$ for all $x, y \in G_1$, then there exists a homomorphism $H : G_1 \to G_2$ with $d(h(x), H(x)) < \epsilon$ for all $x \in G_1$?

Hyers [2] gave the first affirmative partial answer to the question of Ulam regarding Banach spaces. It was further generalized via excellent results obtained by a number of authors [3–12].

Xu *et al.* [13] achieved the general solution of the quintic functional equation

$$f(x + 3y) - 5f(x + 2y) + 10f(x + y) - 10f(x) \tag{8.1}$$
$$+ 5f(x - y) - f(x - 2y) = 120f(y)$$

and the sextic functional equation

$$f(x + 3y) - 6f(x + 2y) + 15f(x + y) - 20f(x) \tag{8.2}$$
$$+ 15f(x - y) - 6f(x - 2y) + f(x - 3y) = 720f(y)$$

and investigated the stability of (8.1) and (8.2) in quasi-β-normed spaces via fixed point method. Since $f(x) = cx^5$ is a solution of (8.1), we say that it is quintic

Mathematical Analysis and Applications: Selected Topics, First Edition.
Edited by Michael Ruzhansky, Hemen Dutta, and Ravi P. Agarwal.
© 2018 John Wiley & Sons, Inc. Published 2018 by John Wiley & Sons, Inc.

functional equation. Similarly, $f(x) = cx^6$ is a solution of (8.2), we say that it is sextic functional equation. Every solution of the quintic or sextic functional equation is said to be a quintic or sextic mapping, respectively.

Mohammad *et al.* [14] proved the stability for the approximately quintic and sextic mappings on the probabilistic normed spaces.

Cho *et al.* [15] investigate the generalized Hyers–Ulam–Rassias stability problem in quasi-β-normed spaces and then the stability by using a subadditive function for the quintic function $f : X \to Y$ such that

$$2f(2x + y) + 2f(2x - y) + f(x + 2y) + f(x - 2y) \tag{8.3}$$
$$= 20[f(x + y) + f(x - y)] + 90f(x).$$

In 2010, Xu *et al.* [13] introduced and obtained the general solution of the quintic and sextic functional equations

$$f(x + 3y) - 5f(x + 2y) + 10f(x + y) - 10f(x) \tag{8.4}$$
$$+ 5f(x - y) - f(x - 2y) = 120f(y)$$

and

$$f(x + 3y) - 6f(x + 2y) + 15f(x + y) - 20f(x) + 15f(x - y)$$
$$- 6f(x - 2y) + f(x - 3y) = 720f(y) \tag{8.5}$$

and investigated the generalized Ulam–Hyers stability in quasi-β-normed spaces via fixed point method.

Also, Xu *et al.* [16] introduced and discussed the general solution and generalized Ulam–Hyers stability of septic and octic functional equations

$$f(x + 4y) - 7f(x + 3y) + 21f(x + 2y) - 35f(x + y) + 35f(x)$$
$$- 21f(x - y) + 7f(x - 2y) - f(x - 3y) = 5040f(y) \tag{8.6}$$

and

$$f(x + 4y) - 8f(x + 3y) + 28f(x + 2y) - 56f(x + y) + 70f(x)$$
$$- 56f(x - y) + 28f(x - 2y) - 8f(x - 3y) + f(x - 4y) = 40320f(y) \tag{8.7}$$

in quasi-β-normed spaces, respectively.

Very recently the authors Pasupathi *et al.* [17] introduced and achieved the general solution of a new n-dimensional quintic functional equation of the form

$$n[f(nx + y) + f(nx - y)] + f(x + ny) + f(x - ny) \tag{8.8}$$
$$= (n^4 + n^2)[f(x + y) + f(x - y)] + 2(n^6 - n^4 - n^2 + 1)f(x)$$

and a new n-dimensional sextic functional equation of the form

$$f(nx + y) + f(nx - y) + f(x + ny) + f(x - ny) \tag{8.9}$$
$$= (n^4 + n^2)[f(x + y) + f(x - y)] + 2(n^6 - n^4 - n^2 + 1)[f(x) + f(y)]$$

with any real $n \in \mathbb{R}-\{0, \pm1\}$ and investigated the Hyers–Ulam stability, Hyers–Ulam–Rassias stability and generalized Hyers–Ulam–Rassias stability for the quintic and sextic functional equations in Felbin's type fuzzy normed linear spaces. Also authors gave counter examples for the Hyers–Ulam–Rassias stability of quintic and sextic functional equations for some cases.

In this chapter, we present the general solution and generalized Ulam–Hyers stability of the following Euler–Lagrange–Jensen general $(a, b; k = a + b)$-sextic functional equation

$$f(ax + by) + f(bx + ay) + (a - b)^6 \left[f\left(\frac{ax - by}{a - b}\right) + f\left(\frac{bx - ay}{b - a}\right) \right] \quad (8.10)$$
$$= 64(ab)^2(a^2 + b^2) \left[f\left(\frac{x + y}{2}\right) + f\left(\frac{x - y}{2}\right) \right]$$
$$+ 2(a^2 - b^2)(a^4 - b^4)[f(x) + f(y)]$$

where $a \neq b$, such that $k \in R; k = a + b \neq 0, \pm 1$ in a Banach space (BS), Felbin's type fuzzy normed space (FFNS), and Intuitionistic fuzzy normed space (IFNS) by using the standard direct and fixed point method.

Now, we present the following theorem by Margolis and Diaz [18] for the fixed point theory.

Theorem 8.1 (Margolis and Diaz [18]) Suppose that for a complete generalized metric space (Ω, d) and a strictly contractive mapping $T : \Omega \to \Omega$ with Lipschitz constant L. Then, for each given $x \in \Omega$, either

$$d(T^n x, T^{n+1} x) = \infty, \quad \forall \, n \geq 0,$$

or there exists a natural number n_0 such that the following properties hold:

(FP1) $d(T^n x, T^{n+1} x) < \infty$ for all $n \geq n_0$.
(FP2) The sequence $(T^n x)$ is convergent to a fixed to a fixed point y^* of T.
(FP3) y^* is the unique fixed point of T in the set $\Delta = \{y \in \Omega : d(T^{n_0} x, y) < \infty\}$.
(FP4) $d(y^*, y) \leq \frac{1}{1-L} d(y, Ty)$ for all $y \in \Delta$.

In Section 8.2, the general solution of (8.10) is provided.

In Sections 8.3–8.5, the generalized Ulam–Hyers stability of (8.10) is discussed in BS, FFNS, and IFNS, respectively, using both direct and fixed point methods.

8.2 General Solution of Euler–Lagrange–Jensen General $(a, b; k = a + b)$-Sextic Functional Equation

In this section, the general solution of the sextic functional equation (8.10) is given. For this, let us consider A and B be real vector spaces.

Theorem 8.2 If $f : A \to B$ be a mapping satisfying (8.10) for all $x, y \in A$ then f is sextic.

Proof: Letting (x, y) by $(0, 0)$ in (8.10), one obtains

$$2f(0)[1 - (a - b)^6 - 62(ab)^2(a^2 + b^2) - 2(a^2 + b^2)(a^4 + b^4 - a^2b^2)] = 0,$$
$$(8.11)$$

where

$$1 - (a - b)^6 - 62(ab)^2(a^2 + b^2) - 2(a^2 + b^2)(a^4 + b^4 - a^2b^2) \neq 0. \quad (8.12)$$

From (8.11) and (8.12), we obtain

$$f(0) = 0. \tag{8.13}$$

Replacing (x, y) by (x, x) in (8.10), we get

$$f(ax + bx) + f(bx + ax) + (a - b)^6 \left[f\left(\frac{ax - bx}{a - b}\right) + f\left(\frac{bx - ax}{b - a}\right) \right]$$
$$= 64(ab)^2(a^2 + b^2)[f(x)] + 4(a^2 - b^2)(a^4 - b^4)f(x),$$
$$2f((a + b)x) = [64a^4b^2 + 64a^2b^4 + 4a^6 - 4a^2b^4 - 4b^2a^4 + 4b^6 - 2(a - b)^6]f(x)$$
$$= 2[a^6 + 6a^5b + 15a^4b^2 + 20a^3b^3 + 15a^2b^4 + 6ab^5 + b^6]f(x)$$
$$= 2(a + b)^6 f(x) \tag{8.14}$$

for all $x \in A$. By assuming $k \in R; k = a + b \neq 0, \pm 1$ in (8.14), we obtain

$$f(kx) = k^6 f(x) \tag{8.15}$$

for all $x \in A$, $k \in R; k = a + b \neq 0, \pm 1$. Hence, f is a sextic function. This completes the proof of the theorem. $\qquad\square$

8.3 Stability Results in Banach Space

In this section, we investigate the generalized Ulam–Hyers stability of the Euler–Lagrange–Jensen general $(a, b; k = a + b)$-sextic functional equation (8.10) in Banach space using direct and fixed point methods.

Throughout this section, let us consider $\mathcal{G}$ be a normed space and $\mathcal{H}$ be a Banach space. Define a mapping $Df_6 : \mathcal{G} \to \mathcal{H}$ by

$$df_6(x, y) = f(ax + by) + f(bx + ay) + (a - b)^6 \left[f\left(\frac{ax - by}{a - b}\right) + f\left(\frac{bx - ay}{b - a}\right) \right]$$

$$- 64(ab)^2(a^2 + b^2) \left[f\left(\frac{x + y}{2}\right) + f\left(\frac{x - y}{2}\right) \right] - 2(a^2 - b^2)(a^4 - b^4)[f(x) + f(y)],$$

where $a \neq b$, such that $k \in R; k = a + b \neq 0, \pm 1$.

8.3.1 Banach Space: Direct Method

Theorem 8.3 Let $q = \pm 1$ and $\zeta : \mathcal{G}^2 \to [0, \infty)$ be a function such that

$$\lim_{l \to \infty} \frac{\zeta(k^{lq}x, k^{lq}y)}{k^{6lq}} = 0 \tag{8.16}$$

for all $x, y \in \mathcal{G}$. Let $Df_6 : \mathcal{G} \to \mathcal{H}$ be a function satisfying the inequality

$$\|Df_6(x, y)\| \leq \zeta(x, y) \tag{8.17}$$

for all $x, y \in \mathcal{G}$. Then there exists a unique sextic function $S : \mathcal{G} \to \mathcal{H}$ which satisfies (8.10) and

$$\|f(x) - S(x)\| \leq \frac{1}{2k^6} \sum_{p=\frac{1-q}{2}}^{\infty} \frac{\zeta(k^{pq}x, k^{pq}x)}{k^{6pq}} \tag{8.18}$$

where $S(x)$ is defined by

$$S(x) = \lim_{p \to \infty} \frac{f(k^{pq}x)}{k^{pq}} \tag{8.19}$$

for all $x \in \mathcal{G}$.

Proof: **Case (i):** Assume $q = 1$. Setting (x, y) by (x, x) in (8.17), we get

$$\left\| f(ax + bx) + f(bx + ax) + (a - b)^6 \left[f\left(\frac{ax - bx}{a - b} \right) + f\left(\frac{bx - ax}{b - a} \right) \right] \right.$$

$$\left. - 64(ab)^2(a^2 + b^2)[f(x) + f(0)] - 2(a^2 - b^2)(a^4 - b^4)[2f(x)] \right\| \leq \zeta(x, x)$$

$$\left\| 2f((a + b)x) + 2f(x)[(a - b)^6 - 32(ab)^2(a^2 + b^2) \right.$$

$$\left. - 2(a^2 - b^2)(a^4 - b^4)] \right\| \leq \zeta(x, x)$$

$$\left\| 2f((a + b)x) - 2f(x)[a^6 - 6a^5 b + 15a^4 b^2 \right.$$

$$\left. + 20a^3 b^3 + 15a^2 b^4 + 6ab^5 + b^6] \right\| \leq \zeta(x, x)$$

$$\left\| 2f((a + b)x) - 2f(x)(a + b)^6 \right\| \leq \zeta(x, x)$$

$$\|f(kx) - k^6 f(x)\| \leq \frac{1}{2}\zeta(x, x) \tag{8.20}$$

for all $x \in \mathcal{G}$. It follows from (8.20) that

$$\left\| \frac{f(kx)}{k^6} - f(x) \right\| \leq \frac{\zeta(x, x)}{2k^6} \tag{8.21}$$

for all $x \in \mathcal{G}$. Replacing x by kx and dividing by k^6 in (8.21), we have

$$\left\| \frac{f(k^2 x)}{k^{12}} - \frac{f(kx)}{k^6} \right\| \leq \frac{\zeta(kx, kx)}{2k^{12}} \tag{8.22}$$

for all $x \in \mathcal{G}$. From (8.21) and (8.22), we obtain

$$\left\| \frac{f(k^2 x)}{k^{12}} - f(x) \right\| \leq \left\| \frac{f(k^2 x)}{k^{12}} - \frac{f(kx)}{k^6} \right\| + \left\| \frac{f(kx)}{k^6} - f(x) \right\|$$

$$\leq \frac{1}{2k^6} \left[\zeta(x, x) + \frac{\zeta(kx, kx)}{k^6} \right] \tag{8.23}$$

for all $x \in \mathcal{G}$. Generalizing, for a positive integer l, we reach

$$\left\| \frac{f(k^l x)}{k^{6l}} - f(x) \right\| \leq \frac{1}{2k^6} \sum_{p=0}^{l-1} \frac{\zeta(k^p x, k^p x)}{k^{6p}} \tag{8.24}$$

for all $x \in \mathcal{G}$. To prove the convergence of the sequence

$$\left\{ \frac{f(k^l x)}{k^{6l}} \right\},$$

replacing x by $k^m x$ and dividing by k^{6m} in (8.24), for any $l, m > 0$, we get

$$\left\| \frac{f(k^{l+m} x)}{k^{6(l+m)}} - \frac{f(k^m x)}{k^{6m}} \right\| = \frac{1}{k^{6m}} \left\| \frac{f(k^l \cdot k^m x)}{k^{6l}} - f(k^m x) \right\|$$

$$\leq \frac{1}{k^{6m}} \frac{1}{2k^6} \sum_{p=0}^{l-1} \frac{\zeta(k^p \cdot k^m x, \, k^p \cdot k^m x)}{k^{6p}}$$

$$\leq \frac{1}{2k^6} \sum_{p=0}^{\infty} \frac{\zeta(k^{p+m} x, \, k^{p+m} x)}{k^{6(p+m)}}$$

$$\rightarrow 0 \quad as \; m \rightarrow \infty$$

for all $x \in \mathcal{G}$. Thus, it follows that a sequence $\left\{ \dfrac{f(k^l x)}{k^{6l}} \right\}$ is a Cauchy in $\mathcal{H}$ and so it converges. Therefore, we see that a mapping $S(x) : \mathcal{G} \rightarrow \mathcal{H}$ defined by

$$S(x) = \lim_{l \to \infty} \frac{f(k^l x)}{k^{6l}}$$

is well defined for all $x \in \mathcal{G}$. In order to show that S satisfies (8.10), replacing (x, y) by $(k^l x, k^l y)$ and dividing by k^{6l} in (8.17), we have

$$\| S(x, y) \| = \lim_{l \to \infty} \frac{1}{k^{6l}} \left\| Df_6(k^l x, k^l y) \right\| \leq \lim_{l \to \infty} \frac{1}{k^{6l}} \zeta(k^l x, k^l y)$$

for all $x, y \in \mathcal{G}$ and so the mapping S is sextic. Taking the limit as l approaches to infinity in (8.24), we find that the mapping S is a sextic mapping satisfying the inequality (8.18) near the approximate mapping $f : \mathcal{G} \rightarrow \mathcal{H}$ of (8.10). Hence, S satisfies (8.10), for all $x, y \in \mathcal{G}$.

To prove that S is unique, we assume now that there is S' as another sextic mapping satisfying (8.10) and the inequality (8.18). Then, it follows easily that

$$S(k^l x) = k^{6l} S(x), \quad S'(k^l x) = k^{6l} S'(x)$$

for all $x \in \mathcal{G}$ and all $l \in \mathbb{N}$. Thus,

$$\| S(x) - S'(x) \| = \frac{1}{k^{6l}} \left\| S(k^l x) - S'(k^l x) \right\|$$

$$\leq \frac{1}{k^{6l}} \left\{ \left\| S(k^l x) - f(k^l x) \right\| + \left\| f(k^l x) - S'(k^l x) \right\| \right\}$$

$$\leq \frac{1}{k^6} \sum_{p=0}^{\infty} \frac{\zeta\left(k^{p+l} x, \, k^{p+l} x \right)}{k^{6(p+l)}}$$

for all $x \in \mathcal{G}$. Therefore, as $l \to \infty$, in the above inequality, one establishes

$$S(x) - S'(x) = 0$$

for all $x \in \mathcal{G}$, completing the proof of the claimed uniqueness of S. Hence, the theorem holds for $q = 1$.

Case (ii): Assume $q = -1$.

Now replacing x by $\dfrac{x}{k}$ in (8.20), we get

$$\left\| f(x) - k^6 f\left(\frac{x}{k}\right) \right\| \le \frac{1}{2} \varsigma\left(\frac{x}{k}, \frac{x}{k}\right) \tag{8.25}$$

for all $x \in \mathcal{G}$. The rest of the proof is similar to that of case $q = 1$. Hence, for $q = -1$ also the theorem holds. This completes the proof of the theorem. □

The following corollary is an immediate consequence of Theorem 8.3 concerning the stability of Euler–Lagrange–Jensen general $(a, b; k = a + b)$-sextic functional equation (8.10).

Corollary 8.1 Let $Df_6 : \mathcal{G} \to \mathcal{H}$ be a mapping. If there exist real numbers ϑ and σ such that

$$\|Df_6(x, y)\| \le \begin{cases} \vartheta, & \\ \vartheta\{||x||^\sigma + ||y||^\sigma\}, & \sigma \ne 6; \\ \vartheta||x||^\sigma ||y||^\sigma, & 2\sigma \ne 6; \\ \vartheta\{||x||^\sigma ||y||^\sigma + \{||x||^{2\sigma} + ||y||^{2\sigma}\}\}, & 2\sigma \ne 6; \end{cases} \tag{8.26}$$

for all $x, y \in \mathcal{G}$, then there exists a unique sextic function $S : \mathcal{G} \to \mathcal{H}$ such that

$$\|f(x) - S(x)\| \le \begin{cases} \dfrac{\vartheta}{2|k^6 - 1|}, & \\ \dfrac{\vartheta||x||^\sigma}{|k^6 - k^\sigma|}, & \\ \dfrac{\vartheta||x||^{2\sigma}}{2|k^6 - k^{2\sigma}|}, & \\ \dfrac{3\vartheta||x||^{2\sigma}}{2|k^6 - k^{2\sigma}|} \end{cases} \tag{8.27}$$

for all $x \in \mathcal{G}$.

8.3.2 Banach Space: Fixed Point Method

Throughout this section, let $\mathcal{X}$ be a normed space and $\mathcal{Y}$ be a Banach space. Define a mapping $Df_6 : \mathcal{X} \to \mathcal{Y}$ by

$$Df_6(x, y) = f(ax + by) + f(bx + ay) + (a - b)^6 \left[f\left(\frac{ax - by}{a - b}\right) + f\left(\frac{bx - ay}{b - a}\right) \right]$$
$$- 64(ab)^2(a^2 + b^2) \left[f\left(\frac{x + y}{2}\right) + f\left(\frac{x - y}{2}\right) \right] - 2(a^2 - b^2)(a^4 - b^4)[f(x) + f(y)],$$

where $a \neq b$, such that $k \in R; k = a + b \neq 0, \pm 1$. Using Theorem 8.1, we obtain the Hyers–Ulam stability of Euler–Lagrange–Jensen general $(a, b; k = a + b)$-sextic functional equation (8.10).

Theorem 8.4 Let $Df_6 : \mathcal{X} \to \mathcal{Y}$ be a mapping for which there exists a function $\zeta : \mathcal{X}^2 \to [0, \infty)$ with the condition

$$\lim_{n \to \infty} \frac{1}{\hbar_i^{6n}} \zeta(\hbar_i^n x, \hbar_i^n y) = 0 \tag{8.28}$$

where

$$\hbar_i = \begin{cases} k & \text{if } i = 0, \\ \frac{1}{k} & \text{if } i = 1 \end{cases} \tag{8.29}$$

such that the functional inequality

$$\|Df_6(x, y)\| \leq \zeta(x, y) \tag{8.30}$$

holds for all $x, y \in \mathcal{X}$. Assume that there exists $L = L(i)$ such that the function

$$x \to T(x, x) = \frac{1}{2} \zeta\left(\frac{x}{k}, \frac{x}{k}\right)$$

with the property

$$\frac{1}{\hbar_i^6} T(\hbar_i x, \hbar_i x) = L\, T(x, x) \tag{8.31}$$

for all $x \in \mathcal{X}$. Then, there exists a unique sextic mapping $S : \mathcal{X} \to \mathcal{Y}$ satisfying the functional equation (8.10) and

$$\|f(x) - S(x)\| \leq \left(\frac{L^{1-i}}{1 - L}\right) T(x, x) \tag{8.32}$$

for all $x \in \mathcal{X}$.

Proof: Consider the set

$$\Lambda = \{h/h : \mathcal{X} \to \mathcal{Y},\ h(0) = 0\}$$

and introduce the generalized metric on Λ,

$$\inf\{\rho \in (0, \infty) :\ \|h(x) - g(x)\| \leq \rho\, T(x, x),\ x \in \mathcal{X}\}. \tag{8.33}$$

It is easy to see that (8.33) is complete with respect to the defined metric. Define $J : \Lambda \to \Lambda$ by

$$Jh(x) = \frac{1}{\hbar_i^6} h(\hbar_i x)$$

for all $x \in \mathcal{X}$. Now, from (8.33) and $h, g \in \Lambda$, we arrive

$\inf\{\rho \in (0, \infty) : \|h(x) - g(x)\| \le \rho\, T(x, x),\ x \in \mathcal{X}\}$ or

$\inf\left\{\rho \in (0, \infty) : \left\|\dfrac{1}{\hbar_i^6}h(\hbar_i x) - \dfrac{1}{\hbar_i^6}g(\hbar_i x)\right\| \le \dfrac{\rho}{\hbar_i^6}T(\hbar_i x, \hbar_i x),\ x \in \mathcal{X}\right\}$ or

$\inf\left\{L\rho \in (0, \infty) : \left\|\dfrac{1}{\hbar_i^6}h(\hbar_i x) - \dfrac{1}{\hbar_i^6}g(\hbar_i x)\right\| \le L\rho T(x, x),\ x \in \mathcal{X}\right\}$ or

$\inf\{L\rho \in (0, \infty) : \|Jh(x) - Jg(x)\| \le L\rho T(x, x),\ x \in \mathcal{X}\}.$

This implies J is a strictly contractive mapping on Λ with Lipschitz constant L. It follows from (8.33), (8.20), and (8.31) for the case $i = 0$, we reach

$$\inf\left\{1 \in (0, \infty) : \|f(kx) - k^6 f(x)\| \le \tfrac{1}{2}\zeta(x, x), x \in \mathcal{X}\right\} \quad \text{or} \qquad (8.34)$$

$$\inf\left\{1 \in (0, \infty) : \left\|\dfrac{f(kx)}{k^6} - f(x)\right\| \le \dfrac{1}{2}\dfrac{1}{k^6}\zeta(x, x),\ x \in \mathcal{X}\right\} \quad \text{or}$$

$\inf\{L \in (0, \infty) : \|Jf(x) - f(x)\| \le L\, T(x, x),\ x \in \mathcal{X}\}$ or

$\inf\{L^1 \in (0, \infty) : \|Jf(x) - f(x)\| \le L\, T(x, x),\ x \in \mathcal{X}\}$ or

$$\inf\{L^{1-0} \in (0, \infty) : \|Jf(x) - f(x)\| \le L\, T(x, x),\ x \in \mathcal{X}\}. \qquad (8.35)$$

Again replacing $x = \frac{x}{k}$ in (8.34) and (8.31) for the case $i = 1$, we get

$$\inf\left\{1 \in (0, \infty) : \left\|f(x) - k^6 f\left(\dfrac{x}{k}\right)\right\| \le \dfrac{1}{2}\zeta\left(\dfrac{x}{k}, \dfrac{x}{k}\right),\ x \in \mathcal{X}\right\} \quad \text{or}$$

$\inf\{1 \in (0, \infty) : \|f(x) - Jf(x)\| \le T(x, x),\ x \in \mathcal{X}\}$ or

$\inf\{L^0 \in (0, \infty) : \|f(x) - Jf(x)\| \le T(x, x),\ x \in \mathcal{X}\}$ or

$$\inf\{L^{1-1} \in (0, \infty) : \|f(x) - Jf(x)\| \le T(x, x),\ x \in \mathcal{X}\}. \qquad (8.36)$$

Thus, from (8.35) and (8.36), we arrive

$$\inf\{L^{1-i} \in (0, \infty) : \|f(x) - Jf(x)\| \le L^{1-i} T(x, x),\ x \in \mathcal{X}\}. \qquad (8.37)$$

Hence, property (FP1) holds. It follows from property (FP2) that there exists a fixed point S of J in Λ such that

$$S(x) = \lim_{n \to \infty} \dfrac{1}{\hbar_i^{6n}} f(\hbar_i^n x) \qquad (8.38)$$

for all $x \in \mathcal{X}$. In order to show that S satisfies (8.10), replacing (x, y) by $(\hbar_i^n x, \hbar_i^n y)$ and dividing by $\hbar_i^{6n}$ in (8.30), we have

$$\|S_6(x, y)\| = \lim_{n \to \infty} \dfrac{1}{\hbar_i^{6n}}\|Df_6(\hbar_i^n x, \hbar_i^n y)\| \le \lim_{n \to \infty} \dfrac{1}{\hbar_i^{6n}}\zeta(\hbar_i^n x, \hbar_i^n y) = 0$$

for all $x, y \in \mathcal{X}$, and so the mapping S is sextic. That is, S satisfies the functional equation (8.10). By property (FP3), S is the unique fixed point of J in the set

$$\Delta = \{S \in \Lambda : d(f, S) < \infty\}$$

such that

$$\inf\{\rho \in (0, \infty) : \|f(x) - S(x)\| \leq \rho T(x, x), x \in \mathcal{X}\}.$$

Finally by property (FP4), we obtain

$$\|f(x) - S(x)\| \leq \|f(x) - Jf(x)\|,$$

implying

$$\|f(x) - S(x)\| \leq \frac{L^{1-i}}{1 - L},$$

which yields

$$\inf\left\{\frac{L^{1-i}}{1 - L} \in (0, \infty) : \|f(x) - S(x)\| \leq \left(\frac{L^{1-i}}{1 - L}\right) T(x, x), x \in \mathcal{X}\right\}.$$

This completes the proof of the theorem. □

The following corollary is an immediate consequence of Theorem 8.4 concerning the stability of (8.10).

Corollary 8.2 Let $Df_6 : \mathcal{X} \to \mathcal{Y}$ be a mapping. If there exist real numbers ϑ and σ such that

$$\|Df_6(x, y)\| \leq \begin{cases} \vartheta, & \\ \vartheta\{\|x\|^{\sigma} + \|y\|^{\sigma}\}, & \sigma \neq 6, \\ \vartheta\|x\|^{\sigma}\|y\|^{\sigma}, & 2\sigma \neq 6, \\ \vartheta\{\|x\|^{\sigma}\|y\|^{\sigma} + \{\|x\|^{2\sigma} + \|y\|^{2\sigma}\}\}, & 2\sigma \neq 6, \end{cases} \quad (8.39)$$

for all $x, y \in \mathcal{X}$, then there exists a unique sextic function $S : \mathcal{X} \to \mathcal{Y}$ such that

$$\|f(x) - S(x)\| \leq \begin{cases} \dfrac{\vartheta}{2|k^6 - 1|}, & \\ \dfrac{\vartheta}{|k^6 - k^{\sigma}|}\|x\|^{\sigma}, & \\ \dfrac{\vartheta}{2|k^6 - k^{2\sigma}|}\|x\|^{2\sigma}, & \\ \dfrac{3\vartheta}{2|k^6 - k^{2\sigma}|}\|x\|^{2\sigma} & \end{cases} \quad (8.40)$$

for all $x \in \mathcal{X}$.

Proof: Let

$$\zeta(x, y) = \begin{cases} \vartheta, \\ \vartheta\{||x||^\sigma + ||y||^\sigma\} \\ \vartheta||x||^\sigma||y||^\sigma \\ \vartheta\{||x||^\sigma||y||^\sigma + \{||x||^{2\sigma} + ||y||^{2\sigma}\}\} \end{cases}$$

for all $x, y \in \mathcal{X}$. Now

$$\frac{1}{\hbar_i^{6n}} \zeta(\hbar_i^n x, \hbar_i^n y) = \begin{cases} \dfrac{\vartheta}{\hbar_i^{6n}}, & \to 0 \text{ as } n \to \infty, \\[2ex] \dfrac{\vartheta}{\hbar_i^{6n}}\{||\hbar_i^n x||^\sigma + ||\hbar_i^n y||^\sigma\}, & \to 0 \text{ as } n \to \infty, \\[2ex] \dfrac{\vartheta}{\hbar_i^{6n}}||\hbar_i^n x||^\sigma \; ||\hbar_i^n y||^\sigma, & \to 0 \text{ as } n \to \infty, \\[2ex] \dfrac{\vartheta}{\hbar_i^{6n}}\{||\hbar_i^n x||^\sigma \; ||\hbar_i^n y||^\sigma + \{||\hbar_i^n x||^{2\sigma} + ||\hbar_i^n y||^{2\sigma}\}\} & \to 0 \text{ as } n \to \infty. \end{cases}$$

Thus, (8.28) holds. But, we have

$$T(x, x) = \frac{1}{2}\zeta\left(\frac{x}{k}, \frac{x}{k}\right)$$

has the property

$$\frac{1}{\hbar_i^6} T(\hbar_i x, \hbar_i x) = L \; T(x, x)$$

for all $x \in \mathcal{X}$. Hence,

$$T(x, x) = \begin{cases} \dfrac{\vartheta}{2}, \\[2ex] \dfrac{\vartheta}{k^\sigma}||x||^\sigma, \\[2ex] \dfrac{\vartheta}{2k^{2\sigma}}||x||^{2\sigma}, \\[2ex] \dfrac{3\vartheta}{2k^{2\sigma}}||x||^{2\sigma} \end{cases} \tag{8.41}$$

for all $x \in \mathcal{X}$. Now, similarly by (8.41), we prove

$$\frac{1}{\hbar_i^6} T(\hbar_i x, \hbar_i x) = \begin{cases} \hbar_i^{-6}\dfrac{\vartheta}{2}, \\[2ex] \hbar_i^{\sigma-6}\dfrac{\vartheta}{k^\sigma}||x||^\sigma, \\[2ex] \hbar_i^{2\sigma-6}\dfrac{\vartheta}{2k^{2\sigma}}||x||^{2\sigma}, \\[2ex] \hbar_i^{2\sigma-6}\dfrac{3\vartheta}{2k^{2\sigma}}||x||^{2\sigma}. \end{cases}$$

Hence, the inequality (8.32) holds for

(i) $L = \hbar_i^{-6}$ if $i = 0$ and $L = \frac{1}{\hbar_i^{-6}}$ if $i = 1$,

(ii) $L = \hbar_i^{\sigma-6}$ for $\sigma < 6$ if $i = 0$ and $L = \frac{1}{\hbar_i^{\sigma-6}}$ for $\sigma > 6$ if $i = 1$,

(iii) $L = \hbar_i^{2\sigma-6}$ for $2\sigma < 6$ if $i = 0$ and $L = \frac{1}{\hbar_i^{2\sigma-6}}$ for $2\sigma > 6$ if $i = 1$,

(iv) $L = \hbar_i^{2\sigma-6}$ for $2\sigma < 6$ if $i = 0$ and $L = \frac{1}{\hbar_i^{2\sigma-6}}$ for $2\sigma > 6$ if $i = 1$.

Now, from (8.32), we prove the following cases for condition (*i*).

$$L = \hbar_i^{-6}, i = 0 \qquad\qquad L = \frac{1}{\hbar_i^{-6}}, i = 1$$
$$L = k^{-6}, i = 0 \qquad\qquad L = \frac{1}{k^{-6}}, i = 1$$
$$L = k^{-6}, i = 0 \qquad\qquad L = k^6, i = 1$$
$$\|f(x) - S(x)\| \qquad\qquad \|f(x) - S(x)\|$$
$$\leq \left(\frac{L^{1-i}}{1-L}\right) T(x,x) \qquad \leq \left(\frac{L^{1-i}}{1-L}\right) T(x,x)$$
$$= \left(\frac{k^{-6}}{1-k^{-6}}\right)\frac{\vartheta}{2} \qquad\quad = \left(\frac{1}{1-L}\right)\frac{\vartheta}{2}$$
$$= \left(\frac{1}{k^6-1}\right)\frac{\vartheta}{2} \qquad\quad = \left(\frac{1}{1-k^6}\right)\frac{\vartheta}{2}$$

Also, from (8.32), we prove the following cases for condition (*ii*):

$$L = \hbar_i^{\sigma-6},\ \sigma < 6,\ i = 0 \qquad L = \frac{1}{\hbar_i^{\sigma-6}},\ \sigma > 6,\ i = 1$$
$$L = k^{\sigma-6},\ \sigma < 6,\ i = 0 \qquad L = \frac{1}{k^{\sigma-6}},\ \sigma > 6,\ i = 1$$
$$L = k^{\sigma-6},\ \sigma < 6,\ i = 0 \qquad L = k^{6-\sigma},\ \sigma > 6,\ i = 1$$
$$\|f(x) - S(x)\| \qquad\qquad \|f(x) - S(x)\|$$
$$\leq \left(\frac{L^{1-i}}{1-L}\right) T(x,x) \qquad \leq \left(\frac{L^{1-i}}{1-L}\right) T(x,x)$$
$$= \left(\frac{L}{1-L}\right) T(x,x) \qquad = \left(\frac{1}{1-L}\right) T(x,x)$$
$$= \left(\frac{k^{\sigma-6}}{1-k^{\sigma-6}}\right)\frac{\vartheta}{k^\sigma}\|x\|^\sigma \qquad = \left(\frac{1}{1-k^{6-\sigma}}\right)\frac{\vartheta}{k^\sigma}\|x\|^\sigma$$
$$= \left(\frac{\vartheta}{k^6-k^\sigma}\right)\|x\|^\sigma \qquad\quad = \left(\frac{\vartheta}{k^\sigma-k^6}\right)\|x\|^\sigma$$

Again, from (8.32), we prove the following cases for condition (*iii*):

$$L = \hbar_i^{2\sigma-6},\ 2\sigma < 6,\ i = 0 \qquad L = \frac{1}{\hbar_i^{2\sigma-6}},\ 2\sigma > 6,\ i = 1$$
$$L = k^{2\sigma-6},\ 2\sigma < 6,\ i = 0 \qquad L = \frac{1}{k^{2\sigma-6}},\ 2\sigma > 6,\ i = 1$$
$$L = k^{2\sigma-6},\ 2\sigma < 6,\ i = 0 \qquad L = k^{6-2\sigma},\ 2\sigma > 6,\ i = 1$$
$$\|f(x) - S(x)\| \qquad\qquad \|f(x) - S(x)\|$$
$$\leq \left(\frac{L^{1-i}}{1-L}\right) T(x,x) \qquad \leq \left(\frac{L^{1-i}}{1-L}\right) T(x,x)$$
$$= \left(\frac{k^{2\sigma-6}}{1-k^{2\sigma-6}}\right)\frac{\vartheta}{2k^{2\sigma}}\|x\|^{2\sigma} \qquad = \left(\frac{1}{1-k^{6-2\sigma}}\right)\frac{\vartheta}{2k^{2\sigma}}\|x\|^{2\sigma}$$
$$= \left(\frac{1}{k^6-k^{2\sigma}}\right)\frac{\vartheta}{2}\|x\|^{2\sigma} \qquad = \left(\frac{1}{k^{2\sigma}-k^6}\right)\frac{\vartheta}{2}\|x\|^{2\sigma}$$
$$= \left(\frac{\vartheta}{2(k^6-k^{2\sigma})}\right)\|x\|^{2\sigma} \qquad = \left(\frac{\vartheta}{2(k^{2\sigma}-k^6)}\right)\|x\|^{2\sigma}$$

Again, from (8.32), we prove the following cases for condition (*iv*):

$$L = \hbar_i^{2\sigma-6}, \ 2\sigma < 6, \ i = 0 \qquad L = \frac{1}{\hbar_i^{2\sigma-6}}, \ 2\sigma > 6, \ i = 1$$

$$L = k^{2\sigma-6}, \ 2\sigma < 6, \ i = 0 \qquad L = \frac{1}{k^{2\sigma-6}}, \ 2\sigma > 6, \ i = 1$$

$$L = k^{2\sigma-6}, \ 2\sigma < 6, \ i = 0 \qquad L = k^{6-2\sigma}, \ 2\sigma > 6, \ i = 1$$

$$\|f(x) - S(x)\| \qquad\qquad \|f(x) - S(x)\|$$

$$\leq \left(\frac{L^{1-i}}{1-L}\right) T(x, x) \qquad\qquad \leq \left(\frac{L^{1-i}}{1-L}\right) T(x, x)$$

$$= \left(\frac{k^{2\sigma-6}}{1-k^{2\sigma-6}}\right) \frac{3\vartheta}{2k^{2\sigma}} \|x\|^{2\sigma} \qquad = \left(\frac{1}{1-k^{6-2\sigma}}\right) \frac{3\vartheta}{2k^{2\sigma}} \|x\|^{2\sigma}$$

$$= \left(\frac{3\vartheta}{2(k^6 - k^{2\sigma})}\right) \|x\|^{2\sigma} \qquad = \left(\frac{3\vartheta}{2(k^{2\sigma} - k^6)}\right) \|x\|^{2\sigma}.$$

Hence, the proof is complete. $\qquad\qquad\qquad\qquad\qquad\qquad\qquad\square$

8.4 Stability Results in Felbin's Type Spaces

In this section, we investigate the generalized Ulam–Hyers stability of the functional equation (8.10) in Felbin's type spaces using direct and fixed point methods.

Now, we recall the basic definitions and notations in Felbin's type spaces in Grantner *et al.* [19], Lowen [20], Hoehle [21], Kaleva and Seikkala [22–24], Felbin [25], and Xiao and Zhu [26, 27].

Definition 8.1 (**Xiao and Zhu [26]**) Let X be a real linear space, L and $\mathbb{R}$ (resp. left and right norms) be symmetric and nondecreasing mappings from $[0, 1] \times [0, 1] \to [0, 1]$ satisfying $L(0, 0) = 0, \mathbb{R}(1, 1) = 1$. Then $\|\cdot\|$ is called a fuzzy norm and $(X, \|\cdot\|, L, \mathbb{R})$ is a fuzzy normed linear space (abbreviated to FNLS) if the mapping $\|\cdot\| : X \to F^*(R)$ satisfies the following axioms, where $[\|x\|]_\alpha = [\|x\|_\alpha^-, \|x\|_\alpha^+]$ for $x \in X$ and $\alpha \in (0, 1]$:

(A1) $\|x\| = 0$ if and only if $x = 0$,

(A2) $\|rx\| = |r| \odot \|x\|$ for all $x \in X$ and $r \in (-\infty, \infty)$,

(A3) for all $x, y \in X$:

    (A3L) if $s \leq \|x\|_1^-, t \leq \|y\|_1^-$ and $s + t \leq \|x + y\|_1^-$,
        then $\|x + y\|(s + t) \geq L(\|x\|(s), \|y\|(t))$,

    (A3R) if $s \geq \|x\|_1^-, t \geq \|y\|_1^-$ and $s + t \geq \|x + y\|_1^-$,
        then $\|x + y\|(s + t) \leq L(\|x\|(s), \|y\|(t))$.

Lemma 8.1 [27] Let $(X, \|\cdot\|, L, R)$ be an FNLS, and suppose that

(R1) $R(a, b) \leq \max(a, b)$,

(R2) $\forall \alpha \in (0, 1], \exists \beta(0, \alpha]$ such that $R(\beta, y) \leq \alpha$ for all $y \in (0, \alpha)$,

(R3) $\lim_{a \to 0^+} R(a, a) = 0$.

Then, $(R1) \Rightarrow (R2) \Rightarrow (R3)$, but not conversely.

Theorem 8.5 **(Sadeqi *et al.* [28])** Let $(X, || \cdot ||, L, R)$ be an FNLS and $\lim_{a \to 0^+} R(a, a) = 0$. Then, $(X, || \cdot ||, L, R)$ is a Hausdorff topological vector space, whose neighborhood base of origin is $\{N(\epsilon, \alpha) : \epsilon > 0, \alpha \in (0, 1]\}$, where $N(\epsilon, \alpha) = \{x : ||x||_\alpha^+ \leq \epsilon\}$.

Definition 8.2 Let $(X, || \cdot ||, L, R)$ be an FNLS. A sequence $\{x_n\}_{n=1}^\infty \subseteq X$ converges to $x \in X$, if $\lim_{n \to \infty} ||x_n - x||_\alpha^+$, for every $\alpha \in (0, 1]$ denoted by $\lim_{n \to \infty} x_n = x$.

Definition 8.3 Let $(X, || \cdot ||, L, R)$ be an FNLS. A sequence $\{x_n\}_{n=1}^\infty \subseteq X$ is called a Cauchy sequence if $\lim_{m,n \to \infty} ||x_m - x_n||_\alpha^+ = 0$ for every $\alpha \in (0, 1]$.

Definition 8.4 Let $(X, || \cdot ||, L, R)$ be an FNLS. A subset $A \subseteq X$ is said to be complete if every Cauchy sequence in A, converges in A. The fuzzy normed space $(X, || \cdot ||, L, R)$ is said to be a fuzzy Banach space if it is complete.

Throughout this section, let G be a linear space and $(H, |.|, L, R)$ be a fuzzy Banach space satisfying (R2). Define a mapping $Df_6 : G \to H$ by

$$Df_6(x, y) = f(ax + by) + f(bx + ay) + (a - b)^6 \left[f\left(\frac{ax - by}{a - b}\right) + f\left(\frac{bx - ay}{b - a}\right) \right]$$

$$- 64(ab)^2(a^2 + b^2) \left[f\left(\frac{x + y}{2}\right) + f\left(\frac{x - y}{2}\right) \right] - 2(a^2 - b^2)(a^4 - b^4)[f(x) + f(y)],$$

where $a \neq b$, such that $k \in R; k = a + b \neq 0, \pm 1$.

The proofs of the following theorems and corollaries are similar to that of Theorems 8.3 and 8.4 and Corollaries 8.1 and 8.2. Hence, the details of the proofs are omitted.

8.4.1 Felbin's Type Spaces: Direct Method

Theorem 8.6 Let $q = \pm 1$ and $\zeta : G^2 \to F^*(R)$ be a function such that

$$\lim_{l \to \infty} \frac{\zeta(k^{lq}x, k^{lq}y)_\alpha^+}{k^{6lq}} = 0 \tag{8.42}$$

for all $x, y \in G$. Let $Df_6 : G \to H$ be a function satisfying the inequality

$$||Df_6(x, y)||_\alpha^+ \leq \zeta(x, y)_\alpha^+ \tag{8.43}$$

for all $x, y \in G$. Then, there exists a unique sextic function $S : G \to H$ which satisfies Euler–Lagrange–Jensen general $(a, b; k = a + b)$-sextic functional equation (8.10) and

$$\|f(x) - S(x)\|_\alpha^+ \leq \frac{1}{2k^6} \odot \sum_{p=\frac{1-q}{2}}^{\infty} \frac{\zeta(k^{pq}x, k^{pq}x)_\alpha^+}{k^{6pq}} \tag{8.44}$$

where $S(x)$ is defined by

$$S(x) = \lim_{p\to\infty} \frac{f(k^{pq}x)}{k^{pq}} \tag{8.45}$$

for all $x \in G$.

Corollary 8.3 Let $Df_6 : G \to H$ be a mapping. If there exist real numbers ϑ and σ such that

$$\|Df_6(x,y)\|_\alpha^+ \leq \begin{cases} \vartheta, \\ \vartheta \otimes \{\|x\|^\sigma \oplus \|y\|^\sigma\}, & \sigma \neq 6; \\ \vartheta \otimes \|x\|^\sigma \otimes \|y\|^\sigma, & 2\sigma \neq 6; \\ \vartheta \otimes \{\|x\|^\sigma \otimes \|y\|^\sigma + \{\|x\|^{2\sigma} \oplus \|y\|^{2\sigma}\}\}, & 2\sigma \neq 6 \end{cases} \tag{8.46}$$

for all $x, y \in G$, then there exists a unique sextic function $S : G \to H$ such that

$$\|f(x) - S(x)\|_\alpha^+ \leq \begin{cases} \dfrac{\vartheta}{2|k^6 - 1|}, \\ \dfrac{\vartheta(\|x\|^\sigma)_\alpha^+}{|k^6 - k^\sigma|}, \\ \dfrac{\vartheta(\|x\|^{2\sigma})_\alpha^+}{2|k^6 - k^{2\sigma}|}, \\ \dfrac{3\vartheta(\|x\|^{2\sigma})_\alpha^+}{2|k^6 - k^{2\sigma}|} \end{cases} \tag{8.47}$$

for all $x \in G$.

8.4.2 Felbin's Type Spaces: Fixed Point Method

Theorem 8.7 Let $Df_6 : G \to H$ be a mapping for which there exists a function $\zeta : G^2 \to F^*(R)$ with the condition

$$\lim_{n\to\infty} \frac{1}{\hbar_i^{6n}} \zeta(\hbar_i^n x, \hbar_i^n y)_\alpha^+ = 0 \tag{8.48}$$

where

$$\hbar_i = \begin{cases} k & \text{if } i = 0, \\ \frac{1}{k} & \text{if } i = 1 \end{cases} \tag{8.49}$$

such that the functional inequality

$$\|Df_6(x,y)\|_\alpha^+ \leq \zeta(x,y)_\alpha^+ \tag{8.50}$$

holds for all $x, y \in G$. Assume that there exists $L = L(i)$ such that the function

$$x \to T(x,x)_\alpha^+ = \frac{1}{2}\zeta\left(\frac{x}{k}, \frac{x}{k}\right)_\alpha^+$$

with the property

$$\frac{1}{\hbar_i^6} \odot T(\hbar_i x, \hbar_i x) = L \odot T(x, x)_\alpha^+ \tag{8.51}$$

for all $x \in G$. Then, there exists a unique sextic mapping $S : G \to H$ satisfying the functional equation (8.10) and

$$\|f(x) - S(x)\| \;\; \leq \;\; \left(\frac{L^{1-i}}{1 - L}\right) T(x, x)_\alpha^+ \tag{8.52}$$

for all $x \in G$.

Corollary 8.4 Let $Df_6 : G \to H$ be a mapping. If there exist real numbers ϑ and σ such that

$$\|Df_6(x, y)\|_\alpha^+ \;\; \leq \;\; \begin{cases} \vartheta, & \\ \vartheta \otimes \{\|x\|^\sigma \oplus \|y\|^\sigma\}, & \sigma \neq 6; \\ \vartheta \otimes \|x\|^\sigma \otimes \|y\|^\sigma, & 2\sigma \neq 6; \\ \vartheta \otimes \{\|x\|^\sigma \otimes \|y\|^\sigma + \{\|x\|^{2\sigma} \oplus \|y\|^{2\sigma}\}\}, & 2\sigma \neq 6; \end{cases} \tag{8.53}$$

for all $x, y \in G$, then there exists a unique sextic function $S : G \to H$ such that

$$\|f(x) - S(x)\|_\alpha^+ \;\; \leq \;\; \begin{cases} \dfrac{\vartheta}{2|k^6 - 1|}, & \\ \dfrac{\vartheta}{|k^6 - k^\sigma|}(\|x\|^\sigma)_\alpha^+, & \\ \dfrac{\vartheta}{2|k^6 - k^{2\sigma}|}(\|x\|^{2\sigma})_\alpha^+, & \\ \dfrac{3\vartheta}{2|k^6 - k^{2\sigma}|}(\|x\|^{2\sigma})_\alpha^+ & \end{cases} \tag{8.54}$$

for all $x \in G$.

8.5 Intuitionistic Fuzzy Normed Space: Stability Results

In this section, we investigate the generalized Ulam–Hyers stability of Euler–Lagrange–Jensen general $(a, b; k = a + b)$-sextic functional equation (8.10) in IFNS using direct and fixed point methods.

Now, we recall the basic definitions and notations in IFNS.

Definition 8.5 A binary operation $* : [0, 1] \times [0, 1] \to [0, 1]$ is said to be continuous t-norm if $*$ satisfies the following conditions:

(1) $*$ is commutative and associative;
(2) $*$ is continuous;
(3) $a * 1 = a$ for all $a \in [0, 1]$;
(4) $a * b \leq c * d$ whenever $a \leq c$ and $b \leq d$ for all $a, b, c, d \in [0, 1]$.

Definition 8.6 A binary operation $\diamond : [0,1] \times [0,1] \to [0,1]$ is said to be continuous t-conorm if $\diamond$ satisfies the following conditions:

$(1')$ $\diamond$ is commutative and associative;
$(2')$ $\diamond$ is continuous;
$(3')$ $a \diamond 0 = a$ for all $a \in [0,1]$;
$(4')$ $a \diamond b \le c \diamond d$ whenever $a \le c$ and $b \le d$ for all $a, b, c, d \in [0,1]$.

Using the notions of continuous t-norm and t-conorm, Saadati and Park [29] introduced the concept of IFNS as follows:

Definition 8.7 The five-tuple $(X, \mu, \nu, *, \diamond)$ is said to be an IFNS if X is a vector space, $* 1$ is a continuous t-norm, $\diamond$ is a continuous t-conorm, and μ, ν are fuzzy sets on $X \times (0, \infty)$ satisfying the following conditions. For every $x, y \in X$ and $s, t > 0$

(IFN1) $\mu(x, t) + \nu(x, t) \le 1$,
(IFN2) $\mu(x, t) > 0$,
(IFN3) $\mu(x, t) = 1$, if and only if $x = 0$.
(IFN4) $\mu(\alpha x, t) = \mu\left(x, \frac{t}{\alpha}\right)$ for each $\alpha \ne 0$,
(IFN5) $\mu(x, t) * \mu(y, s) \le \mu(x + y, t + s)$,
(IFN6) $\mu(x, \cdot) : (0, \infty) \to [0, 1]$ is continuous,
(IFN7) $\lim_{t \to \infty} \mu(x, t) = 1$ and $\lim_{t \to 0} \mu(x, t) = 0$,
(IFN8) $\nu(x, t) < 1$,
(IFN9) $\nu(x, t) = 0$, if and only if $x = 0$.
(IFN10) $\nu(\alpha x, t) = \nu\left(x, \frac{t}{\alpha}\right)$ for each $\alpha \ne 0$,
(IFN11) $\nu(x, t) \diamond \nu(y, s) \ge \nu(x + y, t + s)$,
(IFN12) $\nu(x, \cdot) : (0, \infty) \to [0, 1]$ is continuous,
(IFN13) $\lim_{t \to \infty} \nu(x, t) = 0$ and $\lim_{t \to 0} \nu(x, t) = 1$.

In this case, (μ, ν) is called an IFNS.

Example 8.1 Let $(X, \|\cdot\|)$ be a normed space. Let $a * b = ab$ and $a \diamond b = \min\{a + b, 1\}$ for all $a, b \in [0, 1]$. For all $x \in X$ and every $t > 0$, consider

$$\mu(x, t) = \begin{cases} \frac{t}{t + \|x\|} & \text{if } t > 0; \\ 0 & \text{if } t \le 0; \end{cases} \quad \text{and} \quad \nu(x, t) = \begin{cases} \frac{\|x\|}{t + \|x\|} & \text{if } t > 0; \\ 0 & \text{if } t \le 0. \end{cases}$$

Then $(X, \mu, \nu, *, \diamond)$ is an IFN-space.

The concepts of convergence and Cauchy sequences in an IFNS are studied in [29].

Definition 8.8 Let $(X, \mu, \nu, *, \diamond)$ be an IFN-space. Then, the sequence $x = \{x_k\}$ is said to be *intuitionistic fuzzy convergent* to a point $L \in X$ if

$$\lim \quad \mu(x_k - L, t) = 1 \quad \text{and} \quad \lim \quad v(x_k - L, t) = 0$$

for all $t > 0$. In this case, we write

$$x_k \xrightarrow{IF} L \quad \text{as} \quad k \to \infty$$

Definition 8.9 Let $(X, \mu, v, *, \diamond)$ be an IFN-space. Then, $x = \{x_k\}$ is said to be *intuitionistic fuzzy Cauchy sequence* if

$$\mu(x_{k+p} - x_k, t) = 1 \quad \text{and} \quad v(x_{k+p} - x_k, t) = 0$$

for all $t > 0$, and $p = 1, 2, \dots$.

Definition 8.10 Let $(X, \mu, v, *, \diamond)$ be an IFN-space. Then, $(X, \mu, v, *, \diamond)$ is said to be *complete* if every intuitionistic fuzzy Cauchy sequence in $(X, \mu, v, *, \diamond)$ is intuitionistic fuzzy convergent $(X, \mu, v, *, \diamond)$.

Hereafter, throughout this section, assume that X be a linear space, (Z, μ', v') be an intuitionistic fuzzy space and (Y, μ, v) be a intuitionistic fuzzy Banach space.

Now, we use the following notation for a given mapping $Df_6 : X \to Y$ such that

$$Df_6(x, y) = f(ax + by) + f(bx + ay) + (a - b)^6 \left[f\left(\frac{ax - by}{a - b}\right) + f\left(\frac{bx - ay}{b - a}\right) \right]$$

$$- 64(ab)^2(a^2 + b^2) \left[f\left(\frac{x + y}{2}\right) + f\left(\frac{x - y}{2}\right) \right] - 2(a^2 - b^2)(a^4 - b^4)[f(x) + f(y)],$$

where $a \neq b$, such that $k \in R$; $k = a + b \neq 0, \pm 1$ and for all $x, y \in X$.

8.5.1 IFNS: Direct Method

Theorem 8.8 Let $\tau \in \{1, -1\}$. Let $\zeta : X \times X \to Z$ be a function such that for some $0 < \left(\frac{p}{k}\right)^{\tau} < 1$,

$$\left. \begin{array}{l} \mu'(\zeta(k^{n\tau}x, k^{n\tau}y), r) \geq \mu'(p^{n\tau}\zeta(x, y), r) \\[2mm] v'(\zeta(k^{n\tau}x, k^{n\tau}y), r) \leq v'(p^{n\tau}\zeta(x, y), r) \end{array} \right\} \tag{8.55}$$

for all $x \in X$ and all $r > 0$ and

$$\left. \begin{array}{l} \lim_{n \to \infty} \mu'(\zeta(k^{\tau n}x, k^{\tau n}y), k^{\tau n}r) = 1 \\[2mm] \lim_{n \to \infty} v'(\zeta(k^{\tau n}x, k^{\tau n}y), k^{\tau n}r) = 0 \end{array} \right\} \tag{8.56}$$

for all $x, y \in X$ and all $r > 0$. Let $Df_6 : X \to Y$ be a function satisfies the inequality

$$\left.\begin{array}{l} \mu(Df_6(x,y),r) \geq \mu'(\zeta(x,y),r) \\[2mm] v(Df_6(x,y),r) \leq v'(\zeta(x,y),r) \end{array}\right\} \tag{8.57}$$

for all $x,y \in X$ and all $r > 0$. Then there exists a unique sextic mapping $S : X \to Y$ satisfying (8.10) and

$$\left.\begin{array}{l} \mu(f(x) - S(x),r) \geq \mu'(\zeta_\mu(x,x), \ |k^6 - p|2r) \\[2mm] v(f(x) - S(x),r) \leq v'(\zeta_v(x,x), \ |k^6 - p|2r) \end{array}\right\} \tag{8.58}$$

for all $x \in X$ and all $r > 0$.

Proof: Let $\tau = 1$. Replacing (x,y) by (x,x) in (8.57), we get

$$\mu\left(f(ax+bx)+f(bx+ax)+(a-b)^6\left[f\left(\frac{ax-bx}{a-b}\right)+f\left(\frac{bx-ax}{b-a}\right)\right]\right.$$
$$\left. - 64(ab)^2(a^2+b^2)f(x) - 2(a^2-b^2)(a^4-b^4)[2f(x)], r\right) \geq \mu'(\zeta(x,x),r),$$

$$\mu(2f((a+b)x)+2(a-b)^6 f(x) - 64(ab)^2(a^2+b^2)f(x)$$
$$- 4(a^2-b^2)(a^4-b^4)f(x),r) \geq \mu'(\zeta(x,x),r),$$

$$\mu(2f((a+b)x) - 2f(x)(a+b)^6, r) \geq \mu'(\zeta(x,x),r),$$

$$\mu\left(f(kx) - k^6 f(x), \frac{r}{2}\right) \geq \mu'(\zeta(x,x),r) \tag{8.59}$$

for all $x \in X$ and all $r > 0$ and $k = a + b$. Similarly, we prove

$$v\left(f(kx) - k^6 f(x), \frac{r}{2}\right) \geq v'(\zeta(x,x),r), \tag{8.60}$$

for all $x \in X$ and all $r > 0$. It follows from (8.59) and (8.60) and using (IFN4), we arrive

$$\left.\begin{array}{l} \mu\left(\frac{f(kx)}{k^6} - f(x), \frac{r}{2k^6}\right) \geq \mu'(\zeta(x,x),r) \\[3mm] v\left(\frac{f(kx)}{k^6} - f(x), \frac{r}{2k^6}\right) \leq v'(\zeta(x,x),r) \end{array}\right\} \tag{8.61}$$

for all $x \in X$ and all $r > 0$. Replacing x by $k^n x$ in (8.61), we have

$$\left.\begin{array}{l} \mu\left(\frac{f(k^{n+1}x)}{k^6} - f(k^n x), \frac{r}{2k^6}\right) \geq \mu'(\zeta(k^n x, k^n x),r) \\[3mm] v\left(\frac{f(k^{n+1}x)}{k^6} - f(k^n x), \frac{r}{2k^6}\right) \leq v'(\zeta(k^n x, k^n x),r) \end{array}\right\} \tag{8.62}$$

for all $x \in X$ and all $r > 0$. It is easy to verify from (8.62) and (IFN4) that

$$\left.\begin{array}{l} \mu\left(\dfrac{f(k^{n+1}x)}{k^{6(n+1)}} - \dfrac{f(k^n x)}{k^{6n}}, \dfrac{r}{2k^6 \cdot k^{6n}}\right) \geq \mu'\left(\zeta(x,x), \dfrac{r}{p^n}\right) \\[4mm] \nu\left(\dfrac{f(k^{n+1}x)}{k^{6(n+1)}} - \dfrac{f(k^n x)}{k^{6n}}, \dfrac{r}{2k^6 \cdot k^{6n}}\right) \leq \nu'\left(\zeta(x,x), \dfrac{r}{p^n}\right) \end{array}\right\} \tag{8.63}$$

for all $x \in X$ and all $r > 0$. Replacing r by $p^n r$ in (8.63), we get

$$\left.\begin{array}{l} \mu\left(\dfrac{f(k^{n+1}x)}{k^{6(n+1)}} - \dfrac{f(k^n x)}{k^{6n}}, \dfrac{p^n r}{2k^6 \cdot k^{6n}}\right) \geq \mu'(\zeta(x,x), r) \\[4mm] \nu\left(\dfrac{f(k^{n+1}x)}{k^{6(n+1)}} - \dfrac{f(k^n x)}{k^{6n}}, \dfrac{p^n r}{2k^6 \cdot k^{6n}}\right) \leq \nu'(\zeta(x,x), r) \end{array}\right\} \tag{8.64}$$

for all $x \in X$ and all $r > 0$. It is easy to see that

$$\frac{f(k^n x)}{k^{6n}} - f(x) = \sum_{i=0}^{n-1} \frac{f(k^{i+1}x)}{k^{6(i+1)}} - \frac{f(k^i x)}{k^{6i}} \tag{8.65}$$

for all $x \in X$. From (8.64) and (8.65), we have

$$\left.\begin{array}{l} \mu\left(\dfrac{f(k^n x)}{k^{6n}} - f(x), \sum_{i=0}^{n-1} \dfrac{p^i r}{2k^6 \cdot k^{6i}}\right) = \mu\left(\sum_{i=0}^{n-1} \dfrac{f(k^{i+1}x)}{k^{6(i+1)}} - \dfrac{f(k^i x)}{k^{6i}}, \sum_{i=0}^{n-1} \dfrac{p^i r}{2k^6 \cdot k^{6i}}\right) \\[4mm] \nu\left(\dfrac{f(k^n x)}{k^{6n}} - f(x), \sum_{i=0}^{n-1} \dfrac{p^i r}{2k^6 \cdot k^{6i}}\right) = \nu\left(\sum_{i=0}^{n-1} \dfrac{f(k^{i+1}x)}{k^{6(i+1)}} - \dfrac{f(k^i x)}{k^{6i}}, \sum_{i=0}^{n-1} \dfrac{p^i r}{2k^6 \cdot k^{6i}}\right) \end{array}\right\} \tag{8.66}$$

for all $x \in X$ and all $r > 0$. From (8.64) and (8.67), we have

$$\left.\begin{array}{l} \mu\left(\dfrac{f(k^n x)}{k^{6n}} - f(x), \sum_{i=0}^{n-1} \dfrac{p^i r}{2k^6 \cdot k^{6i}}\right) \geq \prod_{i=0}^{n-1} \mu\left(\dfrac{f(k^{i+1}x)}{k^{6(i+1)}} - \dfrac{f(k^i x)}{k^{6i}}, \dfrac{p^i r}{2k^6 \cdot k^{6i}}\right) \\[4mm] \nu\left(\dfrac{f(k^n x)}{k^{6n}} - f(x), \sum_{i=0}^{n-1} \dfrac{p^i r}{2k^6 \cdot k^{6i}}\right) \leq \coprod_{i=0}^{n-1} \nu\left(\dfrac{f(k^{i+1}x)}{k^{6(i+1)}} - \dfrac{f(k^i x)}{k^{6i}}, \dfrac{p^i r}{2k^6 \cdot k^{6i}}\right) \end{array}\right\} \tag{8.67}$$

where

$$\prod_{i=0}^{n-1} a_j = a_1 * a_2 * \cdots * a_n \quad \text{and} \quad \coprod_{i=0}^{n-1} b_j = b_1 \diamond b_2 \diamond \cdots \diamond b_n$$

for all $x \in X$ and all $r > 0$. Hence,

$$\left.\begin{aligned}
\mu\left(\frac{f(k^n x)}{k^{6n}} - f(x), \sum_{i=0}^{n-1} \frac{p^i r}{2k^6 \cdot k^{6i}}\right) &\geq \prod_{i=0}^{n-1} \mu'(\zeta(x,x), r) = \mu'(\zeta(x,x), r) \\
v\left(\frac{f(k^n x)}{k^{6n}} - f(x), \sum_{i=0}^{n-1} \frac{p^i r}{2k^6 \cdot k^{6i}}\right) &\leq \coprod_{i=0}^{n-1} v'(\zeta(x,x), r) = v'(\zeta(x,x), r)
\end{aligned}\right\}$$

(8.68)

for all $x \in X$ and all $r > 0$. Replacing x by $k^m x$ in (8.68) and using (8.56), ($IFN4$), we obtain

$$\left.\begin{aligned}
\mu&\left(\frac{f(k^{n+m} x)}{k^{6(n+m)}} - \frac{f(k^m x)}{k^{6m}}, \sum_{i=0}^{n-1} \frac{p^i r}{2k^6 \cdot k^{6(i+m)}}\right) \geq \mu'(\zeta(k^m x, k^m x), r) \\
&= \mu'\left(\zeta(x,x), \frac{r}{p^m}\right) \\
v&\left(\frac{f(k^{n+m} x)}{k^{6(n+m)}} - \frac{f(k^m x)}{k^{6m}}, \sum_{i=0}^{n-1} \frac{p^i r}{2k^6 \cdot k^{6(i+m)}}\right) \leq v'(\zeta(k^m x, k^m x), r) \\
&= v'\left(\zeta(x,x), \frac{r}{p^m}\right)
\end{aligned}\right\}$$

(8.69)

for all $x \in X$ and all $r > 0$ and all $m, n \geq 0$. Replacing r by $p^m r$ in (8.69), we get

$$\left.\begin{aligned}
\mu\left(\frac{f(k^{n+m} x)}{k^{6(n+m)}} - \frac{f(k^m x)}{k^{6m}}, \sum_{i=0}^{n-1} \frac{p^{i+m} r}{2k^6 \cdot k^{6(i+m)}}\right) &\geq \mu'(\zeta(x,x), r) \\
v\left(\frac{f(k^{n+m} x)}{k^{6(n+m)}} - \frac{f(k^m x)}{k^{6m}}, \sum_{i=0}^{n-1} \frac{p^{i+m} r}{2k^6 \cdot k^{6(i+m)}}\right) &\leq v'(\zeta(x,x), r)
\end{aligned}\right\}$$

(8.70)

for all $x \in X$ and all $r > 0$ and all $m, n \geq 0$. It follows from (8.70), that

$$\left.\begin{aligned}
\mu\left(\frac{f(k^{n+m} x)}{k^{6(n+m)}} - \frac{f(k^m x)}{k^{6m}}, r\right) &\geq \mu'\left(\zeta(x,x), \frac{r}{\sum_{i=m}^{n-1} \frac{p^i}{2k^6 \cdot k^{6i}}}\right) \\
v\left(\frac{f(k^{n+m} x)}{k^{6(n+m)}} - \frac{f(k^m x)}{k^{6m}}, r\right) &\leq v'\left(\zeta(x,x), \frac{r}{\sum_{i=m}^{n-1} \frac{p^i}{2k^6 \cdot k^{6i}}}\right)
\end{aligned}\right\}$$

(8.71)

holds for all $x \in X$ and all $r > 0$ and all $m, n \geq 0$. Since $0 < p < 6$ and $\sum_{i=0}^{n} \left(\frac{p}{6}\right)^i < \infty$. The Cauchy criterion for convergence in IFNS shows that $\left\{\frac{f(k^n x)}{k^{6n}}\right\}$ is a Cauchy sequence in (Y, μ, v). Since (Y, μ, v) is a complete IFN-space this sequence converges to some point $S(x) \in Y$. So, one can define the mapping $S : X \to Y$ by $\left\{\frac{f(k^n x)}{k^{6n}}\right\}$

$$\lim_{n \to \infty} \mu \left(\frac{f(k^n x)}{k^{6n}} - S(x), r \right) = 1, \lim_{n \to \infty} \nu \left(\frac{f(k^n x)}{k^{6n}} - S(x), r \right) = 0$$

for all $x \in X$ and all $r > 0$. Hence,

$$\frac{f(k^n x)}{k^{6n}} \xrightarrow{IF} S(x), \text{ as } n \to \infty.$$

Letting $m = 0$ in (8.71), we arrive

$$\left. \begin{aligned} \mu \left(\frac{f(k^n x)}{k^{6n}} - f(x), r \right) &\geq \mu' \left(\zeta(x, x), \frac{r}{\sum_{i=0}^{n-1} \frac{p^i}{2k^6 \cdot k^{6i}}} \right) \\ \nu \left(\frac{f(k^n x)}{k^{6n}} - f(x), r \right) &\leq \nu' \left(\zeta(x, x), \frac{r}{\sum_{i=0}^{n-1} \frac{p^i}{2k^6 \cdot k^{6i}}} \right) \end{aligned} \right\} \tag{8.72}$$

for all $x \in X$ and all $r > 0$. Letting n tend to infinity in (8.72), we have

$$\left. \begin{aligned} \mu(S(x) - f(x), r) &\geq \mu'(\zeta(x, x), 2r|k^6 - p|) \\ \nu(S(x) - f(x), r) &\leq \nu'(\zeta(x, x), 2r|k^6 - p|) \end{aligned} \right\} \tag{8.73}$$

for all $x \in X$ and all $r > 0$. To prove S satisfies (8.10), replacing (x, y) by $(k^n x, k^n y)$ in (8.57), we obtain

$$\left. \begin{aligned} \mu \left(\frac{1}{k^{6n}} Df_6(k^n x, k^n y), r \right) &\geq \mu'(\zeta(k^n x, k^n y), k^{6n} r) \\ \nu \left(\frac{1}{k^{6n}} Df_6(k^n x, k^n y), r \right) &\leq \nu'(\zeta(k^n x, k^n y), k^{6n} r) \end{aligned} \right\} \tag{8.74}$$

for all $x \in X$ and all $r > 0$. Now,

$$\mu \left(S(ax + by) + S(bx + ay) + (a - b)^6 \left[S \left(\frac{ax - by}{a - b} \right) + S \left(\frac{bx - ay}{b - a} \right) \right] \right.$$
$$- 64(ab)^2 (a^2 + b^2) \left[S \left(\frac{x + y}{2} \right) + S \left(\frac{x - y}{2} \right) \right] \tag{8.75}$$
$$\left. - 2(a^2 - b^2)(a^4 - b^4)[S(x) + S(y)], r \right)$$
$$\geq \mu \left(S(ax + by) - \frac{1}{k^{6n}} f(k^n (ax + by)), \frac{r}{9} \right) *$$
$$\mu \left(S(bx + ay) - \frac{1}{k^{6n}} f(k^n (bx + ay)), \frac{r}{9} \right) *$$
$$\mu \left((a - b)^6 \left[S \left(\frac{ax - by}{a - b} \right) - \frac{1}{k^{6n}} f \left(k^n \left(\frac{ax - by}{a - b} \right) \right) \right], \frac{r}{9} \right) *$$
$$\mu \left((a - b)^6 \left[S \left(\frac{bx - ay}{b - a} \right) - \frac{1}{k^{6n}} f \left(k^n \left(\frac{bx - ay}{b - a} \right) \right) \right], \frac{r}{9} \right) *$$
$$\mu \left(-64(ab)^2 (a^2 + b^2) \left[S \left(\frac{x + y}{2} \right) - \frac{1}{k^{6n}} f \left(k^n \left(\frac{x + y}{2} \right) \right) \right], \frac{r}{9} \right) *$$

$$\mu\left(-64(ab)^2(a^2+b^2)\left[S\left(\frac{x-y}{2}\right)-\frac{1}{k^{6n}}f\left(k^n\left(\frac{x-y}{2}\right)\right)\right],\frac{r}{9}\right)*$$

$$\mu\left(-2(a^2-b^2)(a^4-b^4)\left(S(x)-\frac{1}{k^{6n}}f(k^n x)\right),\frac{r}{9}\right)*$$

$$\mu\left(-2(a^2-b^2)(a^4-b^4)\left(S(y)-\frac{1}{k^{6n}}f(k^n y)\right),\frac{r}{9}\right)*$$

$$\mu\left(\frac{1}{k^{6n}}Df_6(k^n x,k^n y),\frac{r}{9}\right).$$

$$\nu\left(S(ax+by)+S(bx+ay)+(a-b)^6\left[S\left(\frac{ax-by}{a-b}\right)+S\left(\frac{bx-ay}{b-a}\right)\right]\right.$$

$$-64(ab)^2(a^2+b^2)\left[S\left(\frac{x+y}{2}\right)+S\left(\frac{x-y}{2}\right)\right] \tag{8.76}$$

$$-2(a^2-b^2)(a^4-b^4)[S(x)+S(y)],r)$$

$$\geq\nu\left(S(ax+by)-\frac{1}{k^{6n}}f(k^n(ax+by)),\frac{r}{9}\right)\diamond$$

$$\nu\left(S(bx+ay)-\frac{1}{k^{6n}}f(k^n(bx+ay)),\frac{r}{9}\right)\diamond$$

$$\nu\left((a-b)^6\left[S\left(\frac{ax-by}{a-b}\right)-\frac{1}{k^{6n}}f\left(k^n\left(\frac{ax-by}{a-b}\right)\right)\right],\frac{r}{9}\right)\diamond$$

$$\nu\left((a-b)^6\left[S\left(\frac{bx-ay}{b-a}\right)-\frac{1}{k^{6n}}f\left(k^n\left(\frac{bx-ay}{b-a}\right)\right)\right],\frac{r}{9}\right)\diamond$$

$$\nu\left(-64(ab)^2(a^2+b^2)\left[S\left(\frac{x+y}{2}\right)-\frac{1}{k^{6n}}f\left(k^n\left(\frac{x+y}{2}\right)\right)\right],\frac{r}{9}\right)\diamond$$

$$\nu\left(-64(ab)^2(a^2+b^2)\left[S\left(\frac{x-y}{2}\right)-\frac{1}{k^{6n}}f\left(k^n\left(\frac{x-y}{2}\right)\right)\right],\frac{r}{9}\right)\diamond$$

$$\nu\left(-2(a^2-b^2)(a^4-b^4)\left(S(x)-\frac{1}{k^{6n}}f(k^n x)\right),\frac{r}{9}\right)\diamond$$

$$\nu\left(-2(a^2-b^2)(a^4-b^4)\left(S(y)-\frac{1}{k^{6n}}f(k^n y)\right),\frac{r}{9}\right)\diamond$$

$$\nu\left(\frac{1}{k^{6n}}Df_6(k^n x,k^n y),\frac{r}{9}\right)$$

for all $x,y\in X$ and all $r>0$. Since

$$\left.\begin{array}{l}\lim\limits_{n\to\infty}\mu\left(\frac{1}{k^{6n}}Df_6(k^n x,k^n y),\frac{r}{9}\right)=1\\[2mm]\lim\limits_{n\to\infty}\nu\left(\frac{1}{k^{6n}}Df_6(k^n x,k^n y),\frac{r}{9}\right)=0\end{array}\right\} \tag{8.77}$$

for all $x\in X$ and all $r>0$. Letting $n\to\infty$ in (8.75) and (8.76) and using (8.77), we observe that S fulfills (8.10). Therefore, S is a sextic mapping.

In order to prove $S(x)$ is unique, let $S'(x)$ be another sextic functional equation satisfying (8.10) and (8.58). Hence,

$$\mu(S(x) - S'(x), r) = \mu\left(\frac{S(k^n x)}{k^{6n}} - \frac{S'(k^n x)}{k^{6n}}, r\right)$$

$$\geq \mu\left(S(k^n x) - f(k^n x), \frac{r.k^{6n}}{2}\right) * \mu\left(f(k^n x) - S'(k^n x), \frac{r.k^{6n}}{2}\right)$$

$$\geq \mu'\left(\zeta(k^n x, k^n x), \frac{2r\, k^{6n}|k^6 - p|}{2}\right)$$

$$\geq \mu'\left(\zeta(x, x), \frac{r\, k^{6n}|k^6 - p|}{p^n}\right)$$

$$v(S(x) - S'(x), r) = v\left(\frac{S(k^n x)}{k^{6n}} - \frac{S'(k^n x)}{k^{6n}}, r\right)$$

$$\leq v\left(S(k^n x) - f(k^n x), \frac{r.k^{6n}}{2}\right) \diamond v\left(f(k^n x) - S'(k^n x), \frac{r.k^{6n}}{2}\right)$$

$$\leq v'\left(\zeta(k^n x, k^n x), \frac{2r\, k^{6n}|k^6 - p|}{2}\right)$$

$$\leq v'\left(\zeta(x, x), \frac{r\, k^{6n}|k^6 - p|}{p^n}\right)$$

for all $x \in X$ and all $r > 0$. Since

$$\lim_{n \to \infty} \frac{r\, k^{6n}|k^6 - p|}{p^n} = \infty,$$

we obtain

$$\left.\begin{aligned} \lim_{n \to \infty} \mu'\left(\zeta_\mu(x, x), \frac{r\, k^{6n}|k^6 - p|}{p^n}\right) &= 1 \\ \lim_{n \to \infty} v'\left(\zeta_\mu(x, x), \frac{r\, k^{6n}|k^6 - p|}{p^n}\right) &= 0 \end{aligned}\right\}$$

for all $x \in X$ and all $r > 0$. Thus,

$$\left.\begin{aligned} \mu(S(x) - S'(x), r) &= 1 \\ v(S(x) - S'(x), r) &= 0 \end{aligned}\right\}$$

for all $x \in X$ and all $r > 0$. Hence, $S(x) = S'(x)$. Therefore, $S(x)$ is unique.

For $\tau = -1$, we can prove the similar stability result. This completes the proof of the theorem. $\qquad\square$

The following corollary is an immediate consequence of Theorem 8.8, regarding the stability of Euler-Lagrange–Jensen general $(a, b; k = a + b)$-sextic functional equation (8.10).

Corollary 8.5 Suppose that a function $Df_6 : X \to Y$ satisfies the inequality

$$\mu(Df_6(x,y),r) \geq \begin{cases} \mu(\lambda, r), \\ \mu(\lambda(||x||^s + ||y||^s), r), \\ \mu(\lambda \ ||x||^s ||y||^s, r), \\ \mu(\lambda\{||x||^s \ ||y||^s + (||x||^{2s} + ||y||^{2s})\}, r), \\ \nu(\lambda, r), \\ \nu(\lambda(||x||^s + ||y||^s), r), \\ \nu(\lambda \ ||x||^s ||y||^s, r), \\ \nu(\lambda\{||x||^s \ ||y||^s + (||x||^{2s} + ||y||^{2s})\}, r), \end{cases} \quad (8.78)$$

for all $x, y \in X$ and all $r > 0$, where λ, s are constants with $\lambda > 0$. Then there exists a unique sextic mapping $S : X \to Y$ such that

$$\mu(f(x) - S(x), r) \geq \begin{cases} \mu(\lambda, |k^6 - p|2r), \\ \mu(2\lambda||x||^s, |k^6 - p|2r), \\ \mu(\lambda \ ||x||^{2s}, |k^6 - p|2r), \\ \mu(3\lambda||x||^{2s}, |k^6 - p|2r), \\ \nu(\lambda, |k^6 - p|2r), \\ \nu(2\lambda \ ||x||^s, |k^6 - p|2r), \\ \nu(\lambda \ ||x||^{2s}, |k^6 - p|2r), \\ \nu(3\lambda \ ||x||^{2s}, |k^6 - p|2r), \end{cases} \quad (8.79)$$

for all $x \in X$ and all $r > 0$.

8.5.2 IFNS: Fixed Point Method

Theorem 8.9 Let $Df_6 : X \to Y$ be a mapping for which there exist a function $\zeta : X \times X \to Z$ with the condition

$$\left. \begin{aligned} \lim_{n\to\infty} \mu'(\zeta(\chi_i^n x, \chi_i^n y), \chi^{6n} r) = 1 \\ \\ \lim_{n\to\infty} \nu'(\zeta(\chi_i^n x, \chi_i^n y), \chi^{6n} r) = 0 \end{aligned} \right\} \quad (8.80)$$

for all $x, y \in X$ and all $r > 0$, where

$$\chi_i = \begin{cases} k & \text{if } i = 0 \\ \frac{1}{k} & \text{if } i = 1 \end{cases} \quad (8.81)$$

and satisfying the functional inequality

$$\left. \begin{aligned} \mu(Df_6(x,y),r) \geq \mu'(\zeta(x,y),r) \\ \\ \nu(Df_6(x,y),r) \leq \nu'(\zeta(x,y),r) \end{aligned} \right\} \quad (8.82)$$

for all $x, y \in X$ and all $r > 0$. If there exists $L = L(i) > 0$ such that the function

$$x \to \rho(x) = \frac{1}{2}\zeta\left(\frac{x}{k}, \frac{x}{k}\right),$$

has the property

$$\left.\begin{aligned}
\mu\left(L^{\frac{\rho(\chi_i x)}{\chi_i}}, r\right) &= \mu(\rho(x), r) \\
v\left(L^{\frac{\rho(\chi_i x)}{\chi_i}}, r\right) &= v(\rho(x), r)
\end{aligned}\right\} \tag{8.83}$$

for all $x \in X$ and all $r > 0$, then there exists a unique sextic function $S : X \rightarrow Y$ satisfying the functional equation (8.10) and

$$\left.\begin{aligned}
\mu(f(x) - S(x), r) &\geq \mu'\left(\rho(x), \frac{L^{1-i}}{1-L}r\right) \\
v(f(x) - S(x), r) &\leq v'\left(\rho(x), \frac{L^{1-i}}{1-L}r\right)
\end{aligned}\right\} \tag{8.84}$$

for all $x \in X$ and all $r > 0$.

Proof: Consider the set

$$\Lambda = \{h/h : X \rightarrow Y, \ h(0) = 0\}$$

And introduce the generalized metric on Λ,

$$\left.\begin{aligned}
\inf\{K \in (0, \infty) &: \mu(h(x) - g(x), r) \geq \mu'(\rho(x), Kr), x \in X\} \\
\inf\{K \in (0, \infty) &: v(h(x) - g(x), r) \leq v'(\rho(x), Kr), x \in X\}
\end{aligned}\right\} \tag{8.85}$$

It is easy to see that (8.85) is complete with respect to the defined metric. Define $J : \Lambda \rightarrow \Lambda$ by

$$Jh(x) = \frac{1}{\hbar_i^6}h(\hbar_i x),$$

for all $x \in \mathcal{X}$. The rest of the proof is similar to that of Theorem 8.4. $\square$

The following corollary is an immediate consequence of Theorem 8.9, regarding the stability of Euler–Lagrange–Jensen general $(a, b; k = a + b)$-sextic functional equation (8.10).

Corollary 8.6 Suppose that a function $Df_6 : X \rightarrow Y$ satisfies the inequality

$$\left.\begin{aligned}
\mu(Df_6(x, y), r) &\geq \begin{cases} \mu(\lambda, r), \\ \mu(\lambda(||x||^s + ||y||^s), r), \\ \mu(\lambda\,||x||^s||y||^s, r), \\ \mu(\lambda\{||x||^s\,||y||^s + (||x||^{2s} + ||y||^{2s})\}, r), \end{cases} \\
v(Df_6(x, y), r) &\leq \begin{cases} v(\lambda, r), \\ v(\lambda(||x||^s + ||y||^s), r), \\ v(\lambda\,||x||^s||y||^s, r), \\ v(\lambda\{||x||^s\,||y||^s + (||x||^{2s} + ||y||^{2s})\}, r), \end{cases}
\end{aligned}\right\} \tag{8.86}$$

for all $x, y \in X$ and all $r > 0$, where λ, s are constants with $\lambda > 0$. Then there exists a unique sextic mapping $S : X \rightarrow Y$ such that

$$\mu(f(x) - S(x), r) \geq \begin{cases} \mu\left(\frac{\lambda}{2}, |k^6 - p|2r\right), \\ \mu(\lambda||x||^s, |k^6 - p|2r), \\ \mu\left(\frac{\lambda}{2}||x||^{2s}, |k^6 - p|2r\right), \\ \mu\left(\frac{3\lambda}{2}||x||^{2s}, |k^6 - p|2r\right), \end{cases}$$

$$v(f(x) - S(x), r) \leq \begin{cases} v\left(\frac{\lambda}{2}, |k^6 - p|2r\right), \\ v(\lambda||x||^s, |k^6 - p|2r), \\ v\left(\frac{\lambda}{2}||x||^{2s}, |k^6 - p|2r\right), \\ v\left(\frac{3\lambda}{2}||x||^{2s}, |k^6 - p|2r\right), \end{cases}$$

(8.87)

for all $x \in X$ and all $r > 0$.

References

1 Ulam, S.M. (1964) *Problems in Modern Mathematics*, Science Editions, John Wiley & Sons, Inc., New York.

2 Hyers, D.H. (1941) On the stability of the linear functional equation. *Proc. Natl. Acad. Sci. U.S.A.*, **27**, 222–224.

3 Aoki, T. (1950) On the stability of the linear transformation in Banach spaces. *J. Math. Soc. Jpn.*, **2**, 64–66.

4 Arunkumar, M., Rassias, M.J., and Zhang, Y. (2012) Ulam - Hyers stability of a 2- variable AC - mixed type functional equation: direct and fixed point methods. *J. Mod. Math. Front.*, **1** (3), 10–26.

5 Gavruta, P. (1994) A generalization of the Hyers-Ulam-Rassias stability of approximately additive mappings. *J. Math. Anal. Appl.*, **184**, 431–436.

6 Abasalt, B., Narasimman, P., Ravi, K., and Shojaee, B. (2015) Mixed type of additive and quintic functional equations. *Ann. Math. Silesianae*, **29**, 35–50. doi: 10.1515/amsil-2015-0004.

7 Pasupathi, N. and Rangappan, A. (2016) Ulam-J Rassias stability of additive functional equation in digital logic circuits. *Math. Sci. Lett.*, **5** (2), 153–159.

8 Narasimman, P. (2016) Solution and fuzzy stability of a mixed type functional equation. *Adv. Pure Appl. Math.*, **7** (1), 29–39.

9 Bodaghi, A., Narasimman, P., Rassias, J.M., and Ravi, K. (2016) Ulam stability of the reciprocal functional equation in non-archimedean fields. *Acta Math. Univ. Comenianae*, **LXXXV** (1), 113–124.

10 Pasupathi, N. and Rassias, J.M. (2016) Ulam-Hyers stabilities of a generalized composite functional equation innon-Archimedean spaces. *Adv. Pure Appl. Math.*, **7** (4), 249–257. doi: 10.1515/apam-2016-0023.

11 Rassias, J.M. (1982) On approximately of approximately linear mappings by linear mappings. *J. Funct. Anal.*, **46**, 126–130.

12 Rassias, Th.M. (1978) On the stability of the linear mapping in Banach spaces. *Proc. Am. Math. Soc.*, **72**, 297–300.

13 Xu, T.Z., Rassias, J.M., Rassias, M.J., and Xu, W.X. (2010) A fixed point approach to the stability of quintic and sextic functional equations in quasi-β-normed spaces. *J. Inequal. Appl.*, **2010** Art. ID 423231, 123.

14 Mohammad, B.G., Majani, H., and Gordji, M.E. (2012) Approximately quintic and sextic mappings on the probabilistic normed spaces. *Bull. Korean Math. Soc.*, **49** (2), 339–352.

15 Cho, I.G., Kang, D., and Koh, H. (2010) Stability problems of quintic mappings in quasi-β-normed spaces. *J. Inequal. Appl.*, **2010**, Article ID 368981, 9p., doi: 10.1155/2010/368981.

16 Xu, T.Z. and Rassias, J.M. (2013) Approximate septic and octic mappings in quasi-β-normed spaces. *J. Comput. Anal. Appl.*, **15** (6), 1110–1119. copyright 2013 Eudoxus Press, LLC.

17 Pasupathi, N., Rassias, J.M., and Ravi, K. (2016) AN n-dimensional quintic and Sextic functional equations and its stabilities in Felbin's type spaces. *Georgian Math. J.*, **23** (1), 121–137. doi: 10.1515/gmj-2015-0039.

18 Margolis, B. and Diaz, J.B. (1968) A fixed point theorem of the alternative for contractions on a generalized complete metric space. *Bull. Am. Math. Soc.*, **126**, 305–309.

19 Gantner, T., Steinlage, R., and Warren, R. (1978) Compactness in fuzzy topological spaces. *J. Math. Anal. Appl.*, **62**, 547–562.

20 Lowen, R. (1996) *Fuzzy Set Theory, (Ch. 5 : Fuzzy Real Numbers)*, Kluwer, Dordrecht.

21 Hoehle, U. (1987) Fuzzy real numbers as Dedekind cuts with respect to a multiple-valued logic. *Fuzzy Sets Syst.*, **24**, 263–278.

22 Kaleva, O. and Seikkala, S. (1984) On fuzzy metric spaces. *Fuzzy Sets Syst.*, **12**, 215–229.

23 Kaleva, O. (1985) The completion of fuzzy metric spaces. *J. Math. Anal. Appl.*, **109**, 194–198.

24 Kaleva, O. (2008) A comment on the completion of fuzzy metric spaces. *Fuzzy Sets Syst.*, **159** (16), 2190–2192.

25 Felbin, C. (1992) Finite dimensional fuzzy normed linear spaces. *Fuzzy Sets Syst.*, **48**, 239–248.

26 Xiao, J. and Zhu, X. (2002) On linearly topological structure and property of fuzzy normed linear space. *Fuzzy Sets Syst.*, **125**, 153–161.

27 Xiao, J. and Zhu, X. (2004) Topological degree theory and fixed point theorems in fuzzy normed space. *Fuzzy Sets Syst.*, **147**, 437–452.

28 Sadeqi, I., Moradlou, F., and Salehi, M. (2013) On approximate Cauchy equation in Felbin's type fuzzy normed linear spaces. *Iran. J. Fuzzy Syst.*, **10** (3), 51–63.

29 Saadati, R. and Park, J.H. (2006) On the intuitionistic fuzzy topological spaces. *Chaos, Solitons Fractals*, **27**, 3313–3344.

9

A Note on the Split Common Fixed Point Problem and its Variant Forms

Adem Kiliçman and L.B. Mohammed

Department of Mathematics, Faculty of Science, University Putra Malaysia, 43400 Serdang, Selangor, Malaysia

2010 AMS Subject Classification 58E35; 47H09, 47J25

9.1   Introduction

Functional analysis is an abstract branch of mathematics that originated from classical analysis. The impetus came from linear algebra, problems related to ordinary and partial differential equations, calculus of variations, approximation theory, integral equations, and so on. Functional analysis can be defined as the study of certain topological–algebraic structures and of the methods by which the knowledge of these structures can be applied to analytic problems [1].

Fixed-point theory (FPT) is one of the most powerful and fruitful tools of modern mathematics and may be considered a core subject of nonlinear analysis. It has been a nourishing area of research for many mathematicians. The origins of the theory, which date to the later part of the nineteenth century, rest in the use of successive approximations to establish the existence and uniqueness of the solutions, particularly to differential equations, for example, see [1–7] and references therein.

The classical importance of FPT in functional analysis is due to its usefulness in the theory of ordinary and partial differential equations. The existence or construction of a solution to a differential equation often reduces to the existence or location of a fixed point for an operator defined on a subset of a space of functions. FPT had also been used to determine the existence of periodic solutions for functional differential equations when solutions are already known to exist, for example, see [8–11] and references therein.

Related to the FPT, we have the split common fixed-point problems (SCFPP). The SCFPP was introduced and studied by Censor and Segal [12] as a generalization of many existing problems in nonlinear sciences, both pure and applied. Moreover, Censor and Segal [12] had shown that the problem of fixed point,

Mathematical Analysis and Applications: Selected Topics, First Edition.
Edited by Michael Ruzhansky, Hemen Dutta, and Ravi P. Agarwal.

convex feasibility, multiple-set split feasibility, split feasibility, and much more can be studied more conveniently as SCFPP. The results and conclusions that are true for the SCFPP continue to hold for these problems, and it shows the significance and range of applicability of the SCFPP. One of the important applications of SCFPP can be seen in intensity modulation radiation therapy (IMRT), for more details, see Censor *et al.* [13].

This research work falls within the general area of "Nonlinear Functional Analysis," an area with the vast amount of applicability in the recent years, as such becoming the object of an increasing amount of study. We focus on an important topic within this area "*A note on the split common fixed-point problem and its variant forms.*" In this regard, we discuss the SCFPP and its variant forms. We show that already known problems are special cases of the SCFPP. We use approximation methods to suggest different iterative algorithms for solving SCFPP and its variant forms. In the end, we give the convergence results of these algorithms.

9.2 Basic Concepts and Definitions

9.2.1 Introduction

In this section, we give some definitions and basic results. We start from the definition of vector space and end with some results from Hilbert spaces. Those results that are commonly used in all the chapters are given in this section, and those results that are relevant to a particular chapter are provided at the beginning of each chapter. In short, this section works as a foundation for the structure of this chapter.

9.2.2 Vector Space

Vector space play a vital role in many branches of mathematics. In fact, in various practical (and theoretical) problems, we have a set V whose elements may be vectors in three-dimensional spaces, or sequences of numbers, or functions; and these elements can be added and multiplied by constants (numbers) in a natural way, the result being again an element of V. Such concrete situations suggest the concept of a vector space as defined below. The definition will involve a general field $\mathbb{F}$, but in functional analysis, $\mathbb{F}$ will be $\mathbb{R}$ or $\mathbb{C}$. The elements of $\mathbb{F}$ are called scalars, while in this chapter they will be real or complex numbers.

Definition 9.1 A vector space over a field $\mathbb{F}$ is a nonempty set denoted by V together with addition (+) and scalar multiplication ($\cdot$) satisfies the following conditions:

i) $x + y = y + x$, for all $x, y \in V$;

ii) $x + (y + w) = (x + y) + w$, for all $x, y, w \in V$;

iii) there exists a vector denoted by θ such that $x + \theta = x$, for all $x \in V$;

iv) for all $x \in V$, there exists a unique vector denoted by $(-x)$ such that $x + (-x) = \theta$;

v) $\alpha.(\beta.x) = (\alpha.\beta).x$, for all $\alpha, \beta \in \mathbb{F}$ and $x \in V$;

vi) $\alpha.(x + y) = \alpha.x + \alpha.y$, for all $x, y \in V$ and $\alpha \in \mathbb{F}$;

vii) $(\alpha + \beta).x = \alpha.x + \beta.x$, for all $\alpha, \beta \in \mathbb{F}$ and $x \in V$;

viii) there exists $1 \in \mathbb{F}$ such that $1.x = x$, $\forall x \in V$.

Remark 9.1 From now we will drop the dot ($\cdot$) in the scalar multiplication and denote $\alpha.\beta$ as $\alpha\beta$.

Let $v_1, v_2, v_3, \ldots, v_n \in V$ and $\alpha_1, \alpha_2, \alpha_3, \ldots, \alpha_n$ be scalars. Consider the equation:

$$\alpha_1 v_1 + \alpha_2 v_2 + \alpha_3 v_3 + \cdots + \alpha_n v_n = 0. \tag{9.1}$$

Trivially, $\alpha_1 = \alpha_2 = \alpha_3 = \cdots = \alpha_n = 0$ solves (9.1). If it is possible to have the solution of (9.1) with at least one of the $\alpha_i's$ non zero, then the vectors $v_1, v_2, v_3, \ldots, v_n$ are called "**Linearly Dependent**" otherwise they are called "**Linearly Independent.**"

If $\mathbb{M} \subseteq V$ consist of a linearly independent set of vectors; we say that $\mathbb{M}$ is a linearly independent set.

Definition 9.2 Span of $\mathbb{M}$ (Span$\mathbb{M}$) is defined as the set of all linear combination of $\mathbb{M}$, that is, Span$\mathbb{M} = \{\alpha_1 v_1 + \alpha_2 v_2 + \alpha_3 v_3 + \cdots, v_1, v_2, v_3, \ldots \in V$, where $\alpha_1, \alpha_2, \ldots$ are scalars$\}$.

Definition 9.3 Let $\mathbb{M} \subseteq V$. $\mathbb{M}$ is said to be basis for the space V, if

i) $\mathbb{M}$ is a linearly independent set,

ii) Span$\mathbb{M} = V$.

Definition 9.4 Let V be a vector space, the dimension of V (dim V) is the number of vectors of the basis of V. V is of finite dimension if its dimension is finite. Otherwise, it is said to be of infinite-dimensional space.

Definition 9.5 Let C be a subset of V. C is said to be convex, if for all $x, y \in C$, $\gamma \in [0, 1]$, $(1 - \gamma)x + \gamma y \in C$. In general, for all $x_1, x_2, x_3, \ldots, x_n \in C$ and for $\gamma_j \geq 0$ such that $\sum_{j=1}^{n} \gamma_j = 1$, the combination $\sum_{j=1}^{n} \gamma_j x_j \in C$ is called the convex combination.

Definition 9.6 A mapping $T : V_1 \to V_2$ is said to be linear, if $\forall u, v \in V_1$ and α, β scalars,

$$T(\alpha u + \beta v) = \alpha T(u) + \beta T(v).$$

Limits (of convergent sequences), differentiation, and integration are examples of a linear map.

Remark 9.2 If, in Definition 9.6, the linear space V_2 is replaced by a scalar field $\mathbb{F}$, then the linear map T is called linear functional on V_1.

9.2.3 Hilbert Space and its Properties

Definition 9.7 Let Y be a linear space. An inner product on Y is a function $\langle .,. \rangle : Y \times Y \to \mathbb{F}$ such that the following conditions are satisfies:

i) $\langle y,y \rangle \geq 0 \; \forall y \in Y$;
ii) $\langle y,y \rangle = 0$ if $y = 0, \forall y \in Y$;
iii) $\langle y,z \rangle = \overline{\langle z,y \rangle}, \forall y, z \in Y$, where the "bar" indicates the complex conjugation;
iv) $\langle \alpha x + \beta y, z \rangle = \overline{\alpha}\langle x,z \rangle + \overline{\beta}\langle y,z \rangle$, for all $x, y, z \in Y$ and $\alpha, \beta \in \mathbb{C}$.

Remark 9.3 The pair $(Y, \langle .,. \rangle)$ is called an inner product space. We shall simply write Y for the inner product space $(Y, \langle .,. \rangle)$ when the inner product $\langle .,. \rangle$ is known. Furthermore, if Y is a real vector space, then condition (iii) above reduces to $\langle x,z \rangle = \langle z,x \rangle$ (Symmetry).

Definition 9.8 Let Y be a linear space over $\mathbb{F}$ ($\mathbb{R}$ or $\mathbb{C}$). A norm on Y is a real-valued function $\|.\| : Y \to \mathbb{R}$ such that the following conditions are satisfies:

i) $\|x\| \geq 0, \forall x \in Y$;
ii) $\|x\| = 0$ if $x = 0, \forall x \in Y$;
iii) $\|\alpha x\| = |\alpha| \|x\|, \forall x \in Y$ and $\alpha \in \mathbb{R}$;
iv) $\|x + z\| \leq \|x\| + \|z\|, \forall x, z \in Y$.

Remark 9.4 A linear space Y with a norm defined on it, that is, $(Y, \|.\|)$ is called a normed linear space. If Y is a normed linear space, the norm $\|.\|$ always induces a metric d on Y given by $d(z,x) = \|z - x\|$ for each $x, z \in Y$, with this, (Y, d) become a metric space. For a quick review of metric space, the reader may consult Dunford *et al.* [14].

Lemma 9.1 Let Y be an inner product space. For arbitrary $x, z \in Y$,

$$|\langle x,z \rangle|^2 \leq \langle x,x \rangle \langle z,z \rangle. \tag{9.2}$$

If x and z are linearly dependent, then (9.2) reduces to

$$|\langle x,z \rangle|^2 = \langle x,x \rangle \langle z,z \rangle.$$

This lemma is known as Cauchy–Schwartz Inequality. For more details about the proof, refer Chidume [15].

Lemma 9.2   A mapping $\|.\| : Y \to \mathbb{R}$ defined by

$$\|x\| = \sqrt{\langle x, x \rangle}, \quad \forall x \in Y$$

is a norm on Y.

Remark 9.5   As the consequence of Lemma 9.2, (9.2) reduces to the following inequality:

$$|\langle x, z \rangle| \leq \|x\| \|z\|, \quad \forall x, z \in Y.$$

Definition 9.9   A sequence $\{y_n\}$ in a normed linear space Y is said to converge to $y \in Y$, if $\forall \epsilon > 0$, there exists $N_\epsilon \in \mathbb{N}$, such that $\|y_n - y\| < \epsilon$, $\forall n \geq N_\epsilon$. The vector $y \in Y$ is called the limit of the sequence $\{y_n\}$ and is written as $\lim_{n \to \infty} y_n = y$ or $y_n \to y$, as $n \to \infty$.

Definition 9.10   A sequence $\{y_n\}$ in a normed linear space Y is said to converge weakly to $y \in Y$, if $\lim h(y_n) = h(y)$, for all $y \in Y^*$ where Y^* denote the dual space of Y.

Next, we give some results regards to the weak convergence of a sequence. For more details about the proof, see [15].

Lemma 9.3   Let $\{y_n\} \subseteq E$ (Banach space). Then the following results are satisfies:

i) $y_n \rightharpoonup y \Leftrightarrow h(y_n) \to h(y)$ for each $h \in E^*$;
ii) $y_n \to y \Leftrightarrow y_n \rightharpoonup y$;
iii) $y_n \rightharpoonup y \Leftrightarrow \{y_n\}$ is bounded and

$$\|y\| \leq \lim_{n \to \infty} \inf \|y_n\|;$$

iv) $y_n \rightharpoonup y$ (in E), $h_n \to h$ (in E^*) $\Leftrightarrow h_n(y_n) \to h(y)$ (in $\mathbb{R}$).

Remark 9.6   Lemma 9.3 (ii) shows that strong convergence implies weak convergence. However, the converse may not necessarily be true, that is, in an infinite-dimensional space, weak convergence does not always imply strong convergence, while they are the same if the dimension is finite. For the example of weak convergence, which is not strong convergence, see [15].

Definition 9.11   Let C be a subset of H. A sequence $\{y_n\}$ in H is said to be Fejer monotone, if

$$\|y_{n+1} - z\| \leq \|y_n - z\|, \quad \forall n \geq 1, \ z \in C.$$

Definition 9.12   A sequence $\{y_n\}$ in a normed linear space Y is said to be Cauchy, if $\forall \epsilon > 0$, $\exists N_\epsilon \in \mathbb{N}$ such that $\|y_n - y_m\| < \epsilon$, $\forall n, m \geq N_\epsilon$.

Definition 9.13   A normed linear space Y is said to be complete if and only if every Cauchy sequence in Y converges.

Remark 9.7   With respect to the norm defined in Lemma 9.2, we can define the Cauchy sequence in an inner product space Y. A sequence $\{y_n\}$ in Y is said to be Cauchy if and only if $\langle y_n - y_m, y_n - y_m \rangle^{1/2} := \|y_n - y_m\| \to 0$ as $n, m \to \infty$.

Definition 9.14   An inner product space Y is said to be complete if and only if every Cauchy sequence converges.

Definition 9.15   A complete inner product space is called a "Hilbert Space" and that of normed linear space is known as a "Banach Space."

9.2.4   Bounded Linear Map and its Properties

Definition 9.16   Let $T : H \to H$ be a linear map. T is said to be bounded, if there exists a constant $M \geq 0$ such that

$$\|T(y)\| \leq M\|y\|, \quad \forall y \in H.$$

Next, we give some results of a linear map that are continuous. For more details about the proof, see [15].

Lemma 9.4   Let X and Y be normed linear spaces and $T : X \to Y$ be a linear operator. Then the following results are equivalent:

i) T is continuous;
ii) T is continuous at the origin, that is, if $\{x_n\}$ is a sequence in X such that

$$\lim_{n \to \infty} x_n = 0, \quad \text{then} \quad \lim_{n \to \infty} Tx_n = 0 \text{ in } Y;$$

iii) T is Lipschitz, that is, in the sense that there exists $M \geq 0$ such that

$$\|Tx\| \leq M \|x\|, \quad \forall x \in X;$$

iv) $T(\Delta)$ is bounded (in the sense that there exists $M \geq 0$ such that $\|Tx\| \leq M$ for all $x \in \Delta$, where $\Delta := \{x \in X : \|x\| \leq 1\}$).

Remark 9.8   In the light of Lemma 9.4, we have that a linear map $T : X \to Y$ is continuous if and only if it is bounded.

Definition 9.17 Let $A : H \to H$ be a bounded linear map. Define a mapping $A^* : H \to H$ by

$$\langle Ay, z \rangle = \langle y, A^* z \rangle, \quad \forall y, \ z \in H.$$

The mapping A^* is called the adjoint of A.

The following results are fundamental for the adjoint operator on Hilbert space. For the proof, see [15].

Lemma 9.5 Let $A : H \to H$ be a bounded linear map with its adjoint A^*. Then the following hold:

  i) $(A^*)^* = A$;
 ii) $\|A\| = \|A^*\|$;
iii) $\|A^* A\| = \|A\|^2$.

9.2.5 Some Nonlinear Operators

Let $T : H \to H$ be a map A point $x \in H$ is called a **fixed point** of T provided $Tx = x$. We denote the set of fixed point of T by $Fix(T)$, that is

$$Fix(T) = \{ x \in H : Tx = x \}.$$

The $Fix(T)$ is closed and convex, for more details, see [16].

  T is said to be $\eta-$**strongly monotone**, if there exists a constant $\eta > 0$ such that

$$\langle Tx - Ty, x - y \rangle \geq \eta \|x - y\|, \quad \forall x, \ y \in H,$$

and it is said to be **contraction**, if

$$\| Tx - Tz \| \leq k \, \|x - z\|, \quad \forall x, \ z \in H, \tag{9.3}$$

where $k \in (0, 1)$.

Remark 9.9 If $T : H \to H$ is a contraction mapping with coefficient $k \in (0, 1)$, then $(I - T)$ is $(1 - k)-$strongly monotone, that is

$$\langle (I - T)w - (I - T)z, \ w - z \rangle \geq (1 - k)\|w - z\|^2, \quad \forall w, \ z \in H.$$

Proof:

$$\langle (I - T)w - (I - T)z, \ w - z \rangle = \langle w - z, \ w - z \rangle + \langle Tz - Tw, \ w - z \rangle$$
$$= \langle w - z, \ w - z \rangle - \langle Tw - Tz, \ w - z \rangle. \tag{9.4}$$

On the other hand,

$$\langle Tw - Tz,\ w - z\rangle \le \|Tw - Tz\|\|w - z\|$$
$$\le k\,\|w - z\|^2 \quad \text{since T is a contraction mapping.}$$
$$(9.5)$$

By (9.4) and (9.5), we deduce that

$$\langle (I - T)w - (I - T)z,\ w - z\rangle \ge (1 - k)\|w - z\|^2.$$

This complete the proof. □

Equation (9.3) reduces to the following equation as $k = 1$.

$$\|Tx - Tz\| \le \|x - z\|, \quad \forall x,\ z \in H.$$

This is known as nonexpansive mapping. As a generalization of nonexpansive mapping, we have *asymptotically nonexpansive* [17], this mapping is defined as:

$$\|T^n x - T^n z\| \le k_n\|x - z\|, \quad \forall n \ge 1 \text{ and } x,\ z \in H,$$

where $k_n \subset [1, \infty)$ such that $\lim_{n\to\infty} k_n = 1$.

The map T is said to be **total asymptotically nonexpansive** [18], if

$$\|T^n x - T^n z\|^2 \le \|x - z\|^2 + v_n\eta(\|x - z\|) + \mu_n, \quad \forall n \ge 1 \text{ and } x,\ z \in H.$$

where $\{v_n\}$ and $\{\mu_n\}$ are sequences in $[0, \infty)$ such that $\lim_{n\to\infty} v_n = 0$, $\lim_{n\to\infty} \mu_n = 0$, and $\eta : \mathfrak{R}^+ \to \mathfrak{R}^+$ is a strictly increasing continuous function with $\eta(0) = 0$. This class of mapping generalizes the class of nonexpansive and asymptotically nonexpansive mappings (for more details, see [19, 20]). Moreover, it is said to be $(k, \{\mu_n\}, \{\xi_n\}, \phi)$- total asymptotically strict pseudocontraction, if there exists a constant $k \in [0, 1)$, $\mu_n \subset [0, \infty)$, $\xi_n \subset [0, \infty)$ with $\mu_n \to 0$ and $\xi_n \to 0$ as $n \to \infty$, and continuous strictly increasing function $\phi : [0, \infty) \to [0, \infty)$ with $\phi(0) = 0$ such that

$$\|T^n x - T^n y\|^2 \le \|x - y\|^2 + k\|(I - T^n)x - (I - T^n)y\|^2$$
$$+ \mu_n\phi(\|x - y\|) + \xi_n, \quad \forall x,\ y \in H.$$

T is said to be **strictly pseudocontractive** [21], if

$$\|Tx - Tz\|^2 \le \|x - z\|^2 + k\|(I - T)x - (I - T)z\|^2, \quad \forall x,\ z \in H,$$

where $k \in [0, 1)$. Moreover, it is said to **pseudocontractive** if

$$\|Tx - Tz\|^2 \le \|x - z\|^2 + \|(I - T)x - (I - T)z\|^2, \quad \forall x,\ z \in H.$$

It is obvious that all nonexpansive mappings and strictly pseudocontractive mappings are pseudocontractive mappings but the converse does not hold.

T is said to be **quasi-nonexpansive** [22], if $Fix(T) \ne \emptyset$ and

$$\|Tx - z\| \le \|x - z\|, \forall x \in H \text{ and } z \in Fix(T).$$

This is equivalent to

$$2\langle x - Tx, z - Tx \rangle \leq \|Tx - x\|^2, \ \forall x \in H \quad \text{and} \quad z \in Fix(T). \tag{9.6}$$

Remark 9.10 Every nonexpansive mapping with $Fix(T) \neq \emptyset$ is a quasi-nonexpansive; however, the converse may not necessarily be true. Thus, the class of quasi-nonexpansive mapping generalizes the class of nonexpansive mapping.

The following is an example of a quasi-nonexpansive mapping, which is not nonexpansive mapping, for more details, see [23].

Example 9.1 Let $H = \mathbb{R}$, defined $T : Q := [0, \infty) \to \mathbb{R}$ by

$$Ty = \frac{y^2 + 2}{1 + y}, \quad \text{for all } y \in Q.$$

T is said to be k−demicontractive, if

$$\|Ty - z\|^2 \leq \|y - z\|^2 + k\|Ty - z\|^2, \quad \forall y \in H \quad \text{and} \quad z \in Fix(T), \tag{9.7}$$

where $k \in [0, 1)$. Trivially, the class of demicontractive mapping generalizes the class of quasi-nonexpansive mapping for $k \geq 0$.

The following is an example of a demicontractive mapping, which is not quasi-nonexpansive mapping, for more details, see [24] and references therein.

Example 9.2 Define a map $T : l_2 \to l_2$ by

$$T(x_1, x_2, x_3, \ldots) = -\frac{5}{2}(x_1, x_2, x_3, \ldots), \text{ for all } (x_1, x_2, x_3, \ldots) \in l_2.$$

Remark 9.11 If $k = -1$, then (9.7) reduces to

$$\|Ty - z\|^2 \leq \|y - z\|^2 - \|Ty - y\|^2, \ \forall y \in H \quad \text{and} \quad z \in Fix(T).$$

This is known as *firmly quasi-nonexpansive mapping*. Every strictly pseudo-contractive mapping with $Fix(T) \neq \emptyset$ is a demicontractive mapping; however, the converse may not necessarily be true. Thus, the class of demicontractive mapping is more general than the class of strictly pseudocontractive mapping.

The following is an example of demicontractive mapping, which is not strictly pseudocontractive mapping, for more details, see [21].

Example 9.3 Let $C = [-1, 1]$ be a subset of a real Hilbert space H. Define T on C by

$$T(x) = \begin{cases} \frac{2}{3}x \sin\left(\frac{1}{x}\right), & \text{if } x \neq 0, \\ 0, & \text{if } x = 0. \end{cases}$$

Clearly, 0 is the only fixed point of T. For $x \in C$, we have

$$|Tx - 0|^2 = |Tx|^2$$

$$= \left| \frac{2}{3} x \sin \left(\frac{1}{x} \right) \right|^2$$

$$\leq \left| \frac{2x}{3} \right|^2$$

$$\leq |x|^2$$

$$\leq |x - 0|^2 + k|Tx - x|^2, \quad \text{for any } k < 1.$$

Thus, T is demicontractive mapping. Next, we see that T is not strictly pseudo-contractive mapping. Let $x = \frac{2}{\pi}$ and $z = \frac{2}{3\pi}$, then $|Tx - Tz|^2 = \frac{256}{81\pi^2}$. However,

$$|x - z|^2 + |(I - T)x - (I - T)z|^2 = \frac{160}{81\pi^2}.$$

T is said to be *asymptotically quasi-nonexpansive*, if $Fix(T) \neq \emptyset$ such that for each $n \geq 1$,

$$\|T^n x - z\|^2 \leq t_n \|x - z\|^2, \quad \forall z \in Fix(T) \text{ and } x \in H,$$

where $\{t_n\} \subseteq [1, \infty)$ with $\lim_{n \to \infty} t_n = 1$. It is clear from this definition that every asymptotically nonexpansive mapping with $Fix(T) \neq \emptyset$ is asymptotically quasi-nonexpansive mapping.

In addition, T is said to be $(\{r_n\}, \{k_n\}, \eta)$-*total quasi-asymptotically nonexpansive mapping*, if

$$\|T^n y - z\|^2 \leq \|y - z\|^2 + r_n \eta(\|y - z\|)$$
$$+ k_n, \quad \forall n \geq 1, \ z \in Fix(T) \text{ and } y \in H,$$

where $\{r_n\}, \{k_n\}$ are sequences in $[0, \infty)$ such that $\lim_{n \to \infty} r_n = 0$, $\lim_{n \to \infty} k_n = 0$ and $\eta : \mathfrak{R}^+ \to \mathfrak{R}^+$ is a strictly continuous function with $\eta(0) = 0$. This class of mapping, generalizes the class of; quasi-nonexpansive, asymptotically quasi-nonexpansive and total asymptotically nonexpansive mapping.

T is said to be K-Lipschitzian, if

$$\|Ty - Tz\| \leq K \|y - z\|, \quad \forall y, z \in H.$$

It is said to be uniformly K-Lipschitzian, if

$$\|T^n y - T^n z\| \leq K \|y - z\|, \quad \forall y, z \in H.$$

Definition 9.18 A mapping $T : H \to H$ is said to be class$-\tau$ operator, if

$$\langle z - Ty, y - Ty \rangle \leq 0, \quad \forall z \in Fix(T) \text{ and } y \in H.$$

It is important to note that, class$-\tau$ operator is also called directed operator [12, 25], separating operator [26], or cutter operator [27].

Definition 9.19 A self mapping T on H_1 is said to be semicompact if for any bounded sequence $\{x_n\} \subset H$ with $(I - T)x_n$ converges strongly to 0, there exists a subsequence say $\{x_{n_k}\}$ of $\{x_n\}$ such that $\{x_{n_k}\}$ converges strongly to x.

Definition 9.20 A self mapping T on C is said to be demiclosed, if for any sequence $\{y_n\}$ in C such that $y_n \rightharpoonup y$ and if the sequence $Ty_n \to z$, then $Ty = z$.

Remark 9.12 In Definition 9.20, if $z = 0$, the zero vector in C, then T is called demiclosed at zero, for more details, see [28].

Lemma 9.6 If a self mapping T on C is a nonexpansive mapping, then T is demiclosed at zero [16].

Lemma 9.7 If a self mapping T on C is a $k-$strictly pseudocontractive, then $(T - I)$ is demiclosed at zero [29].

Lemma 9.8 Let C be a subset of H_1, and P_C be a metric projection from H_1 onto C. Then $\forall y \in C$ and $x \in H_1$,

$$\|x - P_C(x)\|^2 \leq \|y - x\|^2 - \|y - P_C(x)\|^2.$$

For the proof of this lemma, see [30].

Lemma 9.9 For each $x, y \in H_1$, the following results hold.

i) $\|x + y\|^2 = \|x\|^2 + 2\langle x, y \rangle + \|y\|^2$,
ii) $\|\alpha x + (1 - \alpha)y\|^2 = \alpha\|x\|^2 + (1 - \alpha)\|y\|^2 - \alpha(1 - \alpha)\|x - y\|^2, \quad \forall \alpha \in [0, 1].$

For the proof of this lemma, see [29].

Lemma 9.10 Let $\{a_n\}$ be a sequence of nonnegative real number such that

$$a_{n+1} \leq (1 - \gamma_n)a_n + \sigma_n, \quad n \geq 0,$$

where γ_n is a sequence in $(0, 1)$ and σ_n is a sequence of real number such that;

i) $\lim_{n \to \infty} \gamma_n = 0$ and $\sum \gamma_n = \infty$;
ii) $\lim_{n \to \infty} \frac{\sigma_n}{\gamma_n} \leq 0$ or $\sum |\sigma_n| < \infty$. Then $\lim_{n \to \infty} a_n = 0.$

For the proof, see [31].

Lemma 9.11 Let $\{x_n\}, \{y_n\}, \{z_n\}$ be sequences of nonnegative real numbers satisfying

$$x_{n+1} \leq (1 + z_n)x_n + y_n.$$

If $\sum z_n < \infty$ and $\sum y_n < \infty$, then $\lim_{n \to \infty} x_n$ exist.

For the proof of this lemma, see [32].

Lemma 9.12 Let $\{x_n\}$ be a Fejer monotone with respect to C, then the following are satisfied:

i) $x_n \rightharpoonup x^* \in C$ if and only if $\omega_\omega \subset C$;
ii) $\{P_C x_n\}$ converges strongly to some vector in C;
iii) if $x_n \rightharpoonup x^* \in C$, then $x^* = \lim\limits_{n \to \infty} P_C x_n$.

For the proof, see [33].

9.2.6   Problem Formulation

The SCFPP is formulated as follows:

$$\text{Find } x^* \in C := \bigcap_{i=1}^{N} Fix(T_i) \text{ such that } Ax^* \in Q := \bigcap_{j=1}^{M} Fix(G_j). \qquad (9.8)$$

In this chapter, we consider $T_i : H_1 \to H_1$, for $i = 1, 2, 3, \dots, N$ and $G_j : H_2 \to H_2$, for $j = 1, 2, 3, \dots, M$, to be total quasi-asymptotically nonexpansive and or demicontractive mappings.

We denote the solution set of SCFPP (9.8) by

$$\Gamma = \{x^* \in C \text{ such that } Ax^* \in Q\}. \qquad (9.9)$$

In sequel, we assume that $\Gamma \neq \emptyset$.

9.2.7   Preliminary Results

A Banach space E satisfies *Opial's condition* [34], if for any sequence $\{x_n\}$ in E such that $x_n \rightharpoonup x$, as $n \to \infty$ implies that

$$\lim_{n \to \infty} \inf \|x_n - x\| < \lim_{n \to \infty} \inf \|x_n - y\|, \quad \forall y \in E, \, y \neq x.$$

The above equation it is said to have *Kadec–Klee property* [34], if for any sequence $\{x_n\}$ in E such that $x_n \rightharpoonup x$ and $\|x_n\| \to \|x\|$, as $n \to \infty$ implies that

$$x_n \to x \text{ and as } n \to \infty.$$

Remark 9.13 Each Hilbert space satisfies the Opial and Kadec–Klee's properties.

The following lemma was taken from Wang *et al.* [35], and we include the proof here for the sake of completeness.

Lemma 9.13 Let $G : H_1 \to H_1$ be a $(\{v_n\}, \{\mu_n\}, \xi)$-total quasi-asymptotically nonexpansive mapping with $Fix(G) \neq \emptyset$. Then, for each $y \in Fix(G)$, $x \in H_1$ and $n \geq 1$, the following inequalities are equivalent:

$$\|G^n x - y\|^2 \leq \|x - y\|^2 + v_n \xi(\|x - y\|) + \mu_n; \tag{9.10}$$

$$2\langle x - G^n x, x - y \rangle \geq \|G^n x - x\|^2 - v_n \xi(\|x - y\|) - \mu_n; \tag{9.11}$$

$$2\langle x - G^n x, y - G^n x \rangle \leq \|G^n x - x\|^2 + v_n \xi(\|x - y\|) + \mu_n. \tag{9.12}$$

Proof: $(i) \Rightarrow (ii)$

$$\|G^n x - y\|^2 = \|G^n x - x + x - y\|^2$$
$$= \|G^n x - x\|^2 + 2\langle G^n x - x, x - y \rangle + \|x - y\|^2,$$

this imply that

$$2\langle G^n x - x, x - y \rangle = \|G^n x - y\|^2 - \|G^n x - x\|^2 - \|x - y\|^2$$
$$\leq \|x - y\|^2 + v_n \xi(\|x - y\|) + \mu_n - \|G^n x - x\|^2 - \|x - y\|^2.$$

Thus, we deduce that

$$2\langle x - G^n x, x - y \rangle \geq \|G^n x - x\|^2 - v_n \xi(\|x - y\|) - \mu_n.$$

$(ii) \Rightarrow (iii)$

$$\langle x - G^n x, x - y \rangle = \langle x - G^n x, x - G^n x + G^n x - y \rangle$$
$$= \langle x - G^n x, x - G^n x \rangle + \langle x - G^n x, G^n x - y \rangle.$$

This tends to imply that

$$\langle x - G^n x, G^n x - y \rangle = -\|x - G^n x\|^2 + \langle x - G^n x, x - y \rangle$$
$$\geq -\|x - G^n x\|^2 + \frac{1}{2}\|G^n x - x\|^2 - \frac{1}{2}v_n \xi(\|x - y\|) - \frac{1}{2}\mu_n.$$

Thus, we deduce that

$$2\langle x - G^n x, y - G^n x \rangle \leq \|x - G^n x\|^2 + v_n \xi(\|x - y\|) + \mu_n.$$

$(iii) \Rightarrow (i)$

$$2\langle x - G^n x, y - G^n x \rangle \leq \|G^n x - x\|^2 + v_n \xi(\|x - y\|) + \mu_n$$
$$= \|G^n x - y\|^2 + 2\langle G^n x - y, y - x \rangle + \|x - y\|^2$$
$$+ v_n \xi(\|x - y\|) + \mu_n,$$

thus, we deduce that

$$\|G^n x - y\|^2 \leq \|x - y\|^2 + v_n \xi(\|x - y\|) + \mu_n.$$

And thus completes the proof.

Lemma 9.14 [36] Let $P_C : H \to C$ be a metric projection such that

$$\langle x_n - x^*, x_n - P_C x_n \rangle \leq 0.$$

Then for each $n \geq 1$,

$$\|P_C x_n - x_n\| \leq \|P_C x_n - x^*\|, \quad \forall x^* \in C.$$

Proof: Let $x^* \in C$, then

$$
\begin{aligned}
\|x_n - P_C x_n\|^2 &= \|x_n - x^* + x^* - P_C x_n\|^2 \\
&= \|x_n - x^*\|^2 + \|x^* - P_C x_n\|^2 + 2\langle x_n - x^*, x^* - P_C x_n \rangle \\
&= \|x_n - x^*\|^2 + \|x^* - P_C x_n\|^2 + 2\langle x_n - x^*, x^* - x_n + x_n - P_C x_n \rangle \\
&= \|x_n - x^*\|^2 + \|x^* - P_C x_n\|^2 - 2\|x_n - x^*\|^2 \\
&\quad + 2\langle x_n - x^*, x_n - P_C x_n \rangle \\
&= \|x^* - P_C x_n\|^2 - \|x_n - x^*\|^2 + 2\langle x_n - x^*, x_n - P_C x_n \rangle \\
&\leq \|x^* - P_C x_n\|^2.
\end{aligned}
$$

Thus, we conclude that

$$\|x_n - P_C x_n\| \leq \|x^* - P_C x_n\|. \qquad \square$$

9.2.8 Strong Convergence for the Split Common Fixed-Point Problems for Total Quasi-Asymptotically Nonexpansive Mappings

Theorem 9.1 Let $G : H_1 \to H_1$, $T : H_2 \to H_2$ be $(\{v_{n_1}\}, \{\mu_{n_1}\}, \xi_1)$, $(\{v_{n_2}\}, \{\mu_{n_2}\}, \xi_2)$-total quasi-asymptotically nonexpansive mappings and uniformly L_1, L_2-Lipschitzian continuous mappings such that $(G - I)$ and $(T - I)$ are demiclosed at zero. Let $A : H_1 \to H_2$ be a bounded linear operator with its adjoint A^*. In addition, let M and M^* be positive constants such that $\xi(k) \leq \xi(M) + M^* k^2$, $\forall k \geq 0$. Assume that $\Gamma \neq \emptyset$, and let P_Γ be the metric projection of H_1 onto Γ satisfying

$$\langle x_n - x^*, x_n - P_\Gamma x_n \rangle \leq 0.$$

Define a sequence $\{x_n\}$ by

$$
\begin{cases}
x_0 \in H_1, \\
u_n = x_n + \gamma A^* (T^n - I) A x_n, \\
x_{n+1} = \alpha_n u_n + (1 - \alpha_n) G^n u_n, \quad \forall n \geq 0,
\end{cases}
\tag{9.13}
$$

where $\gamma, L, \{v_n\}, \{\mu_n\}, \{\xi_n\}$, and $\{\alpha_n\}$ satisfy the following conditions:

i) $0 < k < \alpha_n < 1$, $\gamma \in (0, \frac{1}{L^*})$ with $L^* = \|AA^*\|$ and $L = \max\{L_1, L_2\}$;

ii) $v_n = \max \{v_{n_1}, v_{n_2}\}$, $\mu_n = \max \{\mu_{n_1}, \mu_{n_2}\}$ and $\xi = \max \{\xi_1, \xi_2\}$.

Then, $x_n \to x^* \in \Gamma$.

Proof: To show that $x_n \to x^*$, as $n \to \infty$, it suffices to show that

$$x_n \rightharpoonup x^* \quad \text{and} \quad \|x_n\| \to \|x^*\|, \ as \ n \to \infty.$$

We divided the proof into five steps as follows:

Step 1. In this step, we show that for each $x^* \in \Gamma$, the following limit exists.

$$\lim_{n\to\infty} \|x_n - x^*\| = \lim_{n\to\infty} \|u_n - x^*\|. \tag{9.14}$$

Now, let $x^* \in \Gamma$. By (9.13) and Lemma 9.13, we have

$$\begin{aligned}
\|x_{n+1} - x^*\|^2 &= \|\alpha_n u_n + (1 - \alpha_n)G^n u_n - x^*\|^2 \\
&= \|\alpha_n(u_n - G^n u_n)\|^2 + 2\alpha_n\langle u_n - G^n u_n, G^n u_n - x^*\rangle + \|G^n u_n - x^*\|^2 \\
&= \alpha_n^2\|u_n - G^n u_n\|^2 + 2\alpha_n\langle u_n - x^* + x^* - G^n u_n, G^n u_n - x^*\rangle \\
&\quad + \|G^n u_n - x^*\|^2 \\
&= \alpha_n^2\|u_n - G^n u_n\|^2 + 2\alpha_n\langle u_n - x^*, G^n u_n - x^*\rangle \\
&\quad + (1 - 2\alpha_n)\|G^n u_n - x^*\|^2 \\
&= \alpha_n^2\|u_n - G^n u_n\|^2 + 2\alpha_n\langle u_n - x^*, G^n u_n - u_n + u_n - x^*\rangle \\
&\quad + (1 - 2\alpha_n)\|G^n u_n - x^*\|^2 \\
&= \alpha_n^2\|u_n - G^n u_n\|^2 + 2\alpha_n\langle u_n - x^*, G^n u_n - u_n\rangle \\
&\quad + 2\alpha_n\langle u_n - x^*, u_n - x^*\rangle + (1 - 2\alpha_n)\|G^n u_n - x^*\|^2 \\
&\leq -\alpha_n(1 - \alpha_n)\|u_n - G^n u_n\|^2 + 2\alpha_n\|u_n - x^*\|^2 + \alpha_n v_n\xi(\|u_n - x^*\|) \\
&\quad + \alpha_n\mu_n + (1 - 2\alpha_n)(\|u_n - x^*\|^2 + v_n\xi(\|u_n - x^*\|) + \mu_n) \\
&\leq -\alpha_n(1 - \alpha_n)\|u_n - G^n u_n\|^2 + \|u_n - x^*\|^2 \\
&\quad + (1 - \alpha_n)(v_n\xi(\|u_n - x^*\|) + \mu_n) \\
&= -\alpha_n(1 - \alpha_n)\|u_n - G^n u_n\|^2 + (1 + (1 - \alpha_n)v_n M^*)\|u_n - x^*\|^2 \\
&\quad + (1 - \alpha_n)(v_n\xi(M) + \mu_n). \tag{9.15}
\end{aligned}$$

On the other hand,

$$\begin{aligned}
\|u_n - x^*\|^2 &= \|x_n - x^* + \gamma A^*(T^n - I)Ax_n\|^2 \\
&= \|x_n - x^*\|^2 + \gamma^2\|A^*(T^n - I)Ax_n\|^2 \\
&\quad + 2\gamma\langle x_n - x^*, A^*(T^n - I)Ax_n\rangle \tag{9.16}
\end{aligned}$$

and

$$\begin{aligned}
\gamma^2\|A^*(T^n - I)Ax_n\|^2 &= \gamma^2\langle A^*(T^n - I)Ax_n, A^*(T^n - I)Ax_n\rangle \\
&= \gamma^2\langle AA^*(T^n - I)Ax_n, (T^n - I)Ax_n\rangle \\
&\leq \gamma^2 L^*\|(T^n - I)Ax_n\|^2. \tag{9.17}
\end{aligned}$$

By Lemma 9.13, it follows that

$$
\begin{aligned}
2\gamma\langle x_n - x^*, A^*(T^n - I)Ax_n\rangle &= 2\gamma\langle Ax_n - T^nAx_n + T^nAx_n - Ax^*, T^nAx_n - Ax_n\rangle \\
&= 2\gamma\langle T^nAx_n - Ax^*, T^nAx_n - Ax_n\rangle \\
&\quad - 2\gamma\|(T^n - I)Ax_n\|^2 \\
&\leq \gamma v_n M^* L^*\|x_n - x^*\|^2 + \gamma(v_n\xi(M) + \mu_n) \\
&\quad - \gamma\|(T^n - I)Ax_n\|^2.
\end{aligned}
\tag{9.18}
$$

Substituting (9.17) and (9.18) into (9.16), we obtain that

$$
\begin{aligned}
\|u_n - x^*\|^2 &\leq (1 + \gamma v_n M^* L^*)\|x_n - x^*\|^2 \\
&\quad - \gamma(1 - \gamma L)\|(T^n - I)Ax_n\|^2 + \gamma(v_n\xi(M) + \mu_n).
\end{aligned}
\tag{9.19}
$$

By (9.19) and (9.15), we deduce that

$$
\begin{aligned}
\|x_{n+1} - x^*\|^2 &\leq (1 + (1 - \alpha_n)v_n M^*)((1 + \gamma v_n M^* L^*)\|x_n - x^*\|^2 \\
&\quad - \gamma(1 - \gamma L^*)\|(T^n - I)Ax_n\|^2 + \gamma(v_n\xi(M) + \mu_n)) \\
&\quad - \alpha_n(1 - \alpha_n)\|x_n - G^n u_n\|^2 + (1 - \alpha_n)(v_n\xi(M) + \mu_n) \\
&= (1 + (1 - \alpha_n)v_n M^*)(1 + \gamma v_n M^* L)\|x_n - x^*\|^2 \\
&\quad - \gamma(1 - \gamma L^*)(1 + (1 - \alpha_n)v_n M^*)\|(T^n - I)Ax_n\|^2 \\
&\quad - \alpha_n(1 - \alpha_n)\|x_n - G^n u_n\|^2 + (1 + (1 - \alpha_n)v_n M^*)\gamma(v_n\xi(M) + \mu_n) \\
&\quad + (1 - \alpha_n)(v_n\xi(M) + \mu_n) \\
&\leq (1 + (1 - \alpha_n)v_n M^*)(1 + \gamma v_n M^* L)\|x_n - x^*\|^2 \\
&\quad - \gamma(1 - \gamma L^*)\|(T^n - I)Ax_n\|^2 - \alpha_n(1 - \alpha_n)\|x_n - G^n u_n\|^2 \\
&\quad + (1 + (1 - \alpha_n)v_n M^*)\gamma(v_n\xi(M) + \mu_n) \\
&\quad + (1 - \alpha_n)(v_n\xi(M) + \mu_n).
\end{aligned}
\tag{9.20}
$$

Thus, we deduce that

$$
\begin{aligned}
\|x_{n+1} - x^*\|^2 &\leq (1 + \gamma v_n M^* L^* + (1 - \alpha_n)v_n M^*(1 + \gamma v_n M^* L^*))\|x_n - x^*\|^2 \\
&\quad + (1 + (1 - \alpha_n)v_n M^*)\gamma(v_n\xi(M) + \mu_n) + (1 - \alpha_n)(v_n\xi(M) + \mu_n).
\end{aligned}
$$

This implies that

$$
\|x_{n+1} - x^*\|^2 \leq (1 + \beta_n)\|x_n - x^*\|^2 + \eta_n,
\tag{9.21}
$$

where $\beta_n = \gamma v_n M^* L^* + (1 - \alpha_n)v_n M^*(1 + \gamma v_n M^* L^*)$ and
$\eta_n = (1 + (1 - \alpha_n)v_n M^*)\gamma(v_n\xi(M) + \mu_n) + (1 - \alpha_n)(v_n\xi(M) + \mu_n)$.

Clearly, $\sum \beta_n < \infty$ and $\sum \eta_n < \infty$. Moreover, $\beta_n \to 0$ and $\eta_n \to 0$. Hence, by Lemma 9.11, we conclude that $\lim_{n\to\infty}\|x_n - x^*\|$ exist.

We now prove that for each $x^* \in \Gamma$, $\lim_{n\to\infty}\|u_n - x^*\|$ exist.

By (9.20), we deduce that

$$\gamma(1 - \gamma L^*)\|(T^n - I)Ax_n\|^2 \le \|x_n - x^*\|^2 - \|x_{n+1} - x^*\|^2 + \beta_n\|x_n - x^*\|^2 + \eta_n, \tag{9.22}$$

and

$$\alpha_n(1 - \alpha_n)\|u_n - G^n u_n\|^2 \le \|x_n - x^*\|^2 - \|x_{n+1} - x^*\|^2 + \beta_n\|x_n - x^*\|^2 + \eta_n. \tag{9.23}$$

Thus, as $n \to \infty$, we deduce from (9.22) and (9.23) that

$$\lim_{n\to\infty}\|u_n - G^n u_n\| = 0 \quad \text{and} \quad \lim_{n\to\infty}\|Ax_n - T^n Ax_n\| = 0. \tag{9.24}$$

Given (9.19), (9.24), and the fact that $\lim_{n\to\infty}\|x_n - x^*\|$ exists, we obtain that

$$\lim_{n\to\infty}\|u_n - x^*\| \text{ exist.}$$

Moreover, by (9.15) and (9.19), we deduce that

$$\|x_{n+1} - x^*\|^2 \le (1 + (1 - \alpha_n)v_n M^*)\|u_n - x^*\|^2 + (1 - \alpha_n)(v_n \xi(M) + \mu_n), \tag{9.25}$$

and

$$\|u_n - x^*\|^2 \le (1 + \gamma v_n M^* L^*)\|x_n - x^*\|^2 + \gamma(v_n \xi(M) + \mu_n). \tag{9.26}$$

The fact that $\lim_{n\to\infty}\|x_n - x^*\|$ and $\lim_{n\to\infty}\|u_n - x^*\|$ exists, it follows from (9.25) and (9.26) that

$$\lim_{n\to\infty}\|u_n - x^*\| = \lim_{n\to\infty}\|x_n - x^*\|.$$

Step 2. In this step, we show that

$$\lim_{n\to\infty}\|x_{n+1} - x_n\| = 0 \quad \text{and} \quad \lim_{n\to\infty}\|u_{n+1} - u_n\| = 0. \tag{9.27}$$

By (9.13), we have that

$$\begin{aligned}
\|x_{n+1} - x_n\| &= \|\alpha_n u_n + (1 - \alpha_n)G^n u_n - x_n\| \\
&= \|(1 - \alpha_n)(G^n u_n - u_n) + u_n - x_n\| \\
&= \|(1 - \alpha_n)(G^n u_n - u_n) + A^*(T^n - I)Ax_n\|.
\end{aligned} \tag{9.28}$$

In view of (9.24), we deduce from (9.28) that

$$\lim_{n\to\infty}\|x_{n+1} - x_n\| = 0. \tag{9.29}$$

On the other hand,

$$\begin{aligned}
\|u_{n+1} - u_n\| &= \|(I + \gamma A^*(T^{n+1} - I)A)x_{n+1} + (I + \gamma A^*(T^n - I)A)x_n\| \\
&= \|x_{n+1} - x_n + \gamma A^*(T^{n+1} - I)Ax_{n+1} - \gamma A^*(T^n - I)Ax_n\|.
\end{aligned}$$

Thus, by (9.24) and (9.29), we obtain that

$$\lim_{n\to\infty} \|u_{n+1} - u_n\| = 0.$$

Step 3. In this step, we show that

$$\lim_{n\to\infty} \|u_n - Gu_n\| = 0 \quad \text{and} \quad \lim_{n\to\infty} \|Ax_n - Tx_n\| = 0. \tag{9.30}$$

The fact that $\lim_{n\to\infty} \|u_n - G^n u_n\| = 0$, $\lim_{n\to\infty} \|u_{n+1} - u_n\| = 0$, and G is uniformly L-Lipschitzian mapping, we have that

$$\begin{aligned}
\|u_n - Gu_n\| &\leq \|u_n - G^n u_n\| + \|Gu_n - G^n u_n\| \\
&\leq \|u_n - G^n u_n\| + L\|u_n - G^{n-1} u_n\| \\
&\leq \|u_n - G^n u_n\| + L\|G^{n-1} u_n - G^{n-1} u_{n-1}\| \\
&\quad + L\|u_n - G^{n-1} u_{n-1}\| \\
&\leq \|u_n - G^n u_n\| + L^2\|u_n - u_{n-1}\| \\
&\quad + L\|u_n - u_{n-1} + u_{n-1} - G^{n-1} u_{n-1}\| \\
&\leq \|u_n - G^n u_n\| + L(L+1)\|u_n - u_{n-1}\| + L\|u_{n-1} - G^{n-1} u_{n-1}\|.
\end{aligned}$$

Thus, as $n \to \infty$, we have that

$$\lim_{n\to\infty} \|u_n - Gu_n\| = 0.$$

Similarly, from the fact that, $\lim_{n\to\infty} \|Ax_n - T^n Ax_n\| = 0$, $\lim_{n\to\infty} \|x_{n+1} - x_n\| = 0$, and T is uniformly L-Lipschitzian mapping, we deduce that

$$\lim_{n\to\infty} \|Ax_n - TAx_n\| = 0.$$

Step 4. In this step, we show that

$$x_n \rightharpoonup x^* \quad \text{and} \quad u_n \rightharpoonup x^* \text{as } n \to \infty. \tag{9.31}$$

Since $\{u_n\}$ is bounded, then there exists a subsequence $u_{n_i} \subset u_n$ such that

$$u_{n_i} \rightharpoonup x^*, \text{ as } i \to \infty. \tag{9.32}$$

By (9.30) and (9.32), we have that

$$\lim_{i\to\infty} \|u_{n_i} - Gu_{n_i}\| = 0. \tag{9.33}$$

From (9.32), (9.33), and the fact that $(G - I)$ is demiclosed at zero, we have that $x^* \in Fix(G)$. By (9.13), (9.32), and the fact $\lim_{n\to\infty} \|Ax_n - T^n Ax_n\| = 0$, we deduce that

$$x_{n_i} = u_{n_i} - \gamma A^*(T^{n_i} - I)Ax_{n_i} \rightharpoonup x^*.$$

By the definition of A, we get

$$Ax_{n_i} \rightharpoonup Ax^* \text{ as } i \to \infty. \tag{9.34}$$

In view of (9.30), we get

$$\lim_{i\to\infty}\|Ax_{n_i} - TAx_{n_i}\| = 0. \tag{9.35}$$

From (9.34), (9.35), and the fact that $(T - I)$ is demiclosed at zero, we have that $Ax^* \in Fix(T)$. Thus, $x^* \in Fix(G)$ and $Ax^* \in Fix(T)$, and this implies that $x^* \in \Gamma$.

Now, we show that x^* is unique. Suppose to the contrary that there exists another subsequence $u_{n_j} \subset u_n$ such that $u_{n_j} \to y^* \in \Gamma$ with $x^* \neq y^*$. By opial's property of Hilbert space, we have that

$$\lim_{j\to\infty}\inf\|u_{n_j} - x^*\| < \lim_{j\to\infty}\inf\|u_{n_j} - y^*\|$$
$$= \lim_{n\to\infty}\inf\|u_n - y^*\|$$
$$= \lim_{j\to\infty}\inf\|u_{n_j} - y^*\|$$
$$< \lim_{j\to\infty}\inf\|u_{n_j} - x^*\|$$
$$= \lim_{n\to\infty}\inf\|u_n - x^*\|$$
$$= \lim_{j\to\infty}\inf\|u_{n_j} - x^*\|.$$

Thus, we have

$$\lim_{j\to\infty}\inf\|u_{n_j} - x^*\| < \lim_{j\to\infty}\inf\|u_{n_j} - x^*\|.$$

This is a contradiction, therefore, $u_n \to x^*$. By using (9.13) and (9.24), we have

$$x_n = u_n - \gamma A^*(T^n - I)Ax_n \to x^*, \text{ as } n \to \infty.$$

Step 5. In this step, we show that

$$\|x_n\| \to \|x^*\|, \text{ as } n \to \infty. \tag{9.36}$$

To show this, it suffices to show that $\|x_{n+1}\| \to \|x^*\|$ as $n \to \infty$.

By (9.21), Lemmas 9.14 and 9.12, and the fact that $\beta_n \to 0$ and $\eta_n \to 0$, we have

$$\big|\|x_{n+1}\| - \|x^*\|\big|^2 \leq \|x_{n+1} - x^*\|^2$$
$$\leq (1 + \beta_n)\|x_n - x^*\|^2 + \eta_n$$
$$= \|x_n - x^*\|^2 + \beta_n\|x_n - x^*\|^2 + \eta_n$$
$$= \|x_n - P_\Gamma x_n + P_\Gamma x_n - x^*\|^2 + \beta_n\|x_n - x^*\|^2 + \eta_n$$
$$\leq 4\|P_\Gamma x_n - x^*\|^2 + \beta_n\|x_n - x^*\|^2 + \eta_n.$$

Thus, as $n \to \infty$, we have that

$$\lim_{n\to\infty}\big|\|x_{n+1}\| - \|x^*\|\big|^2 \leq 4 \lim_{n\to\infty}\|P_\Gamma x_n - x^*\|^2 + \lim_{n\to\infty}\beta_n\|x_n - x^*\|^2 + \lim_{n\to\infty}(\eta_n).$$

Moreover, this implies that

$$\lim_{n\to\infty}\big|\|x_{n+1}\| - \|x^*\|\big| = 0.$$

By (9.31) and (9.36), we conclude that $x_n \to x^*$, as $n \to \infty$. □

9.2.9 Strong Convergence for the Split Common Fixed-Point Problems for Demicontractive Mappings

In this section, we considered an algorithm for solving the SCFPP for demicontractive mappings without any prior information on the norm on the bounded linear operator and established the strong convergence results of the proposed algorithm. In the end, we provide some special cases of our suggested methods.

Theorem 9.2 Let $U : H_1 \to H_1$ and $T : H_2 \to H_2$ be k_1, k_2–demicontractive mappings such that $(U - I)$ and $(T - I)$ are demiclosed at zero, $A : H_1 \to H_2$ be a bounded linear operator with its adjoint A^*. Assume that $\Gamma \neq \emptyset$ and let P_Γ be a metric projection from H_1 onto Γ satisfying

$$\langle x_n - x^*, x_n - P_\Gamma x_n \rangle \leq 0.$$

Define $\{x_n\}$ by

$$\begin{cases} x_0 \in H_1 \text{ is arbitrary chosen,} \\ u_n = x_n + \rho_n A^*(T - I)Ax_n, \\ x_{n+1} = (1 - \alpha_n)u_n + \alpha_n U u_n, \ \forall n \geq 0, \end{cases} \tag{9.37}$$

where $0 < c < \alpha_n < 1 - k$, with $k := \max\{k_1, k_2\}$, and

$$\rho_n = \begin{cases} \dfrac{(1 - k)\|(I - T)Ax_n\|^2}{2\|A^*(I - T)Ax_n\|^2}, & TAx_n \neq Ax_n, \\ 0, & \text{otherwise.} \end{cases} \tag{9.38}$$

Then, $x_n \to x^* \in \Gamma$.

Proof: To show that $x_n \to x^*$, it suffices to show $x_n \rightharpoonup x^*$ and $\|x_n\| \to \|x^*\|$. We divided the proof into four steps as follows.

Step 1. In this step, we show that $\{x_n\}$ is a Fejer monotone. This is divided into two cases.
Case 1. If $\rho_n = 0$ and Case 2. If $\rho_n \neq 0$.
Now, let $x^* \in \Gamma$.
Case 1. If $\rho_n = 0$. The fact that U is demicontractive, we have

$$\begin{aligned} \|x_{n+1} - x^*\|^2 &= \|x_n - x^* + \alpha_n(Ux_n - x_n)\|^2 \\ &= \|x_n - x^*\|^2 + 2\alpha_n\langle x_n - x^*, Ux_n - x_n \rangle + \alpha_n^2\|Ux_n - x_n\|^2 \\ &\leq \|x_n - x^*\|^2 + \alpha_n(k - 1)\|Ux_n - x_n\|^2 + \alpha_n^2\|Ux_n - x_n\|^2 \\ &\leq \|x_n - x^*\|^2 - \alpha_n(1 - k - \alpha_n)\|Ux_n - x_n\|^2. \end{aligned} \tag{9.39}$$

The fact that $0 < \alpha_n < 1 - k$, it follows from (9.39) that $\{x_n\}$ is Fejer monotone.

Case 2. If $\rho_n \neq 0$. Since U and T are demicontractive mappings, we have

$$
\begin{aligned}
\|x_{n+1} - x^*\|^2 &= \|u_n - \alpha_n u_n + \alpha_n U u_n - x^*\|^2 \\
&= \|u_n - x^*\|^2 + 2\alpha_n \langle u_n - x^*, U u_n - u_n \rangle + \alpha_n^2 \|U u_n - u_n\|^2 \\
&\leq \|u_n - x^*\|^2 - \alpha_n(1-k)\|U u_n - u_n\|^2 + \alpha_n^2 \|U u_n - u_n\|^2 \\
&\leq \|u_n - x^*\|^2 - \alpha_n(1 - k - \alpha_n)\|U u_n - u_n\|^2.
\end{aligned} \tag{9.40}
$$

On the other hand,

$$
\begin{aligned}
\|u_n - x^*\|^2 &= \|x_n + \rho_n A^*(T-I)Ax_n - x^*\|^2 \\
&= \|x_n - x^*\|^2 + 2\rho_n \langle TAx_n - Ax_n, Ax_n - Ax^* \rangle \\
&\quad + \rho_n^2 \|A^*(I-T)Ax_n\|^2 \\
&\leq \|x_n - x^*\|^2 - \rho_n(1-k)\|(T-I)Ax_n\|^2 + \rho_n^2 \|A^*(I-T)Ax_n\|^2 \\
&= \|x_n - x^*\|^2 - \frac{(1-k)^2\|(I-T)Ax_n\|^2}{2\|A^*(I-T)Ax_n\|^2}\|(T-I)Ax_n\|^2 \\
&\quad + \frac{(1-k)^2\|(I-T)Ax_n\|^4}{4\|A^*(I-T)Ax_n\|^4}\|A^*(I-T)Ax_n\|^2 \\
&= \|x_n - x^*\|^2 - \frac{(1-k)^2}{4}\frac{\|(T-I)Ax_n\|^4}{\|A^*(T-I)Ax_n\|^2}.
\end{aligned} \tag{9.41}
$$

Substituting (9.41) into (9.40), we deduce that

$$
\begin{aligned}
\|x_{n+1} - x^*\|^2 &\leq \|x_n - x^*\|^2 - \frac{(1-k)^2\|(T-I)Ax_n\|^4}{4\|A^*(T-I)Ax_n\|^2} \\
&\quad - \alpha_n(1 - k - \alpha_n)\|U u_n - u_n\|^2.
\end{aligned} \tag{9.42}
$$

Thus, $\{x_n\}$ is Fejér monotone. Therefore, $\lim_{n\to\infty} \|x_n - x^*\|$ exist.

Step 2. In this step, we show that

$$
\lim_{n\to\infty} \|(I-T)Ax_n\| = 0 \quad \text{and} \quad \lim_{n\to\infty} \|(I-U)x_n\| = 0. \tag{9.43}
$$

Case 1. If $\rho_n = 0$. By (9.38), we see that $\lim_{n\to\infty} \|(I-T)Ax_n\| = 0$. In addition, by (9.39) and the fact $\lim_{n\to\infty} \|x_n - x^*\|$ exist, it follows that $\lim_{n\to\infty} \|(I-U)x_n\| = 0$.

Case 2. If $\rho_n \neq 0$. The fact $\lim_{n\to\infty} \|x_n - x^*\|$ exist, by (9.42), we deduce that

$$
\lim_{n\to\infty} \left(\frac{(1-k)^2\|(T-I)Ax_n\|^4}{4\|A^*(T-I)Ax_n\|^2} \right) \leq \lim_{n\to\infty} (\|x_n - x^*\| - \|x_{n+1} - x^*\|) = 0 \tag{9.44}
$$

and $\lim_{n\to\infty} \|(I-U)u_n\| \leq \lim_{n\to\infty} \left(\dfrac{\|x_n - x^*\| - \|x_{n+1} - x^*\|}{c(1 - k - \alpha_n)} \right) = 0.$ (9.45)

By (9.44), we have that

$$
\lim_{n\to\infty} \left(\frac{\|(T-I)Ax_n\|^2}{\|A^*(T-I)Ax_n\|} \right) = 0. \tag{9.46}
$$

On the other hand,

$$\|(T - I)Ax_n\| = \|A\| \frac{\|TAx_n - Ax_n\|^2}{\|A\|\|TAx_n - Ax_n\|}$$
$$\leq \|A\| \frac{\|TAx_n - Ax_n\|^2}{\|A^*(T - I)Ax_n\|}.$$

Thus, by (9.46), we deduce that

$$\lim_{n\to\infty} \|(T - I)Ax_n\| = 0.$$

Since

$$\rho_n\|A^*(T - I)Ax_n\| = \|u_n - x_n\|$$
$$= \frac{(1 - k)\|(T - I)Ax\|^2}{2\|A^*(T - I)Ax_n\|}.$$

Thus, by (9.46), we have

$$\lim_{n\to\infty} \rho_n\|A^*(T - I)Ax_n\| = 0. \tag{9.47}$$

By (9.37), we have that

$$\|(U - I)x_n\| = \|(U - I)u_n - (U - I)\rho_n A^*(T - I)Ax_n\|$$
$$\leq \|(U - I)u_n\| + \|(U - I)\rho_n A^*(T - I)Ax_n\|. \tag{9.48}$$

Given (9.45), (9.47), and the fact that U is bounded. It follows from (9.48) that

$$\lim_{n\to\infty} \|(I - U)x_n\| = 0.$$

Hence, in both case, (9.43) hold.

Step 3. In this step, we show that

$$x_n \rightharpoonup x^*, \text{ as } n \to \infty. \tag{9.49}$$

To show this, it suffices to show that $\omega_\omega \subseteq \Gamma$ (see Lemma 9.12 (i)).

Now, let $q \in \omega_\omega$, this implies that, there exists $\{x_{n_j}\}$ of $\{x_n\}$ such that $x_{n_j} \rightharpoonup q$. Since $\lim_{j\to\infty}\|Ux_{n_j} - x_{n_j}\| = 0$, together with the demiclosed of $(U - I)$ at zero, we conclude that, $q \in Fix(U)$.

On the other hand, since A is bounded, we have that $Ax_{n_j} \rightharpoonup Aq$. By (9.43) and together with the demiclosed of $(T - I)$ at zero, we have that $Aq \in Fix(T)$. Thus, $q \in \Gamma$, this implies that $\omega_\omega \subseteq \Gamma$. Hence, by Lemma 9.12, we conclude that $x_n \rightharpoonup x^*$, as $n \to \infty$.

Step 4. In this step, we show that

$$\|x_n\| \to \|x^*\|, \text{ as } n \to \infty. \tag{9.50}$$

To show (9.50), it suffices to show that $\|x_{n+1}\| \to \|x^*\|$.

By Lemma 9.9 and the fact that $\{x_n\}$ is a Fejer monotone, we have

$$\left| \|x_{n+1}\| - \|x^*\| \right|^2 \leq \|x_{n+1} - x^*\|^2$$

$$\leq \|x_n - x^*\|^2$$

$$= \|x_n - P_\Gamma x_n + P_\Gamma x_n - x^*\|^2$$

$$\leq 4\|P_\Gamma x_n - x^*\|^2. \tag{9.51}$$

Thus, we deduce that $\|x_{n+1}\| \to \|x^*\|$. By (9.49) and (9.50), we conclude that $x_n \to x^*$ as $n \to \infty$. □

Corollary 9.1 Let $G : H_1 \to H_1$ and $T : H_2 \to H_2$ be (k_{n_1}, k_{n_2})- quasi-asymptotically nonexpansive mappings such that $(G - I)$ and $(T - I)$ are demiclosed at zero, and $A : H_1 \to H_2$ be a bounded linear operator with its adjoint A^*. In addition, let $L^* = \|AA^*\|$, M and M^* be positive constants such that $\xi(k) \leq \xi(M) + M^* k^2, \forall k \geq 0$. Assume that, $\Gamma \neq \emptyset$, and let P_Γ be a metric projection of H_1 onto Γ satisfying

$$\langle x_n - x^*, x_n - P_\Gamma x_n \rangle \leq 0.$$

Define $\{x_n\}$ by

$$\begin{cases} x_0 \in H_1, \\ u_n = x_n + \gamma A^*(T^n - I)Ax_n, \\ x_{n+1} = \alpha_n u_n + (1 - \alpha_n)G^n u_n, \ \forall n \geq 0, \end{cases} \tag{9.52}$$

where $\alpha_n \subset (0, 1)$, $\gamma \in (0, \frac{1}{L^*})$, $L = \max\{L_1, L_2\}$ and $k_n = \max\{k_{n_1}, k_{n_2}\}$. Then $x_n \to x^* \in \Gamma$.

Proof: G and T are $(\{v_n\}, \{\mu_n\}, \xi)$- total quasi-asymptotically nonexpansive mappings with $\{v_n\} = \{k_n - 1\}$, $\mu_n = 0$ and $\xi(k) = k^2, \forall k \geq 0$. Moreover, G and T are uniformly k_{n_1}, k_{n_2}- Lipschitzian mappings. Therefore, all the conditions in Theorem 9.13 are satisfied. Hence, the conclusion of this corollary follows directly from Theorem 9.13. □

Corollary 9.2 Let $G : H_1 \to H_1$ and $T : H_2 \to H_2$ be two quasi-nonexpansive mappings such that $(G - I)$ and $(T - I)$ are demiclosed at zero. Moreover, let A be a bounded linear operator with its adjoint A^*. Assume that, $\Gamma \neq \emptyset$, and let P_Γ be a metric projection of H onto Γ satisfying

$$\langle x_n - x^*, x_n - P_\Gamma x_n \rangle \leq 0.$$

Define $\{x_n\}$ by

$$\begin{cases} x_0 \in H_1; \\ u_n = x_n + \gamma A^*(T - I)Ax_n; \\ x_{n+1} = \alpha_n u_n + (1 - \alpha_n)Gu_n, \ \forall n \geq 0, \end{cases} \qquad (9.53)$$

where $\{\alpha_n\} \subset (0, 1)$ and $\gamma \in (0, \frac{1}{L^*})$ with $L^* = \|AA^*\|$. Then, $x_n \to x^* \in \Gamma$.

Proof: G and T are (1)– quasi-asymptotically nonexpansive mappings. Moreover, G and T are uniformly 1– Lipschitzian mappings. Therefore, all the conditions of Corollary 9.1 are satisfied. Hence, the conclusions of this corollary follow directly from Corollary 9.1. □

Corollary 9.3 Let $H_1, H_2, A, A^*, P_\Gamma$ and $\{x_n\}$ be as in Theorem 9.2. In addition, let $U : H_1 \to H_1$ and $T : H_2 \to H_2$ be quasi-nonexpansive mappings such that $(U - I)$ and $(T - I)$ are demiclosed at zero. Assume that $\Gamma \neq \emptyset$. Then, $x_n \to x^* \in \Gamma$.

Proof: Since T is quasi-nonexpansive, clearly T is 0-demicontractive. Hence, all the hypothesis of Theorem 9.2 are satisfied. Therefore, the proof of this corollary follows trivially from Theorem 9.2. □

Corollary 9.4 Let $H_1, H_2, A, A^*, P_\Gamma$, and $\{x_n\}$ be as in Theorem 9.2. In addition, let $U : H_1 \to H_1$ and $T : H_2 \to H_2$ be firmly quasi-nonexpansive mappings such that $(U - I)$ and $(T - I)$ are demiclosed at zero. Moreover, assume that $\Gamma \neq \emptyset$. Then, $x_n \to x^* \in \Gamma$.

Corollary 9.5 Let $H_1, H_2, A, A^*, P_\Gamma$ and $\{x_n\}$ be as in Theorem 9.2. In addition, let $U : H_1 \to H_1$ and $T : H_2 \to H_2$ be directed operators such that $(U - I)$ and $(T - I)$ are demiclosed at zero and assume that $\Gamma \neq \emptyset$. Then, $x_n \to x^* \in \Gamma$.

Proof: Since T is directed operator, clearly T is (-1)–demicontractive. Hence, all the hypothesis of Theorem 9.2 are satisfied. Therefore, the proof of this corollary follows trivially from Theorem 9.2. □

9.2.10 Application to Variational Inequality Problems

Let $T : C \to H_1$ be a nonlinear mapping. The variational inequality problem with respect to C consists as finding a vector $x^* \in C$ such that

$$\langle Tx^*, x - x^* \rangle \geq 0, \quad \forall x \in C. \qquad (9.54)$$

We denote the solution set of Variational Inequality Problem (9.54) by $VI(T, C)$. It is easy to see that

$$\text{find } x^* \in VI(T, C) \text{ if and only if } x^* \in Fix(P_C(I - \beta T)), \qquad (9.55)$$

where P_C is the metric projection from H_1 onto C and β is a positive constant.

Let $Q := Fix(P_C(I - \beta T))$ (the fixed point set of $P_C(I - \beta T)$) and $A = I$ (the identity operator on H_1), then (9.54) can be written as;

$$\text{find } x^* \in C \text{ such that } Ax^* \in Q. \tag{9.56}$$

9.2.11 On Synchronal Algorithms for Fixed and Variational Inequality Problems in Hilbert Spaces

The aim of this section is to expand the general approximation method proposed by Tian and Di [37] to the class of $(k, \{\mu_n\}, \{\xi_n\}, \phi)$- total asymptotically strict pseudocontraction and uniformly M-Lipschitzian mappings to solve the fixed-point problem (FPP) as well as variational inequality problem in the frame work of Hilbert space. The results presented in this chapter extend, improve, and generalize several known results in the literature.

9.2.12 Preliminaries

In the sequel, we shall make use of the following lemmas in proving the main results of this section.

Lemma 9.15 [38] Let H be a Hilbert space, there hold the following identities;

i) $\|x - y\|^2 = \|x\|^2 - \|y\|^2 - 2\langle x - y, y\rangle$, $\forall x, y \in H$;
ii) $\|tx + (1 - t)y\|^2 = t\|x\|^2 + (1 - t)\|y\|^2 - t(1 - t)\|x - y\|^2$, $\forall t \in [0, 1]$ and $x, y \in H$;
iii) if $\{x_n\}$ is a sequence in H such that $x_n \rightharpoonup z$, then

$$\lim_{n\to\infty} \sup \|x_n - y\|^2 = \lim_{n\to\infty} \sup \|x_n - z\|^2 + \|z - y\|^2, \forall y \in H.$$

Lemma 9.16 [39] Let C be a nonempty, closed, convex subset of a real Hilbert space H and let $T : C \to C$ be a $(k, \{\mu_n\}, \{\xi_n\}, \phi)$- total asymptotically strict pseudocontraction mapping and uniformly L-Lipschitzian. Then, $I - T$ is demiclosed at zero in the sense that if $\{x_n\}$ is a sequence in C such that $x_n \rightharpoonup x^*$, and $\lim_{n\to\infty} \sup \|(T^n - I)x_n\| = 0$, then $(T - I)x^* = 0$.

Lemma 9.17 [37] Assume that $\{a_n\}$ is a sequence of nonnegative real number such that

$$a_{n+1} \le (1 - \gamma_n)a_n + \sigma_n, \quad n \ge 0,$$

where γ_n is a sequence in $(0, 1)$ and σ_n is a sequence of real number such that;

i) $\lim\limits_{n\to\infty} \gamma_n = 0$ and $\sum \gamma_n = \infty$;

ii) $\lim\limits_{n\to\infty} \frac{\sigma_n}{\gamma_n} \leq 0$ or $\sum |\sigma_n| < \infty$. Then $\lim\limits_{n\to\infty} a_n = 0$.

Lemma 9.18 [37] Let $F : H \to H$ be a η- strongly monotone and L-Lipschitzian operator with $L > 0$ and $\eta > 0$. Assume that $0 < \mu < \frac{2\eta}{L^2}$, $\tau = \mu \left(\eta - \frac{L^2\mu}{2} \right)$, and $0 < t < 1$. Then,

$$\|(I - \mu t F)x - (I - \mu t F)y\| \leq (1 - \tau t)\|x - y\|.$$

Lemma 9.19 [40] Let $S : C \to H$ be a uniformly L-Lipschitzian mapping with $L \in (0, 1]$. Define $T : C \to H$ by $T^{\beta_n}x = \beta_n x + (1 - \beta_n)S^n x$ with $\beta_n \in (0, 1)$ and $\forall x \in C$. Then T^{β_n} is nonexpansive and $Fix(T^{\beta_n}) = Fix(S^n)$.

Proof: Let $x, y \in C$, from Lemma (2.1(ii)), we have

$$
\begin{aligned}
\|T^{\beta_n}x - T^{\beta_n}y\|^2 &= \| \beta_n(x - y) + (1 - \beta_n)(S^n x - S^n y)\|^2 \\
&= \beta_n \|x - y\|^2 + (1 - \beta_n)\|S^n x - S^n y\|^2 \\
&\quad - \beta_n(1 - \beta_n)\|(x - y) - (S^n x - S^n y)\|^2 \\
&\leq \beta_n \|x - y\|^2 + (1 - \beta_n)\|S^n x - S^n y\|^2 \\
&\leq (L^2 + \beta_n(1 - L^2))\|x - y\|^2,
\end{aligned}
$$

since $L \in (0, 1]$ and $\beta_n \in (0, 1)$, it follow that, T^{β_n} is nonexpansive, and it is not difficult to see that $Fix(T^{\beta_n}) = Fix(S^n)$. □

Lemma 9.20 [41] Let H be a real Hilbert space, $f : H \to H$ be a contraction with coefficient $0 < \alpha < 1$ and $F : H \to H$ be a L-Lipschitzian continuous operator and η-strongly monotone operator with $L > 0$ and $\eta > 0$. Then for $0 < \gamma < \frac{\mu\eta}{\alpha}$,

$$\langle x - y, (\mu F - \gamma f)x - (\mu F - \gamma f)y \rangle \geq (\mu\eta - \gamma\alpha)\|x - y\|^2.$$

Theorem 9.3 Let $T : H \to H$ be a $(k, \{\mu_n\}, \{\xi_n\}, \phi)$- total asymptotically strict pseudocontraction mapping and uniformly M-Lipschitzian with $\phi(t) = t^2, \forall t \geq 0$, and $M \in (0, 1]$. Assume that $Fix(T^n) \neq \emptyset$, and let f be a contraction with coefficient $\beta \in (0, 1)$, $G : H \to H$ be a η-strongly monotone and L-Lipschitzian operator with $L > 0$ and $\eta > 0$, respectively. Assume that $0 < \gamma < \mu \left(\eta - \frac{\mu L^2}{2} \right) /\beta = \frac{\tau}{\beta}$ and let $x_0 \in H$ be chosen arbitrarily, $\{\alpha_n\}$ and $\{\beta_n\}$ be two sequences in $(0, 1)$ satisfying the following conditions:

$$
\begin{cases}
\text{(i) } \lim_{n \to \infty} \alpha_n = 0 \text{ and } \sum \alpha_n = \infty; \\
\text{(ii) } \sum |\alpha_{n+1} - \alpha_n| < \infty, \sum |\beta_{n+1} - \beta_n| < \infty \text{ and } \sum |\beta_n| < \infty; \\
\text{(iii) } 0 \leq k \leq \beta_n < a < 1, \forall n \geq 0.
\end{cases}
\tag{9.57}
$$

Let $\{x_n\}$ be a sequence defined by

$$
\begin{cases}
T^{\beta_n} = \beta_n I + (1 - \beta_n) T^n; \\
x_{n+1} = \alpha_n \gamma f(x_n) + (I - \alpha_n \mu G) T^{\beta_n} x_n,
\end{cases}
\tag{9.58}
$$

then $\{x_n\}$ converges strongly to a common fixed of T^n, which solve the variational inequality problem

$$
\langle (\gamma f - \mu G) x^*, x - x^* \rangle \leq 0, \quad \forall x \in Fix(T^n).
\tag{9.59}
$$

Proof: The proof is divided into five steps as follows.

Step 1. In this step, we show that

$$
T^{\beta_n} \text{ is nonexpansive and } Fix(T^{\beta_n}) = Fix(T^n).
\tag{9.60}
$$

The proof follows directly from Lemma (9.19).

Step 2. In this step, we show that

$$
\{x_n\}, \{T^n x_n\}, \{f(x_n)\} \quad \text{and} \quad \{GT^n x_n\} \text{ are all bounded.}
\tag{9.61}
$$

Let $x^* \in Fix(T^n)$, from (9.58) and Lemma (9.18), and the fact that f is a contraction, we have

$$
\begin{aligned}
\|x_{n+1} - x^*\| &= \|\alpha_n \gamma f(x_n) + (I - \alpha_n \mu G) T^{\beta_n} x_n - x^*\| \\
&= \|\alpha_n (\gamma f(x_n) - \mu G x^*) + (I - \alpha_n \mu G) T^{\beta_n} x_n - (I - \alpha_n \mu G) x^*\| \\
&\leq (1 - \alpha_n \tau)\|x_n - x^*\| + \alpha_n \|\gamma (f(x_n) - f(x^*)) + \gamma f(x^*) - \mu G x^*)\| \\
&\leq (1 - \alpha_n (\tau - \gamma \beta))\|x_n - x^*\| + \alpha_n \|\gamma f(x^*) - \mu G x^*)\| \\
&\leq \max \left\{ \|x_n - x^*\|, \frac{\|\gamma f(x^*) - \mu G x^*)\|}{(\tau - \gamma \beta)} \right\}.
\end{aligned}
$$

By using induction, we have

$$
\|x_{n+1} - x^*\| \leq \max \left\{ \|x_0 - x^*\|, \frac{\|\gamma f(x^*) - \mu G x^*)\|}{(\tau - \gamma \beta)} \right\}.
\tag{9.62}
$$

Hence $\{x_n\}$ is bounded, and also

$$
\begin{aligned}
\|T^n x_n - x^*\|^2 &\leq \|x_n - x^*\|^2 + k\|x_n - x^* - (T^n x_n - x^*)\|^2 + \mu_n \phi(\|x_n - x^*\|) + \xi_n \\
&= \|x_n - x^*\|^2 + k\|x_n - x^*\|^2 + k\|T^n x_n - x^*\|^2 \\
&\quad + 2k\|x_n - x^*\|\|T^n x_n - x^*\| + \mu_n \|x_n - x^*\|^2 + \xi_n \\
&\leq (1 + k + \mu_n)\|x_n - x^*\|^2 + 2k\|x_n - x^*\|\|T^n x_n - x^*\| \\
&\quad + k\|T^n x_n - x^*\|^2 + \xi_n.
\end{aligned}
\tag{9.63}
$$

From (9.63), we deduce that

$$(1 - k)\|T^n x_n - x^*\|^2 - 2k\|x_n - x^*\|$$
$$\|T^n x_n - x^*\| - (1 + k + \mu_n)\|x_n - x^*\|^2 - \xi_n \leq 0.$$

This implies that

$$\|T^n x_n - x^*\| \leq \frac{k\|x_n - x^*\|}{(1 - k)}$$

$$+ \frac{\sqrt{4k^2\|x_n - x^*\|^2 + 4(1 - k)\{(1 + k + \mu_n)\|x_n - x^*\|^2 + \xi_n\}}}{2(1 - k)}$$

$$= \frac{k\|x_n - x^*\| + \sqrt{(1 + (1 - k)\mu_n)\|x_n - x^*\|^2 + (1 - k)\xi_n}}{(1 - k)}$$

$$\leq \frac{k\|x_n - x^*\| + (1 + (1 - k)\mu_n)\|x_n - x^*\|^2 + (1 - k)\xi_n}{(1 - k)}$$

$$\|T^n x_n - x^*\| \leq M^*, \tag{9.64}$$

where M^* is chosen arbitrarily such that

$$\sup \left(\frac{k\|x_n - x^*\| + (1 + (1 - k)\mu_n)\|x_n - x^*\|^2 + (1 - k)\xi_n}{(1 - k)} \right) \leq M^*.$$

It follows from (9.64) that $\{T^n x_n\}$ is bounded. Since G is L-Lipschitzian, f is contraction, and the fact that $\{x_n\}$ *and* $\{T^n x_n\}$ are bounded, it is easy to see that $\{GT^n x_n\}$ and $\{f(x_n)\}$ are also bounded.

Step 3. In this step, we show that

$$\lim_{n \to \infty} \|x_{n+1} - x_n\| = 0. \tag{9.65}$$

Now,

$$\|x_{n+2} - x_{n+1}\| = (\alpha_{n+1}\gamma f(x_{n+1}) + (I - \alpha_{n+1}\mu G)T^{\beta_{n+1}} x_{n+1})$$
$$- (\alpha_n \gamma f(x_n) + (I - \alpha_n \mu G)T^{\beta_n} x_n)$$
$$= \alpha_{n+1}\gamma(f(x_{n+1}) - f(x_n)) + (\alpha_{n+1} - \alpha_n)\gamma f(x_n)$$
$$+ (I - \alpha_{n+1}\mu G)T^{\beta_{n+1}} x_{n+1} - (I - \alpha_{n+1}\mu G)T^{\beta_n} x_n$$
$$+ (\alpha_n - \alpha_{n+1})\mu GT^{\beta_n} x_n,$$

this turn to implies that

$$\|x_{n+2} - x_{n+1}\| \leq \alpha_{n+1}\gamma\beta\|x_{n+1} - x_n\| + (1 - \alpha_{n+1}\tau)\|T^{\beta_{n+1}} x_{n+1} - T^{\beta_n} x_n\|$$
$$+ |\alpha_{n+1} - \alpha_n|(\gamma\|f(x_n)\| + \mu\|GT^{\beta_n} x_n\|)$$
$$\leq \alpha_{n+1}\gamma\beta\|x_{n+1} - x_n\| + (1 - \alpha_{n+1}\tau)\|T^{\beta_{n+1}} x_{n+1} - T^{\beta_n} x_n\|$$
$$+ |\alpha_{n+1} - \alpha_n|N_1, \tag{9.66}$$

where N_1 is chosen arbitrarily so that $\sup_{n \geq 1}(\gamma\|f(x_n)\| + \mu\|GT^{\beta_n} x_n\|) \leq N_1$.

On the other hand,

$$\|T^{\beta_{n+1}}x_{n+1} - T^{\beta_n}x_n\| \leq \|T^{\beta_{n+1}}x_{n+1} - T^{\beta_{n+1}}x_n\| + \|T^{\beta_{n+1}}x_n - T^{\beta_n}x_n\|$$

$$\leq \|x_{n+1} - x_n\| + |\beta_{n+1} - \beta_n|\|x_n\| + |\beta_{n+1}|\|T^{n+1}x_n\|$$

$$+ |\beta_n|\|T^n x_n\|$$

$$\leq \|x_{n+1} - x_n\| + |\beta_{n+1} - \beta_n|N_2 + |\beta_{n+1}|N_3 + |\beta_n|N_4,$$

$$(9.67)$$

where $N_{2,3,4}$ satisfy the following relations:

$$N_2 \geq \sup_{n\geq 1}\|x_n\|, \quad N_3 \geq \sup_{n\geq 1}\|T^{n+1}x_n\|, \quad \| \text{ and } \quad N_4 \geq \sup_{n\geq 1}\|T^n x_n\|,$$

respectively.

Now substituting (9.67) into (9.66) yields

$$\|x_{n+2} - x_{n+1}\| \leq \alpha_{n+1}\gamma\beta\|x_{n+1} - x_n\| + (1 - \alpha_{n+1}\tau)(\|x_{n+1} - x_n\|$$

$$+ |\beta_{n+1} - \beta_n|N_2 + |\beta_{n+1}|N_3 + |\beta_n|N_4) + |\alpha_{n+1} - \alpha_n|N_1$$

$$= (1 + \alpha_{n+1}(\gamma\beta - \tau))\|x_{n+1} - x_n\| + |\alpha_{n+1} - \alpha_n|N_1$$

$$+ (1 - \alpha_{n+1}\tau)(|\beta_{n+1} - \beta_n|N_2 + |\beta_{n+1}|N_3 + |\beta_n|N_4)$$

$$\leq (1 - \alpha_{n+1}(\tau - \gamma\beta))\|x_{n+1} - x_n\| + (1 - \alpha_{n+1}\tau)$$

$$(|\beta_{n+1} - \beta_n| + |\beta_{n+1}| + |\beta_n| + |\alpha_{n+1} - \alpha_n|)N_5,$$

where N_5 choosing appropriately such that $N_5 \geq \max\{N_1, N_2, N_3, N_4\}$. By Lemma (2.3) and (ii), it follows that

$$\lim_{n\to\infty}\|x_{n+1} - x_n\| = 0.$$

From (9.58), we have,

$$\|x_{n+1} - T^{\beta_n}x_n\| = \|\alpha_n\gamma f(x_n) + (I - \alpha_n\mu G)T^{\beta_n}x_n - T^{\beta_n}x_n\|$$

$$\leq \alpha_n\|\gamma f(x_n) - \mu G T^{\beta_n}x_n\| \to 0.$$

On the other hand,

$$\|x_{n+1} - T^{\beta_n}x_n\| = \|x_{n+1} - (\beta_n + (1 - \beta_n)T^n)x_n\|$$

$$= \|(x_{n+1} - x_n) + (1 - \beta_n)(x_n - T^n x_n)\|$$

$$\geq (1 - \beta_n)\|x_n - T^n x_n\| - \|x_{n+1} - x_n\|,$$

this implies that

$$\|x_n - T^n x_n\| \leq \frac{\|x_{n+1} - T^{\beta_n}x_n\| + \|x_{n+1} - x_n\|}{(1 - \beta_n)}$$

$$\leq \frac{\|x_{n+1} - T^{\beta_n}x_n\| + \|x_{n+1} - x_n\|}{(1 - a)} \to 0.$$

From the boundedness of $\{x_n\}$, we deduce that $\{x_n\}$ converges weakly. Now assume that $x_n \rightharpoonup p$, by Lemma (2.2) and the fact that $\|x_n - T^n x_n\| \to 0$, we obtain $p \in Fix(T^n)$. So, we have

$$\omega_\omega(x_n) \subset Fix(T^n). \tag{9.68}$$

By Lemma (2.6), it follows that $(\gamma f - \mu G)$ is strongly monotone, so the variational inequality (9.59) has a unique solution $x^* \in Fix(T^n)$.

Step 4. In this step, we show that

$$\lim_{n\to\infty} \sup \langle (\gamma f - \mu G)x^*, x_n - x^* \rangle \le 0. \tag{9.69}$$

The fact that $\{x_n\}$ is bounded, we have $\{x_{n_i}\} \subset \{x_n\}$ such that

$$\lim_{n\to\infty} \sup \langle (\gamma f - \mu G)x^*, x_n - x^* \rangle = \lim_{i\to\infty} \sup \langle (\gamma f - \mu G)x^*, x_{n_i} - x^* \rangle \le 0.$$

Suppose without loss of generality that $x_{n_i} \rightharpoonup x$, from (9.68), it follows that $x \in Fix(T^n)$. Since x^* is the unique solution of (9.58), implies that

$$\lim_{n\to\infty} \sup \langle (\gamma f - \mu G)x^*, x_n - x^* \rangle = \lim_{i\to\infty} \sup \langle (\gamma f - \mu G)x^*, x_{n_i} - x^* \rangle.$$
$$= \langle (\gamma f - \mu G)x^*, x - x^* \rangle \le 0.$$

Step 5. In this step, we show that

$$\lim_{n\to\infty} \|x_n - x^*\| = 0. \tag{9.70}$$

By Lemma (2.4) and the fact that f is a contraction, we have

$$\begin{aligned}
\|x_{n+1} - x^*\|^2 &= \|\alpha_n(\gamma f(x_n) - \mu Gx^*) + (I - \alpha_n \mu G)T^{\beta_n}x_n - (I - \alpha_n \mu G)x^*\|^2 \\
&\le \|(I - \alpha_n \mu G)T^{\beta_n}x_n - (I - \alpha_n \mu G)x^*\|^2 \\
&\quad + 2\alpha_n \langle \gamma f(x_n) - \mu Gx^*, x_{n+1} - x^* \rangle \\
&\le (1 - \alpha_n \tau)^2 \|x_n - x^*\|^2 + 2\alpha_n \gamma \langle f(x_n) - f(x^*), x_{n+1} - x^* \rangle \\
&\quad + 2\alpha_n \langle \gamma f(x^*) - \mu Gx^*, x_{n+1} - x^* \rangle \\
&\le (1 - \alpha_n \tau)^2 \|x_n - x^*\|^2 + 2\alpha_n \beta \gamma \|x_n - x^*\| \|x_{n+1} - x^*\| \\
&\quad + 2\alpha_n \langle \gamma f(x^*) - \mu Gx^*, x_{n+1} - x^* \rangle \\
&\le (1 - \alpha_n \tau)^2 \|x_n - x^*\|^2 + \alpha_n \beta \gamma (\|x_n - x^*\|^2 + \|x_{n+1} - x^*\|^2) \\
&\quad + 2\alpha_n \langle \gamma f(x^*) - \mu Gx^*, x_{n+1} - x^* \rangle,
\end{aligned}$$

this implies that

$$\begin{aligned}
\|x_{n+1} - x^*\|^2 &\le \frac{((1 - \alpha_n \tau)^2 + \alpha_n \beta \gamma)\|x_n - x^*\|^2}{(1 - \alpha_n \gamma \beta)} + \frac{2\alpha_n \langle \gamma f(x^*) - \mu Gx^*, x_{n+1} - x^* \rangle}{(1 - \alpha_n \gamma \beta)} \\
&\le (1 - (2\tau - \gamma \beta)\alpha_n)\|x_n - x^*\|^2 + \frac{(\alpha_n \tau)^2}{(1 - \alpha_n \gamma \beta)}\|x_n - x^*\|^2 \\
&\quad + \frac{2\alpha_n \langle \gamma f(x^*) - \mu Gx^*, x_{n+1} - x^* \rangle}{(1 - \alpha_n \gamma \beta)},
\end{aligned}$$

and this implies that

$$\|x_{n+1} - x^*\|^2 \le (1 - \gamma_n)\|x_n - x^*\|^2 + \sigma_n,$$

where

$$\gamma_n := (2\tau - \gamma\beta)\alpha_n \quad \text{and}$$

$$\sigma_n := \frac{\alpha_n}{(1 - \alpha_n\gamma\beta)}(\alpha_n\tau^2\|x_n - x^*\|^2 + 2\langle \gamma f(x^*) - \mu Gx^*, x_{n+1} - x^*\rangle).$$

From (3.1 (i)), it follows that

$$\lim_{n\to\infty} \gamma_n = 0,$$

$$\sum \gamma_n = \infty,$$

$$\frac{\sigma_n}{\gamma_n} = \frac{1}{(2\tau-\gamma\beta)(1 - \alpha_n\gamma\beta)}(\alpha_n\tau^2\|x_n - x^*\|^2 + 2\langle \gamma f(x^*) - \mu Gx^*, x_{n+1} - x^*\rangle).$$

Thus $\lim_{n\to\infty} \dfrac{\sigma_n}{\gamma_n} \le 0$.

Hence by Lemma (2.3), it follows that $x_n \to x^*$ as $n \to \infty$. $\qquad\square$

Corollary 9.6 Let B be a unit ball is a real Hilbert space l_2, and let the mapping $T : B \to B$ be defined by

$$T : (x_1, x_2, x_3, \dots) \to (0, x_1^2, a_2x_2, a_3x_3, \dots), (x_1, x_2, x_3, \dots) \in B,$$

where $\{a_i\}$ is a sequence in $(0, 1)$ such that $\displaystyle\prod_{i=2}^{\infty}(a_i) = \frac{1}{2}$. Let, $f, G, \gamma, \{\alpha_n\}, \{\beta_n\}$ be as in Theorem (3.1). Then the sequence $\{x_n\}$ define by Algorithm (9.58), converges strongly to a common fixed point of T^n which solve the variational inequality Problem (3.3).

Proof: By Example (1.1), it follows that T is $(k, \{\mu\}, \{\xi_n\}, \phi)$-total asymptotically strict pseudocontraction mapping and uniformly M-Lipschitzian with $M = 2\displaystyle\prod_{i=2}^{n}(a_i)$. Hence, the conclusion of this corollary, follows directly from Theorem (3.1). $\qquad\square$

Corollary 9.7 Let H be a real Hilbert space and $T : H \to H$ be a $(k, \{k_n\})$- asymptotically strict pseudocontraction mapping and uniformly M-Lipschitzian with $M \in (0, 1]$. Assume that $\text{Fix}(T^n) \ne \emptyset$, and let f, G, γ $\{\alpha_n\}$ and $\{\beta_n\}$ be as in Theorem (3.1). Then, the sequence $\{x_n\}$ generated by Algorithm (9.58), converges strongly to a common fixed point of T^n which solve the variational inequality Problem (9.59).

Corollary 9.8 [41] Let the sequence $\{x_n\}$ be generated by the mapping

$$x_{n+1} = \alpha_n\gamma f(x_n) + (I - \mu\alpha_n F)Tx_n,$$

where T is nonexpansive, α_n is a sequence in $(0,1)$ satisfying the following conditions:

$$
\begin{cases}
\text{(i)} \quad \lim_{n\to\infty} \alpha_n = 0, \quad \sum \alpha_n = \infty; \\
\text{(ii)} \quad \sum |\alpha_{n+1} - \alpha_n| < \infty, \quad \sum |\beta_{n+1} - \beta_n| < \infty; \\
\text{(iii)} \quad 0 \le \max_i k_i \le \beta_n < a < 1, \ \forall n \ge 0.
\end{cases} \tag{9.71}
$$

It was proved in [41] that $\{x_n\}$ converged strongly to the common fixed-point x^* of T, which is the solution of variational inequality problem

$$
\langle (\gamma f - \mu F)x^*, x - x^* \rangle \le 0, \quad \forall x \in Fix(T). \tag{9.72}
$$

Proof: Take $n = 1$, $k = \mu_n = \xi_n = 0$ and $F = G$ in Theorem (3.1). Therefore, all the conditions in Theorem (3.1) are satisfied. Hence the conclusion of this corollary follows directly from Theorem (3.1). □

Corollary 9.9 [42] Let the sequence $\{x_n\}$ be generated by

$$
x_{n+1} = \alpha_n \gamma f(x_n) + (I - \alpha_n A) T x_n,
$$

where T is nonexpansive and the sequence $\alpha_n \subset (0, 1)$ satisfies the conditions in (9.57). Then it was proved in [42] that $\{x_n\}$ converged strongly to x^* which solve the variational inequality

$$
\langle (\gamma f - A)x^*, x - x^* \rangle \le 0, \quad \forall x \in Fix(T). \tag{9.73}
$$

Proof: Take $n = 1$, $\mu_n = \xi_n = 0$ and $\mu = 1$ and $G = A$ in Theorem (3.1). Therefore, all the conditions in Theorem (3.1) are satisfied. Hence the conclusion of this corollary follows directly from Theorem (3.1). □

Corollary 9.10 [43] Let the sequence $\{x_n\}$ be generated by

$$
x_{n+1} = T x_n - \mu \lambda_n F(T x_n),
$$

where T is nonexpansive mapping on H, F is L-Lipschitzian and η-strongly monotone with $L > 0, \eta > 0$, and $0 < \mu < \frac{2\eta}{L^2}$, if the sequence $\lambda_n \subset (0, 1)$ satisfies the following conditions:

$$
\begin{cases}
\text{(i)} \quad \lim_{n\to\infty} \lambda_n = 0, \sum \lambda_n = \infty; \\
\text{(ii) either } \sum |\lambda_{n+1} - \lambda_n| = 0 \quad \text{or} \quad \lim_{n\to\infty} \frac{\lambda_{n+1}}{\lambda_n} = 1.
\end{cases} \tag{9.74}
$$

Then, it was proved by Butnariu *et al.* [43] that $\{x_n\}$ converged strongly to the unique solution of the variational inequality

$$
\langle Fx^*, x - x^* \rangle \ge 0, \quad \forall x \in Fix(T). \tag{9.75}
$$

Proof: Take $n = 1$, $k = \mu_n = \xi_n = 0$ and also take $\gamma = 0$, $\beta_n = 0$, and $G = F$. Therefore, all the conditions in Theorem (3.1) are satisfied. Hence the result follows directly from Theorem (3.1). □

9.3 A Note on the Split Equality Fixed-Point Problems in Hilbert Spaces

In this section, we propose the split feasibility and fixed-point equality problems (SFFPEP) and split common fixed-point equality problems (SCFPEP). Furthermore, we formulate and analyze the algorithms for solving these problems for the class of quasi-nonexpansive mappings in Hilbert spaces. In the end, we study the convergence results of the proposed algorithms.

9.3.1 Problem Formulation

The SFFPEP formulated as follows:

$$\text{Find } x^* \in C \cap Fix(U) \text{ and } y^* \in Q \cap Fix(T) \text{ such that } Ax^* = By^*. \quad (9.76)$$

While the split common fixed point equality problems (in short, SCFPEP) obtained as follows:

$$\text{Find } x^* \in \bigcap_{i=1}^{N} Fix(U_i) \quad \text{and} \quad y^* \in \bigcap_{j=1}^{M} Fix(T_j) \text{ such that } Ax^* = By^*,$$

$$(9.77)$$

where $U_{i=1} : H_1 \rightarrow H_1, i = 1, 2, 3, \ldots, N$, and $T_{j=1} : H_2 \rightarrow H_2, j = 1, 2, 3, \ldots, M$, are quasi-nonexpansive mappings with $Fix(U_i) \neq \emptyset$ and $Fix(T_j) \neq \emptyset$, respectively, $A : H_1 \rightarrow H_3$ and $B : H_2 \rightarrow H_3$ are bounded linear operators.

Note that if $C := Fix(U)$, $Q := Fix(T)$, $H_2 = H_3$ and $B = I$. Then Problem (9.76) reduces to the following problems:

$$\text{Find } x^* \in C \quad \text{and} \quad y^* \in Q \text{ such that } Ax^* = By^*, \quad (9.78)$$

and

$$x^* \in C \cap Fix(U) \text{ such that } Ax^* \in Q \cap Fix(T). \quad (9.79)$$

Equations (9.78) and (9.79) are called the split equality fixed-point problems (SEFPP) and split feasibility and fixed-point problems (SFFPP), respectively. In the light of this, it is worth to mention here that the SFFPEP generalizes the split feasibility problem (SFP), SFFPP, and SEFPP. Therefore, the results and conclusions that are true for the SFFPEP continue to hold for these problems (SFP, SFFPP, and SEFPP), and it shows the significance and the range of applicability of the SFFPEP.

Furthermore, Problem (9.77) reduces to Problem (9.8) as $H_2 = H_3$ and $B = I$. This shows that the SCFPEP generalizes the SCFPP. Therefore, the results and conclusions that are true for the SCFPEP continue to hold for the SCFPP.

We denote the solution of sets SFFPEP (9.76) and SCFPEP (9.77) by

$$\Phi = \{x^* \in C \cap Fix(U) \text{ and } y^* \in Q \cap Fix(T) \text{ such that } Ax^* = By^*\},$$

(9.80)

and

$$\Psi = \left\{ x^* \in \bigcap_{i=1}^{N} Fix(U_i) \text{ and } y^* \in \bigcap_{j=1}^{M} Fix(T_j) \text{ such that } Ax^* = By^* \right\},$$

(9.81)

respectively. In sequel, we assume that Φ and Ψ are nonempty.

9.3.2 Preliminaries

In this section, we present some lemmas used in proving our main result.

Lemma 9.21 Let $C \subset H$ and $\{x_n\}$ be a sequence in H such that the following conditions are satisfied:

i) For each $x \in C$, $\lim_{n \to \infty} \|x_n - x\|$ exists,
ii) Any weak-cluster point of the sequence $\{x_n\}$ belongs to C.

Then, there exists $y \in C$ such that $\{x_n\}$ converges weakly to y.

For the proof, see [34] and references therein.

Lemma 9.22 Let $T_i : H \to H$, for $i = 1, 2, 3, \ldots, N$ be N-quasi-nonexpansive mappings. Defined $U = \sum_{i=1}^{N} \delta_i U_{\beta_i}$, where $U_{\beta_i} = (1 - \beta_i)I + \beta_i T_i$, and $\delta_i \in (0, 1)$ such that $\sum_{i=1}^{N} \delta_i = 1$. Then

i) U is a quasi-nonexpansive mapping,
ii) $Fix(U) = \bigcap_{i=1}^{N} Fix(U_{\beta_i}) = \bigcap_{i=1}^{N} Fix(T_i)$,
iii) in addition, if $(T_i - I)$ for $i = 1, 2, 3, \ldots, N$ is demiclosed at zero, then $(U - I)$ is also demiclosed at zero.

For the proof, see [30] and the references therein.

9.3.3 The Split Feasibility and Fixed-Point Equality Problems for Quasi-Nonexpansive Mappings in Hilbert Spaces

To approximate the solution of the SFFPEP (9.80), we make the following assumptions:

(B_1) $U : H_1 \to H_1$ and $T : H_2 \to H_2$ are quasi-nonexpansive mappings with $Fix(U) \neq \emptyset$ and $Fix(T) \neq \emptyset$, respectively.

(B_2) $A : H_1 \to H_3$ and $B : H_2 \to H_3$ are bounded linear operators with their adjoints A^* and B^*, respectively.

(B_3) $(U - I)$ and $(T - I)$ are demiclosed at zero.

(B_4) P_C and P_Q are metric projections of H_1 and H_2 onto C and Q, respectively.

(B_5) For arbitrary $x_1 \in H_1$ and $y_1 \in H_2$, define a sequence $\{(x_n, y_n)\}$ by:

$$
\begin{cases}
z_n = P_C(x_n - \lambda_n A^*(Ax_n - By_n)), \\
w_n = (1 - \beta_n)z_n + \beta_n U(z_n), \\
x_{n+1} = (1 - \alpha_n)z_n + \alpha_n U(w_n), \\
\\
u_n = P_Q(y_n + \lambda_n B^*(Ax_n - By_n)), \\
r_n = (1 - \beta_n)u_n + \beta_n T(u_n), \\
y_{n+1} = (1 - \alpha_n)u_n + \alpha_n T(r_n), \forall n \geq 1,
\end{cases}
\tag{9.82}
$$

where $0 < a < \beta_n < 1$, $0 < b < \alpha_n < 1$, and $\lambda_n \in \left(0, \frac{2}{L_1 + L_2}\right)$, where $L_1 = A^*A$ and $L_2 = B^*B$, respectively.

We are now in the position to state and prove the main result of this chapter.

Theorem 9.4 Suppose that assumption $(B_1) - (B_5)$ are satisfied, also assume that the solution set $\Phi \neq \emptyset$. Then $(x_n, y_n) \rightharpoonup (x^*, y^*) \in \Phi$.

Proof: Let $(x^*, y^*) \in \Phi$. By (9.82), we have

$$
\begin{aligned}
\|x_{n+1} - x^*\|^2 &= \|(1 - \alpha_n)(z_n - x^*) + \alpha_n(Uw_n - x^*)\|^2 \\
&= (1 - \alpha_n)\|z_n - x^*\|^2 + \alpha_n\|Uw_n - x^*\|^2 - \alpha_n(1 - \alpha_n)\|Uw_n - z_n\|^2 \\
&\leq (1 - \alpha_n)\|z_n - x^*\|^2 + \alpha_n\|w_n - x^*\|^2 \\
&\quad - \alpha_n(1 - \alpha_n)\|Uw_n - z_n\|^2.
\end{aligned}
\tag{9.83}
$$

On the other hand,

$$
\begin{aligned}
\|w_n - x^*\|^2 &= \|(1 - \beta_n)(z_n - x^*) + \beta_n(Uz_n - x^*)\|^2 \\
&= (1 - \beta_n)\|z_n - x^*\|^2 + \beta_n\|Uz_n - x^*\|^2 - \beta_n(1 - \beta_n)\|Uz_n - z_n\|^2 \\
&\leq \|z_n - x^*\|^2 - \beta_n(1 - \beta_n)\|Uz_n - z_n\|^2.
\end{aligned}
\tag{9.84}
$$

Substituting (9.84) into (9.83), we have

$$
\begin{aligned}
\|x_{n+1} - x^*\|^2 &\leq (1 - \alpha_n)\|z_n - x^*\|^2 + \alpha_n\|z_n - x^*\|^2 \\
&\quad - \alpha_n\beta_n(1 - \beta_n)\|Uz_n - z_n\|^2 - \alpha_n(1 - \alpha_n)\|Uw_n - z_n\|^2.
\end{aligned}
\tag{9.85}
$$

On the other hand,

$$
\begin{aligned}
\|z_n - x^*\|^2 &= \|P_C(x_n - \lambda_n A^*(Ax_n - By_n)) - P_C(x^*)\|^2 \\
&\leq \|x_n - \lambda_n A^*(Ax_n - By_n) - x^*\|^2
\end{aligned}
$$

$$= \|x_n - x^*\|^2 - 2\lambda_n \langle Ax_n - Ax^*, Ax_n - By_n \rangle$$
$$+ \lambda_n^2 L_1 \|Ax_n - By_n\|^2. \tag{9.86}$$

Substituting (9.86) into (9.85), we have

$$\|x_{n+1} - x^*\|^2 \le \|x_n - x^*\|^2 - 2\lambda_n \langle Ax_n - Ax^*, Ax_n - By_n \rangle$$
$$+ \lambda_n^2 L_1 \|Ax_n - By_n\|^2 - \alpha_n \beta_n (1 - \beta_n) \|U(z_n) - z_n\|^2$$
$$- \alpha_n (1 - \alpha_n) \|Uw_n - z_n\|^2. \tag{9.87}$$

Similarly, the second equation of (9.82) gives

$$\|y_{n+1} - y^*\|^2 \le \|y_n - y^*\|^2 + 2\lambda_n \langle By_n - By^*, Ax_n - By_n \rangle$$
$$+ \lambda_n^2 L_2 \|Ax_n - By_n\|^2 - \alpha_n \beta_n (1 - \beta_n) \|T(u_n) - u_n\|^2$$
$$- \alpha_n (1 - \alpha_n) \|Tr_n - u_n\|^2. \tag{9.88}$$

By (9.87), (9.88), and noticing that $Ax^* = By^*$, we deduce that

$$\|x_{n+1} - x^*\|^2 + \|y_{n+1} - y^*\|^2 \le \|x_n - x^*\|^2 + \|y_n - y^*\|^2 - 2\lambda_n \|Ax_n - By_n\|^2$$
$$+ \lambda_n^2 (L_1 + L_2) \|Ax_n - By_n\|^2$$
$$- \alpha_n \beta_n (1 - \beta_n) \|U(z_n) - z_n\|^2$$
$$- \alpha_n \beta_n (1 - \beta_n) \|T(u_n) - u_n\|^2. \tag{9.89}$$

Thus, we deduce that

$$\Phi_{n+1} \le \Phi_n - \lambda_n (2 - \lambda_n (L_1 + L_2)) \|Ax_n - By_n\|^2$$
$$- \alpha_n \beta_n (1 - \beta_n) \|U(z_n) - z_n\|^2$$
$$- \alpha_n \beta_n (1 - \beta_n) \|T(u_n) - u_n\|^2, \tag{9.90}$$

where

$$\Phi_n := \|x_n - x^*\|^2 + \|y_n - y^*\|^2.$$

Thus, $\{\Phi_n\}$ is a nonincreasing sequence and bounded below by 0, therefore, it converges.

From (9.90) and the fact that $\{\Phi_n\}$ converges, we deduce that

$$\lim_{n \to \infty} \|Ax_n - By_n\| = 0, \tag{9.91}$$

$$\lim_{n \to \infty} \|Uz_n - z_n\| = 0 \quad \text{and} \quad \lim_{n \to \infty} \|Tu_n - u_n\| = 0. \tag{9.92}$$

Furthermore, since $\{\Phi_n\}$ converges, this ensures that $\{x_n\}$ and $\{y_n\}$ also converges. This further implies that $x_n \rightharpoonup x$ and $y_n \rightharpoonup y$ for some $(x, y) \in \Phi$.

Now, $(x, y) \in \Phi$, implies that $x \in C \cap Fix(U)$ and $y \in Q \cap Fix(T)$ such that $Ax = By$. The fact that $x_n \rightharpoonup x$ and $\lim_{n \to \infty} \|Ax_n - By_n\| = 0$ together with

$$z_n = P_C(x_n - \lambda_n A^*(Ax_n - By_n)),$$

we deduce that $z_n \rightharpoonup P_C x$. Since $x \in C$, by projection theorem, we obtain that $P_C x = x$. Hence, $z_n \rightharpoonup x$.

Similarly, The fact that $y_n \rightharpoonup y$ and $\lim_{n \to \infty} \|Ax_n - By_n\| = 0$ together with

$$u_n = P_Q(y_n + \lambda_n B^*(Ax_n - By_n)),$$

we deduce that $u_n \rightharpoonup P_Q y$. Since $y \in Q$, by projection theorem, we obtain that $P_Q y = y$. Hence, $u_n \rightharpoonup y$.

Now, $z_n \rightharpoonup x$, $\lim_{n \to \infty} \|Uz_n - z_n\| = 0$, and together with the demiclosed of $(U - I)$ at zero, we deduce that $Ux = x$, this implies that $x \in Fix(U)$.

On the other hand, $u_n \rightharpoonup y$ and $\lim_{n \to \infty} \|Tu_n - u_n\| = 0$ together with the demiclosed of $(T - I)$ at zero, we deduce that $Ty = y$, this implies that $y \in Fix(T)$.

Since $z_n \rightharpoonup x, u_n \rightharpoonup y$ and the fact that A and B are bounded linear operators, we have

$$Az_n \rightharpoonup Ax \quad \text{and} \quad Bu_n \rightharpoonup By,$$

this implies that

$$Az_n - Bu_n \rightharpoonup Ax - By,$$

which turn to implies that

$$\|Ax - By\| \leq \lim_{n \to \infty} \inf \|Az_n - Bu_n\| = 0,$$

which further implies that $Ax = By$. Noticing that $x \in C, x \in Fix(U), y \in Q$ and $y \in Fix(T)$, we have that $x \in C \cap Fix(U)$ and $y \in Q \cap Fix(T)$. Hence, we conclude that $(x, y) \in \Phi$.

Summing up, we have proved that:

i) for each $(x^*, x^*) \in \Phi$, the $\lim_{n \to \infty} (\|x_n - x^*\|^2 + \|y_n - y^*\|^2)$ exist;

ii) the weak cluster of the sequence (x_n, y_n) belongs to Φ.

Thus, by Lemma (9.21), we conclude that the sequences (x_n, y_n) converge weakly to $(x^*, x^*) \in \Phi$. This completes the proof. □

Theorem 9.5 Suppose that all the hypothesis of Theorem 9.4 is satisfied. In addition, assume that U and T are semicompacts, then $(x_n, y_n) \to (x^*, y^*) \in \Phi$.

Proof: As in the proof of Theorem 9.4, $\{u_n\}$ and $\{z_n\}$ are bounded, by (9.92) and the fact that U and T are semicompacts, then there exists subsequences $\{u_{n_k}\}$ and $\{z_{n_k}\}$ (suppose without loss of generality) of $\{u_n\}$ and $\{z_n\}$ such that $u_{n_k} \to x$ and $z_{n_k} \to y$. Since, $u_n \rightharpoonup x^*$ and $z_n \rightharpoonup y^*$, we have $x = x^*$ and $y = y^*$. By (9.91) and the fact that $u_{n_k} \to x^*$ and $z_{n_k} \to y^*$, we have

$$\lim_{n \to \infty} \|Ax^* - Ay^*\| = \lim_{n \to \infty} \|Au_{n_k} - Bz_{n_k}\| = 0, \tag{9.93}$$

which tends to imply that $Ax^* = Ay^*$. Hence $(x^*, y^*) \in \Phi$. Thus, the iterative algorithm of Theorem 9.4 conveges strongly to the solution of Problem 9.80. □

9.3.4 The Split Common Fixed-Point Equality Problems for Quasi-Nonexpansive Mappings in Hilbert Spaces

To approximate the solution of split common fixed-point equality problems, we make the following assumptions:

(A_1) $T_1, T_2, T_3, \ldots, T_N : H_1 \to H_1$ and $U_1, U_2, U_3, \ldots, U_M : H_2 \to H_2$ are quasi-nonexpansive mappings with $\bigcap_{i=1}^{N} Fix(T_i) \neq \emptyset$ and $\bigcap_{i=1}^{N} Fix(U_j) \neq \emptyset$, respectively.

(A_2) $(T_i - I)$, for $i = 1, 2, 3, \ldots, N$ and $(U_j - I)$, for $i = 1, 2, 3, \ldots, M$ are demiclosed at zero.

(A_3) $A : H_1 \to H_3$ and $B : H_2 \to H_3$ are bounded linear operators with their adjoints A^* and B^*, respectively.

(A_4) For arbitrary $x_1 \in H_1$ and $y_1 \in H_2$, define $\{(x_n, y_n)\}$ by:

$$
\begin{cases}
z_n = x_n - \lambda_n A^*(Ax_n - By_n), \\
w_n = (1 - \beta_n)z_n + \beta_n \sum_{j=1}^{M} \delta_j U_{\gamma_j}(z_n), \\
x_{n+1} = (1 - \alpha_n)z_n + \alpha_n \sum_{j=1}^{M} \delta_j U_{\gamma_j}(w_n), \\
u_n = y_n + \lambda_n B^*(Ax_n - By_n), \\
r_n = (1 - \beta_n)u_n + \beta_n \sum_{i=1}^{N} \lambda_i T_{\tau_i}(u_n), \\
y_{n+1} = (1 - \alpha_n)u_n + \alpha_n \sum_{i=1}^{N} \lambda_i T_{\tau_i}(r_n), \quad \forall n \geq 1,
\end{cases}
\tag{9.94}
$$

where $U_{\gamma_j} = (1 - \gamma_j)I + \gamma_j U_j$ and $\gamma_j \in (0, 1)$, for $j = 1, 2, 3, \ldots, M$, $T_{\tau_i} = (1 - \tau_i)I + \tau_i T_i$, and $\tau_i \in (0, 1)$, for $i = 1, 2, 3, \ldots, N$, $\sum_{j=1}^{M} \delta_j = 1$ and $\sum_{i=1}^{N} \lambda_i = 1$; $0 < a < \beta_n < 1$, $0 < b < \alpha_n < 1$, and $\lambda_n \in \left(0, \frac{2}{L_1+L_2}\right)$ where $L_1 = A^*A$ and $L_2 = B^*B$.

Theorem 9.6 Suppose that conditions $(A_1) - -(A_4)$ above are satisfied, also, assume that the solution set $\Psi \neq \emptyset$. Then, $(x_n, y_n) \rightharpoonup (x^*, y^*) \in \Psi$.

Proof: Let $(x^*, y^*) \in \Psi$ and $U = \sum_{j=1}^{M} \delta_j U_{\gamma_j}$ and $T = \sum_{i=1}^{N} \lambda_i T_{\tau_i}$. By Lemma 9.22, we deduce that U and T are quasi-nonexpansive mappings, $Fix(U) = \bigcap_{j=1}^{M} Fix(U_{\delta_j}) = \bigcap_{j=1}^{M} Fix(U_j)$ and $Fix(T) = \bigcap_{i=1}^{N} Fix(T_{\tau_i}) = \bigcap_{i=1}^{N} Fix(T_i)$, respectively. By Algorithm (9.94), we deduce the following algorithm.

$$
\begin{cases}
z_n = x_n - \lambda_n A^*(Ax_n - By_n), \\
w_n = (1 - \beta_n)z_n + \beta_n U(z_n), \\
x_{n+1} = (1 - \alpha_n)z_n + \alpha_n U(w_n), \\
\\
u_n = y_n + \lambda_n B^*(Ax_n - By_n), \\
r_n = (1 - \beta_n)u_n + \beta_n T(u_n), \\
y_{n+1} = (1 - \alpha_n)u_n + \alpha_n T(r_n), \; \forall n \geq 1.
\end{cases}
\tag{9.95}
$$

Thus, all the hypothesis of Theorem 9.4 is satisfied. Hence the proof of this theorem follows directly from Theorem 9.4. □

Corollary 9.11 Suppose that conditions (B_1) to (B_5) are satisfied and let $\{(x_n, y_n)\}$ be the sequence generated by Algorithm (9.82). Assume that $\Phi \neq \emptyset$, and let U and T be the firmly of quasi-nonexpansive mappings. Then the sequence $\{(x_n, y_n)\}$ generated by Algorithm (9.82) converges weakly to the solution of Problem (9.80).

Corollary 9.12 Suppose that conditions (B_1) to (B_4) are satisfied and let the sequence $\{(x_n, y_n)\}$ be generated by

$$
\begin{cases}
z_n = x_n - \lambda_n A^*(Ax_n - By_n), \\
x_{n+1} = (1 - \alpha_n)z_n + \alpha_n U(z_n), \\
\\
u_n = y_n + \lambda_n B^*(Ax_n - By_n), \\
y_{n+1} = (1 - \alpha_n)u_n + \alpha_n T(y_n), \; \forall n \geq 0,
\end{cases}
\tag{9.96}
$$

where $0 < a < \beta_n < 1$, and $\lambda_n \in \left(0, \frac{2}{L_1 + L_2}\right)$, where $L_1 = A^*A$ and $L_2 = B^*B$. Assume that $\Phi \neq \emptyset$. Then the sequence $\{(x_n, y_n)\}$ generated by Algorithm (9.96) converges weakly to the solution of SEFPP (9.78).

Proof: Trivially, Algorithm (9.82) reduces to Algorithm (9.96) as $\beta = 0$, $P_C = P_Q = I$ and SFFPEP (9.79) reduces to SEFPP (9.78) as $C := Fix(U)$ and $Q := Fix(T)$. Therefore, all the hypothesis of Theorem 9.4 is satisfied. Hence, the proof of this corollary follows directly from Theorem 9.4. □

Corollary 9.13 Suppose that conditions $(A_1) - -(A_4)$ are satisfied, and let the sequence $\{(x_n, y_n)\}$ be defined by Algorithm (9.94). Assume that $\Psi \neq \emptyset$ and let U and T be firmly quasi-nonexpansive mappings, where $U = \sum_{j=1}^{M} \delta_j U_{Y_j}$ and $T = \sum_{i=1}^{N} \lambda_i T_{\tau_i}$. Then $(x_n, y_n) \rightarrow (x^*, x^*) \in \Psi$.

9.4 Numerical Example

In this section, we give the numerical examples that illustrate our theoretical results.

Example 9.4 Let $H_1 = \mathfrak{R}$ with the inner product defined by $\langle x, y \rangle = xy$ for all $x, y \in \mathfrak{R}$ and $\|.\|$ stands for the corresponding norm. Let $C := [0, \infty)$ and $Q := [0, \infty)$. Defined $T : C \to \mathfrak{R}$ and $S : Q \to \mathfrak{R}$ by $Tx = \frac{x^2+5}{1+x}$, $\forall x \in C$ and $Sx = \frac{x+5}{5}$, $\forall x \in Q$. Then T and S are quasi-nonexpansive mappings.

Proof: Trivially, $Fix(T) = 5$ and $Fix(S) = \frac{5}{4}$. □

Now,

$$
\begin{aligned}
|Tx - 5| &= \left| \frac{x^2 + 5}{1 + x} - 5 \right| \\
&= \frac{x}{1 + x} |x - 5| \\
&\leq |x - 5|.
\end{aligned}
$$

On the other hand,

$$
\begin{aligned}
\left| Sx - \frac{5}{4} \right| &= \left| \frac{x + 5}{5} - \frac{5}{4} \right| \\
&= \frac{1}{5} \left| x - \frac{5}{4} \right| \\
&\leq \left| x - \frac{5}{4} \right|.
\end{aligned}
$$

Thus, T and S are quasi-nonexpansive mappings.

Example 9.5 Let $H_1 = \mathfrak{R}$, $H_2 = \mathfrak{R}$, $C := [0, \infty)$, and $Q := [0, \infty)$ be subset of H_1 and H_2, respectively. Defined $T : C \to C$ by $Tx = \frac{x+2}{3}$ $\forall x \in C$, and $U : Q \to Q$ by

$$
Ux = \begin{cases} \dfrac{2x}{x+1}, & \forall x \in (1, +\infty) \\ 0, & \forall x \in [0, 1]. \end{cases} \tag{9.97}
$$

Then, U and T are quasi-nonexpansive mappings.

Proof: Trivially, $Fix(T) = 1$ and $Fix(U) = 1$.
Now,

$$
\begin{aligned}
|Tx - 1| &= \left| \frac{x + 2}{3} - 1 \right| \\
&\leq |x - 1|.
\end{aligned}
$$

In addition,

$$|Ux - 1| = \frac{1}{1 + x}|x - 1|$$
$$\leq |x - 1|.$$

Thus, U and T are quasi-nonexpansive mappings.

The following example is a particular case of Theorem 9.4. □

Example 9.6 Let $H_1 = \mathfrak{R}$ with the inner product defined by $\langle x, y \rangle = xy$ for all $x, y \in \mathfrak{R}$ and $\|.\|$ stands for the corresponding norm. Let $C := [0, \infty)$ and $Q := [0, \infty)$. Defined $U : C \to \mathfrak{R}$ and $T : Q \to \mathfrak{R}$ by $Ux = \frac{x^2 + 5}{1 + x}$, $\forall x \in C$ and $Tx = \frac{x + 5}{5}$, $\forall x \in Q$. In addition, let $P_C = P_Q = I$, $Ax = x$, $By = 4y$, $\lambda_n = 1$, $\alpha_n = \frac{1}{5}$, $\beta_n = \frac{1}{8}$ and $\{(x_n, y_n)\}$ be the sequence generated by Algorithm (9.82). That is

$$\begin{cases} x_0 \in C \text{ and } y_0 \in Q, \\ z_n = P_C(x_n - A^*(x_n - 4y_n)), \\ w_n = \left(1 - \frac{1}{8}\right) z_n + \frac{1}{8} U(z_n), \\ x_{n+1} = \left(1 - \frac{1}{5}\right) z_n + \frac{1}{5} U(w_n), \\ \\ u_n = P_Q(y_n + B^*(x_n - 4y_n)), \\ r_n = \left(1 - \frac{1}{8}\right) u_n + \frac{1}{8} T(u_n), \\ y_{n+1} = \left(1 - \frac{1}{5}\right) u_n + \frac{1}{5} T(r_n), \ \forall n \geq 0. \end{cases} \tag{9.98}$$

Then (x_n, y_n) converges to $(5, 5/4) \in \Psi$.

Proof: By Example 9.4, U and T are quasi-nonexpansive mappings. Clearly, A and B are bounded linear operator on $\mathfrak{R}$ with $A = A^* = 1$ and $B = B^* = 4$, respectively. Furthermore, it is easy to see that $Fix(U) = 5$ and $Fix(T) = \frac{5}{4}$. Hence,

$$\Psi = \{5 \in C \cap Fix(U) \text{ and } 5/4 \in Q \cap Fix(T) \text{ such that } A(5) = B(5/4)\}.$$

Simplifying Algorithm (9.98), we obtain the following algorithm.

$$\begin{cases} x_0 \in C \text{ and } y_0 \in Q, \\ z_n = x_n, \\ w_n = \frac{7}{8} z_n + \frac{1}{8} \left(\frac{z_n^2 + 5}{z_n + 1} \right), \\ x_{n+1} = \frac{4}{5} z_n + \frac{1}{5} \left(\frac{w_n^2 + 5}{w_n + 1} \right), \\ \\ u_n = y_n, \\ r_n = \frac{7}{8} u_n + \frac{1}{8} \left(\frac{u_n + 5}{5} \right), \\ y_{n+1} = \frac{4}{5} u_n + \frac{1}{5} \left(\frac{r_n + 5}{5} \right), \ \forall n \geq 0. \end{cases} \tag{9.99}$$

□

Table 9.1 Shows the numerical values of Example 9.6 Algorithm (9.99), starting with the initial values $x_0 = 10$ and $y_0 = 15$.

n	x_n	y_n
0	10.00000000	15.00000000
1	9.898293685	12.74500000
2	9.797736851	10.85982000
3	9.698337655	9.283809520
⋮	⋮	⋮
248	5.001051418	1.250000002
249	5.001012726	1.250000002
250	5.000975458	1.250000002

Table 9.2 Shows the numerical values of Example 9.6 Algorithm (9.99), starting with the initial values $x_0 = 5$ and $y_0 = 1.25$.

n	x_n	y_n
0	5.000000000	1.250000000
1	5.000000000	1.250000000
2	5.000000000	1.250000000
⋮	⋮	⋮
98	5.000000000	1.250000000
99	5.000000000	1.250000000
100	5.000000000	1.250000000

We used Maple and obtained the numerical values of Algorithm 9.99 and 9.101 in Tables 9.1, 9.2 and Figures 9.1, 9.2 respectively.

The following example is a particular case of Theorem 9.6.

Example 9.7 Let $H_1 = \Re$ and $H_2 = \Re$, $C := [0, \infty)$ and $Q := [0, \infty)$ be subset of H_1 and H_2, respectively. Define $T : C \to C$ by $Tx = \frac{x+2}{3}$ $\forall x \in C$, and $U : Q \to Q$ by

$$Ux = \begin{cases} \frac{2x}{x+1}, & \forall x \in (1, +\infty) \\ 0, & \forall x \in [0, 1]. \end{cases} \tag{9.100}$$

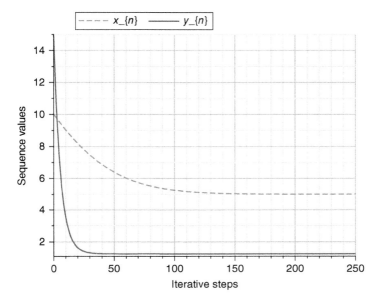

Figure 9.1 Shows the convergence of Example 9.6 Algorithm (9.99), starting with the initial value $x_0 = 10$ and $y_0 = 15$.

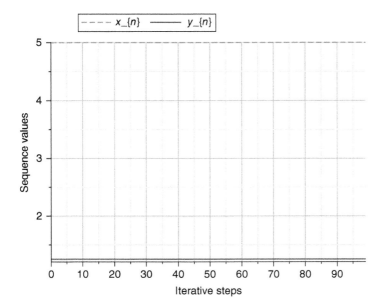

Figure 9.2 Shows the convergence of Example 9.6 Algorithm (9.99), starting with the initial value $x_0 = 5$ and $y_0 = 1.25$.

Table 9.3 Shows the numerical values of Example 9.7 Algorithm (9.101), starting with the initial values $x_0 = -5$. and $y_0 = -5$.

n	x_n	y_n
0	−5.000000000	−5.000000000
1	−4.614625287	−4.874708995
2	−3.892459776	−4.752034296
⋮	⋮	⋮
148	1.513044281	0.7359128532
149	1.497072269	0.7414274772

Table 9.4 Shows the numerical values of Example 9.7 Algorithm (9.101), starting with the initial values $x_0 = 5$ and $y_0 = 5$.

n	x_n	y_n
0	5.000000000	5.000000000
1	4.916472663	4.760850019
2	4.834689530	4.537828465
3	4.754614179	4.329771078
⋮	⋮	⋮
148	1.176058095	1.007392532
149	1.172381679	1.007122340

Let also $\lambda_n = 1$, $Ax = x$, $By = y$, $\gamma_j = \frac{1}{3}$, $\tau_i = \frac{1}{5}$, $\alpha_n = \frac{1}{7}$ and $\beta_n = \frac{1}{9}$. The sequence $\{(x_n, y_n)\}$ defined by Algorithm 9.94 can be written as follows (Tables 9.3, 9.4 and Figures 9.3, 9.4):

$$
\begin{cases}
z_n = x_n - A^*(Ax_n - By_n), \\
w_n = \frac{8}{9}z_n + \frac{1}{9}\left(\frac{2z_n}{3} + \frac{2z_n}{3(z_n+1)}\right), \\
x_{n+1} = \frac{6}{7}z_n + \frac{1}{7}\left(\frac{2w_n}{3} + \frac{2w_n}{3(w_n+1)}\right), \\
\\
u_n = y_n + B^*(Ax_n - By_n), \\
r_n = \frac{8}{9}u_n + \frac{1}{9}\left(\frac{4u_n}{5} + \frac{u_n+2}{15}\right), \\
y_{n+1} = \frac{6}{7}u_n + \frac{1}{7}\left(\frac{4r_n}{5} + \frac{r_n+2}{15}\right), \quad \forall n \geq 1.
\end{cases}
\tag{9.101}
$$

Then (x_n, y_n) converges to $(1, 1) \in \Psi$.

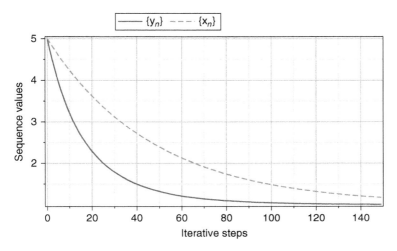

Figure 9.3 Shows the convergence of Example 9.7 Algorithm (9.101), starting with the initial value $x_0 = 5$ and $y_0 = 5$.

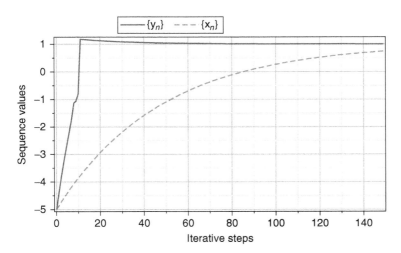

Figure 9.4 Shows the convergence of Example 9.7 Algorithm (9.101), starting with the initial value $x_0 = -5$ and $y_0 = -5$.

Proof: By Example 9.5, U and T are quasi-nonexpansive mappings with $Fix(U) = 1$ and $Fix(T) = 1$, respectively. Clearly, A, B are bounded linear on $\Re$, $A = A^* = 1$ and $B = B^* = 1$. Hence,

$$\Psi = \{1 \in Fix(T) \text{ and } 1 \in Fix(U) \text{ such that } A(1) = B(1)\}.$$

Simplifying Algorithm (9.101), we have

$$
\begin{cases}
z_n = y_n, \\
w_n = \frac{8}{9}z_n + \frac{1}{9}\left(\frac{2z_n}{3} + \frac{2z_n}{3(z_n+1)}\right), \\
x_{n+1} = \frac{6}{7}z_n + \frac{1}{7}\left(\frac{2w_n}{3} + \frac{2w_n}{3(w_n+1)}\right), \\
u_n = x_n, \\
r_n = \frac{8}{9}u_n + \frac{1}{9}\left(\frac{4u_n}{5} + \frac{u_n+2}{15}\right), \\
y_{n+1} = \frac{6}{7}u_n + \frac{1}{7}\left(\frac{4r_n}{5} + \frac{r_n+2}{15}\right), \quad \forall n \geq 1.
\end{cases}
\tag{9.102}
$$

9.5 The Split Feasibility and Fixed Point Problems for Quasi-Nonexpansive Mappings in Hilbert Spaces

In this section, we propose Ishikawa-type extra-gradient algorithms for solving the SFFPP. Under some suitable assumptions imposed on some parameters and operators involved, we prove the strong convergence theorems of these algorithms.

9.5.1 Problem Formulation

The SFFPP required to find a vector

$$
x^* \in C \cap Fix(T_1) \text{ such that } Ax^* \in Q \cap Fix(T_2),
\tag{9.103}
$$

where $T_1 : H_1 \rightarrow H_1$ and $T_2 : H_2 \rightarrow H_2$ are quasi-nonexpansive mappings, and $A : H_1 \rightarrow H_2$ is a bounded linear operator.

We denote the solution set of Problem (9.103) by

$$
\Delta = \{x^* \in C \cap Fix(T_1) \text{ such that } Ax^* \in Q \cap Fix(T_2)\}.
\tag{9.104}
$$

In sequel, we assume that $\Delta \neq \emptyset$.

9.5.2 Preliminary Results

The following well-known results are significant in proving the main result of this chapter.

Lemma 9.23 Let C be a nonempty closed convex subset of a Hilbert space.

i) If G and S are two quasi-nonexpansive mappings on C, then GS is also quasi-nonexpansive mapping.

ii) Let $G_\alpha = (1 - \alpha)I + \alpha G$, where $\alpha \in (0, 1]$ and G is a quasi-nonexpansive mapping on C. Then for all $x \in C$ and $q \in Fix(G)$, G_α is also a quasi-nonexpansive.

iii) Let $\{G_i\}_{i=1}^N : C \to C$ be N-quasi-nonexpansive mappings and $\{\alpha_i\}_{i=1}^N$ be a positive sequence in $(0,1)$ such that $\sum_{i=1}^N \alpha_i = 1$. Suppose that $\{G_i\}_{i=1}^N$ has a fixed point. Then

$$Fix\left(\sum_{i=1}^N \alpha_i G_i\right) = \bigcap_{i=1}^N Fix(G_i).$$

iv) Let $\{G_i\}_{i=1}^N$ and $\{\alpha_i\}_{i=1}^N$ be as in (iii) above. Then $\sum_{i=1}^N \alpha_i G_i$, is a quasi-nonexpansive mapping. Furthermore, if for each $i = 1, 2, 3, \ldots, N$, $G_i - I$ is demiclosed at zero, then $\sum_{i=1}^N \alpha_i G_i - I$ is also.

The proof of (i) follows trivially, for the proof of (ii) see Moudafi [28] while the proof of (iii) and (iv) are deduce from Li and He [30].

9.6 Ishikawa-Type Extra-Gradient Iterative Methods for Quasi-Nonexpansive Mappings in Hilbert Spaces

Theorem 9.7 Let $T : C \to H_1$ and $G : Q \to H_2$ be two quasi-nonexpansive mappings and $A : H_1 \to H_2$ be a bounded linear operator with its adjoint A^*. Assume that $(T - I)$ and $(GP_Q - I)$ are demiclosed at zero, and $\Delta \neq \emptyset$. Define $\{x_n\}$ by

$$\begin{cases} x_0 \in C \text{ chosen arbitrarily,} \\ y_n = P_C(x_n - \gamma_n A^*(I - GP_Q)Ax_n), \\ z_n = P_C(y_n - \gamma_n A^*(I - GP_Q)Ay_n), \\ w_n = (1 - \alpha_n)z_n + \alpha_n T((1 - \beta_n)z_n + \beta_n Tz_n), \\ C_{n+1} = \{z \in C_n : \|w_n - z\|^2 \le \|z_n - z\|^2 \le \|y_n - z\|^2 \le \|x_n - z\|^2\}, \\ x_{n+1} = P_{C_{n+1}}(x_0), \forall n \ge 0, \end{cases}$$

(9.105)

where P is a projection operator, $0 < a < \alpha_n < 1$, $0 < b < \beta_n < 1$, and $0 < c < \gamma_n < \frac{1}{L}$, with $L = \|AA^*\|$. Then $x_n \to x^* \in \Delta$.

Proof: **Step 1.** First, we show that $P_{C_{n+1}}$ is well defined. To show this, it suffices to show that for each $n \ge 0$, C_n is closed and convex. Trivially, C_n is closed.

Next, we show that C_n is convex. To show this, it suffices to show that for each $r_1, r_2 \in C_n$ and $\xi \in (0, 1)$, $\xi r_1 + (1 - \xi)r_2 \in C_n$.

Now, we compute

$$\begin{aligned} \|w_n - \xi r_1 - (1 - \xi)r_2\|^2 &= \|\xi(w_n - r_1) + (1 - \xi)(w_n - r_2)\|^2 \\ &= \xi \|w_n - r_1\|^2 + (1 - \xi) \|w_n - r_2\|^2 \\ &\quad - (1 - \xi)\xi \|r_1 - r_2\|^2 \end{aligned}$$

$$\leq \xi \, \|z_n - r_1\|^2 + (1 - \xi) \, \|z_n - r_2\|^2$$
$$- (1 - \xi)\xi \, \|r_1 - r_2\|^2$$
$$= \|z_n - \xi r_1 - (1 - \xi)r_2\|^2. \tag{9.106}$$

Similarly, we obtain that

$$\|z_n - \xi r_1 - (1 - \xi)r_2\|^2 \leq \|y_n - \xi r_1 - (1 - \xi)r_2\|^2$$
$$\leq \|x_n - \xi r_1 - (1 - \xi)r_2\|^2. \tag{9.107}$$

Thus, for each $r_1, r_2 \in C_n, \xi r_1 + (1 - \xi)r_2 \in C_n$.

Step 2. Here, we show that $\Delta \subset C_n, n \geq 0$.

Let $q \in \Delta$ and $u_n = (1 - \beta_n)z_n + \beta_n T z_n$. The fact that T is quasi-nonexpansive, it follows from (9.105) that

$$\|w_n - q\|^2 = \|(1 - \alpha_n)z_n + \alpha_n T u_n - q\|^2$$
$$= \|(1 - \alpha_n)(z_n - q) + \alpha_n(T u_n - q)\|^2$$
$$= (1 - \alpha_n)\|z_n - q\|^2 + \alpha_n\|T u_n - q\|^2 - \alpha_n(1 - \alpha_n)\|T u_n - z_n\|^2$$
$$\leq (1 - \alpha_n)\|z_n - q\|^2 + \alpha_n\|u_n - q\|^2 - \alpha_n(1 - \alpha_n)\|T u_n - z_n\|^2$$
$$\leq (1 - \alpha_n)\|z_n - q\|^2 + \alpha_n\|(1 - \beta_n)(z_n - q) + \beta_n(T z_n - q)\|^2$$
$$- \alpha_n(1 - \alpha_n)\|T u_n - z_n\|^2$$
$$= (1 - \alpha_n)\|z_n - q\|^2 + \alpha_n(1 - \beta_n)\|z_n - q\|^2 + \alpha_n\beta_n\|T z_n - q\|^2$$
$$- \alpha_n\beta_n(1 - \beta_n)\|T z_n - z_n\|^2 - \alpha_n(1 - \alpha_n)\|T u_n - z_n\|^2$$
$$\leq (1 - \alpha_n)\|z_n - q\|^2 + \alpha_n(1 - \beta_n)\|z_n - q\|^2 + \alpha_n\beta_n\|z_n - q\|^2$$
$$- \alpha_n\beta_n(1 - \beta_n)\|T z_n - z_n\|^2 - \alpha_n(1 - \alpha_n)\|T u_n - z_n\|^2$$
$$\leq \|z_n - q\|^2. \tag{9.108}$$

On the other hand, since G and P_C are both quasi-nonexpansive, by Lemma (9.23), we obtain that GP_C is also quasi-nonexpansive. Thus, we have

$$\|z_n - q\|^2 = \|P_C(y_n - \gamma_n A^*(I - GP_Q)Ay_n) - q\|^2$$
$$\leq \|y_n - \gamma_n A^*(I - GP_Q)Ay_n - q\|^2$$
$$= \|y_n - q\|^2 - 2\gamma_n\langle y_n - q, A^*(I - GP_Q)Ay_n\rangle + \|\gamma_n A^*(I - GP_Q)Ay_n\|^2$$
$$= \|y_n - q\|^2 - 2\gamma_n\langle Ay_n - GP_Q Ay_n + GP_Q Ay_n - Aq, Ay_n - GP_Q Ay_n\rangle$$
$$+ \gamma_n^2 L\|(I - GP_Q)Ay_n\|^2$$
$$= \|y_n - q\|^2 + 2\gamma_n\langle Aq - GP_Q Ay_n, Ay_n - GP_Q Ay_n\rangle$$
$$- \gamma_n(2 - \gamma_n L)\|GP_Q Ay_n - Ay_n\|^2$$
$$\leq \|y_n - q\|^2 - \gamma_n(1 - \gamma_n L)\|GP_Q Ay_n - Ay_n\|^2. \tag{9.109}$$

Following the same way as in the proof of (9.109), we obtain that

$$\|y_n - q\|^2 \le \|x_n - q\|^2. \tag{9.110}$$

Combine Eqs. (9.108)–(9.110), we have

$$\|w_n - q\|^2 \le \|z_n - q\|^2 \le \|y_n - q\|^2 \le \|x_n - q\|^2. \tag{9.111}$$

Thus, we have that $q \in C_n$, this implies that $\Delta \subset C_n$.

Noticing that $\Delta \subset C_{n+1} \subset C_n$ and $x_{n+1} = P_{C_{n+1}}(x_0) \subset C_n$, we have that

$$\|x_{n+1} - x_0\| \le \|q - x_0\|, \quad \forall n \ge 0 \text{ and } q \in \Delta. \tag{9.112}$$

This shows that $\{x_n\}$ is bounded. By Lemma (9.8), we have

$$\|x_{n+1} - x_n\|^2 + \|x_{n+1} - x_0\|^2 = \|P_{C_{n+1}}(x_0) - x_n\|^2 + \|P_{C_{n+1}}(x_0) - x_0\|^2$$
$$\le \|x_n - x_0\|^2. \tag{9.113}$$

This implies that

$$\|x_{n+1} - x_0\| \le \|x_n - x_0\| .$$

Thus, $\{\|x_n - x_0\|\}$ is a nonincreasing sequence and bounded below by zero. Therefore, $\lim_{n \to \infty} \|x_n - x_0\|$ exists.

On the other hand, for each $k > n$, we also obtain that

$$\|x_k - x_n\|^2 + \|x_n - x_0\|^2 = \|P_{C_n}(x_0) - x_k\|^2 + \|P_{C_n}(x_0) - x_0\|^2$$
$$\le \|x_k - x_0\|^2. \tag{9.114}$$

Thus, by (9.114) and the fact that the $\lim_{n \to \infty} \|x_{n+1} - x_0\|$ exist, we obtain that

$$\lim_{k, n \to \infty} \| x_k - x_n \| = 0.$$

This shows that $\{x_n\}$ is Cauchy.

Since $x_{n+1} = P_{C_{n+1}}(x_0) \in C_{n+1} \subset C_n$ and the fact that $\{x_n\}$ is a Cauchy sequence, we deduce that

$$\|y_n - x_n\| \le \|y_n - x_{n+1}\| + \|x_{n+1} - x_n\|$$
$$\le 2 \|x_{n+1} - x_n\|,$$

and

$$\|z_n - x_n\| \le \|z_n - x_{n+1}\| + \|x_{n+1} - x_n\|$$
$$\le 2 \|x_{n+1} - x_n\| .$$

Thus, as $n \to \infty$, we deduce that

$$\lim_{n \to \infty} \|z_n - x_n\| = 0 \quad \text{and} \quad \lim_{n \to \infty} \|y_n - x_n\| = 0. \tag{9.115}$$

On the other hand,

$$\|y_n - z_n\| \le \|y_n - x_n\| + \|x_n - z_n\| .$$

By (9.115), we obtain that

$$\lim_{n\to\infty} \|y_n - z_n\| = 0. \tag{9.116}$$

Similarly, we obtain that

$$\lim_{n\to\infty} \|w_n - x_n\| = 0, \lim_{n\to\infty} \|z_n - y_n\| = 0 \text{ and } \lim_{n\to\infty} \|w_n - z_n\| = 0. \tag{9.117}$$

By (9.109), we obtain that

$$\|GP_Q Ay_n - Ay_n\|^2 \le \frac{\|y_n - x^*\|^2 - \|z_n - x^*\|^2}{\gamma_n(1 - \gamma_n L)}$$
$$\le \frac{\|y_n - z_n\|(\|z_n - y_n\| + 2\|z_n - x^*\|)}{c(1 - \gamma_n L)}.$$

Thus, by (9.116) we deduce that

$$\lim_{n\to\infty} \|GP_Q Ay_n - Ay_n\| = 0.$$

Similarly, by (9.108) and (9.117), we deduce that

$$\lim_{n\to\infty} \|Tz_n - z_n\| = 0.$$

Finally, we show that $x_n \to x^*$.

Since $\{x_n\}$ is Cauchy, we assume that $x_n \to p$. By (9.105), we have that $z_n \to p$, this implies that $z_n \rightharpoonup p$. The fact that the $\lim_{n\to\infty} \|Tz_n - z_n\| = 0$ together with the demiclosed of $(T - I)$ at zero, we deduce that $p \in Fix(T)$.

On the other hand, since $x_n \to p$, implies that $y_n \rightharpoonup p$, by (9.105), we deduce that $p = P_C p$, which implies that $p \in C$; and therefore, we have $p \in C \cap Fix(T)$.

Furthermore, by the definition of A, we have that $Ay_n \to Ap$, this implies that $Ay_n \rightharpoonup Ap$. The fact that the $\lim_{n\to\infty} \|GP_Q Ay_n - Ay_n\| = 0$ together with the demiclosed of $(GP_Q - I)$ at zero, we deduce that $Ap \in Fix(GP_Q)$, this implies that $Ap \in Q \cap Fix(G)$. Hence $p \in \Delta$. This show that $x_n \to x^*$. The proof is complete.

Next, we consider the SFFPP for the class of finite family of quasi-nonexpansive mappings. □

Theorem 9.8 Let $\{T_i\}_{i=1}^M : C \to H_1$ and $\{G_j\}_{j=1}^N : Q \to H_2$ be quasi-nonexpansive mappings with $\bigcap_{i=1}^M Fix(T_i) \ne \emptyset$ and $\bigcap_{j=1}^N Fix(G_j) \ne \emptyset$. In addition, let $A : H_1 \to H_2$ be a bounded linear operator with its adjoint A^*. Assume that $(T_i - I), i = 1, 2, 3, \dots, M$ and $(G_j P_Q - I), j = 1, 2, 3, \dots, N$ are demiclosed at zero, and $\Delta \neq \emptyset$. Define $\{x_n\}$ by

$$
\begin{cases}
x_0 \in C \text{ chosen arbitrarily,} \\
y_n = P_C\left(x_n - \gamma_n A^*\left(I - \sum_{j=1}^{N} \delta_j G_j P_Q\right) A x_n\right), \\
z_n = P_C\left(y_n - \gamma_n A^*\left(I - \sum_{j=1}^{N} \delta_j G_j P_Q\right) A y_n\right), \\
w_n = (1 - \alpha_n)z_n + \alpha_n \sum_{i=1}^{M} \lambda_i T_i\left((1 - \beta_n)z_n + \beta_n \sum_{i=1}^{M} \lambda_i T_i z_n\right), \\
C_{n+1} = \{z \in C_n : \|w_n - z\|^2 \leq \|z_n - z\|^2 \leq \|y_n - z\|^2 \leq \|x_n - z\|^2\}, \\
x_{n+1} = P_{C_{n+1}}(x_0), \ \forall n \geq 0,
\end{cases}
$$

$$(9.118)$$

where P is a projection operator, $0 < a < \alpha_n < 1$, $0 < b < \beta_n < 1$, and $0 < c < \gamma_n < \frac{1}{L}$ with $L = \|AA^*\|$. Then $x_n \to x^* \in \Delta$. $\qquad \square$

Proof: By Lemma 9.23, we deduce that

i) $\sum_{j=1}^{N} \delta_j G_j$ and $\sum_{i=1}^{M} \lambda_i T_i$ are quasi-nonexpansive mappings.
ii) $\sum_{j=1}^{N} \delta_j(G_j - I)$ and $\sum_{i=1}^{M} \lambda_i(T_i - I)$ are demiclosed at zero.
iii) $Fix\left(\sum_{i=1}^{M} \lambda_i T_i\right) = \bigcap_{i=1}^{M} Fix(T_i)$ and $Fix\left(\sum_{j=1}^{N} \delta_j G_j\right) = \bigcap_{j=1}^{N} Fix(G_j)$.

Thus, all the hypothesis of Theorem 9.8 is satisfied. Therefore, the proof of this theorem follows trivially from Theorem 9.7.

As the consequence of Theorem 9.8, we immediately obtain the following corollary.

Corollary 9.14 Let $\{T_i\}_{i=1}^{M} : C \to H_1$ and $\{G_j\}_{j=1}^{N} : Q \to H_2$ be quasi-nonexpansive mappings with $\bigcap_{i=1}^{M} Fix(T_i) \neq \emptyset$ and $\bigcap_{j=1}^{N} Fix(G_j) \neq \emptyset$, respectively. In addition, let $A : H_1 \to H_2$ be a bounded linear operator with its adjoint A^*. Assume that $(T_i - I), i = 1, 2, 3, \dots, M$ and $(G_j P_Q - I), j = 1, 2, 3, \dots, N$ are demiclosed at zero and $\Delta \neq \emptyset$. Define $\{x_n\}$ by

$$
\begin{cases}
x_0 \in C \text{ chosen arbitrarily,} \\
z_n = P_C(x_n - \gamma_n A^*(I - \sum_{j=1}^{N} \delta_j G_j P_Q) A x_n), \\
w_n = (1 - \alpha_n)z_n + \alpha_n \sum_{i=1}^{M} \lambda_i T_i z_n, \\
C_{n+1} = \{z \in C_n : \|w_n - z\|^2 \leq \|z_n - z\|^2 \leq \|x_n - z\|^2\}, \\
x_{n+1} = P_{C_{n+1}}(x_0), \ \forall n \geq 0,
\end{cases}
$$

$$(9.119)$$

where P is a projection operator, $0 < a < \alpha_n < 1$, $0 < b < \beta_n < 1$, and $0 < c < \gamma_n < \frac{1}{L}$ with $L = \|AA^*\|$. Then $x_n \to x^* \in \Delta$.

Proof: In Algorithm (9.118), take $y_n = x_n$ and $\beta_n = 0$, then, Algorithm (9.118) reduces to Algorithm (9.119); therefore, all the hypothesis of Theorem 9.8 is satisfied. Hence, the proof of this corollary follows directly from Theorem 9.8. $\qquad\square$

9.6.1 Application to Split Feasibility Problems

As a special case of Problem (9.104), we give the following theorems for solving SFP and the FPP.

Theorem 9.9 Let $\{T_i\}_{i=1}^M : C \to H_1$ and $\{G_j\}_{j=1}^N : Q \to H_2$ be quasi-nonexpansive mappings with $\bigcap_{i=1}^M Fix(T_i) \neq \emptyset$ and $\bigcap_{j=1}^N Fix(G_j) \neq \emptyset$. In addition, let $A : H_1 \to H_2$ be a bounded linear operator with its adjoint A^*. Assume that $(T_i - I), i = 1, 2, 3, \ldots, M$ and $(G_j - I), j = 1, 2, 3, \ldots, N$ are demiclosed at zero and $\Delta \neq \emptyset$. Define $\{x_n\}$ by

$$
\begin{cases}
x_0 \in C \text{ chosen arbitrarily,} \\
y_n = P_C(x_n - \gamma_n A^*(I - \sum_{j=1}^N \delta_j G_j) A x_n), \\
z_n = P_C(y_n - \gamma_n A^*(I - \sum_{j=1}^N \delta_j G_j) A y_n), \\
w_n = (1 - \alpha_n) z_n + \alpha_n \sum_{i=1}^M \lambda_i T_i \left((1 - \beta_n) z_n + \beta_n \sum_{i=1}^M \lambda_i T_i z_n \right), \\
C_{n+1} = \{z \in C_n : \|w_n - z\|^2 \le \|z_n - z\|^2 \le \|y_n - z\|^2 \le \|x_n - z\|^2\}, \\
x_{n+1} = P_{C_{n+1}}(x_0), \ \forall n \ge 0,
\end{cases}
$$

$$(9.120)$$

where P is a projection operator, $0 < a < \alpha_n < 1$, $0 < b < \beta_n < 1$, and $0 < c < \gamma_n < \frac{1}{L}$ with $L = \|AA^*\|$. Then $x_n \to x^* \in \Omega$.

Proof: In Algorithm (9.118), take $P_Q = I$ (the identity mapping), then, Algorithm (9.118) reduces to Algorithm (9.120), therefore, all the hypothesis of Theorem 9.8 is satisfied. Hence, the proof of this theorem, follows directly from Theorem 9.8. $\qquad\square$

Theorem 9.10 Let $\{T_i\}_{i=1}^M : C \to H_1$ be quasi-nonexpansive mapping with $\bigcap_{i=1}^M Fix(T_i) \neq \emptyset$. Assume that $(T_i - I), i = 1, 2, 3, \ldots, M$ is demiclosed at zero.

Define $\{x_n\}$ by

$$
\begin{cases}
x_0 \in C \text{ chosen arbitrarily,} \\
u_n = (1 - \beta_n)x_n + \beta_n \sum_{i=1}^{M} \lambda_i T_i x_n, \\
w_n = (1 - \alpha_n)u_n + \alpha_n \sum_{i=1}^{M} \lambda_i T_i u_n, \\
C_{n+1} = \{z \in C_n : \|w_n - z\|^2 \leq \|u_n - z\|^2 \leq \|x_n - z\|^2\}, \\
x_{n+1} = P_{C_{n+1}}(x_0), \ \forall n \geq 0,
\end{cases}
\tag{9.121}
$$

where P is a projection operator, $0 < a < \alpha_n < 1$ and $0 < b < \beta_n < 1$. Then $\{x_n\}$ converges strongly to the solution of common fixed point of $\{T_i\}_{i=1}^{M}$.

Proof: In Algorithm (9.118), take $\gamma_n = 0$ and $P_C = I$ (the identity mapping), then, Algorithm (9.118) reduces to Algorithm (9.121), therefore; all the hypothesis of Theorem 9.8 is satisfied. Hence, the proof of this theorem, follows directly from Theorem 9.8.

Remark 9.14 In this section, we have proposed Ishikawa-type extra-gradient methods for solving the SFFPP for the class of quasi-nonexpansive mappings in Hilbert spaces. Under some suitable assumptions imposed on some parameters and operators involved, we proved the strong convergence theorems of these algorithms. Furthermore, as an application, we gave the strong convergence theorem for the SFP. The results presented in this chapter, not only extend the result of Chen *et al.* [44] but also extend, improve and generalize the results of; Takahashi and Toyoda [45], Nadezhkina and Takahashi [46], Ceng *et al.* [47], and Li and He [30] in the following ways:

- The theorem of Chen *et al.* [44] gave the weak convergence results while ours gave the strong convergence results.
- The technique of proving our results is entirely different from that of Chen *et al.* [44]. Furthermore, the algorithms of Chen *et al.* [44] involve the class of nonexpansive mappings while our algorithm includes the class of quasi-nonexpansive mappings, which are more general than that nonexpansive mappings.
- The method for finding the solution of the SFFPP is more general that the method of finding the solution to SFP.
- The theorem of Li and He [30] gave the strong convergence results for the SFP while ours gave the strong convergence for the SFFPP. Furthermore, our algorithms generalize that of Li and He [30]. For instance, in Theorem 9.7 Algorithm (9.105), take $y_n = x_n$ and $\beta_n = 0$, hence, our algorithm reduces to that of Li and He [30] Theorem 2.1 Algorithm 2.1.

- Our theorems gave the strong convergence for the solution of the SFFPP for the class of quasi-nonexpansive mappings, while the results of Ceng *et al.* [48] gave a weak convergence result for the solution of the SFFPP for the class of nonexpansive mappings.
- The SFFPP is a fascinating problem. It generalizes the SFP and FPP. All the results and conclusions that are true for the SFFPP continue to holds for these problems (SFP, FPP), and it shows the significance and the range of applicability of SFFPP.
- The novelty of our theorems gives strong convergence results while the theorem of [45] Nadezhkina and Takahashi [46], Ceng *et al.* [47], and Li and He [30] all gives weak convergence results.

9.7 Conclusion

In this work, we have studied the SCFPP and its applications. We have suggested some algorithms for solving this SCFPP and its variant forms for different classes of nonlinear mappings.

Proceeding systematically in our work, we gave the basic definitions and results from the literature. In addition, we briefly provided an overview of the SCFPP and its variant forms in Section 9.2. In the next section, we have suggested and analyzed iterative algorithms for solving the SCFPP for the class of total quasi-asymptotically nonexpansive mappings in Hilbert spaces. In addition, we gave the strong convergence results of the proposed algorithms. In addition, we considered an algorithm for solving this SCFPP for the class of demicontractive mappings without any prior information on the normed on the bounded linear operator and established the strong convergence results of the proposed algorithm.

As a generalization of the SFP, we proposed Ishikawa-type extra-gradient methods for solving the SFFPP for the class of quasi-nonexpansive mappings in Hilbert spaces. Under some suitable assumptions imposed on some parameters and operators involved, we proved the strong convergence theorems of these algorithms.

In the end, we proposed a new problem called "SFFPEP" and study it for the class of quasi-nonexpansive mappings in Hilbert spaces. We also proposed new iterative methods for solving this SFFPEP and proved the convergence results of the proposed algorithms. In additions, as a generalization of SFFPEP, we consider another problem called "Split Common Fixed Equality Problems (SCFPEP)" and study it for the class of finite family of quasi-nonexpansive mappings in Hilbert spaces. Finally, We suggested some algorithms for solving this SCFPEP and proved the convergence results of the proposed algorithms.

References

1 Rudin, W. (1973) *Functional Analysis*, McGraw-Hill Series in Higher Mathematics, McGraw Hill Book Company, New York.

2 Arino, O., Gautier, S., and Penot, J.P. (1984) A fixed point theorem for sequentially continuous mappings with application to ordinary differential equations. *Funk. Ekvac.*, **27** (3), 273–279.

3 Yamamoto, N. (1998) A numerical verification method for solutions of boundary value problems with local uniqueness by Banach's fixed-point theorem. *SIAM J. Numer. Anal.*, **35** (5), 2004–2013.

4 Taleb, A. and Hanebaly, E. (2000) A fixed point theorem and its application to integral equations in modular function spaces. *Proc. Am. Math. Soc.*, **128** (2), 419–426.

5 Nieto, J.J. and Rodríguez-López, R. (2005) Contractive mapping theorems in partially ordered sets and applications to ordinary differential equations. *Order*, **22** (3), 223–239.

6 Pathak, H.K., Khan, M.S., and Tiwari, R. (2007) A common fixed point theorem and its application to nonlinear integral equations. *Comput. Math. Appl.*, **53** (6), 961–971.

7 Sestelo, R.F. and Pouso, R.L. (2015) Existence of solutions of first-order differential equations via a fixed point theorem for discontinuous operators. *Fixed Point Theory Appl.*, **2015** (1), 220.

8 Chow, S.N. (1974) Existence of periodic solutions of autonomous functional differential equations. *J. Differ. Equ.*, **15** (2), 350–378.

9 Grimmer, R. (1979) Existence of periodic solutions of functional differential equations. *J. Math. Anal. Appl.*, **72** (2), 666–673.

10 Torres, P.J. (2003) Existence of one-signed periodic solutions of some second-order differential equations via a Krasnoselskii fixed point theorem. *J. Differ. Equ.*, **190** (2), 643–662.

11 Kiss, G. and Lessard, J.P. (2012) Computational fixed-point theory for differential delay equations with multiple time lags. *J. Differ. Equ.*, **252** (4), 3093–3115.

12 Censor, Y. and Segal, A. (2009) The split common fixed point problem for directed operators. *J. Convex Anal.*, **16** (2), 587–600.

13 Censor, Y., Bortfeld, T., Martin, B., and Trofimov, A. (2006) A unified approach for inversion problems in intensity-modulated radiation therapy. *Phys. Med. Biol.*, **51** (10), 2353–2365.

14 Dunford, N., Schwartz, J.T., Bade, W.G., and Bartle, R.G. (1971) *Linear Operators*, Wiley-interscience, New York.

15 Chidume, C.E. (2006) *Applicable Functional Analysis*, International Centre for Theoretical Physics Trieste, Italy.

16 Goebel, K. and Kirk, W.A. (1990) *Topics in Metric Fixed Point Theory*, vol. **28**, Cambridge University Press, Cambridge.

17 Goebel, K. and Kirk, W.A. (1972) A fixed point theorem for asymptotically nonexpansive mappings. *Proc. Am. Math. Soc.*, **35** (1), 171–174.

18 Alber, Y., Chidume, C.E., and Zegeye, H. (2006) Approximating fixed points of total asymptotically nonexpansive mappings. *Fixed Point Theory Appl.*, **2006** (1), 1–20.

19 Chidume, C.E. and Ofoedu, E.U. (2007) Approximation of common fixed points for finite families of total asymptotically nonexpansive mappings. *J. Math. Anal. Appl.*, **333** (1), 128–141.

20 Chidume, C.E. and Ofoedu, E.U. (2009) A new iteration process for approximation of common fixed points for finite families of total asymptotically nonexpansive mappings. *Int. J. Math. Math. Sci.*, **2009**, 1–17.

21 Browder, F.E. and Petryshyn, W.V. (1967) Construction of fixed points of nonlinear mappings in Hilbert space. *J. Math. Anal. Appl.*, **20** (2), 197–228.

22 Diaz, J.B. and Metcalf, F.T. (1967) On the structure of the set of subsequential limit points of successive approximations. *Bull. Am. Math. Soc.*, **73** (4), 516–519.

23 He, Z. and Du, W.S. (2012) Nonlinear algorithms approach to split common solution problems. *Fixed Point Theory Appl.*, **2012** (1), 130.

24 Chidume, C.E., Ndambomve, P., and Bello, U.A. (2015) The split equality fixed point problem for demi-contractive mappings. *J. Nonlinear Anal. Optim. Theory Appl.*, **6** (1), 61–69.

25 Zaknoon, M. (2003) Algorithmic developments for the convex feasibility problem, University of Haifa Faculty of Science and Science Education, Department of Mathematics.

26 Cegielski, A. (2010) Generalized relaxations of nonexpansive operators and convex feasibility problems. *Contemp. Math.*, **513**, 111–123.

27 Cegielski, A. and Censor, Y. (2011) Opial-type theorems and the common fixed point problem, in *Fixed-Point Algorithms for Inverse Problems in Science and Engineering*, Springer, New York, pp. 155–183.

28 Moudafi, A. (2011) A note on the split common fixed-point problem for quasi-nonexpansive operators. *Nonlinear Anal. Theory Methods Appl.*, **74** (12), 4083–4087.

29 Acedo, G.L. and Xu, H.K. (2007) Iterative methods for strict pseudo-contractions in Hilbert spaces. *Nonlinear Anal. Theory Methods Appl.*, **67** (7), 2258–2271.

30 Li, R. and He, Z. (2015) A new iterative algorithm for split solution problems of quasi-nonexpansive mappings. *J. Inequal. Appl.*, **2015** (1), 131.

31 Xu, H.K. (2002) Iterative algorithms for nonlinear operators. *J. London Math. Soc.*, **66** (1), 240–256.

32 Tan, K.K. and Xu, H.K. (1993) Approximating fixed points of nonexpansive mappings by the Ishikawa iteration process. *J. Math. Anal. Appl.*, **178** (2), 301–308.

33 Bauschke, H.H. and Borwein, J.M. (1996) On projection algorithms for solving convex feasibility problems. *SIAM Rev.*, **38** (3), 367–426.

34 Opial, Z. (1967) Weak convergence of the sequence of successive approximations for nonexpansive mappings. *Bull. Am. Math. Soc.*, **73** (4), 591–597.

35 Wang, X.R., Chang, S.S., Wang, L., and Zhao, Y.H. (2012) Split feasibility problems for total quasi-asymptotically nonexpansive mappings. *Fixed Point Theory Appl.*, **2012** (1), 151.

36 Mohammed, L.B. and Kiliçman, A. (2015) Strong convergence for the split common fixed-point problem for total quasi-asymptotically nonexpansive mappings in Hilbert space. *Abstr. Appl. Anal.*, **2015**, 1–7.

37 Tian, M. and Di, L. (2011) Synchronal algorithm and cyclic algorithm for fixed point problems and variational inequality problems in Hilbert space. *J. Fixed Point Appl.*, **21**, 1–14.

38 Marino, G. and Xu, H.K. (2007) Weak and strong convergence theorems for strict-pseudocontractions in Hilbert spaces. *J. Math. Anal. Appl.*, **329**, 336–346.

39 Changa, S.S., Joseph Lee, H.W., Chanb, C.K., Wanga, L., and Qin, L.J. (2013) Split feasibility problem for quasi-nonexpansive multi-valued mappings and total asymptotically strict pseudo-contractive mapping. *Appl. Math. Comput.*, **219** (20), 10416–10424.

40 Bulama, L.M. and Kiliçman, A. (2016) On synchronal algorithm for fixed point and variational inequality problems in Hilbert spaces. *SpringerPlus*, **5** (1), 103.

41 Tain, M. (2010) A general iterative algorithm for quasi-nonexpansive mapping in Hilbert space. *J. Nonlinear Anal. Theory Methods Appl.*, **73** (3), 689–694.

42 Marino, G. and Xu, H.K. (2006) A general iterative method for nonexpansive mapping in Hilbert space. *J. Math. Anal. Appl.*, **318** (1), 43–52.

43 Butnariu, D., Reich, S., and Censor, Y. (2001) *Inherently Parallel Algorithms in Feasibility and Optimization and their Application*, Studies in Computational Mathematics, vol. **8**, Elsevier, p. 516. ISBN: 0080508766, 9780080508764.

44 Chen, J.Z., Ceng, L.C., Qiu, Y.Q., and Kong, Z.R. (2015) Extra-gradient methods for solving split feasibility and fixed point problems. *Fixed Point Theory Appl.*, **2015** (1), 192.

45 Takahashi, W. and Toyoda, M. (2003) Weak convergence theorems for nonexpansive mappings and monotone mappings. *J. Optim. Theory Appl.*, **118** (2), 417–428.

46 Nadezhkina, N. and Takahashi, W. (2006) Weak convergence theorem by an extragradient method for nonexpansive mappings and monotone mappings. *J. Optim. Theory Appl.*, **128** (1), 191–201.

47 Ceng, L.C., Petru?el, A., and Yao, J.C. (2013) Relaxed extragradient methods with regularization for general system of variational inequalities with

constraints of split feasibility and fixed point problems. *Abstr. Appl. Anal.,* **2013**, 1–25.

48 Ceng, L.C., Ansari, Q.H., and Yao, J.C. (2012) An extragradient method for solving split feasibility and fixed point problems. *Comput. Math. Appl.,* **64** (4), 633–642.

10

Stabilities and Instabilities of Rational Functional Equations and Euler–Lagrange–Jensen (a, b)-Sextic Functional Equations

John Michael Rassias[1], Krishnan Ravi[2], and Beri V. Senthil Kumar[3]

[1] *Pedagogical Department E.E., Section of Mathematics and Informatics, National and Capodistrian University of Athens, Athens 15342, Greece*
[2] *Department of Mathematics, Sacred Heart College, Tirupattur 635 601, Tamil Nadu, India*
[3] *Department of Mathematics, C. Abdul Hakeem College of Engg. and Tech., Melvisharam 632 509, Tamil Nadu, India*

10.1   Introduction

The idea of functional equation is an elegant and powerful technique used in many fields of mathematics and science. The applications of functional equations are involved in every field, and all fields are promoted from their usage and simplicity. Prior to the development of the differential calculus, the physical processes were studied in terms of functions. There was an input variable x (or several input variables) and an output variable $f(x)$, and $f(x)$ had to satisfy some relations corresponding to some properties of the physical process, which were often known by observation. This leads to some functional equation for f.

Functional equations represent an alternative way of modeling problems in Physics. The interest of modeling physical problems by functional equations is that we do not have to assume the differentiability of the function f. Consequently, the functional equations lead often to other solutions than those given by partial differential equations, and these other solutions can be of interest to physicists.

Functional equations are being used with vigor in ever-increasing numbers to investigate problems in many fields of science. This chapter presents a comprehensive study of the classical topic of functional equations.

In mathematics, a function is denoted by a functional equation in hidden style. The importance of a function (or functions) is (or are) communicated by the equation at some point. For instance, functional equations are used to determine properties of functions.

The subject of functional equations forms a modern branch of mathematics. A functional equation is defined as an equation that has finite number of

Mathematical Analysis and Applications: Selected Topics, First Edition.
Edited by Michael Ruzhansky, Hemen Dutta, and Ravi P. Agarwal.
© 2018 John Wiley & Sons, Inc. Published 2018 by John Wiley & Sons, Inc.

indefinite functions and a finite number of independent variables on its both sides.

If we solve a functional equation, then its solution is a function that fulfills the equation. Functional equations comprise a traditional branch of mathematics offering wide scope for algebraic, analytic, order theoretic, and topological considerations. In earlier days, functional equations in different forms were studied by mathematicians such as Euler (eighteenth century) and Cauchy (nineteenth century).

The most well known among the functional equations is $\phi(x + y) = \phi(x) + \phi(y)$. Its solution in the case of real variables was achieved by Cauchy in the year 1821. This equation is known as Cauchy functional equation. In addition, the properties of the additive Cauchy functional equation are dominant tools in more or less every field of natural and social sciences.

This new theory of functional equation is now rapidly growing. The number of mathematical papers and mathematicians dealing with functional equations is increasing day by day. The investigations of functional equations in other subjects such as differential geometry, iterations and analytic functional, differential equations, number theory, and abstract algebra indicate the growing importance of functional equations. In this way, this theory acquired its own personality. Since analytic methods have already been fatigue to some extent in many branches of mathematics, the mathematicians throughout the world are interested in functional equations. Application of elementary methods helps one to achieve much deeper and more general results than by applying classical methods of mathematical analysis.

10.1.1 Growth of Functional Equations

The theory of functional equation was dealt by mathematicians like Abel, D'Alembert, Babbage, Cauchy, Euler, Gauss, Legendre, Schroder and many other mathematicians. Functional equations in several variables were first used by Oresme during 1347–1350. Tannery, Artin, and Reich also used some other type of functional equations. The monographs by Napier (1614), Kepler (1624), and Galileo (1638) contain more functional equations.

The functional equations were applied in the work done by the great mathematicians Picard (1827), Hilbert (1832), Hardy (1840), Bellman (1855), and Hille (1856). In 1905, Aczel collected all the works previously done by other mathematicians and published them in a monograph. This monograph serves as a resource book for many researchers working in the field of functional equations.

10.1.2 Importance of Functional Equations

Currently, the theory of functional equations is an ever-growing area of mathematics with extensive applications. To model and study problems in mathematical analysis, statistics, physics, biology, behavioral sciences, engineering and information theory, the application of functional equation play a significant role.

For solving many problems in economics, psychology, and sociology, functional equations, and their solutions have become important. The theory of functional equations is relatively new and it contributes to the development of strong tools in present-day mathematics. On the contrary, many mathematical thoughts in different fields have become crucial to the basis of functional equations. Many new applied problems and theories have inspired and encouraged specialists on functional equations to develop new approaches and new methods.

10.1.3 Functional Equations Relevant to Other Fields

Functional equations arise in many fields of mathematics, such as geometry, statistics, measure theory, algebraic geometry, and group theory. There are many fascinating uses of functional equations to study various problems in the field of probability. Solutions of functional equations can be employed in describing joint distributions from conditional distributions. Functional equations find many applications in the study of stochastic process, classical mechanics, astronomy, economics, dynamic programming, game theory, computer graphics, neural networks, digital image processing, coding theory, fuzzy set theory, decision theory, artificial intelligence, cluster analysis, multivalued logic, population ethics, finance, information theory, wireless sensor networks, and many other fields.

10.1.4 Definition of Functional Equation with Examples

A Hungarian mathematician Aczel [1], an outstanding expert in functional equations, defines the functional equation as follows:

Functional equation: Functional equations are equations which have a fixed number of unidentified functions and a limited number of free variables on both sides.

Solutions of functional equation: A solution of a functional equation is a function which satisfies the equation.

Example:

(i) The Cauchy functional equations
- $A(u + v) = A(u) + A(v)$,
- $E(u + v) = E(u)E(v)$,
- $M(uv) = M(u)M(v)$, and
- $L(uv) = L(u) + L(v)$

have solutions $A(u) = ku, E(u) = e^u$, $M(u) = u^c$, and $L(u) = \log u$, respectively.

(ii) $f(u) = cu + k$ is the solution of the functional equation $f\left(\frac{u+v}{2}\right) = \frac{1}{2}[f(u) + f(v)]$.

10.2   Ulam Stability Problem for Functional Equation

The stability of functional equations is an interesting topic that has been dealt for the last six decades. In mathematics, a stipulation in which a slight disturbance in a system does not create a considerable disturbing consequence on that system. An equation is said to be stable if a slightly different solution is close to the exact solution of that equation. In mathematical modeling of physical problems, the deviations in measurements will result with errors and these deviations can be dealt with the stability of equations. Hence, the stability of equations is essential in mathematical models. A stable solution will be sensible in spite of such deviations.

An interesting and eminent talk given by Ulam [2] in 1940, inspired to study the investigation of stability of functional equations. The ensuing question was asked by him pertaining to the stability of homomorphisms in group theory:

Let $\mathcal{X}, \mathcal{Y}$ be a group and a metric group with metric $d(\cdot, \cdot)$, respectively. Suppose $\epsilon > 0$ is a fixed constant. Then, whether there exists a constant $\delta > 0$ so that if a mapping $g : \mathcal{X} \to \mathcal{Y}$ satisfies

$$d(g(uv), g(u)g(v)) < \delta$$

for all $u, v \in \mathcal{X}$, and there exists a homomorphism $a : \mathcal{X} \to \mathcal{Y}$ with the condition

$$d(g(u), a(u)) < \epsilon$$

for all $u \in \mathcal{X}$?

If the answer is confirmative, then the functional equation for homomorphism is said to be stable.

10.2.1 ϵ-Stability of Functional Equation

The foremost answer to the question of Ulam was provided by Hyers [3]. He luminously replied to the question of Ulam by considering $\mathcal{X}$ and $\mathcal{Y}$ as Banach spaces. The result obtained by Hyers is stated in the succeeding celebrated theorem.

Theorem 10.1   **[3]**   Assume that $\mathcal{G}$ and $\mathcal{H}$ are Banach spaces. If a function $g : \mathcal{G} \to \mathcal{H}$ satisfies the inequality

$$\|g(u + v) - g(u) - g(v)\| \leq \epsilon \tag{10.1}$$

for some $\epsilon > 0$ and for all $u, v \in \mathcal{G}$, then the limit

$$\mathcal{A}(u) = \lim_{n \to \infty} \frac{1}{2^n} g(2^n u) \tag{10.2}$$

exists for all $u \in \mathcal{G}$ is the unique additive function such that

$$\|g(u) - \mathcal{A}(u)\| \leq \epsilon \tag{10.3}$$

for all $u \in \mathcal{G}$. In addition, if $g(tu)$ is continuous in t, for all $u \in \mathcal{G}$, then $\mathcal{A}$ is linear.

Based on the above said outcome, one can finalize that the additive functional equation $g(u + v) = g(u) + g(v)$ has **Hyers–Ulam stability or ϵ-stability** on $(\mathcal{X}, \mathcal{Y})$. In the above Theorem 10.1, an additive function $\mathcal{A}$ is created directly from the given function g which also fulfills (10.3) and it is the most dominant technique to investigate the stability of several functional equations. Hyers theorem was indiscriminated by Aoki [4] in 1950 for additive mappings.

10.2.2 Stability Involving Sum of Powers of Norms

After Hyers gave a positive answer to Ulam's question, a huge number of papers have been brought out by publishing in association with various simplifications of Ulam's problem and Hyers theorem.

Since there is no elucidation for the boundedness of Cauchy difference $g(x + y) - g(x) - g(y)$ in the expression of (10.1), in the year 1978, Rassias [5] tried to deteriorate the stipulation for the Cauchy difference and thrived in establishing what is now known to be the Hyers–Ulam–Rassias stability pertinent to the additive Cauchy equation. This jargon is reasonable because the theorem of Rassias has strongly persuaded many mathematicians studying stability problems of functional equation. In fact, Rassias proved the ensuing theorem.

Theorem 10.2 [5] Let X and Y be Banach spaces, let $0 < k < \infty$ and $0 \le \alpha < 1$. If a function $f : X \to Y$ satisfies

$$\|f(x + y) - f(x) - f(y)\| \le k(\|x\|^\alpha + \|y\|^\alpha) \tag{10.4}$$

for all $x, y \in X$, then there is a unique additive mapping $A : X \to Y$ such that

$$\|f(x) - A(x)\| \le \frac{2k}{2 - 2^\alpha} \|x\|^\alpha \tag{10.5}$$

for all $x \in X$. Furthermore, if $f(tx)$ is continuous in t, for all $x \in X$, then A is linear.

Rassias noticed that the proof of this theorem also works for $p < 0$ and asked whether such a theorem can also be proved for $p \ge 1$. Gajda [6] answered the question of Rassias for the case $p > 1$ by a slight modification of the expression in (10.2). His idea to demonstrate the theorem for this case is to replace n by $-n$ in the formula (10.2).

Hence, Theorem 10.2 cannot be extended for the only critical value of $p = 1$. Gajda [6] showed that the theorem is false for $p = 1$ by constructing the succeeding counter-example. Hello
For a fixed $\theta > 0$ and $\mu = \frac{1}{6}\theta$, define a function $f : \mathbb{R} \to \mathbb{R}$ by

$$f(x) = \sum_{n=0}^{\infty} 2^{-n} \phi(2^n x),$$

where the function $\phi : \mathbb{R} \to \mathbb{R}$ is given by

$$\phi(x) = \begin{cases} \mu & \text{for } x \in [1, \infty), \\ \mu x & \text{for } x \in (-1, 1), \\ -\mu & \text{for } x \in (-\infty, -1]. \end{cases}$$

Then, the function f serves as a counter-example for $p = 1$ as presented in the ensuing theorem.

Theorem 10.3 [6] The function f defined above satisfies

$$|f(x + y) - f(x) - f(y)| \leq \theta(|x| + |y|) \tag{10.6}$$

for all $x, y \in \mathbb{R}$, while there is no constant $\delta \geq 0$ and no additive function $A : \mathbb{R} \to \mathbb{R}$ satisfying the condition

$$|f(x) - A(x)| \leq \delta|x| \tag{10.7}$$

for all $x, y \in \mathbb{R}$.

10.2.3 Stability Involving Product of Powers of Norms

In 1982, Rassias [7] provided a further generalization of the result of Hyers and established a theorem using weaker conditions bounded by product of powers of norms, which is presented in the subsequent theorem:

Theorem 10.4 [7] Let $f : X \to Y$ be a mapping from a normed vector space X into a Banach space Y subject to the inequality

$$\|f(x + y) - f(x) - f(y)\| \leq k\|x\|^{\alpha}\|y\|^{\alpha} \tag{10.8}$$

for all $x, y \in X$, where k and α are constants with $k > 0$ and $0 \leq \alpha < \frac{1}{2}$. Then the limit

$$A(x) = \lim_{n \to \infty} \frac{1}{2^n} f(2^n x) \tag{10.9}$$

occurs for all $x \in X$ and $T : X \to Y$ is the unique additive mapping which satisfies

$$\|f(x) - T(x)\| \leq \frac{k}{2 - 2^{2\alpha}}\|x\|^{2\alpha} \tag{10.10}$$

for all $x \in X$. If $\alpha < 0$, then the inequality (10.8) holds for $x, y \neq 0$ and (10.10) for $x \neq 0$. If $\alpha > \frac{1}{2}$, then the inequality (10.8) holds for $x, y \in X$ and the limit

$$T(x) = \lim_{n \to \infty} 2^n f\left(\frac{x}{2^n}\right) \tag{10.11}$$

exists for all $x \in X$ and $T : X \to Y$ is the unique additive mapping which satisfies

$$\|f(x) - T(x)\| \leq \frac{k}{2^{2\alpha} - 2}\|x\|^{2\alpha} \tag{10.12}$$

for all $x \in X$. Moreover, if $f : X \to Y$ is a mapping such that the transformation $t \to f(tx)$ is continuous in $t \in \mathbb{R}$ for every fixed $x \in X$, then T is $\mathbb{R}$-linear mapping.

This type of stability containing product of powers of norms is identified as Ulam–Gavruta–Rassias stability by Bouikhalene and Elquorachi [8], Nakmahachalasint [9, 10], Park and Najati [11], Pietrzyk [12], and Sibaha *et al.* [13].

10.2.4 Stability Involving a General Control Function

In 1994, a further generalization of the Rassias' theorem was obtained by Gavruta [14] who replaced the bounds $k(\|x\|^p + \|y\|^p)$ and $k\|x\|^p\|y\|^p$ by a general control function $\varphi(x, y)$. The ensuing theorem provides his result.

Theorem 10.5 **[14]** Let G and H be an abelian group and a Banach space, respectively, and let $\phi : G^2 \to [0, \infty)$ be a function satisfying

$$\Phi(x, y) = \sum_{k=0}^{\infty} 2^{-k-1} \varphi(2^k x, 2^k y) < \infty \tag{10.13}$$

for all $x, y \in G$. If a function $f : G \to E$ satisfies the inequality

$$\|f(x + y) - f(x) - f(y)\| \leq \phi(x, y) \tag{10.14}$$

for all $x, y \in G$, then there exists a unique additive function $T : G \to H$ with

$$\|f(x) - T(x)\| \leq \Phi(x, x)$$

for all $x \in G$. Moreover, suppose $f(tx)$ is continuous in t, for all $x \in G$, then T is a linear function.

This type of stability is celebrated as **Generalized Hyers–Ulam–Rassias stability**.

10.2.5 Stability Involving Mixed Product–Sum of Powers of Norms

In 2008, Ravi *et al.* [15] investigated the stability of a new quadratic functional equation

$$Q(kx_1 + x_2) + Q(kx_1 - x_2)$$
$$= 2Q(x_1 + x_2) + 2Q(x_1 - x_2) + 2(k^2 - 2)Q(x_1) - 2Q(x_2)$$

for any arbitrary but fixed real constant k with $k \neq 0$; $k \neq \pm 1$; $k \neq \pm\sqrt{2}$ using mixed product-sum of powers of norms in the succeeding theorem.

Theorem 10.6 Let $g : X \to Y$ be a mapping that satisfies the inequality

$$\|g(mu + v) + g(mu - v) - 2g(u + v) - 2g(u - v) - 2(k^2 - 2)g(u) + 2g(v)\|_Y$$
$$\leq \delta\{ \|u\|_X^\alpha \|v\|_X^\alpha + (\|u\|_X^{2\alpha} + \|v\|_X^{2\alpha}) \} \tag{10.15}$$

for all $u, v \in X$ with $u \perp v$, where δ and *alpha* are constants with $\delta, \alpha > 0$ and either $k > 1$; $\alpha < 1$ or $k < 1$; $\alpha > 1$ with $k \neq 0$; $k \neq \pm 1$; $k \neq \pm\sqrt{2}$ and $-1 \neq |k|^{\alpha-1} < 1$. Then the limit

$$Q(u) = \lim_{n\to\infty} \frac{g(m^n u)}{m^{2n}} \tag{10.16}$$

exists for all $u \in X$ and $Q : X \to Y$ is the unique orthogonally Euler–Lagrange quadratic mapping such that

$$\|g(u) - Q(u)\|_Y \leq \frac{\delta}{2|k^2 - k^{2\alpha}|} \|u\|_X^{2\alpha}, \tag{10.17}$$

for all $u \in X$.

The above-mentioned stability is acknowledged as John M. Rassias stability including mixed product–sum of powers of norms by Ravi *et al.* [15, 16].

10.2.6 Application of Ulam Stability Theory

Several mathematicians have noticed interesting applications of the Hyers–Ulam–Rassias stability theory to various types of mathematical problems. Stability theory is useful in the study of nonlinear analysis, particularly in fixed point theory. In nonlinear analysis it is a well-known fact that finding the expression of the asymptotic derivative of a nonlinear operator can be a difficult problem. In this sense, how the Hyers–Ulam–Rassias stability theory can be used to evaluate the asymptotic derivative of some nonlinear operators is explained.

Jung [17] proved the Hyers–Ulam stability for Jensen's equation on a restricted domain and his result is applied for studying an interesting property of additive mappings.

The stability properties of various functional equations are applied in many unrelated fields. For example, Zhou [18] used the stability result of the functional equation $g(x + y) + g(x - y) = 2g(x)$ to show a conjecture of Ditzian about the relationship between the smoothness of a mapping and the degree of its approximation by the related Bernstein polynomials. These stability results are applied in stochastic analysis [19], financial and actuarial mathematics, psychology, and sociology.

10.3 Various Forms of Functional Equations

The functional equation

$$Q(x_1 + x_2) + Q(x_1 - x_2) = 2Q(x_1) + 2Q(x_2) \tag{10.18}$$

is called as a quadratic functional equation since the quadratic function $Q(x) = \alpha x^2$ is a solution of (10.18) [20]. The solution of the quadratic functional equation is known as a quadratic mapping. A quadratic functional equation was used to describe inner product spaces. A square norm on an inner product space fulfills the well-known parallelogram law:

$$\|x_1 + x_2\|^2 + \|x_1 - x_2\|^2 = 2\|x_1\|^2 + 2\|x_2\|^2.$$

The quadratic functional equation (10.18) holds good in the above parallelogram law.

A mapping $Q : X \to Y$ between vector spaces is quadratic if and only if there exists a unique symmetric biadditive mapping $B : X \times X \to Y$ so that $Q(x_1) = B(x_1, x_1)$ for all $x_1 \in X$, where the mapping B is defined by

$$B(x_1, x_2) = \frac{Q(x_1 + x_2) - Q(x_1 - x_2)}{4}, \quad \text{for all } x_1, x_2 \in X.$$

Chang and Kim [21] solved the succeeding quadratic functional equations

$$f(2x_1 + x_2) + f(2x_1 - x_2) = f(x_1 + x_2) + f(x_1 - x_2) + 6f(x_1) \tag{10.19}$$

and

$$f(2x_1 + x_2) + f(x_1 + 2x_2) = 4f(x_1 + x_2) + f(x_1) + f(x_2) \tag{10.20}$$

for their general solutions and obtained their Hyers–Ulam stability.

Park *et al.* [22] solved the following quadratic functional equation

$$f(x_1 + 2x_2) + f(x_1 - 2x_2) = 2f(x_1) + 8f(x_2) \tag{10.21}$$

and proved Hyers–Ulam stability of (10.21).

Rassias [23] introduced Euler–Lagrange type quadratic functional equation of the form

$$f(ax + by) + f(bx - ay) = (a^2 + b^2)(f(x) + f(y)) \tag{10.22}$$

motivated from the following pertinent algebraic equation

$$|ax + by|^2 + |bx - ay|^2 = (a^2 + b^2)(|x|^2 + |y|^2). \tag{10.23}$$

The solution of the functional equation (10.22) is called an Euler–Lagrange quadratic type mapping. In addition, Rassias [23–27] generalized the standard quadratic equation to the following quadratic equation

$$m_1 m_2 |a_1 x_1 + a_2 x_2|^2 + |m_2 a_2 x_1 - m_1 a_1 x_2|^2$$
$$= (m_1 |a_1|^2 + m_2 |a_2|^2)(m_2 |x_1|^2 + m_1 |x_2|^2).$$

He introduced and investigated the general pertinent Euler–Lagrange quadratic mappings. These Euler–Lagrange mappings are named Euler–Lagrange–Rassias mappings, and the corresponding Euler–Lagrange equations are called Euler–Lagrange–Rassias equations [9, 12, 28, 29]. These notions provide a cornerstone in analysis, because of their particular interest in probability theory and stochastic analysis in marrying these fields of research to functional equations via the pioneering introduction of the

Euler–Lagrange–Rassias quadratic weighted means and fundamental mean equations [25, 26, 28].

Quadratic functional equations of various forms are dealt by many mathematicians and one can see [16, 26, 30–40].

Rassias [41] introduced the cubic functional equation

$$f(x_1 + 2x_2) - 3f(x_1 + x_2) + 3f(x_1) - f(x_1 - x_2) = 6f(x_2) \qquad (10.24)$$

and obtained its general solution and proved its Hyers–Ulam stability problem. It is easy to verify that the cubic function $f(x) = ax^3$ satisfies (10.24). Every solution of the cubic functional equation (10.24) is said to be a cubic mapping.

Recently polynomial equations are applied in self-testing, self-correcting, and approximate checking of computer programs that compute polynomials. This concept triggered to study various types of cubic functional equations.

Jun and Kim [42] brought out the solution of the familiar cubic functional equation

$$C(2x_1 + x_2) + C(2x_1 - x_2) = 2C(x_1 + x_2) + 2C(x_1 - x_2) + 12C(x_1) \qquad (10.25)$$

and obtained its generalized Hyers–Ulam–Rassias stability.

A mapping $C : A \rightarrow B$ between vector spaces satisfies (10.25) if and only if there exists a mapping $S : A \times A \times A \rightarrow B$ such that $C(x_1) = S(x_1, x_1, x_1)$, $\forall x_1 \in A$ and S, which is symmetric for each fixed variable and additive for each fixed two variables.

Several other types of cubic functional equations were dealt by many authors in [43–50].

The quartic functional equation

$$Q(x_1 + 2x_2) + Q(x_1 - 2x_2)$$
$$= 4Q(x_1 + x_2) + 4Q(x_1 - x_2) + 6Q(x_1) + 24Q(x_2) \qquad (10.26)$$

was introduced by Rassias [51]. It is easy to find that the function $Q(x) = ax^4$ is a solution of (10.26). The solution of the quartic functional equation is said to be a quartic mapping.

The general solution of (10.26) is obtained without supposing any regularity conditions on the unknown function (refer [51]). The function $Q : \mathbb{R} \rightarrow \mathbb{R}$ is a solution of (10.26) if and only if $Q(x_1) = B(x_1, x_1, x_1, x_1)$, where the function $B : \mathbb{R}^4 \rightarrow \mathbb{R}$ is symmetric and additive in each variable. Since the solution of equation (10.26) is even, we can rewrite (10.26) as

$$Q(2x_1 + x_2) + Q(2x_1 - x_2)$$
$$= 4Q(x_1 + x_2) + 4Q(x_1 - x_2) + 24Q(x_1) - 6Q(x_2). \qquad (10.27)$$

The solution of the quartic functional equation (10.26) is said to be a quartic mapping.

Lee *et al.* [52] gained the general solution and investigated the Hyers–Ulam stability of the quartic functional equation (10.27).

Petapirak and Nakmahachalasint [53] applied a different method to obtain the general solution of the quartic functional equation

$$Q(3x_1 + x_2) + Q(x_1 + 3x_2)$$
$$= 64Q(x_1) + 64Q(x_2) + 24Q(x_1 + x_2) - 6Q(x_1 - x_2) \tag{10.28}$$

by proving the succeeding theorem.

Theorem 10.7 [53] A mapping $Q : A \to B$ between vector spaces satisfies the functional equation (10.28) if and only if there exists a 4-additive mapping $S_4 : A \times A \times A \times A \to B$ which symmetric such that $Q(y) = S_4(y, y, y, y), \forall y \in A$.

The results connected with solution and Hyers–Ulam stability of other type of quartic functional equations are available in [28, 54–57].

In 2010, Xu *et al.* [58] achieved the general solution and proved the stability of the quintic functional equation

$$f(x + 3y) - 5f(x + 2y) + 10f(x + y) - 10f(x) + 5f(x - y)$$
$$- f(x - 2y) = 120f(y) \tag{10.29}$$

and the sextic functional equation

$$f(x + 3y) - 6f(x + 2y) + 15f(x + y) - 20f(x)$$
$$+ 15f(x - y) - 6f(x + 2y) + f(x - 3y) = 720f(y) \tag{10.30}$$

in quasi-β-normed spaces using fixed point method.

Consider the following mixed type functional equation deriving from quadratic and additive functions

$$f(u + 2v) + f(u - 2v) + 4f(u) = 3[f(u + v) + f(u - v)] + f(2v) - 2f(v). \tag{10.31}$$

It is easy to see that the function $f(u) = \alpha u^2 + \beta u$ is a solution of the functional equation (10.31).

Jun and Kim [59] accomplished the solution and investigated Hyers–Ulam stability of the mixed type additive-quadratic functional equation

$$h(x_1 + 2x_2) + 2h(x_1 - x_2) = h(x_1 - 2x_2) + 2h(x_1 + x_2). \tag{10.32}$$

Rassias *et al.* [60] solved for the solution and studied Hyers–Ulam stability of the mixed type of additive–cubic functional equation

$$3A(u + v + w) + A(-u + v + w)$$
$$+ A(u - v + w) + A(u + v - w) + 4[A(u) + A(v) + A(w)]$$
$$= 4[A(u + v) + A(u + w) + A(v + w)]. \tag{10.33}$$

The solution of (10.33) is $A(u) = S(u, u, u) + T(u)$, where S is symmetric for each fixed variable and is additive for fixed two variables, and T is additive. The function $A(u) = \alpha u + \beta u^3$ is a solution of (10.33).

Eshaghi Gordji [61] obtained the general solution of mixed type additive-quartic functional equation

$$g(2x_1 + x_2) + g(2x_1 - x_2)$$
$$= 4g(f(x_1 + x_2) + g(x_1 - x_2)) - \frac{3}{7}(g(2x_2) - 2g(x_2)) + 2g(2x_1) - 8g(x_1)$$

(10.34)

as $g(x_1) = M(x_1, x_1, x_1, x_1) + A(x_1)$, where M is a multi-additive symmetric function and A is an additive function. The function $g(x_1) = \alpha x_1 + \beta x_1^4$ is a solution of (10.34).

Towanlong and Nakmachalasint [62] investigated the general solution of the mixed type quadratic-cubic functional equation

$$f(x_1 + 3x_2) - 3f(x_1 + 2x_2) + 3f(x_1 + x_2) - f(x_1) = 3f(x_2) - 3f(-x_2).$$

(10.35)

It is easy to see that the mixed type function $f(x_1) = c_1 x_1^2 + c_2 x_1^3$ fulfills (10.35).

Eshaghi Gordji et al. [63] solved the following mixed type quadratic-quartic functional equation

$$f(2u + v) + f(2u - v) = 4[f(u + v) + f(u - v)] + 2[f(2u) - 4f(u)] - 6f(v)$$

(10.36)

for its general solution and investigated its Hyers–Ulam stability in random normed spaces. The function $f(u) = \alpha u^2 + \beta u^4$ is a solution of (10.36).

Eshaghi Gordji et al. [64] acquired the general solution and investigated HyersUlam–Rassias stability of the following mixed type cubic–quartic functional equation

$$g(u + 2v) + g(2u - v) = 4[g(u + v) + g(u - v)] - 24g(u) - 6g(u) + 3g(2v)$$

(10.37)

in quasi-Banach spaces. The function $g(u) = \alpha u^3 + \beta u^4$ is a solution of (10.37).

Jun and Kim [65] acquired the general solution of mixed type additive-quadratic-cubic functional equation

$$q(x_1 + 2x_2) + q(x_1 - 2x_2) + 6q(x_1) = 4q(x_1 + x_2) + 4q(x_1 - x_2) \quad (10.38)$$

by proving the following theorem.

Theorem 10.8 **[65]** A function $q : X \to Y$ between vector spaces satisfies the functional equation (10.38) if and only if there exist functions $S : X \times X \times X \to Y$, $M : X \times X \to Y$, $T : X \to Y$ and a constant $k \in Y$ such that $q(u) = S(u, u, u) + M(u, u) + T(u) + k$, $\forall u \in X$, where S is symmetric, for each fixed variable and is additive for fixed two variables, M is symmetric biadditive and T is additive.

It is easy to note that the function $q(x) = \alpha x + \beta x^2 + \gamma x^3$ is a solution of (10.38).

Park [66] obtained the general solution of the mixed type additive-quadratic-quartic functional equation

$$f(x_1 + 2x_2) + f(x_1 - 2x_2) = 2f(x_1 + x_2) + 2f(-x_1 - x_2) + 2f(x_1 - x_2)$$
$$+ 2f(x_2 - x_1) - 4f(-x_1) - 2f(x_1) + f(2x_2)$$
$$+ f(-2x_2) - 4f(x_2) - 4f(-x_2) \quad (10.39)$$

and proved its Hyers–Ulam stability. The function $f(x_1) = ax_1 + bx_1^2 + cx_1^4$ is a solution of (10.39).

Eshaghi Gordji *et al.* [67] solved and studied Hyers–Ulam stability of the mixed type additive–cubic–quartic functional equation

$$11[g(u + 2v) + g(u - 2v)] = 44[g(u + v) + g(u - v) + 12g(3v)]$$
$$- 48g(2v) + 60g(v) - 66g(u). \quad (10.40)$$

The function $g(u) = \alpha u + \beta u^3 + \gamma u^4$ is a solution of (10.40). A lot of mixed type functional equations were dealt by many researchers and there are many interesting results concerning these equations [38, 68–76].

In 1996, Isac and Rassias [77] were the first to provide applications of the stability theory of functional equations for the proof of new fixed point theorems. The stability problems of several various functional equations have been extensively investigated by a number of authors using fixed point methods [22, 76, 78, 79].

Several stability results have recently been obtained for various functional equations and functional inequalities, also for mappings with more general domains and ranges [8, 21, 36, 38, 64, 73, 80–82]. Many research monographs are also available on functional equations, one can see [1, 34, 83–87].

10.4 Preliminaries

In this section, we recall some preliminary notions associated with Banach spaces.

Definition 10.1 A normed space is a vector space $\mathcal{X}$ over a field $\mathbb{K}$ together with a function (the norm) $\|\cdot\| : \mathcal{X} \to \mathbb{R}$ satisfying

(i) $\|u\| = 0$ if and only if $u = 0$;
(ii) $\|\rho u\| = |\rho| \|u\|, \forall u \in \mathcal{X}$ and $\rho \in \mathbb{K}$;
(iii) $\|u + v\| \leq \|u\| + \|v\|$.

Definition 10.2 A sequence $\{u_n\}$ is a Cauchy sequence if for all $\epsilon > 0$, there exists $n(\epsilon)$ such that for all $m, n \geq n(\epsilon)$, we have

$$\|u_m - u_n\| \leq \epsilon.$$

Definition 10.3   A normed space is complete if every Cauchy sequence $\{u_n\}$ converges, that is, there exists $u \in \mathcal{X}$ with

$$\|u_n - u\| \to 0, \quad \text{as } n \to \infty.$$

Definition 10.4   A complete normed space is called a Banach space.

We recall some fundamental notions in association with quasi-β-normed spaces and m-additive symmetric mappings.

Let β be a fixed real number with $0 < \beta \le 1$ and let $\mathbb{K}$ denote either $\mathbb{R}$ or $\mathbb{C}$.

Definition 10.5   Let $\mathcal{X}$ be a linear space over $\mathbb{K}$. A quasi-β-norm $\|\cdot\|$ is a real-valued function on $\mathcal{X}$ satisfying the following conditions:

(i) $\|a\| \ge 0$ for all $a \in \mathcal{X}$ and $\|a\| = 0$ if and only if $a = 0$.

(ii) $\|\eta a\| = |\eta|^\beta \cdot \|a\|$ for all $\eta \in \mathbb{K}$ and all $a \in \mathcal{X}$.

(iii) There is a constant $K \ge 1$ such that

$$\|a + b\| \le K(\|a\| + \|b\|) \quad \text{for all } a, b \in \mathcal{X}.$$

The pair $(\mathcal{X}, \|\cdot\|)$ is called quasi-β-normed space if $\|\cdot\|$ is a quasi-β-norm on $\mathcal{X}$. The smallest possible K is called the modulus of concavity of $\|\cdot\|$.

Definition 10.6   A complete quasi-β-normed space is called a quasi-β-Banach space.

Definition 10.7   A quasi-β-norm $\|\cdot\|$ is called a (β, p)-norm $(0 < p < 1)$ if

$$\|x + y\|^p \le \|x\|^p + \|y\|^p$$

for all $x, y \in \mathcal{X}$. In this case, a quasi-β-Banach space is called a (β, p)-Banach space.

Let us recall basic concepts of non-Archimedean space.

A non-Archimedean field is a field $\mathbb{A}$ equipped with a function (valuation) $|\cdot|$ from $\mathbb{A}$ into $[0, \infty)$ such that for all $r, s \in \mathbb{A}$,

(i) $|r| = 0$ if and only if $r = 0$,

(ii) $|rs| = |r||s|$,

(iii) $|r + s| \le \max\{|r|, |s|\}$.

Clearly $|1| = |-1| = 1$ and $|n| \le 1$ for all $n \in \mathbb{N}$.

We always assume, in addition, that $|\cdot|$ is nontrivial, that is, there subsists an $a_0 \in \mathbb{A}$ such that $|a_0| \ne 0, 1$.

An example of a non-Archimedean valuation is the mapping $|\cdot|$ taking everything but 0 into 1 and $|0| = 0$. This valuation is called trivial. Another example of a non-Archimedean valuation on a field $\mathbb{A}$ is the mapping

$$
|r| =
\begin{cases}
0 & \text{if } r = 0, \\[2mm]
\dfrac{1}{r} & \text{if } r > 0, \\[2mm]
-\dfrac{1}{r} & \text{if } r < 0
\end{cases}
$$

for any $r \in \mathbb{A}$.

Let p be a prime number. For any nonzero rational number $x = p^r \frac{m}{n}$ in which m and n are coprime to the prime number p. Consider the p-adic absolute value $|x|_p = p^{-r}$ on $\mathbb{Q}$. It is easy to check that $|\cdot|$ is a non-Archimedean norm on $\mathbb{Q}$. The completion of $\mathbb{Q}$ with respect to $|\cdot|$ which is denoted by $\mathbb{Q}_p$ is said to be the p-adic number field. Note that if $p > 2$, then $|2^n| = 1$ for all integers n.

10.5 Rational Functional Equations

10.5.1 Reciprocal Type Functional Equation

For the first time, Ravi and Senthil Kumar [88] investigated the generalized Hyers–Ulam stability for the reciprocal type functional equation

$$
r(x + y) = \frac{r(x)r(y)}{r(x) + r(y)}, \tag{10.41}
$$

where $r : X \to \mathbb{R}$ is a mapping with X is the space of nonzero real numbers, $x + y \neq 0$ and $r(x) + r(y) \neq 0$ for all $x, y \in X$.

Definition 10.8 Let X be the space of nonzero real numbers. Then a mapping $r : X \to \mathbb{R}$ defined as $r(x) = \frac{c}{x}$, c being a constant, is called reciprocal, if the reciprocal functional equation (10.41) holds with $x + y \neq 0$, $r(x) \neq -r(y)$ and $r(x) \neq 0$, for all $x, y \in X$.

Note that the mapping r is called reciprocal because the following algebraic identity

$$
\frac{1}{x + y} = \frac{\frac{1}{x}\frac{1}{y}}{\frac{1}{x} + \frac{1}{y}}
$$

holds for all $x, y \in X$.

10.5.2 Solution of Reciprocal Type Functional Equation

In the following theorem, we assume $r(3x) \neq 0$, $r(-x) \neq 0$, $2r(x) + r(y) \neq 0$, $2r(x) + r(-y) \neq 0$, for all $x, y \in \mathbb{R} - \{0\}$ (10.41).

Theorem 10.9 Let r be a real-valued function of a nonzero real variable satisfying the reciprocal type functional equation (10.41). Then, $r(x)$ is of the form $r(x) = \frac{c}{x}$, for all $x \in \mathbb{R} - \{0\}$, where c is a constant.

Proof: Replacing (x, y) by (x, x) in (10.41), we obtain

$$r(2x) = \frac{1}{2}r(x) \tag{10.42}$$

for all $x \in \mathbb{R} - \{0\}$. Similarly, we obtain $r(3x) = \frac{1}{3}r(x)$, for all $x \in \mathbb{R} - \{0\}$. Now, replacing (x, y) by $(x, 2y)$ in (10.41) and using (10.42), we get

$$r(x + 2y) = \frac{r(x)r(y)}{2r(x) + r(y)} \tag{10.43}$$

for all $x, y \in \mathbb{R} - \{0\}$. Substituting $(x, y) = (x, -2y)$ in (10.41) and using (10.42), we obtain

$$r(x - 2y) = \frac{r(x)r(-y)}{2r(x) + r(-y)} \tag{10.44}$$

for all $x, y \in \mathbb{R} - \{0\}$. Equation (10.43) divided by (10.44) gives

$$\frac{r(x + 2y)}{r(x - 2y)} = \frac{r(y)[2r(x) + r(-y)]}{r(-y)[2r(x) + r(y)]} \tag{10.45}$$

for all $x, y \in \mathbb{R} - \{0\}$. Now, replacing (x, y) by $(x, -x)$ in (10.45) to get

$$\frac{r(-x)}{r(3x)} = \frac{r(-x)[2r(x) + r(x)]}{r(3x)[2r(x) + r(-x)]}$$

for all $x \in \mathbb{R} - \{0\}$, which on further simplification yields $r(-x) = -r(x)$, for all $x \in \mathbb{R} - \{0\}$, which shows that r is an odd function. From (10.41) and using induction on a positive integer n, it is easy to show that

$$r\left(\sum_{i=1}^{n} x_i\right) = \frac{\prod_{i=1}^{n} r(x_i)}{\sum_{i=1}^{n} \left(\prod_{j=1, j \neq i}^{n} r(x_j)\right)}. \tag{10.46}$$

Replacing x_i by x for $i = 1, 2, \ldots, n$ in (10.46), we get

$$r(nx) = \frac{1}{n}r(x) \tag{10.47}$$

for $x \in \mathbb{R} - \{0\}$, where n is a positive integer. Replacing x by $\frac{x}{n}$ in (10.47), we obtain

$$r\left(\frac{x}{n}\right) = nr(x) \tag{10.48}$$

for $x \in \mathbb{R} - \{0\}$, where n is a positive integer. Now, replacing x by $-x$ in (10.47) and (10.48) and using oddness of r, we get

$$r(-nx) = -\frac{1}{n}r(x), \tag{10.49}$$

$$r\left(-\frac{x}{n}\right) = -nr(x), \tag{10.50}$$

respectively, for $x \in \mathbb{R} - \{0\}$, where n is a positive integer. When $x = 1$, (10.47) and (10.48) become

$$r(n) = \frac{1}{n}r(1) = \frac{c}{n}, \tag{10.51}$$

$$r\left(\frac{1}{n}\right) = nr(1) = nc, \tag{10.52}$$

respectively, for some constant $c = r(1)$, where n is a positive integer. Similarly, when $x = 1$, (10.49) and (10.50) become

$$r(-n) = -\frac{1}{n}r(1) = -\frac{c}{n}, \tag{10.53}$$

$$r\left(-\frac{1}{n}\right) = -nr(1) = -nc, \tag{10.54}$$

respectively, for some constant $c = r(1)$, where n is a positive integer. Hence, we conclude that

$$r(m) = \frac{1}{m}r(1) = \frac{c}{m} \quad \text{and} \quad r\left(\frac{1}{m}\right) = mr(1) = mc$$

for some constant $c = r(1)$ and $m \in \mathbb{Z} - \{0\}$.

Next, let $k = \frac{m}{n}$ be any rational number, where $m, n \in \mathbb{Z} - \{0\}$. Then (10.47)–(10.50) yield

$$r(kx) = r\left(m.\frac{x}{n}\right) = \frac{1}{m}r\left(\frac{x}{n}\right) = \frac{n}{m}r(x) = \frac{1}{k}r(x) \tag{10.55}$$

for $x \in \mathbb{R} - \{0\}, k \in \mathbb{Q}$. When $x = 1$, (10.55) becomes

$$r(k) = \frac{1}{k}r(1) = \frac{c}{k}, \quad \text{for some constant } c = r(1).$$

Note that $r(1) = c \neq 0$, otherwise it will lead to constant function. Hence, we conclude that

$$r(x) = \frac{c}{x}, \quad \text{for all } x \in \mathbb{R} - \{0\}. \qquad \square$$

10.5.3 Generalized Hyers–Ulam Stability of Reciprocal Type Functional Equation

We assume that $x + y \neq 0$, $g(x) \neq 0$ and $g(x) + g(y) \neq 0$ for all $x, y \in X$ in the following theorems. We also assume that X as the space of nonzero real numbers in the following results (10.41).

Theorem 10.10   Let $g : X \to \mathbb{R}$ be a mapping satisfying

$$|Dg(x, y)| \le \phi(x, y) \tag{10.56}$$

for all $x, y \in X$, where $\phi : X \times X \to \mathbb{R}$ be a given function such that

$$\Phi(x) = \sum_{i=0}^{\infty} \frac{1}{2^i} \phi\left(\frac{x}{2^{i+1}}, \frac{x}{2^{i+1}}\right) \tag{10.57}$$

with the condition

$$\lim_{n \to \infty} \frac{1}{2^n} \phi\left(\frac{x}{2^{n+1}}, \frac{x}{2^{n+1}}\right) = 0 \tag{10.58}$$

holds for all $x \in X$. Then, there exists a unique reciprocal mapping $r : X \to \mathbb{R}$, which satisfies (10.41) and the inequality

$$|g(x) - r(x)| \le \Phi(x) \tag{10.59}$$

for all $x \in X$.

Proof: Replacing (x, y) by $\left(\frac{x}{2}, \frac{x}{2}\right)$ in (10.56), we obtain

$$\left| g(x) - \frac{1}{2} g\left(\frac{x}{2}\right) \right| \le \phi\left(\frac{x}{2}, \frac{x}{2}\right) \tag{10.60}$$

for all $x \in X$. Again replacing x by $\frac{x}{2}$ in (10.60), dividing by 2 and summing the resulting inequality with (10.60), we obtain $\left| g(x) - \frac{1}{2^2} g\left(\frac{x}{2^2}\right) \right| \le \sum_{i=0}^{1} \frac{1}{2^i} \phi\left(\frac{x}{2^{i+1}}, \frac{x}{2^{i+1}}\right)$ for all $x \in X$. Proceeding further and using induction on a positive integer n, we obtain

$$|g(x) - 2^{-n} g(2^{-n} x)| \le \sum_{i=0}^{\infty} \frac{1}{2^i} \phi\left(\frac{x}{2^{i+1}}, \frac{x}{2^{i+1}}\right) \tag{10.61}$$

for all $x \in X$. In order to prove the convergence of the sequence $\{2^{-n} g(2^{-n} x)\}$, replace x by $2^{-p} x$ in (10.61) and divide by 2^p, we find that for $n > p > 0$

$$|2^{-p} g(2^{-p} x) - 2^{-n-p} g(2^{-n-p} x)| = 2^{-p} |g(2^{-p} x) - 2^{-n} g(2^{-n-p} x)|$$

$$\le \sum_{i=0}^{\infty} \frac{1}{2^{p+i}} \phi\left(\frac{x}{2^{p+i+1}}, \frac{x}{2^{p+i+1}}\right)$$

$$\to 0 \text{ as } p \to \infty.$$

Thus, the sequence $\{2^{-n} g(2^{-n} x)\}$ is a Cauchy sequence. Allowing $n \to \infty$ in (10.61), we arrive (10.59) with $r(x) = \lim_{n \to \infty} 2^{-n} g(2^{-n} x)$. To show that r satisfies (10.41), replacing (x, y) by $(2^{-n} x, 2^{-n} y)$ in (10.56) and dividing by 2^n, we obtain

$$2^{-n} |Dg(2^{-n} x, 2^{-n} y)| \le 2^{-n} \phi(2^{-n} x, 2^{-n} y). \tag{10.62}$$

Allowing $n \to \infty$ in (10.62), we see that r satisfies (10.41) for all $x, y \in X$. To prove r is a unique reciprocal mapping satisfying (10.41). Let $r_1 : X \to \mathbb{R}$ be another reciprocal mapping that satisfies (10.41) and the inequality (10.59). Clearly $r_1(2^{-n}x) = 2^n r_1(x)$, $r(2^{-n}x) = 2^n r(x)$ and using (10.59), we arrive

$$|r_1(x) - r(x)| \leq 2^{-n}|r_1(2^{-n}x) - r(2^{-n}x)|$$

$$\leq 2^{-n}(|r_1(2^{-n}x) - g(2^{-n}x)| + |g(2^{-n}x) - r(2^{-n}x)|)$$

$$\leq 2 \sum_{i=0}^{\infty} \frac{1}{2^{n+i}} \phi\left(\frac{x}{2^{n+i+1}}, \frac{x}{2^{n+i+1}}\right)$$

$$\to 0 \text{ as } n \to \infty,$$

which implies that r is unique. This completes the proof of Theorem 10.10. $\square$

Theorem 10.11 Let $g : X \to \mathbb{R}$ be a mapping satisfying (10.56), for all $x, y \in X$, where $\phi : X \times X \to \mathbb{R}$ be a given function such that

$$\Phi(x) = \sum_{i=0}^{\infty} 2^{i+1} \phi(2^i x, 2^i x) \tag{10.63}$$

with the condition

$$\lim_{n \to \infty} 2^{n+1} \phi(2^n x, 2^n x) = 0 \tag{10.64}$$

holds for every $x \in X$. Then, there exists a unique reciprocal mapping $r : X \to \mathbb{R}$ which satisfies (10.41) and the inequality

$$|g(x) - r(x)| \leq \Phi(x) \tag{10.65}$$

for all $x \in X$.

Proof: The proof is obtained by replacing (x, y) by (x, x) in (10.56), multiplying by 2 and proceeding by similar arguments as in Theorem 10.10. $\square$

Corollary 10.1 Let $g : X \to \mathbb{R}$ be a mapping and let there exist real numbers $p \neq -1$ and $c_1 \geq 0$ such that

$$|Dg(x, y)| \leq c_1(|x|^p + |y|^p) \tag{10.66}$$

for all $x, y \in X$. Then there exists a unique reciprocal mapping $r : X \to \mathbb{R}$ satisfying (10.41) and

$$|g(x) - r(x)| \leq \begin{cases} \dfrac{4c_1}{2^{p+1} - 1}|x|^p, & \text{for } p > -1, \\[3mm] \dfrac{4c_1}{1 - 2^{p+1}}|x|^p, & \text{for } p < -1 \end{cases} \tag{10.67}$$

for every $x \in X$.

Proof: The proof follows immediately by taking $\phi(x, y) = c_1(|x|^p + |y|^p)$, for all $x, y \in X$ in Theorems 10.10 and 10.11, respectively. $\square$

Corollary 10.2 Let $g : X \to \mathbb{R}$ be a mapping and let there exist real numbers $a, b : \rho = a + b \neq -1$. Let there exist $c_2 \geq 0$ such that

$$|Dg(x, y)| \leq c_2 |x|^a |y|^b \tag{10.68}$$

for all $x, y \in X$. Then there exists a unique reciprocal mapping $r : X \to \mathbb{R}$ satisfying (10.41) and

$$|g(x) - r(x)| \leq \begin{cases} \dfrac{2c_2}{2^{\rho+1} - 1} |x|^\rho, & \text{for } \rho > -1 \\[2mm] \dfrac{2c_2}{1 - 2^{\rho+1}} |x|^\rho, & \text{for } \rho < -1 \end{cases} \tag{10.69}$$

for every $x \in X$.

Proof: The required results in Corollary 10.2 can be easily derived by considering $\phi(x, y) = c_2 |x|^a |y|^b$, for all $x, y \in X$ in Theorems 10.10 and 10.11, respectively. □

Corollary 10.3 Let $c_3 \geq 0$ and $q \neq -\frac{1}{2}$ be real numbers, and $g : X \to \mathbb{R}$ be a mapping satisfying the functional inequality

$$|Dg(x, y)| \leq c_3(|x|^q |y|^q + (|x|^{2q} + |y|^{2q})) \tag{10.70}$$

for all $x, y \in X$. Then, there exists a unique reciprocal mapping $r : X \to \mathbb{R}$ satisfying (10.41) and

$$|g(x) - r(x)| \leq \begin{cases} \dfrac{6c_3}{2^{2q+1} - 1} |x|^{2q}, & \text{for } q > -\dfrac{1}{2}, \\[2mm] \dfrac{6c_3}{1 - 2^{2q+1}} |x|^{2q}, & \text{for } q < -\dfrac{1}{2} \end{cases} \tag{10.71}$$

for every $x \in X$.

Proof: By choosing $\phi(x, y) = c_3(|x|^q |y|^q + (|x|^{2q} + |y|^{2q}))$, for all $x, y \in X$ in Theorems 10.10 and 10.11, respectively, the proof of Corollary 10.3 is complete. □

10.5.4 Counter-Example

The following example illustrates the fact that the functional equation (10.41) is not stable for $p = -1$ in Corollary 10.1. Consider the function $\varphi : X \to \mathbb{R}$ defined by

$$\varphi(x) = \begin{cases} \dfrac{c_1}{x}, & \text{for } x \in (1, \infty), \\[2mm] c_1, & \text{otherwise,} \end{cases}$$

where $c_1 > 0$ is a constant, and define a mapping $g : X \to \mathbb{R}$ by

$$g(x) = \sum_{n=0}^{\infty} \frac{\varphi(2^{-n}x)}{2^n}, \quad \text{for all } x \in X.$$

The function g defined above serves as a counter-example to prove that the functional equation (10.41) is not stable for $p = -1$ in Corollary 10.1.

Theorem 10.12 Suppose the function g satisfies the inequality

$$|Dg(x, y)| \leq 3c_1(|x|^{-1} + |y|^{-1}) \tag{10.72}$$

for all $x, y \in X$. Therefore, there do not exist a reciprocal mapping $r : X \to \mathbb{R}$ and a constant $\delta > 0$ such that

$$|g(x) - r(x)| \leq \delta|x|^{-1} \tag{10.73}$$

for all $x \in X$.

Proof: $|g(x)| \leq \sum_{n=0}^{\infty} \frac{|\varphi(2^{-n}x)|}{|2^n|} \leq \sum_{n=0}^{\infty} \frac{c_1}{2^n} = 2c_1$. Hence, g is bounded by $2c_1$. If $|x|^{-1} + |y|^{-1} \geq 1$, then the left-hand side of (10.72) is less than $3c_1$. Now, suppose that $0 < |x|^{-1} + |y|^{-1} < 1$. Then there exists a positive integer m such that

$$\frac{1}{2^{m+1}} \leq |x|^{-1} + |y|^{-1} < \frac{1}{2^m}. \tag{10.74}$$

Hence, $|x|^{-1} + |y|^{-1} < \frac{1}{2^m}$ implies

$$2^m|x|^{-1} + 2^m|y|^{-1} < 1$$

$$\text{or } \frac{x}{2^m} > 1, \ \frac{y}{2^m} > 1$$

$$\text{or } \frac{x}{2^{m-1}} > 2 > 1, \ \frac{y}{2^{m-1}} > 2 > 1$$

and consequently

$$\frac{1}{2^{m-1}}(x + y) > 1.$$

Therefore, for each value of $n = 0, 1, 2, \dots, m - 1$, we obtain

$$\frac{1}{2^n}(x), \frac{1}{2^n}(y), \frac{1}{2^n}(x + y) > 1$$

and $D\varphi\left(\frac{1}{2^n}x, \frac{1}{2^n}y\right) = 0$ for $n = 0, 1, 2, \dots, m - 1$. Using (10.74) and the definition of g, we obtain

$$\frac{|Dg(x, y)|}{(|x|^{-1} + |y|^{-1})} \leq \sum_{n=m}^{\infty} \frac{\left|\varphi(2^{-n}(x + y)) - \frac{\varphi(2^{-n}x)\varphi(2^{-n}y)}{\varphi(2^{-n}x) + \varphi(2^{-n}y)}\right|}{2^n(|x|^{-1} + |y|^{-1})}$$

$$\leq \sum_{k=0}^{\infty} \frac{\frac{3}{2}c_1}{2^k 2^m (|x|^{-1} + |y|^{-1})}$$

$$\leq \sum_{k=0}^{\infty} \frac{\frac{3}{2}c_1}{2^k} = \frac{3}{2}c_1 \left(1 - \frac{1}{2}\right)^{-1} = 3c_1, \quad \text{for all } x, y \in X.$$

That is, the inequality (10.72) holds true.

Now, assume that there exists a reciprocal mapping $r : X \to \mathbb{R}$ satisfying (10.73). Therefore, we have

$$|g(x)| \leq (\delta + 1)|x|^{-1}. \tag{10.75}$$

However, we can choose a positive integer p with $pc_1 > \delta + 1$. If $x \in (1, 2^{p-1})$, then $2^{-n}x \in (1, \infty)$ for all $n = 0, 1, 2, \ldots, p-1$ and therefore

$$|g(x)| = \sum_{n=0}^{\infty} \frac{\varphi(2^{-n}x)}{2^n} \geq \sum_{n=0}^{m-1} \frac{\frac{c_1}{2^{-n}x}}{2^n} = \frac{pc_1}{x} > (\delta + 1)x^{-1},$$

which contradicts (10.75). Therefore, the reciprocal type functional equation (10.41) is not stable for $p = -1$ in Corollary 10.1. □

10.5.5 Geometrical Interpretation of Reciprocal Type Functional Equation

Consider a right-angled triangle ABC with "a" and "b" as sides shown in Figure 10.1 (10.41).

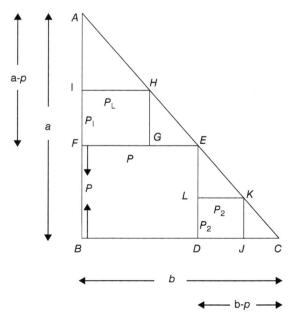

Figure 10.1 Geometrical interpretation of (10.41).

Construct a square BDEF inside the triangle ABC as shown in Figure 10.1 with side "p." Then $AD = a - p$, $FC = b - p$. Now,

$$\text{Area of triangle ABC} = \frac{1}{2}ab. \tag{10.76}$$

From (10.76), it is easy to show that

$$p = \frac{ab}{a+b}. \tag{10.77}$$

Now, construct another two squares FGHI and DJKL with sides "p_1" and "p_2," respectively, as shown in Figure 10.1. Then

$$p_1 = \frac{p(a-p)}{a} = \frac{a^2b}{(a+b)^2} \tag{10.78}$$

and

$$p_2 = \frac{p(b-p)}{b} = \frac{ab^2}{(a+b)^2}. \tag{10.79}$$

Now,

$$p_1 + p_2 = \frac{ab}{(a+b)^2}(a+b) = \frac{ab}{a+b} = p.$$

Therefore,

$$p_1 + p_2 = p. \tag{10.80}$$

In this construction, if we take $a = \frac{1}{x}$, $b = \frac{1}{y}$, then

$$p_1 = \frac{y}{(x+y)^2}, \tag{10.81}$$

$$p_2 = \frac{x}{(x+y)^2} \tag{10.82}$$

and

$$p = \frac{\frac{1}{x}\frac{1}{y}}{\frac{1}{x} + \frac{1}{y}} \tag{10.83}$$

Substituting the relations (10.81)–(10.83) in (10.80), we obtain

$$\frac{1}{x+y} = \frac{\frac{1}{x}\frac{1}{y}}{\frac{1}{x} + \frac{1}{y}}. \tag{10.84}$$

Since $r(x) = \frac{1}{x}$ is a solution of the functional equation (10.41), the property (10.84) is satisfied by the functional equation (10.41). Hence, the functional equation (10.41) holds good in the above geometric construction.

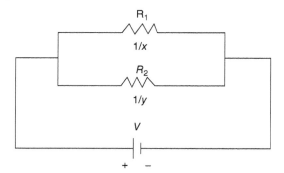

Figure 10.2 Electric circuit connected with two resistors in parallel.

10.5.6 An Application of Equation (10.41) to Electric Circuits

We know, the conductance of any material is reciprocal of its resistance. The parallel circuit is one in which several resistances are connected across one another in such a way that one terminal of each is connected to form a junction point while the remaining ends are also joined to form another junction point.

Consider two resistors R_1 and R_2 with resistances $\frac{1}{x}$ and $\frac{1}{y}$, respectively, are connected in parallel as shown in Figure 10.2.

Then x and y are the conductances of the resistors R_1 and R_2, respectively and hence $x + y$ is the total circuit conductance. To find the equivalent resistance of loads wired in parallel, we use a mathematical formula known as the "reciprocal formula, which is given by

$$\frac{1}{\text{Total circuit conductance}}$$
$$= \text{Total equivalent resistance of the parallel circuit.} \tag{10.85}$$

The reciprocal formula (10.85) satisfies the following algebraic identity:

$$\frac{1}{x+y} = \frac{\frac{1}{x}\frac{1}{y}}{\frac{1}{x} + \frac{1}{y}} \tag{10.86}$$

and hence, the functional equation (10.41) holds good in the above circuit.

10.5.7 Reciprocal-Quadratic Functional Equation

In this section, we introduce a new reciprocal-quadratic functional equation of the form

$$f(x+y) = \frac{f(x)f(y)}{f(x) + f(y) + 2\sqrt{f(x)f(y)}}. \tag{10.87}$$

We obtain the general solution of the functional equation (10.87) and investigate the generalized Hyers–Ulam stability of (10.87).

It is easy to see that the reciprocal-quadratic function $f(x) = \frac{1}{x^2}$ is a solution of the functional equation (10.87). Hence, we can say that every solution of the functional equation (10.87) is a reciprocal-quadratic function.

We recall the following definition and prove the lemmas that will be useful to get our main results. Throughout this paper, we assume that X is the set of nonzero real numbers.

Definition 10.9 [89] A function $f : X \times X \to \mathbb{R}$ is said to be bireciprocal if and only if f satisfies the following functional equations:

$$f(x+y,z) = \frac{f(x,z)f(y,z)}{f(x,z)f(y,z)}$$

$$f(x,y+z) = \frac{f(x,y)f(x,z)}{f(x,y)+f(x,z)}$$

for all $x, y, z \in X$.

For example, consider

$$f(x,y) = \frac{c}{xy} \tag{10.88}$$

for $x, y \in X$, where c is a constant. Then f bireciprocal, since

$$f(x+y,z) = \frac{c}{(x+y)z} = \frac{\frac{c}{xz}\frac{c}{yz}}{\frac{c}{xz}+\frac{c}{yz}} = \frac{f(x,z)f(y,z)}{f(x,z)+f(y,z)}$$

and

$$f(x,y+z) = \frac{c}{x(y+z)} = \frac{\frac{c}{xy}\frac{c}{xz}}{\frac{c}{xy}+\frac{c}{xz}} = \frac{f(x,y)f(x,z)}{f(x,y)+f(x,z)}.$$

Lemma 10.1 If $f : X \to \mathbb{R}$ be a mapping satisfies (10.87) for all $x, y \in X$, then f is reciprocal-quadratic function.

Proof: Setting $y = x$ in (10.87), we have

$$f(2x) = \frac{f(x)f(x)}{2f(x)+2f(x)} = \frac{1}{4}f(x) = \frac{1}{2^2}f(x) \tag{10.89}$$

for all $x \in X$. Further, letting $y = 2x$ in (10.87) and using (10.89) in the resulting equation, one obtains

$$f(3x) = \frac{f(x)f(2x)}{f(x)+f(2x)+2\sqrt{f(x)f(2x)}}$$

$$= \frac{\frac{1}{4}f(x)}{f(x)+\frac{1}{4}f(x)+f(x)} = \frac{1}{9}f(x) = \frac{1}{3^2}f(x)$$

for all $x \in X$. Continuing similarly, for any positive integer n, we assume $f(nx) = \frac{1}{n^2} f(x)$, for all $x \in X$. Now, replacing $y = nx$ in (10.87), we find

$$f((n+1)x) = \frac{f(x)f(nx)}{f(x) + f(nx) + 2\sqrt{f(x)f(nx)}}$$

$$= \frac{\frac{1}{n^2} f(x)}{f(x) + \frac{1}{n^2} f(x) + \frac{2}{n} f(x)}$$

$$= \frac{\frac{1}{n^2} f(x)}{\left(1 + \frac{1}{n^2} + \frac{2}{n}\right)}$$

$$= \frac{1}{(n+1)^2} f(x)$$

for all $x \in X$. Hence, by mathematical induction, we conclude that $f(kx) = \frac{1}{k^2} f(x)$ for all $x \in X$ and any positive integer which shows that f is reciprocal-quadratic function. $\qquad\square$

10.5.8 General Solution of Reciprocal-Quadratic Functional Equation

In this section, we achieve the general solution of reciprocal-quadratic functional equation (10.87).

Theorem 10.13 A mapping $f : X \to \mathbb{R}$ satisfies (10.87) for all $x, y \in X$ if and only if there exists a symmetric bireciprocal function $R : X \times X \to \mathbb{R}$ such that $R(x, y) = \sqrt{f(x)f(y)}$, for all $x, y \in X$.

Proof: Let us assume that there exists a symmetric bireciprocal function $R : X \times X \to \mathbb{R}$ such that

$$R(x, y) = \sqrt{f(x)f(y)}, \quad \text{for all } x, y \in X.$$

If $y = x$, then

$$R(x, x) = \sqrt{f(x)f(x)} = f(x) \tag{10.90}$$

for all $x \in X$. Now, using (10.90), we have

$$f(x + y) = R(x + y, x + y)$$

$$= \frac{R(x, x+y)R(y, x+y)}{R(x, x+y) + R(y, x+y)}$$

$$= \frac{\frac{R(x,x)R(x,y)}{R(x,x)+R(x,y)} \cdot \frac{R(y,x)R(y,y)}{R(y,x)+R(y,y)}}{\frac{R(x,x)R(x,y)}{R(x,x)+R(x,y)} + \frac{R(y,x)R(y,y)}{R(y,x)+R(y,y)}}$$

$$= \frac{R(x,x)R(y,y)R(x,y)R(y,x)}{R(x,x)R(x,y)[R(y,x)+R(y,y)]+R(y,x)R(y,y)[R(x,x)+R(x,y)]}$$

$$= \frac{f(x)f(y)R(x,y)R(y,x)}{R(x,y)\{f(x)[R(y,x)+f(y)]+f(y)[f(x)+R(x,y)]\}}$$

$$= \frac{f(x)f(y)R(y,x)}{f(x)R(y,x)+f(x)f(y)+f(x)f(y)+f(y)R(x,y)}$$

$$= \frac{f(x)f(y)R(y,x)}{2f(x)f(y)+[f(x)+f(y)]R(x,y)}$$

$$= \frac{f(x)f(y)R(y,x)}{2R(x,y)R(y,x)+[f(x)+f(y)]R(x,y)}$$

$$= \frac{f(x)f(y)}{2R(x,y)+f(x)+f(y)}$$

$$= \frac{f(x)f(y)}{f(x)+f(y)+2\sqrt{f(x)f(y)}}$$

for all $x,y \in X$. Hence, f satisfies (10.87) for all $x,y \in X$.

Conversely, if f satisfies (10.87), for all $x,y \in X$, we have to prove there exists a symmetric bireciprocal function defined as $R : X \times X \to \mathbb{R}$ such that $R(x,y) = \sqrt{f(x)f(y)}$ for all $x,y \in X$.

f is well defined. Therefore using equation (10.90), define the value of

$$\sqrt{f(x)f(y)} = \frac{1}{2}\left[\frac{f(x)f(y)}{f(x+y)} - (f(x)+f(y))\right] \tag{10.91}$$

for all $x,y \in X$. The right-hand side of (10.91) is a well-defined expansion and $f(x+y) \neq 0$. Therefore, left-hand side of (10.91) is well defined and it exists. We call

$$R(x,y) = \sqrt{f(x)f(y)} \tag{10.92}$$

for all $x,y \in X$. Hence, there exists $R(x,y)$ for all $x,y \in X$ with $R(x,y) = R(y,x)$, which implies R is symmetric. Now, we prove that R is bireciprocal function.

$$R(x_1+x_2,y) = \sqrt{f(x_1+x_2)f(y)}$$

$$= \sqrt{\frac{f(x_1)f(x_2)f(y)}{f(x_1)+f(x_2)+2\sqrt{f(x_1)f(x_2)}}}$$

$$= \sqrt{\frac{f(x_1)f(x_2)f(y)}{(\sqrt{f(x_1)}+\sqrt{f(x_2)})^2}}$$

$$= \frac{\sqrt{f(x_1)f(x_2)f(y)}}{\sqrt{f(x_1)}+\sqrt{f(x_2)}} \cdot \frac{\sqrt{f(y)}}{\sqrt{f(y)}}$$

$$= \frac{\sqrt{f(x_1)f(x_2)}f(y)}{\sqrt{f(x_1)f(y)}+\sqrt{f(x_2)f(y)}}$$

$$= \frac{1}{\frac{1}{\sqrt{f(x_1)f(y)}} + \frac{1}{\sqrt{f(x_2)f(y)}}}$$

$$= \frac{1}{\frac{1}{R(x_1,y)} + \frac{1}{R(x_2,y)}}$$

$$= \frac{R(x_1,y)R(x_2,y)}{R(x_1,y) + R(x_2,y)}$$

for all $x_1, x_2, y \in X$. Similarly, we have

$$R(x, y_1 + y_2) = \frac{R(x,y_1)R(x,y_2)}{R(x,y_1) + R(x,y_2)}$$

for all $x, y_1, y_2 \in X$. Therefore, R is symmetric bireciprocal function. □

Remark 10.1 A mapping $f : X \to \mathbb{R}$ satisfies (10.87) and $\sqrt{f(x)f(y)} = R(x,y) = R(y,x)$ for all $x, y \in X$. Then we have $R(x,y)R(y,x) = f(x)f(y)$. Since R is symmetric, we obtain

$$R^2(x, y) = f(x)f(y)$$

or $\quad \dfrac{R(x,y)}{f(x)} = \dfrac{f(y)}{R(x,y)}.$

Therefore, $f(x), R(x,y)$ and $f(y)$ are in geometric progression.

10.5.9 Generalized Hyers–Ulam Stability of Reciprocal-Quadratic Functional Equations

In this section, we investigate the generalized Hyers–Ulam stability of the functional equation (10.87) in the setting of real numbers. Moreover, we prove Hyers–Ulam stability, Hyers–Ulam–Rassias stability, Ulam–Gavruta–Rassias stability and J.M. Rassias stability controlled by the mixed product-sum of powers of norms for (10.87).

Theorem 10.14 Let $\phi : X^2 \to \mathbb{R}$ be a function satisfying

$$\sum_{j=0}^{\infty} \frac{1}{4^j} \phi\left(\frac{x}{2^{j+1}}, \frac{y}{2^{j+1}}\right) < +\infty \tag{10.93}$$

for all $x, y \in X$. If a function $f : X \to \mathbb{R}$ satisfies functional inequality

$$|D_f(x,y)| \le \phi(x,y) \tag{10.94}$$

for all $x, y \in X$, then there exists a unique reciprocal-quadratic mapping $r : X \to \mathbb{R}$, which satisfies (10.87) and the inequality

$$|f(x) - r(x)| \le \sum_{j=0}^{\infty} \frac{1}{4^j} \phi\left(\frac{x}{2^{j+1}}, \frac{x}{2^{j+1}}\right) \tag{10.95}$$

for all $x \in X$.

Proof: Switching (x, y) to $\left(\frac{x}{2}, \frac{x}{2}\right)$ in (10.94), we obtain

$$\left| f(x) - \frac{1}{4} f\left(\frac{x}{2}\right) \right| \le \phi\left(\frac{x}{2}, \frac{x}{2}\right) \tag{10.96}$$

for all $x \in X$. Now, plugging x by $\frac{x}{2}$ in (10.96), dividing by 4 and then summing the resulting inequality with (10.96), one finds

$$\left| f(x) - \frac{1}{4^2} f\left(\frac{x}{2^2}\right) \right| \le \sum_{j=0}^{1} \frac{1}{4^j} \phi\left(\frac{x}{2^{j+1}}, \frac{x}{2^{j+1}}\right)$$

for all $x \in X$. Proceeding further and using induction arguments on a positive integer n, we arrive

$$\left| f(x) - \frac{1}{4^n} f\left(\frac{x}{2^n}\right) \right| \le \sum_{j=0}^{n-1} \phi\left(\frac{x}{2^{j+1}}, \frac{x}{2^{j+1}}\right) \tag{10.97}$$

for all $x \in X$. For any positive integer j and $x \in X$, we have

$$\left| \frac{1}{4^{j+1}} f\left(\frac{x}{2^{j+1}}\right) - \frac{1}{4^j} f\left(\frac{x}{2^j}\right) \right| = \frac{1}{4^j} \left| f\left(\frac{x}{2^j}\right) - \frac{1}{4} f\left(\frac{x}{2^{j+1}}\right) \right|$$

$$\le \frac{1}{4^j} \phi\left(\frac{x}{2^{j+1}}, \frac{x}{2^{j+1}}\right).$$

Hence, for any integers n, m with $n > m > 0$, we obtain by using the triangular inequality

$$\left| \frac{1}{4^n} f\left(\frac{x}{2^n}\right) - \frac{1}{4^m} f\left(\frac{x}{2^m}\right) \right|$$

$$= \left| \frac{1}{4^n} f\left(\frac{x}{2^n}\right) - \frac{1}{4^{n-1}} f\left(\frac{x}{2^{n-1}}\right) + \frac{1}{4^{n-1}} f\left(\frac{x}{2^{n-1}}\right) - \cdots \right.$$

$$\left. + \frac{1}{4^{m+1}} f\left(\frac{x}{2^{m+1}}\right) - \frac{1}{4^m} f\left(\frac{x}{2^m}\right) \right|$$

$$\le \frac{1}{4^{n-1}} \phi\left(\frac{x}{2^n}, \frac{x}{2^n}\right) + \cdots + \frac{1}{4^m} \phi\left(\frac{x}{2^{m+1}}, \frac{x}{2^{m+1}}\right)$$

$$\le \sum_{j=m}^{n-1} \frac{1}{4^j} \phi\left(\frac{x}{2^{j+1}}, \frac{x}{2^{j+1}}\right) \tag{10.98}$$

for all $x \in X$. Taking the limit as $n \to +\infty$ in (10.98) and considering (10.93), it follows that the sequence $\left\{ \frac{1}{4^n} f\left(\frac{x}{2^n}\right) \right\}$ is a Cauchy sequence for each $x \in X$. Since $\mathbb{R}$ is complete, we can define $r : X \to \mathbb{R}$ by $r(x) = \lim_{n \to \infty} \frac{1}{4^n} f\left(\frac{x}{2^n}\right)$. To show that r satisfies (10.87), replacing (x, y) by $(2^{-n}x, 2^{-n}y)$ in (10.94) and dividing by 4^n, we obtain

$$|4^{-n} D_f(4^{-n}x, 2^{-n}y)| \le 4^{-n}\phi(2^{-n}x, 2^{-n}y) \tag{10.99}$$

for all $x, y \in X$ and for all positive integer n. Using (10.93) in (10.99), we see that r satisfies (10.87), for all $x, y \in X$. Taking limit $n \to \infty$ in (10.97), we arrive (10.95). Now, it remains to show that r is uniquely defined. Let $r_1 : X \to \mathbb{R}$ be another reciprocal mapping which satisfies (10.87) and the inequality (10.95). Clearly, $r_1(2^{-n}x) = 4^n r_1(x)$, $r(2^{-n}x) = 4^n r(x)$ and using (10.95), we arrive

$$|r_1(x) - r(x)| = 4^{-n}|r_1(2^{-n}x) - r(2^{-n}x)|$$
$$\leq 4^{-n}(|r_1(2^{-n}x) - f(2^{-n}x)| + |f(2^{-n}x) - r(2^{-n}x)|)$$
$$\leq 2\sum_{j=0}^{\infty} \frac{1}{4^{n+j}} \phi\left(\frac{x}{2^{n+j+1}}, \frac{x}{2^{n+j+1}}\right)$$
$$\leq 2\sum_{j=n}^{\infty} \frac{1}{4^j} \phi\left(\frac{x}{2^{j+1}}, \frac{x}{2^{j+1}}\right) \tag{10.100}$$

for all $x \in X$. Allowing $n \to \infty$ in (10.100), we find that r is unique. This completes the proof of Theorem 10.14. $\qquad\square$

Theorem 10.15   Let $\phi : X^2 \to \mathbb{R}$ be a function satisfying

$$\sum_{i=0}^{\infty} 4^i \phi(2^j x, 2^j y) < +\infty \tag{10.101}$$

for all $x_1, x_2, \ldots, x_m \in X$. If a function $f : X \to \mathbb{R}$ satisfies the functional inequality (10.94) for all $x, y \in X$, then there exists a unique reciprocal-quadratic mapping $r : X \to \mathbb{R}$ which satisfies (10.87) and the inequality

$$|r(x) - f(x)| \leq \sum_{j=0}^{\infty} 4^{j+1} \phi(2^j x, 2^j x) \tag{10.102}$$

for all $x \in X$.

Proof: Letting y as x in (10.94) and multiplying by 4, one obtains

$$|4f(2x) - f(x)| \leq 4\phi(x, x) \tag{10.103}$$

for all $x \in X$. Again considering x as $2x$ in (10.103) and multiplying by 4 and then adding the resulting inequality with (10.103), one finds

$$|4^2 f(2^2 x) - f(x)| \leq 4\sum_{j=0}^{1} 4^j \phi(2^j x, 2^j x)$$

for all $x \in X$. Continuing further in the similar manner and using mathematical induction, one can arrive at

$$|4^n f(2^n x) - f(x)| \leq \sum_{j=0}^{n-1} 4^{j+1} \phi(2^j x, 2^j x)$$

for all $x \in X$. The remaining part of the proof is obtained by similar arguments as in Theorem 10.14. □

Corollary 10.4   Let $\epsilon > 0$ be fixed. If $f : X \to \mathbb{R}$ satisfies

$$|D_f(x, y)| \le \epsilon$$

for all $x, y \in E$, then there exists a unique reciprocal-quadratic mapping $r : X \to Y$ such that

$$|f(x) - r(x)| \le \frac{4\epsilon}{3}$$

for all $x, y \in E$.

Proof: Letting $\phi(x, y) = \epsilon$, for all $x, y \in X$ in Theorem 10.14, we arrive at the desired result. □

Corollary 10.5   For any fixed $k_1 \ge 0$ and $\alpha \ne -2$, if $f : X \to \mathbb{R}$ satisfies

$$|D_f(x, y)| \le k_1(|x|^\alpha + |y|^\alpha)$$

for all $x, y \in X$, then there exists a unique reciprocal-quadratic mapping $r : X \to \mathbb{R}$ such that

$$|f(x) - r(x)| \le \begin{cases} \dfrac{8k_1}{2^{\alpha+2} - 1}|x|^\alpha & \text{for } \alpha > -2 \\[2mm] \dfrac{8k_1}{1 - 2^{\alpha+2}}|x|^\alpha & \text{for } \alpha < -2 \end{cases}$$

for all $x \in X$.

Proof: If we choose $\phi(x, y) = k_1(|x|^\alpha + |y|^\alpha)$ for all $x, y \in X$, then by Theorem 10.14, we arrive

$$|f(x) - r(x)| \le \frac{8k_1}{2^{\alpha+2} - 1}|x|^\alpha$$

for all $x \in X$ and $\alpha > -2$ and using Theorem 10.15, we arrive

$$|f(x) - r(x)| \le \frac{8k_1}{1 - 2^{\alpha+2}}|x|^\alpha$$

for all $x \in X$ and $\alpha < -2$. □

Corollary 10.6   Let $f : X \to \mathbb{R}$ be a mapping and there exist α, β with $\lambda = \alpha + \beta$ such that $\lambda \ne -2$. If there exists $k_2 \ge 0$ such that

$$|D_f(x, y)| \le k_2|x|^\alpha |y|^\beta$$

for all $x, y \in X$, then there exists a unique reciprocal-quadratic mapping $r : X \to \mathbb{R}$ satisfying the functional equation (10.87) and

$$|f(x) - r(x)| \leq \begin{cases} \dfrac{4k_2}{2^{\lambda+2} - 1} |x|^\lambda & \text{for } \lambda > -2, \\[3mm] \dfrac{4k_2}{1 - 2^{\lambda+2}} |x|^\lambda & \text{for } \lambda < -2 \end{cases}$$

for all $x \in X$.

Proof: Considering $\phi(x, y) = k_2 |x|^\alpha |y|^\beta$, for all $x, y \in X$, then by Theorem 10.14, we arrive

$$|f(x) - r(x)| \leq \frac{4k_2}{2^{\lambda+2} - 1} |x|^\lambda$$

for all $x \in X$ and $\lambda > -2$ and using Theorem 10.15, we arrive

$$|f(x) - r(x)| \leq \frac{4k_2}{1 - 2^{\lambda+2}} |x|^\lambda$$

for all $x \in X$ and $\lambda < -2$. □

Corollary 10.7 Let $k_3 > 0$ and $\alpha \neq -2$ be real numbers, and $f : X \to \mathbb{R}$ be a mapping satisfying the functional inequality

$$|D_f(x, y)| \leq k_3 \left(|x|^{\alpha/2} |y|^{\alpha/2} + (|x|^\alpha + |y|^\alpha) \right)$$

for all $x, y \in X$. Then, there exists a unique reciprocal-quadratic mapping $r : X \to \mathbb{R}$ satisfying the functional equation (10.87) and

$$|r(x) - f(x)| \leq \begin{cases} \dfrac{12k_3}{2^{\alpha+2} - 1} |x|^\alpha & \text{for } \alpha > -2 \\[3mm] \dfrac{12k_3}{1 - 2^{\alpha+2}} |x|^\alpha & \text{for } \alpha < -2 \end{cases}$$

for all $x \in X$.

Proof: Choosing $\phi(x, y) = k_3(|x|^{\alpha/2} |y|^{\alpha/2} + (|x|^\alpha + |y|^\alpha))$, for all $x, y \in X$, then by Theorem 10.14, we arrive

$$|f(x) - r(x)| \leq \frac{12k_3}{2^{\alpha+2} - 1} |x|^\alpha$$

for all $x \in X$ and $\alpha > -2$ and using Theorem 10.15, we arrive

$$|f(x) - r(x)| \leq \frac{12k_3}{1 - 2^{\alpha+2}} |x|^\alpha$$

for all $x \in X$ and $\alpha < -2$. □

10.5.10 Counter-Examples

In this section, using the idea of the well-known counter-example provided by Gajda [6], we illustrate a counter-example that the functional equation (10.87) is not stable for $\alpha = -2$ in Corollary 10.5.

We consider the function

$$\varphi(x) = \begin{cases} \dfrac{\delta}{x^2}, & \text{for } x \in (1, \infty), \\ \delta, & \text{otherwise,} \end{cases} \tag{10.104}$$

where $\varphi : \mathbb{R} \to \mathbb{R}$. Let $f : \mathbb{R} \to \mathbb{R}$ be defined by

$$f(x) = \sum_{n=0}^{\infty} 4^{-n} \varphi(2^{-n} x) \tag{10.105}$$

for all $x \in \mathbb{R}$. The function f serves as a counter-example for the fact that the functional equation (10.87) is not stable for $\alpha = -2$ in Corollary 10.5 in the following theorem.

Theorem 10.16 If the function $f : X \to \mathbb{R}$ defined in (10.105) satisfies the functional inequality

$$|D_f(x, y)| \le \frac{20\delta}{3} (|x|^{-2} + |y|^{-2}) \tag{10.106}$$

for all $x, y \in X$, then there do not exist a reciprocal-quadratic mapping $r : X \to \mathbb{R}$ and a constant $\mu > 0$ such that

$$|f(x) - r(x)| \le \mu |x|^{-2} \tag{10.107}$$

for all $x \in X$.

Proof: First, we are going to show that f satisfies (10.106).

$$|f(x)| = \left| \sum_{n=0}^{\infty} 4^{-n} \varphi(2^{-n} x) \right| \le \sum_{n=0}^{\infty} \frac{\delta}{4^n} = \frac{4\delta}{3}.$$

Therefore, we see that f is bounded by $\frac{4\delta}{3}$ on $\mathbb{R}$. If $|x|^{-2} + |y|^{-2} \ge 1$, then the left-hand side of (10.106) is less than $\frac{20\delta}{3}$. Now, suppose that $0 < |x|^{-2} + |y|^{-2} < 1$. Then there exists a positive integer k such that

$$\frac{1}{4^{k+1}} \le |x|^{-2} + |y|^{-2} < \frac{1}{4^k}. \tag{10.108}$$

Hence, $|x|^{-2} + |y|^{-2} < \frac{1}{4^k}$ implies

$$4^k (|x|^{-2} + |y|^{-2}) < 1$$
$$\text{or} \quad 4^k x^{-2} < 1, 4^k y^{-2} < 1$$

or $\quad \dfrac{x^2}{4^k} > 1, \dfrac{y^2}{4^k} > 1$

or $\quad \dfrac{x^2}{4^k} > 1 > \dfrac{1}{4}, \dfrac{y^2}{4^k} > 1 > \dfrac{1}{4}$

or $\quad \dfrac{x^2}{4^{k-1}} > 4 > 1, \dfrac{y^2}{4^{k-1}} > 4 > 1$

and consequently

$$\frac{1}{4^{k-1}}(x) > 1, \qquad \frac{1}{4^{k-1}}(y) > 1, \qquad \frac{1}{4^{k-1}}(x+y) > 1.$$

Therefore, for each value of $n = 0, 1, 2, \ldots, k-1$, we obtain

$$\frac{1}{4^n}(x) > 1, \qquad \frac{1}{4^n}(y) > 1, \qquad \frac{1}{4^n}(x+y) > 1$$

and $D_\varphi(4^{-n}x, 4^{-n}y) = 0$ for $n = 0, 1, 2, \ldots, k-1$. Using (10.108) and the definition of f, we obtain

$$|D_f(x, y)|$$

$$= \left| \sum_{n=0}^{\infty} 4^{-n}\varphi(2^{-n}(x+y)) - \frac{\displaystyle\sum_{n=0}^{\infty} 4^{-n}\varphi(2^{-n}x) \sum_{n=0}^{\infty} 4^{-n}\varphi(2^{-n}x)}{\displaystyle\sum_{n=0}^{\infty} 4^{-n}(\varphi(2^{-n}x) + \varphi(2^{-n}y) + 2\sqrt{\varphi(2^{-n}x)\varphi(2^{-n}y)})} \right|$$

$$\leq \left| \sum_{n=k}^{\infty} 4^{-n}\varphi(2^{-n}(x+y)) - \frac{\displaystyle\sum_{n=k}^{\infty} 4^{-n}\varphi(2^{-n}x) \sum_{n=k}^{\infty} 4^{-n}\varphi(2^{-n}x)}{\displaystyle\sum_{n=k}^{\infty} 4^{-n}(\varphi(2^{-n}x) + \varphi(2^{-n}y) + 2\sqrt{\varphi(2^{-n}x)\varphi(2^{-n}y)})} \right|$$

$$\leq \sum_{n=k}^{\infty} \frac{\delta}{4^n} + \frac{1}{4}\sum_{n=k}^{\infty} \frac{\delta}{4^n}$$

$$\leq \frac{5\delta}{4}\frac{1}{4^k}\left(1 - \frac{1}{4}\right)^{-1}$$

$$\leq \frac{5\delta}{3}\frac{1}{4^k}$$

$$\leq \frac{20\delta}{3}\frac{1}{4^{k+1}} \leq \frac{20\delta}{3}(|x|^{-2} + |y|^{-2})$$

for all $x, y \in X$. Therefore, the inequality (10.106) holds true.

We claim that the reciprocal-quadratic functional equation (10.87) is not stable for $\alpha = -2$ in Corollary 10.5.

Assume that there exists a reciprocal-quadratic mapping $r : X \to \mathbb{R}$ satisfying (10.107). Therefore, we have

$$|f(x)| \leq (\mu + 1)|x|^{-2}. \tag{10.109}$$

However, we can choose a positive integer m with $m\delta > \mu + 1$. If $x \in (1, 2^{m-1})$ then $2^{-n}x \in (1, \infty)$ for all $n = 0, 1, 2, \ldots, m - 1$ and therefore,

$$|f(x)| = \sum_{n=0}^{\infty} \frac{\varphi(2^{-n}x)}{4^n} \geq \sum_{n=0}^{m-1} \frac{\frac{4^n\delta}{x}}{4^n} = \frac{m\delta}{x} > (\mu + 1)x^{-2},$$

which contradicts (10.109). Therefore, the reciprocal-quadratic functional equation (10.87) is not stable for $\alpha = -2$ in Corollary 10.5. □

10.5.11 Reciprocal-Cubic and Reciprocal-Quartic Functional Equations

In this section, we introduce the reciprocal-cubic functional equation

$$c(2x + y) + c(x + 2y)$$
$$= \frac{9c(x)c(y)\{c(x) + c(y) + 2c(x)^{1/3}c(y)^{1/3}[c(x)^{1/3} + c(y)^{1/3}]\}}{[2c(x)^{2/3} + 2c(y)^{2/3} + 5c(x)^{1/3}c(y)^{1/3}]^3} \quad (10.110)$$

and the reciprocal-quartic functional equation

$$q(2x + y) + q(2x - y) = \frac{2q(x)q(y)\left[q(x) + 16q(y) + 24\sqrt{q(x)q(y)}\right]}{\left[4\sqrt{q(y)} - \sqrt{q(x)}\right]^4}.$$

$$(10.111)$$

It can be verified that the reciprocal-cubic function $c(x) = \frac{1}{x^3}$ and the reciprocal-quartic function $q(x) = \frac{1}{x^4}$ are solutions of the functional equations (10.110) and (10.111), respectively. Then, we investigate the generalized Hyers–Ulam stability of these new functional equations in the framework of non-Archimedean fields. We extend the results concerning Hyers–Ulam stability, Hyers–Ulam–Rassias stability, and Ulam–Găvruta–Rassias stability controlled by the mixed product-sum of powers of norms for (10.110) and (10.111). We also provide counter-examples that the functional equations (10.110) and (10.111) are not stable for the singular cases.

10.5.12 Hyers–Ulam Stability of Reciprocal-Cubic and Reciprocal-Quartic Functional Equations

In this section, we investigate the generalized Hyers–Ulam stability of (10.110) and (10.111) in non-Archimedean fields. We also establish the results pertaining to Hyers–Ulam stability, Hyers–Ulam–Rassias stability and Ulam–Găvruta–Rassias stability controlled by product–sum of powers of norms.

Theorem 10.17 Let $l \in \{1, -1\}$ be fixed, and let $F : \mathbb{X}^* \times \mathbb{X}^* \longrightarrow \mathbb{Y}^*$ be a mapping such that

$$\lim_{n \to \infty} \left| \frac{1}{27} \right|^{ln} F\left(\frac{x}{3^{ln + \frac{l+1}{2}}}, \frac{y}{3^{ln + \frac{l+1}{2}}} \right) = 0 \quad (10.112)$$

for all $x, y \in \mathbb{X}^*$. Suppose that $c : \mathbb{X}^* \longrightarrow \mathbb{Y}$ is a mapping satisfying the inequality

$$|\Delta_1 c(x, y)| \leq F(x, y) \tag{10.113}$$

for all $x, y \in \mathbb{X}^*$. Then, there exists a unique reciprocal-cubic mapping $C : \mathbb{X}^* \longrightarrow \mathbb{Y}$ such that

$$|c(x) - C(x)| \leq \max \left\{ \left| \frac{1}{27} \right|^{jl + \frac{l-1}{2}} F \left(\frac{x}{3^{jl + \frac{l+1}{2}}}, \frac{x}{3^{jl + \frac{l+1}{2}}} \right) : j \in \mathbb{N} \cup \{0\} \right\} \tag{10.114}$$

for all $x \in \mathbb{X}^*$.

Proof: Interchanging (x, y) by (x, x) in (10.113), we obtain

$$\left| c(x) - \frac{1}{27^l} c \left(\frac{x}{3^l} \right) \right| \leq |27|^{|l-1|/2} F \left(\frac{x}{3^{l+1/2}}, \frac{x}{3^{l+1/2}} \right) \tag{10.115}$$

for all $x \in \mathbb{X}^*$. Replacing x by $\frac{x}{3^{ln}}$ in (10.115) and multiplying by $\left| \frac{1}{27} \right|^{ln}$, we have

$$\left| \frac{1}{27^{ln}} c \left(\frac{x}{3^{ln}} \right) - \frac{1}{27^{(n+1)l}} c \left(\frac{x}{3^{(n+1)l}} \right) \right| \leq \left| \frac{1}{27} \right|^{ln + \frac{l-1}{2}} F \left(\frac{x}{3^{ln + \frac{l+1}{2}}}, \frac{x}{3^{ln + \frac{l+1}{2}}} \right) \tag{10.116}$$

for all $x \in \mathbb{X}^*$. It follows from the relations (10.112) and (10.116) that the sequence $\left\{ \frac{1}{27^{ln}} c \left(\frac{x}{3^{ln}} \right) \right\}$ is Cauchy. Since $\mathbb{Y}$ is complete, this sequence converges to a mapping $C : \mathbb{X}^* \longrightarrow \mathbb{Y}$ defined by

$$C(x) = \lim_{n \to \infty} \frac{1}{27^{ln}} c \left(\frac{x}{3^{ln}} \right). \tag{10.117}$$

On the other hand, for each $x \in \mathbb{X}^*$ and nonnegative integers n, we have

$$\left| \frac{1}{27^{ln}} c \left(\frac{x}{3^{ln}} \right) - c(x) \right|$$

$$= \left| \sum_{j=0}^{n-1} \left\{ \frac{1}{27^{(j+1)l}} c \left(\frac{x}{3^{(j+1)l}} \right) - \frac{1}{27^{jl}} c \left(\frac{x}{3^{jl}} \right) \right\} \right|$$

$$\leq \max \left\{ \left| \frac{1}{27^{(j+1)l}} c \left(\frac{x}{3^{(j+1)l}} \right) - \frac{1}{27^{jl}} c \left(\frac{x}{3^{jl}} \right) \right| : 0 \leq i < n \right\}$$

$$\leq \max \left\{ \left| \frac{1}{27} \right|^{jl + \frac{l-1}{2}} F \left(\frac{x}{3^{jl + \frac{l+1}{2}}}, \frac{x}{3^{jl + \frac{l+1}{2}}} \right) : 0 \leq j < n \right\}. \tag{10.118}$$

Applying (10.112) and letting $n \to \infty$ in the inequality (10.118), we find that the inequality (10.114) holds. Using (10.112), (10.113), and (10.117) for all $x, y \in \mathbb{X}^*$

we have

$$|\Delta_1 C(x,y)| = \lim_{n\to\infty} \left|\frac{1}{27}\right|^{ln} \left|\Delta_1 c\left(\frac{x}{3^{ln}}, \frac{y}{3^{ln}}\right)\right| \leq \lim_{n\to\infty} \left|\frac{1}{27}\right|^{ln} F\left(\frac{x}{3^{ln}}, \frac{y}{3^{ln}}\right) = 0.$$

Thus, the mapping C satisfies (10.110) and hence it is reciprocal-cubic mapping. In order to prove the uniqueness of C, let us consider $C' : \mathbb{X}^* \longrightarrow \mathbb{Y}$ be another reciprocal-cubic mapping satisfying (10.114). Then

$$|C(x) - C'(x)|$$

$$= \lim_{m\to\infty} \left|\frac{1}{27}\right|^{lm} \left|C\left(\frac{x}{3^{lm}}x\right) - C'\left(\frac{x}{3^{lm}}\right)\right|$$

$$\leq \lim_{m\to\infty} \left|\frac{1}{27}\right|^{lm} \max\left\{\left|C\left(\frac{x}{3^{lm}}\right) - c\left(\frac{x}{3^{lm}}\right)\right|, \left|c\left(\frac{x}{3^{lm}}\right) - C'\left(\frac{x}{3^{lm}}\right)\right|\right\}$$

$$\leq \lim_{m\to\infty}\lim_{n\to\infty} \max\left\{\max\left\{\left|\frac{1}{27}\right|^{(j+m)l+\frac{l-1}{2}} F\left(\frac{x}{3^{(j+m)l+\frac{l+1}{2}}}, \frac{x}{3^{(j+m)l+\frac{l+1}{2}}}\right) : m \leq j \leq n+m\right\}\right\}$$

$$= 0$$

for all $x \in \mathbb{X}^*$, which shows that C is unique. This finishes the proof.

The following corollaries are immediate consequence of Theorem 10.17 concerning the stability of (10.110). $\qquad\square$

Corollary 10.8   Let $\epsilon > 0$ be a constant. If $c : \mathbb{X}^* \longrightarrow \mathbb{Y}$ satisfies $|\Delta_1 c(x,y)| \leq \epsilon$ for all $x, y \in \mathbb{X}^*$, then there exists a unique reciprocal-cubic mapping $C : \mathbb{X}^* \longrightarrow \mathbb{Y}$ satisfying (10.110) and $|c(x) - C(x)| \leq \epsilon$ for all $x \in \mathbb{X}^*$.

Proof: Defining $F(x,y) = \epsilon$ and applying Theorem 10.17, we get the desired result. $\qquad\square$

Corollary 10.9   Let $\epsilon \geq 0$ and $r \neq -3$, be fixed constants. If $c : \mathbb{X}^* \longrightarrow \mathbb{Y}$ satisfies $|\Delta_1 c(x,y)| \leq \epsilon(|x|^r + |y|^r)$ for all $x, y \in \mathbb{X}^*$, then there exists a unique reciprocal-cubic mapping $C : \mathbb{X}^* \to \mathbb{Y}$ satisfying (10.110) and

$$|c(x) - C(x)| \leq \begin{cases} \dfrac{|2|\epsilon}{|3|^r}|x|^r, & r > -3, \\[2mm] |2|\epsilon|3|^3|x|^r, & r < -3 \end{cases}$$

for all $x \in \mathbb{X}^*$.

Proof: The result follows immediately from by taking $F(x,y) = \epsilon(|x|^r + |y|^r)$ in Theorem 10.17. $\qquad\square$

Corollary 10.10   Let $c : \mathbb{X}^* \longrightarrow \mathbb{Y}$ be a mapping and let there exist real numbers $p, q, r = p + q \neq -3$ and $\epsilon \geq 0$ such that $|\Delta_1 c(x,y)| \leq \epsilon|x|^p|y|^q$ for all $x, y \in$

X^*. Then, there exists a unique reciprocal-cubic mapping $C : X^* \longrightarrow Y$ satisfying (10.110) and

$$|c(x) - C(x)| \leq \begin{cases} \dfrac{\epsilon}{|3|^r}|x|^r, & r > -3, \\ \epsilon|3|^3||x|^r, & r < -3 \end{cases}$$

for all $x \in X^*$.

Proof: The required results are obtained by choosing $F(x,y) = \epsilon|x|^p|y|^q$, for all $x, y \in X^*$ in Theorem 10.17. $\square$

Corollary 10.11 Let $\epsilon \geq 0$ and $r \neq -3$ be real numbers, and $c : X^* \longrightarrow Y$ be a mapping satisfying the functional inequality

$$|\Delta_1 c(x,y)| \leq \epsilon \left(|x|^{r/2}|y|^{r/2} + (|x|^r + |y|^r) \right)$$

for all $x, y \in X^*$. Then, there exists a unique reciprocal-cubic mapping $C : X^* \longrightarrow Y$ satisfying (10.110) and

$$|c(x) - C(x)| \leq \begin{cases} \dfrac{|3|\epsilon}{|3|^r}|x|^r, & r > -3, \\ |3|\epsilon|3|^3|x|^r, & r < -3 \end{cases}$$

for all $x \in X^*$.

Proof: Considering $F(x,y) = \epsilon \left(|x|^{r/2}|y|^{r/2} + (|x|^r + |y|^r) \right)$ in Theorem 10.17, one can find the result. $\square$

We have the following result which is analogous to Theorem 10.17 for (10.111). We include the proof for the sake of completeness.

Theorem 10.18 Let $l \in \{1, -1\}$ be fixed, and let $G : X^* \times X^* \longrightarrow Y^*$ be a mapping such that

$$\lim_{n \to \infty} \left| \frac{1}{81} \right|^{ln} G\left(\frac{x}{3^{ln + \frac{l+1}{2}}}, \frac{y}{3^{ln + \frac{l+1}{2}}} \right) = 0 \tag{10.119}$$

for all $x, y \in X^*$. Suppose that $q : X^* \longrightarrow Y$ is a mapping satisfying the inequality

$$|\Delta_2 q(x,y)| \leq G(x,y) \tag{10.120}$$

for all $x, y \in X^*$. Then, there exists a unique reciprocal-quartic mapping $Q : X^* \longrightarrow Y$ such that

$$|q(x) - Q(x)| \leq \max\left\{ \left| \frac{1}{81} \right|^{jl + \frac{l-1}{2}} F\left(\frac{x}{3^{jl + \frac{l+1}{2}}}, \frac{x}{3^{jl + \frac{l+1}{2}}} \right) : j \in \mathbb{N} \cup \{0\} \right\} \tag{10.121}$$

for all $x \in \mathbb{X}^*$.

Proof: Replacing (x, y) by (x, x) in (10.120), we obtain

$$\left| q(x) - \frac{1}{81^l} q\left(\frac{x}{3^l}\right) \right| \le |81|^{|l-1|/2} G\left(\frac{x}{3^{(l+1)/2}}, \frac{x}{3^{(l+1)/2}}\right) \tag{10.122}$$

for all $x \in \mathbb{X}^*$. Switching x into $\frac{x}{3^{ln}}$ in (10.122) and multiplying by $\left|\frac{1}{81}\right|^{ln}$, we arrive at

$$\left| \frac{1}{81^{ln}} c\left(\frac{x}{3^{ln}}\right) - \frac{1}{81^{(n+1)l}} c\left(\frac{x}{3^{(n+1)l}}\right) \right| \le \left|\frac{1}{81}\right|^{ln+\frac{l-1}{2}} G\left(\frac{x}{3^{ln+\frac{l+1}{2}}}, \frac{x}{3^{ln+\frac{l+1}{2}}}\right) \tag{10.123}$$

for all $x \in \mathbb{X}^*$. The relations (10.119) and (10.123) imply that $\left\{\frac{1}{81^{ln}} q\left(\frac{x}{3^{ln}}\right)\right\}$ is a Cauchy sequence. Due to the completeness of $\mathbb{Y}$, there is a mapping $Q : \mathbb{X}^* \longrightarrow \mathbb{Y}$ so that

$$Q(x) = \lim_{n \to \infty} \frac{1}{81^{ln}} q\left(\frac{x}{3^{ln}}\right) \tag{10.124}$$

for all $x \in \mathbb{X}^*$. The rest of the proof is similar to the proof of Theorem 10.17. $\square$

Here, we bring some corollaries regarding to the stability of functional equation (10.111) which are direct consequences of Theorem 10.18.

Corollary 10.12 Let $\delta > 0$ be a constant, and let $q : \mathbb{X}^* \longrightarrow \mathbb{Y}$ satisfies $|\Delta_2 q(x, y)| \le \delta$ for all $x, y \in \mathbb{X}^*$. Then, there exists a unique reciprocal-quartic mapping $Q : \mathbb{X}^* \longrightarrow \mathbb{Y}$ satisfying (10.111) and $|q(x) - Q(x)| \le \delta$ for all $x \in \mathbb{X}^*$.

Proof: It is enough to put $G(x, y) = \delta$ in Theorem 10.18. $\square$

Corollary 10.13 Let $\delta \ge 0$ and $\alpha \ne -4$, be fixed constants. If $q : \mathbb{X}^* \longrightarrow \mathbb{Y}$ satisfies $|\Delta_2 q(x, y)| \le \delta(|x|^\alpha + |y|^\alpha)$ for all $x, y \in \mathbb{X}^*$, then there exists a unique reciprocal-quartic mapping $Q : \mathbb{X}^* \longrightarrow \mathbb{Y}$ satisfying (10.111) and

$$|q(x) - Q(x)| \le \begin{cases} \frac{|2|\delta}{|3|^\alpha} |x|^\alpha, & \alpha > -4 \\ |2|\delta|3|^4 |x|^\alpha, & \alpha < -4, \end{cases}$$

for all $x \in \mathbb{X}^*$.

Proof: Considering $G(x, y) = \delta(|x|^\alpha + |y|^\alpha)$, for all $x, y \in \mathbb{X}^*$ in Theorem 10.18, we reach the result. $\square$

Corollary 10.14 Let $q : \mathbb{X}^* \longrightarrow \mathbb{Y}$ be a mapping and let there exist real numbers $a, b, \alpha = a + b \ne -4$ and $\delta \ge 0$ such that

$$|D_2 q(x, y)| \le \delta |x|^a |y|^b$$

for all $x, y \in \mathbb{X}^*$. Then, there exists a unique reciprocal-quartic mapping $Q: \mathbb{X}^* \longrightarrow \mathbb{Y}$ satisfying (10.111) and

$$|q(x) - Q(x)| \leq \begin{cases} \dfrac{\delta}{|3|^\alpha}|x|^\alpha, & \alpha > -4 \\ \delta|3|^4|x|^\alpha, & \alpha < -4, \end{cases}$$

for all $x \in \mathbb{X}^*$.

Proof: Choosing $G(x, y) = \delta|x|^\alpha|y|^\alpha$ in Theorem 10.18, one can derive the desired result. □

Corollary 10.15 Let $\delta \geq 0$ and $\alpha \neq -4$ be real numbers, and $q: \mathbb{X}^* \longrightarrow \mathbb{Y}$ be a mapping satisfying the functional inequality

$$|D_2q(x, y)| \leq \delta \left(|x|^{\alpha/2}|y|^{\alpha/2} + (|x|^\alpha + |y|^\alpha)\right)$$

for all $x, y \in \mathbb{X}^*$. Then, there exists a unique reciprocal-quartic mapping $Q: \mathbb{X}^* \longrightarrow \mathbb{Y}$ satisfying (10.111) and

$$|q(x) - Q(x)| \leq \begin{cases} \dfrac{|3|\delta}{|3|^\alpha}|x|^\alpha, & \alpha > -4 \\ |3|\delta|3|^4|x|^\alpha, & \alpha < -4 \end{cases}$$

for every $x \in \mathbb{X}^*$.

Proof: The proof follows immediately by taking $G(x, y) = \delta(|x|^{\alpha/2}|y|^{\alpha/2} + (|x|^\alpha + |y|^\alpha))$ in Theorem 10.18. □

10.5.13 Counter-Examples

In this section, applying the idea of the well-known counter-example provided by Gajda [6], we illustrate some counter-examples that the functional equations (10.110) and (10.111) are not stable for $r = -3$ in Corollary 10.9 and $\alpha = -4$ in Corollary 10.13, respectively.

Consider the function

$$\varphi(x) = \begin{cases} \dfrac{\delta}{x^3}, & \text{for } x \in (1, \infty), \\ \delta, & \text{otherwise,} \end{cases} \tag{10.125}$$

where $\varphi: \mathbb{R}^* \longrightarrow \mathbb{R}$. Let $f: \mathbb{R}^* \longrightarrow \mathbb{R}$ be defined by

$$f(x) = \sum_{n=0}^{\infty} 27^{-n} \varphi(3^{-n}x) \tag{10.126}$$

for all $x \in \mathbb{R}$. The function f serves as a counter-example for the fact that the functional equation (10.110) is not stable for $r = -3$ in Corollary 10.9 in the next theorem.

Theorem 10.19 If the function $f : \mathbb{R}^* \longrightarrow \mathbb{R}$ defined in (10.126) satisfies the functional inequality

$$|\Delta_1 f(x, y)| \leq \frac{28\delta}{13}(|x|^{-3} + |y|^{-3}) \tag{10.127}$$

for all $x, y \in X$, then there do not exist a reciprocal-cubic mapping $c : \mathbb{R}^* \longrightarrow \mathbb{R}$ and a constant $\mu > 0$ such that

$$|f(x) - c(x)| \leq \mu |x|^{-3} \tag{10.128}$$

for all $x \in \mathbb{R}^*$.

Proof: First, we are going to show that f satisfies (10.127). By computation, we have

$$|f(x)| = \left| \sum_{n=0}^{\infty} 27^{-n} \varphi(3^{-n} x) \right| \leq \sum_{n=0}^{\infty} \frac{\delta}{27^n} = \frac{27\delta}{26}.$$

Therefore, we see that f is bounded by $\frac{27\delta}{26}$ on $\mathbb{R}$. If $|x|^{-3} + |y|^{-3} \geq 1$, then the left-hand side of (10.127) is less than $\frac{28\delta}{13}$. Now, suppose that $0 < |x|^{-3} + |y|^{-3} < 1$. Hence, there exists a positive integer k such that

$$\frac{1}{27^{k+1}} \leq |x|^{-3} + |y|^{-3} < \frac{1}{27^k}. \tag{10.129}$$

Thus, the relation (10.129) necessities $27^k(|x|^{-3} + |y|^{-3}) < 1$, or equivalently; $27^k x^{-3} < 1, 27^k y^{-3} < 1$. So, $\frac{x^3}{27^k} > 1, \frac{y^3}{27^k} > 1$. The last inequalities imply that $\frac{x^3}{27^{k-1}} > 27 > 1, \frac{y^3}{27^{k-1}} > 27 > 1$ and consequently

$$\frac{1}{3^{k-1}}(x) > 1, \frac{1}{3^{k-1}}(y) > 1, \frac{1}{3^{k-1}}(2x + y) > 1, \frac{1}{3^{k-1}}(x + 2y) > 1.$$

Therefore, for each value of $n = 0, 1, 2, \ldots, k - 1$, we obtain

$$\frac{1}{3^n}(x) > 1, \frac{1}{3^n}(y) > 1, \frac{1}{3^n}(2x + y) > 1, \frac{1}{3^n}(x + 2y) > 1.$$

and $\Delta_1 \varphi(3^{-n}x, 3^{-n}y) = 0$ for $n = 0, 1, 2, \ldots, k - 1$. Using (10.129) and the definition of f, we obtain

$$|\Delta_1 f(x, y)| \leq \sum_{n=k}^{\infty} \frac{\delta}{27^n} + \sum_{n=k}^{\infty} \frac{\delta}{27^n} + \frac{54}{729} \sum_{n=k}^{\infty} \frac{\delta}{27^n} \leq 2\delta \sum_{n=k}^{\infty} \frac{1}{27^n} + \frac{2\delta}{27} \sum_{n=k}^{\infty} \frac{1}{27^n}$$

$$\leq \frac{56\delta}{27} \frac{1}{27^k} \left(1 - \frac{1}{27} \right)^{-1} \leq \frac{28\delta}{13} \frac{1}{27^k} \leq \frac{28\delta}{13} \frac{1}{27^{k+1}} \leq \frac{28\delta}{13}(|x|^{-3} + |y|^{-3})$$

for all $x, y \in \mathbb{R}^*$. Therefore, the inequality (10.127) holds. We claim that the reciprocal-cubic functional equation (10.110) is not stable for $r = -3$ in Corollary 10.9. Assume that there exists a reciprocal-cubic mapping $c : \mathbb{R}^* \to \mathbb{R}$ satisfying (10.128). Therefore, we have

$$|f(x)| \leq (\mu + 1)|x|^{-3}. \tag{10.130}$$

However, we can choose a positive integer m with $m\delta > \mu + 1$. If $x \in (1, 3^{m-1})$ then $3^{-n}x \in (1, \infty)$ for all $n = 0, 1, 2, \ldots, m - 1$ and thus,

$$|f(x)| = \sum_{n=0}^{\infty} \frac{\varphi(3^{-n}x)}{27^n} \geq \sum_{n=0}^{m-1} \frac{\frac{27^n \delta}{x^3}}{27^n} = \frac{m\delta}{x^3} > (\mu + 1)x^{-3},$$

which contradicts (10.130). Therefore, the reciprocal-cubic functional equation (10.110) is not stable for $r = -3$ in Corollary 10.9. □

Now, we consider the function

$$\phi(x) = \begin{cases} \dfrac{\lambda}{x^4} & \text{for } x \in (1, \infty) \\ \lambda & \text{otherwise} \end{cases} \tag{10.131}$$

where $\phi : \mathbb{R}^* \longrightarrow \mathbb{R}$. Let $g : \mathbb{R}^* \to \mathbb{R}$ be defined by

$$g(x) = \sum_{n=0}^{\infty} 81^{-n} \phi(3^{-n}x) \tag{10.132}$$

for all $x \in \mathbb{R}$. In analogy with Theorem 10.19, we have the upcoming result which serves as a counter-example for the fact that the functional equation (10.10) is not stable for $\alpha = -4$ in Corollary 10.13. The method of proof is similar, but we include it.

Theorem 10.20   If the function $g : \mathbb{R}^* \longrightarrow \mathbb{R}$ defined in (10.132) satisfies the functional inequality

$$|\Delta_2 g(x, y)| \leq \frac{61\lambda}{20}(|x|^{-4} + |y|^{-4}) \tag{10.133}$$

for all $x, y \in X$, then there do not exist a reciprocal-quartic mapping $q : \mathbb{R}^* \longrightarrow \mathbb{R}$ and a constant $\beta > 0$ such that

$$|g(x) - q(x)| \leq \beta |x|^{-4} \tag{10.134}$$

for all $x \in \mathbb{R}^*$.

Proof: Let us first prove that g satisfies (10.133).

$$|g(x)| = \left| \sum_{n=0}^{\infty} 81^{-n} \phi(3^{-n}x) \right| \leq \sum_{n=0}^{\infty} \frac{\lambda}{81^n} = \frac{81\lambda}{80}.$$

Hence, we find that g is bounded by $\frac{81\lambda}{80}$ on $\mathbb{R}$. If $|x|^{-4} + |y|^{-4} \geq 1$, then the left-hand side of (10.133) is less than $\frac{61\lambda}{20}$. Now, suppose that $0 < |x|^{-4} + |y|^{-4} < 1$. Then, there exists a positive integer m such that

$$\frac{1}{81^{m+1}} \leq |x|^{-4} + |y|^{-4} < \frac{1}{81^m}. \tag{10.135}$$

Similar arguments as in Theorem 10.19, the relation $|x|^{-4} + |y|^{-4} < \frac{1}{81^m}$ implies

$$\frac{1}{3^{m-1}}(x) > 1, \frac{1}{3^{m-1}}(y) > 1, \frac{1}{3^{m-1}}(2x+y) > 1, \frac{1}{3^{m-1}}(2x-y) > 1.$$

Therefore, for any $n = 0, 1, 2, \ldots, m-1$, we get

$$\frac{1}{3^n}(x) > 1, \frac{1}{3^n}(y) > 1, \frac{1}{3^n}(2x+y) > 1, \frac{1}{3^n}(2x-y) > 1.$$

and $\Delta_2\phi(3^{-n}x, 3^{-n}y) = 0$ for $n = 0, 1, 2, \ldots, m-1$. Using (10.135) and the definition of g, we find

$$|\Delta_2 g(x,y)| \leq \sum_{n=m}^{\infty} \frac{\lambda}{81^n} + \sum_{n=m}^{\infty} \frac{\lambda}{81^n} + \frac{82}{81}\sum_{n=k}^{\infty} \frac{\lambda}{81^n} \leq 2\lambda \sum_{n=m}^{\infty} \frac{1}{81^n} + \frac{82\lambda}{81}\sum_{n=m}^{\infty} \frac{1}{81^n}$$

$$\leq \frac{244\lambda}{81}\frac{1}{81^m}\left(1 - \frac{1}{81}\right)^{-1} \leq \frac{244\lambda}{80}\frac{1}{81^m} \leq \frac{244\lambda}{80}\frac{1}{81^{k+1}}$$

$$\leq \frac{61\lambda}{20}(|x|^{-4} + |y|^{-4})$$

for all $x, y \in \mathbb{R}^*$. This shows that the inequality (10.133) holds. Here, we prove that that the reciprocal-quartic functional equation (10.111) is not stable for $\alpha = -4$ in Corollary 10.9. Assume that there exists a reciprocal-quartic mapping $q : \mathbb{R}^* \longrightarrow \mathbb{R}$ satisfying (10.134). Hence,

$$|g(x)| \leq (\beta + 1)|x|^{-4}. \tag{10.136}$$

On the other hand, we can choose a positive integer k with $k\lambda > \beta + 1$. If $x \in (1, 3^{k-1})$ then $3^{-n}x \in (1, \infty)$ for all $n = 0, 1, 2, \ldots, k-1$ and so

$$|g(x)| = \sum_{n=0}^{\infty} \frac{\phi(3^{-n}x)}{81^n} \geq \sum_{n=0}^{k-1} \frac{\frac{81^n\lambda}{x^4}}{81^n} = \frac{k\lambda}{x^4} > (\beta+1)x^{-4},$$

which contradicts (10.136). Therefore, the reciprocal-quartic functional equation (10.2) is not stable for $\alpha = -4$ in Corollary 10.13. $\qquad\square$

10.6   Euler-Lagrange–Jensen $(a, b; k = a + b)$-Sextic Functional Equations

In this section, we introduce a new Euler–Lagrange–Jensen $(a, b; k = a + b)$-sextic functional equation

$$f(ax + by) + f(bx + ay) + (a - b)^6 \left[f\left(\frac{ax - by}{a - b} \right) + f\left(\frac{bx - ay}{b - a} \right) \right]$$

$$= 64(ab)^2(a^2 + b^2) \left[f\left(\frac{x + y}{2} \right) + f\left(\frac{x - y}{2} \right) \right]$$

$$+ 2(a^2 - b^2)(a^4 - b^4)[f(x) + f(y)] \tag{10.137}$$

where $a \neq b$, such that $k \in \mathbb{R}$; $k = a + b \neq 0, \pm 1$ and $\lambda = 1 + (a - b)^6 - 2(a^6 + b^6) - 62(ab)^2(a^2 + b^2) \neq 0$. Then, we investigate the generalized Ulam–Hyers stability of (10.137) in quasi-β-normed spaces using fixed point method. We extend the stability results involving sum of powers of norms, product of powers of norms and mixed product-sum of powers of norms of the above functional equation. We also provide a counter-example to show that the functional equation (10.137) is not stable for singular case. It is easy to see that the function $f(x) = kx^6$ is a solution of (10.137). Hence, we say that it is a sextic functional equation.

10.6.1   Generalized Ulam–Hyers Stability of Euler-Lagrange-Jensen Sextic Functional Equation Using Fixed Point Method

Throughout this section, we assume that $\mathcal{X}$ is a linear space and $\mathcal{Y}$ is a (β, p)-Banach space with (β, p)-norm $\|\cdot\|_{\mathcal{Y}}$. Let K be the modulus of concavity of $\|\cdot\|_{\mathcal{Y}}$. For notational convenience, we define the difference operator for a given mapping $f : \mathcal{X} \to \mathcal{Y}$ as

$$D_s f(x, y) = f(ax + by) + f(bx + ay) + (a - b)^6 \left[f\left(\frac{ax - by}{a - b} \right) + f\left(\frac{bx - ay}{b - a} \right) \right]$$

$$= 64(ab)^2(a^2 + b^2) \left[f\left(\frac{x + y}{2} \right) + f\left(\frac{x - y}{2} \right) \right]$$

$$+ 2(a^2 - b^2)(a^4 - b^4)[f(x) + f(y)]$$

for all $x, y \in \mathcal{X}$.

Lemma 10.2   [58]   Let $j \in \{-1, 1\}$ be fixed, $m, b \in \mathbb{N}$ with $b \geq 2$ and $\Phi : \mathcal{X} \to [0, \infty)$ be a function such that there exists an $L < 1$ with $\Phi(b^j x) \leq b^{jm\beta} L \Phi(x)$ for all $x \in \mathcal{X}$. Let $g : \mathcal{X} \to \mathcal{Y}$ be a mapping satisfying

$$\|g(bx) - b^m g(x)\|_{\mathcal{Y}} \leq \Phi(x) \tag{10.138}$$

for all $x \in \mathcal{X}$, then there exists a uniquely determined mapping $G : \mathcal{X} \to \mathcal{Y}$ such that $G(bx) = b^m G(x)$ and

$$\|g(x) - G(x)\|_{\mathcal{Y}} \leq \frac{1}{b^{m\beta} |1 - L^j|} \Phi(x) \tag{10.139}$$

for all $x \in \mathcal{X}$.

Theorem 10.21   Let $i \in \{-1, 1\}$ be fixed. Let $\phi : \mathcal{X} \times \mathcal{X} \to [0, \infty)$ be a function such that there exists an $L < 1$ with $\phi(k^i x, k^i y) \leq k^{6i\beta} L \phi(x, y)$ for all $x, y \in \mathcal{X}$. Let $f : \mathcal{X} \to \mathcal{Y}$ be a mapping satisfying

$$\|D_s f(x, y)\|_\mathcal{Y} \leq \phi(x, y) \tag{10.140}$$

for all $x, y \in \mathcal{X}$. Then, there exists a unique sextic mapping $S : \mathcal{X} \to \mathcal{Y}$ such that

$$\|f(x) - S(x)\|_\mathcal{Y} \leq \frac{1}{k^{6\beta} |1 - L^i|} \Psi(x) \tag{10.141}$$

for all $x \in \mathcal{X}$, where

$$\Psi(x) = \frac{K}{2^\beta} \left[\phi(x, x) + \frac{32^\beta (ab)^{2\beta} (a^2 + b^2)^\beta}{\lambda^\beta} \phi(0, 0) \right].$$

Proof: Plugging (x, y) into $(0, 0)$ in (10.140), we obtain

$$\|f(0)\|_\mathcal{Y} \leq \frac{1}{2^\beta \lambda^\beta} \phi(0, 0). \tag{10.142}$$

Switching (x, y) to (x, x) in (10.140), one finds

$$\|f(kx) - k^6 f(x) - 32(ab)^2 (a^2 + b^2) f(0)\|_\mathcal{Y} \leq \frac{1}{2^\beta} \phi(x, x) \tag{10.143}$$

for all $x \in \mathcal{X}$. Using (10.142) and (10.143), we arrive

$$\|f(kx) - k^6 f(x)\|_\mathcal{Y} \leq \Psi(x) \tag{10.144}$$

for all $x \in \mathcal{X}$. By Lemma 10.2, there exists a unique mapping $S : \mathcal{X} \to \mathcal{Y}$ such that $S(kx) = k^6 S(x)$ and

$$\|f(x) - S(x)\|_Y \leq \frac{1}{k^{6\beta} |1 - L^i|} \Psi(x)$$

for all $x \in \mathcal{X}$. It remains to show that S is a sextic map. By (10.140), we have

$$\left\| \frac{1}{k^{6in}} D_s f(k^{in} x, k^{in} y) \right\|_\mathcal{Y} \leq k^{-6in\beta} \phi(k^{in} x, k^{in} y)$$

$$\leq k^{-6in\beta} (k^{6i\beta} L)^n \phi(x, y)$$

$$= L^n \phi(x, y)$$

for all $x, y \in \mathcal{X}$ and $n \in \mathbb{N}$. So $\|D_s S(x, y)\|_\mathcal{Y} = 0$ for all $x, y \in \mathcal{X}$. Thus, the mapping $S : \mathcal{X} \to \mathcal{Y}$ is sextic, which completes the proof of theorem.   □

Corollary 10.16   Let $\mathcal{X}$ be a quasi-α-normed space with quasi-α-norm $\|\cdot\|_\mathcal{X}$, and let $\mathcal{Y}$ be a (β, p)-Banach space with (β, p)-norm $\|\cdot\|_\mathcal{Y}$. Let k_1, p be positive numbers with $p \neq \frac{6\beta}{\alpha}$ and $f : \mathcal{X} \to \mathcal{Y}$ be a mapping satisfying

$$\|D_s f(x, y)\|_\mathcal{Y} \leq k_1 (\|x\|_\mathcal{X}^p + \|y\|_\mathcal{X}^p)$$

for all $x, y \in \mathcal{X}$. Then, there exists a unique sextic mapping $S : \mathcal{X} \to \mathcal{Y}$ such that

$$\|f(x) - S(x)\|_{\mathcal{Y}} \leq \begin{cases} \dfrac{k_1 K}{2^{\beta}(k^{6\beta} - k^{p\alpha})} \|x\|_{\mathcal{X}}^{p}, & p \in \left(0, \dfrac{6\beta}{\alpha}\right), \\[4mm] \dfrac{k^{p\alpha} k_1 K}{k^{6\beta} 2^{\beta}(k^{p\alpha} - k^{6\beta})} \|x\|_{\mathcal{X}}^{p}, & p \in \left(\dfrac{6\beta}{\alpha}, \infty\right) \end{cases}$$

for all $x \in \mathcal{X}$.

Proof: The proof is obtained by taking $\phi(x, y) = k_1(\|x\|_{\mathcal{X}}^{p} + \|y\|_{\mathcal{X}}^{p})$, for all $x, y \in \mathcal{X}$ and $L = \frac{k^{p\alpha}}{k^{6\beta}}$ in Theorem 10.21. $\qquad\square$

Corollary 10.17 Let $\mathcal{X}$ be a quasi-α-normed space with quasi-α-norm $\|\cdot\|_{\mathcal{X}}$, and let $\mathcal{Y}$ be a (β, p)-Banach space with (β, p)-norm $\|\cdot\|_{\mathcal{Y}}$. Let k_2, p, q be positive numbers with $\rho = p + q \neq \frac{6\beta}{\alpha}$ and $f : \mathcal{X} \to \mathcal{Y}$ be a mapping satisfying

$$\|D_s f(x, y)\|_Y \leq k_2 \|x\|_{\mathcal{X}}^{p} \|y\|_{\mathcal{X}}^{p}$$

for all $x, y \in X$. Then, there exists a unique sextic mapping $S : \mathcal{X} \to \mathcal{Y}$ such that

$$\|f(x) - S(x)\|_{\mathcal{Y}} \leq \begin{cases} \dfrac{k_2 K}{2^{\beta}(k^{6\beta} - k^{\rho\alpha})} \|x\|_{\mathcal{X}}^{p}, & \rho \in \left(0, \dfrac{6\beta}{\alpha}\right), \\[4mm] \dfrac{k^{\rho\alpha} k_2 K}{k^{6\beta} 2^{\beta}(k^{\rho\alpha} - k^{6\beta})} \|x\|_{\mathcal{X}}^{p}, & \rho \in \left(\dfrac{6\beta}{\alpha}, \infty\right) \end{cases}$$

for all $x \in \mathcal{X}$.

Proof: Letting $\phi(x, y) = k_2 \|x\|_{\mathcal{X}}^{p} \|y\|_{\mathcal{X}}^{q}$, for all $x, y \in \mathcal{X}$ and $L = \frac{k^{\rho\alpha}}{k^{6\beta}}$ in Theorem 10.21, we obtain the required results. $\qquad\square$

Corollary 10.18 Let $\mathcal{X}$ be a quasi-α-normed space with quasi-α-norm $\|\cdot\|_{\mathcal{X}}$, and let $\mathcal{Y}$ be a (β, p)-Banach space with (β, p)-norm $\|\cdot\|_{\mathcal{Y}}$. Let k_3, r be positive numbers $r \neq \frac{3\beta}{\alpha}$ and $f : \mathcal{X} \to \mathcal{Y}$ be a mapping satisfying

$$\|D_s f(x, y)\|_Y \leq k_3 [\|x\|_{\mathcal{X}}^{r} \|x\|_{\mathcal{X}}^{r} + (\|x\|_{\mathcal{X}}^{2r} + \|y\|_{\mathcal{X}}^{2r})]$$

for all $x, y \in \mathcal{X}$. Then there exists a unique sextic mapping $S : \mathcal{X} \to \mathcal{Y}$ such that

$$\|f(x) - S(x)\|_Y \leq \begin{cases} \dfrac{3 k_3 K}{2^{\beta}(k^{6\beta} - k^{2r\alpha})} \|x\|_{\mathcal{X}}^{2r}, & r \in \left(0, \dfrac{3\beta}{\alpha}\right) \\[4mm] \dfrac{3 k^{2r\alpha} k_3 K}{k^{6\beta} 2^{\beta}(k^{2r\alpha} - k^{6\beta})} \|x\|_{\mathcal{X}}^{2r}, & r \in \left(\dfrac{3\beta}{\alpha}, \infty\right) \end{cases}$$

for all $x \in \mathcal{X}$.

Proof: By taking $\varphi(x, y) = k_3[\|x\|_{\mathcal{X}}^r \|y\|_{\mathcal{X}}^r + (\|x\|_{\mathcal{X}}^{2r} + \|y\|_{\mathcal{X}}^{2r})]$, for all $x, y \in \mathcal{X}$ and $L = \frac{k^{2r\alpha}}{k^{6\beta}}$ in Theorem 10.21, we arrive at the desired results. □

10.6.2 Counter-Example

In this section, using the idea of the well-known counter-example provided by Gajda [6], we illustrate a counter-example that the functional equation (10.137) is not stable for $p = \frac{6\beta}{\alpha}$ in Corollary 10.16. We consider the function

$$\varphi(x) = \begin{cases} x^6, & \text{for } |x| < 1, \\ 1, & \text{for } |x| \geq 1, \end{cases} \tag{10.145}$$

where $\varphi : \mathbb{R} \to \mathbb{R}$. Let $f : \mathbb{R} \to \mathbb{R}$ be defined by

$$f(x) = \sum_{n=0}^{\infty} 2^{-6n} \varphi(2^n x) \tag{10.146}$$

for all $x \in \mathbb{R}$. The function f serves as a counter-example for the fact that the functional equation (10.137) is not stable for $p = \frac{6\beta}{\alpha}$ in Corollary 10.16 in the following theorem.

Theorem 10.22 If the function f defined in (10.146) satisfies the functional inequality

$$|D_s f(x, y)| \leq \frac{64^3 \delta}{63}(|x|^6 + |y|^6) \tag{10.147}$$

where $\delta = 2[1 + (a - b)^6 - 2(a^6 + b^6) - 62(ab)^2(a^2 + b^2)] > 0$, for all $x, y \in \mathbb{R}$, then there do not exist a sextic mapping $S : \mathbb{R} \to \mathbb{R}$ and a constant $\epsilon > 0$ such that

$$|f(x) - S(x)| \leq \epsilon |x|^6$$

for all $x \in \mathbb{R}$.

Proof: First, we are going to show that f satisfies (10.147).

$$|f(x)| = \left| \sum_{n=0}^{\infty} 2^{-6n} \varphi(2^n x) \right| \leq \sum_{n=0}^{\infty} \frac{1}{2^{6n}} = \frac{64}{63}.$$

Therefore, we see that f is bounded by $\frac{64}{63}$ on $\mathbb{R}$. If $|x|^6 + |y|^6 = 0$ or $|x|^6 + |y|^6 \geq \frac{1}{64}$, then

$$|D_s f(x, y)| \leq \frac{64\delta}{63} \leq \frac{64^2 \delta}{63}(|x|^6 + |y|^6).$$

Now, suppose that $0 < |x|^6 + |y|^6 < \frac{1}{64}$. Then there exists a nonnegative integer k such that

$$\frac{1}{64^{k+1}} \leq |x|^6 + |y|^6 < \frac{1}{64^k}. \tag{10.148}$$

Hence, $64^k|x|^6 < 1$, $64^k|y|^6 < 1$ and $2^n(ax + by)$, $2^n(bx + ay)$, $2^n\left(\frac{ax-by}{a-b}\right)$, $2^n\left(\frac{bx-ay}{b-a}\right)$, $2^n\left(\frac{x+y}{2}\right)$, $2^n\left(\frac{x-y}{2}\right)$, 2^nx, $2^ny \in (-1, 1)$ for all $n = 0, 1, 2, \ldots, k-1$.

Hence, for $n = 0, 1, 2, \ldots, k-1$,

$$\varphi(2^n(ax + by)) + \varphi(2^n(bx + ay))$$
$$+ (a-b)^6\left[\varphi\left(2^n\left(\frac{ax-by}{a-b}\right)\right) + \varphi\left(2^n\left(\frac{bx-ay}{b-a}\right)\right)\right]$$
$$- 64(ab)^2(a^2+b^2)\left[\varphi\left(2^n\left(\frac{x+y}{2}\right)\right) + \varphi\left(2^n\left(\frac{x-y}{2}\right)\right)\right]$$
$$- 2(a^2-b^2)(a^4-b^4)[\varphi(2^nx) + \varphi(2^ny)] = 0. \tag{10.149}$$

From the definition of f and the inequality (10.148), we obtain that

$$|D_sf(x,y)|$$
$$= \left|\sum_{n=0}^{\infty} 2^{-6n}\varphi(2^n(ax + by)) + \sum_{n=0}^{\infty} 2^{-6n}\varphi(2^n(bx + ay))\right.$$
$$+ (a-b)^6\left[\sum_{n=0}^{\infty} 2^{-6n}\varphi\left(2^n\left(\frac{ax-by}{a-b}\right)\right) + \sum_{n=0}^{\infty} 2^{-6n}\varphi\left(2^n\left(\frac{bx-ay}{b-a}\right)\right)\right]$$
$$- 64(ab)^2(a^2+b^2)\left[\sum_{n=0}^{\infty} 2^{-6n}\varphi\left(2^n\left(\frac{x+y}{2}\right)\right) + \sum_{n=0}^{\infty} 2^{-6n}\varphi\left(2^n\left(\frac{x-y}{2}\right)\right)\right]$$
$$\left. - 2(a^2-b^2)(a^4-b^4)\left[\sum_{n=0}^{\infty} 2^{-6n}\varphi(2^nx) + \sum_{n=0}^{\infty} 2^{-6n}\varphi(2^ny)\right]\right|$$
$$\leq \sum_{n=0}^{\infty} 2^{-6n}\left|\varphi(2^n(ax + by)) + \varphi(2^n(bx + ay))\right.$$
$$+ (a-b)^6\left[\varphi\left(2^n\left(\frac{ax-by}{a-b}\right)\right) + \varphi\left(2^n\left(\frac{bx-ay}{b-a}\right)\right)\right]$$
$$- 64(ab)^2(a^2+b^2)\left[\varphi\left(2^n\left(\frac{x+y}{2}\right)\right) + \varphi\left(2^n\left(\frac{x-y}{2}\right)\right)\right]$$
$$\left. - 2(a^2-b^2)(a^4-b^4)[\varphi(2^nx) + \varphi(2^ny)]\right|$$
$$\leq \sum_{n=0}^{\infty} 2^{-6n}\delta = \frac{2^{6(1-k)}\delta}{63} \leq \frac{64^3\delta}{63}(|x|^6 + |y|^6). \tag{10.150}$$

Therefore, f satisfies (10.147) for all $x, y \in \mathbb{R}$. Now, we claim that the functional equation (10.137) is not stable for $p = \frac{6\beta}{\alpha}$ in Corollary 10.16. Suppose on the contrary that there exists a sextic mapping $S:\mathbb{R} \to \mathbb{R}$ and a constant $\epsilon > 0$ such that

$$|f(x) - S(x)| \leq \epsilon|x|^6 \quad \text{for all } x \in \mathbb{R}.$$

Then, there exists a constant $c \in \mathbb{R}$ such that $S(x) = cx^6$ for all rational numbers x [44]. So we obtain that

$$|f(x)| \leq (\epsilon + |c|)|x|^6 \tag{10.151}$$

for all $x \in \mathbb{Q}$. Let $m \in \mathbb{N}$ with $m + 1 > \epsilon + |c|$. If x is a rational number in $(0, 2^{-m})$, then $2^n x \in (0, 1)$ for all $n = 0, 1, 2, \dots, m$, and for this x, we get

$$f(x) = \sum_{n=0}^{\infty} 2^{-6n} \varphi(2^n x) \geq \sum_{n=0}^{m} 2^{-6n}(2^n x)^6 = (m + 1)x^6 > (\epsilon + |c|)x^6,$$

$$\tag{10.152}$$

which contradicts (10.151). Hence, the functional equation (10.137) is not stable for $p = \frac{6\beta}{\alpha}$ in Corollary 10.16. $\qquad\square$

10.6.3 Generalized Ulam–Hyers Stability of Euler-Lagrange-Jensen Sextic Functional Equation Using Direct Method

In this section, we prove various stabilities associated with Hyers, Rassias, and Gavruta for Euler–Lagrange–Jensen (a, b)-sextic functional equation (10.137) using direct method.

Throughout this section, we assume that $\mathcal{A}$ is a linear space and $\mathcal{B}$ is a (β, p)-Banach space with (β, p)-norm $\|\cdot\|_{\mathcal{B}}$. Let K be the modulus of concavity of $\|\cdot\|_{\mathcal{B}}$.

Theorem 10.23 Let $\phi : \mathcal{A} \times \mathcal{A} \to [0, \infty)$ be a mapping satisfying

$$\sum_{j=0}^{\infty} \left(\frac{K}{\mu^{6\beta}} \right)^{j} \phi(\mu^j x, \mu^j y) < \infty \tag{10.153}$$

for all $x, y \in \mathcal{A}$. Let $f : \mathcal{A} \to \mathcal{B}$ be a mapping with the condition $f(0) = 0$ such that

$$\|D_s f(x, y)\|_{\mathcal{B}} \leq \phi(x, y) \tag{10.154}$$

for all $x, y \in \mathcal{A}$. Then there exists a unique sextic mapping $S : \mathcal{A} \to \mathcal{B}$ satisfying (10.137) and

$$\|f(x) - S(x)\|_{\mathcal{B}} \leq \frac{K}{2^{\beta} \mu^{6\beta}} \sum_{j=0}^{\infty} \left(\frac{K}{\mu^{6\beta}} \right)^{j} \phi(\mu^j x, \mu^j x) \tag{10.155}$$

for all $x \in \mathcal{A}$. The mapping $S(x)$ is defined by

$$S(x) = \lim_{n \to \infty} \frac{1}{\mu^{6n}} f(\mu^n x) \tag{10.156}$$

for all $x \in \mathcal{A}$.

Proof: Switching (x, y) to (x, x) in (10.154) and simplifying further, we obtain

$$\left\| \frac{1}{\mu^6} f(\mu x) - f(x) \right\|_{\mathcal{B}} \leq \frac{1}{2^{\beta} \mu^{6\beta}} \phi(x, x) \tag{10.157}$$

for all $x \in \mathcal{A}$. Now, replacing x by μx, dividing by $\mu^{6\beta}$ in (10.157), we find

$$\left\| \frac{1}{\mu^{12}} f(\mu^2 x) - \frac{1}{\mu^6} f(\mu x) \right\|_B \leq \frac{1}{2^\beta \mu^{12\beta}} \phi(\mu x, \mu x) \tag{10.158}$$

for all $x \in \mathcal{A}$. Combining (10.157) with (10.158) and using triangle inequality and since $K \geq 1$,

$$\left\| \frac{1}{\mu^{12}} f(\mu^2 x) - f(x) \right\|_B \leq \frac{K}{2^\beta \mu^{6\beta}} \sum_{j=0}^{1} \left(\frac{K}{\mu^{6\beta}} \right)^j \phi(\mu^j x, \mu^j x) \tag{10.159}$$

for all $x \in \mathcal{A}$. Using induction arguments on a positive integer n, we arrive

$$\left\| \frac{1}{\mu^{6n}} f(\mu^n x) - f(x) \right\|_B \leq \frac{K}{2^\beta \mu^{6\beta}} \sum_{j=0}^{n-1} \left(\frac{K}{\mu^{6\beta}} \right)^j \phi(\mu^j x, \mu^j x) \tag{10.160}$$

for all $x \in \mathcal{A}$. From (10.157), we obtain

$$\left\| \frac{1}{\mu^{6(j+1)}} f(\mu^{j+1} x) - \frac{1}{\mu^{6j}} f(\mu^j x) \right\|_B \leq \frac{1}{\mu^{6j\beta} 2^\beta \mu^{6\beta}} \phi(\mu^j x, \mu^j x) \tag{10.161}$$

for all $x \in \mathcal{A}$. For $n > m$,

$$\left\| \frac{1}{\mu^{6n}} f(\mu^n x) - \frac{1}{\mu^{6m}} f(\mu^m x) \right\|_B \leq \sum_{j=m}^{n-1} \left\| \frac{1}{\mu^{6(j+1)}} f(\mu^{j+1} x) - \frac{1}{\mu^{6j}} f(\mu^j x) \right\|_B$$

$$\leq \frac{1}{2^\beta \mu^{6\beta}} \sum_{j=m}^{n-1} \frac{1}{\mu^{6j\beta}} \phi(\mu^j x, \mu^j x) \tag{10.162}$$

for all $x \in \mathcal{A}$. The right-hand side of the above inequality (10.162) tends to 0 as $n \to \infty$. Hence, $\left\{ \frac{1}{\mu^{6n}} f(\mu^n x) \right\}$ is a Cauchy sequence in B. Hence, we may define

$$S(x) = \lim_{n \to \infty} \frac{1}{\mu^{6n}} f(\mu^n x)$$

for all $x \in \mathcal{A}$. Since $K \geq 1$, replacing (x, y) by $(\mu^n x, \mu^n y)$ and dividing by $\mu^{6n\beta}$ in (10.154), we have

$$\frac{1}{\mu^{6n\beta}} \| D_s f(\mu^n x, \mu^n y) \|_B \leq \frac{1}{\mu^{6n\beta}} K^n \phi(\mu^n x, \mu^n y) \tag{10.163}$$

for all $x, y \in \mathcal{A}$. By taking $n \to \infty$, the definition of S implies that S satisfies (10.137) for all $x, y \in \mathcal{A}$. Thus, S is a sextic mapping. In addition, the inequality (10.163) implies the inequality (10.155). Now, it remains to show the uniqueness of S. Assume that there exists $S' : \mathcal{A} \to B$ satisfying (10.137) and (10.155). It is easy to show that for all $x \in \mathcal{A}$, $S'(\mu^n x) = \mu^{6n} S'(x)$ and $S(\mu^n x) = \mu^{6n} S(x)$. Then

$$
\begin{aligned}
\|S'(x) - S(x)\|_B &= \left\| \frac{1}{\mu^{6n}} S'(\mu^n x) - \frac{1}{\mu^{6n}} S(\mu^n x) \right\|_B \\
&= \frac{1}{\mu^{6n\beta}} \|S'(\mu^n x) - S(\mu^n x)\|_B \\
&\leq \frac{1}{\mu^{6n\beta}} K(\|S'(\mu^n x) - f(\mu^n x)\|_B + \|f(\mu^n x) - S(\mu^n x)\|_B) \\
&\leq 2K \sum_{j=0}^{\infty} \left(\frac{K}{\mu^{6\beta}} \right)^{n+j} \phi(\mu^{n+j} x, \mu^{n+j} x)
\end{aligned}
$$

for all $x \in \mathcal{A}$. By letting $n \to \infty$, we immediately have the uniqueness of S. $\qquad\square$

Theorem 10.24 Let $\phi : \mathcal{A} \times \mathcal{A} \to [0, \infty)$ be a mapping satisfying

$$
\sum_{j=0}^{\infty} (K\mu^{6\beta})^j \phi \left(\frac{x}{\mu^{j+1}}, \frac{y}{\mu^{j+1}} \right) < \infty \tag{10.164}
$$

for all $x, y \in \mathcal{A}$. Let $f : \mathcal{A} \to \mathcal{B}$ be a mapping with the condition $f(0) = 0$ satisfying (10.154) for all $x, y \in \mathcal{A}$. Then there exists a unique sextic mapping $S : \mathcal{A} \to \mathcal{B}$ satisfying (10.137) and

$$
\|f(x) - S(x)\|_B \leq \frac{K}{2^\beta} \sum_{j=0}^{\infty} (K\mu^{6\beta})^j \phi \left(\frac{x}{\mu^{j+1}}, \frac{x}{\mu^{j+1}} \right) \tag{10.165}
$$

for all $x \in \mathcal{A}$. The mapping $S(x)$ is defined by

$$
S(x) = \lim_{n \to \infty} \mu^{6n} f \left(\frac{x}{\mu^n} \right) \tag{10.166}
$$

for all $x \in \mathcal{A}$.

Proof: Plugging (x, y) into $\left(\frac{x}{\mu}, \frac{x}{\mu} \right)$ in (10.154), we obtain

$$
\left\| f(x) - \mu^6 f \left(\frac{x}{\mu} \right) \right\|_B \leq \frac{1}{2^\beta} \phi \left(\frac{x}{\mu}, \frac{x}{\mu} \right) \tag{10.167}
$$

for all $x \in \mathcal{A}$. Now, substituting x as $\frac{x}{\mu}$, multiplying by μ^6 in (10.167) and summing the resulting inequality with (10.167), we have

$$
\left\| f(x) - \mu^{12} f \left(\frac{x}{\mu^2} \right) \right\|_B \leq \frac{K}{2^\beta} \sum_{j=0}^{1} (K\mu^{6\beta})^j \phi \left(\frac{x}{\mu^{j+1}}, \frac{x}{\mu^{j+1}} \right)
$$

for all $x \in \mathcal{A}$. Using induction arguments on a positive integer n, we conclude that

$$
\left\| f(x) - \mu^{6n} f \left(\frac{x}{\mu^n} \right) \right\|_B \leq \frac{K}{2^\beta} \sum_{j=0}^{n-1} (K\mu^{6\beta})^j \phi \left(\frac{x}{\mu^{j+1}}, \frac{x}{\mu^{j+1}} \right)
$$

for all $x \in \mathcal{A}$. The rest of the proof is obtained by similar arguments as in Theorem 10.23. ☐

Corollary 10.19 Let $\theta \geq 0$ be fixed. If a mapping $f : \mathcal{A} \rightarrow \mathcal{B}$ satisfies the inequality

$$\|D_s f(x, y)\|_B \leq \theta$$

for all $x, y \in \mathcal{A}$, then there exists a unique sextic mapping $S : \mathcal{A} \rightarrow \mathcal{B}$ satisfying (10.137) and

$$\|f(x) - S(x)\|_B \leq \frac{K\theta}{2^\beta(\mu^{6\beta} - K)}$$

for all $x \in \mathcal{A}$.

Proof: Considering $\phi(x, y) = \theta$, for all $x, y \in \mathcal{A}$ in Theorem 10.23, we have

$$\|f(x) - S(x)\|_B \leq \frac{K}{2^\beta \mu^{6\beta}} \sum_{j=0}^{\infty} \left(\frac{K}{\mu^{6\beta}}\right)^j \theta$$

$$\leq \frac{K\theta}{2^\beta \mu^{6\beta}} \left(1 - \frac{K}{\mu^{6\beta}}\right)^{-1}$$

$$\leq \frac{K\theta}{2^\beta(\mu^{6\beta} - K)}$$

for all $x \in \mathcal{A}$. ☐

Corollary 10.20 Let $\theta_1 \geq 0$ be fixed and $r \neq 6\beta$. If a mapping $f : \mathcal{A} \rightarrow \mathcal{B}$ satisfies the inequality

$$\|D_s f(x, y)\|_B \leq \theta_1(\|x\|_{\mathcal{A}}^r + \|y\|_{\mathcal{A}}^r)$$

for all $x, y \in \mathcal{A}$, then there exists a unique sextic mapping $S : \mathcal{A} \rightarrow \mathcal{B}$ satisfying (10.137) and

$$\|f(x) - S(x)\|_B \leq \begin{cases} \dfrac{2\theta_1 K}{2^\beta(\mu^{6\beta} - K\mu^r)}\|x\|_{\mathcal{A}}^r, & \text{for } r < 6\beta, \\[3mm] \dfrac{2\theta_1 K}{2^\beta(\mu^r - K\mu^{6\beta})}\|x\|_{\mathcal{A}}^r, & \text{for } r > 6\beta \end{cases}$$

for all $x \in \mathcal{A}$.

Proof: By choosing $\phi(x, y) = \theta_1(\|x\|_{\mathcal{A}}^r + \|y\|_{\mathcal{A}}^r)$, for all $x, y \in \mathcal{A}$ and $r < 6\beta$ in Theorem 10.23, we obtain

$$\|f(x) - S(x)\|_B \leq \frac{K}{2^\beta \mu^{6\beta}} \sum_{j=0}^{\infty} \frac{2\theta_1 K^j}{\mu^{6j\beta}} \mu^{jr} \|x\|_{\mathcal{A}}^r$$

$$\leq \frac{2\theta_1 K}{2^\beta \mu^{6\beta}} \sum_{j=0}^{\infty} (K\mu^{r-6\beta})^j \|x\|_{\mathcal{A}}^r$$

$$\leq \frac{2\theta_1 K}{2^\beta (\mu^{6\beta} - K\mu^r)} \|x\|_{\mathcal{A}}^r \qquad (10.168)$$

for all $x \in \mathcal{A}$ and $r > 6\beta$ in Theorem 10.24, we have

$$\|f(x) - S(x)\|_B \leq \frac{2\theta_1 K}{2^\beta \mu^r} \sum_{j=0}^{\infty} (K\mu^{6\beta-r})^j \|x\|_{\mathcal{A}}^r$$

$$\leq \frac{2\theta_K}{2^\beta \mu^r} (1 - K\mu^{6\beta-r})^{-1} \|x\|_{\mathcal{A}}^r$$

$$\leq \frac{2\theta_1 K}{2^\beta (\mu^r - K\mu^{6\beta})} \|x\|_{\mathcal{A}}^r \qquad (10.169)$$

for all $x \in \mathcal{A}$. Combining (10.168) and (10.169), we arrive at the required results. □

Corollary 10.21 Let $\theta_2 \geq 0$ be fixed and r, s such that $\gamma = r + s \neq 6\beta$. If a mapping $f : \mathcal{A} \to B$ satisfies the inequality

$$\|D_s f(x, y)\|_B \leq \theta_2 \|x\|_{\mathcal{A}}^r \|y\|_{\mathcal{A}}^s$$

for all $x, y \in \mathcal{A}$, then there exists a unique sextic mapping $S : \mathcal{A} \to B$ satisfying (10.137) and

$$\|f(x) - S(x)\|_B \leq \begin{cases} \dfrac{\theta_2 K}{2^\beta (\mu^{6\beta} - K\mu^\gamma)} \|x\|_{\mathcal{A}}^\gamma, & \text{for } \gamma < 6\beta \\[3mm] \dfrac{\theta_2 K}{2^\beta (\mu^\gamma - K\mu^{6\beta})} \|x\|_{\mathcal{A}}^\gamma, & \text{for } \gamma > 6\beta \end{cases}$$

for all $x \in \mathcal{A}$.

Proof: By replacing $\phi(x, y) = \theta_2 \|x\|_{\mathcal{A}}^r \|y\|_{\mathcal{A}}^s$, for all $x, y \in \mathcal{A}$ and considering $\gamma < 6\beta$ in Theorem 10.23, one have

$$\|f(x) - S(x)\|_B \leq \frac{\theta_2 K}{2^\beta \mu^{6\beta}} \sum_{j=0}^{\infty} (K\mu^{\gamma-6\beta})^j \|x\|_{\mathcal{A}}^\gamma$$

$$\leq \frac{\theta_2 K}{2^\beta \mu^{6\beta}} (1 - K\mu^{\gamma-6\beta})^{-1} \|x\|_{\mathcal{A}}^\gamma$$

$$\leq \frac{\theta_2 K}{2^\beta (\mu^{6\beta} - K\mu^\gamma)} \|x\|_{\mathcal{A}}^\gamma \qquad (10.170)$$

for all $x \in \mathcal{A}$ and assuming $\gamma > 6\beta$ in Theorem 10.24, we arrive

$$\|f(x) - S(x)\|_B \leq \frac{\theta_2 K}{2^\beta} \sum_{j=0}^{\infty} \frac{(K\mu^{6\beta})^j}{\mu^{j\gamma} \mu^\gamma} \|x\|_{\mathcal{A}}^\gamma$$

$$\leq \frac{\theta_2 K}{2^\beta \mu^\gamma} \sum_{j=0}^{\infty} (K\mu^{6\beta-\gamma})^j \|x\|_{\mathcal{A}}^\gamma$$

$$\leq \frac{\theta_2 K}{2^\beta (\mu^\gamma - K\mu^{6\beta})} \|x\|_{\mathcal{A}}^\gamma \tag{10.171}$$

for all $x \in \mathcal{A}$. From (10.170) and (10.171), we obtain the desired results. $\quad\square$

Corollary 10.22 Let $\theta_3 \geq 0$ be fixed and $\gamma \neq 6\beta$. If a mapping $f : \mathcal{A} \to \mathcal{B}$ satisfies the inequality

$$\|D_s f(x, y)\|_B \leq \theta_3 \left(\|x\|_{\mathcal{A}}^{\gamma/2} \|y\|_{\mathcal{A}}^{\gamma/2} + (\|x\|_{\mathcal{A}}^\gamma + \|y\|_{\mathcal{A}}^\gamma) \right)$$

for all $x, y \in \mathcal{A}$, then there exists a unique sextic mapping $S : \mathcal{A} \to \mathcal{B}$ satisfying (10.137) and

$$\|f(x) - S(x)\|_B \leq \begin{cases} \dfrac{3\theta_3 K}{2^\beta (\mu^{6\beta} - K\mu^\gamma)} \|x\|_{\mathcal{A}}^\gamma, & \text{for } \gamma < 6\beta, \\[3mm] \dfrac{3\theta_3 K}{2^\beta (\mu^\gamma - K\mu^{6\beta})} \|x\|_{\mathcal{A}}^\gamma, & \text{for } \gamma > 6\beta \end{cases}$$

for all $x \in \mathcal{A}$.

Proof: By putting $\phi(x, y) = \theta_3(\|x\|_{\mathcal{A}}^{\gamma/2} \|y\|_{\mathcal{A}}^{\gamma/2} + (\|x\|_{\mathcal{A}}^\gamma + \|y\|_{\mathcal{A}}^\gamma))$, for all $x, y \in \mathcal{A}$ and taking $\gamma < 6\beta$ in Theorem 10.23, we get

$$\|f(x) - S(x)\|_B \leq \frac{3\theta_3 K}{2^\beta \mu^{6\beta}} \sum_{j=0}^{\infty} (K\mu^{\gamma-6\beta})^j \|x\|_{\mathcal{A}}^\gamma$$

$$\leq \frac{3\theta_3 K}{2^\beta \mu^{6\beta}} (1 - K\mu^{\gamma-6\beta})^j \|x\|_{\mathcal{A}}^\gamma$$

$$\leq \frac{3\theta_3 K}{2^\beta (\mu^{6\beta} - K\mu^\gamma)} \|x\|_{\mathcal{A}}^\gamma \tag{10.172}$$

for all $x \in \mathcal{A}$ and considering $\gamma > 6\beta$ in Theorem 10.24, we have

$$\|f(x) - S(x)\|_B \leq \frac{3\theta_3 K}{2^\beta \mu^\gamma} \sum_{j=0}^{\infty} (K\mu^{6\beta-\gamma})^j \|x\|_{\mathcal{A}}^\gamma$$

$$\leq \frac{3\theta_3 K}{2^\beta \mu^\gamma} (1 - K\mu^{6\beta-\gamma})^{-1} \|x\|_{\mathcal{A}}^\gamma$$

$$\leq \frac{3\theta_3 K}{2^\beta (\mu^\gamma - K\mu^{6\beta})} \|x\|_{\mathcal{A}}^\gamma \tag{10.173}$$

for all $x \in \mathcal{A}$. Using (10.172) and (10.173), we arrive at the requisite results. $\quad\square$

References

1 Aczel, J. (1966) *Lectures on Functional Equations and their Applications*, vol. **19**, Academic Press, New York.
2 Ulam, S.M. (1960) *Problems in Modern Mathematics*, Rend. Chapter VI, John Wiley & Sons, Inc., New York.
3 Hyers, D.H. (1941) On the stability of the linear functional equation. *Proc. Natl. Acad. Sci. U.S.A.*, **27**, 222–224.
4 Aoki, T. (1950) On the stability of the linear transformation in Banach spaces. *J. Math. Soc. Jpn.*, **2**, 64–66.
5 Rassias, Th.M. (1978) On the stability of the linear mapping in Banach spaces. *Proc. Am. Math. Soc.*, **72**, 297–300.
6 Gajda, Z. (1991) On stability of additive mappings. *Int. J. Math. Math. Sci.*, **14** (3), 431–434.
7 Rassias, J.M. (1982) On approximately of approximately linear mappings by linear mappings. *J. Funct. Anal. U.S.A.*, **46**, 126–130.
8 Bouikhalene, B. and Elquorachi, E. (2007) Ulam-Gavruta-Rassias stability of the Pexider functional equation. *Int. J. Appl. Math. Stat.*, **7**, 7–39.
9 Nakmahachalasint, P. (2007) On the generalized Ulam-Găvrută-Rassias stability of a mixed type linear and Euler-Lagrange-Rassias functional equation. *Int. J. Math. Math. Sci.*, **2007**, 10.
10 Nakmahachalasint, P. (2007) Hyers-Ulam-Rassias and Ulam-Gavruta-Rassias stabilities of additive functional equation in several variables. *Int. J. Math. Math. Sci.*, **2007**, 6. Article ID 13437.
11 Park, C. and Najati, A. (2007) Homomorphisms and derivations in C*-algebras. *Abstr. Appl. Anal.*, **2007**, 12. Article ID 80630.
12 Pietrzyk, A. (2006) Stability of the Euler-Lagrange-Rassias functional equation. *Demonstr. Math.*, **39** (3), 523–530.
13 Sibaha, A., Bouikhalene, B., and Elquorachi, E. (2007) Ulam-Gavruta-Rassias stability for a linear fuctional equation. *Int. J. Appl. Math. Stat.*, **7**, 157–168.
14 Găvrută, P. (1994) A generalization of the Hyers-Ulam-Rassias stability of approximately additive mappings. *J. Math. Anal. Appl.*, **184**, 431–436.
15 Ravi, K., Arunkumar, M., and Rassias, J.M. (2008) Ulam stability for the orthogonally general Euler-Lagrange type functional equation. *Int. J. Math. Stat.*, **3** (A08), 36–46.
16 Ravi, K., Narasimman, P., and Kishore Kumar, R. (2009) Generalized Hyers-Ulam- Rassias stability and J.M. Rassias stability of a quadratic functional equation. *Int. J. Math. Sci. Eng. Appl.*, **3** (2), 79–94.
17 Jung, S.M. (1998) Hyers-Ulam-Rassias stability of Jensen's equation and its application. *Proc. Am. Math. Soc.*, **126** (11), 3137–3143.
18 Zhou, D.X. (1992) On a conjecture of Z. Ditzian. *J. Approx. Theory*, **69**, 167–172.
19 Malliavin, P. (1997) *Stochastic Analysis*, Springer-Verlag, Berlin.

20 Aczél, J. and Dhombres, J. (1989) *Functional Equations in Several Variables*, Cambridge University Press, Cambridge.

21 Chang, I.S. and Kim, H.M. (2002) On the Hyers-Ulam stability of quadratic functional equations. *J. Inequal. Appl. Math.*, **33**, 1–12.

22 Park, C., Lee, J.R., and Shin, D.Y. (2012) Generalized Ulam-Hyers stability of random homomorphisms in random normed algebras associated with the Cauchy functional equation. *Appl. Math. Lett.*, **25**, 200–205.

23 Rassias, J.M. (1992) On the stability of the Euler-Lagrange functional equation. *Chin. J. Math.*, **20**, 185–190.

24 Rassias, J.M. (1994) On the stability of the non-linear Euler-Lagrange functional equation in real normed linear spaces. *J. Math. Phys. Sci.*, **28**, 231–235.

25 Rassias, J.M. (1996) On the stability of the general Euler-Lagrange functional equation. *Demonstr. Math.*, **29**, 755–766.

26 Rassias, J.M. (1998) Solution of the Ulam stability problem for Euler-Lagrange quadratic mappings. *J. Math. Anal. Appl.*, **220**, 613–639.

27 Rassias, J.M. (1999) On the stability of the multi-dimensional Euler-Lagrange functional equation. *J. Indian Math. Soc.*, **66**, 1–9.

28 Koh, H. and Kang, D. (2013) Solution and stability of Euler-Lagrange-Rassias quartic functional equations in various quasi-normed spaces. *Abstr. Appl. Anal.*, **2013**, 8. Article ID 908168.

29 Park, C.G. (2007) Stability of an Euler-Lagrange-Rassias type additive mapping. *Int. J. Appl. Math. Stat.*, **7**, 101–111.

30 Adam, M. (2011) On the stability of some quadratic functional equation. *J. Nonlinear Sci. Appl.*, **4** (1), 50–59.

31 Czerwik, S. (1994) The stability of the quadratic functional equation, in *Stability of Mappings of Hyers-Ulam Type* (eds T.M. Rassias and J. Tabor), Hadronic Press, Palm Harbor, FL, pp. 81–91.

32 Eshaghi Gordji, M. and Khodaei, H. (2009) On the generalized Hyers-Ulam-Rassias stability of quadratic functional equations. *Abstr. Appl. Anal.*, **2009**, 11. Article ID 923476.

33 Jung, S.M. (1999) On the Hyers-Ulam-Rassias stability of a quadratic functional equation. *J. Math. Anal. Appl.*, **232**, 384–393.

34 Kannappan, P. (1995) Quadratic functional equation and inner product spaces. *Results Math.*, **27** (3-4), 368–372.

35 Lee, J.R., An, J.S., and Park, C. (2008) On the stability of quadratic functional equations. *Abstr. Appl. Anal.*, **2008**, 8. Article ID 628178.

36 Movahednia, E. (2013) Fixed point and generalized Hyers-Ulam-Rassias stability of a quadratic functional equation. *J. Math. Comput. Sci.*, **6**, 72–78.

37 Ravi, K., Murali, R., and Arunkumar, M. (2008) The generalized Hyers-Ulam-Rassias stability of a quadratic functional equation. *J. Inequal. Pure Appl. Math.*, **9** (1), 5. Article 20.

38 Ravi, K., Rassias, J.M., Arunkumar, M., and Kodandan, R. (2009) Stability of a generalized mixed type additive, quadratic, cubic and quartic functional equation. *J. Inequal. Pure Appl. Math.*, **10** (4), 1–29.

39 Wang, L. (2010) Intuitionistic fuzzy stability of a quadratic functional equation. *Fixed Point Theory Appl.*, **2010** (1), 7. Article ID 107182.

40 Zivari-Kazempour, A. and Eshaghi Gordji, M. (2012) Generalized Hyers-Ulam stabilities of an Euler-Lagrange-Rassias quadratic functional equation. *Asian-Eur. J. Math.*, **5** (1), doi: 10.1142/S1793557112500143.

41 Rassias, J.M. (2001) Solution of the Ulam stablility problem for cubic mappings. *Glasnik Matematicki. Serija III*, **36** (1), 63–72.

42 Jun, K.W. and Kim, H.M. (2003) On the Hyers-Ulam-Rassias stability of a general cubic functional equation. *Math. Inequal. Appl.*, **6**, 87–95.

43 Eskandani, G.Z., Rassias, J.M., and Gavruta, P. (2011) Generalized Hyers-Ulam stability for a general cubic functional equation in quasi-β-normed spaces. *Asian-Eur. J. Math.*, **4**, 413–425.

44 Jun, K.W. and Kim, H.M. (2007) On the stability of Euler-Lagrange type cubic mappings in quasi-Banach spaces. *J. Math. Anal. Appl.*, **332** (2), 1335–1350.

45 Mursaleen, M. and Mohiuddine, S.A. (2009) On stability of a cubic functional equation in intuitionistic fuzzy normed spaces. *Chaos, Solitons Fractals*, **42**, 2997–3005.

46 Mursaleen, M. and Ansari, K.J. (2013) Stability results in intuitionistic fuzzy normed spaces for a cubic functional equation. *Appl. Math. Inf. Sci.*, **7** (5), 1685–1692.

47 Najati, A. (2007) Hyers-Ulam-Rassias stability of a cubic functional equation. *Bull. Korean Math. Soc.*, **44** (4), 825–840.

48 Najati, A. (2007) The generalized Hyers-Ulam-Rassias stability of a cubic functional equation. *Turk. J. Math.*, **31**, 395–408.

49 Park, K.H. and Jung, Y.S. (2004) Stability of a cubic functional equation on groups. *Bull. Korean Math. Soc.*, **41** (2), 347–357.

50 Ravi, K., Rassias, J.M., and Narasimman, P. (2011) Stability of a cubic functional equation in fuzzy normed space. *J. Appl. Anal. Comput.*, **1** (3), 411–425.

51 Rassias, J.M. (1999) Solution of the Ulam stability problem for quartic mappings. *Glasnic Matematicki. Serija III*, **34** (2), 243–252.

52 Lee, S.H., Im, S.M., and Hwang, I.S. (2005) Quartic functional equations. *J. Math. Anal. Appl.*, **307**, 387–394.

53 Petapirak, M. and Nakmahachalasint, P. (2008) A quartic functional equation and its generalized Hyers-Ulam-Rassias stability. *Thai J. Math.*, (Special Issue), 77–84.

54 Bodaghi, A. (2014) Stability of a quartic functional equation. *Sci. World J.*, **2014**, 9. Article ID 752146.

55 Hengkrawit, C. and Thanyacharoen, A. (2013) A general solution of a quartic functional equation and its stability. *Int. J. Pure Appl. Math.*, **85** (4), 691–706.

56 Chung, J.K. and Sahoo, P.K. (2003) On the general solution of a quartic functional equation. *Bull. Korean Math. Soc.*, **40** (4), 565–576.

57 Saadati, R., Cho, Y.J., and Vahidi, J. (2010) The stability of the quartic functional equation in various spaces. *Comput. Math. Appl.*, **60**, 1994–2002.

58 Xu, T.Z., Rassias, J.M., Rassias, M.J., and Xu, W.X. (2010) A fixed point approach to the stability of quintic and sextic functional equations in quasi-β-normed spaces. *J. Inequal. Appl.*, **2010** (1), 23. Article ID 423231.

59 Jun, K.W. and Kim, H.M. (2005) On the Hyers-Ulam stability of a generalized quadratic and additive functional equation. *Bull. Korean Math. Soc.*, **42** (1), 133–148.

60 Rassias, J.M., Ravi, K., Arunkumar, M., and Senthil Kumar, B.V. (2010) Solution and Ulam stability of a mixed type cubic and additive functional equation, in *Functional equations, Difference Inequalities and Ulam stability Notions*, Chapter 13, Nova Science Publishers, pp. 149–175.

61 Eshaghi Gordji, M. (2010) Stability of a functional equation deriving from quartic and additive functions. *Bull. Korean Math. Soc.*, **47** (3), 491–502.

62 Towanlong, W. and Nakmahachalasint, P. (2008) A quadratic functional equation and its generalized Hyers-Ulam Rassias stability. *Thai J. Math.* Special Issue (Annual Meeting in Mathematics), 85–91.

63 Eshaghi Gordji, M., Savadkouhi, M.B., and Park, C. (2009) Quadratic-quartic functional equations in RN-spaces. *J. Inequal. Appl.*, **2009** (1), 14. Article ID 868423.

64 Eshaghi Gordji, M., Zolfaghari, S., Rassias, J.M., and Savadkouhi, M.B. (2009) Solution and stability of a mixed type cubic and quartic functional equation in quasi-Banach spaces. *Abstr. Appl. Anal.*, **2009**, 1–14. Article ID 417473.

65 Jun, K.W. and Kim, H.M. (2002) The generalized Hyers-Ulam-Rassias stability of a cubic functional equation. *J. Math. Anal. Appl.*, **274**, 867–878.

66 Park, C. (2010) Fuzzy stability of an additive-quadratic-quartic functional equation. *J. Inequal. Appl.*, **2010**, 22. Article ID 253040.

67 Eshaghi Gordji, M., Gharetapeh, S.K., Park, C., and Zolfaghri, S. (2009) Stability of an additive-cubic-quartic functional equation. *Adv. Differ. Equ.*, **2009**, 20. Article 395693.

68 Al-Fhaid, A.S. and Mohiuddine, S.A. (2013) On the Ulam stability of mixed type QA mappings in IFN-spaces. *Adv. Differ. Equ.*, **2013**, 1–16.

69 Cho, Y.J., Eshaghi Gordji, M., and Zolfaghari, S. (2010) Solutions and stability of generalized mixed type QC functional equations inrandom normed spaces. *J. Inequal. Appl.*, **2010**, 16. Article ID 403101 (pages.

70 Ebadian, A. and Zolfaghari, S. (2012) Stability of a mixed additive and cubic functional equation in several variables in non-Archimedean spaces. *Ann. Univ. Ferrara*, **58**, 291–306.

71 Mohamadi, M., Cho, Y.J., Park, C., Vetro, P., and Saadati, R. (2010) Random stability of an additive-quadratic-quartic functional equation. *J. Inequal. Appl.*, **2010**, 18. Article ID 754210.

72 Moradlou, F., Vaezi, H., and Eskandani, G.Z. (2009) Hyers-Ulam-Rassias stability of a quadratic and additive functional equation in quasi-Banach spaces. *Mediterr. J. Math.*, **6**, 233–248.

73 Najati, A. and Park, C. (2013) Cauchy-Jensen additive mappings in quasi-Banach algebras and its applications. *J. Nonlinear Anal. Appl.*, **2013**, 1–16.

74 Park, C., Jo, S.W., and Kho, D.Y. (2009) On the stability of an AQCQ-functional equation. *J. Chungcheong Math. Soc.*, **22** (4), 757–770.

75 Park, C., Azadi Kenary, H., and Rassias, Th.M. (2012) Hyers-Ulam-Rassias stability of the additive-quadratic mappings in non-Archimedean Banach spaces. *J. Inequal. Appl.*, **2012**, 174.

76 Xu, T.Z., Rassias, J.M., and Xu, W.X. (2011) A fixed point approach to the stability of a general mixed additive-cubic functional equation in quasi fuzzy normed spaces. *Int. J. Phys. Sci.*, **6** (2), 313–324.

77 Isac, G. and Rassias, Th.M. (1996) Stability of ψ-additive mappings: applications to nonlinear analysis. *Int. J. Math. Math. Sci.*, **19** (2), 219–228.

78 Cadariu, L. and Radu, V. (2009) Fixed points and stability for functional equations in probabilistic metric and random normed spaces. *Fixed Point Theory Appl.*, **2009**, 18. Article ID 589143.

79 Xu, T.Z., Rassias, J.M., and Xu, W.X. (2010) A fixed point approach to the stability of a general mixed *AQCQ*-functional equation in non-Archimedean normed spaces. *Discrete Dyn. Nat. Soc.*, **2010**, 24. Article ID 812545.

80 Chang, I.S. and Jung, Y.S. (2003) Stability of functional equations deriving from cubic and quadratic functions. *J. Math. Anal. Appl.*, **283**, 491–500.

81 Ghobadipour, N. and Park, C. (2010) Cubic-quartic functional equations in fuzzy normed spaces. *Int. J. Nonlinear Anal. Appl.*, **1**, 12–21.

82 Lee, J.R., Shin, D.Y., and Park, C. (2013) Hyers-Ulam stability of functional equations in matrix normed spaces. *J. Inequal. Appl.*, **2013**, 22.

83 Aczel, J. (1984) *Functional Equations, History, Applications and Theory*, D. Reidel Publishing Company.

84 Alsina, C. (1987) *On the Stability of a Functional Equation*, General Inequalities, vol. **5**, Oberwolfach, Birkhauser, Basel, pp. 263–271.

85 Czerwik, S. (2002) *Functional Equations and Inequalities in Several Variables*, World Scientific Publishing Company, New Jersey, London, Singapore, Hong Kong.

86 Hyers, D.H., Isac, G., and Rassias, Th.M. (1998) *Stability of Functional Equations in Several Variables*, Birkhauser, Basel.

87 Jung, S.M. (2001) *Hyers-Ulam-Rassias Stability of Functional Equations in Mathematical Analysis*, Hardonic press, Palm Harbor, FL.

88 Ravi, K. and Senthil Kumar, B.V. (2010) Ulam-Găvruta-Rassias stability of Rassias reciprocal functional equation. *Global J. Appl. Math. Math. Sci.,* **3** (1-2), 57–79.

89 Ravi, K. and Senthil Kumar, B.V. (2015) Generalized Hyers-Ulam-Rassias stability of a sytem of bi-reciprocal functional equations. *Eur. J. Pure Appl. Math.,* **8** (2), 283–293.

11

Attractor of the Generalized Contractive Iterated Function System

Mujahid Abbas[1,2] and Talat Nazir[3,4]

[1]Department of Mathematics, Government College University, Katchery Road, Lahore 54000, Pakistan
[2]Department of Mathematics, King Abdulaziz University, Jeddah 21589, Saudi Arabia
[3]Department of Mathematics, University of Jeddah, Jeddah 21589, Saudi Arabia
[4]Department of Mathematics, COMSATS Institute of Information Technology, Abbottabad 22060, Pakistan

2010 AMS Subject Classification 47H10, 54E50, 54H25

11.1   Iterated Function System

Iterated function systems (IFS) are based on the mathematical foundations laid by Hutchinson [1]. He showed that the Hutchinson operator constructed with the help of a finite system of contraction mappings defined on an Euclidean space $\mathbb{R}^n$ has closed and bounded subset of $\mathbb{R}^n$ as its fixed point, called attractor of IFS [2]. In this context, fixed point theory plays significant and vital role to help in construction of fractals.

Fixed point theory is studied in an environment created with appropriate mappings satisfying certain conditions. Recently, many researchers have obtained fixed point results for single and multivalued mappings defined on metric spaces.

Banach contraction principle [3], one of the basic and the most widely applied fixed point theorems in all of analysis reads as follows:

Theorem 11.1   Let (X, d) be a complete metric space and $f : X \to X$ a contraction on X with contraction constant $\alpha \in [0, 1)$, that is, for any $x, y \in X$, the following holds:

$$d(fx, fy) \leq \alpha d(x, y). \tag{11.1}$$

Then f has a unique fixed point in X. Furthermore, for any initial guess $x_0 \in X$, the sequence of simple iterates $\{x_0, fx_0, f^2x_0, f^3x_0, \ldots\}$ converges to a fixed point of f.

Mathematical Analysis and Applications: Selected Topics, First Edition.
Edited by Michael Ruzhansky, Hemen Dutta, and Ravi P. Agarwal.
© 2018 John Wiley & Sons, Inc. Published 2018 by John Wiley & Sons, Inc.

Banach contraction principle [3] is of paramount importance in metrical fixed point theory with a wide range of applications, including iterative methods for solving linear, nonlinear, differential, integral, and difference equations. This initiated several researchers to extend and enhance the scope of metric fixed point theory. As a result, Banach contraction principle has been extended either by generalizing the domain of the mapping [4–10] or by extending the contractive condition on the mappings [11–16]. There are certain cases when the range X of a mapping is replaced with a family of sets possessing some topological structure and consequently a single-valued mapping is replaced with a multivalued mapping. Nadler [17] was the first who combined the ideas of multivalued mappings and contractions and hence initiated the study of metric fixed point theory of multivalued operators, see also [18–20]. The fixed point theory of multivalued operators provides important tools and techniques to solve the problems of pure, applied, and computational mathematics, which can be restructured as an inclusion equation for an appropriate multivalued operator.

The purpose of this chapter is to construct a fractal set of iterated function system, a certain finite collection of mappings defined on a metric space, which induce compact-valued mappings defined on a family of compact subsets of a metric space. We prove that Hutchinson operator defined with the help of a finite family of F-contraction mappings on a complete metric space is itself generalized F-contraction mapping on a family of compact subsets of X. We then obtain a final fractal obtained by successive application of a generalized F–Hutchinson operator. A nontrivial example is presented to support the result proved herein.

In what follows, the letters $\mathbb{R}$, $\mathbb{R}^n$, $\mathbb{R}_+$, and $\mathbb{N}$ will denote the set of all real numbers, the set of all n-tuples of real numbers, the set of all positive real numbers, and the set of all natural numbers, respectively.

Definition 11.1   Let (X, d) be a metric space and $C \subseteq X$. Then C is compact if every sequence $\{x_n\}$ in C contains a subsequence having a limit in C. Note that closed and bounded subsets of $\mathbb{R}^n$ are compact. In addition, every finite set in $\mathbb{R}^n$ is compact. On the other hand, $(0, 1] \subset \mathbb{R}$ is not compact as $\{1, \frac{1}{2}, \frac{1}{2^2}, \dots\} \subset (0, 1]$ does not have any convergent subsequence. In addition, $\mathbb{Z} \subset \mathbb{R}$ is not compact.

Let (X, d) be a metric space and $\mathcal{H}(X)$ denotes the set of all nonempty compact subsets of X. For $A, B \in \mathcal{H}(X)$, let

$$H(A, B) = \max\{\sup_{b \in B} \mathrm{d}(b, A), \sup_{a \in A} \mathrm{d}(a, B)\},$$

where $\mathrm{d}(x, B) = \inf\{\mathrm{d}(x, b) : b \in B\}$ is the distance of a point x from the set B. The mapping H is said to be the Pompeiu–Hausdorff metric induced by d. If (X, d) is a complete metric space, then $(\mathcal{H}(X), H)$ is also a complete metric space.

For the sake of completeness, we state and prove the following lemma.

Lemma 11.1   Let (X, d) be a metric space. For all $A, B, C, D \in \mathcal{H}(X)$, the following hold:

(i)   If $B \subseteq C$, then $\sup\limits_{a \in A} d(a, C) \leq \sup\limits_{a \in A} d(a, B)$.

(ii) $\sup\limits_{x \in A \cup B} d(x, C) = \max\{\sup\limits_{a \in A} d(a, C), \sup\limits_{b \in B} d(b, C)\}$.

(iii) $H(A \cup B, C \cup D) \leq \max\{H(A, C), H(B, D)\}$.

Proof: To prove (i): Since $B \subseteq C$, for all $a \in A$, we have

$$d(a, C) = \inf\{d(a, c) : c \in C\}$$
$$\leq \inf\{d(a, b) : b \in B\} = d(a, B),$$

which implies that

$$\sup\limits_{a \in A} d(a, C) \leq \sup\limits_{a \in A} d(a, B).$$

To prove (ii):

$$\sup\limits_{x \in A \cup B} d(x, C) = \sup\{d(x, C) : x \in A \cup B\}$$
$$= \max\{\sup\{d(x, C) : x \in A\}, \sup\{d(x, C) : x \in B\}\}$$
$$= \max\{\sup\limits_{a \in A} d(a, C), \sup\limits_{b \in B} d(b, C)\}.$$

To prove (iii): Note that

$$\sup\limits_{x \in A \cup B} d(x, C \cup D)$$
$$\leq \max\{\sup\limits_{a \in A} d(a, C \cup D), \sup\limits_{b \in B} d(b, C \cup D)\} \quad \text{(by using (ii))}$$
$$\leq \max\{\sup\limits_{a \in A} d(a, C), \sup\limits_{b \in B} d(b, D)\} \quad \text{(by using (i))}$$
$$\leq \max\left\{ \max\{\sup\limits_{a \in A} d(a, C), \sup\limits_{c \in C} d(c, A)\}, \max\{\sup\limits_{b \in B} d(b, D), \sup\limits_{u \in D} d(u, B)\} \right\}$$
$$= \max\{H(A, C), H(B, D)\}.$$

In the similar way, we obtain that

$$\sup\limits_{y \in C \cup D} d(y, A \cup B) \leq \max\{H(A, C), H(B, D)\}.$$

Hence it follows that

$$H(A \cup B, C \cup D) = \max\{ \sup\limits_{x \in A \cup B} d(x, C \cup D), \sup\limits_{y \in C \cup D} d(y, A \cup B)\}$$
$$\leq \max\{H(A, C), H(B, D)\}. \qquad \square$$

Wardowski [21] introduced a new contraction called F-contraction and proved a fixed point result as an interesting generalization of the Banach contraction principle.

Consistent with [21], the following definition and examples are needed.

Let F be the collection of all continuous mappings $F : \mathbb{R}_+ \to \mathbb{R}$ that satisfy the following conditions:

(F_1) F is strictly increasing, that is, for all $\alpha, \beta \in \mathbb{R}_+$ such that $\alpha < \beta$ implies that $F(\alpha) < F(\beta)$.

(F_2) For every sequence $\{\alpha_n\}$ of positive real numbers, $\lim\limits_{n \to \infty} \alpha_n = 0$ and $\lim\limits_{n \to \infty} F(\alpha_n) = -\infty$ are equivalent.

(F_3) There exists $k \in (0, 1)$ such that $\lim\limits_{\alpha \to 0_+} \alpha^k F(\alpha) = 0$.

Definition 11.2 [21] Let (X, d) be a metric space. A self-mapping f on X is called an F-contraction if for any $x, y \in X$, there exists $F \in F$ and $\tau > 0$ such that

$$\tau + F(d(fx, fy)) \leq F(d(x, y)), \tag{11.2}$$

whenever $d(fx, fy) > 0$.

From (F_1) and (11.2), we conclude that

$$d(fx, fy) < d(x, y), \quad \text{for all} \quad x, y \in X, \quad fx \neq fy.$$

Indeed from (11.2), for all $x, y \in X$ with $d(fx, fy) > 0$, we have

$$F(d(fx, fy)) < F(d(x, y)).$$

Since F is strictly increasing (F_1), it follows that

$$d(fx, fy) < d(x, y) \text{ for all } x, y \in X \text{ whenever } fx \neq fy.$$

Thus, every F-contraction mapping is contractive, and in particular, every F-contraction mapping is continuous.

Following examples show that there are varieties of contractive conditions corresponding to different choices of elements in F.

Example 11.1 Let $F : \mathbb{R}_+ \to \mathbb{R}$ be defined by $F(\lambda) = \ln(\lambda)$ for $\lambda > 0$. Then F satisfies (F_1)–(F_3). A mapping $f : X \to X$ satisfying (11.2) is a contraction with contractive factor $e^{-\tau}$, that is,

$$d(fx, fy) \leq e^{-\tau} d(x, y), \quad \text{for all } x, y \in X, \quad fx \neq fy. \tag{11.3}$$

It is clear that for $x, y \in X$ such that $fx = fy$ the inequality $d(fx, fy) \leq e^{-\tau} d(x, y)$ holds.

Example 11.2 If we take $F(\lambda) = \ln(\lambda) + \lambda$, $\lambda > 0$, then F satisfies (F_1)–(F_3) and (11.2) is of the form

$$\frac{d(fx, fy)}{d(x, y)} e^{d(fx, fy) - d(x, y)} \leq e^{-\tau}, \quad \text{for all } x, y \in X, \quad fx \neq fy. \tag{11.4}$$

Example 11.3 Consider $F(\lambda) = -1/\sqrt{\lambda}$ for $\lambda > 0$, then $F \in \mathcal{F}$. In this case, F-contraction mapping f satisfies

$$d(fx, fy) \leq \frac{1}{(1 + \tau\sqrt{d(x, y)})^2} d(x, y), \quad \text{for all } x, y \in X, \ fx \neq fy. \quad (11.5)$$

Note that, the above is a special case of nonlinear contraction of the type $d(fx, fy) \leq \psi(d(x, y))d(x, y)$ for all $x, y \in X, fx \neq fy$. For details, see [11, 15].

Example 11.4 Let $F(\lambda) = \ln(\lambda^2 + \lambda)$, $\lambda > 0$. Then F satisfies (F_1)–(F_3) and the mapping f satisfies the following condition

$$\frac{d(fx, fy)(d(fx, fy) + 1)}{d(x, y)(d(x, y) + 1)} \leq e^{-\tau}, \quad \text{for all } x, y \in X, \ fx \neq fy. \quad (11.6)$$

In all above Examples, conditions (11.3)–(11.6) are satisfied for any $x, y \in X$ with $fx = fy$.

Theorem 11.2 [21] Let (X, d) be a complete metric space and $f : X \to X$ an F-contraction mapping. Then f has a unique fixed point in X and for every x_0 in X, a sequence of iterates $\{x_0, fx_0, f^2x_0, \ldots\}$ converges to the fixed point of f.

Theorem 11.3 Let (X, d) be a metric space and $f : X \to X$ an F contraction. Then

(1) f maps elements in $\mathcal{H}(X)$ to elements in $\mathcal{H}(X)$.
(2) If for any $A \in \mathcal{H}(X)$,

$$f(A) = \{f(x) : x \in A\}.$$

Then $f : \mathcal{H}(X) \to \mathcal{H}(X)$ is a F-contraction mapping on $(\mathcal{H}(X), H)$.

Proof: As F-contraction mapping is continuous. The image of a compact subset under $f : X \to X$ is compact, that is, $A \in \mathcal{H}(X)$ implies $f(A) \in \mathcal{H}(X)$.

To prove (2): Let $A, B \in \mathcal{H}(X)$ with $H(f(A), f(B)) \neq \emptyset$. Since $f : X \to X$ is F contraction, we obtain that

$$0 < d(fx, fy) < d(x, y) \quad \text{for all } x, y \in X, \ x \neq y.$$

Thus, we have

$$d(fx, f(B)) = \inf_{y \in B} d(fx, fy) < \inf_{y \in B} d(x, y) = d(x, B).$$

In addition,

$$d(fy, f(A)) = \inf_{x \in A} d(fy, fx) < \inf_{x \in A} d(y, x) = d(y, A).$$

Now

$$H(f(A), f(B)) = \max\{\sup_{x \in A} d(fx, f(B)), \sup_{y \in B} d(fy, f(A))\}$$
$$< \max\{\sup_{x \in A} d(x, B), \sup_{y \in B} d(y, A)\} = H(A, B).$$

Since F is strictly increasing,

$$F(H(f(A), f(B))) < F(H(A, B)).$$

Consequently, there exists some $\tau^* > 0$ such that

$$\tau^* + F(H(f(A), f(B))) \leq F(H(A, B)).$$

Hence $f : \mathcal{H}(X) \to \mathcal{H}(X)$ is a F contraction. □

Theorem 11.4 Let (X, d) be a metric space and $\{f_n : n = 1, 2, \ldots, N\}$, a finite family of F-contraction self-mappings on X. Define $T : \mathcal{H}(X) \to \mathcal{H}(X)$ by

$$T(A) = f_1(A) \cup f_2(A) \cup \cdots \cup f_N(A)$$
$$= \cup_{n=1}^{N} f_n(A),$$

for each $A \in \mathcal{H}(X)$. Then T is F contraction on $\mathcal{H}(X)$.

Proof: We demonstrate the claim for $N = 2$. Let $f_1, f_2 : X \to X$ be two F contractions. Take $A, B \in \mathcal{H}(X)$ with $H(T(A), T(B)) \neq 0$. From Lemma 11.1 (iii), it follows that

$$\tau + F(H(T(A), T(B))) = \tau + F(H(f_1(A) \cup f_2(A), f_1(B) \cup f_2(B)))$$
$$\leq \tau + F(\max\{H(f_1(A), f_1(B)), H(f_2(A), f_2(B))\})$$
$$\leq F(H(A, B)). \qquad \square$$

Theorem 11.5 Let (X, d) be a complete metric space and $\{f_n : n = 1, 2, \ldots, N\}$, a finite family of F contractions on X. Define a mapping on $\mathcal{H}(X)$ as

$$T(A) = f_1(A) \cup f_2(A) \cup \cdots \cup f_N(A)$$
$$= \cup_{n=1}^{N} f_n(A),$$

for each $A \in \mathcal{H}(X)$. Then

(1) $T : \mathcal{H}(X) \to \mathcal{H}(X)$;
(2) T has a unique fixed point $U \in \mathcal{H}(X)$, that is $U = T(U) = \cup_{n=1}^{k} f_n(U)$;
(3) for any initial set $A_0 \in \mathcal{H}(X)$, the sequence of compact sets $\{A_0, T(A_0), T^2(A_0), \ldots\}$ converges to a fixed point of T.

Proof: (1) Since each f_i is F contraction, therefore from definition of T and Theorem 11.3 the conclusion follows immediately. (2) From Theorem 11.4, $T : \mathcal{H}(X) \to \mathcal{H}(X)$ is F−contraction. Moreover, the completeness of (X, d) implies that $(\mathcal{H}(X), H)$ is complete. Consequently, (2) and (3) follow from Theorem 11.2. □

Definition 11.3 Let (X, d) be a metric space. A mapping $T : \mathcal{H}(X) \to \mathcal{H}(X)$ is said to be a generalized F contraction if there exists $F \in \mathcal{F}$ and $\tau > 0$ such that for any $A, B \in \mathcal{H}(X)$ with $H(T(A), T(B)) \neq 0$, the following holds:

$$\tau + F(H(T(A), T(B))) \leq F(M_T(A, B)), \qquad (11.7)$$

where $M_T(A, B) = \max\{H(A, B), H(A, T(A)), H(B, T(B)),$
$$\frac{H(A, T(B)) + H(B, T(A))}{2},$$
$$H(T^2(A), T(A)), H(T^2(A), B), H(T^2(A), T(B))\}.$$

The operator T defined above is also called generalized F-Hutchinson operator. Note that if T defined in Theorem 11.5 is F contraction, then it is trivially generalized F contraction and so T is generalized F-Hutchinson operator. The converse does not hold [22].

Definition 11.4 Let X be a metric space. If $f_n : X \to X$, $n = 1, 2, \dots, N$ are F-contraction mappings, then $(X; f_1, f_2, \dots, f_N)$ is called generalized (F-contractive) IFS.

Thus, the generalized IFS consists of a metric space and finite family of F-contraction mappings on X.

Definition 11.5 A nonempty compact set $A \subseteq X$ is said to be an attractor of the generalized F-contractive IFS if

(a) $T(A) = A$ and
(b) there is an open set $V \subseteq X$ such that $A \subseteq V$ and $\lim_{k \to \infty} T^k(B) = A$ for any compact set $B \subseteq V$, where the limit is taken with respect to the Hausdorff metric.

The largest open set V satisfying (b) is called a basin of attraction.

11.2 Generalized *F*-contractive Iterated Function System

We start with the following result. In this result, we prove the existence of fixed point of generalized F-contraction operator T.

Theorem 11.6 Let (X, d) be a complete metric space and $\{X : f_n, n = 1, 2, \dots, k\}$ a generalized IFS. Let $T : \mathcal{H}(X) \to \mathcal{H}(X)$ be defined by

$$T(A) = f_1(A) \cup f_2(A) \cup \cdots \cup f_N(A)$$
$$= \cup_{n=1}^N f_n(A),$$

for each $A \in \mathcal{H}(X)$. If T is a generalized F-Hutchinson operator, then T has a unique fixed point $U \in \mathcal{H}(X)$, that is

$$U = T(U) = \cup_{n=1}^k f_n(U).$$

Moreover, for any initial set $A_0 \in \mathcal{H}(X)$, the sequence of compact sets $\{A_0, T(A_0), T^2(A_0), \dots\}$ converges to a fixed point of T.

Proof: Let A_0 be an arbitrary element in $\mathcal{H}(X)$. If $A_0 = T(A_0)$, then the proof is finished. Therefore, we assume that $A_0 \neq T(A_0)$. Define

$$A_1 = T(A_0),\ A_2 = T(A_1),\ \ldots,\ A_{m+1} = T(A_m)$$

for $m \in \mathbb{N}$.

We may assume that $A_m \neq A_{m+1}$ for all $m \in \mathbb{N}$. If not, then $A_k = A_{k+1}$ for some k implies $A_k = T(A_k)$, and this completes the proof. Take $A_m \neq A_{m+1}$ for all $m \in \mathbb{N}$. From (11.7), we have

$$\tau + F(H(A_{m+1}, A_{m+2})) = \tau + F(H(T(A_m), T(A_{m+1})))$$
$$\leq F(M_T(A_m, A_{m+1})),$$

where

$$M_T(A_m, A_{m+1}) = \max\{H(A_m, A_{m+1}), H(A_m, T(A_m)), H(A_{m+1}, T(A_{m+1})),$$
$$\frac{H(A_m, T(A_{m+1})) + H(A_{m+1}, T(A_m))}{2},$$
$$H(T^2(A_m), T(A_m)), H(T^2(A_m), A_{m+1}),$$
$$H(T^2(A_m), T(A_{m+1}))\}$$
$$= \max\{H(A_m, A_{m+1}), H(A_m, A_{m+1}), H(A_{m+1}, A_{m+2}),$$
$$\frac{H(A_m, A_{m+2}) + H(A_{m+1}, A_{m+1})}{2},$$
$$H(A_{m+2}, A_{m+1}), H(A_{m+2}, A_{m+1}), H(A_{m+2}, A_{m+2})\}$$
$$\leq \max\{H(A_m, A_{m+1}), H(A_{m+1}, A_{m+2}),$$
$$\frac{H(A_m, A_{m+1}) + H(A_{m+1}, A_{m+2})}{2}\}$$
$$= \max\{H(A_m, A_{m+1}), H(A_{m+1}, A_{m+2})\}.$$

Thus, we have

$$\tau + F(H(A_{m+1}, A_{m+2})) \leq F(\max\{H(A_m, A_{m+1}), H(A_{m+1}, A_{m+2})\})$$
$$= F(H(A_m, A_{m+1})),$$

that is,

$$F(H(A_{m+1}, A_{m+2})) \leq F(H(A_m, A_{m+1})) - \tau$$

for all $m \in \mathbb{N}$. Therefore,

$$F(H(A_n, A_{n+1})) \quad \leq \quad F(H(A_{n-1}, A_n)) - \tau$$
$$\leq \quad F(H(A_{n-2}, A_{n-1})) - 2\tau$$
$$\leq \cdots \leq F(H(A_0, A_1)) - n\tau,$$

and we obtain that $\lim\limits_{n \to \infty} F(H(A_n, A_{n+1})) = -\infty$ that together with (F_2) implies that

$$\lim\limits_{n \to \infty} H(A_n, A_{n+1}) = 0.$$

Now by (F_3), there exists $h \in (0,1)$ such that

$$\lim_{n \to \infty} [H(A_n, A_{n+1})]^h F(H(A_n, A_{n+1})) = 0.$$

Thus, we have

$$[H(A_n, A_{n+1})]^h F(H(A_n, A_{n+1})) - [H(A_n, A_{n+1})]^h F(H(A_0, A_{n+1}))$$
$$\leq -n\tau [H(A_n, A_{n+1})]^h \leq 0.$$

On taking limit as $n \to \infty$, we obtain

$$\lim_{n \to \infty} n[H(A_n, A_{n+1})]^h = 0.$$

As $\lim_{n \to \infty} n^{\frac{1}{h}} H(A_n, A_{n+1}) = 0$, so there exists $n_1 \in \mathbb{N}$ such that

$$n^{\frac{1}{h}} H(A_n, A_{n+1}) \leq 1$$

for all $n \geq n_1$. Therefore, we have

$$H(A_n, A_{n+1}) \leq \frac{1}{n^{1/h}}$$

for all $n \geq n_1$. For $m, n \in \mathbb{N}$ with $m > n \geq n_1$, we have

$$H(A_n, A_m) \leq H(A_n, A_{n+1}) + H(A_{n+1}, A_{n+2}) + \cdots + H(A_{m-1}, A_m)$$
$$\leq \sum_{i=n}^{\infty} \frac{1}{i^{1/h}}.$$

By the convergence of the series $\sum_{i=1}^{\infty} \frac{1}{i^{1/h}}$, we get $H(A_n, A_m) \to 0$ as $n, m \to \infty$. Therefore, $\{A_n\}$ is a Cauchy sequence in X. Since $(\mathcal{H}(X), H)$ is complete, we have $A_n \to U$ as $n \to \infty$ for some $U \in \mathcal{H}(X)$.

In order to show that U is the fixed point of T, we contrary assume that Pompeiu–Hausdorff weight assigned to the U and $T(U)$ is not zero. Now

$$\tau + F(H(A_{n+1}, T(U))) = \tau + F(H(T(A_n), T(U))) \leq F(M_T(A_n, U)), \quad (11.8)$$

where

$$M_T(A_n, U) = \max\{H(A_n, U), H(A_n, T(A_n)), H(U, T(U)),$$
$$\frac{H(A_n, T(U)) + H(U, T(A_n))}{2}, H(T^2(A_n), T(A_n)),$$
$$H(T^2(A_n), U), H(T^2(A_n), T(U))\}$$
$$= \max\{H(A_n, U), H(A_n, A_{n+1}), H(U, T(U)),$$
$$\frac{H(A_n, T(U)) + H(U, A_{n+1})}{2}, H(A_{n+2}, A_{n+1}),$$
$$H(A_{n+2}, U), H(A_{n+2}, T(U))\}.$$

Now we consider the following cases:

(1) If $M_T(A_n, U) = H(A_n, U,)$, then on taking limit as $n \to \infty$ in (11.8), we have

$$\tau + F(H(T(U), U)) \leq F(H(U, U)),$$

a contradiction.

(2) When $M_T(A_n, U) = H(A_n, A_{n+1})$, then

$$\tau + F(H(T(U), U)) \leq F(H(U, U)),$$

gives a contradiction.

(3) In case $M_T(A_n, U) = H(U, T(U))$, then on taking limit as $n \to \infty$ in (11.8), we get

$$\tau + F(H(T(U), U)) \leq F(H(U, T(U))),$$

a contradiction.

(4) If $M_T(A_n, U) = \frac{H(A_n, T(U)) + H(U, A_{n+1})}{2}$, then on taking limit as $n \to \infty$, we have

$$\tau + F(H(T(U), U)) \leq F\left(\frac{H(U, T(U)) + H(U, U)}{2}\right)$$

$$= F\left(\frac{H(U, T(U))}{2}\right),$$

a contradiction.

(5) When $M_T(A_n, U) = H(A_{n+2}, A_{n+1})$, then

$$\tau + F(H(T(U), U)) \leq F(H(U, U)),$$

gives a contradiction.

(6) In case $M_T(A_n, U) = H(A_{n+2}, U)$, then on taking limit as $n \to \infty$ in (11.8), we get

$$\tau + F(H(T(U), U)) \leq F(H(U, U)),$$

a contradiction.

(7) Finally if $M_T(A_n, U) = H(A_{n+2}, T(U))$, then on taking limit as $n \to \infty$, we have

$$\tau + F(H(T(U), U)) \leq F(H(U, T(U))),$$

a contradiction.

Thus, U is the fixed point of T.

To show the uniqueness of fixed point of T, assume that U and V are two fixed points of T with $H(U, V)$ is not zero. Since T is a F-contraction map, we obtain that

$$\tau + F(H(U, V)) = \tau + F(H(T(U), T(V)))$$
$$\leq F(\max\{H(U, V), H(U, T(U)), H(V, T(V)),$$
$$\frac{H(U, T(V)) + H(V, T(U))}{2},$$
$$H(T^2(U), U), H(T^2(U), V), H(T^2(U), T(V))\})$$
$$= F(\max\{H(U, V), H(U, U), H(V, V), \frac{H(U, V) + H(V, U)}{2},$$
$$H(U, U), H(U, V), H(U, V)\})$$
$$= F(H(U, V)),$$

a contradiction as $\tau > 0$. Thus, T has a unique fixed point $U \in H(X)$. □

Remark 11.1   In Theorem 11.6, if we take $S(X)$ the collection of all singleton subsets of X, then clearly $S(X) \subseteq H(X)$. Moreover, consider $f_n = f$ for each n, where $f = f_1$ then the mapping T becomes

$$T(x) = f(x).$$

With this setting, we obtain the following fixed point result.

Corollary 11.1   Let (X, d) be a complete metric space and $\{X : f_n, n = 1, 2, \ldots, k\}$ a generalized IFS. Let $f : X \to X$ be a mapping defined as in Remark 2.2. If there exists some $F \in F$ and $\tau > 0$ such that for any $x, y \in H(X)$ with $d(f(x), f(y)) \neq 0$, the following holds:

$$\tau + F(d(fx, fy)) \leq F(M_f(x, y)),$$

where

$$M_T(x, y) = \max\{d(x, y), d(x, fx), d(y, fy), \frac{d(x, fy) + d(y, fx)}{2},$$
$$d(f^2x, y), d(f^2x, fx), d(f^2x, fy)\}.$$

Then f has a unique fixed point $x \in X$, Moreover, for any initial set $x_0 \in X$, the sequence of compact sets $\{x_0, fx_0, f^2x_0, \ldots\}$ converges to a fixed point of f.

Corollary 11.2   Let (X, d) be a complete metric space and $(X; f_n, n = 1, 2, \ldots, k)$ be IFS where each f_i for $i = 1, 2, \ldots, k$ is a contraction self-mapping on X. Then $T : H(X) \to H(X)$ defined in Theorem 2.1 has a unique fixed point in $H(X)$. Furthermore, for any set $A_0 \in H(X)$, the sequence of compact sets $\{A_0, T(A_0), T^2(A_0), \ldots\}$ converges to a fixed point of T.

Proof: It follows from Theorem 1.13 that if each f_i for $i = 1, 2, \ldots, k$ is a contraction mapping on X, then the mapping $T : H(X) \to H(X)$ defined by

$$T(A) = \cup_{n=1}^{k} f_n(A),$$

for all $A \in \mathcal{H}(X)$ is contraction on $\mathcal{H}(X)$. Using Theorem 11.6, the result follows. $\qquad\square$

Corollary 11.3 Let (X, d) be a complete metric space and $(X; f_n, n = 1, 2, \ldots, k)$ an IFS where each f_i for $i = 1, 2, \ldots, k$ is a mapping on X satisfying

$$d(f_i x, f_i y)\, e^{d(f_i x, f_i y) - d(x,y)} \le e^{-\tau} d(x, y), \quad \text{for all } x, y \in X,\ f_i x \ne f_i y,$$

where $\tau > 0$. Then the mapping $T : \mathcal{H}(X) \to \mathcal{H}(X)$ defined in Theorem 2.1 has a unique fixed point in $\mathcal{H}(X)$. Furthermore, for any set $A_0 \in \mathcal{H}(X)$, the sequence of compact sets $\{A_0, T(A_0), T^2(A_0), \ldots\}$ converges to a fixed point of T.

Proof: Take $F(\lambda) = \ln(\lambda) + \lambda$, $\lambda > 0$ in Theorem 11.4, then each mapping f_i for $i = 1, 2, \ldots, k$ on X satisfies

$$d(f_i x, f_i y)\, e^{d(f_i x, f_i y) - d(x,y)} \le e^{-\tau} d(x, y), \quad \text{for all } x, y \in X,\ f_i x \ne f_i y,$$

where $\tau > 0$. Again from Theorem 11.4, the mapping $T : \mathcal{H}(X) \to \mathcal{H}(X)$ defined by

$$T(A) = \cup_{n=1}^{k} f_n(A),$$

for all $A \in \mathcal{H}(X)$ satisfies

$$H(T(A), T(B)) e^{H(T(A), T(B)) - H(A,B)} \le e^{-\tau} H(A, B),$$

for all $A, B \in \mathcal{H}(X)$, $H(T(A), T(B)) \ne 0$. Using Theorem 11.6, the result follows. $\qquad\square$

Corollary 11.4 Let (X, d) be a complete metric space and $(X; f_n, n = 1, 2, \ldots, k)$ be IFS such that each f_i for $i = 1, 2, \ldots, k$ is a mapping on X satisfying

$$d(f_i x, f_i y)(d(f_i x, f_i y) + 1) \le e^{-\tau} d(x, y)(d(x, y) + 1), \quad \text{for all } x, y \in X,\ f_i x \ne f_i y,$$

where $\tau > 0$. Then the mapping $T : \mathcal{H}(X) \to \mathcal{H}(X)$ defined in Theorem 11.6 has a unique fixed point in $\mathcal{H}(X)$. Furthermore, for any set $A_0 \in \mathcal{H}(X)$, the sequence of compact sets $\{A_0, T(A_0), T^2(A_0), \ldots\}$ converges to a fixed point of T.

Proof: By taking $F(\lambda) = \ln(\lambda^2 + \lambda) + \lambda$, $\lambda > 0$ in Theorem 11.4, we obtain that each mapping f_i for $i = 1, 2, \ldots, k$ on X satisfies

$$d(f_i x, f_i y)(d(f_i x, f_i y) + 1)$$
$$\le e^{-\tau} d(x, y)(d(x, y) + 1), \quad \text{for all } x, y \in X,\ f_i x \ne f_i y,$$

where $\tau > 0$. Again it follows from Theorem 11.4 that the mapping $T : \mathcal{H}(X) \to \mathcal{H}(X)$ defined by

$$T(A) = \cup_{n=1}^{k} f_n(A),$$

for all $A \in \mathcal{H}(X)$ satisfies

$$H(T(A), T(B))(H(T(A), T(B)) + 1) \le e^{-\tau} H(A, B)(H(A, B) + 1),$$

for all $A, B \in \mathcal{H}(X)$, $H(T(A), T(B)) \neq 0$. Using Theorem 11.6, the result follows. □

Corollary 11.5 Let (X, d) be a complete metric space and $(X; f_n, n = 1, 2, \ldots, k)$ be IFS such that each f_i for $i = 1, 2, \ldots, k$ is a mapping on X satisfying

$$d(f_i x, f_i y) \leq \frac{1}{(1 + \tau \sqrt{d(x, y)})} d(x, y), \quad \text{for all } x, y \in X, \ f_i x \neq f_i y,$$

where $\tau > 0$. Then the mapping $T : \mathcal{H}(X) \to \mathcal{H}(X)$ defined in Theorem 11.6 has a unique fixed point $\mathcal{H}(X)$. Furthermore, for any set $A_0 \in \mathcal{H}(X)$, the sequence of compact sets $\{A_0, T(A_0), T^2(A_0), \ldots\}$ converges to a fixed point of T.

Proof: Take $F(\lambda) = -1/\sqrt{\lambda}$, $\lambda > 0$ in Theorem 11.4, then each mapping f_i for $i = 1, 2, \ldots, k$ on X satisfies

$$d(f_i x, f_i y) \leq \frac{1}{(1 + \tau \sqrt{d(x, y)})^2} d(x, y), \quad \text{for all } x, y \in X, \ f_i x \neq f_i y,$$

where $\tau > 0$. Again it follows from Theorem 11.4 that the mapping $T : \mathcal{H}(X) \to \mathcal{H}(X)$ defined by

$$T(A) = \cup_{n=1}^{k} f_n(A),$$

for all $A \in \mathcal{H}(X)$ satisfies

$$H(T(A), T(B)) \leq \frac{1}{(1 + \tau \sqrt{H(A, B)})^2} H(A, B),$$

for all $A, B \in \mathcal{H}(X)$, $H(T(A), T(B)) \neq 0$. Using Theorem 11.6, the result follows. □

Example 11.5 Let $X = [0, 1] \times [0, 1]$ and d be a Euclidean metric on X. Define $f_1, f_2 : X \to X$ as

$$f_1(x, y) = \left(\frac{1}{x + 1}, \frac{y}{y + 1} \right) \quad \text{and}$$

$$f_2(x, y) = \left(\frac{\sin x}{\sin x + 1}, \frac{1}{\sin y + 1} \right).$$

Note that, for all $\mathbf{x} = (x_1, y_1), \mathbf{y} = (x_2, y_2) \in X$ with $\mathbf{x} \neq \mathbf{y}$,

$$d(f_1(\mathbf{x}), f_1(\mathbf{y}))$$

$$= d\left(\left(\frac{1}{x_1 + 1}, \frac{y_1}{y_1 + 1} \right), \left(\frac{1}{x_2 + 1}, \frac{y_2}{y_2 + 1} \right) \right)$$

$$= \sqrt{\frac{(x_1 - x_2)^2}{(x_1 + 1)^2(x_2 + 1)^2} + \frac{(y_1 - y_2)^2}{(y_1 + 1)^2(y_2 + 1)^2}}$$

$$< \sqrt{(x_1 - x_2)^2 + (y_1 - y_2)^2}$$

$$= d((x_1, y_1), (x_2, y_2))$$
$$= d(\mathbf{x}, \mathbf{y}).$$

In addition,

$$d(f_2(\mathbf{x}), f_2(\mathbf{y}))$$
$$= d\left(\left(\frac{\sin x_1}{\sin x_1 + 1}, \frac{1}{\sin y_1 + 1}\right), \left(\frac{\sin x_2}{\sin x_2 + 1}, \frac{1}{\sin y_2 + 1}\right)\right)$$
$$= \sqrt{\frac{(\sin x_1 - \sin x_2)^2}{(\sin x_1 + 1)^2(\sin x_2 + 1)^2} + \frac{(\sin y_1 - \sin y_2)^2}{(\sin y_1 + 1)^2(\sin y_2 + 1)^2}}$$
$$< \sqrt{(\sin x_1 - \sin x_2)^2 + (\sin y_1 - \sin y_2)^2}$$
$$\leq \sqrt{(x_1 - x_2)^2 + (y_1 - y_2)^2}$$
$$= d((x_1, y_1), (x_2, y_2))$$
$$= d(\mathbf{x}, \mathbf{y}).$$

Now there exists $\tau > 0$ such that

$$d(f_1(\mathbf{x}), f_1(\mathbf{y}))\left(1 + \tau\sqrt{d(\mathbf{x}, \mathbf{y})}\right)^2 \leq d(\mathbf{x}, \mathbf{y}) \text{ and }$$

$$d(f_2(\mathbf{x}), f_2(\mathbf{y}))\left(1 + \tau\sqrt{d(\mathbf{x}, \mathbf{y})}\right)^2 \leq d(\mathbf{x}, \mathbf{y})$$

are satisfied. Consider the IFS $\{\mathbb{R}^2; f_1, f_2\}$ with mapping $T : \mathcal{H}([0, 1]^2) \to \mathcal{H}([0, 1]^2)$ given as

$$T(A) = f_1(A) \cup f_2(A).$$

For all $A, B \in \mathcal{H}([0, 1]^2)$ with $H(T(A), T(B)) \neq 0$, by Theorem 1.10,

$$H(T(A), T(B))\left(1 + \tau\sqrt{H(A, B)}\right)^2 \leq H(A, B)$$

holds. Furthermore, we can analyze the convergence of T to the attractor of IFS in the Figure 11.1.

11.3 Iterated Function System in *b*-Metric Space

The concept of metric has been generalized further in one to many ways. The concept of a *b*-metric space was introduced by Czerwik [23]. Since then, several papers have been published on the fixed point theory of various classes of single-valued and multivalued operators in *b*-metric space [23–33].

In this section, we construct a fractal set of IFS, a certain finite collection of mappings defined on a *b*-metric space, which induce compact-valued mappings defined on a family of compact subsets of a *b*-metric space. We prove that Hutchinson operator defined with the help of a finite family of generalized *F*-contraction mappings on a complete *b*-metric space is itself

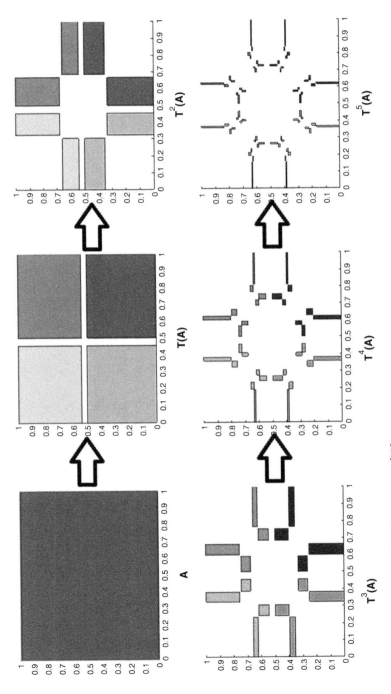

Figure 11.1 Convergence of T to the attractor of IFS.

generalized F-contraction mapping on a family of compact subsets of X. Then in the next section, we obtain a final fractal obtained by successive application of a generalized F-Hutchinson operator in b-metric space.

Definition 11.6 Let X be a nonempty set and $b \geq 1$ a given real number. A function $d : X \times X \to \mathbb{R}_+$ is said to be a b-metric if for any $x, y, z \in X$, the following conditions hold:

(b$_1$) $d(x, y) = 0$ if and only if $x = y$,
(b$_2$) $d(x, y) = d(y, x)$,
(b$_3$) $d(x, y) \leq b(d(x, z) + d(z, y))$,

The pair (X, d) is called a b-metric space with parameter $b \geq 1$.

If $b = 1$, then b-metric space is a metric spaces. But the converse does not hold in general [23, 24, 28].

Example 11.6 [34] Let (X, d) be a metric space, and $\rho(x, y) = (d(x, y))^p$, where $p > 1$ is a real number. Then ρ is a b-metric with $b = 2^{p-1}$.

Obviously conditions (b$_1$) and (b$_2$) of above definition are satisfied. If $1 < p < \infty$, then the convexity of the function $f(x) = x^p$ ($x > 0$) implies

$$\left(\frac{a+b}{2} \right)^p \leq \frac{1}{2}(a^p + b^p),$$

and hence, $(a + b)^p \leq 2^{p-1}(a^p + b^p)$ holds. Thus, for each $x, y, z \in X$, we obtain

$$\begin{aligned} \rho(x, y) = (d(x, y))^p &\leq (d(x, z) + d(z, y))^p \\ &\leq 2^{p-1}((d(x, z))^p + (d(z, y))^p) \\ &= 2^{p-1}(\rho(x, z) + \rho(z, y)). \end{aligned}$$

Therefore, condition (b$_3$) of the above definition is satisfied and ρ is a b-metric.

If $X = \mathbb{R}$ (set of real numbers) and $d(x, y) = |x - y|$ is the usual metric, then $\rho(x, y) = (x - y)^2$ is a b-metric on $\mathbb{R}$ with $b = 2$, but is not a metric on $\mathbb{R}$.

Definition 11.7 [27] Let (X, d) be a b-metric space. Then a subset $C \subset X$ is called:

(i) closed if and only if for each sequence $\{x_n\}$ in C that converges to an element x, we have $x \in C$ (i.e., $C = \overline{C}$).
(ii) compact if and only if for every sequence of elements of C there exists a subsequence that converges to an element of C.
(iii) bounded if and only if $\delta(C) := \sup\{d(x, y) : x, y \in C\} < \infty$.

If (X, d) is a compete b-metric space, then the Pompeiu–Hausdorff metric space $(\mathcal{H}(X), H)$ induced by d is also a complete b -metric space.

For the sake of completeness, we state the following Lemma hold in b-metric space [35].

Lemma 11.2 Let (X, d) be a b-metric space. For all $A, B, C, D \in \mathcal{H}(X)$, the following hold:

(i) If $B \subseteq C$, then $\sup_{a \in A} d(a, C) \leq \sup_{a \in A} d(a, B)$.

(ii) $\sup_{x \in A \cup B} d(x, C) = \max\{\sup_{a \in A} d(a, C), \sup_{b \in B} d(b, C)\}$.

(iii) $H(A \cup B, C \cup D) \leq \max\{H(A, C), H(B, D)\}$.

The following lemmas from [23, 30, 31] will be needed in the sequel to prove the main result of the paper.

Lemma 11.3 Let (X, d) be a b-metric space and $CB(X)$ denotes the set of all nonempty closed and bounded subsets of X. For $x, y \in X$ and $A, B \in CB(X)$, the following statements hold:

(1) $(CB(X), H)$ is a b-metric space.
(2) $d(x, B) \leq H(A, B)$ for all $x \in A$.
(3) $d(x, A) \leq b(d(x, y) + d(y, A))$.
(4) For $h > 1$ and $\acute{a} \in A$, there is a $\acute{b} \in B$ such that $d(\acute{a}, \acute{b}) \leq hH(A, B)$.
(5) For every $h > 0$ and $\acute{a} \in A$, there is a $\acute{b} \in B$ such that $d(\acute{a}, \acute{b}) \leq H(A, B) + h$.
(6) For every $\lambda > 0$ and $\tilde{a} \in A$, there is a $\tilde{b} \in B$ such that $d(\tilde{a}, \tilde{b}) \leq \lambda$.
(7) For every $\lambda > 0$ and $\tilde{a} \in A$, there is a $\tilde{b} \in B$ such that $d(\tilde{a}, \tilde{b}) \leq \lambda$ implies $H(A, B) \leq \lambda$.
(8) $d(x, A) = 0$ if and only if $x \in \overline{A} = A$.
(9) For $\{x_n\} \subseteq X$,

$$d(x_0, x_n) \leq bd(x_0, x_1) + \cdots + b^{n-1}d(x_{n-2}, x_{n-1}) + b^{n-1}d(x_{n-1}, x_n).$$

Definition 11.8 Let (X, d) be a b-metric space. A sequence $\{x_n\}$ in X is called:

(i) Cauchy if and only if for $\varepsilon > 0$, there exists $n(\varepsilon) \in \mathbb{N}$ such that for each $n, m \geq n(\varepsilon)$, we have $d(x_n, x_m) < \varepsilon$.

(ii) Convergent if and only if there exists $x \in X$ such that for all $\varepsilon > 0$ there exists $n(\varepsilon) \in \mathbb{N}$ such that for all $n \geq n(\varepsilon)$, we have $d(x_n, x) < \varepsilon$. In this case, we write $\lim_{n \to \infty} x_n = x$.

It is known that a sequence $\{x_n\}$ in b-metric space X is Cauchy if and only if $\lim_{n \to \infty} d(x_n, x_{n+p}) = 0$ for all $p \in \mathbb{N}$. A sequence $\{x_n\}$ is convergent to $x \in X$ if and only if $\lim_{n \to \infty} d(x_n, x) = 0$. A b-metric space (X, d) is said to be complete if every Cauchy sequence in X is convergent in X.

An *et al.* [24] studied the topological properties of b-metric spaces and stated the following assertions:

(c_1) In a b-metric space (X, d), d is not necessarily continuous in each variable.

(c_2) In a b-metric space (X, d), if d is continuous in one variable then d is continuous in other variable.

(c_3) An open ball in b-metric space (X, d) is not necessarily an open set. An open ball is open if d is continuous in one variable.

Wardowski [21] introduced another generalized contraction called F contraction and proved a fixed point result as an interesting generalization of the Banach contraction principle in complete metric space [36].

Let Υ be the set of all mapping $\tau : \mathbb{R}_+ \to \mathbb{R}_+$ that satisfying $\lim \inf_{t \to 0} \tau(t) > 0$ for all $t \geq 0$.

Definition 11.9 Let (X, d) be a b-metric space. A self-mapping f on X is called a generalized F contraction if for any $x, y \in X$, there exists $F \in F$ and $\tau \in \Upsilon$ such that

$$\tau(d(x, y)) + F(d(fx, fy)) \leq F(d(x, y)), \tag{11.9}$$

whenever $d(fx, fy) > 0$.

Theorem 11.7 Let (X, d) be a b-metric space and $f : X \to X$ an generalized F contraction. Then

(1) f maps elements in $\mathcal{H}(X)$ to elements in $\mathcal{H}(X)$.
(2) if for any $A \in \mathcal{H}(X)$,

$$f(A) = \{f(x) : x \in A\}.$$

Then $f : \mathcal{H}(X) \to \mathcal{H}(X)$ is a generalized F-contraction mapping on $(\mathcal{H}(X), H)$.

Proof: As generalized F-contractive mapping is continuous and the image of a compact subset under $f : X \to X$ is compact, so we obtain $A \in \mathcal{H}(X)$ implies $f(A) \in \mathcal{H}(X)$.

To prove (2): Let $A, B \in \mathcal{H}(X)$ with $H(f(A), f(B)) \neq \emptyset$. Since $f : X \to X$ is a generalized F contraction, we obtain

$$0 < d(fx, fy) < d(x, y) \quad \text{for all } x, y \in X, \ x \neq y.$$

Thus, we have

$$d(fx, f(B)) = \inf_{y \in B} d(fx, fy) < \inf_{y \in B} d(x, y) = d(x, B).$$

In addition,

$$d(fy, f(A)) = \inf_{x \in A} d(fy, fx) < \inf_{x \in A} d(y, x) = d(y, A).$$

Now

$$H(f(A), f(B)) = \max\{\sup_{x \in A} d(fx, f(B)), \sup_{y \in B} d(fy, f(A))\}$$

$$< \max\{\sup_{x \in A} d(x, B), \sup_{y \in B} d(y, A)\} = H(A, B).$$

By strictly increasing of F implies

$$F(H(f(A), f(B))) < F(H(A, B)).$$

Consequently, there exists a function $\tau : \mathbb{R}_+ \to \mathbb{R}_+$ with $\lim \inf_{t \to 0} \tau(t) > 0$ for all $t \geq 0$ such that

$$\tau(H(A, B)) + F(H(f(A), f(B))) \leq F(H(A, B)).$$

Hence $f : \mathcal{H}(X) \to \mathcal{H}(X)$ is a generalized F contraction. $\qquad \square$

Theorem 11.8　Let (X, d) be a b-metric space and $\{ f_n : n = 1, 2, \dots, N \}$ a finite family of generalized F-contraction self-mappings on X. Define $T : \mathcal{H}(X) \to \mathcal{H}(X)$ by

$$T(A) = f_1(A) \cup f_2(A) \cup \cdots \cup f_N(A)$$
$$= \cup_{n=1}^{N} f_n(A),$$

for each $A \in \mathcal{H}(X)$. Then T is a generalized F contraction on $\mathcal{H}(X)$.

Proof: We demonstrate the claim for $N = 2$. Let $f_1, f_2 : X \to X$ be two F contractions. Take $A, B \in \mathcal{H}(X)$ with $H(T(A), T(B)) \neq 0$. From Lemma 11.2 (iii), it follows that

$$\tau(H(A, B)) + F(H(T(A), T(B)))$$
$$= \tau(H(A, B)) + F(H(f_1(A) \cup f_2(A), f_1(B) \cup f_2(B)))$$
$$\leq \tau(H(A, B)) + F(\max\{H(f_1(A), f_1(B)), H(f_2(A), f_2(B))\})$$
$$\leq F(H(A, B)). \qquad \square$$

Definition 11.10　Let (X, d) be a metric space. A mapping $T : \mathcal{H}(X) \to \mathcal{H}(X)$ is said to be a Ciric type generalized F contraction if for $F \in F$ and $\tau \in \Upsilon$ such that for any $A, B \in \mathcal{H}(X)$ with $H(T(A), T(B)) \neq 0$, the following holds:

$$\tau(M_T(A, B)) + F(H(T(A), T(B))) \leq F(M_T(A, B)), \qquad (11.10)$$

where

$$M_T(A, B) = \max\{H(A, B), H(A, T(A)), H(B, T(B)), \frac{H(A, T(B)) + H(B, T(A))}{2b},$$
$$H(T^2(A), T(A)), H(T^2(A), B), H(T^2(A), T(B))\}.$$

Theorem 11.9　Let (X, d) be a b-metric space and $\{ f_n : n = 1, 2, \dots, N \}$ a finite sequence of generalized F-contraction mappings on X. If $T : \mathcal{H}(X) \to \mathcal{H}(X)$ is defined by

$$T(A) = f_1(A) \cup f_2(A) \cup \cdots \cup f_N(A)$$
$$= \cup_{n=1}^{N} f_n(A),$$

for each $A \in \mathcal{H}(X)$. Then T is a Ciric type generalized F -contraction mapping on $\mathcal{H}(X)$.

Proof: Using Theorem 11.8 with property (F_1), the result follows. $\square$

An operator T in above theorem is called Ciric type generalized F-Hutchinson operator.

11.4 Generalized F-Contractive Iterated Function System in b-Metric Space

In this section, we established various results of generalized F-contractive IFS in the setup of b-metric Space. First, we prove the following result.

Theorem 11.10 Let (X, d) be a complete b-metric space and $\{X; f_n, n = 1, 2, \ldots, k\}$ a generalized F-contractive IFS. Then following hold:

a) A mapping $T : \mathcal{H}(X) \to \mathcal{H}(X)$ defined by

$$T(A) = \cup_{n=1}^k f_n(A),$$

is Ciric type generalized F-Hutchinson operator.
b) Operator T has a unique fixed point $U \in \mathcal{H}(X)$, that is

$$U = T(U) = \cup_{n=1}^k f_n(U).$$

c) For any initial set $A_0 \in \mathcal{H}(X)$, the sequence of compact sets $\{A_0, T(A_0), T^2(A_0), \ldots\}$ converges to a fixed point of T.

Proof: Part (a) follows from Theorem 11.9. For parts (b) and (c), we proceed as follows: Let A_0 be an arbitrary element in $\mathcal{H}(X)$. If $A_0 = T(A_0)$, then the proof is finished. Therefore, we assume that $A_0 \neq T(A_0)$. Define

$$A_1 = T(A_0), \ A_2 = T(A_1), \ldots, A_{m+1} = T(A_m)$$

for $m \in \mathbb{N}$.

We may assume that $A_m \neq A_{m+1}$ for all $m \in \mathbb{N}$. If not, then $A_k = A_{k+1}$ for some k implies $A_k = T(A_k)$ and this completes the proof. Take $A_m \neq A_{m+1}$ for all $m \in \mathbb{N}$. From (11.10), we have

$$\begin{aligned} &\tau(M_T(A_m, A_{m+1})) + F(H(A_{m+1}, A_{m+2})) \\ &= \tau(M_T(A_m, A_{m+1})) + F(H(T(A_m), T(A_{m+1}))) \\ &\leq F(M_T(A_m, A_{m+1})), \end{aligned}$$

where

$$\begin{aligned} M_T(A_m, A_{m+1}) = \max\{ &H(A_m, A_{m+1}), H(A_m, T(A_m)), H(A_{m+1}, T(A_{m+1})), \\ &\frac{H(A_m, T(A_{m+1})) + H(A_{m+1}, T(A_m))}{2b}, \\ &H(T^2(A_m), T(A_m)), H(T^2(A_m), A_{m+1}), H(T^2(A_m), T(A_{m+1}))\} \end{aligned}$$

$$= \max\{H(A_m, A_{m+1}), H(A_m, A_{m+1}), H(A_{m+1}, A_{m+2}),$$
$$\frac{H(A_m, A_{m+2}) + H(A_{m+1}, A_{m+1})}{2b},$$
$$H(A_{m+2}, A_{m+1}), H(A_{m+2}, A_{m+1}), H(A_{m+2}, A_{m+2})\}$$
$$= \max\{H(A_m, A_{m+1}), H(A_{m+1}, A_{m+2})\}.$$

In case $M_T(A_m, A_{m+1}) = H(A_{m+1}, A_{m+2})$, we have

$$F(H(A_{m+1}, A_{m+2})) \leq F(H(A_{m+1}, A_{m+2})) - \tau(H(A_{m+1}, A_{m+2})),$$

a contradiction as $\tau(H(A_{m+1}, A_{m+2})) > 0$. Therefore, $M_T(A_m, A_{m+1}) = H(A_m, A_{m+1})$ and we have

$$F(H(A_{m+1}, A_{m+2})) \leq F(H(A_m, A_{m+1})) - \tau(H(A_m, A_{m+1}))$$
$$< F(H(A_m, A_{m+1})).$$

Thus, $\{H(A_{m+1}, A_{m+2})\}$ is decreasing and hence convergent. We now show that $\lim_{m \to \infty} H(A_{m+1}, A_{m+2}) = 0$. By property of τ, there exists $c > 0$ with $n_0 \in \mathbb{N}$ such that $\tau(H(A_m, A_{m+1})) > c$ for all $m \geq n_0$. Note that

$$F(H(A_{m+1}, A_{m+2})) \leq F(H(A_m, A_{m+1})) - \tau(H(A_m, A_{m+1}))$$
$$\leq F(H(A_{m-1}, A_m)) - \tau(H(A_{m-1}, A_m)) - \tau(H(A_m, A_{m+1}))$$
$$\leq \cdots \leq H(A_0, A_1) - [\tau(H(A_0, A_1)) + \tau(H(A_1, A_2))$$
$$+ \cdots + \tau(H(A_m, A_{m+1}))]$$
$$\leq F(H(A_0, A_1)) - n_0,$$

gives $\lim_{m \to \infty} F(H(A_{m+1}, A_{m+2})) = -\infty$, which together with (F_2) implies that $\lim_{m \to \infty} H(A_{m+1}, A_{m+2}) = 0$. By (F_3), there exists $h \in (0, 1)$ such that

$$\lim_{n \to \infty} [H(A_{m+1}, A_{m+2})]^h F(H(A_{m+1}, A_{m+2})) = 0.$$

Thus, we have

$$[H(A_m, A_{m+1})]^h F(H(A_m, A_{m+1})) - [H(A_m, A_{m+1})]^h F(H(A_0, A_1))$$
$$\leq [H(A_m, A_{m+1})]^h (f(H(A_0, A_1) - n_0)) - [H(A_m, A_{m+1})]^h F(H(A_0, A_1))$$
$$\leq -n_0 [H(A_m, A_{m+1})]^h \leq 0.$$

On taking limit as $n \to \infty$, we obtain that $\lim_{m \to \infty} m[H(A_{m+1}, A_{m+2})]^h = 0$. Hence $\lim_{m \to \infty} m^{\frac{1}{h}} H(A_{m+1}, A_{m+2}) = 0$. There exists $n_1 \in \mathbb{N}$ such that $m^{\frac{1}{h}} H(A_{m+1}, A_{m+2}) \leq 1$ for all $m \geq n_1$ and hence $H(A_{m+1}, A_{m+2}) \leq \frac{1}{m^{1/h}}$ for all $m \geq n_1$. For $m, n \in \mathbb{N}$ with $m > n \geq n_1$, we have

$$H(A_n, A_m) \leq H(A_n, A_{n+1}) + H(A_{n+1}, A_{n+2}) + \cdots + H(A_{m-1}, A_m)$$
$$\leq \sum_{i=n}^{\infty} \frac{1}{i^{1/h}}.$$

By the convergence of the series $\sum_{i=1}^{\infty} \frac{1}{i^{1/h}}$, we get $H(A_n, A_m) \to 0$ as $n, m \to \infty$. Therefore, $\{A_n\}$ is a Cauchy sequence in X. Since $(\mathcal{H}(X), d)$ is complete, we have $A_n \to U$ as $n \to \infty$ for some $U \in \mathcal{H}(X)$.

In order to show that U is the fixed point of T, we contrary assume that Pompeiu–Hausdorff weight assign to the U and $T(U)$ is not zero. Now

$$\tau(M_T(A_n, U)) + F(H(A_{n+1}, T(U)))$$
$$= \tau + F(H(T(A_n), T(U))) \leq F(M_T(A_n, U)), \qquad (11.11)$$

where

$$M_T(A_n, U) = \max\{H(A_n, U), H(A_n, T(A_n)), H(U, T(U)),$$
$$\frac{H(A_n, T(U)) + H(U, T(A_n))}{2b}, H(T^2(A_n)),$$
$$T(A_n)), H(T^2(A_n), U), H(T^2(A_n), T(U))\}$$
$$= \max\{H(A_n, U), H(A_n, A_{n+1}), H(U, T(U)),$$
$$\frac{H(A_n, T(U)) + H(U, A_{n+1})}{2b}, H(A_{n+2}, A_{n+1}),$$
$$H(A_{n+2}, U), H(A_{n+2}, T(U))\}\}.$$

Now we consider the following cases:

(1) If $M_T(A_n, U) = H(A_n, U)$, then on taking lower limit as $n \to \infty$ in (11.11), we have

$$\lim_{n \to \infty} \inf \ \tau(H(A_n, U)) + F(H(T(U), U)) \leq F(H(U, U)),$$

a contradiction as $\lim \inf_{t \to 0} \tau(t) > 0$ for all $t \geq 0$.

(2) When $M_T(A_n, U) = H(A_n, A_{n+1})$, then by taking lower limit as $n \to \infty$, we obtain

$$\lim_{n \to \infty} \inf \ \tau(H(A_n, A_{n+1})) + F(H(T(U), U)) \leq F(H(U, U)),$$

which gives a contradiction.

(3) In case $M_T(A_n, U) = H(U, T(U))$, then we get

$$\tau(H(U, T(U))) + F(H(T(U), U)) \leq F(H(U, T(U))),$$

a contradiction as $\tau(H(U, T(U))) > 0$.

(4) If $M_T(A_n, U) = \frac{H(A_n, T(U)) + H(U, A_{n+1})}{2b}$, then on taking lower limit as $n \to \infty$, we have

$$\lim_{n \to \infty} \inf \ \tau \left(\frac{H(A_n, T(U)) + H(U, A_{n+1})}{2b} \right) + F(H(T(U), U))$$
$$\leq F \left(\frac{H(U, T(U)) + H(U, U)}{2b} \right)$$
$$= F \left(\frac{H(U, T(U))}{2b} \right),$$

a contradiction as F is strictly increasing map.

(5) When $M_T(A_n, U) = H(A_{n+2}, A_{n+1})$, then

$$\lim_{n\to\infty} \inf \tau(H(A_{n+2}, A_{n+1})) + F(H(T(U), U)) \le F(H(U, U))$$

gives a contradiction.

(6) In case $M_T(A_n, U) = H(A_{n+2}, U)$, then on taking lower limit as $n \to \infty$ in (11.11), we get

$$\lim_{n\to\infty} \inf \tau(H(A_{n+2}, U)) + F(H(T(U), U)) \le F(H(U, U)),$$

a contradiction.

(7) Finally if $M_T(A_n, U) = H(A_{n+2}, T(U))$, then on taking lower limit as $n \to \infty$, we have

$$\lim_{n\to\infty} \inf \tau(H(A_{n+2}, T(U))) + F(H(T(U), U)) \le F(H(U, T(U))),$$

a contradiction.

Thus, U is the fixed point of T.

To show the uniqueness of fixed point of T, assume that U and V are two fixed points of T with $H(U, V)$ is not zero. Since T is a F-contraction map, we obtain that

$$\tau(M_T(U, V)) + F(H(U, V)) = \tau(M_T(U, V)) + F(H(T(U), T(V)))$$
$$\le F(M_T(U, V)),$$

where

$$M_T(U, V) = \max\{H(U, V), H(U, T(U)), H(V, T(V)),$$
$$\frac{H(U, T(V)) + H(V, T(U))}{2b}, H(T^2(U), U),$$
$$H(T^2(U), V), H(T^2(U), T(V))\}$$
$$= \max\{H(U, V), H(U, U), H(V, V), \frac{H(U, V) + H(V, U)}{2b},$$
$$H(U, U), H(U, V), H(U, V)\}$$
$$= H(U, V),$$

that is,

$$\tau(H(U, V)) + F(H(U, V)) \le F(H(U, V)),$$

a contradiction as $\tau(H(U, V)) > 0$. Thus, T has a unique fixed point $U \in \mathcal{H}(X)$. $\square$

Remark 11.2 In Theorem 11.10, if we take $S(X)$ the collection of all singleton subsets of X, then clearly $S(X) \subseteq \mathcal{H}(X)$. Moreover, consider $f_n = f$ for each n, where $f = f_i$ for any $i \in \{1, 2, 3, \dots, k\}$, then the mapping T becomes

$$T(x) = f(x).$$

With this setting, we obtain the following fixed point result.

Corollary 11.6   Let (X, d) be a complete b-metric space and $\{X : f_n, n = 1, 2, \ldots, k\}$ a generalized IFS. Let $f : X \to X$ be a mapping defined as in Remark 11.2. If there exist some $F \in \mathcal{F}$ and $\tau \in \Upsilon$ such that for any $x, y \in H(X)$ with $d(f(x), f(y)) \neq 0$, the following holds:

$$\tau(M_f(x, y)) + F(d(fx, fy)) \leq F(M_f(x, y)),$$

where

$$M_T(x, y) = \max\{d(x, y), d(x, fx), d(y, fy), \frac{d(x, fy) + d(y, fx)}{2b},$$
$$d(f^2x, y), d(f^2x, fx), d(f^2x, fy)\}.$$

Then f has a unique fixed point in X. Moreover, for any initial set $x_0 \in X$, the sequence of compact sets $\{x_0, fx_0, f^2x_0, \ldots\}$ converges to a fixed point of f.

Corollary 11.7   Let (X, d) be a complete b-metric space and $(X; f_n, n = 1, 2, \ldots, k)$ be IFS where each f_i for $i = 1, 2, \ldots, k$ is a contraction self-mapping on X. Then $T : H(X) \to H(X)$ defined in Theorem 11.10 has a unique fixed point in H (X). Furthermore, for any set $A_0 \in H(X)$, the sequence of compact sets $\{A_0, T(A_0), T^2(A_0), \ldots\}$ converges to a fixed point of T.

Proof: It follows from Theorem 11.8 that if each f_i for $i = 1, 2, \ldots, k$ is a contraction mapping on X, then the mapping $T : H(X) \to H(X)$ defined by

$$T(A) = \cup_{n=1}^k f_n(A),$$

is contraction on $H(X)$. Using Theorem 11.10, the result follows.   □

Corollary 11.8   Let (X, d) be a complete b-metric space and $(X; f_n, n = 1, 2, \ldots, k)$ an IFS. Suppose that each f_i for $i = 1, 2, \ldots, k$ is a mapping on X satisfying

$$d(f_ix, f_iy) \, e^{d(f_ix, f_iy) - d(x, y)} \leq e^{-\tau(d(x, y))} d(x, y),$$

for all $x, y \in X$, $f_ix \neq f_iy$, where $\tau \in \Upsilon$. Then the mapping $T : H(X) \to H(X)$ defined in Theorem 11.10 has a unique fixed point in $H(X)$. Furthermore, for any set $A_0 \in H(X)$, the sequence of compact sets $\{A_0, T(A_0), T^2(A_0), \ldots\}$ converges to a fixed point of T.

Proof: Take $F(\lambda) = \ln(\lambda) + \lambda$, $\lambda > 0$ in Theorem 11.8, then each mapping f_i for $i = 1, 2, \ldots, k$ on X satisfies

$$d(f_ix, f_iy) \, e^{d(f_ix, f_iy) - d(x, y)} \leq e^{-\tau(d(x, y))} d(x, y),$$

for all $x, y \in X$, $f_ix \neq f_iy$, where $\tau \in \Upsilon$. Again from Theorem 11.8, the mapping $T : H(X) \to H(X)$ defined by

$$T(A) = \cup_{n=1}^k f_n(A)$$

satisfies

$$H(T(A), T(B))e^{H(T(A),T(B))-H(A,B)} \leq e^{-\tau}H(A, B),$$

for all $A, B \in \mathcal{H}(X)$, $H(T(A), T(B)) \neq 0$. Using Theorem 11.10, the result follows. □

Corollary 11.9 Let (X, d) be a complete b-metric space and $(X; f_n, n = 1, 2, \ldots, k)$ be IFS. Suppose that each f_i for $i = 1, 2, \ldots, k$ is a mapping on X satisfying

$$d(f_i x, f_i y)(d(f_i x, f_i y) + 1) \leq e^{-\tau(d(x,y))}d(x, y)(d(x, y) + 1),$$

for all $x, y \in X$, $f_i x \neq f_i y$, where $\tau \in \Upsilon$. Then the mapping $T : \mathcal{H}(X) \to \mathcal{H}(X)$ defined in Theorem 11.10 has a unique fixed point in $\mathcal{H}(X)$. Furthermore, for any set $A_0 \in \mathcal{H}(X)$, the sequence of compact sets $\{A_0, T(A_0), T^2(A_0), \ldots\}$ converges to a fixed point of T.

Proof: By taking $F(\lambda) = \ln(\lambda^2 + \lambda) + \lambda$, $\lambda > 0$ in Theorem 11.8, we obtain that each mapping f_i for $i = 1, 2, \ldots, k$ on X satisfies

$$d(f_i x, f_i y)(d(f_i x, f_i y) + 1) \leq e^{-\tau(d(x,y))}d(x, y)(d(x, y) + 1),$$

for all $x, y \in X$, $f_i x \neq f_i y$, where $\tau \in \Upsilon$. Again it follows from Theorem 11.8 that the mapping $T : \mathcal{H}(X) \to \mathcal{H}(X)$ defined by

$$T(A) = \cup_{n=1}^{k} f_n(A),$$

satisfies

$$H(T(A), T(B))(H(T(A), T(B)) + 1) \leq e^{-\tau(H(A,B))}H(A, B)(H(A, B) + 1),$$

for all $A, B \in \mathcal{H}(X)$, $H(T(A), T(B)) \neq 0$. Using Theorem 11.10, the result follows. □

Corollary 11.10 Let (X, d) be a complete b-metric space and $(X; f_n, n = 1, 2, \ldots, k)$ be IFS. Suppose that each f_i for $i = 1, 2, \ldots, k$ is a mapping on X satisfying

$$d(f_i x, f_i y) \leq \frac{1}{(1 + \tau(d(x, y))\sqrt{d(x, y)}}d(x, y),$$

for all $x, y \in X$, $f_i x \neq f_i y$, where $\tau \in \Upsilon$. Then the mapping $T : \mathcal{H}(X) \to \mathcal{H}(X)$ defined in Theorem 11.8 has a unique fixed point $\mathcal{H}(X)$. Furthermore, for any set $A_0 \in \mathcal{H}(X)$, the sequence of compact sets $\{A_0, T(A_0), T^2(A_0), \ldots\}$ converges to a fixed point of T.

Proof: Take $F(\lambda) = -1/\sqrt{\lambda}$, $\lambda > 0$ in Theorem 11.8, then each mapping f_i for $i = 1, 2, \ldots, k$ on X satisfies

$$d(f_i x, f_i y) \leq \frac{1}{(1 + \tau(d(x, y))\sqrt{d(x, y)})^2}d(x, y), \quad \text{for all } x, y \in X, f_i x \neq f_i y,$$

where $\tau \in \Upsilon$. Again it follows from Theorem 11.8 that the mapping $T : \mathcal{H}(X) \to \mathcal{H}(X)$ defined by

$$T(A) = \cup_{n=1}^{k} f_n(A)$$

satisfies

$$H(T(A), T(B)) \leq \frac{1}{(1 + \tau(H(A,B))\sqrt{H(A,B)})^2} H(A,B),$$

for all $A, B \in \mathcal{H}(X)$, $H(T(A), T(B)) \neq 0$. Using Theorem 11.10, the result follows. $\qquad\qquad\square$

References

1 Hutchinson, J. (1981) Fractals and self-similarity. *Indiana Univ. J. Math.*, **30** (5), 713–747.

2 Barnsley, M.F. (1993) *Fractals Everywhere*, 2nd edn, Academic Press, San Diego, CA.

3 Banach, S. (1922) Sur les opérations dans les ensembles abstraits et leur applications aux équations intégrales. *Fund. Math.*, **3**, 133–181.

4 Abdeljawad, T. (2011) Fixed points for generalized weakly contractive mappings in partial metric spaces. *Math. Comput. Modell.*, **54**, 2923–2927.

5 Arandjelović, I., Kadelburg, Z., and Radenović, S. (2011) Boyd-Wong-type common fixed point results in cone metric spaces. *Appl. Math. Comput.*, **217**, 7167–7171.

6 Huang, L.G. and Zhang, X. (2007) Cone metric spaces and fixed point theorems of contractive maps. *J. Math. Anal. Appl.*, **332**, 1467–1475.

7 Illic, D., Abbas, M., and Nazir, T. (2015) Iterative approximation of fixed points of Presić operators on partial metric spaces. *Math. Nachr.*, 1–13. doi: 10.1002/mana.201400235.

8 Kadelburg, Z., Radenović, S., and Rakočević, V. (2009) Remarks on "Quasi-contraction on a cone metric space. *Appl. Math. Lett.*, **22**, 1674–1679.

9 Shukla, S. and Radenovic, S. (2014) PreÅ¡ic-Boyd-Wong type results in ordered metric spaces. *Int. J. Anal. Appl.*, **5** (2), 154–166.

10 Tarafdar, E. (1974) An approach to fixed-point theorems on uniform spaces. *Trans. Am. Math. Soc.*, **191**, 209–225.

11 Boyd, D.W. and Wong, J.S.W. (1969) On nonlinear contractions. *Proc. Am. Math. Soc.*, **20**, 458–464.

12 Edelstein, M. (1961) An extension of Banach's contraction principle. *Proc. Am. Math. Soc.*, **12**, 7–10.

13 Kirk, W.A. (2003) Fixed points of asymptotic contractions. *J. Math. Anal. Appl.*, **277**, 645–650.

14 Meir, A. and Keeler, E. (1969) A theorem on contraction mappings. *J. Math. Anal. Appl.*, **28**, 326–329.

15 Rakotch, E. (1962) A note on contractive mappings. *Proc. Am. Math. Soc.*, **13**, 459–465.

16 Shukla, S., Radenovic, S., and Kadelburg, Z. (2014) Some fixed point theorems for ordered F-generalized contractions in 0-orbitally complete partial spaces. *Theor. Appl. Math. Comput. Sci.*, **4** (11), 87–98.

17 Nadler, S.B. (1969) Multivalued contraction mappings. *Pac. J. Math.*, **30**, 475–488.

18 Abbas, M., Alfuraidan, M.R., and Nazir, T. (2016) Common fixed points of multivalued F-contractions on metric spaces with a directed graph. *Carpathian J. Math.*, **32**, 1–12.

19 Assad, N.A. and Kirk, W.A. (1972) Fixed point theorems for setvalued mappings of contractive type. *Pac. J. Math.*, **43**, 533–562.

20 Sgroi, M. and Vetro, C. (2013) Multi-valued F-Contractions and the solution of certain functional and integral equations. *Filomat*, **27** (7), 1259–1268.

21 Wardowski, D. (2012) Fixed points of new type of contractive mappings in complete metric spaces. *Fixed Point Theory Appl.*, **2012** (1), 94.

22 Wardowski, D. and Dung, N.V. (2014) Fixed points of $F-$ contractions on complete metric spaces. *Demontratio Math.*, **47** (1), 146–155.

23 Czerwik, S. (1993) Contraction mappings in b-metric spaces. *Acta Math. Inform. Univ. Ostrav.*, **1**, 5–11.

24 An, T.V., Tuyen, L.Q., and Dung, N.V. (2015) Stone-type theorem on b-metric spaces and applications. *Topol. Appl.*, **185-186**, 50–64.

25 Aydi, H., Bota, M.F., Karapınar, E., and Mitrović, S. (2012) A fixed point theorem for set-valued quasicontractions in b-metric spaces. *Fixed Point Theory Appl.*, **2012**, Article ID 88.

26 Boriceanu, M., Petruşel, A., and Rus, I.A. (2010) Fixed point theorems for some multivalued generalized contractions in b-metric spaces. *Int. J. Math. Stat.*, **6**, 65–76.

27 Boriceanu, M., Bota, M., and Petruşel, A. (2010) Multivalued fractals in b-metric spaces. *Cent. Eur. J. Math.*, **8** (2), 367–377.

28 Ćirić, Lj., Abbas, M., Rajović, M., and Ali, B. (2012) Suzuki type fixed point theorems for generalized multi-valued mappings on a set endowed with two b-metrics. *Appl. Math. Comput.*, **219**, 1712–1723.

29 Chifu, C. and Petruşel, G. (2014) Fixed points for multivalued contractions in b-metric spaces with applications to fractals. *Taiwan J. Math.*, **18** (5), 1365–1375.

30 Czerwik, S., Dlutek, K., and Singh, S.L. (1997) Round-off stability of iteration procedures for operators in b-metric spaces. *J. Nat. Phys. Sci.*, **11**, 87–94.

31 Czerwik, S. (1998) Nonlinear set-valued contraction mappings in b-metric spaces. *Atti Semin. Mat. Fis. Univ. Modena*, **46**, 263–276.

32 Kutbi, M.A., Karapinar, E., Ahmad, J., and Azam, A. (2014) Some fixed point results for multi-valued mappings in *b*-metric spaces. *J. Inequal. Appl.*, **2014**, 126.

33 Roshan, J.R., Hussain, N., Sedghi, S., and Shobkolaei, N. (2015) Suzuki-type fixed point results in *b*-metric spaces. *Math. Sci.*, **9**, 153–160.

34 Roshan, J.R., Parvaneh, V., and Altun, I. (2014) Some coincidence point results in ordered *b*-metric spaces and applications in a system of integral equations. *Appl. Math. Comput.*, **226**, 725–737.

35 Nazir, T., Silvestrov, S., and Abbas, M. (2016) Fractals of generalized *F*-Hutchinson operator. *Waves, Wavelets Fractals Adv. Anal.*, **2**, 29–40.

36 Klim, D. and Wardowsk, D. (2015) Fixed points of dynamic processes of set-valued *F*-contractions and application to functional equations. *Fixed Point Theory Appl.*, **2015** (22), 9.

12

Regular and Rapid Variations and Some Applications

Ljubiša D.R. Kočinac[1], Dragan Djurčić[2], and Jelena V. Manojlović[1]

[1] Faculty of Sciences and Mathematics, Department of Mathematics, University of Niš, 18000 Niš, Serbia
[2] Faculty of Technical Sciences, Department of Mathematics, University of Kragujevac, 34000 Čačak, Serbia

2010 AMS Subject Classification primary: 26A12, 40A05; secondary: 34A34, 34C11, 34E05, 91A05

12.1 Introduction and Historical Background

In 1826, Abel proved the following result, which is known nowadays as Abel's direct theorem.

Theorem A **(Abel)** Let $(a_k)_{k \in \mathbb{N}}$ be a sequence of real numbers such that the series $\sum_{k=0}^{\infty} a_k$ converges to S. Then, the Abel sum

$$\lim_{x \to 1_-} \sum_{k=0}^{\infty} a_k x^k$$

converges to S, provided $\sum_{k=0}^{\infty} a_k x^k$ is convergent for $|x| < 1$.

In 1897, Tauber proved the inverse of Abel's theorem.

Theorem B **(Tauber)** If the Abel sum

$$\lim_{x \to 1_-} \sum_{k=0}^{\infty} a_k x^k$$

converges to S, and $\sum_{k=0}^{\infty} a_k x^k$ is convergent for $|x| < 1$, then $\sum_{k=0}^{\infty} a_k$ converges to S if the Tauber condition

$$a_k = o\left(\frac{1}{k}\right), \quad k \to \infty$$

is satisfied.

Mathematical Analysis and Applications: Selected Topics, First Edition.
Edited by Michael Ruzhansky, Hemen Dutta, and Ravi P. Agarwal.
© 2018 John Wiley & Sons, Inc. Published 2018 by John Wiley & Sons, Inc.

The Tauber theorem was improved in 1911 by Littlewood who proved the following theorem.

Theorem C (Littlewood) Tauber's theorem holds if the Tauber condition is replaced with the condition

$$a_k = O\left(\frac{1}{k}\right), \quad k \to \infty.$$

This Littlewood's theorem was the starting point of a new branch of mathematical analysis – Tauberian theory.

In 1914, Hardy and Littlewood published a new theorem as a generalization of the Littlewood theorem with a very complicated and very long proof.

Theorem D (Hardy–Littlewood) Let $a_k \geq 0$ for each $k \in \omega$ and let $\sum\limits_{k=0}^{\infty} a_k x^k$ converge for $|x| < 1$. If

$$\lim_{x \to 1_-} (1 - x) \sum_{k=0}^{\infty} a_k x^k = 1,$$

then

$$\lim_{k \to \infty} \frac{1}{k} \sum_{i=0}^{k} a_i = 1.$$

Studying in the 1930s Tauberian theory, in particular working on a simplification of the work of Hardy and Littlewood, J. Karamata initiated investigation in asymptotic analysis of divergent processes, nowadays known as *Karamata's theory of regular variation* (see [1–5], and also [6–10]).

In 1970, de Haan [11] defined and investigated rapid variation and so stimulated further development in asymptotic analysis. The theory of rapid variability is conjugate with the theory of slow (and regular) variability; these theories are conjugated, for example, through generalized inverse [12]. The book by Bingham *et al.* [6] is a nice exposition of Karamata theory and the theory of rapid variability [10, 11], while Hardy's book [13] is a general textbook for the theory of divergent sequences.

These kinds of variability are related to real functions and sequences from the set $\mathbb{S}$ of positive real sequences. Usually, a class C_f of functions has the corresponding partner class C_s of sequences defined in such a way that the restriction $\varphi \upharpoonright \mathbb{N}$ of a function $\varphi \in C_f$ belongs to C_s [6, 10, 14–22].

The theory of regular and rapid variability has many applications in different branches of mathematics: differential and difference equations, in particular in description of asymptotic properties of solutions of these equations, time scales theory, dynamic equations, q-calculus, probability theory, number theory, and so on (see, for instance, [23–28] and Sections 12.4 and 12.5).

12.2   Regular Variation

A function $\varphi : [a, \infty) \to (0, \infty)$, $a > 0$, is called *regularly varying* [6] if it is measurable and satisfies the following asymptotic condition:

$$\lim_{t \to \infty} \frac{\varphi(\lambda t)}{\varphi(t)} = g(\lambda) < \infty, \qquad \lambda > 0. \tag{12.1}$$

If $g(\lambda) = 1$ for each $\lambda > 0$, then φ is said to be *slowly varying*.

The function $g(\lambda)$ is equal to λ^ρ for some $\rho \in \mathbb{R}$ called the *index of variability* of φ.

The class of regularly varying (resp. slowly varying) functions we denote by RV_f (resp. SV_f), while the class of regularly varying functions of index of variability ρ is denoted by $RV_{f,\rho}$.

A fundamental result on regularly varying functions, which will be used especially in Section 12.5, is summarized in the following proposition.

Proposition 12.1

(i) $\varphi \in RV_{f,\rho}$ if and only if $\varphi(t) = t^\rho \ell(t)$ with $\ell \in SV_f$;

(ii) (Representation theorem) $\varphi \in RV_{f,\rho}$ if and only if $\varphi(t)$ is represented in the form

$$\varphi(t) = c(t) \exp \left(\int_{t_0}^{t} \frac{\delta(s)}{s} \, ds \right), \qquad t \geq t_0,$$

for some $t_0 > 0$ and for some measurable functions $c(t)$ and $\delta(t)$ such that

$$\lim_{t \to \infty} c(t) = c_0 \in (0, \infty) \quad \text{and} \quad \lim_{t \to \infty} \delta(t) = \rho.$$

(iii) If $\varphi_1 \in RV_{f,\sigma_1}$, $\varphi_2 \in RV_{f,\sigma_2}$, then $\varphi_1 \varphi_2 \in RV_{f,\sigma_1 + \sigma_2}$, $\varphi_1^\alpha \in RV_{f,\alpha\sigma_1}$) for any $\alpha \in \mathbb{R}$. Moreover, $\varphi_1 \circ \varphi_2 \in RV_{f,\sigma_1 \sigma_2}$ if $\varphi_2(t) \to \infty$, as $t \to \infty$.

(iv) If $\varphi(t) \sim t^\alpha \ell(t)$ as $t \to \infty$ with $\ell \in SV_f$, then φ is a regularly varying function of index α, that is, $\varphi(t) = t^\alpha \ell^*(t)$, $\ell^* \in SV_f$, where in general $\ell^*(t) \neq \ell(t)$, but $\ell^*(t) \sim \ell(t)$ as $t \to \infty$.

(v) Let φ be a positive, continuously differentiable on $[0, \infty)$ and such that

$$\lim_{t \to \infty} \frac{t \varphi'(t)}{\varphi(t)} = 0.$$

Then, φ is slowly varying.

(vi) Regularly varying function of index $\sigma \neq 0$ is almost monotone.

(vii) For the function $\varphi \in RV_{f,\alpha}$, $\alpha > 0$, there exists $g \in RV_{f,1/\alpha}$ such that

$$\varphi(g(t)) \sim g(\varphi(t)) \sim t \quad \text{as} \quad t \to \infty.$$

Here, and in what follows, the symbol $\sim$ is used to denote the strong asymptotic equivalence of two positive functions g, h defined in a neighborhood of infinity:

$$g(t) \sim h(t), \quad t \to \infty \quad \Leftrightarrow \quad \lim_{t \to \infty} \frac{g(t)}{h(t)} = 1.$$

If for $\varphi(t) = t^\rho \ell(t)$ (Proposition 12.1(i)), the slowly varying part ℓ tends to some positive constant as $t \to \infty$, it is called a *trivial slowly varying* one denoted by $\ell \in \mathrm{tr} - \mathcal{SV}_f$, while the function φ is called a *trivial regularly varying function of index* ρ, denoted by $\varphi \in \mathrm{tr} - \mathrm{RV}_{f,\rho}$. Otherwise ℓ is called a *nontrivial slowly varying function*, denoted by $\ell \in \mathrm{ntr} - \mathrm{SV}_f$, and φ is called a *nontrivial* $\mathrm{RV}_{f,\rho}$ *function*, denoted by $\varphi \in \mathrm{ntr} - \mathrm{RV}_{f,\rho}$.

A measurable function $\varphi : (0, a) \to (0, \infty)$ is said to be *regularly varying at zero of index* $\rho \in \mathbb{R}$ if $f\left(\frac{1}{t}\right)$ is regularly varying, that is, if

$$\lim_{t \to 0+} \frac{\varphi(\lambda t)}{\varphi(t)} = \lambda^\rho \quad \text{for all} \quad \lambda > 0. \tag{12.2}$$

A sequence $\mathbf{x} = (x_n)_{n \in \mathbb{N}}$ of positive real numbers is said to be *regularly varying* if for each $\lambda > 0$ it satisfied

$$k_{\mathbf{x}}(\lambda) := \lim_{n \to \infty} \frac{x_{[\lambda n]}}{x_n} < \infty. \tag{12.3}$$

It is known that the limit function $k_{\mathbf{x}}(\lambda)$ is of the form λ^ρ for some $\rho \in \mathbb{R}$ [6]; ρ is called the *index of variability* of $\mathbf{x}$.

If $\rho = 0$, then $\mathbf{x}$ is said to be *slowly varying*.

By RV_s and SV_s we denote the class of regularly varying sequences and the class of slowly varying sequences, respectively. $\mathrm{RV}_{s,\rho}$ is the class of regularly varying sequences of index of variability ρ, and $\mathrm{RV}_{s,+} := \bigcup_{\rho > 0} \mathrm{RV}_{s,\rho}$.

Bojanić and Seneta [29] (see also [30]) unified the theory of slow variability by the following theorem.

Theorem 12.1 For a sequence $\mathbf{x} = (x_n)_{n \in \mathbb{N}}$ in $\mathbb{S}$ the following are equivalent:

(a) $\mathbf{x}$ is slowly varying;
(b) The function $\varphi_{\mathbf{x}}$ defined by $\varphi_{\mathbf{x}}(t) = x_{[t]}$, $t \geq 1$, is slowly varying.

In the sequel, we will see that similar theorems (called Bojanić–Galambos–Seneta type theorems) hold for several other classes of sequences and functions.

12.2.1 The Class $\mathrm{Tr}(\mathrm{RV}_s)$

Definition 12.1 [31, 32] A sequence $\mathbf{x} = (x_n)_{n \in \mathbb{N}} \in \mathbb{S}$ is in the class $\mathrm{Tr}(\mathrm{RV}_s)$ of *translationally regularly varying sequences* if for each $\lambda \in \mathbb{R}$,

$$\lim_{n \to \infty} \frac{x_{[n+\lambda]}}{x_n} = r(\lambda) < \infty. \tag{12.4}$$

Theorem 12.2 [31] If $\mathbf{x} \in \mathrm{Tr}(\mathrm{RV}_s)$, then $r(\lambda) = e^{\rho[\lambda]}$, for some $\rho \in \mathbb{R}$.

Proof: Let $\mathbf{x} = (x_n)_{n \in \mathbb{N}} \in \text{Tr}(\text{RV}_s)$. Then, for each $\lambda \in \mathbb{R}$,

$$\frac{1}{r(\lambda)} = \lim_{n \to \infty} \frac{x_n}{x_{[n+\lambda]}} = \lim_{n \to \infty} \frac{x_n}{x_{n+[\lambda]}} = \lim_{k \to \infty} \frac{x_{k-[\lambda]}}{x_k} = \lim_{k \to \infty} \frac{x_{[k-[\lambda]]}}{x_k} = r(-[\lambda]).$$

Since, for each $\lambda \in \mathbb{R}$, $0 < r(-[\lambda]) < \infty$, we have $r(\lambda) > 0$ for each $\lambda \in \mathbb{R}$.

We prove now that $r(\lambda) = e^{\rho[\lambda]}$. For $\lambda = 0$ the statement is trivially true. Suppose $\lambda > 0$. We have

$$r(\lambda) = \lim_{n \to \infty} \frac{x_{[n+\lambda]}}{x_n} = \lim_{n \to \infty} \frac{x_{n+[\lambda]}}{x_n}$$

$$= \begin{cases} \lim_{n \to \infty} \left(\dfrac{x_{n+1}}{x_n} \cdot \dfrac{x_{n+2}}{x_{n+1}} \cdots \dfrac{x_{n+[\lambda]}}{x_{n+[\lambda]-1}} \right), & \text{for } \lambda \geq 1, \\ 1, & \text{for } 0 < \lambda < 1, \end{cases}$$

$$= (r(1))^{[\lambda]}.$$

By the first part of the proof $r(1) > 0$, so that (for the considered $\lambda > 0$) we have

$$r(\lambda) = e^{[\lambda] \cdot \ln(r(1))}.$$

If we set $\rho = \ln(r(1)) \in \mathbb{R}$, then for $\lambda > 0$ we obtain $r(\lambda) = e^{\rho[\lambda]}$.

Let now $\lambda < 0$. Then, we have

$$r(\lambda) = \frac{1}{\displaystyle\lim_{n \to \infty} \frac{x_n}{x_{[n+\lambda]}}} = \frac{1}{\displaystyle\lim_{n \to \infty} \frac{x_n}{x_{n+[\lambda]}}}$$

$$= \frac{1}{\displaystyle\lim_{n \to \infty} \left(\dfrac{x_{n+[\lambda]+1}}{x_{n+[\lambda]}} \cdots \dfrac{x_n}{x_{n-1}} \right)}$$

$$= \frac{1}{(r(1))^{-[\lambda]}} = (r(1))^{[\lambda]} = e^{\rho[\lambda]}.$$

This completes the proof of the theorem. $\square$

Remark 12.1 It is easy to see that the function $r(\lambda)$ from the previous theorem satisfies:

1. $r(\lambda) = r([\lambda])$ for each $\lambda \in \mathbb{R}$; this means that this function is a step-function on $\mathbb{R}$, with platforms on the intervals $[k, k+1)$, $k \in \mathbb{Z}$.
2. $r(-k) = (r(k))^{-1}$ for each $k \in \{0, 1, 2, \dots\}$.
3. $r(\lambda) = 1$ for $\lambda \in [0, 1)$, and $r(-\lambda + 1) = (r(\lambda))^{-1}$ for each $\lambda \in (1, \infty) \setminus \mathbb{N}$.

For a sequence $\mathbf{x} = (x_n)_{n \in \mathbb{N}}$ in $\text{Tr}(\text{RV}_s)$, the number ρ from Theorem 12.2 is called the *index of variability* of $\mathbf{x}$.

By $\text{Tr}(\text{RV}_{s,\rho})$ we denote the set of all sequences in $\text{Tr}(\text{RV}_s)$ of index of variability ρ.

We prove now a representation theorem for the class $\text{Tr}(\text{RV}_{s,\rho})$.

Theorem 12.3   A sequence $(x_n)_{n \in \mathbb{N}} \in \mathbb{S}$ belongs to the class $\mathrm{Tr}(\mathrm{RV}_{s,\rho})$, $\rho \in \mathbb{R}$, if and only if

$$x_n = x_1 \cdot \exp\left(\sum_{i=1}^{n-1} a_i\right), \quad n \geq 2, \tag{12.5}$$

for some real sequence $(a_n)_{n \in \mathbb{N}}$ with $\lim_{n \to \infty} a_n = \rho$, and $x_1 > 0$.

Proof: ($\Rightarrow$) If $\mathbf{x} = (x_n)_{n \in \mathbb{N}}$ is in $\mathrm{Tr}(\mathrm{RV}_{\rho,s})$, then by Theorem 12.2

$$\lim_{n \to \infty} \frac{x_{n+1}}{x_n} = r(1) = e^\rho < \infty,$$

which means that there is a sequence $(c_n)_{n \in \mathbb{N}} \in \mathbb{S}$ such that $\lim_{n \to \infty} c_n = r(1)$ and $\frac{x_{n+1}}{x_n} = c_n$, $n \in \mathbb{N}$. So for $n \geq 1$ we have

$$x_{n+1} = c_n \cdot x_n = c_n \cdot c_{n-1} \cdots c_1 \cdot x_1.$$

If we put $a_i = \ln c_i$, $i \in \mathbb{N}$, then $\lim_{n \to \infty} a_n = \rho$ and for each $n \in \mathbb{N}$

$$x_{n+1} = x_1 \cdot \exp\left(\sum_{i=1}^{n} a_i\right),$$

that is, for each $n \geq 2$

$$x_n = x_1 \cdot \exp\left(\sum_{i=1}^{n-1} a_i\right),$$

where $\lim_{n \to \infty} a_n = \rho$.

($\Leftarrow$) From (12.5) it follows

$$r = \lim_{n \to \infty} \frac{x_{n+1}}{x_n} = \lim_{n \to \infty} e^{a_n} = e^\rho,$$

so that for each $\lambda \in \mathbb{R}$

$$\lim_{n \to \infty} \frac{x_{[n+\lambda]}}{x_n} = r^{[\lambda]} = e^{\rho \cdot [\lambda]}.$$

This means $(x_n)_{n \in \mathbb{N}} \in \mathrm{Tr}(\mathrm{RV}_{s,\rho})$.   □

12.2.2 Classes of Sequences Related to $\mathrm{Tr}(\mathrm{RV}_s)$

Let $\mathbf{x} = (x_n)_{n \in \mathbb{N}}$ be a sequence in $\mathbb{S}$. For each $k \in \mathbb{N}$ define a new sequence $\mathbf{V}^{(k)}(\mathbf{x}) = (V_n^{(k)}(\mathbf{x}))_{n \in \mathbb{N}}$ inductively by

$$V_n^{(1)}(\mathbf{x}) := \frac{x_{n+1}}{x_n}, \quad n \in \mathbb{N};$$

$$V_n^{(k+1)}(\mathbf{x}) := \frac{V_{n+1}^{(k)}(\mathbf{x})}{V_n^{(k)}(\mathbf{x})}, \quad n \in \mathbb{N}.$$

The sequence $\mathbf{V}^{(k)}(\mathbf{x})$ we call the *quotient sequence of* $\mathbf{x}$ *of order* k. We also put $\mathbf{V}^{(0)}(\mathbf{x}) = \mathbf{x}$ [18].

By a direct calculation, we obtain that for $k \geq 2$

$$V_n^{(k)}(\mathbf{x}) = \frac{x_{n+k}^{s_{k+1}^{(k)}} \cdot x_{n+k-2}^{s_{k-1}^{(k)}} \cdots x_n^{s_1^{(k)}}}{x_{n+k-1}^{s_k^{(k)}} \cdot x_{n+k-3}^{s_{k-2}^{(k)}} \cdots x_{n+1}^{s_2^{(k)}}}, \qquad \text{if } k \text{ is even,} \tag{12.6}$$

and

$$V_n^{(k)}(\mathbf{x}) = \frac{x_{n+k}^{s_{k+1}^{(k)}} \cdot x_{n+k-2}^{s_{k-1}^{(k)}} \cdots x_{n+1}^{s_2^{(k)}}}{x_{n+k-1}^{s_k^{(k)}} \cdot x_{n+k-3}^{s_{k-2}^{(k)}} \cdots x_n^{s_1^{(k)}}}, \qquad \text{if } k \text{ is odd,} \tag{12.7}$$

where

$$s_1^{(i)} = 1, \quad s_{i+1}^{(i)} = 1 \quad \text{for } 1 \leq i \leq k; \quad s_i^{(j)} = s_i^{(j-1)} + s_{i-1}^{(j-1)} \quad \text{for } 2 \leq i, j \leq k.$$

We define now the quotient speed of a sequence in $\mathbb{S}$.

Definition 12.2 Let k be a natural number. A sequence $\mathbf{x} = (x_n)_{n \in \mathbb{N}}$ has finite *kth quotient speed* if there is a real number $v \geq 1$ such that

$$v = \lim_{n \to \infty} V_n^{(k)}(\mathbf{x}).$$

In this case we write $v_k(\mathbf{x}) = v$.

The kth quotient speed of a sequence $\mathbf{x} = (x_n)_{n \in \mathbb{N}}$ is ∞ if

$$\lim_{n \to \infty} V_n^{(k)}(\mathbf{x}) = \infty.$$

Proposition 12.2 Let $\mathbf{x} \in \mathbb{S}$. For all $k \in \mathbb{N}$ we have:

(a) If $v_k(\mathbf{x}) = \infty$, then $v_i(\mathbf{x}) = \infty$ for $1 \leq i < k$.
(b) If $v_k(\mathbf{x}) < \infty$, then $v_i(\mathbf{x}) < \infty$ for $i > k$.

Therefore, the class of sequences having finite quotient speed v_1 coincides with the class $\mathrm{Tr}(\mathrm{RV}_s)$ of translationally regularly varying sequences.

Example 12.1

(1) The sequence $\mathbf{x}$ defined by $x_n = 2^n$, $n \in \mathbb{N}$, satisfies $v_1(\mathbf{x}) = 2$.
(2) The sequences $\mathbf{x}$ and $\mathbf{y}$ defined by $x_n = n^n$ and $y_n = n!$, $n \in \mathbb{N}$, have finite second quotient speed $v_2(\mathbf{x}) = v_2(\mathbf{y}) = 1$ (while $v_1(\mathbf{x}) = v_1(\mathbf{y}) = \infty$).
(3) The sequence $x_n = \prod_{i=1}^{n} i!$ has finite third quotient speed $v_3(\mathbf{x})$ (equal 1), and $v_2(\mathbf{x}) = \infty$.
(4) For each $k \in \mathbb{N}$, the sequence $x_n = 2^{n^k}$, $n \in \mathbb{N}$, has finite quotient speed $v_k(\mathbf{x})$.

The previous consideration suggests to define the following classes related to the class $\text{Tr}(\text{RV}_s)$.

Define

$$\text{Tr}^{(1)}(\text{RV}_s) = \text{Tr}(\text{RV}_s)$$

and for each $k \geq 2$

$$\text{Tr}^{(k)}(\text{RV}_s) = \{\mathbf{x} \in \mathbb{S} : \upsilon_k(\mathbf{x}) < \infty\}.$$

Note that for each $k \in \mathbb{N}$ the class $\text{Tr}^{(k)}(\text{RV}_s)$ is nonempty and it holds

$$\text{Tr}(\text{RV}_s) = \text{Tr}^{(1)}(\text{RV}_s) \subsetneqq \text{Tr}^{(2)}(\text{RV}_s) \subsetneqq \cdots \text{Tr}^{(k)}(\text{RV}_s) \subsetneqq \cdots \tag{12.8}$$

Observe that for each $k \in \mathbb{N}$, the sequence $x_n = 2^{n^{k+1}}$, $n \in \mathbb{N}$, belongs to $\text{Tr}^{(k+1)}(\text{RV}_s) \backslash \text{Tr}^{(k)}(\text{RV}_s)$.

12.2.3 The Class ORV$_s$ and Seneta Sequences

Definition 12.3 [33] A function $\varphi : [a, \infty) \to (0, \infty)$, $a > 0$, is said to be $\mathcal{O}$-*regularly varying* if it is measurable and for each $\lambda > 0$ satisfies the condition

$$\overline{k}_\varphi(\lambda) := \limsup_{t \to \infty} \frac{\varphi(\lambda t)}{\varphi(t)} < \infty. \tag{12.9}$$

The function $\overline{k}_\varphi(\lambda)$, $\lambda > 0$, is called the *index function of* φ. If this function is continuous we say that φ is $\mathcal{O}$-regularly varying with a continuous index function.

The class of all these functions is denoted by ORV$_f$.

Definition 12.4 [33] A sequence $\mathbf{x} = (x_n)_{n \in \mathbb{N}}$ is said to be $\mathcal{O}$-*regularly varying* for each $\lambda > 0$ satisfies the condition

$$\overline{k}_\mathbf{x}(\lambda) := \limsup_{n \to \infty} \frac{x_{[\lambda n]}}{x_n} < \infty. \tag{12.10}$$

The class of these sequences is denoted by ORV$_s$.

The classes ORV$_f$ and ORV$_s$ have been investigated in [14, 34–36].

Theorem 12.4 [14, Theorems 1 and 3] Let $\mathbf{x} = (x_n)_{n \in \mathbb{N}}$ be a sequence of positive real numbers. Then the following assertions are equivalent:

(a) $\mathbf{x} \in \text{ORV}_s$;
(b) $\varphi_\mathbf{x}(t) = x_{[t]} \in \text{ORV}_f$ on the interval $[1, \infty)$;

(c) **x** is represented in the form

$$x_n = \exp\left\{\mu_n + \sum_{k=1}^{n} \frac{\delta_k}{k}\right\},$$

where $(\mu_n)_{n\in\mathbb{N}}$ and $(\delta_n)_{n\in\mathbb{N}}$ are bounded sequences.

Theorem 12.5 [14, Theorem 4] If a sequence **x** is in the class ORV_s, then the index function $k_\mathbf{x}$ is in ORV_f.

Let $\beta \geq 1$. Denote by $SO_{s,\beta}$ the class of all positive sequences $\mathbf{x} = (x_n)_{n\in\mathbb{N}}$ satisfying $\overline{k}_\mathbf{x}(\lambda) \leq \beta$ for each $\lambda > 0$. Sequences from the class $SO_s := \bigcup_{\beta\geq 1} SO_{s,\beta}$ are called *Seneta sequences*.

They form a proper subclass of the class ORV_s of $\mathcal{O}$-regularly varying sequences.

Theorem 12.6 [37] Let $\mathbf{x} = (x_n)_{n\in\mathbb{N}}$ be a sequence of positive real numbers. Then the following assertions are equivalent:

(a) **x** is a Seneta sequence (from the class $SO_{s,\beta}$);
(b) the function $\varphi_\mathbf{x}(t) = x_{[t]}$, $t > 0$, is a Seneta function (from the class $SO_{f,\beta}$);
(c) there is a Seneta function φ on $[a, \infty)$, $a > 0$, such that $\varphi(n) = x_n$ for each $n \geq [a] + 1$;
(d) **x** is represented in the form

$$x_n = \exp\left\{\mu_n + \sum_{k=1}^{n} \frac{\delta_k}{k}\right\}, \quad n \geq n_0 \text{ for some } n_0 \in \mathbb{N},$$

where $(\mu_n)_{n\in\mathbb{N}}$ is a bounded sequence, and $(\delta_n)_{n\in\mathbb{N}}$ is a sequence converging to 0 as $n \to \infty$;

12.3 Rapid Variation

Definition 12.5 [11] (see also [6]) A function $\varphi : [a, \infty) \to (0, \infty)$, $a > 0$, is said to be *rapidly varying of index of variability* ∞ if it is measurable and satisfies the asymptotic condition

$$\lim_{t\to\infty} \frac{\varphi(\lambda t)}{\varphi(t)} = \infty, \quad \lambda > 1. \tag{12.11}$$

The class of rapidly varying functions of index ∞ we denote by $RV_{f,\infty}$.

12.3.1 Some Properties of Rapidly Varying Functions

We give now a few results on rapidly varying functions obtained quite recently. In the following theorem, operator $\underline{D}$ is the *lower Dini derivative* defined by

$$\underline{D}g(t) = \lim_{y \to t} \inf \frac{g(y) - g(t)}{y - t}, \quad \text{for } g : \mathbb{R} \to \mathbb{R}, \ t \in \mathbb{R}.$$

Theorem 12.7 [22] For a function $\varphi : (0, \infty) \to (0, \infty)$ the following assertions are equivalent:

(a) $\varphi \in R_{f,\infty}$;
(b) there is a nondecreasing, absolutely continuous function $g : \mathbb{R} \to \mathbb{R}$ such that $\lim_{t \to \infty} \underline{D}g(t) = \infty$ and there is a measurable function $j : (0, \infty) \to (0, \infty)$ such that $j(t) \sim t$ for $t \to \infty$, so that

$$\varphi(t) = \exp(g(\ln(j(t))))$$

for all $t > 0$;
(c) there are measurable functions $j : (0, \infty) \to (0, \infty)$ and $h : (0, \infty) \to [0, \infty)$, such that $\lim_{t \to \infty} h(t) = \infty$ and $j(t) \sim t$ for $t \to \infty$, for which

$$\varphi(t) = \exp\left\{ c + \int_0^{j(t)} h(u)\frac{du}{u} \right\}$$

for all $t > 0$ and for some $c \in \mathbb{R}$.

Theorem 12.8 [22] Let $\varphi : (0, \infty) \to (0, \infty)$ be a measurable function. Then $\varphi \in R_{f,\infty}$ if and only if for all $\alpha > 0$ there is a measurable function $j_\alpha : (0, \infty) \to (0, \infty)$ such that $j_\alpha(t) \sim t$, for $t \to \infty$, and there is a nondecreasing function $k_\alpha : (0, \infty) \to (0, \infty)$, so that

$$\varphi(t) = t^\alpha \cdot k_\alpha(j_\alpha(t)), \quad \text{for } t > 0.$$

Let $F^{(\infty)}$ be the set of all functions $\varphi : (0, \infty) \to (0, \infty)$, which are bounded on $(0, \alpha)$ for each $\alpha \in (0, \infty)$, and $\lim \sup_{t \to \infty} \varphi(t) = \infty$ [2]. For any $\varphi \in F^{(\infty)}$, the (positive, nondecreasing, and unbounded) function $\varphi^\leftarrow$ defined by

$$\varphi^\leftarrow(y) := \inf\{x > 0 : \varphi(t) > y\} \tag{12.12}$$

on (b, ∞), where $b = \inf\{\varphi(z) : z \in (0, \infty)\}$, is the *generalized inverse* of φ [6]. Further, let

$$F^\infty = \{\varphi \in F^{(\infty)} : \lim \inf_{t \to \infty} \varphi(t) = \infty\}.$$

For any $\varphi \in F^\infty$ we consider the next four positive and nondecreasing functions:

1. $\varphi^{\leftarrow \ell}(y) := \inf\{t > 0 : \varphi(t) > y\}, \quad y > b;$

2. $\varphi^{\leftarrow u}(y) := \sup\{t > 0 : \varphi(t) \le y\}, \quad y > b$;
3. $\overline{\varphi}(t) := \sup\{\varphi(z) : z \le t\}, \quad t > 0$;
4. $\underline{\varphi}(t) := \inf\{\varphi(z) : z \ge t\}, \quad t > 0$.

Theorem 12.9 [20] Let $\varphi \in \mathcal{F}^\infty$ be a measurable function. The following assertions are equivalent:

(a) $\varphi \in R_{f,\infty}$;

(b) $\lim\limits_{t \to \infty} \inf\limits_{\lambda > L} \frac{\varphi(t)}{\varphi\left(\frac{t}{\lambda}\right)} = \infty$ for each $L > 1$;

(c) $\lim\limits_{t \to \infty} \frac{\overline{\varphi}(\lambda t)}{\underline{\varphi}(t)} = \infty$ for each $\lambda > 1$;

(d) $\lim\limits_{t \to \infty} \frac{\varphi^{\leftarrow u}(\lambda t)}{\varphi^{\leftarrow \ell}(t)} = 1$ for each $l\lambda > 1$.

Theorem 12.10 (38, Theorem 2.5) Let $p.q \in [0, \infty)$ be such that $\frac{1}{p} + \frac{1}{q} = 1$. If $\varphi \in R_{f,\infty}$ and $\psi \in RV_{f,\rho}$ are locally bounded on $[0, \infty)$, then

$$\int_0^y \varphi(t)\psi(t) \, dt = o\left(\left(\int_0^y \varphi(t)^p \, dt\right)^{1/p} \cdot \left(\int_0^y \psi(t)^q \, dt\right)^{1/q}\right), \quad y \to \infty.$$

Theorem 12.11 (38, Corollary 2.1) If $\varphi \in R_{f,\infty}$ and $p > 1$, then

$$\frac{f(1) + f(2) + \cdots + f(n)}{n} = o\left(\left(\frac{f(1)^p + f(2)^p + \cdots + f(n)^p}{n}\right)^{1/p}\right), \quad n \to \infty.$$

A function $\varphi : [a, \infty) \to (0, \infty), a > 0$, is said to be *rapidly varying of index of variability* $-\infty$ [11] (see also [6]) if it is measurable and for each $\lambda > 1$ satisfies

$$\lim_{t \to \infty} \frac{\varphi(\lambda t)}{\varphi(t)} = 0. \tag{12.13}$$

$R_{f,-\infty}$ denotes the class of rapidly varying functions of index $-\infty$.

Definition 12.6 A sequence $\mathbf{x} = (x_n)_{n \in \mathbb{N}} \in \mathbb{S}$ is *rapidly varying* (of index of variability ∞) if the following asymptotic condition is satisfied:

$$\lim_{n \to \infty} \frac{x_{[\lambda n]}}{x_n} = 0, \quad 0 < \lambda < 1, \tag{12.14}$$

or equivalently

$$\lim_{n \to \infty} \frac{x_{[\lambda n]}}{x_n} = \infty, \quad \lambda > 1. \tag{12.15}$$

$R_{s,\infty}$ denotes the class of rapidly varying sequences [16].

A sequence $\mathbf{x} = (x_n)_{n\in\mathbb{N}} \in \mathbb{S}$ is said to belong to the class $R_{s,-\infty}$ of *rapidly vary-ing sequences of index of variability* $-\infty$ if for each $\lambda > 1$ the following condition is satisfied:

$$\lim_{n\to\infty} \frac{x_{[\lambda n]}}{x_n} = 0. \tag{12.16}$$

It is easy to see that the following two facts hold:

Fact 1. A function φ belongs to the class $R_{f,\infty}$ if and only if the function $\psi = \frac{1}{\varphi}$ belongs to the class $R_{f,\infty}$.

Fact 2. A sequence $(x_n)_{n\in\mathbb{N}}$ is in the class $R_{s,\infty}$ if and only if the sequence $(1/x_n)_{n\in\mathbb{N}}$ is in the class $R_{s,-\infty}$.

The following theorem, which allowed a unified study of rapidly varying sequences and rapidly varying functions, was shown in [16].

Theorem 12.12 For a sequence $\mathbf{x} = (x_n)_{n\in\mathbb{N}}$ in $\mathbb{S}$ the following are equivalent:

(a) $\mathbf{x}$ belongs to the class $R_{s,\infty}$;
(b) The function $\varphi_\mathbf{x}$ defined by $\varphi_\mathbf{x}(t) = x_{[t]}$, $t \geq 1$, is in the class $R_{f,\infty}$.

Then, we have the following results that are parallel to Theorem 12.13.

Theorem 12.13 For a sequence $\mathbf{x} = (x_n)_{n\in\mathbb{N}}$ in $\mathbb{S}$ the following are equivalent:

(a) $\mathbf{x}$ belongs to the class $R_{s,-\infty}$;
(b) the function $\varphi_\mathbf{x}$ defined by $\varphi_\mathbf{x}(t) = x_{[t]}$, $t \geq 1$, is in the class $R_{f,-\infty}$;
(c) $\lim_{n\to\infty} \frac{x_{[\lambda n]}}{x_n} = \infty$, $0 < \lambda < 1$.

Theorem 12.14 If a sequence $(x_n)_{n\in\mathbb{N}}$ belongs to $R_{s,-\infty}$, then $\lim_{n\to\infty} x_n = 0$.

12.3.2 The Class ARV$_s$

Definition 12.7 ([17, 39] for functions) A sequence $\mathbf{x} = (x_n)_{n\in\mathbb{N}} \in \mathbb{S}$ is said to belong to the *class* ARV$_s$ if for each $\lambda > 1$ the following condition is satisfied:

$$\underline{k}_\mathbf{x}(\lambda) := \liminf_{n\to\infty} \frac{x_{[\lambda n]}}{x_n} > 1. \tag{12.17}$$

Notice that (12.17) is equivalent to the statement: for each $\lambda > 1$ there is $n_0 = n_0(\lambda)$ such that for each $n \geq n_0$

$$\frac{x_{[\lambda n]}}{x_n} \geq \frac{(\underline{k}_\mathbf{x}(\lambda) - 1)}{2} + 1 \tag{12.18}$$

and also to the statement: for each $\lambda > 1$ there are $n_0 = n_0(\lambda)$ and $c(\lambda) > 1$ such that

$$x_{[\lambda n]} \geq c(\lambda) \cdot x_n, \tag{12.19}$$

for every $n \geq n_0$.

Let $\mathbf{x} = (x_n)_{n \in \mathbb{N}}$ be a strictly increasing, unbounded sequence from $\mathbb{S}$. The *numerical function of* $\mathbf{x}$ is the function $\delta_{\mathbf{x}} : [x_1, \infty) \to \mathbb{N}$ defined by [8]

$$\delta_{\mathbf{x}}(t) := \max\{n \in \mathbb{N} : x_n \leq t\}.$$

The numerical function of a sequence is an important characteristic of divergent sequences [8].

A strictly increasing and unbounded sequence $\mathbf{x} = (x_n)_{n \in \mathbb{N}} \in \mathbb{S}$ belongs to ARV_s if and only if the numerical function $\delta_{\mathbf{x}}$ of $\mathbf{x}$ is an $\mathcal{O}$-regularly varying function with a continuous index function [36].

Observe that we have

$$\mathrm{RV}_{s,+} \subsetneq \mathrm{ARV}_s \text{ and } \mathrm{R}_{s,\infty} \subsetneq \mathrm{ARV}_s.$$

The following two lemmas [17] describe important properties of the class ARV_s.

Lemma 12.1 Each sequence $\mathbf{x} = (x_n)_{n \in \mathbb{N}} \in \mathrm{ARV}_s$ contains a subsequence belonging to ARV_s.

Proof: Let p be a fixed prime number. Since $\mathbf{x} \in \mathrm{ARV}_s$ and $p > 1$, there are $c = c(p)$ and $n_0 \in \mathbb{N}$ such that $x_{pn} \geq c \cdot x_n$ for all $n \geq n_0$; without loss of generality one can suppose that $n_0 = 1$. We prove that the subsequence $(y_k)_{k \in \mathbb{N}} := (x_{p^k})_{k \in \mathbb{N}}$ of $\mathbf{x}$ belongs to the class ARV_s; in fact, we show a little bit more: this subsequence is in the class $\mathrm{R}_{s,\infty}$ which is a subclass of ARV_s.

As $p^k > n_0$, $k \in \mathbb{N}$, by the above we have $x_{p \cdot p^k} \geq c \cdot x_{p^k}$, that is, $\frac{y_{k+1}}{y_k} \geq c > 1$. Therefore,

$$\liminf_{k \to \infty} \frac{y_{k+1}}{y_k} \geq c.$$

Thus, for $\lambda > 1$, we have

$$\liminf_{k \to \infty} \frac{y_{[\lambda k]}}{y_k} = \liminf_{k \to \infty} \frac{y_{[\lambda k]}}{y_{[\lambda k]-1}} \cdots \frac{y_{k+1}}{y_k},$$

and because on the right side of this equality there are $[\lambda k] - k$ factors, and $[\lambda k] - k > (\lambda - 1)k - 1$, one obtains

$$\liminf_{k \to \infty} \frac{y_{[\lambda k]}}{y_k} > \liminf_{k \to \infty} c^{(\lambda-1)k-1} = \infty.$$

This means that $(y_k)_{k \in \mathbb{N}} \in \mathrm{R}_{s,\infty} \subset \mathrm{ARV}_s$, which completes the proof. $\square$

Lemma 12.2 Every sequence from the class ARV_s contains a subsequence divergent to ∞.

Proof: Let $\mathbf{x} = (x_n)_{n\in\mathbb{N}} \in ARV_s$. Consider its subsequence $(x_{2^n})_{n\in\mathbb{N}\cup\{0\}}$. For each $\lambda > 1$ there are $n = n_0(\lambda)$ and $c(\lambda) > 1$ such that $x_{[\lambda n]} \geq c(\lambda) \cdot x_n$ for all $n \geq n_0$; one can suppose $n_0 = 1$ (otherwise we apply a similar procedure to the subsequence $(x_{2^n \cdot n_0})_{n\in\mathbb{N}\cup\{0\}}$). Therefore, for $n \geq n_0$ we have

$$x_{2^n} \geq c(2) \cdot x_{2^{n-1}} \geq \cdots \geq (c(2))^n \cdot x_1$$

and thus

$$\liminf_{n\to\infty} x_{2^n} \geq \liminf_{n\to\infty}(x_1 \cdot (c(2))^n) = \infty,$$

that is, $(x_{2^n})_{n\in\mathbb{N}}$ is the required subsequence. $\square$

12.3.3 The Class $KR_{s,\infty}$

In this section, we study an important subclass of the class $R_{\infty,s}$, that we denote by $KR_{s,\infty}$.

For a sequence $\mathbf{x} = (x_n)_{n\in\mathbb{N}}$ of positive real numbers the *lower Matuszewska index* $d(\mathbf{x})$ is defined as the supremum of all $d \in \mathbb{R}$ such that for each $\Lambda > 1$

$$\frac{x_{[\lambda n]}}{x_n} \geq \lambda^d(1 + o(1)), \quad (n \to \infty) \tag{12.20}$$

holds uniformly (with respect to λ) on the segment $[1, \Lambda]$ (compare with the definition of lower Matuszewska index for functions [6, p. 68]). The sequence $\mathbf{x}$ belongs to the *class* $KR_{s,\infty}$ if $d(\mathbf{x}) = \infty$.

Let us mention that in a similar way one defines the *lower Matuszewska index* $d(\varphi)$ of a measurable function $\varphi : [a, \infty) \to (0, \infty)$. The class of all measurable functions whose lower Matuszewska index is ∞ is denoted by $KR_{f,\infty}$. This class of functions has very important asymptotic properties (see [6] in this connection). By a result from [6] we have $KR_{f,\infty} \subsetneq R_{f,\infty}$.

The importance of the class $KR_{s,\infty}$ follows, among others, from the Seneta–de Haan theorem [6, Theorem 2.4.7] that gives nice relations between classes SV_f and $KR_{f,\infty}$ under the generalized inverse, and from the fact that there is a connection between rapidly varying functions and their cumulative maximum functions from the class $KR_{f,\infty}$ (see [6, p. 87], in particular Proposition 2.4.6).

Theorem 12.15 For a sequence $\mathbf{x} = (x_n)_{n\in\mathbb{N}}$ of positive real numbers the following are equivalent:

(a) $\mathbf{x} \in KR_{s,\infty}$;
(b) for each $d \in \mathbb{R}$ it holds $\liminf_{n\to\infty} \inf_{\lambda \geq 1} \frac{x_{[\lambda n]}}{\lambda^d x_n} \geq 1$;

Proof: (a) $\Rightarrow$ (b) From $d(\mathbf{x}) = \infty$, it follows that for every $d \in \mathbb{R}$, every $\Lambda > 1$, and sufficiently large n we have $\frac{x_{[\lambda n]}}{x_n} \geq \lambda^d(1 + o(1))$, where $\lambda \in [1, \Lambda]$ is an arbitrary fixed element. For the same d, λ, Λ, for sufficiently large n we have $\inf_{\lambda \in [1, \Lambda]} \frac{x_{[\lambda n]}}{\lambda^d x_n} \geq 1 + o(1)$. In other words, for each $\varepsilon > 0$ there is $n_0 = n_0(\varepsilon) \in \mathbb{N}$ such that $\inf_{\lambda \in [1, \Lambda]} \frac{x_{[\lambda n]}}{\lambda^d x_n} \geq 1 - \varepsilon$ for each $n \geq n_0$. Because the last inequality is true for each $\Lambda > 1$, it follows that (for the same d) for each $\lambda \geq 1$ we have $\inf_{\lambda \geq 1} \frac{x_{[\lambda n]}}{\lambda^d x_n} \geq 1 - \varepsilon$. As ε was arbitrary (b) follows.

(b) $\Rightarrow$ (a) Suppose that for an arbitrarily fixed $d \in \mathbb{R}$, $\lim \inf_{n \to \infty} \inf_{\lambda \geq 1} \frac{x_{[\lambda n]}}{\lambda^d x_n} \geq 1$ is satisfied. Then for the same d and each $\varepsilon > 0$ there exists $n_0 = n_0(\varepsilon) \in \mathbb{N}$ such that $\inf_{\lambda \geq 1} \frac{x_{[\lambda n]}}{\lambda^d x_n} \geq 1 - \varepsilon$ for each $n \geq n_0$. In other words, for the same d, ε, n_0, and for each $\lambda \geq 1$, especially for $\lambda \in [1, \Lambda]$, $\Lambda > 1$ an arbitrary real number, it holds $\frac{x_{[\lambda n]}}{x_n} \geq \lambda^d(1 - \varepsilon)$ for each $n \geq n_0$. This means that for each $\Lambda > 1$ we have $\frac{x_{[\lambda n]}}{x_n} \geq \lambda^d(1 + o(1))$ uniformly with respect to $\lambda \in [1, \Lambda]$ for $n \to \infty$. Since d is arbitrary, (a) follows. $\square$

The next statement is a result of the Galambos–Bojanić–Seneta type [6, 19, 29, 30, 37, 40].

Theorem 12.16 For a sequence $\mathbf{x} = (x_n)_{n \in \mathbb{N}}$ of positive real numbers the following are equivalent:

(a) $\mathbf{x} \in \mathsf{KR}_{s,\infty}$.
(b) The function $\varphi_{\mathbf{x}}(t) = x_{[t]}$, $t \geq 1$, belongs to the class $\mathsf{KR}_{f,\infty}$.
(c) For each $n \geq 1$, $x_n = \exp\left\{ p_n + q_n + \sum_{k=1}^{n} \frac{r_k}{k} \right\}$, where $(p_n)_{n \in \mathbb{N}}$ is a nondecreasing sequence of real numbers, $(q_n)_{n \in \mathbb{N}}$ and $(r_n)_{n \in \mathbb{N}}$ are real sequences with $\lim_{n \to \infty} q_n = 0$, $\lim_{n \to \infty} r_n = \infty$.

Proof: (a) $\Rightarrow$ (b) Let $\mathbf{x} = (x_n)_{n \in \mathbb{N}} \in \mathsf{KR}_{s,\infty}$. Then by Theorem 12.15 we have $\lim \inf_{n \to \infty} \inf_{\lambda \geq 1} \frac{x_{[\lambda n]}}{\lambda^d x_n} \geq 1$ for each $d \in \mathbb{R}$. This means that for the same d and each $\varepsilon > 0$ there is $n_0 = n_0(d, \varepsilon) \in \mathbb{N}$ such that $\inf_{\lambda \geq 1} \frac{x_{[\lambda t]}}{\lambda^d x_{[t]}} \geq 1 - \varepsilon$ for each $t \geq n_0$ (≥ 1). Therefore, for the same d, ε, n_0 it is true

$$\inf_{\lambda \geq 1} \frac{x_{[\lambda t]}}{\lambda^d x_{[t]}} = \inf_{\lambda \geq 1} \frac{x_{[\frac{t}{[t]} \cdot [t] \cdot \lambda]}}{\lambda^d x_{[t]}} \geq \inf_{\lambda \geq 1} \frac{x_{[[t] \cdot \lambda]}}{\lambda^d x_{[t]}} \geq 1 - \varepsilon,$$

that is, (for this d)

$$\lim \inf_{t \to \infty} \inf_{\lambda \geq 1} \frac{x_{[\lambda t]}}{\lambda^d x_{[t]}} \geq 1.$$

By [6, Proposition 2.4.3(ii)], it follows that the function $x_{[t]}$ belongs to the class $\mathsf{KR}_{f,\infty}$.

(b) $\Rightarrow$ (c) Let $\mathbf{x} = (x_n)_{n \in \mathbb{N}}$ be a sequence satisfying (2), that is, the function $\varphi_{\mathbf{x}}(t) = x_{[t]}, t \geq 1$, is in the class $\mathsf{KR}_{\infty f}$. By [6, Theorem 2.4.5], for each n we have

$$x_n = \exp \left\{ p_n + q_n + \int_1^n \frac{\alpha(t)}{t} \, dt \right\},$$

where $(p_n)_{n \in \mathbb{N}}$ is a nondecreasing real sequence, $(q_n)_{n \in \mathbb{N}}$ is a sequence with $\lim_{n \to \infty} q_n = 0$, and $\alpha : [1, \infty) \to \mathbb{R}$ is a measurable function with $\lim_{t \to \infty} \alpha(t) = \infty$. Then

$$c_n = \exp \left\{ p_n + q_n + \sum_{k=1}^{n_1} \left(\int_k^{k+1} \frac{\alpha(t)}{t} \, dt \right) + \sum_{k=n_1+1}^{n-1} \left(\int_k^{k+1} \frac{\alpha(t)}{t} \, dt \right) \right\},$$

where $n_1 \in \mathbb{N}$ is such that $\inf_{t \geq n_1+1} \alpha(t) \geq A > 0$ for some given A. Define

$$r_k = 0 \text{ for } k = 1; \quad r_k = k \int_{k-1}^k \frac{\alpha(t)}{t} \, dt \text{ for } k \geq 2.$$

For each $k \geq n_1 + 2$ (and $k \leq n$) the following is satisfied:

$$r_k = k \int_{k-1}^k \frac{\alpha(t)}{t} \, dt \geq k \cdot \ln \left(\frac{k}{k-1} \right) \cdot \inf_{t \geq n_1+1} \alpha(t)$$

$$\geq k \cdot \ln \left(\frac{k}{k-1} \right) \cdot A \geq A/2.$$

So, $\lim_{n \to \infty} r_n = \infty$ and (c) follows.

(c) $\Rightarrow$ (a) Let for each $n \in \mathbb{N}$, $x_n = \exp \left\{ p_n + q_n + \sum_{k=1}^n \frac{r_k}{k} \right\}$, where $(p_n)_{n \in \mathbb{N}}$ is a nondecreasing real sequence, and $(q_n)_{n \in \mathbb{N}}$ and $(r_n)_{n \in \mathbb{N}}$ are real sequences with $\lim_{n \to \infty} q_n = 0$ and $\lim_{n \to \infty} r_n = \infty$, respectively. We prove that $\mathbf{x} = (x_n)_{n \in \mathbb{N}} \in \mathsf{KR}_{s,\infty}$. Consider two cases: $d \leq 0$ and $d > 0$.

$d \leq 0$: In this case we have

$$\inf_{\lambda \geq 1} \frac{x_{[\lambda n]}}{\lambda^d x_n} = \min \left\{ 1, \inf_{\lambda > 1} \frac{x_{[\lambda n]}}{\lambda^d x_n} \right\}.$$

Consider

$$\inf_{\lambda > 1} \frac{x_{[\lambda n]}}{\lambda^d x_n} = \inf_{\lambda > 1} \{ \lambda^{-d} \exp\{ (p_{[\lambda n]} - p_n) + (q_{[\lambda n]} - q_n) + I(n, \lambda) \} \},$$

where

$$I(n, \lambda) = \begin{cases} 0, & \text{if } \lambda \in (1, 1 + 1/n), \\ \displaystyle\sum_{k=n+1}^{[\lambda n]} \frac{r_k}{k}, & \text{if } \lambda \in [1 + 1/n, \infty). \end{cases}$$

It holds $\inf_{\lambda > 1} I(n, \lambda) \geq 0$, $\inf_{\lambda > 1} (p_{[\lambda n]} - p_n) \geq 0$, and $\inf_{\lambda > 1} \lambda^{-d} = 1$. Since $\lim_{n \to \infty} q_n = 0$ it follows that the sequence $(q_n)_{n \in \mathbb{N}}$ is a Cauchy sequence and that for each $\varepsilon > 0$ there is $n_1 = n_1(\varepsilon) \in \mathbb{N}$ such that $|q_{[\lambda n]} - q_n| \le \varepsilon$ for each $\lambda > 1$ and each $n \ge n_1$. This means that for each $n \ge n_1$ we have

$$\inf_{\lambda > 1}\{\lambda^{-d}\exp\{(p_{[\lambda n]} - p_n) + (q_{[\lambda n]} - q_n) + I(n, \lambda)\}\}$$

$$\ge \inf_{\lambda > 1}\lambda^{-d} \cdot \inf_{\lambda > 1}\exp\{p_{[\lambda n]} - p_n\} \cdot \inf_{\lambda > 1}\exp\{q_{[\lambda n]} - q_n\}$$

$$\cdot \inf_{\lambda > 1}\exp\{I(n, \lambda)\} \ge e^{-\varepsilon} > 1 - \varepsilon,$$

so that $\liminf_{n \to \infty}\inf_{\lambda \ge 1}\frac{x_{[\lambda n]}}{\lambda^d x_n} \ge 1$, that is, $\mathbf{x} \in KR_{\infty,s}$.

$d > 0$: Following the same arguments as in case $d \le 0$, it is sufficient to consider only $\inf_{\lambda \ge 1}\left\{\frac{\exp\{I(n,\lambda)\}}{\lambda^d}\right\}$, that is, because $\inf_{\lambda \ge 1}\frac{x_{[\lambda n]}}{\lambda^d x_n} = \min\{1, \inf_{\lambda > 1}\frac{x_{[\lambda n]}}{\lambda^d x_n}\}$, it suffices to consider only $\inf_{\lambda > 1}\left\{\frac{\exp\{I(n,\lambda)\}}{\lambda^d}\right\}$. For a given $n \in \mathbb{N}$ we have $\inf_{\lambda \in (1,1+1/n)}\frac{e^0}{\lambda^d} = \frac{1}{(1+1/n)^d}$. Since

$$\liminf_{n \to \infty}\inf_{\lambda > 1}\left\{\frac{\exp\{I(n, \lambda)\}}{\lambda^d}\right\}$$

$$= \liminf_{n \to \infty}\min\left\{\inf_{\lambda \in (1,1+1/n)}\left\{\frac{\exp\{I(n, \lambda)\}}{\lambda^d}\right\}, \inf_{\lambda \ge 1+1/n}\left\{\frac{\exp\{I(n, \lambda)\}}{\lambda^d}\right\}\right\}$$

and since

$$\liminf_{n \to \infty}\inf_{\lambda \in (1,1+1/n)}\left\{\frac{\exp\{I(n, \lambda)\}}{\lambda^d}\right\} = \lim_{n \to \infty}\frac{1}{(1+1/n)^d} = 1,$$

we have to consider only $\liminf_{n \to \infty}\inf_{\lambda \ge 1+1/n}\left\{\frac{\exp\{I(n,\lambda)\}}{\lambda^d}\right\}$. We have

$$\inf_{\lambda \ge 1+1/n}\left\{\frac{\exp\{I(n, \lambda)\}}{\lambda^d}\right\} = \inf_{\lambda \ge 1+1/n}\left\{\exp\left\{-d\ln\lambda + \frac{r_{n+1}}{n+1} + \cdots \frac{r_{[\lambda n]}}{[\lambda n]}\right\}\right\},$$

and if we choose $n_2 \in \mathbb{N}$ so that $r_n \ge 2d$ for all $n \ge n_2$, then for the same n the following is true:

$$\inf_{\lambda \ge 1+1/n}\left\{\exp\left\{-d\ln\lambda + \frac{r_{n+1}}{n+1} + \cdots \frac{r_{[\lambda n]}}{[\lambda n]}\right\}\right\}$$

$$\ge \inf_{\lambda \ge 1+1/n}\{\exp\{-d\ln\lambda + 2d(\ln([\lambda n] + 1) - \ln(n+1))\}\}$$

$$\ge \inf_{\lambda \ge 1+1/n}\left\{\exp\left\{d(-\ln\lambda + 2\ln\frac{\lambda n}{n+1})\right\}\right\}$$

$$= \inf_{\lambda \ge 1+1/n}\left\{\exp\left\{d\ln\frac{\lambda}{(1+1/n)^2}\right\}\right\} = \exp\left\{d\ln\frac{1}{1+1/n}\right\}.$$

Therefore, we obtain

$$\liminf_{n \to \infty}\inf_{\lambda \ge 1+1/n}\left\{\frac{\exp\{I(n, \lambda)\}}{\lambda^d}\right\} \ge \lim_{n \to \infty}\exp\left\{d\ln\frac{1}{1+1/n}\right\} = 1.$$

This completes the proof. $\qquad\square$

Theorem 12.17 If $\mathbf{x} = (x_n)_{n\in\mathbb{N}}$ is a nondecreasing sequence in $R_{s,\infty}$, then $\mathbf{x} \in KR_{s,\infty}$.

The following two theorems give interesting properties of sequences from the class $KR_{s,\infty}$.

For a sequence $\mathbf{x} = (x_n)_{n\in\mathbb{N}} \in \mathbb{S}$, the sequence $\bar{\mathbf{x}} = (\bar{x}_n)_{n\in\mathbb{N}}$ is defined by $\bar{x}_n = \max\{x_k : 1 \le k \le n\}, n \in \mathbb{N}$.

Theorem 12.18 If a sequence $\mathbf{x} = (x_n)_{n\in\mathbb{N}}$ belongs to the class $R_{s,\infty}$, then the sequence $\bar{\mathbf{x}}$ belongs to the class $KR_{s,\infty}$.

Proof: The sequence $\bar{x}_n$ is a nondecreasing sequence of positive real numbers, and for each $\lambda > 1$ satisfies

$$\liminf_{n\to\infty} \frac{\bar{x}_{[\lambda n]}}{\bar{x}_n} = \liminf_{n\to\infty} \frac{\max\{x_k : 1 \le k \le [\lambda n]\}}{\max\{x_k : 1 \le k \le n\}} \ge \liminf_{n\to\infty} \frac{x_{[\lambda n]}}{x_{k^*(n)}},$$

where $k^*(n)$ is the index for which x_k is maximal. It follows

$$\liminf_{n\to\infty} \frac{\bar{x}_{[\lambda n]}}{\bar{x}_n} \ge \liminf_{n\to\infty} \frac{x_{[\lambda n]}}{x_{k^*(n)}} \ge \liminf_{n\to\infty} \frac{x_{[\lambda n]}}{x_n} \cdot \liminf_{n\to\infty} \frac{x_n}{x_{n-1}} \cdots \liminf_{n\to\infty} \frac{x_{k^*(n)+1}}{x_{k^*(n)}} = \infty$$

as the first factor here is equal to ∞, and all other factors are ≥ 1 (because $\mathbf{x} \in R_{s,\infty}$). So, $\bar{\mathbf{x}} \in R_{s,\infty}$, and since it is nondecreasing, by Theorem 12.17, $\bar{\mathbf{x}} \in KR_{s,\infty}$. □

Theorem 12.19 If a sequence $\mathbf{x} = (x_n)_{n\in\mathbb{N}}$ belongs to the class $KR_{s,\infty}$, then $\mathbf{x} \backsim \bar{\mathbf{x}}$.

Proof: Since $\mathbf{x} \in KR_{s,\infty}$, by Theorem 12.16 the function $g(\mathbf{x}) = \mathbf{x}_{[t]}$, $t \ge 1$, is in the class $KR_{f,\infty}$. Define now the function

$$\mathbf{c}_{1[t]} = \begin{cases} 1, & \text{if } t \in [0,1), \\ \mathbf{c}_{[t]}, & \text{if } t \in [1,\infty). \end{cases}$$

By a result in [6, p. 87] we have

$$\mathbf{c}_{[t]} \backsim \mathbf{c}_{1[t]} \backsim \bar{\mathbf{c}}_{1[t]} = \sup\{\mathbf{c}_{1[y]} : 0 \le y \le [t]\}$$
$$= \max\{\mathbf{c}_{[y]} : 1 \le y \le [t]\}, \quad (t \to \infty),$$

so that for $t = n \in \mathbb{N}$

$$x_n \backsim \max\{\mathbf{x}_{[y]} : 1 \le y \le n\}, \quad (n \to \infty).$$

It follows that for $n \to \infty$ we have $x_n \backsim \max\{x_k : 1 \le k \le n\}$, that is, $x_n \backsim \bar{x}_n$ for $n \to \infty$. □

The previous equivalence relation need not be true for sequences from $R_{s,\infty} \backslash KR_{s,\infty}$.

Let $\mathbf{x} \in R_{\infty,s} \backslash KR_{\infty,s}$. Assume, by contradiction, that $\bar{x}_n \backsim x_n$ for $n \to \infty$. It follows $x_n = \beta(n)\bar{x}_n$ for $n \to \infty$, where $\lim_{n\to\infty}\beta(n) = 1$. Take $\beta(n) = e^{\delta(n)}$

with $\lim_{n\to\infty}\delta(n) = 0$. By Theorem 12.18 the sequence $\bar{\mathbf{x}}$ belongs to $KR_{s,\infty}$, so that, by Theorem 12.16, $\bar{x}_n = \exp\left\{p_n + q_n + \sum_{k=1}^{n}\frac{r_k}{k}\right\}$, $n \in \mathbb{N}$, where $(p_n)_{n\in\mathbb{N}}$ is a nondecreasing real sequence, $\lim_{n\to\infty}q_n = 0$, $\lim_{n\to\infty}r_n = \infty$. Again by Theorem 12.16, one obtains $\mathbf{x} \in KR_{s,\infty}$ and we have a contradiction.

12.3.4 The Class $Tr(R_{s,\infty})$

Definition 12.8 [18, 19] A sequence $\mathbf{x} = (x_n)_{n\in\mathbb{N}} \in \mathbb{S}$ is in the class $Tr(R_{s,\infty})$ of *translationally rapidly varying sequences* if for each $\lambda \geq 1$, the following condition holds:

$$\lim_{n\to\infty}\frac{x_{[n+\lambda]}}{x_n} = \infty. \tag{12.21}$$

Some important divergent sequences are translationally rapidly varying. For example, the sequence $(x_n)_{n\in\mathbb{N}}$, $x_n = n!$, $n \in \mathbb{N}$, belongs to the class $Tr(R_{s,\infty})$.

Theorem 12.20 The following hold:

(1) $Tr(R_{s,\infty}) \subsetneqq R_{\infty,s}$;
(2) A sequence $(x_n)_{n\in\mathbb{N}}$ belongs to $Tr(R_{s,\infty})$ if and only if

$$\lim_{n\to\infty}\frac{x_{[n+\lambda]}}{x_n} = \begin{cases} 1, & \text{if } 0 \leq \lambda < 1, \\ 0, & \text{if } \lambda < 0. \end{cases}$$

Proof: (1) Let $\mathbf{x} = (x_n)_{n\in\mathbb{N}}$ be a sequence from $Tr(R_{s,\infty})$. Then $\lim_{n\to\infty}\frac{x_{n+1}}{x_n} = \infty$. Hence there is $n_0 \in \mathbb{N}$ such that for each $n \geq n_0$, $x_{n+1} > 2 \cdot x_n$. Since for $\lambda > 1$

$$\liminf_{n\to\infty}\frac{x_{[\lambda n]}}{x_n} = \liminf_{n\to\infty}\left(\frac{x_{n+1}}{x_n}\cdots\frac{x_{[\lambda n]}}{x_{[\lambda n]-1}}\right)$$
$$\geq \liminf_{n\to\infty}2^{\lambda n-1-n} = \infty,$$

we have $(x_n)_{n\in\mathbb{N}} \in R_{s,\infty}$. So, $Tr(R_{s,\infty}) \subset R_{s,\infty}$. Since $Tr(RV_{s,1}) \subsetneqq R_{s,\infty}$ and $Tr(RV_{s,1}) \cap Tr(R_{s,\infty}) = \emptyset$, one concludes that (1) is true.

(2) Let $\mathbf{x} = (x_n)_{n\in\mathbb{N}} \in Tr(R_{s,\infty})$. The statement is obviously true for $\lambda \in [0,1)$. Suppose $\lambda < 0$. Then

$$\lim_{n\to\infty}\frac{x_{[n+\lambda]}}{x_n} = \lim_{n\to\infty}\frac{x_{n+[\lambda]}}{x_n} = \lim_{n\to\infty}\frac{1}{\frac{x_n}{x_{n+[\lambda]}}} = \lim_{k\to\infty}\frac{1}{\frac{x_{k-[\lambda]}}{x_k}} = 0,$$

because for $\lambda < 0$, $-[\lambda] \geq 1$.

The converse is easily shown.

Theorem 12.21 The following hold:

(1) $KR_{s,\infty} \subsetneq R_{s,\infty}$;
(2) $Tr(R_{s,\infty}) \subsetneq KR_{s,\infty}$.

Proof: (1) Assume $\mathbf{x} = (x_n)_{n \in \mathbb{N}} \in KR_{s,\infty}$. The for any given $d \in \mathbb{R}$ we have

$$\liminf_{n \to \infty} \inf_{\lambda \geq 1} \frac{x_{[\lambda n]}}{\lambda^d x_n} \geq 1,$$

so that

$$\liminf_{n \to \infty} \inf_{\lambda > 1} \frac{x_{[\lambda n]}}{\lambda^d x_n} \geq 1.$$

It follows from here that for an arbitrarily fixed $\lambda_0 > 1$ it is true

$$\liminf_{n \to \infty} \frac{x_{[\lambda_0 n]}}{\lambda_0^d x_n} \geq 1$$

and thus

$$\liminf_{n \to \infty} \frac{x_{[\lambda_0 n]}}{x_n} \geq \lambda_0^d.$$

This means $\mathbf{x} \in R_{s,\infty}$, as d was arbitrary.

In [15] it was proved that there is a sequence in $R_{s,\infty} \setminus KR_{s,\infty}$.

(2) Let us prove $Tr(R_{s,\infty}) \subsetneq KR_{s,\infty}$. Let $\mathbf{x} = (x_n)_{n \in \mathbb{N}}$ be a sequence in $Tr(R_{s,\infty})$. Then we have $\lim_{n \to \infty} \frac{x_{n+1}}{x_n} = \infty$, that is, there is $n_0 \in \mathbb{N}$ such that $x_{n+1} > x_n$ for all $n \geq n_0$. Consider the sequence $\mathbf{x}^* = (x_n^*)_{n \in \mathbb{N}}$ defined by

$$x_n^* = \begin{cases} x_n, & \text{if } n \geq n_0, \\ 1, & \text{if } 1 \leq n \leq n_0 - 1. \end{cases}$$

This sequence is nondecreasing, and by Theorem 12.17 it belongs to $KR_{s,\infty}$ (because $Tr(R_{s,\infty}) \subsetneq R_{s,\infty}$ by Theorem 12.20).

It follows that for any given $d \in \mathbb{R}$, $\liminf_{n \to \infty} \inf_{\lambda \geq 1} \frac{x_{[\lambda n]}^*}{\lambda^d x_n^*} \geq 1$ holds. Therefore, for the same d and a given $\varepsilon > 0$ there is $n_1 = n_1(\varepsilon, d) \in \mathbb{N}$ such that $\inf_{\lambda \geq 1} \frac{x_{[\lambda n]}^*}{\lambda^d x_n^*} \geq 1 - \varepsilon$ for each $n \geq n_1$. This implies $\inf_{\lambda \geq 1} \frac{x_{[\lambda n]}}{\lambda^d x_n} \geq 1 - \varepsilon$ for all $n \geq \max\{n_0, n_1\}$, that is, $\mathbf{x} \in KR_{s,\infty}$. So, $Tr(R_{s,\infty}) \subseteq KR_{s,\infty}$.

It is easy to check that the sequence $\mathbf{x} = (x_n)_{n \in \mathbb{N}}$ defined by $x_n = e^{\sqrt{n}}$ is in $KR_{s,\infty} \setminus Tr(R_{s,\infty})$. □

12.3.5 Subclasses of $Tr(R_{s,\infty})$

In Section 12.2.2, we defined kth quotient speed of divergence. We consider now subclasses of the class $Tr(R_{s,\infty})$ on the basis of this notion.

Proposition 12.3 Let $\mathbf{x} = (x_n)_{n \in \mathbb{N}} \in \mathbb{S}$ and let $k \in \mathbb{N}$. If $v_{k+1}(\mathbf{x}) = \infty$, then $v_k(\mathbf{x}) = \infty$.

Proof: If $v_{k+1}(\mathbf{x}) = \infty$, then there is $n_0 \in \mathbb{N}$ such that $V_{n+1}^{(k)}(\mathbf{x}) \geq V_n^{(k)}(\mathbf{x})$ for each $n \geq n_0$. Therefore, there is the limit $\lim_{n \to \infty} V_n^{(k)}(\mathbf{x})$ in $\overline{\mathbb{R}}$. This limit cannot be finite, because otherwise we would have $\lim_{n \to \infty} V_n^{(k+1)}(\mathbf{x}) = 1$. $\quad\square$

Corollary 12.1

(1) If for a sequence $\mathbf{x} = (x_n)_{n \in \mathbb{N}}$ it holds $v_k(\mathbf{x}) = \infty$, $k \in \mathbb{N}$, then there is $n_0 \in \mathbb{N}$ such that $x_{n+1} > x_n$ for $n \geq n_0$ and $\lim_{n \to \infty} x_n = \infty$.
(2) Let $\mathbf{x} \in \mathbb{S}$. For all $k \in \mathbb{N}$ we have: if $v_k(\mathbf{x}) = \infty$, then $v_i(\mathbf{x}) = \infty$ for $1 \leq i < k$.

Therefore, sequences having infinite quotient speed v_1 are in the class $\mathsf{Tr}(\mathsf{R}_{s,\infty})$ of translationally rapidly varying sequences.

Note that translationally rapidly varying sequences can have both finite and infinite kth quotient speed (see Example 12.1). The second quotient speed $v_2(\mathbf{x})$ of the translationally rapidly varying sequence $x_n = n!$, $n \in \mathbb{N}$, is equal 1. The quotient speed $v_k(\mathbf{x})$ of the translationally rapidly varying sequence $x_n = 2^{2^n}$, $n \in \mathbb{N}$, is ∞ for each $k \in \mathbb{N}$.

The previous facts suggests to define the following subclasses of the class $\mathsf{Tr}(\mathsf{R}_{s,\infty})$.
Put

$$\mathsf{Tr}^{(1)}(\mathsf{R}_{s,\infty}) = \mathsf{Tr}(\mathsf{R}_{s,\infty})$$

and for each $k \geq 2$

$$\mathsf{Tr}^{(k)}(\mathsf{R}_{s,\infty}) = \{x \in \mathbb{S} : v_k(x) = \infty\}.$$

Also let

$$\mathsf{Tr}^{(\infty)}(\mathsf{R}_{s,\infty}) = \bigcap_{k=1}^{\infty} \mathsf{Tr}^{(k)}(\mathsf{R}_{s,\infty}).$$

Note that for each $k \in \mathbb{N} \cup \{\infty\}$ the class $\mathsf{Tr}^{(k)}(\mathsf{R}_{s,\infty})$ is nonempty and it holds

$$\mathsf{Tr}^{(\infty)}(\mathsf{R}_{s,\infty}) \subsetneqq \cdots \subsetneqq \mathsf{Tr}^{(k)}(\mathsf{R}_{s,\infty}) \cdots \subsetneqq \mathsf{Tr}^{(2)}(\mathsf{R}_{s,\infty}) \subsetneqq \mathsf{Tr}^{(1)}(\mathsf{R}_{s,\infty}) \subsetneqq \mathsf{R}_{s,\infty}. \tag{12.22}$$

Example 12.2

(1) For each $k \in \mathbb{N}$, the sequence $x_n = 2^{n^{k+1}}$, $n \in \mathbb{N}$, belongs to $\mathsf{Tr}^{(k)}(\mathsf{R}_{s,\infty}) \setminus \mathsf{Tr}^{(k+1)}(\mathsf{R}_{s,\infty})$.
(2) The sequences $x_n = 2^{2^n}$, $n \in \mathbb{N}$, and $y_n = e^{n!}$, $n \in \mathbb{N}$, belong to the class $\mathsf{Tr}^{(\infty)}(\mathsf{R}_{s,\infty})$.

The following two theorems give exponential representations of sequences from the classes $\mathsf{Tr}^{(1)}(\mathsf{R}_{s,\infty})$ and $\mathsf{Tr}^{(2)}(\mathsf{R}_{s,\infty})$.

Theorem 12.22 A sequence $\mathbf{x} = (x_n)_{n \in \mathbb{N}} \in \mathbb{S}$ belongs to the class $\text{Tr}^{(1)}(R_{s,\infty})$ if and only if

$$x_n = x_1 \cdot \exp\left(\sum_{i=1}^{n-1} a_i\right), \quad n \geq 2, \tag{12.23}$$

for some real sequence $(a_n)_{n \in \mathbb{N}}$ with $\lim_{n \to \infty} a_n = \infty$, and arbitrary $x_1 > 0$.

Theorem 12.23 A sequence $\mathbf{x} = (x_n)_{n \in \mathbb{N}} \in \mathbb{S}$ belongs to the class $\text{Tr}^{(2)}(R_{s,\infty})$ if and only if

$$x_n = x_2 \cdot \left(\frac{x_2}{x_1}\right)^{n-2} \cdot \exp\left(\sum_{i=1}^{n-2}(n-i-1)a_i\right), \quad n \geq 3, \tag{12.24}$$

for some real sequence $(a_n)_{n \in \mathbb{N}}$ with $\lim_{n \to \infty} a_n = \infty$, and arbitrary $x_1, x_2 > 0$.

Proof: ($\Rightarrow$) If $\mathbf{x} = (x_n)_{n \in \mathbb{N}}$ is in $\text{Tr}^{(2)}(R_{s,\infty})$, then

$$\lim_{n \to \infty} V_n^{(2)}(x) = \lim_{n \to \infty} \frac{V_{n+1}^{(1)}(x)}{V_n^{(1)}(x)} = \infty,$$

which means that there is a sequence $(c_n)_{n \in \mathbb{N}} \in \mathbb{S}$ such that $\lim_{n \to \infty} c_n = \infty$ and for each $n \in \mathbb{N}$

$$V_{n+1}^{(1)}(x) = c_n \cdot V_n^{(1)}(x) = c_n \cdot c_{n-1} \cdots c_1 \cdot V_1^{(1)}(x) = c_n \cdot c_{n-1} \cdots c_1 \cdot \frac{x_2}{x_1}$$

$$= \frac{x_2}{x_1} \cdot \exp\left(\sum_{i=1}^{n} \ln c_i\right) = \frac{x_2}{x_1} \cdot \exp\left(\sum_{i=1}^{n} a_i\right),$$

where $\lim_{n \to \infty} a_i = \infty$. So, for every $n \geq 2$ we have

$$V_n^{(1)}(x) = \frac{x_2}{x_1} \cdot \exp\left(\sum_{i=1}^{n-1} a_i\right);$$

from here we get

$$x_{n+1} = \frac{x_2}{x_1} \cdot \exp\left(\sum_{i=1}^{n-1} a_i\right) \cdot x_n$$

$$= \frac{x_2}{x_1} \cdot \exp\left(\sum_{i=1}^{n-1} a_i\right) \cdots \frac{x_2}{x_1} \cdot \exp\left(\sum_{i=1}^{1} a_i\right) \cdot x_2$$

$$= x_2 \cdot \left(\frac{x_2}{x_1}\right)^{n-1} \cdot \exp\left(\sum_{i=1}^{n-1}(n-i)a_i\right).$$

Therefore, for each $n \geq 3$ we have

$$x_n = x_2 \cdot \left(\frac{x_2}{x_1}\right)^{n-2} \cdot \exp\left(\sum_{i=1}^{n-2}(n-i-1)a_i\right) \quad \text{and} \quad \lim_{n\to\infty} a_n = \infty.$$

($\Leftarrow$) From (12.24) it follows

$$\lim_{n\to\infty} V_n^{(2)} = \lim_{n\to\infty} \frac{V_{n+1}^{(1)}(x)}{V_n^{(1)}(x)} = \lim_{n\to\infty} e^{a_n} = \infty,$$

which means $(x_n)_{n\in\mathbb{N}} \in \mathrm{Tr}^{(2)}(\mathsf{R}_{s,\infty})$. $\qquad\square$

12.3.6   The Class Γ_s

A function $\varphi : \mathbb{R} \to (0, \infty)$ belongs to the class Γ_f (see [41]; also [6]) if it is nondecreasing, continuous from the right, and if there is a measurable function $g : \mathbb{R} \to (0, \infty)$, called the *auxiliary function* of φ, such that for each $\lambda \in \mathbb{R}$,

$$\lim_{t\to\infty} \frac{\varphi(t + \lambda g(t))}{\varphi(t)} = e^\lambda. \tag{12.25}$$

Fact 1.   If $\varphi \in \Gamma_f$, then for each $A \in (0, \infty)$, the restriction of φ on $[A, \infty)$ belongs to the class $\mathsf{R}_{f,\infty}$ of rapidly varying functions [6].

Fact 2.   If $\Gamma_{f,+}$ is the set of functions which are restrictions of functions from Γ_f on some interval $[A, \infty)$, $A > 0$, then [6, Proposition 2.4.4]

$$\Gamma_{f,+} \subsetneqq \mathsf{KR}_{f,\infty} \subsetneqq \mathsf{R}_{f,\infty}.$$

We define now an important class of sequences corresponding to the function class Γ_f, contained in the class of rapidly varying sequences (see (12.27) for some clarifications concerning this definition).

Definition 12.9   A sequence $\mathbf{x} = (x_n)_{n\in\mathbb{N}} \in \mathbb{S}$ is in the *class* Γ_s if it is nondecreasing and there is a sequence $(\alpha_n)_{n\in\mathbb{N}}$ of positive real numbers, called the *auxiliary sequence* of $\mathbf{x}$, such that for each $\lambda \in \mathbb{R}$,

$$\lim_{n\to\infty} \frac{x_{\lfloor n+\lambda\alpha_n\rfloor}}{x_n} = e^{\lfloor\lambda\rfloor}. \tag{12.26}$$

Theorem 12.24 $\Gamma_s \subsetneqq \mathsf{R}_{s,\infty}$.

Proof: Let $\mathbf{x} = (x_n)_{n\in\mathbb{N}}$ belong to the class Γ_s and let $(\alpha_n)_{n\in\mathbb{N}}$ be the auxiliary sequence of $\mathbf{x}$. Then the sequence $(x_n)_{n\in\mathbb{N}}$ is nondecreasing and $\lim_{n\to\infty} x_n = \infty$; otherwise (12.26) would not be satisfied (say for $\lambda = 1$). Therefore, for every $\lambda \in \mathbb{R}$, $\lim_{n\to\infty}(n + \lambda\alpha_n) = \lim_{n\to\infty} n(1 + \lambda\frac{\alpha_n}{n}) = \infty$.

Claim. $\lim_{n\to\infty} \frac{a_n}{n} = \lim\sup_{n\to\infty} \frac{a_n}{n} = 0$.

Suppose, on the contrary, that $\lim\sup_{n\to\infty} \frac{a_n}{n} = H > 0$. Consider two cases: 1. $H < \infty$, and 2. $H = \infty$.

1. There is an increasing mapping $k : \mathbb{N} \to \mathbb{N}$ such that $\lim_{n\to\infty} \frac{a_{k(n)}}{k(n)} = H$. For sufficiently large n we have $\frac{a_{k(n)}}{k(n)} \geq \frac{H}{2}$. So, for $\lambda = -\frac{3}{H}$ and n sufficiently large we have that $n + \lambda a_n$ does not tend to ∞ as $n \to \infty$, which is a contradiction.

2. Again, there is an increasing mapping $k : \mathbb{N} \to \mathbb{N}$ such that $\lim_{n\to\infty} \frac{a_{k(n)}}{k(n)} = H$. For sufficiently large n we have $\frac{a_{k(n)}}{k(n)} \geq 2$. So, for $\lambda = -1$ and n sufficiently large we have that $n + \lambda a_n$ does not tend to ∞ as $n \to \infty$, which is again a contradiction.

This completes the proof of Claim.

Therefore, for each $\lambda > 1$ we have $(\lambda - 1) \cdot \frac{n}{a_n} \to \infty$ as $n \to \infty$, so that

$$\liminf_{n\to\infty} \frac{x_{[\lambda n]}}{x_n} = \liminf_{n\to\infty} \frac{x_{[n+(\lambda-1)\frac{n}{a_n}\cdot a_n]}}{x_n} \geq \liminf_{n\to\infty} \frac{x_{[n+Aa_n]}}{x_n} = e^{[A]},$$

for each $A > 0$. If here $A \to \infty$, we have

$$\liminf_{n\to\infty} \frac{x_{[\lambda n]}}{x_n} = \lim_{n\to\infty} \frac{x_{[\lambda n]}}{x_n} = \infty.$$

By [16, Corollary 2.2] we conclude that $\mathbf{x} \in R_{\infty,s}$. □

Example 12.3 The class Γ_s is a proper subclass of $R_{s,\infty}$.

(1) The reader can easily check that the sequence $\mathbf{x} = (x_n)_{n\in\mathbb{N}}$ defined by

$$x_n = \begin{cases} e^{n+2}, & \text{if } n \text{ is odd}, \\ e^n, & \text{if } n \text{ is even} \end{cases}$$

belongs to $R_{s,\infty} \backslash \Gamma_s$.

(2) The sequence $\mathbf{x} = (x_n)_{n\in\mathbb{N}}$ defined by $x_n = e^{n/2}$, $n \in \mathbb{N}$ belongs to $R_{s,\infty}$.

We prove $\mathbf{x} \notin \Gamma_s$. Suppose $\mathbf{x} \in \Gamma_s$. There is a sequence $(\alpha_n)_{n\in\mathbb{N}}$, the auxiliary sequence of $\mathbf{x}$, so that for each $\lambda \in \mathbb{R}$ it holds

$$\lim_{n\to\infty} \frac{e^{\frac{1}{2}(n+[\alpha_n\lambda])}}{e^{1/2n}} = e^{[\lambda]},$$

that is,

$$\lim_{n\to\infty} e^{\frac{1}{2}[\alpha_n\lambda]} = e^{[\lambda]}.$$

The last equality is equivalent to the fact that for each $\lambda \in \mathbb{R}$

$$\lim_{n\to\infty} \frac{1}{2}[\alpha_n\lambda] = [\lambda].$$

Let $\lambda \in [0, 1)$. Then $\lim_{n\to\infty} \frac{1}{2}[\alpha_n \lambda] = 0$, that is, for sufficiently large n we have $0 \leq \alpha_n \cdot \lambda < 1$. In particular, for $\lambda = \frac{2}{3}$ there is $n_0 = n_0(\frac{2}{3}) \in \mathbb{N}$ such that for each $n \geq n_0$,

$$0 < \alpha_n < \frac{3}{2}.$$

Let now $\lambda \in [1, 2)$. Then the fact $\lim_{n\to\infty} \frac{1}{2}[\alpha_n \lambda] = [\lambda]$ is equivalent to

$$2[\lambda] = \lim_{n\to\infty}[\alpha_n \cdot \lambda],$$

which implies $2[\lambda] \leq [\frac{3}{2}\lambda]$. We get a contradiction, since the last inequality does not hold for $\lambda \in [1, \frac{4}{3})$.

By definitions of the classes Γ_s and $\mathrm{Tr}(\mathrm{RV}_s)$ and Theorem 12.2 we directly have the following:

Corollary 12.2   If $\mathbf{x} = (x_n)_{n\in\mathbb{N}}$ is a nondecreasing sequence from $\mathrm{Tr}(\mathrm{RV}_{s,1})$, then $\mathbf{x} \in \Gamma_s$.

Let us consider now a class of sequences which is closely related to the class Γ_s.

For a nondecreasing sequence $\mathbf{x} = (x_n)_{n\in\mathbb{N}} \in \mathbb{S}$ we say that belongs to the *class* Γ_s^* if there is a sequence $(\alpha_n)_{n\in\mathbb{N}} \in \mathbb{S}$ (called the *auxiliary sequence* for $\mathbf{x}$) such that for each $\lambda \in \mathbb{R}$ the following holds:

$$\lim_{n\to\infty} \frac{x_{[n+\alpha_n\lambda]}}{x_n} = e^\lambda. \tag{12.27}$$

The class Γ_s^* is nonempty. The sequence $\mathbf{x} = (x_n)_{n\in\mathbb{N}}$ defined by $x_n = e^{\sqrt{n}}$, $n \in \mathbb{N}$, is in this class. Indeed, if we take $\alpha_n = 2\sqrt{n}$, $n \in \mathbb{N}$, then for each $\lambda \in \mathbb{R}$ we have

$$\lim_{n\to\infty} \frac{x_{[n+\alpha_n\lambda]}}{x_n} = e^\lambda.$$

Notice that, on the other hand, the sequence $\mathbf{y} = (y_n)_{n\in\mathbb{N}}$, $y_n = e^n$, $n \in \mathbb{N}$, does not belong to the class Γ_s^*, but belongs to Γ_s.

Let us note also that $\Gamma_s^* \subset \mathrm{R}_{s,\infty}$.

We do not know if it is true $\Gamma_s \subsetneqq \Gamma_s^*$ or $\Gamma_s^* \subsetneqq \Gamma_s$.

12.4   Applications to Selection Principles

In this section, we demonstrate interesting and nice relationships between theory of regular and rapid variations of sequences and selection principles theory (related also with game theory and Ramsey theory). This connections were studied in the series of papers [15–19, 31, 42–44] (see also the survey paper [45]).

The theory of selection principles has its roots in the works 1920s of [46–52], and others. A systematic study of this subject began in the paper [53]. For survey papers concerning selection principles and their interplay with game theory (and Ramsey theory) we refer the reader to [54–60]

Let $\mathcal{A}$ and $\mathcal{B}$ be sets whose elements are families of subsets of an infinite set X. Then [53] $S_1(\mathcal{A}, \mathcal{B})$ denotes the selection principle:

> For each sequence $(A_n : n \in \mathbb{N})$ of elements of $\mathcal{A}$ there is a sequence $(b_n : n \in \mathbb{N})$ such that for each n, $b_n \in A_n$ and $\{b_n : n \in \mathbb{N}\}$ is an element of $\mathcal{B}$.

In this paper $\mathcal{A}$ and $\mathcal{B}$ will be certain subclasses of the class $\mathbb{S}$.

The symbol $G_1(\mathcal{A}, \mathcal{B})$ denotes the infinitely long game for two players, ONE and TWO, who play a round for each $n \in \mathbb{N}$. In the nth round ONE chooses a set $A_n \in \mathcal{A}$, and TWO responds by choosing an element $b_n \in A_n$. TWO wins a play $(A_1, b_1; \ldots ; A_n, b_n; \ldots)$ if $\{b_n : n \in \mathbb{N}\} \in \mathcal{B}$; otherwise, ONE wins.

It is evident that if TWO has a winning strategy (or, more general, if ONE does not have a winning strategy) in the game $G_1(\mathcal{A}, \mathcal{B})$, then the selection hypothesis $S_1(\mathcal{A}, \mathcal{B})$ is true. The converse implication is not always true.

A strategy σ for the player TWO is a *coding strategy* (see [61] where this concept was introduced) if TWO remembers only the most recent move by ONE and by TWO before deciding how to play the next move. More precisely the moves of TWO are: $b_1 = \sigma(A_1, \emptyset)$; $b_n = \sigma(A_n, b_{n-1})$, $n \geq 2$.

In [62] new selection principles were introduced and studied (see also [63] for a detail exposition). $\mathcal{A}$ and $\mathcal{B}$ are as above.

Definition 12.10 The symbol $\alpha_i(\mathcal{A}, \mathcal{B})$, $i = 1, 2, 3, 4$, denotes the following selection hypothesis. For each sequence $(A_n : n \in \mathbb{N})$ of infinite elements of $\mathcal{A}$ there is an element $B \in \mathcal{B}$ such that:

$\alpha_1(\mathcal{A}, \mathcal{B})$: for each $n \in \mathbb{N}$ the set $A_n \backslash B$ is finite;
$\alpha_2(\mathcal{A}, \mathcal{B})$: for each $n \in \mathbb{N}$ the set $A_n \cap B$ is infinite;
$\alpha_3(\mathcal{A}, \mathcal{B})$: for infinitely many $n \in \mathbb{N}$ the set $A_n \cap B$ is infinite;
$\alpha_4(\mathcal{A}, \mathcal{B})$: for infinitely many $n \in \mathbb{N}$ the set $A_n \cap B$ is nonempty.

Evidently,

$$\alpha_1(\mathcal{A}, \mathcal{B}) \Rightarrow \alpha_2(\mathcal{A}, \mathcal{B}) \Rightarrow \alpha_3(\mathcal{A}, \mathcal{B}) \Rightarrow \alpha_4(\mathcal{A}, \mathcal{B}) \Leftarrow S_1(\mathcal{A}, \mathcal{B}).$$

Observe that all above selection principles are monotone in the second coordinate $\mathcal{B}$ and antimonotone in the first coordinate $\mathcal{A}$. For this reason it is natural to find the widest collection $\mathcal{A}$ and the smallest collection $\mathcal{B}$ so that a selection principle is satisfied.

12.4.1 First Results

The first results concerning applications of selection principles and games in asymptotic analysis of divergent processes were obtained in [16, 43].

In [16] we showed that the class $R_{s,\infty}$ has nice selection properties.

Theorem 12.25 The class $R_{s,\infty}$, satisfies each of the following (equivalent) principles:

$$S_1(R_{s,\infty}, R_{s,\infty}); \alpha_2(R_{s,\infty}, R_{s,\infty}); \alpha_3(R_{s,\infty}, R_{s,\infty}); \alpha_4(R_{s,\infty}, R_{s,\infty}).$$

In [17] this result was improved, changing the first coordinate $R_{s,\infty}$ by a wider class ARV_s.

In fact, it was proved the following theorem:

Theorem 12.26 The player TWO has a winning coding strategy in the game $G_1(ARV_s, R_{s,\infty})$.

whose consequence is the following improvement of Theorem 12.25.

Corollary 12.3 The (equivalent) selection principles $S_1(ARV_s, R_{s,\infty})$, $\alpha_2(ARV_s, R_{s,\infty})$, $\alpha_3(ARV_s, R_{s,\infty})$ and $\alpha_4(ARV_s, R_{s,\infty})$ are satisfied.

12.4.2 Improvements

The following two theorems and their corollaries give a significant improvement of the previous three results (because $Tr(R_{s,\infty}) \subsetneqq R_{s,\infty}$).

Theorem 12.27 The player TWO has a winning coding strategy in the game $G_1(ARV_s, Tr(R_{s,\infty}))$.

Proof: A strategy σ for TWO will be defined as follows. Assume that in the first round ONE plays the sequence $\mathbf{x_1} = (x_{1,m})_{m \in \mathbb{N}}$ from ARV_s. TWO responds by choosing $\sigma(\mathbf{x_1}, \emptyset) = x_{1,m_1} = y_1$ – any element in $\mathbf{x_1}$. If in the second round ONE has played $\mathbf{x_2} = (x_{2,m})_{m \in \mathbb{N}} \in ARV_s$, then TWO picks $x_{2,m_2} \in \mathbf{x_2}$ such that $x_{2,m_2} > 2 \cdot y_1$ (which is possible by Lemma 12.2) and responds by $\sigma(\mathbf{x_2}, y_1) = x_{2,m_2} = y_2$. Let in the nth round ONE play $\mathbf{x_n} = (x_{n,m})_{m \in \mathbb{N}}$; then TWO finds $x_{n,m_n} \in \mathbf{x_n}$ such that $x_{n,m_n} > n \cdot y_{n-1}$ and plays $\sigma(\mathbf{x_n}, y_{n-1}) = x_{n,m_n} = y_n$. And so on.

We claim that $\mathbf{y} = (y_n)_{n \in \mathbb{N}}$ is translationally rapidly varying. Indeed, for each $\lambda \geq 1$ we have

$$\liminf_{n \to \infty} \frac{y_{[n+\lambda]}}{y_n} \geq \liminf_{n \to \infty} (n+1)^{[\lambda]} = \infty.$$

that is, $\mathbf{y} \in Tr(R_{s,\infty})$. $\qquad \square$

Corollary 12.4   The selection principle $S_1(ARV_s, Tr(R_{s,\infty}))$ is satisfied.

Further significant improvement is given in the following theorem.

Theorem 12.28   The following selection properties are equivalent (and all are satisfied):

(1) $S_1(ARV_s, Tr^{(2)}(R_{s,\infty}))$;
(2) $\alpha_2(ARV_s, Tr^{(2)}(R_{s,\infty}))$;
(3) $\alpha_3(ARV_s, Tr^{(2)}(R_{s,\infty}))$;
(4) $\alpha_4(ARV_s, Tr^{(2)}(R_{s,\infty}))$.

Proof: Because (2) $\Rightarrow$ (3) and (3) $\Rightarrow$ (4) are obvious, we should prove only (1) $\Rightarrow$ (2) and (4) $\Rightarrow$ (1).

(1) $\Rightarrow$ (2): Let $(\mathbf{x}_n : n \in \mathbb{N})$ be a sequence of elements of ARV_s. By using Lemma 12.1 applied to the sequence $p_1 < p_2 < \cdots p_k < \cdots$ of prime numbers, for each $n \in \mathbb{N}$ consider a sequence $(\mathbf{x}_{n,m} : m \in \mathbb{N})$ of pairwise disjoint subsequences of $\mathbf{x}_n$ each belonging to ARV_s. Apply (1) to the sequence $(\mathbf{x}_{n,m} : n, m \in \mathbb{N})$; there is a sequence $(y_{n,m})_{n,m\in\mathbb{N}}$ such that for each $(n, m) \in \mathbb{N} \times \mathbb{N}$, $y_{n,m} \in \mathbf{x}_{n,m}$ and $\mathbf{y} := (y_{n,m})_{n,m\in\mathbb{N}} \in Tr^{(2)}(R_{s,\infty})$. It is understood that for each $n \in \mathbb{N}$ the set $\mathbf{x}_n \cap \mathbf{y}$ is infinite. In other words, $\mathbf{y}$ is a selector for the sequence $(\mathbf{x}_n : n \in \mathbb{N})$ witnessing that $\alpha_2(ARV_s, Tr^{(2)}(R_{s,\infty}))$ holds.

(4) $\Rightarrow$ (1): Suppose that $\alpha_4(ARV_s, Tr^{(2)}(R_{s,\infty}))$ is satisfied and let $(\mathbf{x}_n : n \in \mathbb{N})$ be a sequence of elements of ARV_s. Suppose that for every n, $\mathbf{x}_n = (x_{n,m})_{m\in\mathbb{N}}$. By (4) there is an increasing sequence $n_1 < n_2 < \cdots$ in $\mathbb{N}$ and a sequence $\mathbf{y} = (x_{n_i,m_i})_{i\in\mathbb{N}} \in Tr^{(2)}(R_{s,\infty})$ such that for each $i \in \mathbb{N}$, $x_{n_i,m_i} \in \mathbf{x}_{n_i}$. By Lemma 12.1 and Corollary 12.1 one may suppose that for each $i \geq 1$, x_{n_i,m_i} is large enough so that for each k with $n_{i-1} < k \leq n_i$, we can pick an element $x_{k,m_k} \in \mathbf{x}_k$ in such a way that $x_{k,m_k} > \frac{k \cdot (x_{k-1,m_{k-1}})^2}{x_{k-2,m_{k-2}}}$. (In other words, we can insert, in an appropriate way, $n_i - n_{i-1}$ new elements between $x_{n_{i-1},m_{i-1}}$ and x_{n_i,m_i}.) The elements chosen in this way (together with elements x_{n_i,m_i}) witness that selection principle $S_1(ARV_s, Tr^{(2)}(R_{s,\infty}))$ is true. The proof of this fact is obtained by a direct calculation. □

Because by (12.22), $Tr^{(2)(R_{s,\infty})} \subsetneq Tr(R_{s,\infty})$, and $Tr(R_{s,\infty}) \subsetneq KR_{s,\infty}$ (Theorem 12.21), we have the following corollary.

Corollary 12.5   The following (equivalent) selection properties are satisfied:

(1) $S_1(ARV_{s,\infty}, KR_{s,\infty})$;
(2) $\alpha_2(ARV_{s,\infty}, KR_{s,\infty})$;
(3) $\alpha_3(ARV_{s,\infty}, KR_{s,\infty})$;
(4) $\alpha_4(ARV_{s,\infty}, KR_{s,\infty})$.

At the end of this section we prove that under additional assumptions the player TWO has a winning strategy in a modified game of the G_1-type. In what follows $\uparrow \mathrm{Tr}(RV_{s,0})$ denotes the class of strictly increasing sequences from $\mathrm{Tr}(RV_{s,0})$, and $u \uparrow \mathrm{Tr}(RV_{s,0})$ denotes the class of unbounded sequences from $\uparrow \mathrm{Tr}(RV_{s,0})$.

For two sequences $\mathbf{x} = (x_n)_{n \in \mathbb{N}}$ and $\mathbf{y} = (y_n)_{n \in \mathbb{N}}$ in $\mathbb{S}$ we write $\mathbf{x} \prec \mathbf{y}$ if and only if $x_n \leq y_n$ for each $n \in \mathbb{N}$. The relation $\prec$ is an order relation on $\mathbb{S}$.

Further, for $\mathbf{x} = (x_n)_{n \in \mathbb{N}} \in \mathbb{S}$ we define the *sequence of singularity of* $\mathbf{x}$ *in the sense of d'Alembert*, denoted by $\mathbf{K}(\mathbf{x}) = (K_n(\mathbf{x}))_{n \in \mathbb{N}}$, by

$$K_n(\mathbf{x}) = \frac{x_{n+1}}{x_n} - 1, \quad n \in \mathbb{N}.$$

The importance of the sequence $\mathbf{K}(\mathbf{x})$ for asymptotic behavior of the sequence $\mathbf{x}$ is well known in the literature [6, 29].

Fact A. If a sequence $\mathbf{x}$ is strictly increasing, then $\mathbf{K}(\mathbf{x}) \in \mathbb{S}$.

Fact B. A sequence $\mathbf{x}$ belongs to the class $\mathrm{Tr}(RV_{s,0})$ if and only if $\lim_{n \to \infty} K_n(\mathbf{x}) = 0$.

For $\mathcal{A}$ and $\mathcal{B}$ infinite subclasses of $\mathbb{S}$, the symbol $G_1^p(\mathcal{A}, \mathcal{B})$ denotes a modification of the game $G_1(\mathcal{A}, \mathcal{B})$: in the nth round ONE chooses a sequence $\mathbf{x}_n \in \mathcal{A}$ so that

(i) $\mathbf{x}_n \prec \mathbf{x}_{n-1}$;
(ii) $\mathbf{K}(\mathbf{x}_n) \prec \mathbf{K}(\mathbf{x}_{n-1})$,

while the other rules are the same as in $G_1(\mathcal{A}, \mathcal{B})$.

Theorem 12.29 TWO has a winning strategy in the game $G_1^p(u \uparrow \mathrm{Tr}(RV_{s,0}), \uparrow \mathrm{Tr}(RV_{s,0}))$.

Proof: Fix $\varepsilon > 0$.

Round 1: Let the first move of ONE be a sequence $\mathbf{x}_1 = (x_{1,m})_{m \in \mathbb{N}} \in u \uparrow \mathrm{Tr}(RV_{s,0})$. Set

$$m_1 = m_1(\varepsilon) = \min\{m \in \mathbb{N} : K_m(\mathbf{x}_1) \leq \varepsilon\}.$$

TWO picks

$$y_1 = x_{1,m_1} = x_{1,m_1^*}.$$

Round 2: ONE chooses a sequence $\mathbf{x}_2 = (x_{2,m})_{m \in \mathbb{N}}$ in $u \uparrow \mathrm{Tr}(RV_{s,0})$ such that $\mathbf{x}_2 \prec \mathbf{x}_1$ and $\mathbf{K}(\mathbf{x}_2) \prec \mathbf{K}(\mathbf{x}_1)$. Let

$$m_2 = m_2(\varepsilon) = \min\{m \in \mathbb{N} : K_m(\mathbf{x}_2) \leq \varepsilon\}.$$

Then $m_2(\varepsilon) \leq m_1(\varepsilon)$. There is $m_2^* \in \mathbb{N}$ such that $x_{2,m_2^*-1} \leq y_1 < x_{2,m_2^*}$. Otherwise, we have the following two possibilities:

1. for each $m \in \mathbb{N}$, $y_1 \geq x_{2,m}$; it is impossible because the sequence $\mathbf{x}_2$ is unbounded.

2. for each $m \in \mathbb{N}$, $y_1 < x_{2,m}$. However, we have $x_{2,m_1} \leq x_{1,m_1} = y_1$.
   TWO chooses $y_2 = x_{2,m_2^*} \in \mathbf{x}_2$. Note $m_2^* > m_1^*$.

   Round 3: ONE chooses a sequence $\mathbf{x}_3 = (x_{3,m})_{m \in \mathbb{N}} \in \mathsf{u} \uparrow \mathrm{Tr}(\mathrm{RV}_{s,0})$ satisfying $\mathbf{x}_3 \prec \mathbf{x}_2$ and $K(\mathbf{x}_3) \prec K(\mathbf{x}_2)$. Denote

$$m_3 = m_3(\varepsilon) = \min\{m \in \mathbb{N} : K_m(\mathbf{x}_3) \leq \varepsilon\}.$$

Then $m_3(\varepsilon) \leq m_2(\varepsilon)$. There is $m_3^* > m_2^*$ such that $x_{3,m_3^*-1} \leq y_2 < x_{3,m_3^*}$. TWO chooses $y_3 = x_{3,m_3^*} \in \mathbf{x}_3$.

And so on.

During the play

$$\mathbf{x}_1, y_1; \mathbf{x}_2, y_2; \mathbf{x}_3, y_3; \cdots$$

one obtains the sequence of elements chosen by TWO

$$y_1, y_2, y_3, \cdots$$

This sequence is obviously strictly increasing. It remains to show that it belongs to $\mathrm{Tr}(\mathrm{RV}_{s,0})$.

Let $k \geq 2$. Then $m_1(\varepsilon) \leq m_1(\varepsilon/k)$ and $\Delta_1(k) = m_1(\varepsilon/k) - m_1(\varepsilon) + 1$ is a (finite) natural number. Also

$$1 < \frac{y_{\Delta_1(k)+1}}{y_{\Delta_1(k)}} \leq 1 + \frac{\varepsilon}{k}.$$

Since the sequence $\mathbf{y} = (y_n)_{n \in \mathbb{N}}$ has the property that for some $h \in \mathbb{N} \cup \{0\}$ and every $j \in \{\Delta_1(k), \cdots, \Delta_1(k) + h = \Delta_1(k+1)\}$ it holds $K_j(\mathbf{y}) \leq \frac{\varepsilon}{k}$, we conclude

$$\lim_{n \to \infty} \frac{y_{n+1}}{y_n} = 1,$$

that is, $\mathbf{y} \in \mathrm{Tr}(\mathrm{RV}_{s,0})$. $\qquad\square$

Corollary 12.6 The selection principle $S_1^p(\mathsf{u} \uparrow \mathrm{Tr}(\mathrm{RV}_{s,0}), \uparrow \mathrm{Tr}(\mathrm{RV}_{s,0}))$ corresponding to the game $G_1^p(\mathsf{u} \uparrow \mathrm{Tr}(\mathrm{RV}_{s,0}), \uparrow \mathrm{Tr}(\mathrm{RV}_{s,0}))$ is true.

Example 12.4 Conditions (i) and (ii) in the definition of the game are essential. A counterexample is the sequence $(\mathbf{x}_n : n \in \mathbb{N})$ of sequences $\mathbf{x_n} = (x_{n,m})_{m \in \mathbb{N}}$ from $\mathsf{u} \uparrow \mathrm{Tr}(\mathrm{RV}_{s,0})$ defined by

$$x_{n,m} = e^n \cdot \ln(m+1), \quad n, m \in \mathbb{N}.$$

Now we give an application of the class of rapidly varying sequences of index of variability $-\infty$ in selection principles theory.

Definition 12.11 [17] Let $A \in [0, \infty)$. A sequence $\mathbf{x} = (x_n)_{n \in \mathbb{N}} \in S$ converging to A is said to *converge rapidly* to A if the Landau–Hurwicz sequence $(w_n(\mathbf{x}))_{n \in \mathbb{N}}$ of $\mathbf{x}$ defined by

$$w_n(\mathbf{x}) = \sup\{|x_m - x_k| : m \geq n, k \geq n\}, \quad n \in \mathbb{N}$$

belongs to de Haan's class $R_{s,-\infty}$ of rapidly varying sequences of index of variability $-\infty$.

For example, if $A \in [0, \infty)$, then each sequence $\mathbf{x_n}$ defined by $\mathbf{x_n} = ((\frac{1}{n}e^{-m} + A)_{m\in\mathbb{N}}$, $n \in \mathbb{N}$, converges rapidly to A.

For $A \in [0, \infty)$ let

$$[A]_{R_{s,-\infty}} = \{\mathbf{x} = (x_n)_{n\in\mathbb{N}} \in \mathbb{S} : \mathbf{x} \text{ converges rapidly to } A\}.$$

Theorem 12.30   Let $A \in [0, \infty)$. Then $S_1([A]_{R_{s,-\infty}}, [A]_{R_{s,-\infty}})$ is satisfied.

Proof: Let $(\mathbf{x}_n = (x_{n,m})_{m\in\mathbb{N}} : n \in \mathbb{N})$, be a sequence of elements of $[A]_{R_{s,-\infty}}$. Construct a sequence $\mathbf{y} = (y_n)_{n\in\mathbb{N}}$ in the following way:

(a) $y_1 = x_{1,m}$, where $m \in \mathbb{N}$ is arbitrary (and fixed) such that $x_{1,m} \neq A$;
(b) for $n \geq 2$, $n \in \mathbb{N}$, $y_n = x_{n,m}$, where m is chosen in such a way that

$$m \geq \tilde{m}_n := \min\{m \in \mathbb{N} : |x_{n,m} - A| \leq \frac{1}{4}|y_{n-1} - A|, x_{n,m} \neq A\}.$$

From the construction of $\mathbf{y}$ it is easy to see that $\mathbf{y} \in \mathbb{S}$, $\mathbf{x} \cap \mathbf{x}_n = y_n$ for every $n \in \mathbb{N}$, and that $\mathbf{y}$ converges to A.

Claim 1. The sequence $(|y_n - A|)_{n\in\mathbb{N}}$ belongs to $R_{s,-\infty}$.
Since

$$|y_n - A| - \frac{1}{4}|y_n - A| \leq w_n(\mathbf{y}) \leq |y_n - A| + \frac{1}{4}|y_n - A|, \quad n \in \mathbb{N},$$

we have

$$\frac{3}{4}|y_n - A| \leq w_n(\mathbf{y}) \leq \frac{5}{4}|y_n - A|.$$

This means that for $n \in \mathbb{N}$ it holds

$$w_n(\mathbf{y}) = \theta_n|y_n - A|,$$

where $\frac{3}{4} \leq \theta_n \leq \frac{5}{4}$ for $n \in \mathbb{N}$.

For every $n \in \mathbb{N}$, there is no $m_{0n} \in \mathbb{N}$, such that $x_{n,m} = A$, for $m \geq m_{0n}$, because of the assumption $(w_n(x_n)) \in R_{s,-\infty}$.

The sequence $(|y_n - A|)_{n\in\mathbb{N}}$ belongs to $\mathbb{S}$ and for every $n \in \mathbb{N}$ it holds $|y_{n+1} - A| \leq \frac{1}{4}|y_n - A|$.

Therefore, for each $\lambda > 1$ we have

$$\limsup_{n\to\infty} \frac{|y_{[\lambda n]} - A|}{|y_n - A|} \leq \limsup_{n\to\infty} \left(\frac{1}{4}\right)^{[\lambda n]-n} \leq \limsup_{n\to\infty} \left(\frac{1}{4}\right)^{(\lambda-1)n} = 0.$$

This just means $(|y_n - A|)_{n\in\mathbb{N}} \in R_{s,-\infty}$.

Claim 2. $(w_n(\mathbf{y})) \in R_{s,-\infty}$.

By the above for each $\lambda > 1$, we have

$$
\limsup_{n\to\infty} \frac{w_{[\lambda n]}(\mathbf{y})}{w_n(\mathbf{y})} = \limsup_{n\to\infty} \left(\frac{\theta_{[\lambda n]}}{\theta_n} \cdot \frac{|y_{[\lambda n]} - A|}{|y_n - A|} \right)
$$

$$
\leq \limsup_{n\to\infty} \frac{\theta_{[\lambda n]}}{\theta_n} \cdot \limsup_{n\to\infty} \frac{|y_{[\lambda n]} - A|}{|y_n - A|}
$$

$$
\leq \frac{5}{3} \limsup_{n\to\infty} \frac{|y_{[\lambda n]} - A|}{|y_n - A|} = \frac{5}{3} \cdot 0 = 0,
$$

and thus $(w_n(\mathbf{y})) \in R_{s,-\infty}$.

So, $\mathbf{y} \in [A]_{R_{s,-\infty}}$, and the theorem is shown. □

Remark 12.2 The proof of the previous theorem actually says that the player TWO has a winning strategy in the game $G_1([A]_{R_{s,-\infty}}, [A]_{R_{s,-\infty}})$.

12.4.3 When ONE has a Winning Strategy?

In the previous section, we considered mainly the classes of sequences for which the player TWO has a winning strategy in a game. It is natural to ask for which classes of sequences the player ONE has a winning strategy.

Theorem 12.31 For any $\rho > 0$ the player ONE has a winning strategy in the game $G_1(\mathrm{Tr}(\mathrm{RV}_{s,\rho}), \mathrm{Tr}(\mathrm{RV}_s))$.

Proof: Let σ be a strategy for ONE. Suppose that the first move of ONE is the sequence $\sigma(\emptyset) = \mathbf{x}_1 = (x_{1,m})_{m\in\mathbb{N}}$ belonging to $\mathrm{Tr}(\mathrm{RV}_{s,\rho})$. Let TWO's response be $y_1 = x_{1,m_1} \in \mathbf{x}_1$. Then ONE looks at x_{1,m_1} and picks a new sequence $\mathbf{x}_2 = (x_{2,m})_{m\in\mathbb{N}} \in \mathrm{Tr}(\mathrm{RV}_{s,\rho})$ such that all its elements are bigger than $2 \cdot y_1$. (This is possible, passing if necessary to subsequences, because $\mathbf{x}_2 \in \mathrm{Tr}(\mathrm{RV}_{s,\rho}) \subsetneq R_{s,\infty}$, and thus $\mathbf{x}_2$ is unbounded.) Let $y_2 = x_{2,m_2} \in \mathbf{x}_2$ be the choice of TWO. In the nth round ONE looks at y_{n-1}, TWO's choice in the $(n-1)$th round, and chooses a sequence $\mathbf{x}_n = (x_{n,m})_{m\in\mathbb{N}} \in \mathrm{Tr}(\mathrm{RV}_{s,\rho})$ such that all its elements are bigger than $n \cdot y_{n-1}$. And so on.

The sequence $(y_n)_{n\in\mathbb{N}}$ obtained during the play

$$
\mathbf{x}_1, y_1; \mathbf{x}_2, y_2; \dots, \mathbf{x}_n, y_n; \dots
$$

for each $\lambda \geq 1$ satisfies

$$
\liminf_{n\to\infty} \frac{y_{[n+\lambda]}}{y_n} = \liminf_{n\to\infty} \frac{y_{n+[\lambda]}}{y_n} = \liminf_{n\to\infty} \left(\frac{y_{n+1}}{y_n} \cdots \frac{y_{n+[\lambda]}}{y_{n+[\lambda]-1}} \right)
$$

$$
\geq \liminf_{n\to\infty} (n+1)^{[\lambda]} = \infty.
$$

Therefore, for $\lambda \geq 1$, $\lim_{n\to\infty} \frac{y_{[n+\lambda]}}{y_n} = \infty$ and thus $(y_n)_{n\in\mathbb{N}} \notin \mathrm{Tr}(\mathrm{RV}_s)$. □

Corollary 12.7 For any $\rho > 0$ and $\delta \in \mathbb{R}$ ONE has a winning strategy in the game $G_1(\mathrm{Tr}(\mathrm{RV}_{s,\rho}), \mathrm{Tr}(\mathrm{RV}_{s,\delta}))$.

Corollary 12.8 ONE has a winning strategy in the game $G_1(\mathrm{ARV}_s, \mathrm{Tr}(\mathrm{RV}_s))$ (and thus also in $G_1(\mathrm{R}_{s,\infty}, \mathrm{Tr}(\mathrm{RV}_s))$ and $G_1(\Gamma_s, \mathrm{Tr}(\mathrm{RV}_s))$).

Similarly to the proof of Theorem 12.31 one proves the following statement.

Theorem 12.32 For any $\rho > 0$ the player ONE has a winning strategy in the game $G_1(\mathrm{RV}_{s,\rho}, \mathrm{RV}_s)$.

Proof: Let σ be a strategy for ONE. Let in the first round ONE choose a sequence $\sigma(\emptyset) = \mathbf{x_1} = (x_{1,m})_{m\in\mathbb{N}}$ from $\mathrm{RV}_{s,\rho}$, and let TWO's response be $y_1 = x_{1,m_1} \in \mathbf{x_1}$. Then ONE chooses a new sequence $\mathbf{x_2} = (x_{2,m})_{m\in\mathbb{N}} \in \mathrm{RV}_{s,\rho}$ such that all its elements are bigger than $2 \cdot y_1$. (Note that for any $\rho > 0$, $\mathrm{RV}_{s,\rho} \not\subseteq \mathrm{ARV}_s$ [17], which implies that sequences from $\mathrm{RV}_{s,\rho}$, $\rho > 0$, are unbounded.) Let $y_2 = x_{2,m_2} \in \mathbf{x_2}$ be the choice of TWO. In the nth round ONE looks at y_{n-1}, chosen by TWO in the $(n-1)$th round, and chooses a sequence $\mathbf{x_n} = (x_{n,m})_{m\in\mathbb{N}} \in \mathrm{RV}_{s,\rho}$ such that all its elements are bigger than $2 \cdot y_{n-1}$. And so on.

For each $\lambda > 1$, we have

$$\lim_{n\to\infty} \frac{y_{[\lambda n]}}{y_n} = \lim_{n\to\infty} \left(\frac{y_{[\lambda n]}}{y_{[\lambda n]-1}} \cdots \frac{y_{n+1}}{y_n} \right)$$
$$> \lim_{n\to\infty} 2^{[\lambda n]-n} > \lim_{n\to\infty} 2^{(\lambda-1)n-1} = \infty,$$

and therefore $(y_n)_{n\in\mathbb{N}} \in \mathrm{R}_{s,\infty}$, that is, $(y_n)_{n\in\mathbb{N}} \notin \mathrm{RV}_s$. $\qquad\square$

Corollary 12.9 ONE has a winning strategy in the games $G_1(\mathrm{ARV}_s, \mathrm{RV}_s)$ (in particular, in $G_1(\mathrm{ARV}_s, \mathrm{SV}_s)$), $G_1(\mathrm{R}_{s,\infty} \mathrm{RV}_s)$, $G_1(\Gamma_s, \mathrm{RV}_s)$, and for $\rho > 0$ in $G_1(\mathrm{Tr}(\mathrm{RV}_{s,\rho}), \mathrm{RV}_s)$.

In [31, Theorem 3.14] it was shown.

Theorem 12.33 For any $\rho > 0$ the player ONE has a winning strategy in the game $G_1(\mathrm{Tr}(\mathrm{RV}_{s,\rho}), \mathrm{Tr}(\mathrm{RV}_s))$.

In contrast to this result, we show now the following game-theoretic result for the case $\rho = 0$. The two-person game we consider is denoted by $G^*_{fin}(\mathcal{A}, \mathcal{B})$, and the rules are the following: in the nth round, $n \geq 2$, ONE chooses an element $A_n \in \mathcal{A}$ and TWO responds by choosing a finite set $B_n \subset A_{n-1} \cup A_n$. TWO wins a play $A_1, B_1; \ldots; A_n, B_n; \ldots$ if $\bigcup_{n\in\mathbb{N}} B_n \in \mathcal{B}$; otherwise ONE wins [17].

Theorem 12.34 TWO has a winning strategy in the game $G^*_{fin}(u \uparrow \mathrm{Tr}(\mathrm{RV}_{s,0}),$ $u \uparrow \mathrm{Tr}(\mathrm{RV}_{s,0}))$.

Proof: Let $\varepsilon > 0$ be arbitrary and fixed.

Round 1: Let the first move of ONE be a sequence $\mathbf{x}_1 = (x_{1,m})_{m \in \mathbb{N}} \in$ u ↑ Tr(RV$_{s,0}$). There is $m_1 \in \mathbb{N}$ such that for each $m \geq m_1$

$$1 < \frac{x_{1,m+1}}{x_{1,m}} < 1 + \varepsilon.$$

Two picks

$$F_1 = \{x_{1,m_1}\}.$$

Round 2: ONE chooses a sequence $\mathbf{x}_2 = (x_{2,m})_{m \in \mathbb{N}}$ in u ↑ Tr(RV$_{s,0}$). There are $m_1^*, m_2 \in \mathbb{N}$ such that:

1. $m_1^* > m_1, m_2 > m_1$;
2. $1 < \frac{x_{2,m+1}}{x_{2,m}} < 1 + \varepsilon/2$ for each $m \geq m_2$;
3. $2x_{1,m_1} < x_{1,m_1^*} \leq x_{2,m_2} \leq x_{1,m_1^*+1}$.

Note that from here we have

$$1 \leq \frac{x_{2,m_2}}{x_{1,m_1^*}} \leq \frac{x_{1,m_1^*+1}}{x_{1,m_1^*}} < 1 + \varepsilon.$$

TWO takes

$$F_2 = \{x_{1,m_1+1}, \ldots, x_{1,m_1^*}\} \cup \{x_{2,m_2}\}.$$

Round 3: ONE chooses a sequence $\mathbf{x}_3 = (x_{3,m})_{m \in \mathbb{N}} \in$ u ↑ Tr(RV$_{s,0}$). There are $m_2^*, m_3 \in \mathbb{N}$ such that:

1. $m_2^* > m_2, m_3 > m_2$,
2. $1 < \frac{x_{3,m+1}}{x_{3,m}} < 1 + \varepsilon/3$, for each $m \geq m_3$,
3. $2x_{2,m_2} < x_{2,m_2^*} \leq x_{3,m_3} \leq x_{2,m_2^*+1}$.

Notice that

$$1 \leq \frac{x_{3,m_3}}{x_{2,m_2^*}} \leq \frac{x_{2,m_2^*+1}}{x_{2,m_2^*}} < 1 + \varepsilon/2.$$

TWO chooses

$$F_3 = \{x_{2,m_2+1}, \ldots, x_{2,m_2^*}\} \cup \{x_{3,m_3}\}.$$

And so on.

During the play

$$\mathbf{x}_1, F_1; \mathbf{x}_2, F_2; \mathbf{x}_3, F_3; \ldots$$

one obtains the sequence (of elements chosen by TWO)

$$x_{1,m_1}; x_{1,m_1+1}, \ldots, x_{1,m_1^*}, x_{2,m_2}; x_{2,m_2+1}, \ldots, x_{2,m_2^*}, x_{3,m_3}; \ldots \qquad (12.28)$$

This sequence is obviously increasing and, by construction, it is in $\mathrm{Tr}(\mathrm{RV}_{s,0})$. On the other hand, we have

$$x_{2,m_2} > 2x_{1,m_1}; x_{3,m_3} > 2x_{2,m_2} > 2^2 x_{1,m_1}; \cdots ; x_{k,m_k} > 2^{k-1} x_{1,m_1}; \cdots ,$$

so that $x_{k,m_k} \to \infty$ as $k \to \infty$, and as the sequence (11) is increasing it also diverges to ∞. $\qquad\qquad\square$

12.5 Applications to Differential Equations

The objective of this section is to make a detailed survey of the recent progress in the study of the existence and the asymptotic behavior of positive solutions of the Thomas–Fermi differential equation

(A) $$x'' = q(t)x^\gamma,$$

and next the more general equation of the form

(B) $$(p(t)x')' = q(t)x^\gamma,$$

where p, q are continuous regularly varying function on $[a, \infty)$, $a > 0$ and $\gamma > 0$. Equation (A) (or (B)) is called *sublinear* or *superlinear* according as $\gamma < 1$ or $\gamma > 1$. Our aim is to provide comprehensive overview of our present knowledge of the asymptotic analysis of positive solutions of these equation in both sublinear and superlinear case, placing emphasis on some new results giving a complete answer to the three important questions: Are all solutions of regularly varying? What are necessary and sufficient conditions for the existence of such solutions? Is it possible to determine the precise asymptotic formulas for such solutions?

Investigation of the equation of type (A) was inspired by the classical Thomas–Fermi atomic model described by the following nonlinear singular boundary value problem

$$x'' = \frac{1}{\sqrt{t}} x^{3/2}, \quad x(0) = 1, \quad x(\infty) = 0,$$

[64, 65].

The study of equation (A) (in fact of differential equations in general) in the framework of regular variation is initiated by V.G. Avakumović in [66]. For some physical reasons only solutions decreasing to zero of superlinear equation (A) were of interest in [66]. Later on, results on the decreasing solutions of (A) for the superlinear case were further developed in [67–69], while increasing solutions were studied recently in [70]. Sublinear case of (A) has been considered in [71–78]. This section is designed to present a survey of the main results developed in the papers listed above.

A comprehensive survey of results on the asymptotic analysis of ordinary differential equations in the framework of regular variation up to 2000 can be found in the monograph [24].

Emden–Fowler type of differential equation (A) (with $q(t) < 0$) in the framework of regular variation is treated, for example, in [79–84], and high-order nonlinear differential equations, for example, in [85–88]. For the use of the discrete Karamata theory in difference equations, see [89].

12.5.1 The Existence of all Solutions of (A)

If a solution x of (A) exists on an interval of the form $[t_x, \infty)$, $t_x \geq a$, and is eventually nontrivial, then it is called *proper*. A nontrivial solution which is not proper is called *singular*. Further, a singular solution is classified into two types.

Definition 12.12   (i) A solution x of (A) defined on $[t_0, \infty)$ is said to be extinct at a finite time t_1 (type (S_1)) if there exists $t_1 > t_0$ such that

$$x(t) \neq 0 \quad \text{on } [t_0, t_1) \text{ and } x(t) \equiv 0 \quad \text{on } [t_1, \infty).$$

(ii) A solution x of (A) defined on $[t_0, \infty)$ is said to blow up at a finite time t_1 (type (S_2)) if there exists $t_1 > t_0$ such that

$$x(t) \neq 0 \quad \text{on } [t_0, t_1) \text{ and } \lim_{t \to t_1 - 0} |x(t)| = \lim_{t \to t_1 - 0} |x'(t)| = \infty.$$

For the existence of singular solutions we have the following result:

Theorem 12.35   (i) Superlinear equation (A) has solutions of type (S_2), but has no solutions of type (S_1).

(ii) Sublinear equation (A) has solutions of type (S_1), but has no solutions of type (S_2).

Proof: Claim (i) follows from [90, Theorems 2.1, 2.9]. Claim (ii) follows from [90, Theorems 3.1 and 3.9]. □

It is known that all proper solutions of (A) are nonoscillatory and eventually strictly monotone [91]. If x satisfies (A), then so does $-x$, and so considering the equation (A) in the framework of regular variation, we focus our attention on positive proper solutions of (A). Each positive proper solution satisfies one of four different features:

- all possible positive decreasing solutions fall into following two types

(I) $$\lim_{t \to \infty} x(t) = 0, \quad \lim_{t \to \infty} x'(t) = 0$$

(II) $$\lim_{t \to \infty} x(t) = \text{const} > 0, \quad \lim_{t \to \infty} x'(t) = 0$$

- all possible positive increasing solutions fall into following two types

(III) $$\lim_{t \to \infty} x(t) = \infty, \quad \lim_{t \to \infty} \frac{x(t)}{t} = \text{const} > 0,$$

(IV) $$\lim_{t \to \infty} x(t) = \infty, \quad \lim_{t \to \infty} \frac{x(t)}{t} = \infty.$$

A solution of type (I), (II), (III), or (IV) is called respectively *strongly decreasing, asymptotically constant,* asymptotically linear and *strongly increasing* solution of (A).

The existence in the above four classes is described by the convergence or divergence of the two integrals

$$I = \int_a^\infty t q(t) \, \mathrm{d}t, \quad J = \int_a^\infty t^\gamma q(t) \, \mathrm{d}t.$$

It is known that the existence of solutions of types (II) and (III) can be fully characterized in both superlinear and sublinear case.

Proposition 12.4 (i) Equation (A) either superlinear or sublinear, possesses a positive solution x satisfying (II) if and only if $I < \infty$.
(ii) Equation (A) either superlinear or sublinear, possesses a positive solution x satisfying (III) if and only if $J < \infty$.

Proof: Claim (i) follows from [90, Theorems 2.3 and 3.6] and claim (ii) follows from [90, Theorems 2.4 and 3.7]. □

As regards the existence of strongly decreasing and strongly increasing solutions, the problem of establishing necessary and sufficient conditions turns out to be extremely difficult to solve in some cases. In fact, the existence of strongly decreasing solutions in superlinear case and strongly increasing solutions in sublinear case is completely characterized, while for the existence of strongly decreasing solutions in sublinear case and strongly increasing solutions in superlinear case, only necessary or sufficient conditions are known.

Proposition 12.5

(i) Superlinear equation (A) possesses a positive solution of type (I) if and only if $I = \infty$.
(ii) Superlinear equation (A) possesses a positive solution of type (IV) if $J < \infty$.
(iii) Superlinear equation (A) does not possess positive solutions of type (IV) if

$$\liminf_{t \to \infty} t^{\gamma+1} q(t) > 0.$$

(iv) Sublinear equation (A) possesses a positive solution of type (I) if $I < \infty$.

(v) Sublinear equation (A) does not possess positive solutions of type (I) if

$$\liminf_{t \to \infty} t^2 q(t) > 0.$$

(vi) Sublinear equation (A) possesses a positive solution of type (IV) if and only if $J = \infty$.

Proof: Claim (i)–(vi) follow, respectively, from [90, Theorems 2.2, 2.5, 2.6, 3.2, 3.3, 3.8]. □

While the asymptotic behavior (as $t \to \infty$) of asymptotically constant and asymptotically linear solutions is reasonably clear, this is not the case of the other two types of solutions for which determination of precise asymptotic formula is not an easy problem. At the beginning of the research in this area, assuming that coefficients $q(t) \sim t^\sigma$, $t \to \infty$, Kamo and Naito [92, 93] showed that, under some specific assumptions on σ, strongly increasing and strongly decreasing solutions have the form $x(t) \sim k \, t^\rho$, where ρ is constant depending on σ and γ.

Considering regularly varying functions as a (nontrivial) extension of functions asymptotically equivalent to power ones, natural question arises: How about an extension in the sense that the coefficient in the equation (A) is a regularly varying function? Such study of asymptotic of solutions of differential equations via regular variation was initiated in the seminal paper of V.G. Avakumović [66] and about 30 years later extended and developed in [67]. Avakumović showed that assuming that coefficient q is regularly varying of certain index all decreasing solutions of (A) are regularly varying with precise asymptotic behavior as $t \to \infty$. Initiated by Avakumović's paper asymptotic analysis of differential equations in the framework of regularly varying functions (or Karamata functions) means considering equation (A) with regularly varying coefficient q and also more generally nonlinear equation with regularly varying function $\phi(x)$ instead of x^γ.

12.5.2 Superlinear Thomas–Fermi Equation (A)

The first paper connecting regular variation and differential equations was Avakumović [66] in 1947.

Theorem 12.36 (Avakumović [66]) Let $q : [a, \infty) \to \mathbb{R}$ be regularly varying function of index $\sigma > -2$, then any positive solutions x of superlinear equation (A) tending to zero is regularly varying and satisfies

$$x(t) \sim \left(\frac{(\gamma - 1)^2}{(\sigma + \gamma + 1)(\sigma + 2)} \, t^2 q(t) \right)^{-\frac{1}{\gamma - 1}}, \quad t \to \infty.$$

His method of proof is rather involved and make use, in addition to several artifices, of an elementary Tauberian theorem. In 1991, Geluk [94] presented a simple and elegant proof of Theorem 12.36 using results on smoothly varying functions proved meanwile by Balkema, Haan, and himself (for the proof of Theorem 12.36 see also [24, Theorem 3.2]).

However, Avakumović's paper did not attract much attention and regularly varying functions were totally distant from the theory of DE at that time, until the investigation of Marić and Tomić [67] in 1976. Neither Avakumović nor Geluk consider the border case $\sigma = -2$ when the solutions tending to zero may still exists. Therefore, Marić and Tomić [67–69] considering in fact the more general equation

$$x'' = q(t)\phi(x),$$

with ϕ be a regularly varying function at zero of index $\gamma > 1$ proved the following (for the proof see also [24, Theorems 3.4 and 3.5]):

Theorem 12.37 Let q be regularly varying function of index $\sigma \geq -2$. For every positive solution x tending to zero as $t \to \infty$ of superlinear equation (A) there holds:

1. If $\sigma > -2$ solution x is regularly varying of index $\rho = \frac{\sigma+2}{1-\gamma} < 0$. All such decreasing solutions of (A) have one and the same asymptotic behavior

$$x(t) \sim \left[\frac{t^2 q(t)}{\rho(\rho-1)}\right]^{1/1-\gamma} \quad \text{as} \quad t \to \infty. \tag{12.29}$$

2. If $\sigma = -2$ solution x is slowly varying. All such decreasing solutions of (A) have one and the same asymptotic behavior

$$x(t) \sim \left((\gamma-1)\int_a^t sq(s)\,ds\right)^{1/1-\gamma}, \quad t \to \infty. \tag{12.30}$$

Now, we will give an answer to the question which naturally arises: Is the requirement $\sigma \geq -2$ necessary for the superlinear equation (A) to have a regularly varying solution of negative index or a slowly varying solution? The answer is the affirmative as the following lemma shows.

Lemma 12.3 Let $q \in RV_{f,\sigma}$, $q(t) = t^\sigma \ell(t)$, $\ell \in SV_f$.

(i) If equation (A) has a regularly varying solution of index $\rho < 0$, then $\sigma > -2$.
(ii) If equation (A) has a nontrivial slowly varying solution, then

$$\sigma = -2, \quad \text{and} \quad \int_a^\infty sq(s)\,ds = \infty. \tag{12.31}$$

Proof: (i) Let $x \in RV_{f,\rho}$, $x(t) = t^\rho \xi(t)$, $\xi \in SV_f$, with $\rho < 0$, be a solution of (A) on $[T, \infty)$. Since $x'(t) \to 0$, $t \to \infty$, integrating (A) from t to ∞, we have

$$-x'(t) = \int_t^\infty q(s)x(s)^\gamma \, ds = \int_t^\infty s^{\sigma + \rho\gamma} l(s)\xi(s)^\gamma \, ds, \quad t \geq T. \tag{12.32}$$

The convergence of the last integral implies $\sigma + \rho\gamma \leq -1$. However, the possibility $\sigma + \rho\gamma = -1$ is excluded. If fact, if this is the case, then (12.32) reduces to

$$-x'(t) = \int_t^\infty s^{-1} l(s)\xi(s)^\gamma \, ds \in SV_f,$$

which is impossible, because taking that $\lim_{t\to\infty} x(t) = c \in [0, \infty)$ the left-hand side is integrable on $[T, \infty)$, while the right side is a slowly varying function and thus it is not integrable on any neighborhood of infinity. Thus, we have $\sigma + \rho\gamma < -1$. Then, by Karamata's integration theorem [6, Proposition 1.5.10], from (12.32) we get

$$-x'(t) \sim \frac{t^{\sigma + \rho\gamma + 1} \ell(t)\xi(t)^\gamma}{-(\sigma + \rho\gamma + 1)}, \quad t \to \infty. \tag{12.33}$$

From the integrability of the left-hand side of (12.33) on $[T, \infty)$ we have $\sigma + \rho\gamma + 1 \leq -1$. If $\sigma + \rho\gamma = -2$, then (12.33) reduces to

$$-x'(t) \sim t^{-1}\ell(t)\xi(t)^\gamma, \quad t \to \infty.$$

Integration of the last relation on $[t, \infty)$ yields

$$x(t) \sim \int_t^\infty \frac{\ell(s)\xi(s)^\gamma}{s} \, ds, \quad t \to \infty,$$

which implies that $x \in SV_f$, that is, $\rho = 0$, an impossibility. Therefore, we must have $\sigma + \rho\gamma < -2$, in which case, integrating (12.33) from t to ∞ with application of Karamata's integration theorem shows that

$$x(t) \sim \frac{t^{\sigma + \rho\gamma + 2} \ell(t)\xi(t)^\gamma}{(\sigma + \rho\gamma + 1)(\sigma + \rho\gamma + 2)} = \frac{t^{\sigma + 2} \ell(t)x(t)^\gamma}{(\sigma + \rho\gamma + 1)(\sigma + \rho\gamma + 2)}, \quad t \to \infty,$$

or

$$x(t) \sim ((\sigma + \rho\gamma + 1)(\sigma + \rho\gamma + 2))^{1/\gamma - 1} t^{-\frac{\sigma + 2}{\gamma - 1}} \ell(t)^{-\frac{1}{\gamma - 1}}, \quad t \to \infty.$$

This shows that x is regularly varying of index $\rho = -(\sigma + 2)/(\gamma - 1) < 0$. Using this value of ρ, we see from $\sigma + \rho\gamma < -2$ that $\sigma > -2$. Moreover, since $(\sigma + \rho\gamma + 1)(\sigma + \rho\gamma + 2) = (\rho - 1)\rho$, the asymptotic behavior of x is given by (12.29).

(ii) Let $x \in ntr - SV_f$ be a solution of (A) on $[t_0, \infty)$. Then $x(t) \to 0$ and $x'(t) \to 0$ as $t \to \infty$. From (A) we have

$$-x'(t) = \int_t^\infty q(s)x(s)^\gamma \, ds = \int_t^\infty s^\sigma l(s)x(s)^\gamma \, ds, \quad t \geq t_0, \tag{12.34}$$

implying that $\sigma \leq -1$. If $\sigma = -1$, the right side of (12.34) is a slowly varying function, which is not integrable on $[t_0, \infty)$. This contradicts the integrability of the left side of (12.34) and accordingly, it must be $\sigma < -1$. In this case from (12.34) by application of Karamata's integration theorem it follows that

$$-x'(t) \sim \frac{t^{\sigma+1}\ell(t)x(t)^\gamma}{-(\sigma+1)}, \quad t \to \infty, \tag{12.35}$$

which by integration on $[t, \infty)$ yields

$$x(t) \sim \int_t^\infty \frac{s^{\sigma+1}l(s)}{-(\sigma+1)}\, ds, \quad t \to \infty.$$

This means that $\sigma + 1 \le -1$. We claim that the possibility that $\sigma < -2$ is not allowed. In fact, we rewrite (12.35) as

$$-x(t)^{-\gamma}x'(t) \sim \frac{t^{\sigma+1}\ell(t)}{-(\sigma+1)}, \quad t \to \infty. \tag{12.36}$$

Thus, if $\sigma < -2$ the right-hand side of (12.36) is integrable on $[t_0, \infty)$, implying that $x(t)^{1-\gamma}$ tends to a finite limit as $t \to \infty$, which is contradiction. Therefore, it must be $\sigma = -2$ and (12.36) becomes

$$-x(t)^{-\gamma}x'(t) \sim t\, q(t) = tq(t), \quad t \to \infty. \tag{12.37}$$

Since $x(t)^{1-\gamma} \to \infty$, $t \to \infty$ the right-hand side of (12.37) is not integrable on $[t_0, \infty)$ implying $\int_a^\infty tq(t)\, dt = \infty$. If we integrate (12.37) on $[t_0, t]$ we get

$$x(t)^{1-\gamma} \sim (\gamma - 1)\int_{t_0}^t s\, q(s)\, ds, \quad t \to \infty,$$

which yields (12.30). $\qquad\square$

Combining Theorem 12.37 with Lemma 12.3 we have the following result.

Theorem 12.38 Let $q \in \mathrm{RV}_{\mathrm{f},\sigma}$, $\sigma \in \mathbb{R}$.

1. All strongly decreasing solutions x of superlinear equation (A) are regularly varying of index $\rho < 0$, with $\rho = \frac{\sigma+2}{1-\gamma}$, if and only if $\sigma > -2$. All such solutions have the exact asymptotic behavior given by (12.29).
2. All strongly decreasing solutions x of superlinear equation (A) are nontrivial slowly varying if and only if (12.31) is satisfied. For all such solutions (12.30) holds.

Results of the same type for all increasing solutions of superlinear equation (A) have been obtained by Kusano *et al.* [70].

Theorem 12.39 Let $q \in \mathrm{RV}_{\mathrm{f},\sigma}$, $\sigma \in \mathbb{R}$. Then, all increasing solutions x of superlinear equation (A) such that $x(t)/t \to \infty$ as $t \to \infty$ are:

1. Regularly varying of index $\rho > 1$ with $\rho = \frac{\sigma+2}{1-\gamma}$, if and only if $\sigma < -\gamma - 1$, and all such solutions have the exact asymptotic behavior given by (12.29).

2. Nontrivial regularly varying of index 1 if and only if

$$\sigma = -\gamma - 1, \quad \text{and} \quad \int_a^\infty s^\gamma q(s) \, ds < \infty,$$

in which case any such solution the exact asymptotic behavior given by

$$x(t) \sim t\left((\gamma - 1)\int_t^\infty s^\gamma q(s) \, ds\right)^{1/1-\gamma}, \quad t \to \infty.$$

Proof: See [70, Theorem 2.2]. □

12.5.3 Sublinear Thomas–Fermi Equation (A)

Sublinear Thomas–Fermi equation (A) has been considered first by Kusano *et al.* [74, 75] and later on in [71–73, 76–78] by other authors. Considering equation (A) with regularly varying coefficient necessary and sufficient conditions for the existence of two types of strongly increasing regularly varying solutions and two types of strongly decreasing regularly varying solutions have been obtained and precise asymptotic formulas have been derived for such solutions.

Theorem 12.40 Suppose that $q \in \mathrm{RV}_{f,\sigma}$.

(i) Sublinear equation (A) possesses strongly decreasing regularly varying solutions of index $\rho < 0$ if and only if $\sigma < -2$, in which case ρ is given by

$$\rho = \frac{\sigma + 2}{1 - \gamma}. \tag{12.38}$$

All such solutions have one and the same asymptotic behavior

$$x(t) \sim \left[\frac{t^2 q(t)}{\rho(\rho - 1)}\right]^{1/1-\gamma} \quad \text{as } t \to \infty. \tag{12.39}$$

(ii) Sublinear equation (A) possesses a nontrivial slowly varying solution if and only if

$$\sigma = -2 \quad \text{and} \quad \int_a^\infty tq(t) < \infty, \tag{12.40}$$

in which case any such solution has one and the same asymptotic behavior

$$x(t) \sim \left[(1 - \gamma)\int_t^\infty sq(s) \, ds\right]^{1/1-\gamma}, \quad t \to \infty. \tag{12.41}$$

(iii) Sublinear equation (A) possesses strongly increasing regularly varying solutions of index $\rho > 1$ if and only if $\sigma > -\gamma - 1$, in which case ρ is given by (12.38) and any such solution x has one and the same asymptotic behavior given by (12.39).

(iv) Sublinear equation (A) possesses a nontrivial regularly varying solution of index 1 if and only if

$$\sigma = -\gamma - 1 \quad \text{and} \quad \int_a^\infty t^\gamma q(t)\, dt = \infty. \tag{12.42}$$

in which case any such solution has one and the same asymptotic behavior

$$x(t) \sim t\left((1-\gamma)\int_a^t s^\gamma q(s)\, ds\right)^{1/1-\gamma}, \quad t \to \infty. \tag{12.43}$$

Proof: Claim (i) follows from [76, Theorem 2.1] and [72, Theorem 5.1].
Claim (ii) follows from [76, Theorem 2.3] and [75, Theorem 2.4].
Claim (iii) follows from [76, Theorem 2.1] and [72, Theorem 5.2].
Claim (iv) follows from [73, Theorem 3.2] and [75, Theorem 3.4]. □

In comparison with superlinear case, the answer to the question of whether all solutions are regularly varying assuming that q is regularly varying has not been given in these papers. However, Matucci and Řehák [77] and Řehák [78] partially solve this problem recently. They proved more general results for positive decreasing solutions of a system of two coupled nonlinear second-order equations of Thomas–Fermi type in [77] and for positive increasing solutions of a cyclic system of n nonlinear differential equations of Thomas–Fermi type in [90]. The above-mentioned systems includes, as special cases, nonlinear scalar differential equation of type (A) and so applications of results from [77, 78] gives improvement of Theorem 12.40 by giving a positive answer to the above question in the case $\sigma < -2$ and $\sigma > -\gamma - 1$. To complete the story, we will adapt proofs in [77, 78] and presented them in Theorems 12.41 and 12.43. However, we note that in neither one of these two papers border cases $\sigma = -2$ and $\sigma = -\gamma - 1$ have not been treated, so the answer to the above question in these cases is still an open problem, which we will work out here in Theorems 12.42 and 12.44.

Throughout proofs all minimizing constants will be denoted by the same letter m and all majorizing ones by M.

Theorem 12.41 Suppose that $q \in \mathrm{RV}_{f,\sigma}$, $\sigma < -2$. All strongly decreasing solutions x of sublinear equation (A) are regularly varying of index ρ given by (12.38).

Proof: Let $q \in \mathrm{RV}_{f,\sigma}$, $q(t) = t^\sigma \ell(t)$, $\ell \in \mathrm{SV}_f$, $\sigma < -2$. First, we show that for each strongly decreasing solution x there exist positive constants m, M such that

$$m\, t^\rho \ell(t)^{1/1-\gamma} \leq x(t) \leq M\, t^\rho \ell(t)^{1/1-\gamma}. \tag{12.44}$$

Since $x'(\infty) = x(\infty) = 0$, integrating (A) twice first from t to ∞ we have

$$-x'(t) = \int_t^\infty q(s)x(s)^\gamma \, ds, \quad x(t) = \int_t^\infty \int_s^\infty q(r)x(r)^\gamma \, dr \, ds, \quad t \geq T,$$

which using that x is decreasing implies

$$-x'(t) \leq x(t)^\gamma \int_t^\infty q(s) \, ds, \quad x(t) \leq x(t)^\gamma \int_t^\infty \int_s^\infty q(r) \, dr \, ds, \quad t \geq T.$$
$$\text{(12.45)}$$

Because $\sigma < -2$, application of Karamata's integration theorem to the both integrals in (12.45) yields that there exists $M > 0$ such that

$$-x'(t) \leq M \, x(t)^\gamma t^{\sigma+1} \ell(t), \quad x(t) \leq M \, x(t)^\gamma t^{\sigma+2} \ell(t). \tag{12.46}$$

Second inequality in (12.46) implies directly the right-hand side inequality in (12.44).

Next, we prove the left-hand side inequality in (12.44). Setting $w(t) = x(t)|x'(t)|$ and

$$v = \frac{\gamma+1}{\gamma+3}, \quad \mu = \frac{2}{\gamma+3}, \quad \kappa = \frac{1-\gamma}{\gamma+3} \tag{12.47}$$

an application of Young's inequality gives

$$-w'(t) = w(t) \left(\frac{q(t)x(t)^\gamma}{|x'(t)|} + \frac{|x'(t)|}{x(t)} \right) \geq \frac{w(t)}{\mu^\mu v^v} \left(\frac{q(t)x(t)^\gamma}{|x'(t)|} \right)^\mu \left(\frac{|x'(t)|}{x(t)} \right)^v$$
$$= \frac{w(t)}{\mu^\mu v^v} x(t)^{\gamma\mu - v} |x'(t)|^{v-\mu} q(t)^\mu.$$

Since, $\gamma\mu - v = v - \mu = -\kappa$, we get

$$-w'(t) \geq m \, w(t)^{1-\kappa} q(t)^\mu. \tag{12.48}$$

After dividing (12.48) with $w(t)^{1-\kappa}$, using $\kappa > 0$ and $w(\infty) = 0$, by integration on $[t, \infty]$ we obtain

$$w(t)^\kappa \geq m \int_t^\infty q(s)^\mu \, ds = m \int_t^\infty s^{\sigma\mu} \ell(s)^\mu \, ds \quad \text{for} \quad m > 0. \tag{12.49}$$

Since $-\frac{\gamma+3}{2} > -2$, assumption $\sigma < -2$ implies $\sigma\mu + 1 < 0$. Thus, application of Karamata's integration theorem on the right-hand side of the previous inequality together with the first inequality in (12.46) gives

$$x(t)^\kappa \geq m \, (-x'(t))^{-\kappa} t^{\sigma\mu+1} \ell(t)^\mu \geq m \, x(t)^{-\kappa\gamma} t^{\sigma\mu+1-(\sigma+1)\kappa} \ell(t)^{\mu-\kappa}. \tag{12.50}$$

Using (12.47), we have

$$\sigma\mu + 1 - (\sigma+1)\kappa = \rho\kappa(\gamma+1), \quad \frac{\mu-\kappa}{\kappa(\gamma+1)} = \frac{1}{1-\gamma} \tag{12.51}$$

so that from (12.50) we get the left-hand side inequality in (12.44).

It remains to prove that solutions satisfying (12.44) are regularly varying of index $\rho = \frac{\sigma+2}{1-\gamma}$. We define the function

$$X(t) = t^\rho \ell(t)^{1/1-\gamma}(\rho(\rho-1))^{-\frac{1}{1-\gamma}}, \quad \ell \in SV_f. \tag{12.52}$$

It is a matter of straightforward computation, with application of Karamata's integration theorem, to verify that X satisfies integral asymptotic relation

$$\int_t^\infty \int_s^\infty q(r)X(r)^\gamma \, dr \, ds \sim X(t), \quad t \to \infty \tag{12.53}$$

Put

$$k = \liminf_{t\to\infty} \frac{x(t)}{X(t)}, \quad K = \limsup_{t\to\infty} \frac{x(t)}{X(t)}. \tag{12.54}$$

and

$$J(t) = \int_t^\infty \int_s^\infty q(r)X(r)^\gamma \, dr \, ds, \quad t \geq T,$$

In view of (12.44) it is clear that $0 < k \leq K < \infty$. Application of generalized L'Hospital's rule [95] two times gives

$$k = \liminf_{t\to\infty} \frac{x(t)}{X(t)} = \liminf_{t\to\infty} \frac{x(t)}{J(t)} \geq \liminf_{t\to\infty} \frac{x'(t)}{J'(t)} = \liminf_{t\to\infty} \frac{\int_t^\infty q(s)x(s)^\gamma \, ds}{\int_t^\infty q(s)X(s)^\gamma \, ds}$$

$$\geq \liminf_{t\to\infty} \frac{x(t)^\gamma}{X(t)^\gamma} = \left(\liminf_{t\to\infty} \frac{x(t)}{X(t)}\right)^\gamma = k^\gamma.$$

It follows that $k \geq k^\gamma$, implying that $k \geq 1$ because $\gamma < 1$. Similarly, we are led to the inequality $K \leq K^\gamma$, which implies that $K \leq 1$. Thus we conclude that $k = K = 1$, that is, $x(t) \sim X(t)$, $t \to \infty$, which yields that x is a regularly varying function of index ρ. □

Theorem 12.42 Suppose that $q \in RV_{f,\sigma}$ satisfies (12.40). All strongly decreasing solutions x of sublinear equation (A) are slowly varying.

Proof: Let $q \in RV_{f,-2}$, $q(t) = t^{-2}\ell(t)$, $\ell \in SV_f$. First, we show that for each strongly decreasing solution x there exist positive constants m, M such that

$$m\left(\int_t^\infty s^{-1}\ell(s) \, ds\right)^{1/1-\gamma} \leq x(t) \leq M\left(\int_t^\infty s^{-1}\ell(s) \, ds\right)^{1/1-\gamma}. \tag{12.55}$$

Integrating (A) twice first from t to ∞, applying Karamata's integration theorem and using that x is decreasing gives

$$-x'(t) \leq x(t)^\gamma t^{-1}\ell(t), \tag{12.56}$$

and

$$x(t) \leq x(t)^\gamma \int_t^\infty s^{-1}\ell(s) \, ds, \quad t \geq T,$$

implying the right-hand side inequality in (12.55).

To prove the left-hand side inequality in (12.55), first, note that in view of Proposition 12.1 (vi), there exist numbers p, r, ($r < -\sigma < p$) such that

$$t^p q(t) \text{ almost increases and } t^r q(t) \text{ almost decreases.} \tag{12.57}$$

Bearing in mind x decreases, by integrating on both sides of (A) over (t, kt) with an arbitrary fixed $k > 1$, in view of (12.57), one obtains for $t \geq T$,

$$-x'(t) \geq m t^p q(t) x(kt)^\gamma \int_t^{kt} s^{-p} \, ds,$$

which leads to

$$-x'(t) \geq m t q(t) x(kt)^\gamma, \quad t \geq T. \tag{12.58}$$

On the other hand, by multiplying on both sides of (A) by $-x'(t)$, integrating over (t, kt) and using (12.57), one obtains for any fixed $k > 1$ and $t \geq T$

$$x'(t)^2 \geq m t^p q(t) \int_t^{kt} s^{-p} x(s)^\gamma (-x'(s)) \, ds,$$

implying that

$$-x'(t) \geq m \, (q(t) x(t)^{\gamma+1})^{1/2} \left[1 - \left(\frac{x(kt)}{x(t)} \right)^{\gamma+1} \right]^{1/2}. \tag{12.59}$$

From (12.58) and (12.59) we shall derive the following inequality, holding for all $t \geq T$

$$-x'(t) \geq m t q(t) x(t)^\gamma. \tag{12.60}$$

Obviously, the behavior of the quotient $0 < x(kt)/x(t) < 1$ is essential in that. For, if, for example, $\limsup_{t \to \infty} x(kt)/x(t) = 1$, inequality (12.59) is useless. Therefore consider the following alternative:

Take a fixed $k > 1$, and an arbitrary fixed α such that $0 < \alpha < 1$. There holds: Either

$$\frac{x(kt)}{x(t)} \geq \alpha \tag{12.61}$$

for all t belonging to some intervals $\bar{I}_n$, $n \geq 1$ which might be all ultimately neighboring when $\bigcup_{n=1}^{\infty} \bar{I}_n = [T, \infty)$ for some $T \geq a$, or

$$\frac{x(kt)}{x(t)} < \alpha \tag{12.62}$$

for all to belonging to some intervals $\underline{I}_n$, $n \geq 1$, which again might be all ultimately neighboring when $\bigcup_{n=1}^{\infty} \underline{I}_n = [T, \infty)$ for some $T \geq a$.

In general, due to the continuity of x, one has

$$\bigcup_{n \geq 1} (\underline{I}_n \cup \overline{I}_n) = [T, \infty). \tag{12.63}$$

Now, if (12.61) holds, inequality (12.58) gives (12.60) for all $t \in \overline{I}_n$.

However, if all $\overline{I}_n$ are ultimately neighboring then $\underline{I}_n$ do not exist and so (12.60) holds for all $t \geq T$.

If, on the other hand, (12.62) holds, choose a sequence $\{t_n\}$, $n \geq 1$ of arbitrary points $t_n \in \underline{I}_n$ so that (12.62) holds for $t = t_n$. But then, because of [24, Lemma 3.1], there exist numbers $0 < \mu < 1$ and $0 < \alpha' < 1$ such that $x(kt)/x(t) < \alpha'$ for all $t \in [\mu t_n, t_n]$. Hence, from (12.59) and the preceding inequality, one obtains

$$-x'(t) \geq m(q(t)x(t)^{\gamma+1})^{1/2}, \quad t \in [\mu t_n, t_n], \tag{12.64}$$

so after dividing by $x(t)^{\gamma+1/2}$ and integrating over $[\mu t_n, t_n]$, since $\gamma < 1$, we get

$$x(\mu t_n)^{1-\gamma/2} \geq m \int_{\mu t_n}^{t_n} (t^p q(t))^{1/2} t^{-p/2} \, dt \geq m \, (t_n^p q(\mu t_n))^{1/2}$$

$$\times \int_{\mu t_n}^{t_n} t^{-p/2} \, dt \geq m t_n q(\mu t_n)^{1/2}, \tag{12.65}$$

which multiplying by $q(\mu t_n)^{1/2} x(\mu t_n)^{\gamma}$ gives

$$(q(\mu t_n)x(\mu t_n)^{\gamma+1})^{1/2} \geq m t_n q(\mu t_n)x(\mu t_n)^{\gamma}. \tag{12.66}$$

Since t_n is arbitrary in $\underline{I}_n$, inequalities (12.64) and (12.66) together give (12.60) for all $t \in \underline{I}_n$.

Again, if all $\underline{I}_n$ are ultimately neighboring, then $\overline{I}_n$ do not exist, t_n is arbitrary in $[T, \infty)$ and (12.60) holds for all $t \geq T$. Finally, if both sequences of considered intervals exist, then again (12.60) holds for all $t \geq T$ due to (12.63).

To conclude the proof divide (12.60) by $x(kt)^{\gamma}$, integrate over $(t/k, \infty)$ to obtain for $t \geq T$

$$x(t)^{1-\gamma} \geq m \int_t^\infty sq(s) \, ds = m \int_t^\infty s^{-1}\ell(s) \, ds,$$

which because $1 - \gamma > 0$ is the same as the left-hand side of inequality (12.55).

It remains to prove that solutions satisfying (12.55) are slowly varying. Therefore, in view of (12.56) and (12.55) we have

$$0 \leq t \frac{-x'(t)}{x(t)} \leq Mx(t)^{\gamma-1}\ell(t) \leq M\ell(t)\left(\int_t^\infty s^{-1}\ell(s) \, ds\right)^{-1} \tag{12.67}$$

Since, by Karamata's integration theorem,

$$\lim_{t \to \infty} \ell(t)\left(\int_t^\infty s^{-1}\ell(s) \, ds\right)^{-1} = 0,$$

we conclude that

$$\lim_{t \to \infty} t \frac{x'(t)}{x(t)} = 0.$$

Thus, $x \in SV_f$ by Karamata's integration theorem. □

Theorem 12.43 Suppose that $q \in RV_{f,\sigma}$, $\sigma > -\gamma - 1$. All strongly increasing solutions x of sublinear equation (A) are regularly varying of index ρ given by (12.38).

Proof: Let $q \in RV_f$, $q(t) = t^\sigma \ell(t)$, $\ell \in SV_f$, $\sigma > -\gamma - 1$. First, we show that for each strongly increasing solution x there exist positive constants m, M such that

$$m \, t^\rho \ell(t)^{1/1-\gamma} \leq x(t) \leq M \, t^\rho \ell(t)^{1/1-\gamma}. \tag{12.68}$$

for all large t. Using that $x(t) \to \infty$, $t \to \infty$ we have

$$x(t) \sim \int_T^t x'(s) \, ds, \quad t \geq T,$$

which since x' is increasing gives

$$x(t) \leq t \, x'(t), \quad t \geq T. \tag{12.69}$$

Integration of (A) from T to t, since $x'(t) \to \infty$, $t \to \infty$, in view of (12.69), gives

$$x'(t) \sim \int_T^t q(s)x(s)^\gamma \, ds \leq x'(t)^\gamma \int_T^t q(s)s^\gamma \, ds.$$

Using $\sigma + \gamma > -1$ application of Karamata's integration theorem to the above integral yields that there exists $M > 0$ such that

$$x'(t)^{1-\gamma} \leq M \, t^{\gamma+1} q(t) = M \, t^{\gamma+\sigma+1} \ell(t) \tag{12.70}$$

which together with (12.69) implies the right-hand side inequality in (12.68).

Next, we prove the left-hand side inequality in (12.68). Setting $w(t) = x(t)x'(t)$ and v, μ, κ as in (12.47), application of Young's inequality gives

$$w'(t) = w(t) \left(\frac{q(t)x(t)^\gamma}{x'(t)} + \frac{x'(t)}{x(t)} \right) \geq \frac{w(t)}{\mu^\mu v^v} \left(\frac{q(t)x(t)^\gamma}{x'(t)} \right)^\mu \left(\frac{x'(t)}{x(t)} \right)^v$$

$$= \frac{w(t)}{\mu^\mu v^v} x(t)^{\gamma\mu-v} x'(t)^{v-\mu} q(t)^\mu = \frac{1}{\mu^\mu v^v} w(t)^{1-\kappa} q(t)^\mu.$$

and integration on $[T, t]$ implies

$$w(t)^\kappa \geq m \int_T^t q(s)^\mu \, ds. \tag{12.71}$$

Since $-\frac{\gamma+3}{2} < -\gamma - 1$, assumption $\sigma > -\gamma - 1$ implies $\sigma\mu + 1 > 0$. Thus, application of Karamata's integration theorem on the right-hand side of the previous inequality together with (12.70) gives

$$x(t)^\kappa \geq m\, x'(t)^{-\kappa}\, t^{\sigma\mu+1}\ell(t)^\mu \geq m\, t^{\sigma\mu+1-\kappa\frac{\gamma+\sigma+1}{1-\gamma}}\ell(t)^{\mu-\frac{\kappa}{1-\gamma}}, \qquad (12.72)$$

for some $m > 0$. Using (12.47) we have

$$\sigma\mu + 1 - \kappa\frac{\gamma+\sigma+1}{1-\gamma} = \frac{\sigma+2}{\gamma+3} = \kappa\,\rho, \qquad \mu - \frac{\kappa}{1-\gamma} = \frac{1}{\gamma+3},$$

so that from (12.72) we get the left-hand side inequality in (12.68).

To prove that solutions satisfying (12.55) are regularly varying of index $\rho = \frac{\sigma+2}{1-\gamma}$, we define the function $X(t)$ with (12.52) and with application of Karamata's integration theorem verify that X satisfies the integral asymptotic relation

$$\int_T^t \int_T^s q(r)X(r)^\gamma \, dr \, ds \sim X(t), \quad t \to \infty. \qquad (12.73)$$

Put k, K as in (12.54) and in view of (12.55) it is clear that $0 < k \leq K < \infty$. Application of L'Hospital's rule gives $k \geq k^\gamma$ and $K \leq K^\gamma$, implying that $k \geq 1$ and $K \leq 1$. Thus we conclude that $k = K = 1$, that is, $x(t) \sim X(t), t \to \infty$, which yields that x is a regularly varying function of index ρ. $\qquad\square$

Theorem 12.44 Suppose that $q \in RV_{f,\sigma}$ satisfies (12.42). All strongly increasing solutions x of sublinear equation (A) are regularly varying of index 1.

Proof: Let $q \in RV_{f,-\gamma-1}$, $q(t) = t^{-\gamma-1}\ell(t)$, $\ell \in SV_f$. First, we show that for each strongly increasing solution x there exist positive constants m, M such that

$$m\, t\left(\int_T^t s^{-1}\ell(s) \, ds\right)^{1/1-\gamma} \leq x(t) \leq Mt\left(\int_T^t s^{-1}\ell(s) \, ds\right)^{1/1-\gamma}. \qquad (12.74)$$

Integration of (A) from T to t, since $x'(t) \to \infty, t \to \infty$, in view of (12.69), gives

$$x'(t)^{1-\gamma} \leq M\int_T^t s^{-1}\ell(s) \, ds \in SV_f,$$

which together with (12.69) and application of Karamata's integration theorem, implies the right-hand side inequality in (12.74).

To prove the left-hand side inequality in (12.74) we perform the substitution $x(t) = ty(t)$ in (A) and obtain the following differential equation for y:

(C) $\qquad\qquad\qquad (t^2 y'(t))' = t^{\gamma+1}q(t)y(t)^\gamma,$

Obviously, y increases and $y(t) \to \infty$, as $t \to \infty$. Clearly, in order to prove the left-hand side inequality in (12.74) it suffices to prove that y satisfies inequalities

$$y(t) \geq m\left(\int_T^t s^{-1}\ell(s) \, ds\right)^{1/1-\gamma}, \quad t \geq T. \qquad (12.75)$$

Bearing in mind y increases, by integrating on both sides of (C) over (t, kt) with an arbitrary fixed $k > 1$, in view of (12.57), one obtains for $t \geq T$,

$$y'(kt) \geq mt^r q(kt) y(t)^\gamma \int_t^{kt} s^{\gamma+1-r}\, ds,$$

which leads to

$$y'(kt) \geq mt^\gamma q(kt) y(t)^\gamma, \quad t \geq T. \tag{12.76}$$

On the other hand, by multiplying on both sides of (C) by $t^2 y'(t)$, integrating over (t, kt) and using that the function $s^{r+\gamma+3} q(s)$ is almost decreasing for some r, one obtains for any fixed $k > 1$ and $t \geq T$

$$y'(kt)^2 \geq mt^{\gamma-1-r} q(kt) t^r \int_t^{kt} s^{-r} y(s)^\gamma y'(s)\, ds,$$

implying that

$$y'(kt) \geq m\, (t^{\gamma-1} q(kt) y(kt)^{\gamma+1})^{1/2} \left\{ 1 - \left(\frac{y(t)}{y(kt)} \right)^{\gamma+1} \right\}^{1/2}. \tag{12.77}$$

From (12.76) and (12.77) we shall derive the following inequality, holding for all $t \geq T$

$$y'(kt) \geq mt^\gamma q(kt) y(kt)^\gamma. \tag{12.78}$$

Obviously, the behavior of the quotient $0 < y(t)/y(kt) < 1$ is essential in that. For, if, for example, $\limsup_{t\to\infty} y(t)/y(kt) = 1$, inequality (12.77) is useless. Therefore consider the following alternative:

Take a fixed $k > 1$ and an arbitrary fixed α such that $0 < \alpha < 1$. There holds: Either

$$\frac{y(t)}{y(kt)} \geq \alpha \tag{12.79}$$

for all t belonging to some intervals $\bar{I}_n$, $n \geq 1$ which might be all ultimately neighboring when $\bigcup_{n=1}^{\infty} \bar{I}_n = [T, \infty)$ for some $T \geq a$, or

$$\frac{y(t)}{y(kt)} < \alpha \tag{12.80}$$

for all to belonging to some intervals $\underline{I}_n$, $n \geq 1$, which again might be all ultimately neighboring when $\bigcup_{n=1}^{\infty} \underline{I}_n = [T, \infty)$ for some $T \geq a$.

In general, due to the continuity of y, one has

$$\bigcup_{n\geq 1} (\underline{I}_n \cup \bar{I}_n) = [T, \infty). \tag{12.81}$$

Now, if (12.79) holds, inequality (12.76) gives (12.78) for all $t \in \bar{I}_n$.

However, if all $\overline{I}_n$ are ultimately neighboring then $\underline{I}_n$ do not exist and so (12.78) holds for all $t \geq T$.

If, on the other hand, (12.80) holds, choose a sequence $\{t_n\}, n \geq 1$ of arbitrary points $t_n \in \underline{I}_n$ so that (12.80) holds for $t = t_n$. But then, because of Lemma [70, Lemma 1.1, Remark 1.1], there exists $0 < \alpha' < 1$ such that $y(t)/y(kt) < \alpha'$ for all $t \in [t_n, kt_n]$. Hence, from (12.77) and the preceding inequality, one obtains

$$y'(kt) \geq m(t^{\gamma-1}q(kt)y(kt)^{\gamma+1})^{1/2}, \quad t \in [t_n, kt_n], \tag{12.82}$$

so after dividing by $y(kt)^{\gamma+1/2}$ and integrating over $[t_n, kt_n]$, since $\gamma > 1$, we get

$$y(kt_n)^{1-\gamma/2} \geq m \int_{t_n}^{kt_n} (t^{\gamma-1}q(kt))^{1/2} \, dt. \tag{12.83}$$

Using (12.57) for the integral on the right-hand side of (12.65) we have

$$\int_{t_n}^{kt_n} (t^{\gamma-1}q(kt))^{1/2} \, dt \geq m \, (t_n^r q(kt_n))^{1/2} \int_{t_n}^{kt_n} t^{\gamma-1-r/2} \, dt$$
$$\geq m(t_n^{\gamma+1}q(kt_n))^{1/2},$$

which together with (12.83) gives

$$(t_n^{\gamma-1}q(kt_n)y(kt_n)^{\gamma+1})^{1/2} \geq t_n^{\gamma}q(kt_n)y(kt_n)^{\gamma}. \tag{12.84}$$

Since t_n is arbitrary in $\underline{I}_n$, inequalities (12.82) and (12.84) together give (12.78) for all $t \in \underline{I}_n$.

Again, if all $\underline{I}_n$ are ultimately neighboring, then $\overline{I}_n$ do not exist, t_n is arbitrary in $[T, \infty)$ and (12.78) holds for all $t \geq T$. Finally, if both sequences of considered intervals exist, then again (12.78) holds for all $t \geq T$ due to (12.81).

At this point we observe that one could not use such a procedure with the intervals $\underline{I}_n$ instead of $[t_i, kt_i]$, since the former may tend to 0 when $n \to \infty$.

To conclude the proof divide (12.78) by $y(kt)^\gamma$, integrate over $[T/k, t/k]$ to obtain for $t \geq T$

$$y(t)^{1-\gamma} \geq m \int_T^t s^\gamma q(s) \, ds = m \int_T^t s^{-1}\ell(s) \, ds,$$

which because $1 - \gamma > 0$ is the same as the right-hand side of inequality (12.75) implying the left-hand side inequality in (12.74) for x.

It remains to prove that solutions satisfying (12.74) are $\mathrm{RV}_{\mathrm{f},1}$. Therefore, in view of (12.69) and (12.55) we have

$$0 \leq t\frac{x''(t)}{x'(t)} \leq t^2 q(t)x(t)^{\gamma-1} = t^{1-\gamma}\ell(t)x(t)^{\gamma-1} \leq M\ell(t)\left(\int_T^t s^{-1}\ell(s) \, ds\right)^{-1}, \tag{12.85}$$

which by Karamata's integration theorem yields

$$\lim_{t\to\infty} t\frac{x''(t)}{x'(t)} = 0.$$

Thus, $x' \in SV_f$ and by application of Karamata's integration theorem we get

$$x(t) \sim \int_T^t x'(s) \, ds \sim t x'(t) \in RV_{f,1}, \quad t \to \infty$$

implying that $x \in RV_{f,1}$. $\qquad\qquad\qquad\square$

Combining Theorem 12.40 with Theorems 12.41–12.44, we have the following results for sublinear equation (A).

Theorem 12.45 Let $q \in RV_{f,\sigma}, \sigma \in \mathbb{R}$. Then, all increasing solutions x of sublinear equation (A) such that $x(t)/t \to \infty$ as $t \to \infty$ are:

1. Regularly varying of index $\rho > 1$ with $\rho = \frac{\sigma+2}{1-\gamma}$, if and only if $\sigma > -\gamma - 1$, in which case any such solution has one and the same asymptotic behavior given by (12.39).
2. Nontrivial regularly varying of index 1 if and only if (12.42) holds, in which case any such solution has one and the same asymptotic behavior given by (12.43).

Theorem 12.46 Let $q \in RV_{f,\sigma}, \sigma \in \mathbb{R}$. Then, all decreasing solutions x of sublinear equation (A) such that $x(t) \to 0$ as $t \to \infty$ are:

1. Regularly varying of index $\rho < 0$ with $\rho = \frac{\sigma+2}{1-\gamma}$, if and only if $\sigma < -2$, in which case any such solution has one and the same asymptotic behavior given by (12.39).
2. Nontrivial slowly varying if and only if (12.40) holds, in which case any such solution has one and the same asymptotic behavior given by (12.41).

12.5.4 A Generalization

This section is devoted to the asymptotic analysis of positive solutions of the more general equation (B), where p, q are continuous on $[a, \infty)$. Any positive solution x of (B), continuable and eventually different from zero, belong to one of the classes:

$$\mathbb{M}^+ = \{x \text{ solution of (B)} \ : \ x'(t) > 0 \text{ for large } t\},$$
$$\mathbb{M}^- = \{x \text{ solution of (B)} \ : \ x'(t) < 0 \text{ for large } t\}.$$

The class $\mathbb{M}^+$ can be further divided into mutually disjoint subclasses:

$$\mathbb{M}_C^+ = \{x \in \mathbb{M}^+ \ : \ \lim_{t\to\infty} x(t) = d_x, \quad 0 < d_x < \infty\},$$
$$\mathbb{M}_{\infty,C}^+ = \{x \in \mathbb{M}^+ \ : \ \lim_{t\to\infty} x(t) = \infty, \quad \lim_{t\to\infty} p(t)x'(t) = c_x, \quad 0 < c_x < \infty\},$$
$$\mathbb{M}_{\infty,\infty}^+ = \{x \in \mathbb{M}^+ \ : \ \lim_{t\to\infty} x(t) = \infty, \quad \lim_{t\to\infty} p(t)x'(t) = \infty\},$$

while the class $\mathbb{M}^-$ is divided into the next three disjoint subclasses:

$$\mathbb{M}_C^- = \{x \in \mathbb{M}^- : \lim_{t \to \infty} x(t) = d_x, \quad 0 < d_x < \infty\},$$

$$\mathbb{M}_{0,C}^- = \{x \in \mathbb{M}^- : \lim_{t \to \infty} x(t) = 0, \quad \lim_{t \to \infty} p(t)x'(t) = -c_x, \quad 0 < c_x < \infty\},$$

$$\mathbb{M}_{0,0}^- = \{x \in \mathbb{M}^- : \lim_{t \to \infty} x(t) = 0, \quad \lim_{t \to \infty} p(t)x'(t) = 0\},$$

It is known that the following two cases

$$I_p = \int_a^\infty \frac{dt}{p(t)} = \infty, \quad I_p = \int_a^\infty \frac{dt}{p(t)} < \infty,$$

give a different aspect on the existence and asymptotic behavior of positive solutions of (B). The behavior of solutions x belonging in classes $\mathbb{M}_C^+, \mathbb{M}_{\infty,C}^+, \mathbb{M}_C^-, \mathbb{M}_{0,C}^-$ is obvious (for the existence of such solutions see [91, 96, 97, 98]), since it is easy to see that if $x \in \mathbb{M}_{0,C}^-$, then

$$x(t) \sim c_x\, \pi(t), \quad t \to \infty, \quad \text{with } \pi(t) = \int_t^\infty \frac{ds}{p(s)}$$

while if $x \in \mathbb{M}_{\infty,C}^+$, then

$$x(t) \sim c_x\, P(t), \quad t \to \infty \text{ with } P(t) = \int_a^t \frac{ds}{p(s)}.$$

Therefore, we shall consider increasing solutions in the class $\mathbb{M}_{\infty,\infty}^+$ and decreasing solutions in the class $\mathbb{M}_{0,0}^-$, which are usually called *strongly increasing solutions* and *strongly decreasing solutions* of (B). In order to derive the precise asymptotic behavior of these solutions of equation (B) we assume in addition that

$$p \in RV_{f,\mu}, \quad q \in RV_{f,\nu}, \quad \mu, \nu \in \mathbb{R}.$$

Observe that $I_p = \infty$ $[I_p < \infty]$, may hold for $\mu \le 1$ $[\mu \ge 1]$, respectively. Since regularly varying function p with $\mu = 1$ may either satisfies $I_p = \infty$ or $I_p < \infty$, our attention is focused on the cases where either $\mu < 1$ or $\mu > 1$.

Case $I_p = \infty$. We make the change of variables $(t, x) \leftrightarrow (\tau, X)$

$$\tau = P(t) \text{ with } P(t) = \int_a^t \frac{ds}{p(s)}, \quad X(\tau) = x(t). \tag{12.86}$$

Thus, Equation (B) is transformed into

$$\ddot{X} = Q(\tau)X^\gamma, \quad \text{with } Q(\tau) = p(t(\tau))q(t(\tau)), \tag{12.87}$$

with use of $\dot{X} = dX/d\tau$.

Proposition 12.5, in view of (12.86), ensures the existence of solution in classes $\mathbb{M}_{\infty,\infty}^+$, $\mathbb{M}_{0,0}^-$, depending on the convergence or divergence of integrals:

$$Z_1 = \int_a^\infty P(u)q(u)\, du, \quad W_1 = \int_a^\infty P(u)^\gamma q(u)\, du.$$

Proposition 12.6 If $\gamma > 1$:

(i) $\mathbb{M}_{0,0}^{-} \neq \emptyset$ if and only if $Z_1 = \infty$.

(ii) If $W_1 < \infty$, then $\mathbb{M}_{\infty,\infty}^{+} \neq \emptyset$.

If $\gamma < 1$:

(iii) If $Z_1 < \infty$, then $\mathbb{M}_{0,0}^{-} \neq \emptyset$.

(ii) $\mathbb{M}_{\infty,\infty}^{+} \neq \emptyset$ if and only if $W_1 = \infty$.

An application of Karamata's integration theorem to (12.86) with $p(t) = t^\mu L(t)$, $L \in SV_f$, gives for $t \to \infty$

$$\tau = P(t) \sim \frac{t^{1-\mu} L(t)^{-1}}{1-\mu} = \frac{t p(t)^{-1}}{1-\mu} \tag{12.88}$$

and so, by Proposition 12.1(iv), $P \in RV_{f,1-\mu}$. Thus, since $1 - \mu > 0$, by Proposition 12.1 (vii) one has $t \sim P^{-1}(\tau)$, $P^{-1} \in RV_{f,(1-\mu)^{-1}}$ (here P^{-1} denotes the asymptotic inverse of P). Hence, by Proposition 12.1(iii)

$$p \circ P^{-1} \in RV_{f,\frac{\mu}{1-\mu}}, \quad q \circ P^{-1} \in RV_{f,\frac{\nu}{1-\mu}}, \quad Q \in RV_{f,\frac{\mu+\nu}{1-\mu}}.$$

Thus, an application of Theorems 12.38 and 12.39 and Theorems 12.45 and 12.46 to (12.87) with $\sigma = \frac{\mu+\nu}{1-\mu}$ and then by substitution (12.86), with use of (12.88), to return to the original variable t one obtains:

Theorem 12.47 Suppose that $p \in RV_{f,\mu}$ with $\mu < 1$ and $q \in RV_{f,\nu}$. All strongly decreasing solutions x of superlinear equation (B) are:

(i) Regularly varying of index $\rho < 0$ with ρ given by

$$\rho = \frac{\mu - \nu - 2}{\gamma - 1}, \tag{12.89}$$

if and only if $\nu > \mu - 2$, and the asymptotic behavior of any such solution is determined by

$$x(t) \sim \left[\frac{t^2 p(t)^{-1} q(t)}{\rho(\rho - 1 + \mu)} \right]^{-\frac{1}{\gamma-1}}, \quad t \to \infty. \tag{12.90}$$

(ii) Nontrivial slowly varying if and only if

$$\nu = \mu - 2 \quad \text{and} \quad \int_a^\infty P(s) q(s) \, ds = \infty,$$

and all such solutions satisfy

$$x(t) \sim \left((\gamma - 1) \int_a^t P(u) q(u) \, du \right)^{1/1-\gamma}, \quad t \to \infty.$$

Theorem 12.48 Suppose that $p \in RV_{f,\mu}$ with $\mu < 1$ and $q \in RV_{f,\nu}$. All strongly increasing solutions x of superlinear equation (B) are:

(i) Regularly varying of index $\rho > 1 - \mu$, with ρ given by (12.89), if and only if $v < \gamma(\mu - 1) - 1$ and the asymptotic behavior of any such solution is determined by (12.90).

(ii) Nontrivial $RV_{f,1-\mu}$ if and only if

$$v = \gamma(\mu - 1) - 1 \quad \text{and} \quad \int_a^\infty P(s)^\gamma q(s) \, ds < \infty,$$

and all such solutions satisfy

$$x(t) \sim P(t)\left((\gamma - 1) \int_t^\infty P(u)^\gamma q(u) \, du \right)^{1/1-\gamma}, \quad t \to \infty.$$

Theorem 12.49   Let $p \in RV_{f,\mu}$ with $\mu < 1$ and $q \in RV_{f,v}$. Then, all strongly decreasing solutions x of sublinear equation (B) are:

(i) Regularly varying of index $\rho < 0$, with ρ given by (12.89), if and only if $v < \mu - 2$, in which case any such solution has one and the same asymptotic behavior given by (12.90).

(ii) Nontrivial slowly varying if and only if

$$v = \mu - 2 \quad \text{and} \quad \int_a^\infty P(s)q(s) \, ds < \infty,$$

in which case any such solution has one and the same asymptotic behavior given by

$$x(t) \sim \left((1 - \gamma) \int_t^\infty P(u)q(u) \, du \right)^{1/1-\gamma}, \quad t \to \infty.$$

Theorem 12.50   Let $p \in RV_{f,\mu}$ with $\mu < 1$ and $q \in RV_{f,v}$. Then, all strongly increasing solutions x of sublinear equation (B) are:

(i) Regularly varying of index $\rho > 1 - \mu$, with ρ given by (12.89), if and only if $v > \gamma(\mu - 1) - 1$, in which case any such solution has one and the same asymptotic behavior given by (12.90).

(ii) Nontrivial $RV_{f,1-\mu}$ if and only if

$$v = \gamma(\mu - 1) - 1 \quad \text{and} \quad \int_a^\infty P(s)^\gamma q(s) \, ds = \infty,$$

in which case any such solution has one and the same asymptotic behavior given by

$$x(t) \sim P(t)\left((1 - \gamma) \int_a^t P(s)^\gamma q(s) \, ds \right)^{1/1-\gamma}, \quad t \to \infty.$$

Case $I_p < \infty$. We make the change of variables $(t, x) \leftrightarrow (\tau, X)$

$$\tau = \frac{1}{\pi(t)} \quad \text{with} \quad \pi(t) = \int_t^\infty \frac{ds}{p(s)}, \quad X(\tau) = \tau x(t). \tag{12.91}$$

Thus equation (B) is transformed into

$$\ddot{X} = Q(\tau)X^\gamma, \quad \text{with} \quad Q(\tau) = \frac{p(t(\tau))q(t(\tau))}{\tau^{\gamma+3}}, \tag{12.92}$$

with the use of $\dot{X} = dX/d\tau$.

The existence of solution in classes $\mathbb{M}^+_{\infty,\infty}$, $\mathbb{M}^-_{0,0}$, in this case, depends on the convergence or divergence of integrals:

$$Z_2 = \int_a^\infty \pi(u)q(u)\, du, \quad W_2 = \int_a^\infty \pi(u)^\gamma q(u)\, du.$$

Proposition 12.7 If $\gamma > 1$:

(i) $\mathbb{M}^-_{0,0} \neq \emptyset$ if and only if $W_2 = \infty$.
(ii) If $Z_2 < \infty$, then $\mathbb{M}^+_{\infty,\infty} \neq \emptyset$.
   If $\gamma < 1$:
(iii) If $W_2 < \infty$, then $\mathbb{M}^-_{0,0} \neq \emptyset$.
(ii) $\mathbb{M}^+_{\infty,\infty} \neq \emptyset$ if and only if $Z_2 = \infty$.

Similarly as in the previous case one gets:

$$\tau = \frac{1}{\pi(t)} \sim \frac{(\mu - 1)p(t)}{t}, \quad t \to \infty,$$

implying that $1/\pi \in \mathrm{RV}_{f,\mu-1}$, and

$$Q \in \mathrm{RV}_{f,\frac{\mu+\nu}{\mu-1}-\gamma-3}.$$

Then, an application of Theorems 12.38 and 12.39 and Theorems 12.45 and 12.46 to the equation (12.92), with $\sigma = \frac{\mu+\nu}{\mu-1} - \gamma - 3$ and returning to the original variable t as in the previous case, leads to the following:

Theorem 12.51 Suppose that $p \in \mathrm{RV}_{f,\mu}$ with $\mu > 1$ and $q \in \mathrm{RV}_{f,\nu}$. All strongly decreasing solutions x of superlinear equation (B) are:

(i) Regularly varying of index $\rho < 0$, with ρ given by (12.89), if and only if $\nu > \gamma(\mu - 1) - 1$, and the asymptotic behavior of any such solution is determined by (12.90).
(ii) Nontrivial slowly varying if and only if

$$\nu = \gamma(\mu - 1) - 1 \quad \text{and} \quad \int_a^\infty \pi(s)^\gamma q(s)\, ds = \infty,$$

and all such solutions satisfy

$$x(t) \sim \left((\gamma - 1) \int_a^t \pi(u)^\gamma q(u) \, du \right)^{1/1-\gamma}, \quad t \to \infty.$$

Theorem 12.52   Suppose that $p \in RV_{f,\mu}$ with $\mu > 1$ and $q \in RV_{f,\nu}$. All strongly increasing solutions of superlinear equation (B) are:

(i) Regularly varying of index $\rho > \mu - 1$ if and only if $\nu < \mu - 2$, in which case ρ is given by (12.89) and the asymptotic behavior of any such solution is determined by the formula (12.90).
(ii) Nontrivial $RV_{f,\mu-1}$ if and only if

$$\nu = \mu - 2 \quad \text{and} \quad \int_a^\infty \pi(s)q(s) \, ds < \infty,$$

and all such solutions x satisfy

$$x(t) \sim \pi(t)^{-1} \left((\gamma - 1) \int_t^\infty \pi(s)q(s) \, ds \right)^{1/1-\gamma}, \quad t \to \infty.$$

Theorem 12.53   Let $p \in RV_{f,\mu}$ with $\mu > 1$ and $q \in RV_{f,\nu}$. Then, all strongly decreasing solutions x of sublinear equation (B) are:

(i) Regularly varying of index $\rho < 0$, with ρ given by (12.89), if and only if $\nu < \gamma(\mu - 1) - 1$, in which case any such solution has one and the same asymptotic behavior given by (12.90).
(ii) Nontrivial slowly varying if and only if

$$\nu = \gamma(\mu - 1) - 1 \quad \text{and} \quad \int_a^\infty \pi(s)^\gamma q(s) \, ds < \infty,$$

in which case any such solution has one and the same asymptotic behavior given by

$$x(t) \sim \left((1 - \gamma) \int_t^\infty \pi(u)^\gamma q(u) \, du \right)^{1/1-\gamma}, \quad t \to \infty.$$

Theorem 12.54   Let $p \in RV_{f,\mu}$ with $\mu > 1$ and $q \in RV_{f,\nu}$. Then, all strongly increasing solutions x of sublinear equation (B) are:

(i) Regularly varying of index $\rho > \mu - 1$, with ρ given by (12.89), if and only if $\nu > \mu - 2$, in which case any such solution has one and the same asymptotic behavior given by (12.90).

(ii) Nontrivial regularly varying of index $\mu - 1$ if and only if

$$v = \mu - 2 \quad \text{and} \quad \int_a^\infty \pi(s)q(s)\, \mathrm{d}s = \infty,$$

in which case any such solution has one and the same asymptotic behavior given by

$$x(t) \sim \pi(t)^{-1}\left((1-\gamma)\int_a^t \pi(s)q(s)\, \mathrm{d}s\right)^{1/1-\gamma}, \quad t \to \infty.$$

References

1 Karamata, J. (1930) Sur certains "Tauberian theorems" de G.H. Hardy et Littlewood. *Mathematica (Cluj)*, **3**, 33–48.

2 Karamata, J. (1930) Sur un mode de croissance régulière des fonctions. *Mathematica (Cluj)*, **4**, 38–53.

3 Karamata, J. (1930) Über die Hardy-Littlewoodschen Umkehrungen des Abelschen Stetigkeitsätzes. *Math. Z.*, **32**, 319–320.

4 Karamata, J. (1931) Neuer Beweis und Verallgemeinerung der Tauberschen Sätze, welche die Laplacesche un Stieltjessche Transformation betreffen. *J. Reine Angew. Math.*, **164**, 27–39.

5 Karamata, J. (1933) Sur un mode de croissance régulière. Théorèmes fondamenteaux. *Bull. Soc. Math. France*, **61**, 55–62.

6 Bingham, N.H., Goldie, C.M., and Teugels, J.L. (1987) *Regular Variation*, Cambridge University Press, Cambridge.

7 Geluk, J.H. and de Haan, L. (1987) *Regular Variation, Extensions and Tauberian Theorems*, vol. **40**, CWI Tracts, Amsterdam.

8 Karamata, J. (1949) *Theory and Practice of the Stieltjes Integral*, vol. **CLIV**, Serbian Academy of Sciences and Arts, Institute of Mathematics, Belgrade (In Serbian).

9 Korevaar, J. (2004) *Tauberian Theory. A Century of Developments*, Grundl. Math. Wiss., vol. **329**, Springer-Verlag, Berlin.

10 Seneta, E. ((1976)) *Regularly Varying Functions*, Lecture Notes in Mathematics, vol. **508**, Springer-Verlag, Berlin, Heidelberg, New York.

11 de Haan, L. (1970) *On Regular Variations and its Applications to the Weak Convergence of Sample Extremes*, Mathematical Centre Tracts, vol. **32**, Mathematisch Centrum, Amsterdam.

12 Djurčić, D. and Torgašev, A. (2007) Some asymptotic relations for the generalized inverse. *J. Math. Anal. Appl.*, **325**, 1397–1402.

13 Hardy, H. (1949) *Divergent Series*, Oxford University Press.

14 Djurčić, D. and Božin, V. (1997) A proof of a S. Aljančić hypothesis on O-regularly varying sequences. *Publ. Inst. Math. (Beograd)*, **62** (76), 46–52.

15 Djurčić, D., Elez, N., and Kočinac, Lj.D.R. (2015) On a subclass of the class of rapidly varying sequences. *Appl. Math. Comput.*, **251**, 626–632.

16 Djurčić, D., Kočinac, Lj.D.R., and Žižović, M.R. (2007) Some properties of rapidly varying sequences. *J. Math. Anal. Appl.*, **327**, 1297–1306.

17 Djurčić, D., Kočinac, Lj.D.R., and Žižović, M.R. (2008) Rapidly varying sequences and rapid convergence. *Topolo. Appl.*, **155**, 2143–2149.

18 Djurčić, D., Kočinac, Lj.D.R., and Žižović, M.R. (2009) A few remarks on divergent sequences: rates of divergence. *J. Math. Anal. Appl.*, **360**, 588–598.

19 Djurčić, D., Kočinac, Lj.D.R., and Žižović, M.R. (2010) A few remarks on divergent sequences: rates of divergence II. *J. Math. Anal. Appl.*, **327**, 705–709.

20 Elez, N. and Djurčić, D. (2013) Some properties of rapidly varying functions. *J. Math. Anal. Appl.*, **401**, 888–895.

21 Elez, N. and Djurčić, D. (2014) Rapid variability and Karamata's integral theorem. *Filomat*, **28**, 487–492.

22 Elez, N. and Djurčić, D. (2015) Representation and characterization of rapidly varying functions. *Hacettepe J. Math. Stat.*, **44**, 317–322.

23 Došlý, O. and Řehák, P. (2005) *Half-Linear Differential Equations*, Elsevier, North Holland.

24 Marić, V. (2000) *Regular Variation and Differential Equations*, Lecture Notes Mathematics, vol. **1726**, Springer-Verlag, Berlin.

25 Matucci, S. and Řehák, P. (2008) Rapidly varying sequences and second-order difference equations. *J. Differ. Equ. Appl.*, **14**, 17–30.

26 Matucci, S. and Řehák, P. (2009) Rapidly varying decreasing solutions of half-linear difference equations. *Math. Comput. Modell.*, **49**, 1692–1699.

27 Řehák, P. (2014) *Nonlinear Differential Equations in the Framework of Regular Variation*, AMathNet, Brno.

28 Vítovec, J. (2010) Theory of rapid variation on time scales with applications to dynamic equations. *Arch. Math. (Brno)*, **46**, 263–284.

29 Bojanić, R. and Seneta, E. (1973) A unified theory of regularly varying sequences. *Math. Z.*, **134**, 91–106.

30 Galambos, J. and Seneta, E. (1973) Regularly varying sequences. *Proc. Am. Math. Soc.*, **41**, 110–116.

31 Djurčić, D., Kočinac, Lj.D.R., and Žižović, M.R. (2008) Classes of sequences of real numbers, games and selection properties. *Topol. Appl.*, **156**, 46–55.

32 Tasković, M. (2003) Fundamental facts on translationally $\mathcal{O}$-regularly varying functions. *Math. Moravica*, **7**, 107–152.

33 Avakumović, V.G. (1936) Über einen $\mathcal{O}$-inversionssatz. *Bull. Int. Acad. Youg. Sci.*, **29–30**, 107–117.

34 Aljančić, S. and Arandjelović, D. (1977) $\mathcal{O}$-regularly varying functions. *Publ. Inst. Math. (Beograd)*, **22** (36), 5–22.

35 Arandjelović, D. (1990) $\mathcal{O}$-regularly variation and uniform convergence. *Publ. Inst. Math. (Beograd)*, **48** (62), 25–40.

36 Djurčić, D. (1998) $\mathcal{O}$-regularly varying functions and strong asymptotic equivalence. *J. Math. Anal. Appl.*, **220**, 451–461.

37 Djurčić, D. and Torgašev, A. (2006) On the Seneta sequences. *Acta Math. Sin.*, **22**, 689–692.

38 Elez, N. and Vladičić, V. (2015) Integral properties of rapidly and regularly varying functions. *Publ. Inst. Math. (N.S.)*, **98** (112), 91–96.

39 Matuszewska, W. (1964) On a generalization of regularly increasing functions. *Stud. Math.*, **24**, 271–279.

40 Djurčić, D. and Torgašev, A. (2009) A theorem of Galambos-Bojanić-Seneta type. *Abstr. Appl. Anal.*, **2009**, 6. Article ID 360794.

41 de Haan, L. (1974) Equivalence classes of regularly varying functions. *Stoch. Process. Appl.*, **2**, 243–259.

42 Djurčić, D., Kočinac, Lj.D.R., and Žižović, M.R. (2007) On increasing rapidly varying sequences. *Note Mat.*, **27** (Suppl. 1), 55–63.

43 Djurčić, D., Kočinac, Lj.D.R., and Žižović, M.R. (2007) Relations between sequences and selection properties. *Abstr. Appl. Anal.*, **2007**, 8. Article ID 43081.

44 Djurčić, D., Kočinac, Lj.D.R., and Žižović, M.R. (2012) On the class $\mathbb{S}_0$ of real sequences. *Appl. Math. Lett.*, **25**, 1296–1298.

45 Djurčić, D., Kočinac, Lj.D.R., and Žižović, M.R. (2006) On selection principles and games in divergent processes, in *Selection Principles and Covering Properties in Topology, Quaderni di Matematica*, Dipartimento di Matematica, vol. **18** (ed. Lj.D.R. Kočinac), Seconda Universita di Napoli, Caserta, pp. 133–155.

46 Borel, E. (1919) Sur la classification des ensembles de mesure nulle, *Bull. Soc. Math. France*, **47**, 97–125.

47 Menger, K. (1924) Einige Überdeckungssätze derPunktmengenlehre, *Sitzungsberichte Abt. 2a, Mathematik,Astronomie, Physik, Meteorologie und Mechanik*, vol. **133**, Wiener Akademie, Wien, pp. 421–444.

48 Hurewicz, W. (1925) Über die Verallgemeinerg des Borelschen Theorems. *Math. Z.*, **24**, 401–425.

49 Hurewicz, W. (1927) Über Folgen stetiger Funktionen. *Fund. Math.*, **9**, 193–204.

50 Rothberger, F. (1938) Eine Verschärfung der Eigenschaft C. *Fund. Math.*, **30**, 50–55.

51 Sierpiński, W. (1928) Sur un ensemble nondénombrable, donc toute image continue est de mesure nulle. *Fund. Math.*, **11**, 301–304.

52 Sierpiński, W. (1926) Sur un problème de K. Menger. *Fund. Math.*, **8**, 223–224.

53 Scheepers, M. (1996) Combinatorics of open covers I: Ramsey theory. *Topol. Appl.*, **69**, 31–62.

54 Kočinac, Lj.D.R. (2003) Generalized Ramsey theory and topological properties: a survey. *Rend. Sem. Mat. Messina, Ser. II*, **25**, 119–132.

55 Kočinac, Lj.D.R. (2004) Selected results on selection principles, in *Proceedings of the 3rd Seminar Geometry and Topology* (ed. Sh. Rezapour), Azarbaijan University of Tarbiat Moallem, Tabriz, Iran, pp. 71–104.

56 Kočinac, Lj.D.R. (2006) Some covering properties in topological and uniform spaces. *Proc. Steklov Inst. Math.*, **252**, 122–137.

57 Kočinac, Lj.D.R. (2015) Star selection principles: a survey. *Khayyam J. Math.*, **1**, 82–106.

58 Sakai, M. and Scheepers, M. (2014) The combinatorics of open covers, in *Recent Progress in General Topology III* (eds K.P. Hart, J. van Mill, and P. Simon), Atlantis Press, pp. 751–800.

59 Scheepers, M. (2003/2004) Selection principles and covering properties in topology. *Note Mat.*, **22** (2), 3–41.

60 Tsaban, B. (2006) Some new directions in infinite-combinatorial topology, in *Topics in Set Theory and its Applications* (eds J. Bagaria and S. Todorvcevic), Birkhäuser, pp. 225–255.

61 Scheepers, M. (1991) Meager-nowhere dense games (II): coding strategies. *Proc. Am. Math. Soc.*, **112**, 1107–1115.

62 Kočinac, Lj.D.R. (2008) Selection principles related to α_i-properties. *Taiwan. J. Math.*, **12**, 561–571.

63 Kočinac, Lj.D.R. (2011) On the α_i-selection principles and games. *Contemp. Math.*, **533**, 107–124.

64 Thomas, L.H. (1927) The calculation of atomic fields. *Proc. Cambridge Philos. Soc.*, **23**, 542–548.

65 Fermi, E. (1927) Un metodo statistico per la determinazione di alcune proprietá dell'atomo. *Rend. Accad. Naz. Lincei*, **6**, 602–607.

66 Avakumović, V.G. (1947) Sur l'équation différentielle de Thomas-Fermi. *Publ. Inst. Math. (Beograd)*, **1**, 101–113.

67 Marić, V. and Tomić, M. (1976) Asymptotic properties of solutions of the equation $y'' = f(x)\varphi(y)$. *Math. Z.*, **149**, 261–266.

68 Marić, V. and Tomić, M. (1977) Regular variation and asymptotic properties of solutions of nonlinear differential equations. *Publ. Inst. Math. (Beograd)*, **21** (35), 119–129.

69 Marić, V. and Tomić, M. (1980) Asymptotic of solutions of second order Thomas-Fermi equation. *J. Differ. Equ.*, **35**, 36–40.

70 Kusano, T., Manojlović, J., and Marić, V. (2014) Increasing solutions of Thomas-Fermi type differential equations - the superlinear case. *Nonlinear Anal.*, **108**, 114–127.

71 Jaroš, J. and Kusano, T. (2013) Slowly varying solutions of a class of first order systems of nonlinear differential equations. *Acta Math. Univ. Comenianae*, **82**, 265–284.

72 Jaroš, J. and Kusano, T. (2013) Existence and precise asymptotic behavior of strongly monotone solutions of systems of nonlinear differential equations. *Differ. Equ. Appl.*, **5**, 185–204.

73 Kusano, T., Manojlović, J., and Marić, V. (2011) Increasing solutions of Thomas-Fermi type differential equations - the sublinear case. *Bull. Acad. Serbe Sci. Arts, Classe Sci. Math. Nat., Sci. Math.*, **CXLIII** (36), 21–36.

490 | *Mathematical Analysis and Applications*

74 Kusano, T., Marić, V., and Tanigawa, T. (2009) Regularly varying solutions of generalized Thomas-Fermi equations. *Bull. Acad. Serbe Sci. Arts, Classe Sci. Math. Nat., Sci. Math.*, **CXLIII** (34), 43–73.

75 Kusano, T., Marić, V., and Tanigawa, T. (2012) An asymptotic analysis of positive solutions of generalized Thomas-Fermi differential equations - the sub-half-linear case. *Nonlinear Anal.*, **75**, 2474–2485.

76 Manojlović, J.V. and Marić, V. (2012) An asymptotic analysis of positive solutions of Thomas-Fermi type sublinear differential equations. *Mem. Differ. Equ. Math. Phys.*, **57**, 75–94.

77 Matucci, S. and Řehák, P. (2014) Asymptotics of decreasing solutions of coupled p-Laplacian systems in the framework of regular variation. *Anal. Mat. Pura Appl.*, **193**, 837–858.

78 Řehák, P. (2013) Asymptotic behavior of increasing solutions to a system of n nonlinear differential equations. *Nonlinear Anal.*, **77**, 45–58.

79 Jaroš, J., Kusano, T., and Manojlović, J. (2013) Asymptotic analysis of positive solutions of generalized Emden-Fowler differential equations in the framework of regular variation. *Cent. Eur. J. Math.*, **11**, 2215–2233.

80 Kusano, T. and Manojlović, J.V. (2011) Asymptotic analysis of Emden-Fowler differential equations in the framework of regular variation. *Anal. Mat. Pura Appl.*, **190**, 619–644.

81 Kusano, T. and Manojlović, J.V. (2011) Precise asymptotic behavior of solutions of the sublinear Emden-Fowler differential equation, Appl. *Math. Comput.*, **217**, 4382–4396.

82 Kusano, T. and Manojlović, J.V. (2011) Asymptotic behavior of positive solutions of sublinear differential equations of Emden-Fowler type. *Comput. Math. Appl.*, **62**, 551–565.

83 Kusano, T., Manojlović, J.V., and Milošević, J. (2013) Intermediate solutions of second order quasilinear ordinary differential equations in the framework of regular variation. *Appl. Math. Comput.*, **219**, 8178–8191.

84 Milošević, J. and Manojlović, J.V. (2015) Positive decreasing solutions of second order quasilinear ordinary differential equations in the framework of regular variation. *Filomat*, **29**, 1995–2010.

85 Kusano, T. and Manojlović, J.V. (2012) Asymptotic behavior of positive solutions of odd order Emden-Fowler type differential equations in the framework of regular variation. *Electron. J. Qual. Theory Differ. Equ.*, **45**, 1–23.

86 Kusano, T. and Manojlović, J.V. (2013) Precise asymptotic behavior of intermediate solutions of even order nonlinear differential equation in the framework of regular variation. *Moscow Math. J.*, **13**, 649–666.

87 Kusano, T., Manojlović, J.V., and Milošević, J. (2014) Intermediate solutions of fourth order quasilinear differential equations in the framework of regular variation. *Appl. Math. Comput.*, **248**, 246–272.

88 Milošević, J. and Manojlović, J.V. (2015) Asymptotic analysis of fourth order quasilinear differential equations in the framework of regular variation. *Taiwan. J. Math.*, **19**, 1415–1456.

89 Agarwal, R.P. and Manojlović, J.V. (2013) On the existence and the asymptotic behavior of nonoscillatory solutions of second order quasilinear difference equations. *Funk. Ekvac.*, **56**, 81–109.

90 Mizukami, M., Naito, M., and Usami, H. (2002) Asymptotic behavior of solutions of a class of second order quasilinear ordinary differential equations. *Hiroshima Math. J.*, **32**, 51–78.

91 Cecchi, M., Došlá, Z., and Marini, M. (2001) On nonoscillatory solutions of differential equations with p–Laplacian. *Adv. Math. Sci. Appl.*, **11**, 419–436.

92 Kamo, K. and Usami, H. (2000) Asymptotic forms of positive solutions of second-order quasilinear ordinary differential equations. *Adv. Math. Sci. Appl.*, **10**, 673–688.

93 Kamo, K. and Usami, H. (2001) Asymptotic forms of positive solutions of secondorder quasilinear ordinary differential equations with sub-homogeneity. *Hiroshima Math. J.*, **31**, 35–49.

94 Geluk, J.L. (1991) Note on a theorem of Avakumović. *Proc. Am. Math. Soc.*, **112**, 429–431.

95 Haupt, O. and Aumann, G. (1938) *Differential- und Integralrechnung*, Walter de Gruyter, Berlin.

96 Cecchi, M., Došlá, Z., and Marini, M. (2000) On the dynamics of the generalized Emden-Fowler equations. *Georgian Math. J.*, **7**, 269–282.

97 Cecchi, M., Došlá, Z., and Marini, M. (2006) Integral conditions for nonoscillation of second order nonlinear differential equations. *Nonlinear Anal.*, **64**, 1278–1289.

98 Kvinikadze, G. (1998) On strongly increasing solutions of system of nonlinear ordinary differential equations. *Trudy Inst. Prikl. Mat. I. Vekua*, **43**, 222–227.

13

n-Inner Products, n-Norms, and Angles Between Two Subspaces

Hendra Gunawan

Department of Mathematics, Bandung Institute of Technology, Bandung 40132, Indonesia

MSC (2010) 46B05, 46B20, 46B99, 47H10, 54H25

13.1   Introduction

Assuming that the reader is familiar with the notion of inner products and norms, that are usually defined on a vector space, we begin our discussion with the definitions of 2-inner products and 2-norms, which have attracted many researchers in the area of functional analysis around three decades ago. For simplicity, we assume that our vector space is real.

Let X be a real vector space of dimension $d \geq 2$. A *2-inner product* on X is a function $\langle \cdot, \cdot | \cdot \rangle : X \times X \times X \to \mathbb{R}$ satisfying the following properties:

(1) $\langle x, x | y \rangle \geq 0$ for all $x, y \in X$; $\langle x, x | y \rangle = 0$ iff x and y are linearly dependent;
(2) $\langle x, x | y \rangle = \langle y, y | x \rangle$ for all $x, y \in X$;
(3) $\langle x, y | z \rangle = \langle y, x | z \rangle$ for all $x, y, z \in X$;
(4) $\langle \alpha x + \beta x', y | z \rangle = \alpha \langle x, y | z \rangle + \beta \langle x', y | z \rangle$ for all $x, x', y, z \in X$ and $\alpha, \beta \in \mathbb{R}$.

The pair $(X, \langle \cdot, \cdot | \cdot \rangle)$ is called a *2-inner product space* [8, 9].

Meanwhile, a *2-norm* on X is a function $\| \cdot, \cdot \| : X \times X \to \mathbb{R}$ satisfying the following properties:

(1) $\| x, y \| = 0$ iff x and y are linearly dependent;
(2) $\| x, y \| = \| y, x \|$ for all $x, y \in X$;
(3) $\| x, \alpha y \| = |\alpha| \ \| x, y \|$ for all $x, y \in X$ and $\alpha \in \mathbb{R}$;
(4) $\| x, y + z \| \leq \| x, y \| + \| x, z \|$ for all $x, y, z \in X$.

The pair $(X, \| \cdot, \cdot \|)$ is called a *2-normed space* [10].

If X is equipped with an inner product $\langle \cdot, \cdot \rangle$, we know that we can define a norm $\| \cdot \|$ on X by $\| x \| := \langle x, x \rangle^{\frac{1}{2}}$. One of its properties is that it satisfies *the*

Mathematical Analysis and Applications: Selected Topics, First Edition.
Edited by Michael Ruzhansky, Hemen Dutta, and Ravi P. Agarwal.
© 2018 John Wiley & Sons, Inc. Published 2018 by John Wiley & Sons, Inc.

triangle inequality:

$$\|x + y\| \leq \|x\| + \|y\|,$$

which is easy to prove once we establish *the Cauchy–Schwarz inequality*:

$$\langle x, y \rangle^2 \leq \|x\|^2 \|y\|^2.$$

There are several ways to prove the Cauchy–Schwarz inequality. We may rewrite it as a determinantal inequality involving a 2×2 *Gram matrix*

$$\begin{vmatrix} \langle x, x \rangle & \langle x, y \rangle \\ \langle y, x \rangle & \langle y, y \rangle \end{vmatrix} \geq 0,$$

and conclude that the Cauchy–Schwarz inequality holds since the matrix is positive semidefinite (see [11, pp. 407–408] for Gram matrices). We also note that the equality holds if and only if x and y are linearly dependent.

On an inner product space $(X, \langle \cdot, \cdot \rangle)$, we can also define a 2-inner product $\langle \cdot, \cdot | \cdot \rangle$ on X by

$$\langle x, y | z \rangle := \begin{vmatrix} \langle x, y \rangle & \langle x, z \rangle \\ \langle z, y \rangle & \langle z, z \rangle \end{vmatrix}$$

and an induced 2-norm $\| \cdot, \cdot \|$ on X given by $\|x, y\| := \langle x, x | y \rangle^{\frac{1}{2}}$, that is

$$\| x, y \| = \begin{vmatrix} \langle x, x \rangle & \langle x, y \rangle \\ \langle y, x \rangle & \langle y, y \rangle \end{vmatrix}^{\frac{1}{2}}.$$

Let us verify that the above formula defines a 2-norm on X. As usual, the properties (N1), (N2) and (N3) are easy to check. To verify the property (N4) or the triangle inequality, it suffices to prove the Cauchy–Schwarz inequality

$$\langle x, y | z \rangle^2 \leq \|x, z\|^2 \|y, z\|^2.$$

But, again, we may rewrite the inequality as

$$\begin{vmatrix} \langle x, x | z \rangle & \langle x, y | z \rangle \\ \langle y, x | z \rangle & \langle y, y | z \rangle \end{vmatrix} \geq 0.$$

We note that the matrix is positive semidefinite: that is, for any $\alpha, \beta \in \mathbb{R}$, we have

$$[\alpha \ \beta] \begin{bmatrix} \langle x, x | z \rangle & \langle x, y | z \rangle \\ \langle y, x | z \rangle & \langle y, y | z \rangle \end{bmatrix} \begin{bmatrix} \alpha \\ \beta \end{bmatrix} = \langle \alpha x + \beta y, \alpha x + \beta y | z \rangle \geq 0.$$

This tells us that the Cauchy–Schwarz inequality holds.

Alternatively, one may observe that, under the assumption that $x \neq 0$, the Cauchy–Schwarz inequality

$$\begin{vmatrix} \langle x, y \rangle & \langle x, z \rangle \\ \langle z, y \rangle & \langle z, z \rangle \end{vmatrix}^2 \leq \begin{vmatrix} \langle x, x \rangle & \langle x, z \rangle \\ \langle z, x \rangle & \langle z, z \rangle \end{vmatrix} \begin{vmatrix} \langle y, y \rangle & \langle y, z \rangle \\ \langle z, y \rangle & \langle z, z \rangle \end{vmatrix}$$

is equivalent to

$$\begin{vmatrix} \langle x,x \rangle & \langle x,y \rangle & \langle x,z \rangle \\ \langle y,x \rangle & \langle y,y \rangle & \langle y,z \rangle \\ \langle z,x \rangle & \langle z,y \rangle & \langle z,z \rangle \end{vmatrix} \geq 0.$$

Since the matrix is positive semidefinite, the inequality holds. Moreover, we also see that the equality holds iff the three vectors x, y, and z are linearly dependent.

The above observation on inner product spaces and 2-inner product spaces suggests how the concept of inner products and 2-inner products, as well as norms and 2-norms, can be generalized to that of n-inner products and n-norms for any $n \in \mathbb{N}$. We shall discuss some results on n-inner product spaces and n-normed spaces, including the topology and the notion of orthogonality in n-normed spaces. Related to an n-inner product and its induced n-norm, we have the Cauchy–Schwarz inequality which is closely related to the angle between two n-dimensional subspaces intersecting on an $(n-1)$-dimensional subspace. In the last part of this note, we shall also discuss the notion of angles between two subspaces of an inner product space in general.

13.2 *n*-Inner Product Spaces and *n*-Normed Spaces

Let $n \in \mathbb{N}$ and X be a real vector space of dimension $d \geq n$. The mapping $\langle \cdot, \cdot | \cdot, \dots, \cdot \rangle : X^{n+1} \to \mathbb{R}$ which satisfies the following five properties:

(I1) $\langle x_1, x_1 | x_2, \dots, x_n \rangle \geq 0$; $\langle x_1, x_1 | x_2, \dots, x_n \rangle = 0$ iff $x_1, x_2, \dots, x_n$ are linearly dependent,

(I2) $\langle x_1, x_1 | x_2, \dots, x_n \rangle = \langle x_{i_1}, x_{i_1} | x_{i_2}, \dots, x_{i_n} \rangle$ for any permutation $(i_1, \dots, i_n)$ of $(1, \dots, n)$,

(I3) $\langle x_0, x_1 | x_2, \dots, x_n \rangle = \langle x_1, x_0 | x_2, \dots, x_n \rangle$,

(I4) $\langle \alpha x_0, x_1 | x_2, \dots, x_n \rangle = \alpha \langle x_0, x_1 | x_2, \dots, x_n \rangle$,

(I5) $\langle x_0 + x_0^*, x_1 | x_2, \dots, x_n \rangle = \langle x_0, x_1 | x_2, \dots, x_n \rangle + \langle x_0^*, x_1 | x_2, \dots, x_n \rangle$,

is called an n-inner product on X [1, 12].

Along with the n-inner product, we can also define $\|\cdot, \dots, \cdot\|$ on X^n by

$$\|x_1, \dots, x_n\| := \langle x_1, x_1 | x_2, \dots, x_n \rangle^{1/2}.$$

This mapping is called an *n-norm* on X [2–4]. It satisfies the following four properties:

(N1) $\|x_1, \dots, x_n\| = 0$ iff $x_1, \dots, x_n$ are linearly dependent,

(N2) $\|x_1, \dots, x_n\|$ is invariant under permutation,

(N3) $\|\alpha x_1, x_2, \dots, x_n\| = |\alpha| \, \|x_1, x_2, \dots, x_n\|$,

(N4) $\|x_0 + x_1, x_2, \dots, x_n\| \leq \|x_0, x_2, \dots, x_n\| + \|x_1, x_2, \dots, x_n\|$.

As usual, the first three properties are easy to see. To prove the last property or the triangle inequality, we need to establish the Cauchy–Schwarz inequality.

Proposition 13.1 (The Cauchy–Schwarz Inequality) For all $x_0, x_1, \ldots,$ $x_n \in X$, we have

$$\langle x_0, x_1 | x_2, \ldots, x_n \rangle^2 \le \| x_0, x_2, \ldots, x_n \|^2 \| x_1, x_2, \ldots, x_n \|^2; \tag{13.1}$$

and the equality holds iff $x_0, x_1, \ldots, x_n$ are linearly dependent.

Proof: First observe that the inequality may be rewritten as

$$\begin{vmatrix} \langle x_0, x_0 | x_2, \ldots, x_n \rangle & \langle x_0, x_1 | x_2, \ldots, x_n \rangle \\ \langle x_1, x_0 | x_2, \ldots, x_n \rangle & \langle x_1, x_1 | x_2, \ldots, x_n \rangle \end{vmatrix} \ge 0,$$

which obviously holds since the matrix is positive semidefinite. Next suppose that we have the equality

$$\begin{vmatrix} \langle x_0, x_0 | x_2, \ldots, x_n \rangle & \langle x_0, x_1 | x_2, \ldots, x_n \rangle \\ \langle x_1, x_0 | x_2, \ldots, x_n \rangle & \langle x_1, x_1 | x_2, \ldots, x_n \rangle \end{vmatrix} = 0,$$

If $\langle x_0, x_0 | x_2, \ldots, x_n \rangle = 0$ or $\langle x_1, x_1 | x_2, \ldots, x_n \rangle = 0$, then $x_0, x_1, \ldots, x_n$ are linearly dependent. Otherwise, there exists a $\beta \ne 0$ such that

$$\langle x_0, x_1 | x_2, \ldots, x_n \rangle = \beta \langle x_0, x_0 | x_2, \ldots, x_n \rangle \quad \text{and}$$
$$\langle x_1, x_1 | x_2, \ldots, x_n \rangle = \beta \langle x_1, x_0 | x_2, \ldots, x_n \rangle.$$

Hence

$$\langle x_0, \beta x_0 - x_1 | x_2, \ldots, x_n \rangle = 0 \quad \text{and} \quad \langle x_1, \beta x_0 - x_1 | x_2, \ldots, x_n \rangle = 0,$$

and so

$$\langle \beta x_0 - x_1, \beta x_0 - x_1 | x_2, \ldots, x_n \rangle = 0.$$

But this implies that $\beta x_0 - x_1, x_2, \ldots, x_n$ are linearly dependent, and so are $x_0, x_1, x_2, \ldots, x_n$.

Conversely, suppose that $x_0, x_1, x_2, \ldots, x_n$ are linearly dependent. If $x_2, \ldots, x_n$ are linearly dependent, then the right-hand side of (13.1) is equal to zero and so is the left-hand side. Now suppose that $x_2, \ldots, x_n$ are linearly independent. Since the equation

$$\alpha x_0 + \beta x_1 + \gamma_2 x_2 + \cdots + \gamma_n x_n = 0$$

has a non-trivial solution, we must have α or $\beta \ne 0$. Without loss of generality, assume that $\alpha \ne 0$, so that

$$x_0 = a_1 x_1 + a_2 x_2 + \cdots + a_n x_n$$

for some $a_1, a_2, \ldots, a_n \in \mathbb{R}$.

Next we note that, for each $k = 2, \ldots, n$, we have $\langle x_k, x_k | x_2, \ldots, x_n \rangle = 0$ and consequently

$$\langle x_0, x_k | x_2, \ldots, x_n \rangle^2 \leq \langle x_0, x_0 | x_2, \ldots, x_n \rangle \langle x_k, x_k | x_2, \ldots, x_n \rangle = 0,$$

implying that $\langle x_0, x_k | x_2, \ldots, x_n \rangle = 0$. The same is true when x_0 is replaced by x_1. Hence

$$\langle x_0, x_0 | x_2, \ldots, x_n \rangle = \langle a_1 x_1, a_1 x_1 | x_2, \ldots, x_n \rangle = a_1^2 \langle x_1, x_1 | x_2, \ldots, x_n \rangle$$

and

$$\langle x_0, x_1 | x_2, \ldots, x_n \rangle = \langle a_1 x_1, x_1 | x_2, \ldots, x_n \rangle = a_1 \langle x_1, x_1 | x_2, \ldots, x_n \rangle,$$

and therefore the equality follows. □

If $(X, \langle \cdot, \cdot \rangle)$ is a real inner product space of dimension $d \geq n$, we may define the *standard n-inner product* and the *standard n-norm* to be the following mappings:

$$\langle x_0, x_1 | x_2, \ldots, x_n \rangle_S := \begin{vmatrix} \langle x_0, x_1 \rangle & \langle x_0, x_2 \rangle & \cdots & \langle x_0, x_n \rangle \\ \langle x_2, x_1 \rangle & \langle x_2, x_2 \rangle & \cdots & \langle x_2, x_n \rangle \\ \vdots & \vdots & \ddots & \vdots \\ \langle x_n, x_1 \rangle & \langle x_n, x_2 \rangle & \cdots & \langle x_n, x_n \rangle \end{vmatrix}$$

and

$$\|x_1, \ldots, x_n\|_S := \langle x_1, x_1 | x_2, \ldots, x_n \rangle_S^{1/2}$$

$$= \begin{vmatrix} \langle x_1, x_1 \rangle & \cdots & \langle x_1, x_n \rangle \\ \vdots & \ddots & \vdots \\ \langle x_n, x_1 \rangle & \cdots & \langle x_n, x_n \rangle \end{vmatrix}^{\frac{1}{2}}.$$

Accordingly we call $(X, \langle \cdot, \cdot | \cdot, \ldots, \cdot \rangle_S)$ and $(X, \|\cdot, \ldots, \cdot\|_S)$ the *standard n-inner product space* and the *standard n-normed space* [13].

Note that $\langle \cdot, \cdot | \cdot, \ldots, \cdot \rangle_S$ is a mapping on X^{n+1} while $\|\cdot, \ldots, \cdot\|_S$ is a mapping on X^n. For $n = 1$, we know that $\| \cdot \|_S$ is a norm, and $\|x_1\|_S$ is the length of x_1. For $n = 2$, $\|\cdot, \cdot\|_S$ defines a 2-norm, and $\|x_1, x_2\|_S$ gives us the area of the parallelogram spanned by x_1 and x_2. For $n = 3$, one may apply the Gram–Schmidt process to observe that $\|x_1, x_2, x_3\|_S$ is nothing but the volume of the parallelepiped spanned by x_1, x_2, and x_3. In general, $\|x_1, \ldots, x_n\|_S$ represents the volume of the n-dimensional parallelepiped spanned by $x_1, \ldots, x_n$ in X. We find this geometric interpretation useful not only for the standard n-normed spaces but also for arbitrary n-normed spaces. In any n-normed space $(X, \|\cdot, \ldots, \cdot\|)$, one may observe that

$$\|x_1, x_2, \ldots, x_n\| = \|x_1 + \alpha_2 x_2 + \cdots + \alpha_n x_n, x_2, \ldots, x_n\|$$

for every $x_1, x_2, \ldots, x_n \in X$ and $\alpha_2, \ldots, \alpha_n \in \mathbb{R}$. Geometrically, this means that the volume of the parallelepiped is unchanged if we alter one of the vectors by a linear combination of the other vectors.

As indicated in our introductory observation, we have the following result for standard n-inner product spaces.

Proposition 13.2 In the standard n-inner product space, the Cauchy–Schwarz inequality (13.1) is equivalent to

$$
\begin{vmatrix}
\langle x_0, x_0 \rangle & \langle x_0, x_1 \rangle & \cdots & \langle x_0, x_n \rangle \\
\langle x_1, x_0 \rangle & \langle x_1, x_1 \rangle & \cdots & \langle x_1, x_n \rangle \\
\vdots & \vdots & \ddots & \vdots \\
\langle x_n, x_0 \rangle & \langle x_n, x_1 \rangle & \cdots & \langle x_n, x_n \rangle
\end{vmatrix} \geq 0.
\tag{13.2}
$$

Remark 13.1 (1) The determinant on the left hand side of (13.2) is clearly nonnegative since it is the volume of the $(n + 1)$-dimensional parallelepiped spanned by $x_0, x_1, \ldots, x_n$. Moreover, we also see that the equality holds if and only if $x_0, x_1, \ldots, x_n$ are linearly dependent (see [14, 15]).

(2) The proof employs some facts about symmetric matrices. For 2×2 matrices $A = [a_{ij}]$, we have $|A| = a_{11}a_{22} - a_{12}a_{21}$. Particularly, when $a_{12} = a_{21}$, we have $|A| = a_{11}a_{22} - a_{12}^2$, and so $|A| \geq 0$ is equivalent to $a_{12}^2 \leq a_{11}a_{22}$. Now let $A = [a_{ij}]$ be an $N \times N$ matrix ($N \geq 3$) such that the determinants of the sub-matrices $A_k = [a_{ij}]_{i,j=k,\ldots,N}$ ($k = 3, \ldots, N$) are all non-zero. Then one may observe that

$$
|A_3| \, |A| = |M_{2,2}| \, |M_{1,1}| - |M_{2,1}| \, |M_{1,2}|
$$

where M_{ij} denotes the $(N - 1) \times (N - 1)$ matrix obtained from A by deleting the i-th row and j-th column (i.e., M_{ij} is the minor corresponding to a_{ij}). In particular, if A is symmetric, then

$$
|A_3| \, |A| = |M_{2,2}| \, |M_{1,1}| - |M_{2,1}|^2.
$$

See [5] for the details.

In an n-normed space $(X, \|\cdot, \ldots, \cdot\|)$, we have the following fact:

Fact 1. Let $(X, \|\cdot, \ldots, \cdot\|)$ be an n-normed space. If the dimension of X is equal to n, then for every $x_0, x_1, \ldots, x_n \in X$ one of the following equalities holds:

$$
\|x_0 + x_1, x_2, \ldots, x_n\| = \|x_0, x_2, \ldots, x_n\| + \|x_1, x_2, \ldots, x_n\|
$$

or

$$
\|x_0 - x_1, x_2, \ldots, x_n\| = \|x_0, x_2, \ldots, x_n\| + \|x_1, x_2, \ldots, x_n\|.
$$

Proof: Since the dimension of X is n, the vectors $x_0, x_1, x_2, \ldots, x_n$ must be linearly dependent. As usual, we may assume that $x_0 = a_1 x_1 + a_2 x_2 + \cdots + a_n x_n$ for some $a_1, a_2, \ldots, a_n \in \mathbb{R}$. If $a_1 \geq 0$, then we have

$$
\begin{aligned}
\|x_0 + x_1, x_2, \ldots, x_n\| &= \|(a_1 + 1)x_1, x_2, \ldots, x_n\| \\
&= (a_1 + 1)\,\|x_1, x_2, \ldots, x_n\| \\
&= \|a_1 x_1, x_2, \ldots, x_n\| + \|x_1, x_2, \ldots, x_n\| \\
&= \|x_0, x_2, \ldots, x_n\| + \|x_1, x_2, \ldots, x_n\|.
\end{aligned}
$$

If $a_1 < 0$, then we have

$$
\begin{aligned}
\|x_0 - x_1, x_2, \ldots, x_n\| &= \|(a_1 - 1)x_1, x_2, \ldots, x_n\| \\
&= (1 - a_1)\,\|x_1, x_2, \ldots, x_n\| \\
&= \|x_1, x_2, \ldots, x_n\| + \|a_1 x_1, x_2, \ldots, x_n\| \\
&= \|x_1, x_2, \ldots, x_n\| + \|x_0, x_2, \ldots, x_n\|.
\end{aligned}
$$

Therefore one of the two equalities must hold. $\qquad\square$

13.2.1   Topology in *n*-Normed Spaces

A sequence $(x(k))$ in an n-normed space $(X, \|\cdot, \ldots, \cdot\|)$ is said to be *convergent* to an $x \in X$ iff $\lim\limits_{k\to\infty} \|x(k) - x, x_2, \ldots, x_n\| = 0$ for all $x_2, \ldots, x_n \in X$. In such a case, we write $\lim\limits_{k\to\infty} x(k) = x$ and call x a *limit* of $(x(k))$.

Fact 2.   If $\lim\limits_{k\to\infty} x(k)$ exists, it is unique.

Proof: Suppose that we have a sequence $(x(k))$ that converges to x and y, with $x \neq y$. Choose $x_2, \ldots, x_n$ such that $\|x - y, x_2, \ldots, x_n\| \neq 0$. Now, letting $\epsilon = \frac{1}{2}\|x - y, x_2, \ldots, x_n\|$, there exists $K \in \mathbb{N}$ such that for $k \geq K$ we have $\|x(k) - x, x_2, \ldots, x_n\| < \epsilon$ and $\|x(k) - y, x_2, \ldots, x_n\| < \epsilon$. But then, by the triangle inequality, we get

$$
\begin{aligned}
\|x - y, x_2, \ldots, x_n\| &\leq \|x(k) - x, x_2, \ldots, x_n\| + \|x(k) - y, x_2, \ldots, x_n\| \\
&< 2\epsilon = \|x - y, x_2, \ldots, x_n\|,
\end{aligned}
$$

which is impossible. $\qquad\square$

Although the above definition of the convergence of a sequence works fine, we find that the condition requiring the limit must be 0 for *all* $x_2, \ldots, x_n \in X$ is rather too much. Some researchers weaken the condition by requiring that the limit is 0 for *some* (fixed vectors) $x_2, \ldots, x_n \in X$. However, this might be "too loose." For instance, using this requirement, the sequence $(x(k))$ that has terms $x_2, 2x_n, x_2, 3x_n, x_2, 4x_n, \ldots$, is convergent to 0, because $\|x(k), x_2, \ldots, x_n\| = 0$ for every $k \in \mathbb{N}$. We do not want this to happen. So the problem is to find the sets of vectors $x_2, \dots, x_n$ that we can use to define that a sequence is convergent to a limit such that, when X is also equipped with a norm, the limit coincides with the limit in norm. The problem is solved, in many cases, by choosing any linearly independent set $\{a_1, \dots, a_n\}$ of n vectors in X and requiring that $\lim_{k \to \infty} \|x(k) - x, a_{i_2}, \dots, a_{i_n}\| = 0$ for all $\{i_2, \dots, i_n\} \subset \{1, \dots, n\}$. Thus, instead of checking infinitely many limits, we only need to verify n limits only [13].

We observe that the uniqueness of the limit of a convergent sequence still holds because of the following fact:

Fact 3. Let $(X, \|\cdot, \dots, \cdot\|)$ be an n-normed space and $\{a_1, \dots, a_n\}$ be a set of n linearly independent vectors in X. Then $\|x, a_{i_2}, \dots, a_{i_n}\| = 0$ for every $\{i_2, \dots, i_n\} \subset \{1, \dots, n\}$ if and only if $x = 0$.

Many results in normed spaces, such as fixed point theorems, have analogs in n-normed spaces. Many researchers proved fixed point theorems and many other results in an n-normed space by imitating the proof in normed spaces (see for instance [16–18]). We found that we could obtain the same results differently. Using a set of n linearly independent vectors, we realize that the n-normed space can actually be equipped with a norm derived from the n-norm.

Fact 4. Let $(X, \|\cdot, \dots, \cdot\|)$ be an n-normed space and $\{a_1, \dots, a_n\}$ be a set of n linearly independent vectors in X. Then

$$\|x\| := \sum_{\{i_2, \dots, i_n\} \subset \{1, \dots, n\}} \|x, a_{i_2}, \dots, a_{i_n}\|$$

defines a norm on X.

Using the above fact, we could prove fixed point theorems and many other results by exploiting our knowledge of normed spaces (see for instance [13, 19, 20, 21]).

13.3 Orthogonality in n-Normed Spaces

Several notions of orthogonality in a normed space have been developed by many authors. For example, the following definitions of Pythagorean, isosceles, and the Birkhoff-James orthogonality in a (real) normed space $(X, \|\cdot\|)$ are known:

P-orthogonality: x is P-orthogonal to y (denoted by $x \perp_P y$) if only if

$$\|x + y\|^2 = \|x\|^2 + \|y\|^2.$$

I-orthogonality: x is I-orthogonal to y (denoted by $x \perp_I y$) if only if

$$\|x + y\| = \|x - y\|.$$

BJ-orthogonality: x is BJ-orthogonal to y (denoted by $x \perp_{BJ} y$) if only if

$$\|x + \alpha y\| \geq \|x\| \quad \text{for every } \alpha \in \mathbb{R}.$$

If X is equipped with an inner product $\langle \cdot, \cdot \rangle$, then one may observe that $x \perp_P y$, $x \perp_I y$, and $x \perp_{BJ} y$ are all equivalent to the condition that $\langle x, y \rangle = 0$, for which we have the usual orthogonality $x \perp y$. However, in a normed space which is not an inner product space, one does not imply another. For further properties of these notions of orthogonality, see, for example, [22]. Related results may be found in [23–28].

Many researchers have extended these notions of orthogonality to 2-normed spaces (see, for example, [29, 30, 31]). As the notions of orthogonality in normed spaces are inspired by that in inner product spaces, the notions of orthogonality in 2-normed spaces are also connected to that in 2-inner product spaces. In [32], it is shown that the "standard" definition of orthogonality in a 2-inner product space $(X, \langle \cdot, \cdot | \cdot \rangle)$, where $\dim(X) \geq 3$, is the following:

G-orthogonality (in 2-normed spaces): x is *G-orthogonal to* y, denoted by $x \perp_G y$, if and only if there exists a subspace V of X with $\text{codim}(V) = 1$ such that $\langle x, y | z \rangle = 0$ for all $z \in V$.

We say that this definition is "standard" because when X is a standard 2-inner product space, that is, when X is actually equipped with an inner product $\langle \cdot, \cdot \rangle$ and the 2-inner product

$$\langle x, y | z \rangle_S := \begin{vmatrix} \langle x, y \rangle & \langle x, z \rangle \\ \langle z, y \rangle & \langle z, z \rangle \end{vmatrix},$$

we have $x \perp_G y$ if and only if $x \perp y$ (see [32]). The above definition of G-orthogonality also improves the one proposed by Cho and Kim [29]. (As shown in [32], Cho and Kim's definition of orthogonality in 2-inner product spaces are void.)

Accordingly, we define P-, I-, and BJ-orthogonality in a 2-normed space $(X, \|\cdot, \cdot\|)$ of dimension 3 or higher as follows:

Definition A (*P-, I-, and BJ-orthogonality* in 2-normed spaces)

(a) $x \perp_P y$ if only if there exists a subspace V of X with $\text{codim}(V) = 1$ such that

$$\|x + y, z\|^2 = \|x, z\|^2 + \|y, z\|^2 \quad \text{for every } z \in V;$$

(b) $x \perp_I y$ if only if there exists a subspace V of X with $\text{codim}(V) = 1$ such that

$$\|x + y, z\| = \|x - y, z\| \quad \text{for every } z \in V;$$

(c) $x \perp_{BJ} y$ if only if there exists a subspace V of X with $\text{codim}(V) = 1$ such that

$$\|x + \alpha y, z\| \geq \|x, z\| \quad \text{for every } z \in V \text{ and } \alpha \in \mathbb{R}.$$

These notions of orthogonality are extended to n-normed spaces in [6]. We note that in an n-inner product space $(X, \langle \cdot, \cdot | \cdot, \dots, \cdot \rangle)$, which is also equipped with the induced n-normed $\|\cdot, \dots, \cdot\|$, we have *the parallelogram law*:

$$\|x + y, x_2, \dots, x_n\|^2 + \|x - y, x_2, \dots, x_n\|^2$$
$$= 2\|x, x_2, \dots, x_n\|^2 + 2\|y, x_2, \dots, x_n\|^2,$$

and *the polarization identity*:

$$4\langle x, y | x_2, \dots, x_n \rangle = \|x + y, x_2, \dots, x_n\|^2 - \|x - y, x_2, \dots, x_n\|^2.$$

The theorem below shows that in the standard n-inner product space we cannot define G-orthogonality between x and y by the condition that

$$\langle x, y | x_2, \dots, x_n \rangle_S = 0 \quad \text{for all } x_2, \dots, x_n \notin \text{span}\{x, y\}$$

as suggested by [29] and [33].

Theorem 13.1 Let $(X, \langle \cdot, \cdot | \cdot, \dots, \cdot \rangle_S)$ be a standard n-inner product space of dimension $n + 1$ or higher. Then, the condition that $\langle x, y | x_2, \dots, x_n \rangle_S = 0$ for all $x_2, \dots, x_n \notin \text{span}\{x, y\}$ is satisfied only by $x = 0$ or $y = 0$.

Proof: We shall prove the theorem through its contraposition. Suppose that $x \neq 0$ and $y \neq 0$. Our task is to show that there exist $x_2, \dots, x_n \notin \text{span}\{x, y\}$ such that $\langle x, y | x_2, \dots, x_n \rangle_S \neq 0$. To do so, we consider several cases.

Case 1. If x and y are linearly dependent, that is, $y = kx$ for $k \neq 0$, then

$$\langle x, y | x_2, \dots, x_n \rangle_S = k\langle x, x | x_2, \dots, x_n \rangle_S = k\|x, x_2, \dots, x_n\|_S^2.$$

Now choose $x_2, \dots, x_n \notin \text{span}\{x\}$ such that $\{x, x_2, \dots, x_n\}$ is linearly independent. Then, we have

$$\langle x, y | x_2, \dots, x_n \rangle_S = k\|x, x_2, \dots, x_n\|_S^2 \neq 0.$$

Case 2. If x and y are linearly independent, then we consider the following two subcases.

Case 2a. If $x \not\perp y$ or $\langle x, y \rangle \neq 0$, then we may choose an orthonormal sequence $x_2, \dots, x_n \in \{x, y\}^\perp$, so that

$$\langle x, y | x_2, \dots, x_n \rangle_S = \langle x, y \rangle \|x_2\|^2 \cdots \|x_n\|^2 = \langle x, y \rangle \neq 0.$$

Case 2b. If $x \perp y$ or $\langle x, y \rangle = 0$, then we may choose a nonzero vector $x_2 = x + y + u \notin \text{span}\{x, y\}$ where u is a fixed nonzero vector perpendicular to $\mathrm{span}\{x, y\}$, and nonzero vectors $x_3, \dots, x_n \notin \mathrm{span}\{x, y\}$ where $x_3 \perp \mathrm{span}\{x, y\}$ and $x_i \perp \mathrm{span}\{x, y, x_3, \dots, x_{i-1}\}$ for $i = 4, \dots, n$. Hence, we have

$$\langle x, y | x_2, \dots, x_n \rangle_S = \begin{vmatrix} 0 & \langle x, x_2 \rangle & 0 & \dots & 0 \\ \langle x_2, y \rangle & \langle x_2, x_2 \rangle & \langle x_2, x_3 \rangle & \dots & \langle x_2, x_n \rangle \\ 0 & \langle x_3, x_2 \rangle & \langle x_3, x_3 \rangle & \dots & 0 \\ \vdots & \vdots & \vdots & \ddots & \vdots \\ 0 & \langle x_n, x_2 \rangle & 0 & \dots & \langle x_n, x_n \rangle \end{vmatrix}$$

$$= -\langle x, x_2 \rangle \langle x_2, y \rangle \|x_3\|^2 \dots \|x_n\|^2$$

$$= -\|x\|^2 \|y\|^2 \|x_3\|^2 \dots \|x_n\|^2 \neq 0.$$

Thus, in any case, we can always find $x_2, \dots, x_n \notin \mathrm{span}\{x, y\}$ such that $\langle x, y | x_2, \dots, x_n \rangle_S \neq 0$. The proof is therefore complete. □

13.3.1   G-, P-, I-, and BJ- Orthogonality

As in 2-inner product spaces and 2-normed spaces, we define the notions of G-orthogonality in n-inner product spaces and P-, I-, and BJ-orthogonality in n-normed spaces as follows. For the rest of this section, we assume that X is a vector space of dimension $n + 1$ or higher.

Definition B   (*G-orthogonality* in n-normed spaces) Let $(X, \langle \cdot, \cdot | \cdot, \dots, \cdot \rangle)$ be an n-inner product space. For $x, y \in X$, we say that x is *G-orthogonal to* y and write $x \perp_G y$ if and only if there exists a subspace V of X with $\mathrm{codim}(V) = 1$ such that

$$\langle x, y | x_2, \dots, x_n \rangle = 0 \quad \text{for every } x_2, \dots, x_n \in V.$$

Definition C   (*P-, I-,* and *BJ-orthogonality* in n-normed spaces) Let $(X, \|\cdot, \dots, \cdot\|)$ be an n-normed space. For $x, y \in X$, we define

(a) $x \perp_P y$ if only if there exists a subspace V of X with $\mathrm{codim}(V) = 1$ such that

$$\|x + y, x_2, \dots, x_n\|^2 = \|x, x_2, \dots, x_n\|^2 + \|y, x_2, \dots, x_n\|^2$$
$$\text{for every } x_2, \dots, x_n \in V;$$

(b) $x \perp_I y$ if only if there exists a subspace V of X with $\mathrm{codim}(V) = 1$ such that

$$\|x + y, x_2, \dots, x_n\| = \|x - y, x_2, \dots, x_n\| \quad \text{for every } x_2, \dots, x_n \in V;$$

(c) $x \perp_{BJ} y$ if only if there exists a subspace V of X with $\mathrm{codim}(V) = 1$ such that

$$\|x + \alpha y, x_2, \dots, x_n\| \geq \|x, x_2, \dots, x_n\|$$
for every $x_2, \dots, x_n \in V$ and $\alpha \in \mathbb{R}$.

Theorem 13.2 In an n-inner product space $(X, \langle \cdot, \cdot | \cdot, \ldots, \cdot \rangle)$, P-, I-, and BJ-orthogonality are equivalent to G-orthogonality.

Proof: The proof follows from the properties of the n-inner product. For instance, to prove that $x \perp_{BJ} y$ implies $x \perp_G y$, we take a subspace V of X with $\operatorname{codim}(V) = 1$ such that

$$\|x + \alpha y, x_2, \ldots, x_n\| \geq \|x, x_2, \ldots, x_n\|$$

for every $x_2, \ldots, x_n \in V$ and $\alpha \in \mathbb{R}$. Now fix $x_2, \ldots, x_n \in V$ for the moment. Then, squaring both sides, we get

$$2\alpha \langle x, y | x_2, \ldots, x_n \rangle + \alpha^2 \|y, x_2, \ldots, x_n\|^2 \geq 0$$

for every $\alpha \in \mathbb{R}$. This forces us to have $\langle x, y | x_2, \ldots, x_n \rangle^2 \leq 0$ or $\langle x, y | x_2, \ldots, x_n \rangle = 0$. Since this is true for every $x_2, \ldots, x_n$, we conclude that $x \perp_G y$. $\qquad \square$

Theorem 13.3 Let $(X, \langle \cdot, \cdot | \cdot, \ldots, \cdot \rangle)$ be an n-inner product space. Then,

(a) $x \perp_G x$ if and only if $x = 0$;
(b) $x \perp_G y$ if and only if $y \perp_G x$;
(c) if $x \perp_G y$, then $\alpha x \perp_G \beta y$ for every $\alpha, \beta \in \mathbb{R}$.

Proof: The proof follows directly from the definition of G-orthogonality and the properties of the n-inner product. For instance, to prove the "only if" part of (a), we suppose that $x \perp_G x$ and $x \neq 0$. Then, there exists a subspace V of X with $\operatorname{codim}(V) = 1$ such that $\langle x, x | x_2, \ldots, x_n \rangle = \|x, x_2, \ldots, x_n\|^2 = 0$ for every $x_2, \ldots, x_n \in V$. But, since $\dim(V) \geq n$, we can choose $x_2, \ldots, x_n \in V$ such that $\{x, x_2, \ldots, x_n\}$ is linearly independent. Hence $\|x, x_2, \ldots, x_n\| > 0$, which is a contradiction. $\qquad \square$

The next theorem states that in a standard n-inner product space, G-orthogonality is equivalent to the usual orthogonality (with respect to the inner product).

Theorem 13.4 Let $(X, \langle \cdot, \cdot | \cdot, \ldots, \cdot \rangle_S)$ be a standard n-inner product space. Then, $x \perp_G y$ if and only if $x \perp y$.

Proof: First we prove the necessary condition. Suppose that $x \perp y$, that is, $\langle x, y \rangle = 0$, and that $x, y \neq 0$. Then we can choose $V = \{x\}^\perp$. Clearly V is a subspace of X with $\operatorname{codim}(V) = 1$. Now, for every $x_2, x_3, \ldots, x_n \in V$, we have

$$\langle x, y | x_2, \ldots, x_n \rangle_S = 0.$$

(Alternatively, one may choose $V = \{y\}^\perp$ and obtain the same result.) This shows that $x \perp_G y$.

For the sufficient condition, suppose that $x \not\perp y$, that is, $\langle x, y \rangle \neq 0$. Clearly x and y are nonzero vectors. To show that $x \not\perp_G y$, let V be an arbitrary subspace of X with $\mathrm{codim}(V) = 1$. Since $\dim(V \cap \{x\}^\perp) \geq n - 1$, there must exist an orthonormal sequence $x_2, \ldots, x_n \in V$ such that $\langle x, x_i \rangle = 0$ for every $i = 2, \ldots, n$. Accordingly, we have

$$\langle x, y | x_2, \ldots, x_n \rangle_S = \langle x, y \rangle \|x_2\|^2 \cdots \|x_n\|^2 = \langle x, y \rangle \neq 0.$$

This completes the proof of the theorem. $\qquad\square$

13.3.2 Remarks on the n-Dimensional Case

Let $(X, \|\cdot, \ldots, \cdot\|_S)$ be a standard n-normed space of dimension n. With respect to the G-orthogonality defined earlier, we find that two arbitrary vectors x and y are G-orthogonal to each other. Given $x, y \neq 0$, just take V to be an $n - 1$-dimensional subspace of X such that $x \in V$. Now, for every $x_2, \ldots, x_n \in V$, the set $\{x, x_2, \ldots, x_n\}$ is linearly dependent. Then, supposing that x_n is a linear combination of $x, x_2, \ldots, x_{n-1}$, we get

$$\langle x, y | x_2, \ldots, x_n \rangle_S = \begin{vmatrix} \langle x, y \rangle & \langle x, x_2 \rangle & \cdots & \langle x, x_n \rangle \\ \langle x_2, y \rangle & \langle x_2, x_2 \rangle & \cdots & \langle x_2, x_n \rangle \\ \vdots & \vdots & \ddots & \vdots \\ \langle x_n, y \rangle & \langle x_n, x_2 \rangle & \cdots & \langle x_n, x_n \rangle \end{vmatrix}$$

$$= \begin{vmatrix} \langle x, y \rangle & \langle x, x_2 \rangle & \cdots & \langle x, x_n \rangle \\ \langle x_2, y \rangle & \langle x_2, x_2 \rangle & \cdots & \langle x_2, x_n \rangle \\ \vdots & \vdots & \ddots & \vdots \\ 0 & 0 & \cdots & 0 \end{vmatrix} = 0.$$

(The same also happens if we take V to be an $n - 1$-dimensional subspace of X such that $y \in V$.) This fact is of course undesirable.

To define orthogonality in n-normed spaces of dimension n in general, it seems that we have to use a different way. In light of [13], we can actually derive a norm from the n-norm and then define P-, I-, and BJ-orthogonality using this norm. In the standard case, the derived norm can be obtained from the n-norm in such a way that it coincides with the existing one (see [34]). Therefore, P-, I-, and BJ-orthogonality defined by using the derived norm will coincide with the usual orthogonality (with respect to the existing inner product).

13.4 Angles Between Two Subspaces

Given two subspaces of a vector space, how can we define the angle between them? In statistics, the angle is used as a measure of dependency of one set of random variables on another [35]. The notion of angles between two subspaces of the Euclidean space $\mathbb{R}^d$ has been studied by many researchers since the 1950s or even earlier [36]. Research results on angles between subspaces and related topics can be found in, for examples, [37–41]. Particularly, in [40], Risteski and Trenčevski introduced a more geometrical definition of angle between two subspaces of $\mathbb{R}^d$ and explained its connection with the so-called canonical (or principal) angles. Their definition of the angle is refined in [7], which we shall discuss here.

Let $(X, \langle \cdot, \cdot \rangle)$ be a real inner product space. Given two nonzero, finite-dimensional, subspaces U and V of X with $\dim(U) \leq \dim(V)$, we wish to have a formula for the angle between U and V that can be viewed, in some sense, as a natural generalization of the "usual" definition of the angle at least in two cases, namely (a) between a 1-dimensional subspace and a q-dimensional subspace of X, and (b) between two p-dimensional subspaces intersecting on a common $(p-1)$-dimensional subspace of X.

Before we go further, let us elaborate what we mean by the "usual" definition of the angle in the above two cases.

(a) If $U = \text{span}\{u\}$ is a 1-dimensional subspace and $V = \text{span}\{v_1, \dots, v_q\}$ is a q-dimensional subspace of X, then the angle θ between U and V is defined by

$$\cos^2\theta = \frac{\langle u, u_V \rangle^2}{\|u\|^2 \|u_V\|^2} \tag{13.3}$$

where u_V denotes the (orthogonal) projection of u on V and $\| \cdot \| = \langle \cdot, \cdot \rangle^{\frac{1}{2}}$ denotes the induced norm on X. (Throughout this note, we shall always take θ to be in the interval $[0, \frac{\pi}{2}]$.)

(b) If $U = \text{span}\{u, w_2, \dots, w_p\}$ and $V = \text{span}\{v, w_2, \dots, w_p\}$ are p-dimensional subspaces of X that intersects on $(p-1)$-dimensional subspace $W = \text{span}\{w_2, \dots, w_p\}$ with $p \geq 2$, then the angle θ between U and V may be defined by

$$\cos^2\theta = \frac{\langle u_W^\perp, v_W^\perp \rangle^2}{\|u_W^\perp\|^2 \|v_W^\perp\|^2} \tag{13.4}$$

where $u_W^\perp$ and $v_W^\perp$ are the orthogonal complement of u and v, respectively, on W.

We observe the common property among these two formulas. In (a), if we write $u = u_V + u_V^\perp$ where $u_V^\perp$ is the orthogonal complement of u on V, then (13.3) amounts to

$$\cos\theta = \frac{\|u_V\|}{\|u\|},$$

which tells us that the value of $\cos\theta$ is equal to the ratio between the length of the projection of u on V and the length of u.

Similarly, in (b), the reader may verify that the value of $\cos \theta$ in (13.4) is equal to the ratio between the volume of the p-dimensional parallelepiped spanned by the projection of $u, w_2, \ldots, w_p$ on V and the volume of the p-dimensional parallelepiped spanned by $u, w_2, \ldots, w_p$, namely

$$\cos \theta = \frac{\|u_V, w_2, \ldots, w_p\|}{\|u, w_2, \ldots, w_p\|}, \tag{13.5}$$

where u_V is the projection of u on V. (Here $\|\cdot, \ldots, \cdot\|$, without the subscript S, denotes the standard n-norm:

$$\| x_1, x_2, \ldots, x_n \| := \langle x_1, x_1 | x_2, \ldots, x_n \rangle^{\frac{1}{2}},$$

where $\langle \cdot, \cdot | \cdot, \ldots, \cdot \rangle$ is the standard n-inner product on X. We assume that $n \geq 2$. If $n = 1$, the standard 1-inner product is understood as the given inner product, while the standard 1-norm is the induced norm.)

To verify (13.5), we first observe that θ satisfies

$$\cos^2 \theta = \frac{\langle u, v | w_2, \ldots, w_p \rangle^2}{\|u, w_2, \ldots, w_p\|^2 \|v, w_2, \ldots, w_p\|^2}.$$

Indeed, writing $u = u_W + u_W^\perp$ and $v = v_W + v_W^\perp$ (where u_W and v_W are the projection of u and v, respectively, on $W = \operatorname{span}\{w_2, \ldots, w_p\}$), we obtain

$$\frac{\langle u, v | w_2, \ldots, w_p \rangle^2}{\|u, w_2, \ldots, w_p\|^2 \|v, w_2, \ldots, w_p\|^2} = \frac{\langle u_W^\perp, v_W^\perp \rangle^2 \|w_2, \ldots, w_p\|^4}{\|u_W^\perp\|^2 \|v_W^\perp\|^2 \|w_2, \ldots, w_p\|^4}$$

$$= \frac{\langle u_W^\perp, v_W^\perp \rangle^2}{\|u_W^\perp\|^2 \|v_W^\perp\|^2},$$

which we have defined in (13.4) as the value of $\cos^2 \theta$.

Next suppose that $u_V = \alpha v + \sum_{k=2}^p \beta_k w_k$. In particular, the scalar α is given by

$$\alpha = \frac{\langle u, v | w_2, \ldots, w_p \rangle}{\|v, w_2, \ldots, w_p\|^2}.$$

Hence we have

$$\| u_V, w_2, \ldots, w_p \|^2 = \langle u, u_V | w_2, \ldots, w_p \rangle = \alpha \langle u, v | w_2, \ldots, w_p \rangle$$

$$= \frac{\langle u, v | w_2, \ldots, w_p \rangle^2}{\|v, w_2, \ldots, w_p\|^2}.$$

Therefore, we obtain

$$\frac{\|u_V, w_2, \ldots, w_p\|^2}{\|u, w_2, \ldots, w_p\|^2} = \frac{\langle u, v | w_2, \ldots, w_p \rangle^2}{\|u, w_2, \ldots, w_p\|^2 \|v, w_2, \ldots, w_p\|^2} = \cos^2 \theta,$$

as claimed.

Motivated by the above observations, we may define the angle between a p-dimensional subspace $U = \operatorname{span}\{u_1, \ldots, u_p\}$ and a q-dimensional subspace

$V = \text{span}\{v_1, \dots, v_q\}$ (with $p \leq q$) such that the value of its cosine is equal to the ratio between the volume of the p-dimensional parallelepiped spanned by the projection of $u_1, \dots, u_p$ on V and the p-dimensional parallelepiped spanned by $u_1, \dots, u_p$. As we shall see later, the ratio is a number in $[0, 1]$ and is invariant under any change of basis for U and V, so that our definition of the angle makes sense.

Remember that $\langle x_1, x_1 | x_2, \dots, x_n \rangle = \det[\langle x_i, x_j \rangle]$ is nothing but the Gram's determinant generated by $x_1, x_2, \dots, x_n$. Geometrically, being the square root of the Gram's determinant, $\|x_1, \dots, x_n\|$ represents the volume of the n-dimensional parallelepiped spanned by $x_1, \dots, x_n$, where $\|x_1, \dots, x_n\| = 0$ if and only if $x_1, \dots, x_n$ are linearly dependent. Besides the Cauchy–Schwarz inequality for the n-inner product (which we have discussed in the previous section), we also have Hadamard's inequality which states that

$$\|x_1, \dots, x_n\| \leq \|x_1\| \cdots \|x_n\|$$

for every $x_1, \dots, x_n \in X$.

Next we recall that $\langle x_0, x_1 + x_1' | x_2, \dots, x_n \rangle = \langle x_0, x_1 | x_2, \dots, x_n \rangle$ for any linear combination x_1' of $x_2, \dots, x_n$. Thus, for instance, for $i = 0$ and 1, one may write $x_i = x_i^* + x_i^\perp$, where x_i^* is the projection of x_i on $\text{span}\{x_2, \dots, x_n\}$ and $x_i^\perp$ is its orthogonal complement, to get

$$\langle x_0, x_1 | x_2, \dots, x_n \rangle = \langle x_0^\perp, x_1^\perp | x_2, \dots, x_n \rangle = \langle x_0^\perp, x_1^\perp \rangle \|x_2, \dots, x_n\|^2.$$

(Here $\|x_2, \dots, x_n\|$ represents the volume of the $(n-1)$-parallelepiped spanned by $x_2, \dots, x_n$.)

Using the standard n-inner product and n-norm, we can derive an explicit formula for the projection of a vector x on the subspace spanned by $x_1, \dots, x_n$. Let $x^* = \sum_{k=1}^n \alpha_k x_k$ be the projection of x on $\text{span}\{x_1, \dots, x_n\}$. Taking the inner products of x^* and x_l, we get the following system of linear equations:

$$\sum_{k=1}^n \alpha_k \langle x_k, x_l \rangle = \langle x^*, x_l \rangle = \langle x, x_l \rangle, \quad l = 1, \dots, n.$$

By Cramer's rule together with properties of inner products and determinants, we obtain

$$\alpha_k = \frac{\langle x, x_k | x_{i_2(k)}, \dots, x_{i_n(k)} \rangle}{\|x_1, x_2, \dots, x_n\|^2}$$

where $\{i_2(k), \dots, i_n(k)\} = \{1, 2, \dots, n\} \setminus \{k\}, \quad k = 1, 2, \dots, n$.

We are now ready to define the angle between two subspaces. Let $U = \text{span}\{u_1, \dots, u_p\}$ and $V = \text{span}\{v_1, \dots, v_q\}$, both are subspaces of X with $p \leq q$. Using the standard n-norm (with $n = p$), we define the angle θ between U and V by

$$\cos \theta := \frac{\|\text{proj}_V u_1, \dots, \text{proj}_V u_p\|}{\|u_1, \dots, u_p\|}, \tag{13.6}$$

where $\text{proj}_V u_i$'s denote the projection of u_i's on V.

The following proposition convinces us that our definition makes sense.

Proposition 13.3 The ratio on the right hand side of (13.6) is a number in $[0, 1]$ and is independent of the choice of bases for U and V.

Proof: First note that the projection of u_i's on V is independent of the choice of basis for V. Further, the ratio is invariant under any change of basis for U, because projections are linear transformations. Indeed, the ratio is unchanged if we (a) swap u_i and u_j, (b) replace u_i by $u_i + \alpha u_j$, or (c) replace u_i by αu_i with $\alpha \neq 0$. Next, assuming particularly that $\{u_1, \dots, u_p\}$ is orthonormal, we have $\|u_1, \dots, u_p\| = 1$ and, by Hadamard's inequality,

$$\|\text{proj}_V u_1, \dots, \text{proj}_V u_p\| \leq \prod_{i=1}^{p} \|\text{proj}_V u_i\| \leq \prod_{i=1}^{p} \|u_i\| = 1.$$

This convinces us that the ratio is a number in $[0, 1]$, and the proof is complete. $\square$

13.4.1 An Explicit Formula

From (13.6), we can derive a explicit formula for the cosine in terms of $u_1, \dots, u_p$ and $v_1, \dots, v_q$, assuming for the moment that $\{v_1, \dots, v_q\}$ is orthonormal. For each $i = 1, \dots, p$, the projection of u_i on V is given by

$$\text{proj}_V u_i = \langle u_i, v_1 \rangle v_1 + \cdots + \langle u_i, v_q \rangle v_q.$$

So, for $i, j = 1, \dots, p$, we have

$$\langle \text{proj}_V u_i, \text{proj}_V u_j \rangle = \sum_{k=1}^{q} \langle u_i, v_k \rangle \langle u_j, v_k \rangle$$

Hence, we obtain

$$\|\text{proj}_V u_1, \dots, \text{proj}_V u_p\|^2 = \det \left[\sum_{k=1}^{q} \langle u_i, v_k \rangle \langle u_j, v_k \rangle \right] = \det(MM^{\mathsf{T}})$$

where $M := [\langle u_i, v_k \rangle]$ is a $p \times q$ matrix and M^{T} is its transpose (so that MM^{T} is a $p \times p$ matrix). The cosine of the angle θ between U and V is therefore given by the formula

$$\cos^2 \theta = \frac{\det(MM^{\mathsf{T}})}{\det[\langle u_i, u_j \rangle]}. \tag{13.7}$$

If $\{u_1, \dots, u_p\}$ happens to be orthonormal, then the formula (13.7) reduces to

$$\cos^2 \theta = \det(MM^{\mathsf{T}}).$$

Further, if $p = q$, then $\det(MM^{\mathrm{T}}) = \det M. \det M^{\mathrm{T}} = \det^2 M$. Hence, from the last formula, we get $\cos\theta = |\det M|$.

The reader might think that the angle defined by (13.6) coincides with the one formulated by Risteski and Trenčevski ([40, Equation 1.2]). However, it is not the case. Risteski and Trenčevski defined the angle θ between two subspaces $U = \mathrm{span}\{u_1, \dots, u_p\}$ and $V = \mathrm{span}\{v_1, \dots, v_q\}$ with $p \leq q$ by

$$\cos^2\theta := \frac{\det(MM^{\mathrm{T}})}{\det[\langle u_i, u_j \rangle]. \det[\langle v_k, v_l \rangle]}, \tag{13.8}$$

by first "proving" the following inequality ([40, Theorem 1.1]):

$$\det(MM^{\mathrm{T}}) \leq \det[\langle u_i, u_j \rangle]. \det[\langle v_k, v_l \rangle], \tag{13.9}$$

where $M := [\langle u_i, v_k \rangle]$. However, the inequality (13.9) is only true in the case (a) where $p = q$ (for which the inequality reduces to Kurepa's generalization of the Cauchy–Schwarz inequality, see [42]) or (b) where $\{v_1, \dots, v_q\}$ is orthonormal. Accordingly, (13.8) makes sense only in these two cases. In other cases, the value of the expression on the right hand side of (13.8) may be greater than 1.

To show that the inequality (13.9) is false in general, we may just take for example $X = \mathbb{R}^3$ (equipped with the usual inner product), $U = \mathrm{span}\{u\}$ where $u = (1, 0, 0)$, and $V = \mathrm{span}\{v_1, v_2\}$ where $v_1 = (\frac{1}{2}, \frac{1}{2}, 0)$ and $v_2 = (\frac{1}{2}, -\frac{1}{2}, \frac{1}{2})$. According to (13.9), we should have

$$\langle u, v_1 \rangle^2 + \langle u, v_2 \rangle^2 \leq \|u\|^2 \|v_1, v_2\|^2.$$

But we can compute that the left hand side of the inequality is equal to $\frac{1}{2}$, while the right hand side is equal to $\frac{3}{8}$, so that $\frac{1}{2} \leq \frac{3}{8}$, which cannot be true. This example shows that (13.9) is false even in the case where $\{u_1, \dots, u_p\}$ is orthonormal and $\{v_1, \dots, v_q\}$ is orthogonal (which is close to being orthonormal).

Let us consider the case where $p = 1$ and $q = 2$ more closely. For a unit vector u and an orthonormal set $\{v_1, v_2\}$ in X, it follows from our definition of the angle θ between $U = \mathrm{span}\{u\}$ and $V = \mathrm{span}\{v_1, v_2\}$ that

$$\cos^2\theta = \langle u, v_1 \rangle^2 + \langle u, v_2 \rangle^2 \leq 1.$$

Hence, for a nonzero vector u and an orthogonal set $\{v_1, v_2\}$ in X, we have

$$\cos^2\theta = \left\langle \frac{u}{\|u\|}, \frac{v_1}{\|v_1\|} \right\rangle^2 + \left\langle \frac{u}{\|u\|}, \frac{v_2}{\|v_2\|} \right\rangle^2.$$

Thus, for this case, we have

$$\langle u, v_1 \rangle^2 \|v_2\|^2 + \langle u, v_2 \rangle^2 \|v_1\|^2 \leq \|u\|^2 \|v_1, v_2\|^2,$$

where $\|v_1, v_2\|^2 = \|v_1\|^2 \|v_2\|^2$ is the area of the parallelogram spanned by v_1 and v_2.

More generally, let u be a nonzero vector and $\{v_1, v_2\}$ be a linearly independent set in X. To have an explicit formula for the cosine of the angle θ

between $U = \text{span}\{u\}$ and $V = \text{span}\{v_1, v_2\}$ in terms of u, v_1 and v_2, we do the following (avoiding the Gram–Schmidt process). Let u_V be the projection of u on V. Then u_V may be expressed as

$$u_V = \frac{\langle u, v_1 | v_2 \rangle}{\|v_1, v_2\|^2} v_1 + \frac{\langle u, v_2 | v_1 \rangle}{\|v_1, v_2\|^2} v_2,$$

where $\langle \cdot, \cdot | \cdot \rangle$ is the standard 2-inner product introduced earlier. Now write $u = u_V + u_V^\perp$ where $u_V^\perp$ is the orthogonal complement of u on V. Then

$$\cos^2\theta = \frac{\|u_V\|^2}{\|u\|^2} = \frac{\langle u, u_V \rangle}{\|u\|^2} = \frac{\langle u, v_1 \rangle \langle u, v_1 | v_2 \rangle + \langle u, v_2 \rangle \langle u, v_2 | v_1 \rangle}{\|u\|^2 \|v_1, v_2\|^2}.$$

$$(13.10)$$

Consequently, for any nonzero vector u and linearly independent set $\{v_1, v_2\}$, we have the following inequality

$$\langle u, v_1 \rangle \langle u, v_1 | v_2 \rangle + \langle u, v_2 \rangle \langle u, v_2 | v_1 \rangle \leq \|u\|^2 \|v_1, v_2\|^2. \qquad (13.11)$$

Here (13.10) and (13.11) serve as corrections for (13.8) and (13.9) for $p = 1$ and $q = 2$.

The inequality (13.11) may be viewed as a generalized Cauchy–Schwarz inequality. The difference between our approach and Risteski and Trenčevski's is that we derive the inequality as a consequence of the definition of the angle between two subspaces, while Risteski and Trenčevski use the "inequality" to define the angle between two subspaces. As long as $p = q$ or, otherwise, $\{v_1, \dots, v_q\}$ is orthonormal, their definition makes sense and of course agrees with ours.

13.4.2 A More General Formula

A more general formula for the cosine of the angle θ between a p-dimensional subspace $U = \text{span}\{u_1, \dots, u_p\}$ and a q-dimensional subspace $V = \text{span}\{v_1, \dots, v_q\}$ of X for arbitrary $p \leq q$ can be obtained as follows.

For each $i = 1, \dots, p$, the projection of u_i on V may be expressed as

$$\text{proj}_V u_i = \sum_{k=1}^{q} \alpha_{ik} v_k$$

where

$$\alpha_{ik} = \frac{\langle u_i, v_k | v_{i_2(k)}, \dots, v_{i_q(k)} \rangle}{\|v_1, v_2, \dots, v_q\|^2}$$

with $\{i_2(k), \dots, i_q(k)\} = \{1, 2, \dots, q\} \backslash \{k\}$, $\quad k = 1, 2, \dots, q$. Next observe that

$$\langle \text{proj}_V u_i, \text{proj}_V u_j \rangle = \langle u_i, \text{proj}_V u_j \rangle = \sum_{k=1}^{q} \alpha_{jk} \langle u_i, v_k \rangle$$

for $i, j = 1, \ldots, p$. Hence we have

$$
\| \operatorname{proj}_V u_i, \ldots, \operatorname{proj}_V u_p \|^2 =
\begin{vmatrix}
\sum_{k=1}^{q} \alpha_{1k} \langle u_1, v_k \rangle & \cdots & \sum_{k=1}^{q} \alpha_{pk} \langle u_1, v_k \rangle \\
\vdots & \ddots & \vdots \\
\sum_{k=1}^{q} \alpha_{1k} \langle u_p, v_k \rangle & \cdots & \sum_{k=1}^{q} \alpha_{pk} \langle u_p, v_k \rangle
\end{vmatrix}
$$

$$
= \frac{\det(M\tilde{M}^{\mathrm{T}})}{\| v_1, \ldots, v_q \|^{2p}}
$$

where

$$
M := [\langle u_i, v_k \rangle] \quad \text{and} \quad \tilde{M} := [\langle u_i, v_k | v_{i_2(k)}, \ldots, v_{i_q(k)}] \tag{13.12}
$$

with $i_2(k), \ldots, i_q(k)$ as above. (Note that $M\tilde{M}^{\mathrm{T}}$ is a $p \times p$ matrix.) Dividing both sides by $\| u_1, \ldots, u_p \|^2$, we get the following formula for the cosine:

$$
\cos^2 \theta = \frac{\det(M\tilde{M}^{\mathrm{T}})}{\det[\langle u_i, u_j \rangle] \cdot \det^p[\langle v_k, v_l \rangle]}, \tag{13.13}
$$

which serves as a correction for the formula (13.8).

Remark 13.2 Note that if $\{ v_1, \ldots, v_q \}$ is orthonormal, we get the formula (13.7) obtained earlier. As shown in [43], one may observe that the value of $\cos \theta$ is equal to the product of all critical value of $f(u) := \max_{v \in V, \|v\|=1} \langle u, v \rangle$, $u \in U, \|u\| = 1$.

As a consequence of our formula, we have the following generalization of the Cauchy–Schwarz inequality, which can be considered as a correction for (13.8).

Theorem 13.5 For two linearly independent sets $\{ u_1, \ldots, u_p \}$ and $\{ v_1, \ldots, v_q \}$ in X with $p \leq q$, we have the following inequality

$$
\det(M\tilde{M}^{\mathrm{T}}) \leq \det[\langle u_i, u_j \rangle] \cdot \det^p[\langle v_k, v_l \rangle],
$$

where M and $\tilde{M}$ are $p \times q$ matrices given by (13.12). Moreover, the equality holds if and only if the subspace spanned by $\{ u_1, \ldots, u_p \}$ is contained in the subspace spanned by $\{ v_1, \ldots, v_q \}$.

Proof: The inequality follows immediately since the ratio of the left hand side to the right hand side is equal to the square of the cosine of the angle between $U := \operatorname{span}\{ u_1, \ldots, u_p \}$ and $V := \operatorname{span}\{ v_1, \ldots, v_q \}$. Next, if U is contained in V, then the projection of u_i's on V are the u_i's themselves. Hence the equality holds since the cosine is equal to 1. If at least one of u_i's, say u_{i_0}, is not in V, then, assuming that $\{ u_1, \ldots, u_p \}$ and $\{ v_1, \ldots, v_q \}$ are orthonormal, the length of the projection of u_{i_0} on V will be strictly less than 1. In this case the cosine will be less than 1, and accordingly we have a strict inequality. □

Remark 13.3   As the reader might have realized, the formula (13.6) may also be used to define the angle between a finite p-dimensional subspace U and an infinite dimensional subspace V of X, assuming that the ambient space X is infinite dimensional and complete (i.e., X is an infinite dimensional Hilbert space). See [7] for an example.

References

1 Misiak, A. (1989) n-inner product spaces. *Math. Nachr.*, **140**, 299–319.

2 Gähler, S. (1969) Untersuchungen über verallgemeinerte m-metrische Räume. I. *Math. Nachr.*, **40**, 165–189.

3 Gähler, S. (1969) Untersuchungen über verallgemeinerte m-metrische Räume. II. *Math. Nachr.*, **40**, 229–264.

4 Gähler, S. (1970) Untersuchungen über verallgemeinerte m-metrische Räume. III. *Math. Nachr.*, **41**, 23–26.

5 Gunawan, H. (2001) On n-inner products, n-norms, and the Cauchy-Schwarz inequality. *Sci. Math. Jpn.*, **5**, 47–54.

6 Gunawan, H., Kikianty, E., Mashadi, S.G., Gemawati, S., and Sihwaningrum, I. (2006) Orthogonality in n-normed Spaces. Hibah Pekerti I Research Report, https://kamindo.files.wordpress.com/2009/08/gunemags-rev1.pdf (accessed 13 November 2017).

7 Gunawan, H., Neswan, O., and Setya-Budhi, W. (2005) A formula for angles between subspaces of inner product spaces. *Beitr. Algebra Geom.*, **46**, 311–320.

8 Diminnie, C., Gähler, S., and White, A. (1973) 2-inner product spaces. *Demonstratio Math.*, **6**, 525–536.

9 Diminnie, C., Gähler, S., and White, A. (1977) 2-inner product spaces II. *Demonstratio Math.*, **10**, 169–188.

10 Gähler, S. (1965) Lineare 2-normietre Räume. *Math. Nachr.*, **28**, 1–43.

11 Horn, R. and Johnson, C. (1985) *Matrix Analysis*, Cambridge University Press, New York, Cambridge.

12 Misiak, A. (1989) Orthogonality and orthonormality in n-inner product spaces. *Math. Nachr.*, **143**, 249–261.

13 Gunawan, H. and Mashadi, M. (2001) On n-normed spaces. *Int. J. Math. Math. Sci.*, **27**, 631–639.

14 Gantmacher, F.R. (1960) *The Theory of Matrices*, vol. **I**, Chelsea Publishing Company, New York, pp. 247–256.

15 Mitrinović, D.S., Pečarić, J.E., and Fink, A.M. (1993) *Classical and New Inequalities in Analysis*, Kluwer Academic Publishers, Dordrecht, pp. 595–603.

16 Cho, Y.J., Huang, N.J., and Long, X. (1995) Some fixed point theorems for nonlinear mappings in 2-Banach spaces. *Far East J. Math. Sci.*, **3** (2), 125–133.

17 Iseki, K. (1976) Fixed point theorems in Banach spaces. *Math. Semin. Notes, Kobe Univ.*, **2**, 11–13.

18 Khan, M.S. and Khan, M.D. (1993) Involutions with fixed points in 2-Banach spaces. *Int. J. Math. Math. Sci.*, **16** (3), 429–433.

19 Ekariani, S., Gunawan, H., and Idris, M. (2013) A contractive mapping on the *n*-normed space of *p*-summable sequences. *J. Math. Anal.*, **4**, 1–7.

20 Gunawan, H., Neswan, O., and Sukaesih, E. (2015) Fixed point theorems on bounded sets in an *n*-normed space. *J. Math. Anal.*, **6**, 51–58.

21 Gunawan, H. (2001) The space of *p*-summable sequences and its natural *n*-norm. *Bull. Austr. Math. Soc.*, **64**, 137–147.

22 Partington, J.R. (1986) Orthogonality in normed spaces. *Bull. Aust. Math. Soc.*, **33**, 449–455.

23 Alonso, J. (1994) Uniqueness properties of isosceles orthogonality in normed linear spaces. *Ann. Sci. Math. Québec*, **18** (1), 25–38.

24 Desbiens, J. (1987) Caractérisation d'un espace préhilbertien au moyen de la relation d'orthogonalité de Birkhoff-James. *Ann. Sc. Math. Québec*, **11** (2), 295–303.

25 Diminnie, C.R. (1983) A new orthogonality relation for normed linear spaces. *Math. Nachr.*, **114**, 197–203.

26 Guijarro, P. and Tomas, M.S. (1997) Perpendicular bisector and orthogonality. *Arch. Math.*, **69**, 491–496.

27 Miličić, P.M. (1998) On the Gram-Schmidt projection in normed spaces. *Univ. Beograd. Publ. Elektrotehn. Fak. Ser. Mat.*, **9**, 75–78.

28 Şerb, I. (1999) Rectangular modulus, Birkhoff orthogonality and characterizations of inner product spaces. *Comment. Math. Univ. Carolinae*, **40**, 107–119.

29 Cho, Y.J. and Kim, S.S. (1983) Gateaux derivatives and 2-inner product spaces. *Glas. Mat. Ser. III*, **27** (47), 197–203.

30 Khan, A. and Siddiqui, A. (1982) B-orthogonality in 2-normed space. *Bull. Calcutta Math. Soc.*, **74**, 216–222.

31 Cho, Y.J., Diminnie, C.R., Gähler, S., Freese, R.W., and Andalafte, E.Z. (1992) Isosceles orthogonal triple in linear 2-normed spaces. *Math. Nachr.*, **157**, 225–234.

32 Gunawan, H., Mashadi, M., Gemawati, S., Nursupiamin, and Sihwaningrum, I. (2006) Orthogonality in 2-normed spaces revisited. *Univ. Beograd. Publ. Elektrotehn. Fak. Ser. Mat.*, **17**, 76–83.

33 Godini, G. MR0743643 (85g:46028), MathSciNet: Mathematical Reviews on the Web.

34 Gunawan, H. (2002) Inner products on *n*-inner product spaces. *Soochow J. Math.*, **28**, 289–298.

35 Anderson, T.W. (1958) *An Introduction to Multivariate Statistical Analysis*, John Wiley & Sons, Inc., New York.

36 Davies, C. and Kahan, W. (1970) The rotation of eigenvectors by a perturbation. III. *SIAM J. Numer. Anal.*, **7**, 1–46.

37 Fedorov, S. (1997) Angle between subspaces of analytic and antianalytic functions in weighted L_2 space on a boundary of a multiply connected domain, in *Operator Theory, System Theory and Related Topics* (eds D. Alpay and V. Vinnikov), Beer-Sheva/Rehovot, pp. 229–256.

38 Knyazev, A.V. and Argentati, M.E. (2002) Principal angles between subspaces in an A-based scalar product: algorithms and perturbation estimates. *SIAM J. Sci. Comput.*, **23**, 2008–2040.

39 Rakočević, V. and Wimmer, H.K. (2003) A variational characterization of canonical angles between subspaces. *J. Geom.*, **78**, 122–124.

40 Risteski, I.B. and Trenčevski, K.G. (2001) Principal values and principal subspaces of two subspaces of vector spaces with inner product. *Beitr. Algebra Geom.*, **42**, 289–300.

41 Wimmer, H.K. (1999) Canonical angles of unitary spaces and perturbations of direct complements. *Linear Algebra Appl.*, **287**, 373–379.

42 Kurepa, S. (1966) On the Buniakowsky-Cauchy-Schwarz inequality. *Glasnik Mat. Ser. III*, **1** (21), 147–158.

43 Gunawan, H. and Neswan, O. (2005) On angles between subspaces of inner product spaces. *J. Indones. Math. Soc.*, **11**, 129–135.

14

Proximal Fiber Bundles on Nerve Complexes

James F. Peters[1, 2]

[1] *Computational Intelligence Laboratory, University of Manitoba, WPG, MB, R3T 5V6, Canada*
[2] *Department of Mathematics, Faculty of Arts and Sciences, Adiyaman University, 02040 Adiyaman, Turkey*

2010 AMS Subject Classification 55R10, 55U10, 40C05, 40D05

14.1   Brief Introduction

This chapter reflects recent work on proximity spaces [1–3], Edeslbrunner–Harer nerve complexes [2, 4, 5], fiber bundles in physical geometry [6], and related work in computational topology [1, 7, 8]. Classes of all half-lines with the same endpoint are called bundles in space geometry [9, Section 4]. In a descriptive proximal physical geometry, the focus shifts from a spatial to a descriptive source of bundles. In physical geometry, the focus shifts from a spatial to a descriptive source of fibre bundles. In general, a continuous mapping $\pi : E \longrightarrow B$ is called a *projection* by Zisman [10] (also called a *fibre bundle* by Chern [11, Section 6, p. 683]), where E is the total space, and B is the base space. Luke and Mishchenko [12] call E the fibre space. For any $x \in B$, $\pi^{-1}(x) = e \in E$ is a *fiber* of the map π. The triple (E, π, B) is a *fiber space*, provided E and B are topological spaces and π is surjective and continuous [13].

The arrow diagram

$$X \xrightarrow{\ f\ } E$$
$$\downarrow \pi$$
$$B$$

in Figure 14.1 represents a *spatial fiber space* (E, π, B) in which E, B are topological spaces and $\pi : E \longrightarrow B$ is surjective and continuous. BreMiller and Sloyer [13] define (E, π, B) to be a *sheaf*, provided $\pi : E \longrightarrow B$ is a local surjective homeomorphism, that is, $\pi : E \longrightarrow B$ is a continuous, 1-1 mapping on E onto B. For $x \in B$, $\pi^{-1}(x) = e \in E$ is a fiber over x and $\pi^{-1}(B) = E$ is a fiber bundle over B.

Mathematical Analysis and Applications: Selected Topics, First Edition.
Edited by Michael Ruzhansky, Hemen Dutta, and Ravi P. Agarwal.
© 2018 John Wiley & Sons, Inc. Published 2018 by John Wiley & Sons, Inc.

$$X \xrightarrow{\ f\ } E \qquad\qquad \mathrm{Re}A \xrightarrow{\ f\ } \mathscr{R}\mathrm{re}A$$
$$\Big\downarrow \pi \qquad\qquad\qquad\qquad \Big\downarrow \Phi$$
$$B \qquad\qquad\qquad\qquad B \subset \mathbb{R}^n$$

Figure 14.1 Two forms of fiber bundles: spatial: $\pi : E \longrightarrow B$ and descriptive: $\Phi : \mathscr{R}_{\mathrm{re}A} \longrightarrow \mathscr{B} \subset \mathbb{R}^n$.

Let $\mathrm{Re}A$ be a set of regions $\mathrm{re}A$, $\mathscr{R}_{\mathrm{re}A}$ a class of regions, $B \subset \mathbb{R}^n$, a subset in an n-dimensional feature space. In addition, let $f : \mathrm{Re}A \longrightarrow \mathscr{R}_{\mathrm{re}A}$, $\pi : \mathscr{R}_{\mathrm{re}A} \longrightarrow \mathbb{R}^n$ be continuous mappings such that

$$\mathrm{Re}A \xmapsto{\ f\ } \mathscr{R}_{\mathrm{re}A} \xmapsto{\ \pi\ } B \subset \mathbb{R}^n.$$

For region $\mathrm{re}A \in \mathscr{R}_{\mathrm{re}A}$, let $\mathscr{R}_{\mathrm{re}A} \xmapsto{\ \pi\ } \mathbb{R}^n$ is defined by

$$\pi(\mathrm{re}A) = (\varphi_1(\mathrm{re}A), \dots, \varphi_i(\mathrm{re}A), \dots, \varphi_n(\mathrm{re}A)) : \varphi_i(\mathrm{re}A) \in \mathbb{R}.$$

The projection $\pi(\mathrm{re}A)$ (also written $\Phi(\mathrm{re}A)$ in the sequel) is a feature vector that provides a *description* of region $\mathrm{re}A$. In a descriptive proximal physical geometry, f maps a collection of regions $\mathrm{Re}A$ to a class of regions $\mathscr{R}_{\mathrm{re}A}$, which is projected onto $B \subset \mathbb{R}^n$ by Φ. In effect, $\Phi(f(\mathrm{Re}A)) \in \mathbb{R}^n$.

Let $\mathrm{re}X \in \mathscr{R}_{\mathrm{re}A}$ and let $\Phi(\mathrm{re}X)$ be a feature vector that describes region $\mathrm{re}X$ in the class of regions $\mathscr{R}_{\mathrm{re}A}$. The continuous mapping $\Phi : \mathscr{R}_{\mathrm{re}A} \longrightarrow B$ is a *projection* on the class of regions $\mathscr{R}_{\mathrm{re}A}$ onto B such that $(\mathscr{R}_{\mathrm{re}A}, \Phi, B)$ is a *descriptive fiber space* and the set of feature vectors $B \subset \mathbb{R}^n$ is the *descriptive base space*. For x in B, $\Phi^{-1}(x) \in \mathscr{R}_{\mathrm{re}A}$ is a fiber over x. The leads to a descriptive fiber bundle which is a BreMiller–Sloyer sheaf for a class of regions. In particular, we consider $\mathrm{re}A$ to be a collection of filled triangles (called a Vietoris–Rips complex) that covers a planar region. Subsets of $\mathrm{re}A$ that have a vertex in common are called nerve complexes. In the sequel, we consider fiber bundles that are projection mappings on nerve complexes (fibers) onto a stitched-together set of nerves called nervous system complexes. Because of our interest in the closeness of nerve complexes, such fiber bundles are considered in the context of proximity spaces.

14.2 Preliminaries

This section briefly introduces nerve complexes, proximities, and region sewing.

14.2.1 Nerve Complexes and Nerve Spokes

Borsuk [14] was one of the first to suggest studying sequences of plane shapes in his theory of shapes.

Borsuk also observed that every polytope can be decomposed X into a finite sum of elementary simplexes, which he called brics. A *polytope* is the intersection of finitely many closed half spaces [15]. This leads to a simplicial complex K covered by simplexes $\Delta_1, \ldots, \Delta_n$ (filled triangles) such that the nerve of the decomposition is the same as K [16]. Briefly, a *geometric simplicial complex* (denoted by $\Delta(S)$ or simply by Δ) is the convex hull of a set of points S, that is, the smallest convex set containing S. Geometric simplexes in this paper are restricted to vertices (0-simplexes), line segments (1-simplexes), and filled triangles (2-simplexes) in the Euclidean plane, since our main interest is in the extraction of features of simplexes superimposed on planar digital images. In this paper, we consider only what is known as a Vietoris–Rips complex, which is a collection of 2-simplices determined by subsets of 3 points in a set of points in the Euclidean plane [7]. An important form of simplicial complex is a collection of simplexes called a nerve.

A planar simplicial complex K is a *Alexandroff–Borsuk nerve*, provided the simplexes in K have nonempty intersection (called the nucleus of the nerve). A nerve of a simplicial complex K (denoted by $\mathrm{Nrv}K$) in the triangulation of a plane region is defined by $\mathrm{Nrv}K = \{\Delta \subseteq K : \bigcap \Delta \neq \emptyset\}$ (Nerve complex). In other words, the simplexes in a nerve have proximity to each other, since they share the nucleus. The *nucleus* of a nerve complex is a vertex common to the 2-simplexes in a nerve. Triangulation of point clouds in the plane provides a straightforward basis for the study of nerve complexes. A *spoke A* (denoted by $\mathrm{sk}A$) on a nerve complex is a 2-simplex in the nerve.

Example 14.1 Let X be a planar triangulated region containing a nerve complex $\mathrm{Nrv}K$. Each filled triangle in $\mathrm{Nrv}K$ is a spoke. For example, *skA* in Figure 14.2 is a spoke in $\mathrm{Nrv}K$.

The study of nerves was introduced by Alexandroff [17], elaborated by Borsuk [16], Leray [18], and a number of others such as Adamaszek *et al.* [19], de Verdière *et al.* [20], Edelsbrunner and Harer [21], and more recently by Adamaszek *et al.* [22].

Figure 14.2 NrvK.

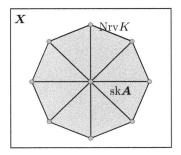

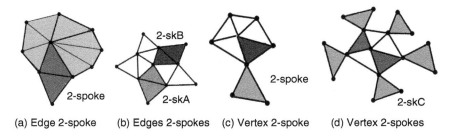

Figure 14.3 Two forms of nerve 2-spokes.

An *n*-**nerve spoke** (denoted by $n - \mathrm{sk}H, n \geq 1$) is a collection of connected 2-simplexes that includes a 2-simplex contained in nerve complex. Each nerve spoke extends outward from the nucleus of a nerve Nrv$K$, giving rise to 1-spokes (2-simplexes within a nerve Nrv$K$), 2-spokes (a 2-simlex that has either a vertex or an edge in common with a 1-spoke in Nrv$K$), 3-spokes (a 2-simlex that has either a vertex or an edge in common with a 2-spoke in Nrv$K$),..., and *n*-spokes (a 2-simlex that has either a vertex or an edge in common with an $(n - 1)$-spoke in Nrv$K$). The focus here is on 2-spokes.

Example 14.2 (**Nerve complex 2-spokes**) Two forms of 2-spokes are shown in Figure 14.3. Let Nrv$K$ be a planar nerve. An edge-based 2-spoke consists of a nerve Nrv$K$ 1-spoke that has an edge in common with a nonnerve 1-spoke (see, e.g., Figure 14.3(a) a). A complete collection of edge-based 2-spokes on Nrv$K$ is shown in Figure 14.3(b)b.

An vertex-based 2-spoke consists of a nerve Nrv$K$ 1-spoke that has a vertex in common with a nonnerve 1-spoke (see, e.g., Figure 14.3(c)a). A complete collection of vertex-based 2-spokes on Nrv$K$ is shown in Figure 14.3(d)b.

A **nervous system complex** K (denoted by Nrvsys$K$) is collection of nerves that share a k-spoke.

Example 14.3 (**Nervous system complex**) Nerve complexes Nrv$A$ and Nrv$B$ in Figure 14.4 have a 2-spoke in common. Hence,

$$\mathrm{Nrvsys}K = \mathrm{Nrv}A \cup \mathrm{Nrv}B$$

is an example of a nervous system complex. In the case, the 2-spoke sk$H$ defined by

$$\mathrm{sk}H = \Delta A \cup \Delta B,$$

which is formed by the pair 1-complexes ΔA, ΔB shown in Figure 14.5. □

Figure 14.4 Nrvsys*K*.

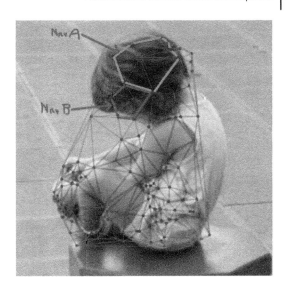

Figure 14.5 Strongly near filled triangles.

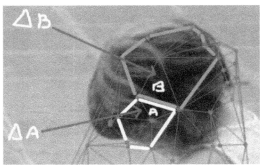

14.2.2   Descriptions and Proximities

This section briefly introduces two basic types of proximities, namely, traditional *spatial proximity* and the more recent *descriptive proximity* in the study of computational proximity [2]. Nonempty sets that have *spatial proximity* are close to each other, either asymptotically or with common points. Sets with points in common are strongly proximal. Nonempty sets that have *descriptive proximity* are close, provided the sets contain one or more elements that have matching descriptions. A commonplace example of descriptive proximity is a pair of paintings that have matching parts such as matching facial characteristics, matching eye, hair, skin color, or matching nose, mouth, ear shape. Each of these proximities has a strong form. A *strong proximity* embodies a special form of tightly twisted nextness of nonempty sets. In simple terms, this means sets that share elements, have strong proximity.

Example 14.4   From Figure 14.4, the pair of nerves NrvA, NrvB exhibit strong proximity, since there is a 2-spoke that overlaps the nerves. To see this, consider triangles ΔA and ΔB in Figure 14.5 and let skE be a 2-spoke in NrvA defined by skE $= \Delta A \cup \Delta B$. Similarly, NrvB has a 2-spoke skH, also defined by skH $= \Delta A \cup \Delta B$. In other words, nerves NrvA, NrvB overlap due to their 2-spokes skE, skH, respectively. In effect, ΔA, ΔB are strongly near, since these triangles belong to overlapping 2-spokes.

Proximities are nearness relations. In other words, a *proximity* between nonempty sets is a mathematical expression that specifies the closeness of the sets. A *proximity space* results from endowing a nonempty set with one or more proximities. Typically, a proximity space is endowed with a common proximity such as the proximities from Čech [23], Efremovič [24], Lodato [25], and Wallman [26], or the more recent descriptive proximity [27–29].

A pair of nonempty sets in a proximity space are *near* (*close to each other*), provided the sets have one or more points in common or each set contains one or more points that are sufficiently close to each other. Let X be a nonempty set, $A, B, C \subset X$. E. Čech [23] introduced axioms for the simplest form of proximity δ_C, which satisfies

Čech Proximity Axioms [23, Section 2.5, p. 439]

(P1) $\emptyset \, \delta A, \forall A \subset X$.
(P2) $A \delta B \Rightarrow B \delta A$.
(P3) $A \cap B \neq \emptyset \Rightarrow A \delta B$.
(P4) $A \delta (B \cup C) \Leftrightarrow A \delta B$ or $A \delta C$.

The Lodato proximity δ_L satisfies the Čech proximity axioms and axiom (P5).

Lodato Proximity Axiom [25]

(P5) $A \delta_L B$ and $\{b\} \delta_L C$ for each $b \in B \Rightarrow A \delta_L C$.

We can associate a topology with the space (X, δ) by considering as closed sets those sets that coincide with their own closure.

Nonempty sets A and B in a topological space X equipped with the proximity $\overset{\wedge\wedge}{\delta}$ are *strongly near* [*strongly contacted*] (denoted $A \overset{\wedge\wedge}{\delta} B$), provided the sets have at least one point in common. The strong contact relation $\overset{\wedge\wedge}{\delta}$ was introduced in [30] and axiomatized in [31], [32, Section 6 Appendix].

Strong Proximity [33, Section 1.2] (see, also, [2, Section 1.5], [30, 34]).

Let X be a topological space, $A, B, C \subset X$ and $x \in X$. The relation $\overset{\wedge\wedge}{\delta}$ on the family of subsets 2^X is a *strong proximity*, provided it satisfies the following axioms:

(snN0) $\emptyset \overset{\wedge\wedge}{\delta} A, \forall A \subset X$, and $X \overset{\wedge\wedge}{\delta} A, \forall A \subset X$.

(snN1) $A \overset{\wedge\wedge}{\delta} B \Leftrightarrow B \overset{\wedge\wedge}{\delta} A$.

(snN2) $A \overset{\wedge\wedge}{\delta} B$ implies $A \cap B \neq \emptyset$.

(snN3) If $\{B_i\}_{i \in I}$ is an arbitrary family of subsets of X and $A \overset{\wedge\wedge}{\delta} B_{i^*}$ for some $i^* \in I$ such that $\text{int}(B_{i^*}) \neq \emptyset$, then $A \overset{\wedge\wedge}{\delta} (\bigcup_{i \in I} B_i)$

(snN4) $\text{int } A \cap \text{ int } B \neq \emptyset \Rightarrow A \overset{\wedge\wedge}{\delta} B$.

When we write $A \overset{\wedge\wedge}{\delta} B$, we read A is *strongly near B* (*A strongly contacts B*). The notation $A \overset{\wedge\wedge}{\not\delta} B$ reads A is not strongly near B (A does not *strongly contact B*). For each *strong proximity* (*strong contact*), we assume the following relations:

(snN5) $x \in \text{int}(A) \Rightarrow x \overset{\wedge\wedge}{\delta} A$

(snN6) $\{x\} \overset{\wedge\wedge}{\delta} \{y\} \Leftrightarrow x = y$

For strong proximity of the nonempty intersection of interiors, we have that $A \overset{\wedge\wedge}{\delta} B \Leftrightarrow \text{int } A \cap \text{int } B \neq \emptyset$ or either A or B is equal to X, provided A and B are not singletons; if $A = \{x\}$, then $x \in \text{int}(B)$, and if B too is a singleton, then $x = y$. It turns out that if $A \subset X$ is an open set, then each point that belongs to A is strongly near A. The bottom line is that strongly near sets always share points, which is another way of saying that sets with strong contact have nonempty intersection. Let δ denote a traditional proximity relation [35].

Proposition 14.1 Let $\text{Nrv}A$, $\text{Nrv}B$ be nerve complexes in a triangulated space X. $\text{Nrv}A \overset{\wedge\wedge}{\delta} \text{Nrv}B$, if and only if 2-spoke $\text{sk}E \in \text{Nrv}A \cap \text{Nrv}B$ for some 2-spoke common to the pair of nerves.

Corollary 14.1 Let $\text{Nrv}A$ and $\text{Nrv}B$ be nerve complexes in a triangulated space X. A 3-spoke $\text{sk}H \in \text{Nrv}A \cup \text{Nrv}B$ for some 3-spoke common to the pair of nerves.

Proof: Immediate from Proposition 14.1 and the definition of a 3-spoke. □

14.2.3 Descriptive Proximities

In the run-up to a close look at extracting features of triangulated image objects, we first consider descriptive proximities, fully covered in [36] and briefly introduced, here. There are two basic types of *object features*, namely, *object characteristic*, and *object location*. For example, an object characteristic of a picture point is color. Descriptive proximities resulted from the introduction of the descriptive intersection pairs of nonempty sets.

> **Descriptive intersection [29] and [37, Section 4.3, p. 84].**

(Φ) $\Phi(A) = \{\Phi(x) \in \mathbb{R}^n : x \in A\}$, set of feature vectors.

($\underset{\Phi}{\cap}$) $A \underset{\Phi}{\cap} B = \{x \in A \cup B : \Phi(x) \in \Phi(A) \& \in \Phi(x) \in \Phi(B)\}$.

The descriptive proximity δ_Φ was introduced in [27–29]. Let $\Phi(x)$ be a feature vector for $x \in X$, a nonempty set of nonabstract points such as picture points. $A\,\delta_\Phi\,B$ reads A is descriptively near B, provided $\Phi(x) = \Phi(y)$ for at least one pair of points, $x \in A, y \in B$. The proximity δ in the Čech, Efremovič, and Wallman proximities is replaced by δ_Φ. Then swapping out δ with δ_Φ in each of the Lodato axioms defines a descriptive Lodato proximity that satisfies the following axioms.

> **Descriptive Lodato Axioms [1, Section 4.15.2]**

(dP0) $\emptyset\,\delta_\Phi\,A, \forall A \subset X$.
(dP1) $A\delta_\Phi B \Leftrightarrow B\delta_\Phi A$.
(dP2) $A \underset{\Phi}{\cap} B \neq \emptyset \Rightarrow A\delta_\Phi B$.
(dP3) $A\,\delta_\Phi\,(B \cup C) \Leftrightarrow A\,\delta_\Phi\,B$ or $A\,\delta_\Phi\,C$.
(dP4) $A\delta_\Phi B$ and $\{b\}\delta_\Phi C$ for each $b \in B \Rightarrow A\delta_\Phi C$.

Nonempty sets A, B in a proximity space X are *strongly near* (denoted $A \overset{\wedge\wedge}{\delta} B$), provided the sets share points. Strong proximity $\overset{\wedge\wedge}{\delta}$ was introduced in [33, Section 2] and completely axiomatized in [31] (see, also, [32, Section 6 Appendix]).

Proposition 14.2 Let (X, δ_Φ) be a descriptive proximity space, $A, B \subset X$. Then $A\,\delta_\Phi\,B \Rightarrow A \underset{\Phi}{\cap} B \neq \emptyset$.

Proof: $A\,\delta_\Phi\,B \Leftrightarrow$ there is at least one $x \in A, y \in B$ such that $\Phi(x) = \Phi(y)$ (by definition of $A\,\delta_\Phi\,B$). Hence, $A \underset{\Phi}{\cap} B \neq \emptyset$. $\qquad\square$

Next, consider a proximal form of a Száz relator [38]. A *proximal relator* $\mathcal{R}$ is a set of relations on a nonempty set X [39]. The pair $(X, \mathcal{R})$ is a proximal relator space. The connection between $\overset{\wedge\wedge}{\delta}$ and δ is summarized in Proposition 14.1.

Lemma 14.1 Let $(X, \{\delta, \delta_\Phi, \overset{\wedge\wedge}{\delta}\})$ be a proximal relator space, $A, B \subset X$. Then

1° $A \overset{\wedge\wedge}{\delta} B \Rightarrow A\delta B$.

2° $A \overset{\wedge\wedge}{\delta} B \Rightarrow A\delta_\Phi B$.

Figure 14.6 sk$A$ $\delta_{\Phi}^{\wedge\wedge}$ sk$B$.

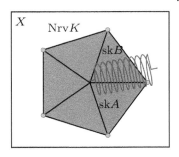

Proof:

1°: From Axiom (snN2), $A \stackrel{\wedge\wedge}{\delta} B$ implies $A \cap B \neq \emptyset$, which implies $A \delta B$ (from Lodato Axiom (P2)).

2°: From 1°, there are $x \in A, y \in B$ common to A and B. Hence, $\Phi(x) = \Phi(y)$, which implies $A \underset{\Phi}{\cap} B \neq \emptyset$. Then, from the descriptive Lodato Axiom (dP2), $A \underset{\Phi}{\cap} B \neq \emptyset \Rightarrow A \delta_{\Phi} B$. This gives the desired result. $\square$

Theorem 14.1 Let $(X, \{\delta_{\Phi}, \stackrel{\wedge\wedge}{\delta}\})$ be a proximal relator triangulated space, $\mathrm{Nrv}A, \mathrm{Nrv}B \subset 2^{X}$. Then

1° $\mathrm{Nrv}A \stackrel{\wedge\wedge}{\delta} \mathrm{Nrv}B$ implies $\mathrm{Nrv}A \delta_{\Phi} \mathrm{Nrv}B$.

2° A 2-spoke sk$A \in \mathrm{Nrv}A \cap \mathrm{Nrv}B$ implies if sk$A \in \mathrm{Nrv}A \underset{\Phi}{\cap} \mathrm{Nrv}B$.

3° A 2-spoke sk$A \in \mathrm{Nrv}A \cap \mathrm{Nrv}B$ implies $\mathrm{Nrv}A \delta_{\Phi} \mathrm{Nrv}B$.

Proof:

1°: Immediate from part 2° in Lemma 14.1.

2°: From Proposition 14.1, sk$A \in \mathrm{Nrv}A \cap \mathrm{Nrv}B$, if and only if $\mathrm{Nrv}A$ $\stackrel{\wedge\wedge}{\delta}$ $\mathrm{Nrv}B$. Consequently, there are members of the 2-spoke sk$A$ common to $\mathrm{Nrv}A, \mathrm{Nrv}B$, which have the same description. Hence, sk$A \in \mathrm{Nrv}A \underset{\Phi}{\cap} \mathrm{Nrv}B$.

3°: Immediate from 2° and Lemma 14.1.

Example 14.5 Let X be a topological space endowed with the strong proximity $\stackrel{\wedge\wedge}{\delta}$ and $A = \{(x, 0) : 0.1 \leq x \leq 1\}, B = \left\{(x, \frac{1}{x}sin(13/x)) : 0.1 \leq x \leq 1\right\}$. In this case, A, B represented by Figure 14.7 are strongly near sets with many points in common.

The descriptive strong proximity $\delta_{\Phi}^{\wedge\wedge}$ is the descriptive counterpart of $\stackrel{\wedge\wedge}{\delta}$.

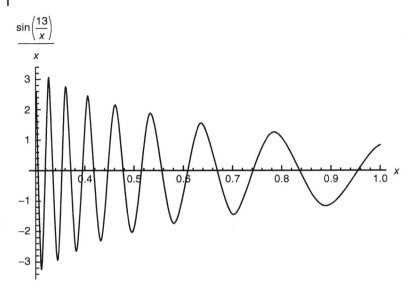

Figure 14.7 Strongly near.

Definition 14.1 Let X be a topological space, $A, B, C \subset X$ and $x \in X$. The relation $\delta_\Phi^{\wedge}$ on the family of subsets 2^X is a *descriptive strong Lodato proximity*, provided it satisfies the following axioms.

Descriptive Strong Lodato proximity [1, Section 4.15.2]

(dsnN0) $\emptyset \; \overset{\emptyset}{\underset{\wedge}{\delta}} \; A, \forall A \subset X$, and $X \, \delta_\Phi^{\wedge} \, A, \forall A \subset X$

(dsnN1) $A \; \delta_\Phi^{\wedge} \; B \Leftrightarrow B \, \delta_\Phi^{\wedge} \, A$

(dsnN2) $A \; \delta_\Phi^{\wedge} \; B \Rightarrow A \underset{\Phi}{\cap} B \neq \emptyset$

(dsnN3) If $\{B_i\}_{i \in I}$ is an arbitrary family of subsets of X and $A \; \delta_\Phi^{\wedge} \; B_{i^*}$ for some $i^* \in I$ such that $\text{int}(B_{i^*}) \neq \emptyset$, then $A \; \delta_\Phi^{\wedge} \; (\bigcup_{i \in I} B_i)$

(dsnN4) $\text{int} A \underset{\Phi}{\cap} \text{int} B \neq \emptyset \Rightarrow A \; \delta_\Phi^{\wedge} \; B$

When we write $A \; \delta_\Phi^{\wedge} \; B$, we read A is *descriptively strongly near B*. The notation $A \; \overset{\emptyset}{\underset{\wedge\Phi}{}} \; B$ reads A is not descriptively strongly near B. For each *descriptive strong proximity*, we assume the following relations:

(dsnN5) $\Phi(x) \in \Phi(\text{int}(A)) \Rightarrow x \, \delta_\Phi^{\wedge} \, A$

(dsnN6) $\{x\} \, \delta_\Phi^{\wedge} \, \{y\} \Leftrightarrow \Phi(x) = \Phi(y)$

So, for example, if we take the strong proximity related to nonempty intersection of interiors, we have that $A \; \delta_\Phi^{\wedge} \; B \Leftrightarrow \text{int} A \underset{\Phi}{\cap} \text{int} B \neq \emptyset$ or either A or B is equal to X, provided A and B are not singletons; if $A = \{x\}$, then $\Phi(x) \in \Phi(\text{int}(B))$, and if B is also a singleton, then $\Phi(x) = \Phi(y)$.

Example 14.6 [**Descriptive strong proximity**] Let X be a triangulated space of picture points represented in Figure 14.5 and let $\Phi : X \longrightarrow \mathbb{R}^n$ be a description of X representing the color of a picture point, where 0 stands for red (r), 1 for brown (b), and 2 for yellow (y). Suppose the range is endowed with the topology given by $\tau = \{\emptyset, \{r,b\}, \{r,b,y\}\}$. Then, $\Delta A \; \delta_\Phi^{\curlywedge} \; \Delta B$, since $\mathrm{int}\Delta A \underset{\Phi}{\cap} \mathrm{int}\Delta B \neq \emptyset$, that is, points in the interior of simplexes $\Delta A, \Delta B$ have matching colors.

Example 14.7 **Nerve spokes with descriptive strong proximity** Let X be a planar triangulated region containing a nerve complex $\mathrm{Nrv}K$, equipped with $\delta_\Phi^{\curlywedge}$. A pair of spokes $\mathrm{sk}A$ and $\mathrm{sk}B$ in a nerve complex $\mathrm{Nrv}K$ is shown in Figure 14.6. $\mathrm{sk}A \; \overset{\curlywedge}{\delta} \; \mathrm{sk}B$, since $\mathrm{sk}A, \mathrm{sk}B$ have common wiring represented by an overlapping coil. Hence, from Lemma 14.1, $\mathrm{sk}A \, \delta_\Phi \, \mathrm{sk}B$. From Axiom (dsnN4), $\mathrm{sk}A \, \delta_\Phi^{\curlywedge} \, \mathrm{sk}B$, since $\mathrm{int}\,\mathrm{sk}A \underset{\Phi}{\cap} \mathrm{int}\,\mathrm{sk}B$.

14.3 Sewing Regions Together

This section introduces a sewing operation. The basic idea is to introduce an edge L between a pair of parallel regions $\mathrm{re}A$, $\mathrm{re}B$ so that $\mathrm{re}A$, $\mathrm{re}B$ are connected and a simplicial complex is constructed. That is, $\mathrm{re}A \; \overset{\curlywedge}{\not\delta} \; \mathrm{re}B$ (the regions do not overlap, even if they are extended indefinitely). Let L have vertices p, q so that $L = \overline{p,q}$ and let $\mathrm{re}A \; \overset{\curlywedge}{\delta} \; p$ (region $\mathrm{re}A$ and p overlap). Further, let $\mathrm{re}A \; \overset{\curlywedge}{\delta} \; q$ (region $\mathrm{re}A$ and q overlap). Then

$$\mathrm{re}A \; \overset{\curlywedge}{\delta} \; L \text{ and } \mathrm{re}B \; \overset{\curlywedge}{\delta} \; L,$$

forming a simplicial complex. In general, nonoverlapping regions are sewn together by joining the regions by one or more edges. Let $k \in \mathbb{N}$ be a natural number. Let $2^{\mathbb{R}^2}$ be a collection of plane regions. From this, the planar mapping

$$sew : 2^{\mathbb{R}^2} \times 2^{\mathbb{R}^2} \times \mathbb{N} \longrightarrow 2^{\mathbb{R}^2}$$

is defined by

$$sew\left(\mathrm{re}A, \, \mathrm{re}B, k\right) = \left\{\mathrm{re}A, \, \mathrm{re}B\right\}$$
$$\cup \bigcup_{i=1}^{k} \left\{a_i \in \mathrm{re}A, \, b_i \in \mathrm{re}B, \{p,q\} \in L : a_i \overset{\curlywedge}{\delta} p \, \& \, b_i \overset{\curlywedge}{\delta} q\right\}$$

By sewing a pair of plane regions together with $sew(\mathrm{re}A, \mathrm{re}B, 1)$, $\mathrm{re}A \; \overset{\curlywedge}{\delta} \; p$, $\mathrm{re}B$ $\overset{\curlywedge}{\delta} \; q$, we mean an edge is added between a subregion $\mathrm{re}a \in \mathrm{re}A$ and vertex p in line $\overline{p,q}$ and a subregion $\mathrm{re}b \in \mathrm{re}B$ and vertex q in line $\overline{p,q}$.

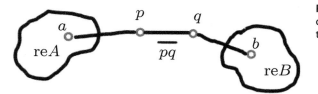

Figure 14.8 Sewing disconnected regions together.

Example 14.8 **(Stitching Regions Together)** Let regions reA, reB contain subregions re a, re b, respectively, as shown in Figure 14.8. Assume that reA, reB are disconnected, that is, there is no path between reA, reB. The sew(reA, reB, $k = 1$) operation transforms reA, reB into connected regions by introducing at least one edge $\overline{p,q}$ so that vertex p is connected to a subregion re a and vertex q is connected to a subregion re b as shown in Figure 14.8.

Lemma 14.2 sew(reA, reB) constructs a simplicial complex.

Proof: sew(reA, reB, k) is defined by a set of edges $\{\overline{p,q}\}$ such that each the vertices of edge are connected to subregions in reA, reB. Without loss of generality, let vertex p in edge $\overline{p,q}$ be connected to subregion re $a \in$ reA and let vertex q in edge $\overline{p,q}$ be connected to subregion re $b \in$ reB. By definition, reA $\overset{\wedge}{\delta}$ p & reA $\overset{\wedge}{\delta}$ q. Hence, reA $\cup \overline{p,q} \cup$ reB is a simplicial complex. If regions reA, reB are connected to $k > 0$ edges as a result of sew(reA, reB, k), then sew(reA, reB, k) constructs a simplicial complex with multiple connecting edges. This gives the desired result. □

Lemma 14.3 Let reA, reB be simplicial complexes. If sew(reA, reB, k), then the resulting region is a simplicial complex.

Proof: From Lemma 14.2, sew(reA, reB, k) is a simplicial complex. Let sew(reA, reB, k) contain a line $\overline{p,q}$ so that reA $\overset{\wedge}{\delta}$ p & reA $\overset{\wedge}{\delta}$ q, that is, vertex p is connected to a subregion re $a \in$ reA and vertex q is connected to a subregion re $b \in$ reB. This gives the desired result. □

Theorem 14.2 Let reX $\in \mathscr{R}_{\text{reA}}$ and let Φ(reX) be a feature vector for region reX in the class of regions $\mathscr{R}_{\text{reA}}$, $x \in \mathbb{R}^n$. Assume that every reX in $\mathscr{R}_{\text{reA}}$ has a unique feature, $\Phi : \mathscr{R}_{\text{reA}} \longrightarrow B \subset \mathbb{R}^n$ is a local surjective homeomorphism in which $\mathscr{R}_{\text{reA}}$, B are topological spaces and $\Phi^{-1}(x) \in \mathscr{R}_{\text{reA}}$ for $x \in B$. Then

1° $(\mathscr{R}_{\text{reA}}, \Phi, B)$ is a BreMiller–Sloyer sheaf.
2° $\Phi^{-1}(B)$ is a descriptive fiber bundle over B.

Proof:
1°: $(\mathscr{R}_{\text{reA}}, \Phi, B)$ is, by definition, a BreMiller–Sloyer sheaf.

2°: Let $\mathrm{re}X \in \mathscr{R}_{\mathrm{re}A}$ and let $x \in \mathbb{R}^n$ be a feature vector that describes $\mathrm{re}X$. Without loss of generality, assume each $\mathrm{re}X$ has a unique shape. $\Phi^{-1}(x) = \mathrm{re}X$, since every $\mathrm{re}X$ has a unique shape in the class of regions $\mathscr{R}_{\mathrm{re}A}$. Hence, $\Phi^{-1}(B) = \mathscr{R}_{\mathrm{re}A}$ is a descriptive fiber bundle over B. $\qquad\square$

Example 14.9 Let $sew(\mathrm{re}A, \mathrm{re}B, k)$ be a simplicial complex (from Lemma 14.2) derived from a pair of disjoint regions $\mathrm{re}A, \mathrm{re}B$, $\mathscr{R}_{sew(\mathrm{re}A,\ \mathrm{re}B,\ k)}$ the class of all planar shapes that are similar to $sew(\mathrm{re}A, \mathrm{re}B, k)$, and B is a set of descriptions of planar shapes. The arrow diagram

$$\{reA,\ \mathrm{re}B\} \xrightarrow{\ \ \mathrm{sew}\ \ } \mathscr{R}_{\mathrm{sew}(\mathrm{re}A,\ \mathrm{re}B,\ k)}$$

$$\downarrow{\scriptstyle\Phi}$$

$$B \subset \mathbb{R}^n$$

represents a descriptive fiber space $(\mathscr{R}_{\mathrm{sew}(\mathrm{re}A,\ \mathrm{re}B,\ k)}, \Phi, B)$ in which $\mathscr{R}_{\mathrm{sew}(\mathrm{re}A,\ \mathrm{re}B,\ k)}$, B are topological spaces and $\Phi : \mathscr{R}_{\mathrm{sew}(\mathrm{re}A,\ \mathrm{re}B,\ k)} \longrightarrow B$ is a homeomorphism. From Theorem 14.2, $(\mathscr{R}_{\mathrm{sew}(\mathrm{re}A,\ \mathrm{re}B,\ k)}, \Phi, B)$ is a BreMiller–Sloyer sheaf and $\Phi^{-1}(B)$ is a descriptive fiber bundle over B.

14.3.1 Sewing Nerves Together with Spokes to Construct a Nervous System Complex

This section introduces a specialized spoke-based sewing operation that results in nervous system complexes. The basic approach is to stitch a pair of neighboring nerve complexes with either one or two 2-spokes. For simplicity, we consider only the union of nerve complexes with an intermediate 2-spoke. By definition, every nerve complex is a collection of 1-spokes. Hence, the union of any 1-spoke adjacent to a nerve 1-spoke is a 2-spoke. This means that adjacent nerve complexes are the union of 2-spokes. In 2D simplicial complex covering a planar region, a 1-spoke between a pair of neighboring nerve complexes (i.e., a pair of nerves separated by a 1-spoke) provide a basis for the construction of a nervous system complex. Let X be a triangulation of a planer region, that is, a Vietoris–Rips complex. Also, 2^X denotes a collection of subsets of X. A **nervous system complex sewing map** (denoted by sewNrv)

$$\mathrm{sewNrv} : 2^X \times 2^X \longrightarrow 2^X \ \ \textbf{(nerve sewing mapping)}$$

is defined by

$$\mathrm{sewNrv}\Big(\mathrm{Nrv}A, \mathrm{Nrv}B\Big) = \mathrm{Nrv}\,A \cup \mathrm{Nrv}\,B \cup$$
$$\Big\{2\mathrm{sk}H,\ 2\mathrm{sk}J \in 2^X : \mathrm{Nrv}A\ \overset{\wedge}{\delta}\ 2\mathrm{sk}H\ \&\ 2\mathrm{sk}J\ \overset{\wedge}{\delta}\ \mathrm{Nrv}B\Big\},$$
$$2\mathrm{sk}H\ \overset{\wedge}{\delta}\ 2\mathrm{sk}J,\ \ i.e.,\ 2\mathrm{sk}H, 2\mathrm{sk}J\ \text{overlap.}$$

Definition 14.2 [**Nervous System Complex**] Let 2^X be a collection of 2-simplexes in a planar Vietoris–Rips complex. The quartet that contains nerve complexes NrvA, NrvB $\in 2^X$ and 2-spokes 2skH, 2skJ $\in 2^X$ is a **nervous system complex** (denoted by NrvsysK), provided

$$\text{Nrvsys}K = \text{sewNrv}(\text{Nrv}A, \text{Nrv}B),$$

that is, NrvA overlaps a simplex 2skH that is strongly near a 2-simplex 2skJ that overlaps NrvB. □

Example 14.10 [**Nervous System Complex**] A sample simplicial complex NrvsysK is shown in Figure 14.9. The pair of 2-spokes 2skH, 2skJ are strongly near, that is, 2skH $\overset{\wedge}{\delta}$ 2skJ, since these 2-spokes have a common edge. Also observe that 2skH contains a 2-simplex, namely, 1skh $\in$ 2skH and 1skh $\in$ NrvA. Hence, NrvA $\overset{\wedge}{\delta}$ 2skH. Similarly, there is a 2-simplex 1skj $\in$ 2skJ and 1skj $\in$ NrvB. Consequently, NrvB $\overset{\wedge}{\delta}$ 2skJ. As a result, NrvsysK = sewNrv(NrvA, NrvB). By definition, NrvsysK is a nervous system complex.

14.4 Some Results for Fiber Bundles

This section presents some results for fiber bundles.

Lemma 14.4 sewNrv(NrvA, NrvB) constructs a Vietoris–Rips complex.

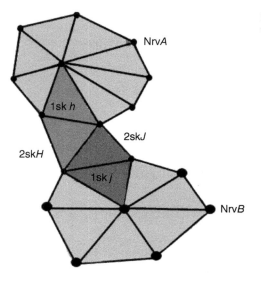

Figure 14.9 Sewing nerve complexes together.

Proof: Recall that a planar Vietoris–Rips complex is a collection of 2-simplexes. Hence, from Definition 14.2, sewNrv(NrvA, NrvB) is a Vietoris–Rips complex.

□

Lemma 14.5 If NrvsysK is a nervous system complex, then the collection of 2-simplexes in NrvsysK is a Vietoris–Rips complex.

Proof: From Lemma 14.4, sewNrv(NrvA, NrvB) is a Vietoris–Rips complex. By definition, NrvsysK = sewNrv(NrvA, NrvB) for a pair of nerve complexes NrvA, NrvB ∈ NrvsysK. Hence, we have the desired result, namely, NrvsysK is a Vietoris–Rips complex.

□

Theorem 14.3 Let sewNrv(NrvA, NrvB) be homeomorphism on a Vietoris–Rips complex E onto a subregion $B \subset E$ with never complexes NrvA, NrvB ∈ 2^E. Then $(E, \text{sewNrv}(\text{Nrv}A, \text{Nrv}B), B)$ is a BreMiller–Sloyer sheaf on a Vietoris–Rips complex.

Proof: From Lemma 14.5, the nervous system constructed by sewNrv(NrvA, NrvB) is Vietoris–Rips complex. By definition, sewNrv(NrvA, NrvB) on E maps onto a subregion $B \subset E$, which is a nervous system complex. Hence, the fiber bundle $(E, \text{sewNrv}(\text{Nrv}A, \text{Nrv}B), B)$ is a BreMiller–Sloyer sheaf.

□

Observe that $A \stackrel{\wedge}{\delta} B$ does not imply $A \stackrel{\delta}{\wedge} B$. In fact, this is the case when the proximity δ is not an *EF-proximity*.

This gives rise to a view of parallelism that does not depend on the parallel lines being straight and does not depend on the lines being cut by a straight line whose interior angles are both right angles.

PG.12 Proximal Parallel Axiom

If lines A, B are extended indefinitely and $A \stackrel{\delta}{\wedge} B$, then A and B are parallel. That is, no part of line A overlaps line B.

Lemma 14.6 Let reA, reB be nonempty regions. If reA and reB are expanded indefinitely and reA $\stackrel{\delta}{\wedge}$ reB, then reA and reB are parallel.

Proof: Partition regions reA and reB into lines that are open sets, that is, each line does not include into border points. Since region is extended indefinitely, each regional line is extended indefinitely. Let line $L_A \in$ reA, line $L_B \in$ reB. $L_A \stackrel{\delta}{\wedge} L_B$, since reA $\stackrel{\delta}{\wedge}$ reB. Consequently, from the **PG.12** Proximal Parallel Axiom, $L_A \| L_B$. Since this holds true for all pairs of regional lines, reA $\|$ reB.

□

Let $\mathscr{R}_{\text{re}A}, \mathscr{R}_{\text{re}A'}$ be a pair of classes of regions. In addition, let $X \in \mathscr{R}_{\text{re}A}$ (a region in class $\mathscr{R}_{\text{re}A}$), $Y \in \mathscr{R}_{\text{re}A'}$ (a region in class $\mathscr{R}_{\text{re}A'}$). Assume $X \stackrel{\delta}{\wedge} Y$

(X and Y are strongly far apart) for each pair of regions in $\mathscr{R}_{reA}, \mathscr{R}_{reA'}$, respectively. Then, from Lemma 14.6, $\mathscr{R}_{reA}, \mathscr{R}_{reA'}$ are parallel classes. That is, $\mathscr{R}_{reA} \| \mathscr{R}_{reA'}$ are parallel, if and only if $X \| Y$ for all $X \in \mathscr{R}_{reA}, Y \in \mathscr{R}_{reA'}$.

Theorem 14.4 Let $B, H \subset \mathbb{R}^n$ be sets of feature vectors that describe regions in the classes of regions $\mathscr{R}_{reA}, \mathscr{R}_{reA'}$, respectively. Assume that every region $reX \in \mathscr{R}_{reA}, reY \in \mathscr{R}_{reA'}$ has a unique feature and that $reX \not\underset{M}{\delta} reY$. Also, let $\Phi_1 : \mathscr{R}_{reA} \longrightarrow B \subset \mathbb{R}^n, \Phi_2 : \mathscr{R}_{reA'} \longrightarrow H \subset \mathbb{R}^n$ be homeomorphisms and let $\Phi^{-1}(B), \Phi^{-1}(H)$ be descriptive fiber bundles. Classes $\mathscr{R}_{reA}, \mathscr{R}_{reA'}$ are parallel if and only if $\Phi^{-1}(B) \| \Phi^{-1}(H)$.

Proof: Since each feature vector in X is unique in its description of a region in $\mathscr{R}_{reA}$, we have $\Phi_1^{-1}(\mathscr{B}) = \mathscr{R}_{reA}$. Similarly, $\Phi_2^{-1}(\mathscr{H}) = \mathscr{R}_{reA'}$. From the fact that $\Phi_1^{-1}(b) \not\underset{M}{\delta} \Phi_1^{-1}(h), b \in B, h \in H$ for each pair of regions, by Lemma 14.6, classes $\mathscr{R}_{reA}, \mathscr{R}_{reA'}$ are parallel. Hence, $\mathscr{R}_{reA} \| \mathscr{R}_{reA} \Leftrightarrow \Phi^{-1}(B) \| \Phi^{-1}(H)$. □

Axiom d.1.4: Descriptively Parallel Regions.
$reA \|_\Phi reB$, if and only if $strA \not\underset{M}{\delta} strB$ and $reA \, \delta_\Phi \, reB$.

The result from Theorem 14.4 is similar to the result for descriptively parallel classes in Theorem 14.5.

Theorem 14.5 Let $X, Y \subset \mathbb{R}^n$ be sets of feature vectors that describe regions in the classes of regions $\mathscr{R}_{reA}, \mathscr{R}_{reB}$, respectively. Assume that every region $reX \in \mathscr{R}_{reA}, reY \in \mathscr{R}_{reB}$ has a unique feature and that $reX \not\underset{M}{\delta} reY$. Let $\Phi_1 : \mathscr{R}_{reA} \longrightarrow \mathbb{R}^n, \Phi_2 : \mathscr{R}_{reB} \longrightarrow \mathbb{R}^n$ be continuous mappings. Classes $\mathscr{R}_{reA}, \mathscr{R}_{reA}$ are descriptively parallel fiber bundles, if and only if $\Phi^{-1}(X) \|_\Phi \Phi^{-1}(Y)$.

Proof: The result follows from Axiom d.14.4 with the proof symmetric with the proof of Theorem 14.4. □

The diagram in Figure 14.10 includes fiber bundles for the derivation of pairs of nerve complexes, nerve complex systems, and their descriptions.

$$
\begin{array}{ccc}
\mathrm{Nrv}A \times \mathrm{Nrv}B & \xrightarrow{\;\text{sewNrv}\;} & \mathrm{Nrv}sysK \\
\downarrow{\scriptstyle \pi_\Phi} & & \downarrow{\scriptstyle \pi'_{\Phi'}} \\
\Phi(\mathrm{Nrv}A) \times \Phi(\mathrm{Nrv}B) & \xrightarrow{\;g\;} & \Phi(\mathrm{Nrv}sysK)
\end{array}
$$

Figure 14.10 Two fiber bundles containing projections π_Φ on a pair of nerve complexes to a pair of descriptions of the nerves and $\pi'_{\Phi'}$ on a nervous system NrvsysK to a description of the nervous system.

Theorem 14.6 Let Nrv$A$, Nrv$B$ be a pair of strongly near nerve complexes in a Vietoris–Rips complex and let sewNrv(Nrv$A$, Nrv$B$) = Nrvsys$K$. Then

1° The diagram in Figure 14.10 is commutative.
2° π_Φ(Nrv$A$, Nrv$B$) describes a Vietoris–Rips complex.
3° $\pi'_{\Phi'}$(Nrvsys$K$) describes a Vietoris–Rips complex.
4° g(Nrv$A$, Nrv$B$) describes the nervous system Nrvsys$K$.
5° An additional mapping h is needed to guarantee that Φ(Nrvsys$K$) includes a description of the 2-spokes used to stitch together the nerve complexes in Nrvsys$K$.

Proof:
1°: Immediate from the fact that $\pi_\Phi \circ g = \pi'_{\Phi'} \circ$ sewNrv.
2°: Since Nrv$A$ $\overset{\wedge\wedge}{\delta}$ Nrv$B$ (by assumption), then either Nrv$A$, Nrv$B$ have a common edge or a common 2-spoke. By definition, Nrv$A$, Nrv$B$ is a Vietoris–Rips complex. Hence, the result follows.
3°: From Lemma 14.5, Nrvsys$K$ is a Vietoris–Rips complex. Hence, $\pi'_{\Phi'}$(Nrvsys$K$) describes a Vietoris–Rips complex.
4°: Since $g(\Phi(\text{Nrv}A), \Phi(\text{Nrv}B)) = \Phi(\text{Nrvsys}K)$, the result follows.
5°: There is no guarantee that $g(\Phi(\text{Nrv}A), \Phi(\text{Nrv}B))$ includes a description of the 2-spokes used to stitch Nrv$A$, Nrv$B$ together. Let 2sk$H$, 2sk$J$ be 2-spokes in Nrv*sys*K. Then let the mapping:

$$\begin{aligned} h : \text{Nrvsys}K \to \Phi\left(\text{Nrv}sysK\right) &\times \Phi\left(2\text{sk}H \overset{\wedge\wedge}{\delta} \text{Nrv}A \in \text{Nrv}sysK\right) \\ &\times \Phi\left(2\text{sk}\,J \overset{\wedge\wedge}{\delta} \text{Nrv}B \in \text{Nrv}sysK\right) \end{aligned}$$

in Figure 14.11 be defined by

$$\begin{aligned} h\left(\text{Nrvsys}K\right) = \Phi\left(\text{Nrv}sysK\right) \cup \Phi\left(2\text{sk}H \overset{\wedge\wedge}{\delta} \text{Nrv}A \in \text{Nrv}sysK\right) \\ \cup \Phi\left(2\text{sk}\,J \overset{\wedge\wedge}{\delta} \text{Nrv}B \in \text{Nrv}sysK\right). \end{aligned}$$

This gives the desired result. □

Figure 14.11 Two fiber bundles containing projections π_Φ on a pair of nerve complexes to a pair of descriptions of the nerves and $\pi'_{\Phi'}$ on a nervous system Nrvsys$K$ to a description of the nervous system.

$$\begin{array}{c} \text{Nrv}sysK \\ \downarrow \pi'_{\Phi'} \\ \Phi(\text{Nrv}sysK) \\ \downarrow h \\ \Phi\left(\text{Nrvsys}K\right) \times \Phi\left(2\text{sk}H \overset{\wedge\wedge}{\delta} \text{Nrv}A \in \text{Nrv}sysK\right) \\ \times \Phi\left(2\text{sk}\,J \overset{\wedge\wedge}{\delta} \text{Nrv}B \in \text{Nrv}sysK\right) \end{array}$$

14.5 Concluding Remarks

A number of forms of fiber bundles have been introduced in this paper. Planar Vietoris–Rips complexes provide a setting for this study. The fulcrum in this study is the introduction of a sewing mapping that makes it possible to stitch Alexandroff–Borsuk nerve complexes together to form nervous system complexes. This is important, since nerve complexes are closed related to the study of shape [40] and have a number of applications such as recent work in descriptive proximity [36], strong proximities in Voronoï tessellations of surfaces [34] and in the detection and classification of shapes in computer vision (see, e.g., [5, Section 1.23]).

References

1 Peters, J.F. (2014) *Topology of Digital Images. Visual Pattern Discovery in Proximity Spaces*, Intelligent Systems Reference Library, vol. **63**, Springer, xv + 411 pp, Zentralblatt MATH Zbl 1295 68010.

2 Peters, J.F. (2016) *Computational Proximity. Excursions in the Topology of Digital Images*, Intelligent Systems Reference Library, vol. **102**, Springer, viii + 445 pp, DOI: 10.1007/978-3-319-30262-1.

3 Peters, J.F. (2016) Proximal relator spaces. *Filomat*, **30** (2), 469–472. MR3497927.

4 Peters, J.F. and İnan, E. (2016) Strongly proximal Edelsbrunner-Harer nerves. *Proc. Jangjeon Math. Soc.*, **19** (3), 563–582. Zbl 06665117.

5 Peters, J.F. (2007) *Foundations of Computer Vision. Computational Geometry, Visual Image Structures and Object Shape Detection*, Intelligent Systems Reference Library, vol. **124**, Springer International Publishing AG, Switzerland, x+370 pp. DOI: 10.1007/978-3-319-52483-2.

6 Peters, J.F. (2016) Two forms of proximal physical geometry. Axioms, sewing regions together, classes of regions, duality and parallel fibre bundles. *Adv. Math. Sci. J.*, **5** (2), 241–268.

7 Chambers, E.W., de Silva, V., Erickson, J., and Ghrist, R. (2010) Vietoris-Rips complexes of planar point sets. *Discrete Comput. Geom.*, **44** (1), 75–90. MR2639819.

8 Edelsbrunner, H. and Harer, J.L. (2010) *Computational Topology. An Introduction*, American Mathematical Society, Providence, RI, xii+241 pp. ISBN: 978-0-8218-4925-5, MR2572029.

9 Brossard, R. (1964) Birkhoff's axioms for space geometry. *Am. Math. Mon.*, **71** (6), 593–606. MR0167872.

10 Zisman, M. (1999) Fibre bundles, fibre maps, in *History of Topology* (ed. I.M. James), North-Holland, Amsterdam, pp. 605–629. MR1721117.

11 Chern, S.-S. (1990) What is geometry. *Am. Math. Mon.*, **97** (8), 679–686. MR1072811.

12 Luke, G. and Mishchenko, A.S. (1998) *Vector Bundles and Their Applications*, Springer-Verlag, Berlin, 254 p.

13 BreMiller, R.S. and Sloyer, C.W. (1967) Fibre spaces and sheaves. *Am. Math. Mon.*, **74** (6), 694–695. MR0214068.

14 Borsuk, K. (1975) *Theory of Shape*, Monografie Matematyczne, Tom **59** (Mathematical Monographs, vol. **59**), PWN—Polish Scientific Publishers. MR0418088; Based on Borsuk, K. (1971) *Theory of Shape*, Lecture Notes Series, No. **28**, Matematisk Institut, Aarhus Universitet, Aarhus. MR0293602.

15 Ziegler, G.M. (1995) *Lectures on Polytopes*, Graduate Texts in Mathematics, vol. **152**, Springer-Verlag, New York, x+370 pp. ISBN: 0-387-94365-X, MR1311028.

16 Borsuk, K. (1948) On the imbedding of systems of compacta in simplicial complexes. *Fund. Math.*, **35**, 217–234.

17 Alexandroff, P. (1928) Über den algemeinen dimensionsbegriff und seine beziehungen zur elementaren geometrischen anschauung. *Math. Ann.*, **98**, 634.

18 Leray, J. (1946) L'anneau d'homologie d'une reprësentation. *C. R. Acad. Sci.*, **222**, 1366–1368.

19 Adamaszek, M., Adams, H., Frick, F., Peterson, C., and Previte-Johnson, C. (2014) Nerve complexes on circular arcs. *arXiv*, **1410** (4336v1), 1–17.

20 de Verdière, E.C., Ginot, G., and Goaoc, X. (2012) Multinerves and Helly Numbers of Acylic Families. Proceedings of the 28th Annual Symposium on Computational Geometry, pp. 209–218.

21 Edelsbrunner, H. and Harer, J.L. (2010) *Computational Topology. An Introduction*, American Mathematical Society, Providence, RI, xii+110 pp. MR2572029.

22 Adamaszek, M., Adams, H., Frick, F., Peterson, C., and Previte-Johnson, C. (2016) Nerve complexes of circular arcs. *Discrete Comput. Geom.*, **56** (2), 251–273. MR3530967.

23 Čech, E. (1966) *Topological Spaces*, John Wiley & Sons, Ltd., London. fr seminar, Brno, 1936-1939; rev. ed. Z. Frolik, M. Katětov.

24 Efremovič, V.A. (1952) The geometry of proximity I. *Mat. Sb. (N.S.)*, **31** (73) (1), 189–200 (in Russian).

25 Lodato, M.W. (1962) On topologically induced generalized proximity relations. PhD thesis. Rutgers University, Supervisor: S. Leader.

26 Wallman, H. (1938) Lattices and topological spaces. *Ann. Math.*, **39** (1), 112–126.

27 Peters, J.F. (2007) Near sets. Special theory about nearness of objects. *Fund. Inf.*, **75**, 407–433. MR2293708.

28 Peters, J.F. (2007) Near sets. General theory about nearness of sets. *Appl. Math. Sci.*, **1** (53), 2609–2629.

29 Peters, J.F. (2013) Near sets: an introduction. *Math. Comput. Sci.*, **7** (1), 3–9. DOI: 10.1007/s11786-013-0149-6, MR3043914.

30 Peters, J.F. (2015) Proximal Delaunay triangulation regions. *Proc. Jangjeon Math. Soc.*, **18** (4), 501–515. MR3444736.

31 Peters, J.F. and Guadagni, C. (2015) Strongly near proximity and hyperspace topology, arXiv **1502**, no. 05913, 1–6.

32 Guadagni, C. (2015) Bornological convergences on local proximity spaces and ω_μ-metric spaces. PhD thesis. Università degli Studi di Salerno, Salerno, Italy, Supervisor: A. Di Concilio, 79 pp.

33 Peters, J.F. (2015) Visibility in proximal Delaunay meshes and strongly near Wallman proximity. *Adv. Math. Sci. J.*, **4** (1), 41–47.

34 Peters, J.F. and Guadagni, C. (2015) Strong proximities on smooth manifolds and Voronoi diagrams. *Adv. Math. Sci. J.*, **4** (2), 91–107. Zbl 1339.54020.

35 Naimpally, S.A. (1970) *Proximity Spaces*, Cambridge University Press, Cambridge, x+128 pp. ISBN: 978-0-521-09183-1.

36 Di Concilio, A., Guadagni, C., Peters, J.F., and Ramanna, S. (2016) Descriptive proximities. Properties and interplay between classical proximities and overlap, arXiv **1609**, no. 06246v1, 1–12.

37 Naimpally, S.A. and Peters, J.F. (2013) *Topology with Applications. Topological Spaces via Near and Far*, World Scientific, Singapore, xv + 277 pp, American Mathematical Society. MR3075111.

38 Száz, A. (1987) Basic tools and mild continuities in relator spaces. *Acta Math. Hungar.*, **50** (3–4), 177–201. MR0918156.

39 Peters, J.F. (2016) Proximal relator spaces. *Filomat*, **30** (2), 469–472. doi: 10.2298/FIL1602469P, MR3497927.

40 Borsuk, K. and Dydak, J. (1981) What is the theory of shape? *Bull. Austr. Math. Soc.*, **22** (2), 161–198. MR0598690.

15

Approximation by Generalizations of Hybrid Baskakov Type Operators Preserving Exponential Functions

Vijay Gupta

Department of Mathematics, Netaji Subhas Institute of Technology, Dwarka, New Delhi 110078, India

AMS Subject Classification 41A25, 41A36

15.1 Introduction

Concerning the convergence of a sequence of linear positive operators, many methods have been considered, but the basic concept which guarantees the convergence of a sequence of linear positive operators is the well-known Korovkin theorem. In addition, the starting theorem concerning the convergence of linear positive operators is the well-known Weierstrass approximation theorem. The one of the most common proofs of the Weierstrass approximation is based on the Bernstein polynomials, which is the basic operators in approximation theory and for $f \in C[0, 1]$ is defined as

$$B_n(f, x) = \sum_{k=0}^{n} \binom{n}{k} x^k (1 - x)^{n-k} f\left(\frac{k}{n}\right), \quad x \in [0, 1]$$

The operators B_n is an important operator in the field of approximation and other related areas. Another extension of this operator is the Baskakov operator, which is based on negative binomial distribution. The Baskakov operators for $f \in C[0, \infty)$ is defined as

$$V_n(f, x) = \sum_{k=0}^{\infty} \binom{n+k-1}{k} \frac{x^k}{(1+x)^{n+k}} f\left(\frac{k}{n}\right), \quad x \in [0, \infty).$$

Another generalization of the Bernstein operator is the Szász–Mirakyan operator defined by

$$S_n(f, x) = \sum_{k=0}^{\infty} e^{-nx} \frac{(nx)^k}{k!} f\left(\frac{k}{n}\right), \quad x \in [0, \infty).$$

The operators defined above, that is, B_n, V_n, and S_n are discretely defined operators and also known as exponential type operators. As such these

Mathematical Analysis and Applications: Selected Topics, First Edition.
Edited by Michael Ruzhansky, Hemen Dutta, and Ravi P. Agarwal.
© 2018 John Wiley & Sons, Inc. Published 2018 by John Wiley & Sons, Inc.

operators may not be utilized to approximate integrable functions. Further generalizations have been discussed for different operators in order to obtain the convergence. Kantorovich [1] was the first, who introduced the modification of the Bernstein polynomials. A more general integral modification of the Bernstein polynomial was given by Durrmeyer [2]. Later Sahai and Prasad [3] and Mazhar and Totik [4] proposed Baskakov–Durrmeyer and Szász–Durrmeyer operators, respectively. In the last two decades, the author considered several problems related to different operators and studied some generalizations of existing operators, we mention some of them as [3]–[24] and so on. In addition, very recently, Gupta *et al.* [25] provided a list of many usual and hybrid Durrmeyer-type operators.

In the year 1970, Boyanov and Veselinov [26] proposed the following general Korovkin-type theorem for the function e^{-kt}, $k = 0, 1, 2$. First, we define the class $C^*[0, \infty)$ discussed in [26], which denoted the linear space of all real-valued continuous functions, defined and continuous on $[0, \infty)$ with the property that $\lim_{x \to \infty} f(x)$ exists and is finite, and endowed with the uniform norm.

Theorem 15.1 [26, theorem 2] If the sequence of linear positive operators $\{A_n(f(t), x)\}$ defined on $C^*[0, \infty)$ satisfies

$$\lim_{n \to \infty} A_n(e^{-kt}, x) = e^{-kx}, \quad k = 0, 1, 2, \text{ uniformly in } [0, \infty),$$

then

$$\lim_{n \to \infty} A_n(f(t), x) = f(x) \text{ uniformly in } [0, \infty) \text{ for each } f \in C^*[0, \infty).$$

In this context, Holhoş [27] extended the studies and established a quantitative result for such exponential functions as follows:

Theorem 15.2 [27, theorem 2.1] Let $f \in C^*[0, \infty)$ and $A_n : C^*[0, \infty) \to C^*[0, \infty)$ be a sequence of positive linear operators. If

$$||A_n 1 - 1||_\infty = a_n,$$
$$||A_n(e^{-t}, x) - e^{-x}||_\infty = b_n,$$
$$||A_n(e^{-2t}, x) - e^{-2x}||_\infty = c_n,$$

where a_n, b_n, and c_n tend to zero for n sufficiently large then we have

$$||A_n f - f||_\infty \leq ||f||_\infty a_n + (2 + a_n) \cdot \omega^* \left(f, \sqrt{a_n + 2b_n + c_n} \right),$$

where the modulus of continuity is defined by

$$\omega^*(f, \delta) = \sup_{\substack{x, t \geq 0 \\ |e^{-x} - e^{-t}| \leq \delta}} |f(x) - f(t)|,$$

for every $\delta \geq 0$ and every function $f \in C^*[0, \infty)$.

In [27], Holhoş also provided some of the well-known operators as applications for the above Theorem 15.2. He used some identities and inequalities to estimate the results (Theorem 15.2), for Baskakov, Szász–Mirakyan, and Chlodowsky-type operators. Very recently, Acar *et al.* [28] considered the Szász–Mirakyan operators, which preserve the function $e^{2ax}, a > 0$, and they estimated the Theorem 15.2 for the modified Szász–Mirakyan operators. They also used the complicated analysis to obtain the result for modified Szász–Mirakyan operators. In this chapter, we discuss the result of the hybrid Szász–Mirakyan operators and their different variants. We use the mathematica software to estimate the results. It provides exact estimates and the complicated analysis and upper bound are avoided here.

15.2 Baskakov–Szász Operators

Gupta and Srivastava [29], defined a new sequence of linear positive operators by combining the well-known Baskakov operators and the Szász–Mirakyan operators for the functions integrable on $[0, \infty)$ and proposed hybrid Baskakov–Szász operators, which are defined as follows:

$$M_n(f, x) = n \sum_{k=0}^{\infty} b_{n,k}(x) \int_0^{\infty} s_{n,k}(t) f(t) \, dt, \tag{15.1}$$

where the Baskakov and Szász–Mirakyan basis functions are, respectively, defined by

$$b_{n,k}(x) = \binom{n+k-1}{k} \frac{x^k}{(1+x)^{n+k}}$$

and

$$s_{n,k}(t) = e^{-nt} \frac{(nt)^k}{k!}.$$

In [30], Gupta and Maheshwari discussed some more approximation properties of Baskakov–Szász operators, and they also established an inverse theorem. Some other results were compiled by Gupta and Agarwal in the recent book [31]. Some other generalized form of Baskakov–Szász operator was discussed by Gupta and Malik [32]. Let $f(t) = e^{At}, A \in \mathbb{R}$, then we have

$$n \int_0^{\infty} e^{-nt} \frac{(nt)^k}{k!} e^{At} \, dt = \left(\frac{n}{n-A} \right)^{k+1}. \tag{15.2}$$

Thus, for the operators defined by (15.1), using (15.2) and by the well-known binomial series $\sum_{k=0}^{\infty} \frac{(a)_k}{k!} z^k = (1-z)^{-a}, |z| < 1$, we have

$$M_n(e^{At}, x) = \frac{n}{(n-A)(1+x)^n(n-1)!} \sum_{k=0}^{\infty} \frac{(n+k-1)!}{k!} \left[\frac{nx}{(n-A)(1+x)} \right]^k$$

$$= \frac{n}{(n-A)(1+x)^n(n-1)!} \sum_{k=0}^{\infty} \frac{(n)_k}{k!} \left[\frac{nx}{(n-A)(1+x)} \right]^k$$

$$= \frac{n}{(n-A)(1+x)^n(n-1)!} \left[1 - \frac{nx}{(n-A)(1+x)} \right]^{-n}$$

$$= n(n-A)^{n-1} [n - A(1+x)]^{-n}. \tag{15.3}$$

Here, we observe that $M_n(e^{At}, x)$ may be treated as moment-generating function (m.g.f.) of the operators $M_n f$, which may be utilized to obtain the moments of (15.1). Let $\mu_r(x) = M_n(e_r, x)$, where $e_r(t) = t^r, r \in \mathbb{N} \cup \{0\}$. The moments are given by

$$\mu_r(x) = \left[\frac{\partial^r}{\partial A^r} M_n(e^{At}, x) \right]_{A=0}$$

$$= \left[\frac{\partial^r}{\partial A^r} \{ n(n-A)^{n-1} [n - A(1+x)]^{-n} \} \right]_{A=0}.$$

Using the software Mathematica, we get the expansion of (15.3) in powers of A as follows:

$$M_n(e^{At}, x) = 1 + \left(\frac{1+nx}{n} \right) A + \left(\frac{n^2x^2 + nx^2 + 4nx + 2}{n^2} \right) \frac{A^2}{2!}$$

$$+ \left(\frac{n^3x^3 + 3n^2x^3 + 2nx^3 + 9n^2x^2 + 9nx^2 + 18nx + 6}{n^3} \right) \frac{A^3}{3!}$$

$$+ \frac{\left[\begin{array}{l} n^4x^4 + 6n^3x^4 + 11n^2x^4 + 6nx^4 + 16n^3x^3 + 48n^2x^3 \\ +32nx^3 + 72n^2x^2 + 72nx^2 + 96nx + 24 \end{array} \right]}{n^4} \frac{A^4}{4!}$$

$$+ \frac{\left[\begin{array}{l} n^5x^5 + 10n^4x^5 + 35n^3x^5 + 50n^2x^5 + 24nx^5 + 25n^4x^4 \\ +150n^3x^4 + 275n^2x^4 + 150nx^4 + 200n^3x^3 + 600n^2x^3 \\ +400nx^3 + 600n^2x^2 + 600nx^2 + 600nx + 120 \end{array} \right]}{n^5} \frac{A^5}{5!}$$

$$+ \mathcal{O}(A^6).$$

In particular, first few moments may be obtained as

$$\mu_0(x) = 1,$$
$$\mu_1(x) = \frac{1+nx}{n},$$
$$\mu_2(x) = \frac{n^2x^2 + nx^2 + 4nx + 2}{n^2},$$
$$\mu_3(x) = \frac{n^3x^3 + 3n^2x^3 + 2nx^3 + 9n^2x^2 + 9nx^2 + 18nx + 6}{n^3}, \tag{15.4}$$
$$\mu_4(x) = \frac{\left[\begin{array}{l} n^4x^4 + 6n^3x^4 + 11n^2x^4 + 6nx^4 + 16n^3x^3 + 48n^2x^3 \\ +32nx^3 + 72n^2x^2 + 72nx^2 + 96nx + 24 \end{array} \right]}{n^4}.$$

Lemma 15.1 Applying (15.4), the central moments $\phi_x^m(t) = (t - x)^m$, are given by:

$$M_n(\phi_x^0(t), x) = 1,$$

$$M_n(\phi_x^1(t), x) = \frac{1}{n},$$

$$M_n(\phi_x^2(t), x) = \frac{nx^2 + 2nx + 2}{n^2},$$

$$M_n(\phi_x^3(t), x) = \frac{2nx^3 + 9nx^2 + 12nx + 6}{n^3},$$

$$M_n(\phi_x^4(t), x) = \frac{\left[\begin{array}{c} -19n^2x^4 + 6nx^4 + 12n^2x^3 + 32nx^3 + 12n^2x^2 \\ +72nx^2 + 72nx + 24 \end{array}\right]}{n^4}.$$

Lemma 15.2 Using the software Mathematica, we obtain the series expansion as indicated below.

$$M_n(e^{-t}, x) = n(n + 1)^{n-1} (n + 1 + x)^{-n}$$

$$= e^{-x} + \frac{e^{-x}(x^2 + 2x - 2)}{2n} + \frac{e^{-x}(3x^4 + 4x^3 - 24x^2 - 48x + 24)}{24n^2}$$

$$+ \frac{e^{-x}(x^6 - 2x^5 - 22x^4 - 24x^3 + 72x^2 + 144x - 48)}{48n^3}$$

$$+ \frac{\left[\begin{array}{c} e^{-x}(15x^8 - 120x^7 - 400x^6 + 1248x^5 + 6240x^4 + 5760x^3 \\ -11520x^2 - 23040x + 5760) \end{array}\right]}{5760n^4}$$

$$+ \mathcal{O}\left(\frac{1}{n^5}\right) \tag{15.5}$$

and

$$M_n(e^{-2t}, x) = n(n + 2)^{n-1} [n + 2(1 + x)]^{-n}$$

$$= e^{-2x} + \frac{e^{-2x}[2(1 + x)^2 - 4]}{n}$$

$$+ \frac{2e^{-2x}(3x^4 + 8x^3 - 6x^2 - 24x + 6)}{3n^2}$$

$$+ \frac{4e^{-2x}(x^6 + 2x^5 - 8x^4 - 24x^3 + 36x - 6)}{3n^3}$$

$$+ \frac{\left[\begin{array}{c} 2e^{-2x}(15x^8 - 280x^6 - 504x^5 + 840x^4 \\ +2880x^3 + 720x^2 - 2880x + 360) \end{array}\right]}{45n^4} + \mathcal{O}\left(\frac{1}{n^5}\right).$$

For the operators (15.1), Theorems 15.1 and 15.2 take the following form:

Theorem 15.3 If the linear positive operators (15.1) satisfies

$$\lim_{n \to \infty} M_n(e^{-kt}, x) = e^{-kx}, \quad k = 0, 1, 2, \text{ uniformly in } [0, \infty),$$

then

$$\lim_{n\to\infty} M_n(f(t),x) = f(x) \text{ uniformly in } [0,\infty) \text{ for each } f \in C^*[0,\infty).$$

The proof of this theorem for the Baskakov–Szász operators is immediate, if we apply (15.4) and Lemma 15.2.

Theorem 15.4   Let $f \in C^*[0,\infty)$. If

$$||M_n 1 - 1||_\infty = a_n,$$
$$||M_n(e^{-t},x) - e^{-x}||_\infty = b_n,$$
$$||M_n(e^{-2t},x) - e^{-2x}||_\infty = c_n,$$

where a_n, b_n, and c_n tend to zero for n sufficiently large then we have

$$||M_n f - f||_\infty \le 2 \cdot \omega^*(f, \sqrt{2b_n + c_n}),$$

where

$$b_n = \left|\left| \frac{e^{-x}(x^2 + 2x - 2)}{2n} + \frac{e^{-x}(3x^4 + 4x^3 - 24x^2 - 48x + 24)}{24n^2} \right.\right.$$
$$+ \frac{e^{-x}(x^6 - 2x^5 - 22x^4 - 24x^3 + 72x^2 + 144x - 48)}{48n^3}$$
$$+ \frac{\left[\begin{array}{c} e^{-x}(15x^8 - 120x^7 - 400x^6 + 1248x^5 + 6240x^4 \\ +5760x^3 - 11520x^2 - 23040x + 5760) \end{array}\right]}{5760n^4} \left.\left. + \mathcal{O}\left(\frac{1}{n^5}\right)\right|\right|$$

and

$$c_n = \left|\left| \frac{e^{-2x}[2(1+x)^2 - 4]}{n} + \frac{2e^{-2x}(3x^4 + 8x^3 - 6x^2 - 24x + 6)}{3n^2} \right.\right.$$
$$+ \frac{4e^{-2x}(x^6 + 2x^5 - 8x^4 - 24x^3 + 36x - 6)}{3n^3}$$
$$+ \frac{\left[\begin{array}{c} 2e^{-2x}(15x^8 - 280x^6 - 504x^5 + 840x^4 \\ +2880x^3 + 720x^2 - 2880x + 360) \end{array}\right]}{45n^4} \left.\left. + \mathcal{O}\left(\frac{1}{n^5}\right)\right|\right|.$$

The proof follows using (15.4) and Lemma 15.2.

15.3   Genuine Baskakov–Szász Operators

The operators defined by (15.1) reproduce only the constant functions. The genuine Baskakov–Szász–Mirakyan operators were discussed in [33], which reproduce constant as well as linear functions and are defined as

$$G_n(f,x) = n \sum_{k=1}^{\infty} b_{n,k}(x) \int_0^\infty s_{n,k-1}(t) f(t)\,dt + b_{n,0}(x) f(0), \tag{15.6}$$

where

$$b_{n,k}(x) = \frac{(n)_k}{k!} \frac{x^k}{(1+x)^{n+k}}$$

and

$$s_{n,k-1}(t) = e^{-nt} \frac{(nt)^{k-1}}{(k-1)!}.$$

Let $f(t) = e^{At}$, $A \in \mathbb{R}$, then we have

$$n \int_0^\infty e^{-nt} \frac{(nt)^{k-1}}{(k-1)!} e^{At} \, dt = \left(\frac{n}{n-A}\right)^k. \tag{15.7}$$

Thus, using (15.7) and the well-known binomial series $\sum_{k=0}^\infty \frac{(a)_k}{k!} z^k = (1-z)^{-a}$, $|z| < 1$, we have

$$G_n(e^{At}, x) = \frac{1}{(1+x)^n} \sum_{k=0}^\infty \frac{(n)_k}{k!} \left[\frac{nx}{(n-A)(1+x)}\right]^k$$

$$= \frac{1}{(1+x)^n} \left[1 - \frac{nx}{(n-A)(1+x)}\right]^{-n}$$

$$= (n-A)^n \left[n - A(1+x)\right]^{-n}. \tag{15.8}$$

Here, we observe that $G_n(e^{At}, x)$ may be treated as m.g.f. of the operators $G_n f$, which may be utilized to obtain the moments of (15.1). Let $\mu_{n,r}^G(x) = G_n(e_r, x)$, where $e_r(t) = t^r$, $r \in \mathbb{N} \cup \{0\}$. The moments are given by

$$\mu_{n,r}^G(x) = \left[\frac{\partial^r}{\partial A^r} G_n(e^{At}, x)\right]_{A=0}$$

$$= \left[\frac{\partial^r}{\partial A^r} \{n(n-A)^n \left[n - A(1+x)\right]^{-n}\}\right]_{A=0}.$$

Using the software Mathematica, we get the expansion of (15.8) in powers of A as follows:

$$G_n(e^{At}, x) = 1 + xA + \left(\frac{nx^2 + x^2 + 2x}{n}\right) \frac{A^2}{2!}$$

$$+ \left(\frac{n^2 x^3 + 6x^2 + 6nx^2 + 2x^3 + 3nx^3 + n^2 x^3}{n^2}\right) \frac{A^3}{3!}$$

$$+ \frac{\left[\begin{matrix} n^3 x^4 + 6n^2 x^4 + 11nx^4 + 6x^4 + 12n^2 x^3 + 36nx^3 \\ +24x^3 + 36nx^2 + 36x^2 + 24x \end{matrix}\right]}{n^3} \frac{A^4}{4!} + \mathcal{O}(A^5).$$

In particular, first few moments may be obtained as

$$\mu_{n,0}^G(x) = 1,$$

$$\mu_{n,1}^G(x) = x,$$

$$\mu_{n,2}^G(x) = \frac{nx^2 + x^2 + 2x}{n},$$

$$\mu_{n,3}^G(x) = \frac{n^2x^3 + 6x^2 + 6nx^2 + 2x^3 + 3nx^3 + n^2x^3}{n^2}, \qquad (15.9)$$

$$\mu_{n,4}^G(x) = \frac{\left[\begin{array}{c} n^3x^4 + 6n^2x^4 + 11nx^4 + 6x^4 + 12n^2x^3 + 36nx^3 \\ +24x^3 + 36nx^2 + 36x^2 + 24x \end{array}\right]}{n^3},$$

which provides moments using m.g.f., which is an alternate way to the one found in [29].

Lemma 15.3 Applying (15.9), the central moments $\phi_x^m(t) = (t - x)^m$ are given by:

$$G_n(\phi_x^0(t), x) = 1,$$
$$G_n(\phi_x^1(t), x) = 0,$$
$$G_n(\phi_x^2(t), x) = \frac{x^2 + 2x}{n}.$$

Lemma 15.4 Using the software Mathematica, we obtain the series expansion as indicated below.

$$G_n(e^{-t}, x) = (n + 1)^n(n + 1 + x)^{-n}$$
$$= e^{-x} + \frac{e^{-x}(x(x + 2))}{n} + \frac{e^{-x}x(3x^3 + 4x^2 - 12x - 24)}{24n^2}$$
$$+ \frac{e^{-x}x(x^6 - 2x^5 - 22x^4 - 16x^3 - 2x^4 + x^5)}{48n^3} + \mathcal{O}\left(\frac{1}{n^4}\right) (15.10)$$

and

$$G_n(e^{-2t}, x) = (n + 2)^n[n + 2(1 + x)]^{-n}$$
$$= e^{-2x} + \frac{e^{-2x}2x(2 + x)}{n} + \frac{2e^{-2x}x(3x^3 + 8x^2 + 3x^3)}{3n^2}$$
$$+ \frac{4e^{-2x}x(x^5 + 2x^4 - 5x^3 - 16x^2 - 6x + 12)}{3n^3} + \mathcal{O}\left(\frac{1}{n^4}\right).$$

For the operators (15.6), Theorems 15.1 and 15.2 take the following form:

Theorem 15.5 If the linear positive operators (15.6) satisfies

$$\lim_{n \to \infty} G_n(e^{-kt}, x) = e^{-kx}, \quad k = 0, 1, 2, \text{ uniformly in } [0, \infty),$$

then

$$\lim_{n \to \infty} G_n(f(t), x) = f(x) \text{ uniformly in } [0, \infty) \text{ for each } f \in C^*[0, \infty).$$

Theorem 15.6 Let $f \in C^*[0, \infty)$. If

$$\|G_n 1 - 1\|_\infty = a_n,$$
$$\|G_n(e^{-t}, x) - e^{-x}\|_\infty = b_n,$$
$$\|G_n(e^{-2t}, x) - e^{-2x}\|_\infty = c_n,$$

where a_n, b_n, and c_n tend to zero for n sufficiently large. Then, we have

$$||G_n f - f||_\infty \le 2 \cdot \omega^*(f, \sqrt{2b_n + c_n}),$$

where

$$b_n = \left\| \frac{e^{-x}(x(x+2))}{n} + \frac{e^{-x}x(3x^3 + 4x^2 - 12x - 24)}{24n^2} \right.$$
$$\left. + \frac{e^{-x}x(x^6 - 2x^5 - 22x^4 - 16x^3 - 2x^4 + x^5)}{48n^3} + \mathcal{O}\left(\frac{1}{n^4}\right) \right\|$$

and

$$c_n = \left\| \frac{e^{-2x}2x(2+x)}{n} + \frac{2e^{-2x}x(3x^3 + 8x^2 + 3x^3)}{3n^2} \right.$$
$$\left. + \frac{4e^{-2x}x(x^5 + 2x^4 - 5x^3 - 16x^2 - 6x + 12)}{3n^3} + \mathcal{O}\left(\frac{1}{n^4}\right) \right\|.$$

The proof follows using Lemma 15.4.

15.4 Preservation of e^{Ax}

In [34], King proposed a generalization of the Bernstein polynomials so as to preserve the test function e_2, and he was able to achieve better approximation. Later in [35], Gonska *et al.* discussed some other King type operators. Recently, Acar *et al.* in [28] considered the Szász–Mirakyan operators, which preserve the function e^{2ax} and estimated some approximation properties. Motivated by the recent work that we discusses in this section, the genuine Baskakov–Szász operators, which preserve the test function e^{Ax}. Suppose the operators defined by (15.6) preserve the function e^{Ax}, then we start with

$$\tilde{G}_n(f, x) = n \sum_{k=1}^{\infty} b_{n,k}(a_n(x)) \int_0^\infty s_{n,k-1}(t) f(t) \, \mathrm{d}t + b_{n,0}(x) f(0), \qquad (15.11)$$

where $b_{n,k}(a_n(x)) = \binom{n+k-1}{k} \frac{(a_n(x))^k}{(1+a_n(x))^{n+k}}$ and $s_{n,k}(t) = e^{-nt}\frac{(nt)^k}{k!}$. Using (15.7) and the well-known binomial series $\sum_{k=0}^{\infty} \frac{(a)_k}{k!} z^k = (1-z)^{-a}, |z| < 1$, the modified operators (15.6) takes the following form

$$\tilde{G}_n(e^{At}, x) = \frac{1}{(1+a_n(x))^n} \sum_{k=0}^{\infty} \frac{(n)_k}{k!} \left[\frac{na_n(x)}{(n-A)(1+a_n(x))} \right]^k$$
$$= \frac{1}{(1+a_n(x))^n} \left[1 - \frac{na_n(x)}{(n-A)(1+a_n(x))} \right]^{-n}$$
$$= \left[\frac{n-A}{n-A(1+a_n(x))} \right]^n. \qquad (15.12)$$

As per our assumption that the operators $\tilde{G}_n(e^{At}, x)$ preserve e^{Ax}, we have

$$e^{Ax} = \left[\frac{n - A}{n - A(1 + a_n(x))}\right]^n,$$

which implies that

$$a_n(x) = \frac{n - A}{A}[1 - e^{-Ax/n}]. \tag{15.13}$$

Now the operators (15.11) can be represented as

$$\tilde{G}_n(f, x) = n \sum_{k=1}^{\infty} b_{n,k}\left(\frac{n - A}{A}[1 - e^{-Ax/n}]\right)\int_0^{\infty} s_{n,k-1}(t)\,f(t)\,dt + b_{n,0}(x)f(0),$$

Lemma 15.5 Using the software Mathematica, we obtain the series expansion as indicated below.

$$\tilde{G}_n(e^{-t}, x) = \left[\frac{n + 1}{n + 1 + (n/A - 1)(1 - e^{-Ax/n})}\right]^n$$

$$= e^{-x} + \frac{1}{2n}(1 + A)e^{-x}x(2 + x)$$

$$+ \frac{1}{24n^2}e^{-x}x\left(-24 - 24A - 12x - 12Ax + 4x^2 + 12Ax^2/\right.$$
$$\left. + 8A^2x^2 + 3x^3 + 6Ax^3 + 3A^2x^3\right)$$

$$+ \frac{1}{48n^3}e^{-x}x\left(48 + 48A + 24x + 24Ax - 16x^2 - 48Ax^2 - 40A^2x^2\right.$$
$$- 8A^3x^2 - 16x^3 - 40Ax^3 - 30A^2x^3 - 6A^3x^3 - 2x^4 - 2Ax^4$$
$$\left. + 2A^2x^4 + 2A^3x^4 + x^5 + 3Ax^5 + 3A^2x^5 + A^3x^5\right) + O(n^{-4})$$

and

$$\tilde{G}_n(e^{-2t}, x) = \left[\frac{n + 2}{n + 2 + 2(n/A - 1)(1 - e^{-Ax/n})}\right]^n$$

$$= e^{-2x} + (2 + A)e^{-2x}x(2 + x)\frac{1}{n}$$

$$+ \frac{1}{6n^2}e^{-2x}x\left(-48 - 24A + 12Ax + 6A^2x + 32x^2 + 36Ax^2\right.$$
$$\left. + 10A^2x^2 + 12x^3 + 12Ax^3 + 3A^2x^3\right)$$

$$+ \frac{1}{12n^3}e^{-2x}x\left(192 + 96A - 96x - 144Ax - 48A^2x - 256x^2\right.$$
$$- 288Ax^2 - 88A^2x^2 - 4A^3x^2 - 80x^3 - 64Ax^3 - 2A^2x^3$$
$$+ 5A^3x^3 + 32x^4 + 64Ax^4 + 40A^2x^4 + 8A^3x^4 + 16x^5$$
$$\left. + 24Ax^5 + 12A^2x^5 + 2A^3x^5\right) + O(n^{-4}).$$

Lemma 15.6 In particular, first few moments may be obtained as

$$\mu_{n,0}^{\tilde{G}}(x) = 1,$$
$$\mu_{n,1}^{\tilde{G}}(x) = a_n(x),$$
$$\mu_{n,2}^{\tilde{G}}(x) = \frac{(n+1)(a_n(x))^2 + 2a_n(x)}{n}.$$

Lemma 15.7 Following Lemma 15.3, the central moments $\phi_x^m(t) = (t - x)^m$ are given by:

$$\tilde{G}_n(\phi_x^0(t), x) = 1,$$
$$\tilde{G}_n(\phi_x^1(t), x) = a_n(x) - x,$$
$$\tilde{G}_n(\phi_x^2(t), x) = (a_n(x) - x)^2 + \frac{(a_n(x))^2 + 2a_n(x)}{n}.$$

Further, we have

$$\lim_{n \to \infty} n\tilde{G}_n(\phi_x^1(t), x) = \lim_{n \to \infty} n[a_n(x) - x] = \frac{-Ax(2 + x)}{2},$$
$$\lim_{n \to \infty} n\tilde{G}_n(\phi_x^2(t), x) = \lim_{n \to \infty} n\left[(a_n(x) - x)^2 + \frac{(a_n(x))^2 + 2a_n(x)}{n}\right] = x(2 + x).$$

Theorem 15.7 Let us suppose that $f \in C^*[0, \infty)$. Further, if

$$||\tilde{G}_n 1 - 1||_\infty = a_n,$$
$$||\tilde{G}_n(e^{-t}, x) - e^{-x}||_\infty = b_n,$$
$$||\tilde{G}_n(e^{-2t}, x) - e^{-2x}||_\infty = c_n,$$

where a_n, b_n, and c_n tend to zero for n sufficiently large. Then, we have

$$||\tilde{G}_n f - f||_\infty \leq 2 \cdot \omega^*\left(f, \sqrt{2b_n + c_n}\right),$$

where

$$b_n = \left\|\frac{1}{2n}(1 + A)e^{-x}x(2 + x)\right.$$
$$+ \frac{1}{24n^2}e^{-x}x\left(-24 - 24A - 12x - 12Ax + 4x^2 + 12Ax^2\right.$$
$$+ 8A^2x^2 + 3x^3 + 6Ax^3 + 3A^2x^3\big)$$
$$+ \frac{1}{48n^3}e^{-x}x\left(48 + 48A + 24x + 24Ax - 16x^2 - 48Ax^2\right.$$
$$- 40A^2x^2 - 8A^3x^2 - 16x^3 - 40Ax^3 - 30A^2x^3 - 6A^3x^3$$
$$- 2x^4 - 2Ax^4 + 2A^2x^4 + 2A^3x^4 + x^5 + 3Ax^5 + 3A^2x^5 + A^3x^5\big)$$
$$\left. + O(n^{-4})\right\|$$

and

$$c_n = \left\| (2+A)e^{-2x}x(2+x)\frac{1}{n} \right.$$
$$+ \frac{1}{6n^2}e^{-2x}x\left(-48 - 24A + 12Ax + 6A^2x + 32x^2 + 36Ax^2\right.$$
$$+ 10A^2x^2 + 12x^3 + 12Ax^3 + 3A^2x^3\big)$$
$$+ \frac{1}{12n^3}e^{-2x}x\left(192 + 96A - 96x - 144Ax - 48A^2x - 256x^2 - 288Ax^2\right.$$
$$- 88A^2x^2 - 4A^3x^2 - 80x^3 - 64Ax^3 - 2A^2x^3 + 5A^3x^3 + 32x^4 + 64Ax^4$$
$$\left. + 40A^2x^4 + 8A^3x^4 + 16x^5 + 24Ax^5 + 12A^2x^5 + 2A^3x^5\right) + O(n^{-4}) \Bigg\| .$$

The proof follows using Lemma 15.5.

Theorem 15.8   Let $f, f'' \in C^*[0, \infty)$. Then we have

$$\left| n[\tilde{G}_n(f; x) - f(x)] + \frac{x(2+x)}{2}[Af'(x) - f''(x)] \right|$$
$$\leq \left| n\tilde{G}_n(\phi_x^1(t), x) + \frac{Ax(2+x)}{2} \right| |f'(x)|$$
$$+ \frac{1}{2}\left| n\tilde{G}_n(\phi_x^2(t), x) - \frac{x(2+x)}{2} \right| |f''(x)| + 2\omega^*\left(f''; \frac{1}{\sqrt{n}}\right)$$
$$\left[n\tilde{G}_n(\phi_x^2(t), x) + \sqrt{n^2\tilde{G}_n((e^{-x} - e^{-t})^4; x)}\sqrt{n^2\tilde{G}_n(\phi_x^4(t), x)} \right].$$

Proof: By the Taylor's formula, we can write such that

$$f(t) = f(x) + f'(x)(t - x) + \frac{f''(x)}{2}(t - x)^2 + h(t, x)(t - x)^2,$$

where $h(t, x) := [f''(\eta) - f''(x)]/2$, η lying between x and t, and h is a continuous function that vanishes at zero. Applying the operator $\tilde{G}_n$ to both sides of above equality, we obtain

$$\tilde{G}_n(f; x) - f(x) = f'(x)\tilde{G}_n(\phi_x^1(t), x) + \frac{f''(x)}{2}\tilde{G}_n(\phi_x^2(t), x)$$
$$+ \tilde{G}_n(h(t, x)(t - x)^2; x),$$

also we can write that

$$\left| n[\tilde{G}_n(f; x) - f(x)] + \frac{Ax(2+x)}{2}f'(x) - \frac{x(2+x)}{2}f''(x) \right|$$
$$\leq f'(x)\left[n\tilde{G}_n(\phi_x^1(t), x) + \frac{Ax(2+x)}{2} \right] + \frac{f''(x)}{2}\left[n\tilde{G}_n(\phi_x^1(t), x) - \frac{x(2+x)}{2} \right]$$
$$+ n\tilde{G}_n(|h(t, x)|(t - x)^2; x).$$

To estimate last inequality following [1], we get

$$|h(t,x)| \leq \left(1 + \frac{(e^{-x} - e^{-t})^2}{\delta^2}\right) \omega^*(f''; \delta).$$

Since

$$|h(y,x)| \leq \begin{cases} 2\omega^*(f''; \delta), |e^{-x} - e^{-t}| < \delta \\ 2\frac{(e^{-x} - e^{-t})^2}{\delta^2}\omega^*(f''; \delta), |e^{-x} - e^{-t}| \geq \delta \end{cases}.$$

Therefore, we have

$$|h(t,x)| \leq 2\left(1 + \frac{(e^{-x} - e^{-t})^2}{\delta^2}\right) \omega^*(f''; \delta).$$

Applying the Cauchy–Schwarz inequality, we obtain

$$n\tilde{G}_n(|h(t,x)|(t-x)^2; x) \leq 2n\omega^*(f''; \delta)\tilde{G}_n(\phi_x^2(t), x)$$
$$+ \frac{2n}{\delta^2}\omega^*(f''; \delta)\sqrt{\tilde{G}_n((e^{-x} - e^{-t})^4; x)}\sqrt{\tilde{G}_n(\phi_x^4(t), x)}.$$

Choosing $\delta = \frac{1}{\sqrt{n}}$, we have

$$n\tilde{G}_n(|h(t,x)|(t-x)^2; x)$$
$$\leq 2\omega^*\left(f''; \frac{1}{\sqrt{n}}\right)\left[n\tilde{G}_n(\phi_x^2(t), x) + \sqrt{n^2\tilde{G}_n((e^{-x} - e^{-t})^4; x)}\sqrt{n^2\tilde{G}_n(\phi_x^4(t), x)}\right].$$

Thus, we immediately get

$$\left|n[\tilde{G}_n(f; x) - f(x)] + \frac{Ax(2+x)}{2}f'(x) - \frac{x(2+x)}{2}f''(x)\right|$$
$$\leq \left|n\tilde{G}_n(\phi_x^1(t), x) + 2ax\right||f'(x)| + \frac{1}{2}\left|n\tilde{G}_n(\phi_x^2(t), x) - 2x\right||f''(x)|$$
$$+ 2\omega^*\left(f''; \frac{1}{\sqrt{n}}\right)\left[n\tilde{G}_n(\phi_x^2(t), x) + \sqrt{n^2S_n^*((e^{-x} - e^{-t})^4; x)}\sqrt{n^2\tilde{G}_n(\phi_x^4(t), x)}\right],$$

which was our claim. $\square$

Corollary 15.1 Let $f, f'' \in C^*[0, \infty)$, then for $x \in [0, \infty)$, we have

$$\lim_{n\to\infty} n[\tilde{G}_n(f, x) - f(x)] = \frac{x(2+x)}{2}[Af'(x) + f''(x)].$$

The proof of the above corollary follows from Theorem 15.8.

15.5 Conclusion

It is observed here that for the Baskakov–Szász type operators, the m.g.f. can be obtained, which may further be used to find the moments. Here we estimate some direct results for different variants of Baskakov–Szász operators. In addition, here we used the software Mathematica to obtain the expansions, limits, and moments and so on, which is different from the methods discussed in [27, 28].

References

1 Kantorovich, L.V. (1930) Sur certains dévelopements suivant les polynómes de la forme de S. *Bernstein C. R. Acad. Sci. USSR*, **1**, 563–568.

2 Durrmeyer, J.L. (1967) Une formule d́ inversion de la transformee de Laplace applications á la theórie desmoments. Thése de 3 e cycle. Faculté des Sciences de Í Université de Paris.

3 Sahai, A. and Prasad, G. (1985) On simultaneous approximation by modified Lupas operators. *J. Approx. Theory*, **45**, 122–128.

4 Mazhar, S.M. and Totik, V. (1985) Approximation by Szász operators. *Acta Sci. Math.*, **49**, 257–269.

5 Gupta, V. (1994) A note on modified Baskakov type operators. *Approx. Theory Appl.*, **10** (3), 74–78.

6 Gupta, V. (1995) Global approximation by modified Baskakov type operators. *Publ. Math.*, **39** (2), 263–271.

7 Gupta, V. (1995) Simultaneous approximation by Szász-Durrmeyer operators. *Math. Stud.*, **64** (14), 27–36.

8 Gupta, V. (2002) Rate of convergence on Baskakov Beta Bézier operators for functions of bounded variation. *Int. J. Math. Math. Sci.*, **32** (8), 471–479.

9 Gupta, V. (2002) Inequality estimate on Durrmeyer type operators. *J. Nonlinear Funct. Anal. Appl.*, **7** (4), 481–494.

10 Gupta, V. (2003) Rate of approximation by new sequence of linear positive operators. *Comput. Math. Appl.*, **45** (12), 1895–1904.

11 Gupta, V. (2004) Degree of approximation to function of bounded variation by Bézier variant of MKZ operators. *J. Math. Anal. Appl.*, **289** (1), 292–300.

12 Gupta, V. (2004) The Bezier variant of Kantorovitch operators. *Comput. Math. Appl.*, **47** (2/3), 227–232.

13 Gupta, V. (2004) Rate of convergence of Baskakov-Beta Bézier operators for locally bounded functions. *Turk. J. Math.*, **28** (3), 271–280.

14 Gupta, V. (2005) Some approximation properties for modified Baskakov type operators. *Georgian Math. J.*, **12** (2), 217–228.

15 Gupta, V. (2005) An Estimate on the convergence of Baskakov Bézier operators. *J. Math. Anal. Appl.*, **312**, 280–288.

16 Gupta, V. (2006) Error estimation by mixed summation integral type operators. *J. Math. Anal. Appl.*, **313** (2), 632–641.

17 Gupta, V. (2008) Some approximation properties on q-Durrmeyer operators. *Appl. Math. Comput.*, **197** (1), 172–178.

18 Gupta, V. (2008) Rate of approximation for the Balazs-Kantorovich-Bézier operators. *Appl. Math. Comput.*, **199** (2), 823–827.

19 Gupta, V. (2009) Approximation for modified Baskakov Durrmeyer type operators. *Rocky Mt. J. Math.*, **39** (3), 825–841.

20 Gupta, V. (2009) Certain family of Durrmeyer type operators. *Ann. Univ. Mariae Curie-Sklodowska Lublin-Polonia, Ser. A*, **LXIII**, 109–115.

21 Gupta, V. (2010) A note on q Baskakov-Szász operators. *Lobachevskii J. Math.*, **31** (4), 359–366.

22 Gupta, V. (2013) Approximation properties by Bernstein-Durrmeyer type operators. *Complex Anal. Oper. Theory*, **7**, 363–374.

23 Gupta, V. (2013) Combinations of integral operators. *Appl. Math. Comput.*, **224**, 876–881.

24 Gupta, V. (2015) Direct estimates for a new general family of Durrmeyer type operators. *Boll. Unione Mat. Ital.*, **7**, 279–288.

25 Gupta, V., Rassias, Th.M., and Sinha, J. (2016) A survey on Durrmeyer operators, in *Contributions in Mathematics and Engineering* (eds P.M. Pardalos and T.M. Rassias), Springer International Publishing, Switzerland, pp. 299–312.

26 Boyanov, B.D. and Veselinov, V.M. (1970) A note on the approximation of functions in an infinite interval by linear positive operators. *Bull. Math. Soc. Sci. Math. Roum.*, **14** (62), 9–13.

27 Holhoş, A. (2010) The rate of approximation of functions in an infinite interval by positive linear operators. *Stud. Univ. Babe-Bolyai, Math.*, (2), 133–142.

28 Acar, T., Aral, A., and Gonska, H. (2017) On Szász-Mirakyan operators preserving $e^{2ax}, a > 0,$. *H. Mediterr. J. Math.*, **14** (6). doi: 10.1007/s00009-016-0804-7.

29 Gupta, V. and Srivastava, G.S. (1993) Simultaneous approximation by Baskakov-Szász type operators. *Bull. Math. Soc. Sci. Math. Roum., Nouv. Sr.*, **37** (85), 73–85.

30 Gupta, V. and Maheshwari, P. (2003) On Baskakov-Szász type operators. *Kyungpook Math. J.*, **43** (3), 315–325.

31 Gupta, V. and Agarwal, R.P. (2014) *Convergence Estimates in Approximation Theory*, Springer-Verlag.

32 Gupta, V. and Malik, N. (2016) Direct estimations of new generalized Baskakov-Szász operators. *Publ. Inst. Math., Nouv. Sr.*, **99** (113), 265–279.

33 Agrawal, P.N. and Mohammad, A.J. (2003) Linear combination of a new sequence of linear positive operators. *Rev. Union Mat. Argent.*, **44** (1), 33–41.

34 King, J.P. (2003) Positive linear operators which preserve x^2. *Acta Math. Hungar.*, **99** (3), 203–208.

35 Gonska, H., Pitul, P., and Rasa, I. (2009) General King-type operators. *Result. Math.*, **53** (3-4), 279–296.

16

Well-Posed Minimization Problems via the Theory of Measures of Noncompactness

Józef Banaś and Tomasz Zając

Department of Nonlinear Analysis, Rzeszów University of Technology, Aleja Powstańców Warszawy 8, 35-959 Rzeszów, Poland

16.1   Introduction

The important aim of this chapter is to discuss the well-posed minimization problems in connection with the theory of measures of noncompactness. We show that the above-mentioned theory enables us to create a deep insight into the problem of attaining of infimum by functionals defined on some closed subsets of a metric space.

Investigations of this chapter are continuation of those conducted in the chapter in [1] in which we described the concept of the well-posedness in the context of the theory of measures of noncompactness. While we formulated the basic concepts of our approach and we presented the main theorems characterizing minimization problems that are well-posed in a general sense in [1], here we focus on delivering of several examples of functionals defined in diverse metric and Banach spaces and on discussing the well-posedness of minimization problems for those functionals.

First, we are going to explain how our theory can be applied in the case when we consider concrete functionals in various Banach spaces and when we search the well-posedness of those functionals with respect to various measures of noncompactness. It is worthwhile mentioning that if a minimization problem is well-posed with respect to a measure of noncompactness (or, equivalently, with respect to a kernel of a measure of noncompactness) then we obtain the information that the set of points at which a considered functional attains its infimum forms a compact set belonging to the kernel of a considered measure of noncompactness. Such information allows us to characterize the set of those points in terms of properties that have sets belonging to a kernel of a measure of noncompactness in question.

Our approach presented here was initiated in [1] but the idea of the well-posedness of minimization problems goes back to Tikhonov [2] and was

Mathematical Analysis and Applications: Selected Topics, First Edition.
Edited by Michael Ruzhansky, Hemen Dutta, and Ravi P. Agarwal.

subsequently developed by Levitin and Polyak [3] and then by Furi and Vignoli [4]. We refer also to expository books and monographs, where optimization problems are widely treated [5–9].

Finally, let us point out that the approach to the study of minimization problems via the theory of measures of noncompactness seems to be new and, according to our knowledge, up to now there have appeared only exposition [1] presenting the approach discussed in this chapter. Nevertheless, some basic ideas initiating this theory have been given in paper [10], which was not noticed by mathematical community.

16.2 Minimization Problems and Their Well-Posedness in the Classical Sense

This section is devoted to formulate the minimization problem and to present the classical notation and definitions concerning this problem.

Thus, let us assume that (X, d) is a given metric space and let D be a nonempty subset of the space X. Denote by $\mathbb{R}$ the set of real numbers. We will write $\mathbb{R}_+$ to denote the interval $[0, \infty)$.

Next, let J be a functional acting from D into $\mathbb{R}$. We will always assume that D is a closed subset of the metric space X and J is a lower bounded and lower semicontinuous functional on the set D. Such assumptions ensure that there exists infimum of J on the set D. So, let us write

$$m_J = \inf_{x \in D} J(x).$$

Obviously, m_J is a real number.

The sequence $(x_n) \subset D$ will be called the *minimizing sequence* for the functional J on the set D if

$$\lim_{n \to \infty} J(x_n) = m_J.$$

Observe that if (x_n) is a minimizing sequence of the functional J on the set D then its limit has to be the cluster point of the set D. Thus, if we impose the assumption concerning the closedness of the set D then the limit of a minimizing sequence for J must belong to the set D. It is also worthwhile noticing that the assumption guaranteeing that the minimizing sequence (x_n) of the functional J on D is convergent, implies that (x_n) has always a unique limit, that is, each minimizing sequence of the functional J tends to the same limit x which is the cluster point of the set D.

Indeed, suppose contrary. This means that there exist two minimizing sequences (x_n) and (y_n) for the functional J on the set D such that (x_n) converges to a point x and (y_n) converges to a point y and $x \neq y$. In such a case the sequence $\{x_1, y_1, x_2, y_2, x_3, y_3, \dots\}$ is also a minimizing sequence for J on the set D, which is not convergent.

Now, we recall the definition of the well-posedness of the functional J on the set D given by Tikhonov [2].

Definition 16.1   We will say that the minimization problem for the functional J is *well-posed* on the set D if every minimizing sequence of the functional J in the set D is convergent in the metric space X.

From the above properties concerning the minimization problem for the functional J on D it follows that if the minimization problem for J is *well-posed on the set D in the sense of Tikhonov* (i.e., in the sense of Definition 16.1) then J has exactly one minimum in the set D.

One of the most important topic illustrating the minimization problem is the problem concerning the so-called nearest point in a set D to a given point y. To describe briefly that problem, let us assume that D is a nonempty closed subset of the metric space X and $y \in X$ is a given point such that $y \notin D$.

A point x_0 realizing the distance of y to the set D, that is, a point $x_0 \in D$ such that

$$d(y, x_0) = \text{dist}(y, D) = \inf\{d(y, x): \quad x \in X\}$$

is called the *nearest point* of the set D to the point y.

Let us mention that in the fixed point theory the more general problem than that presented above, called the best proximity point, is considered [11–14].

Observe, that in general a point being the nearest point of the set D to a point y not always exist and, even if it exists, it has not to be determined uniquely. The problem of the existence and uniqueness of nearest points depends on the topology of the metric space X and (if we consider it in the Banach space setting) on the geometry of an underlying Banach space. This problem was thoroughly discussed in [1] and we will not repeat here those considerations.

In what follows, we describe the generalization of the well-posedness mini-mization problems in the classical Tikhonov sense given by Levitin and Polyak [3]. As previously discussed, we will assume that J is a lower bounded and lower semicontinuous functional defined on a closed subset D of a metric space (X, d).

Definition 16.2   We say that the minimization problem for the functional J is *well-posed in the sense of Levitin and Polyak* if each minimizing sequence for the functional J is compact.

The assumption included in the above definition and requiring that each min-imizing sequence for the functional J defined on the set D is compact means that the set of all terms of any minimizing sequence is compact (more precisely: relatively compact). Particularly, this implies that the set of all accumulation (limit) points of any minimizing sequence for the functional J is compact. On the other hand this allows us to deduce that the set of all points of the set D at which the functional J attains its infimum is a compact subset of the metric space X.

Let us observe that if the minimization problem for the functional J is well-posed in the sense of Tikhonov then it is well-posed in the sense of Levitin–Polyak. The inverse implication is not true (cf. [1]). This means that Definition 16.2 creates the essential generalization of Definition 16.1.

Further, we are going to discuss other generalizations of the well-posedness of minimization problems. In order to present suitable definitions, we will need to be familiar with the concept of a measure of noncompactness. We will discuss this concept in the following section.

16.3 Measures of Noncompactness

The concept of a measure of noncompactness was introduced by Kuratowski [15] in 1930. Namely, for a nonempty and bounded subset A of a metric space (X, d) we define the quantity $\alpha(X)$ in the following way:

$$\alpha(A) = \inf \left\{ \varepsilon > 0 : \ A \text{ can be covered by a finite number of sets of diameter smaller than } \varepsilon \right\}.$$
(16.1)

Notice that the quantity $\alpha(A)$ can be defined equivalently as

$$\alpha(A) = \inf \left\{ \varepsilon > 0 : A \subset \cup_{i=1}^{n} A_i, \ A_i \subset X, \text{diam } A_i < \varepsilon \quad (i = 1, 2, \dots, n), \ n \text{ is an arbitrary natural number} \right\},$$

where the symbol diam Y denotes the diameter of the set $Y (Y \subset X)$. The quantity $\alpha(A)$ is nowadays called the *Kuratowski measure of noncompactness* of the set A.

In order to discuss the Kuratowski measure $\alpha(A)$ and also other measures of noncompactness, we introduce first some notation. Namely, if A is a subset of a metric space X then the symbol $\overline{A}$ will denote the closure of the set A. By the symbol $B_0(x, r)$ we will denote the open ball centered at x and with radius r. The closed ball centered at x and with radius r will be denoted by $B(x, r)$.

When X has the structure of a Banach space, we can define the symbol Conv A that denotes the closed convex hull of the set A. Moreover, if A, B are subsets of a Banach space X then we can define the algebraic operations $A + B$ and λA for $\lambda \in \mathbb{R}$.

Apart from this, we will denote by $\mathfrak{M}_X$ the family of all nonempty and bounded subsets of the complete metric space X and by $\mathfrak{N}_X$ its subfamily consisting of all relatively compact sets.

Now, let us observe that the Kuratowski measure of noncompactness α defined by (16.1) is a function acting from $\mathfrak{M}_X$ into $\mathbb{R}_+$ and has the following properties [16]:

(i) $\alpha(A) = 0 \iff A \in \mathfrak{N}_X$,
(ii) $A \subset B \Rightarrow \alpha(A) \leqslant \alpha(B)$,

(iii) $\alpha(\overline{A}) = \alpha(A)$,
(iv) $\alpha(A) \leqslant \operatorname{diam} A$.

The most important property of the function α is contained in the following theorem.

Theorem 16.1 Let (A_n) be a sequence of nonempty, bounded, and closed subsets of the space X such that $A_n \supset A_{n+1}$ for $n = 1, 2, \ldots$ and $\lim_{n \to \infty} \alpha(A_n) = 0$. Then the intersection set A_∞ of the sequence (A_n), that is, the set

$$A_\infty = \bigcap_{n=1}^{\infty} A_n$$

is nonempty and compact.

The property of the function α described by the above theorem is called the *Cantor intersection property* since it is the generalization of the classical Cantor intersection theorem.

When α is considered in a Banach space E, we can prove that it has the following additional properties:

(i) $\alpha(\operatorname{Conv} A) = \alpha(A)$,
(ii) $\alpha(A + B) \leqslant \alpha(A) + \alpha(B)$,
(iii) $\alpha(\lambda A) = |\lambda| A$ for $\lambda \in \mathbb{R}$.

It is worthwhile mentioning that property (v) was proved by Darbo [17] who (with help of the Kuratowski measure of noncompactness α) obtained an important generalization of the classical Schauder fixed point principle.

The concept of the Kuratowski measure of noncompactness was modified in papers [18, 19], where the authors introduced the new quantity being a kind of a measure of noncompactness. Indeed, for a set $A \in \mathfrak{M}_X$ (X is a complete metric space) they defined the so-called *Hausdorff measure of noncompactness* $\chi(A)$ in the following way

$$\chi(A) = \inf\{\varepsilon > 0 : \ A \text{ has a finite } \varepsilon - \text{net in } X\}. \tag{16.2}$$

It may be shown that the function χ has all the above listed properties (i)–(vii) of the function α. Moreover, the function χ has also the Cantor intersection property expressed in Theorem 16.1 (cf. [16, 20]).

On the other hand, it turns out that the Hausdorff measure χ is more convenient than the Kuratowski measure α. There are two essential reasons which justify such an opinion. To present the first reason, we first describe the concept of the Hausdorff distance between sets [21]. To this end, for an arbitrary nonempty subset A of the metric space (X, d) and for a given number $r > 0$ let us denote by $B(A, r)$ the ball centered at the set A and with radius r, which is defined by the equality

$$B(A, r) = \bigcup_{x \in A} B_0(x, r).$$

Next, for nonempty subsets A, B of X let us define:

$$h(A, B) = \inf\{r > 0 : A \subset B(B, r)\}.$$

The number $h(A, B)$ is called the *nonsymmetric Hausdorff distance* of the set A to the set B. Finally, let us define the quantity $H(A, B)$ as follows:

$$H(A, B) = \max\{h(A, B), h(B, A)\}.$$

Obviously, the above quantity is well defined provided we assume that $A, B \in \mathfrak{M}_X$.

The number $H(A, B)$ is called the *Hausdorff distance* of the sets A and B. It can be shown that H is the pseudometric on the family $\mathfrak{M}_X$ and is the metric if we consider it on the family $\mathfrak{M}_X^c$ consisting of all nonempty, bounded and closed subsets of the metric space X. If the space X is complete then the metric space $(\mathfrak{M}_X^c, H)$ is also complete [21].

If $\mathcal{Z}$ is a nonempty subfamily of the family $\mathfrak{M}_X$ then for an arbitrary set $X \in \mathfrak{M}_X$ we denote by $H(X, \mathcal{Z})$ the distance of X to $\mathcal{Z}$ with respect to the Hausdorff distance H, that is,

$$H(X, \mathcal{Z}) = \inf\{H(X, Z) : Z \in \mathcal{Z}\} = \text{dist}(X, \mathcal{Z}).$$

Then, we have the following theorem [16] (cf. also [22]).

Theorem 16.2 Let (X, d) be a complete metric space. Then

$$\chi(A) = H(A, \mathfrak{N}_X)$$

for an arbitrary set $A \in \mathfrak{M}_X$.

The equality from the above theorem has some consequences being important in applications [16, 23].

The second reason showing that the Hausdorff measure χ is more convenient than the Kuratowski measure of noncompactness α is caused by the fact that in some Banach spaces we are able to give formulas expressing the Hausdorff measure of noncompactness χ in a convenient way. On the other hand, there are no such formulas known for the Kuratowski measure α [16, 20].

Now, we are going to provide the mentioned formulas in some Banach spaces.

First, let us consider the Banach space $C([a, b])$ consisting of all real functions $x = x(t)$ defined and continuous on the interval $[a, b]$ and furnished with the standard maximum norm

$$||x|| = \max\{|x(t)| : t \in [a, b]\}.$$

Then, taking into account the Arzéla–Ascoli criterion for compactness in $C([a, b])$ we can express the Hausdorff measure of noncompactness in the space $C([a, b])$ in the way described subsequently below.

Namely, for $x \in C([a, b])$ denote by $\omega(x, \varepsilon)$ the modulus of continuity of the function x defined as

$$\omega(x, \varepsilon) = \sup\{|x(t) - x(s)| : t, s \in [a, b], |t - s| \leqslant \varepsilon\},$$

for $\varepsilon > 0$. Next, for a set $X \in \mathfrak{M}_{C([a,b])}$ let us assume

$$\omega(X, \varepsilon) = \sup\{\omega(x, \varepsilon) : \ x \in X\},$$

$$\omega_0(X) = \lim_{\varepsilon \to 0} \omega(X, \varepsilon).$$

It can be shown [16, 19] that for $X \in \mathfrak{M}_{C([a,b])}$, we have

$$\chi(X) = \frac{1}{2}\omega_0(X). \tag{16.3}$$

Now, let us denote by c_0 the space of all real sequences $x = (x_n)$ converging to zero and endowed with the maximum norm, that is,

$$||x|| = ||(x_n)|| = \max\{|x_n| : \quad n = 1, 2, \dots\}.$$

In order to present the formula expressing the Hausdorff measure χ in c_0 let us fix a set $X \in \mathfrak{M}_{c_0}$. Then we have the following equality [16]

$$\chi(X) = \lim_{n \to \infty} \left\{ \sup_{x \in X} \{\max\{|x_k| : \quad k \geqslant n\}\} \right\}. \tag{16.4}$$

It turns out that the situation is more complicated in the sequence space c consisting of all sequences $x = (x_n)$ converging to proper limits and furnished with the standard supremum norm

$$||x|| = ||(x_n)|| = \sup\{|x_n| : \quad n = 1, 2, \dots\}.$$

Indeed, we do not know a formula expressing the Hausdorff measure of noncompactness χ in this space. However, if for an arbitrary set $X \in \mathfrak{M}_c$ we define the quantity

$$\mu(X) = \lim_{n \to \infty} \left\{ \sup_{x \in X} \{\sup\{|x_p - x_q| : \quad p, q \geqslant n\}\} \right\}, \tag{16.5}$$

then it can be shown that this quantity has all properties (i)–(vii) of the Hausdorff (or Kuratowski) measure of noncompactness and the following inequalities hold [24]

$$\chi(X) \leqslant \mu(X) \leqslant 2\chi(X) \tag{16.6}$$

for $X \in \mathfrak{M}_c$, where $\chi(X)$ denotes the Hausdorff measure of X in the space c. Thus, in the case of the space c we do not know a formula expressing the Hausdorff measure χ but we can provide the formula for the quantity μ being in some sense equivalent to the Hausdorff measure χ.

Taking into account our further considerations such a situation is quite satisfactory.

Further, we provide the formula expressing the Hausdorff measure of noncompactness χ in the Banach sequence space l_p consisting of all sequences $x = (x_n)$ such that $\sum_{n=1}^{\infty} |x_n|^p < \infty$, where p is a fixed number, $1 \leqslant p < \infty$. The space l_p is furnished with the norm

$$||x|| = ||(x_n)|| = \left(\sum_{n=1}^{\infty} |x_n|^p \right)^{1/p}.$$

Thus, let us fix an arbitrary set $X \in \mathfrak{M}_{l_p}$. Then, we have (cf. [16])

$$\chi(X) = \lim_{n \to \infty} \left\{ \sup_{x \in X} \left(\sum_{k=n}^{\infty} |x_k|^p \right)^{1/p} \right\}.$$

Particularly, the Hausdorff measure of noncompactness in the space l_1 can be expressed by the formula

$$\chi(X) = \lim_{n \to \infty} \left\{ \sup_{x \in X} \sum_{k=n}^{\infty} |x_k| \right\}. \tag{16.7}$$

Our survey of Banach spaces in which we know formulas expressing the Hausdorff measure of noncompactness is now complete (cf. [16, 20, 25]). Let us only mention that in the classical Lebesgue space $L^p(a, b)(1 \leqslant p < \infty)$ we do not know a formula for the Hausdorff measure χ but (similarly as in the case of the sequence space c) we can provide the formula expressing a quantity equivalent to the measure χ [16, 26].

It is worthwhile mentioning that the Kuratowski measure of noncompactness α and the Hausdorff measure χ are equivalent in an arbitrary metric space X, that is, for any set $A \in \mathfrak{M}_X$ the following inequalities are satisfied [16]

$$\chi(X) \leqslant \alpha(X) \leqslant 2\chi(X). \tag{16.8}$$

Later on we will discuss more detailed the problem of the equivalence of measures of noncompactness.

As we pointed out above there are Banach spaces in which we do not know formulas expressing the Hausdorff measure of noncompactness χ but only we know formulas expressing quantities equivalent to that measure. But there are Banach spaces in which the situation is even more complicated since in those spaces we do not know convenient criteria for relative compactness of sets. In spaces of such a type we know only sufficient conditions for relative compactness of sets.

The above-described situation forced us to formulate an axiomatic approach to the definition of a measure of noncompactness, which overcomes the above-indicated difficulties and which creates a handy and useful tool being applicable in nonlinear analysis.

The below given axiomatics seems to be an appropriate answer to the above raised requirements. This axiomatics was introduced in [16]. Taking into account our further considerations we present the mentioned axiomatics in a Banach space E. Obviously, we will also discuss the mentioned axiomatics in the case of a complete metric space.

Definition 16.3 The function $\mu : \mathfrak{M}_E \to \mathbb{R}_+$ is said to be a *measure of noncompactness* in the Banach space E if it satisfies the following conditions:

1° The family ker $\mu = \{A \in \mathfrak{M}_E : \mu(A) = 0\}$ in nonempty and ker $\mu \subset \mathfrak{N}_E$
2° $A \subseteq B \Rightarrow \mu(A) \leqslant \mu(B)$
3° $\mu(\overline{A}) = \mu(A)$
4° $\mu(\text{Conv } A) = \mu(A)$
5° $\mu(\lambda A + (1 - \lambda)B) \leqslant \lambda\mu(A) + (1 - \lambda)\mu(B)$ for $\lambda \in [0, 1]$
6° If (A_n) is a sequence of closed sets from $\mathfrak{M}_E$ such that $A_{n+1} \subset A_n$ for $n = 1, 2, \ldots$ and if $\lim_{n\to\infty} \mu(A_n) = 0$ then the intersection set $A_\infty = \bigcap_{n=1}^{\infty} A_n$ is nonempty.

The family ker μ described in axiom 1° is called the *kernel* of the measure of noncompactness μ.

Observe also that in axiom 6°, from the inclusion $A_\infty \subset A_n$ for $n = 1, 2, \ldots$, it follows that $\mu(A_\infty) \leqslant \mu(A_n)$ for $n = 1, 2, \ldots$. This implies that $\mu(A_\infty) = 0$, that is, the set A_∞ is a member of the family ker μ.

Now, we give an equivalent approach to the concept of a measure of noncompactness. In that approach, we will expose the role of the kernel of the measure of noncompactness.

Definition 16.4 A nonempty family $\mathcal{P}$ ($\mathcal{P} \subset \mathfrak{N}_E$) will be called the *kernel (of a measure of noncompactness)* if it satisfies the following conditions:

1° $A \in \mathcal{P} \Rightarrow \overline{A} \in \mathcal{P}$,
2° $A \in \mathcal{P}, B \subset A, B \neq \emptyset \Rightarrow B \in \mathcal{P}$
3° $A \in \mathcal{P} \Rightarrow \text{Conv } A \in \mathcal{P}$,
4° $A, B \in \mathcal{P} \Rightarrow \lambda A + (1 - \lambda)B \in \mathcal{P}$ for $\lambda \in [0, 1]$,
5° $\mathcal{P}^c = \{A \in \mathcal{P} : A = \overline{A}\}$ is closed in $\mathfrak{M}_E^c$ with respect to Hausdorff metric.

Observe that the family $\mathfrak{N}_E$ is a kernel of a measure of noncompactness since it satisfies all axioms 1°–5° of Definition 16.4. Other example of the kernel of a measure of noncompactness may serve the family $\mathfrak{N}_E^0$ formed by all singletons from E.

Now, we provide the definition of a measure of noncompactness connected with the concept of the kernel given in Definition 16.4.

Definition 16.5 Let $\mathcal{P}$ be a given kernel of a measure of noncompactness. The function $\mu : \mathfrak{M}_E \to \mathbb{R}_+$ is said to be a *measure of noncompactness with the kernel $\mathcal{P}$* (ker $\mu = \mathcal{P}$) if it satisfies the following conditions:

1° $\mu(A) = 0 \Leftrightarrow A \in \mathcal{P}$
2° $A \subseteq B \Rightarrow \mu(A) \leqslant \mu(B)$
3° $\mu(\overline{A}) = \mu(A)$
4° $\mu(\text{Conv } A) = \mu(A)$

5° $\mu(\lambda A + (1 - \lambda)B) \leqslant \lambda\mu(A) + (1 - \lambda)\mu(B)$ for $\lambda \in [0, 1]$

6° If (A_n) is a sequence of closed sets from $\mathfrak{M}_E$ such that $A_{n+1} \subset A_n$ for $n = 1, 2, \ldots$, and if $\lim\limits_{n\to\infty} \mu(A_n) = 0$ then the set $A_\infty = \bigcap\limits_{n=1}^{\infty} A_n$ is nonempty.

In what follows we discuss the concepts included in Definitions 16.3–16.5 in the situation when we assume that the underlying space is a complete metric space (with a metric d). In order to formulate suitable definitions we have to delete axioms connected with the algebraic structure of the family $\mathfrak{M}_E$. For example, we formulate counterparts of Definitions 16.3 and 16.4.

Definition 16.6 Let X be a complete metric space. The function $\mu : \mathfrak{M}_X \to \mathbb{R}_+$ is said to be a measure of noncompactness in X if it is subject to the following conditions:

1° The family $\ker \mu = \{A \in \mathfrak{M}_X : \mu(A) = 0\}$ is nonempty and $\ker \mu \subset \mathfrak{N}_X$

2° $A \subseteq B \Rightarrow \mu(A) \leqslant \mu(B)$,

3° $\mu(\overline{A}) = \mu(A)$,

4° If (A_n) is a sequence of closed sets from $\mathfrak{M}_X$ such that $A_{n+1} \subset A_n$ $(n = 1, 2, \ldots)$ and if $\lim\limits_{n\to\infty} \mu(A_n) = 0$ then the intersection set $A_\infty = \bigcap\limits_{n=1}^{\infty}$ is nonempty.

Definition 16.7 A nonempty family $\mathcal{P} \subset \mathfrak{N}_X$ (X is a complete metric space) is called the kernel of a measure of noncompactness if it satisfies the following conditions:

1° $A \in \mathcal{P} \Rightarrow \overline{A} \in \mathcal{P}$

2° $A \in \mathcal{P}, B \subset A, B \neq \emptyset \Rightarrow B \in \mathcal{P}$

3° $\mathcal{P}^c$ is closed in $\mathfrak{M}_X^c$ with respect to the topology generated by the Hausdorff metric.

In applications connected with the theory of measures of noncompactness we consider frequently measures with the same kernel being "almost" the same from the view point of properties of those measures. This leads to the concept of equivalent measures of noncompactness. In the sequel, we will only deal with measures of noncompactness and kernels of measures defined in a Banach space E.

Definition 16.8 Let μ_1, μ_2 be measures of noncompactness defined in the Banach space E with the same kernel $\mathcal{P}$. We say that μ_1, μ_2 are *equivalent* provided there exist two positive constants α, β such that the inequalities

$$\alpha\mu_1(A) \leqslant \mu_2(A) \leqslant \beta\mu_1(A)$$

are satisfied for an arbitrary set $A \in \mathfrak{M}_E$.

It is clear that the relation $\approx$ defined between measures of noncompactness μ_1, μ_2 with the same kernel $\mathcal{P}$ in the following way:

$$\mu_1 \approx \mu_2 \Leftrightarrow \mu_1 \text{ and } \mu_2 \text{ are equivalent}$$

is an equivalence relation. It is not hard to check that the equivalence class $[\mu]$ with respect to the relation $\approx$ has the algebraic structure of the cone. This means that if μ_1 and μ_2 are measures of noncompactness with the same kernel $\mathcal{P}$ belonging to the equivalence class $[\mu]$ then $\mu_1 + \mu_2$ and $c\mu_1 (c > 0)$ belong also to the equivalence class $[\mu]$.

It is an interesting question to investigate if arbitrary measures of noncompactness μ_1 and μ_2 with the same kernel $\mathcal{P}$ are equivalent in the sense of Definition 16.8. The answer to this question is, in general, negative even if we consider measures of noncompactness having "nice" properties. We return to the above-question later on.

In the sequel, we pay our attention to the class of measures of noncompactness having some additional useful properties apart from those included in Definitions 16.3 and 16.5. Thus, assume that μ is a measure of noncompactness in the Banach space E in the sense of Definition 16.3. We will say that μ is *sublinear* if it satisfies the following conditions:

7° $\mu(A + B) \leqslant \mu(A) + \mu(B)$,
8° $\mu(\lambda A) = |\lambda| \mu(A)$ for $\lambda \in \mathbb{R}$.

If a measure μ satisfies the condition

9° $\mu(A \cup B) = \max\{\mu(A), \mu(B)\}$

then, we call it a *measure with maximum property*.

If μ is a measure of noncompactness such that ker $\mu = \mathfrak{N}_E$ then μ is called a *full measure*.

Finally, a full and sublinear measure with maximum property will be called a *regular measure*.

Let us pay attention to the fact that the Kuratowski measure of noncompactness α defined by (16.1) and the Hausdorff measure χ defined by (16.2) are regular measures. In view of (16.8) those measures of noncompactness are equivalent. It was shown in [27] that in some Banach spaces there exist regular measures of noncompactness that are not equivalent to the Hausdorff measure χ. The question of the equivalence of regular measures of noncompactness to the Hausdorff measure was recently studied intensively in [28–30].

It is worthwhile mentioning that the quantity μ defined in the sequence space c by formula (16.5) is a regular measure of noncompactness which, in view of (16.6) is equivalent to the Hausdorff measure χ. Similar assertion can also be formulated with respect to a quantity defined in the Lebesgue space $L^p(a, b)$ and mentioned previously (cf. [16, 26]).

The above facts show that, in general, measures of noncompatness with the same kernel have not to be equivalent.

Finally, we provide a few examples of measures of noncompactness which are not regular. Such measures are mainly constructed in those Banach spaces in which we do not know necessary and sufficient conditions for relative compactness of sets [16]. Our examples will be constructed in the Banach space $BC(\mathbb{R}_+)$ consisting of all functions $x : \mathbb{R}_+ \to \mathbb{R}$ which are continuous and bounded on $\mathbb{R}_+$. This space is endowed with the classical supremum norm, that is,

$$||x|| = \sup\{|x(t)| : t \in \mathbb{R}_+\}.$$

Let us point out that the space $BC(\mathbb{R}_+)$ plays an important role in applications to nonlinear functional integral equations [31].

In order to define the above announced measures of noncompactness let us fix arbitrarily numbers $\varepsilon > 0$ and $T > 0$. Next, take a set $X \in \mathfrak{M}_{BC(\mathbb{R}_+)}$. For a fixed function $x \in X$ let us put

$$\omega^T(x, \varepsilon) = \sup\{|x(t) - x(s)| :   t, s \in [0, T], |t - s| \leqslant \varepsilon\}.$$

The above quantity represents the so-called *modulus of continuity* of the function x on the interval $[0, T]$. Further, let us define

$$\omega^T(X, \varepsilon) = \sup\{\omega^T(x, \varepsilon) : x \in X\}.$$

It is easily seen that the function $\varepsilon \to \omega^T(X, \varepsilon)$ is nondecreasing, so there exists a finite limit $\lim_{\varepsilon \to 0} \omega^T(X, \varepsilon)$. Thus, we can assume

$$\omega_0^T(X) = \lim_{\varepsilon \to 0} \omega^T(X, \varepsilon)$$

and

$$\omega_0(X) = \lim_{T \to \infty} \omega_0^T(X). \tag{16.9}$$

In contrast to the classical space $C([a, b])$ the quantity defined by (16.9) is not a measure of noncompactness in the space $BC(\mathbb{R}_+)$ (cf. [31]). In order to define measures of noncompactness in the space $BC(\mathbb{R}_+)$ with help of the above defined quantity $\omega_0(X)$ we have to add a component related to the behavior of functions from the set X at infinity. Thus, we define the following quantities:

$$a(X) = \lim_{T \to \infty} \left\{ \sup_{x \in X}\{\sup\{|x(t)| : t \geqslant T\}\} \right\},$$

$$b(X) = \lim_{T \to \infty} \left\{ \sup_{x \in X}\{\sup\{|x(t) - x(s)| : t, s \geqslant T\}\} \right\},$$

$$c(X) = \limsup_{t \to \infty} \operatorname{diam} X(t),$$

where $X(t) = \{x(t) : x \in X\}$ and the symbol $\operatorname{diam} X(t)$ denotes the diameter of the set $X(t)$. Next, let us define on the family $\mathfrak{M}_{BC(\mathbb{R}_+)}$ the quantities

μ_a, μ_b, μ_c in the following way:

$$\mu_a(X) = \omega_0(X) + a(X), \tag{16.10}$$
$$\mu_b(X) = \omega_0(X) + b(X), \tag{16.11}$$
$$\mu_c(X) = \omega_0(X) + c(X). \tag{16.12}$$

We can show [31] that the functions μ_a, μ_b, and μ_c defined subsequently by (16.10), (16.11), and (16.12), respectively, are measures of noncompactness: in the space $BC(\mathbb{R}_+)$. Moreover, for an arbitrary set $X \in \mathfrak{M}_{BC(\mathbb{R}_+)}$ the following inequalities are satisfied [31]

$$\chi(X) \leqslant \mu_b(X),$$
$$\chi(X) \leqslant \mu_c(X),$$
$$\mu_b(X) \leqslant 2\mu_a(X), \quad \mu_c(X) \leqslant 2\mu_a(X).$$

It is worthwhile noticing that measures of noncompactness μ_a and μ_b defined by (16.10) and (16.11) are sublinear and have maximum property. On the other hand the measure μ_c defined by (16.12) is sublinear and has no maximum property. Moreover, these three measures of noncompactness are not regular. This fact is a consequence of the below given description of the kernels of those measures.

Thus, let us observe that the kernel ker μ_a consists of all bounded subsets X of the space $BC(\mathbb{R}_+)$ such that functions from X are locally equicontinuous on $\mathbb{R}_+$ and tend to zero at infinity with the same rate. Similarly, the kernel ker μ_b consists of bounded sets X such that functions from X are locally equicontinuous on $\mathbb{R}_+$ and tend to (finite) limits at infinity with the same rate, that means, functions from X tend to limits uniformly with respect to the set X. Finally, the kernel ker μ_c contains all bounded sets X such that functions belonging to X are locally equicontinuous on $\mathbb{R}_+$ and the thickness of the bundle formed by graphs of functions from X tends to zero at infinity. Moreover, let us mention that ker $\mu_a \subset$ ker μ_b, and ker $\mu_a \subset$ ker μ_c but there are no inclusions between ker μ_b and ker μ_c. It can be also shown [31] that the measures μ_a, μ_b, and μ_c are not full. Thus, those three measures of noncompactness are not regular.

16.4 Well-Posed Minimization Problems with Respect to Measures of Noncompactness

In this section, we will continue our discussion given in Section 16.2 and concerning the well-posed minimization problems for functionals defined on a subset of a metric space. Thus, let us assume that (X, d) is a given metric space. For simplicity we will always assume that the space X is complete. Further, assume that D is a nonempty and closed subset of the metric space X and J is a functional defined on D, that is, $J : D \to \mathbb{R}$. Similarly as previously we will assume that J is lower bounded and lower semicontinuous on D. Observe that in view of the imposed assumptions there exists a real number m_J such that $m_J = \inf_{x \in D} J(x)$.

In Section 16.2, we defined the well-posedness of the minimization problem for the functional J on the set D. We discussed both the classical definition of this concept given by Tikhonov (cf. Definition 16.1) and its generalization given by Levitin and Polyak (Definition 16.2). Now, we are going to provide further generalization of the definition of the well-posed minimization problem. We start with the definition given by Furi and Vignoli [4] in which the concept of the Kuratowski measure of noncompactness α is involved (cf. Definition 16.1). In order to formulate that definition, for a given number $\varepsilon > 0$ let us denote:

$$D_\varepsilon = \{x \in D : J(x) \leqslant m_J + \varepsilon\}.$$

Notice that in view of lower semicontinuity of the functional J, we infer that the set D_ε is closed. Apart from this observe that if $0 < \varepsilon_1 < \varepsilon_2$ then $D_{\varepsilon_1} \subset D_{\varepsilon_2}$. In what follows we will always assume that there exists a number $\varepsilon_0 > 0$ such that the set D_{ε_0} is bounded. Obviously in such a case we have that D_ε is bounded whenever $0 < \varepsilon \leqslant \varepsilon_0$. Let us observe that if the set D is bounded then the set D_ε is bounded for each $\varepsilon > 0$.

Now, we present the above-definition of Furi and Vignoli [4].

Definition 16.9 We say that the minimization problem for the functional J is *well-posed on the set D in the sense of Furi and Vignoli* if

$$\lim_{\varepsilon \to 0} \alpha(D_\varepsilon) = 0. \tag{16.13}$$

Let us observe that if the minimization problem is well-posed in the sense of Furi and Vignoli, then it is well-posed in the sense of Levitin and Polyak (cf. Definition 16.2). In order to prove this assertion let us consider the set D_0 defined as the intersection of the family $\{D_\varepsilon\}_{\varepsilon > 0}$, that is,

$$D_0 = \bigcap_{\varepsilon > 0} D_\varepsilon.$$

In view of Theorem 16.1 and the fact that the set D_ε is nonempty, closed, and bounded (for $\varepsilon \leqslant \varepsilon_0$) we deduce that the set D_0 is nonempty and compact. On the other hand, the set D_0 consists of all accumulation points of an arbitrary minimizing sequence of the functional J on the set D. This means that the set of all accumulation points of all minimizing sequences of functional J is compact and the minimization problem for J is well-posed in the sense of Levitin and Polyak.

Let us mention that in mathematical literature concerning optimization problems one can encounter a lot of papers devoted to the relations among well-posed minimization problems (cf. [8, 32–36], among others). Let us also pay attention to the fact the well-posedness for minimization problems is mainly discussed for functionals under some additional constraints.

In addition, let us notice that if any minimization problem which is well-posed for functional J on the set D (with the use of the Kuratowski measure of noncompactness α in Definition 16.9) is also well-posed if we use the Hausdorff measure of noncompactness χ (cf. formula (16.2) defining the measure χ) and vice versa. Even more, in Definition 16.9 we can use any regular measure of noncompactness which is equivalent to the Kuratowski (or Hausdorff) measure of noncompactness (cf. Section 16.3).

In what follows, we are going to present the general approach to the definition of the concept of the well-posedness. That definition was accepted in [1] and it creates the generalization of the definition of the well-posed minimization problem in the sense of Furi nad Vignoli (Definition 16.9).

Thus, let us consider the general situation when J is a lower bounded and lower semicontinuous functional defined on a nonempty and closed subset D of a complete metric space (X, d). We will use the notation introduced before, concerning the sets $D_\varepsilon (\varepsilon > 0)$ and D_0. Moreover, similarly, as previously, we assume that the set D_{ε_0} is bounded for some $\varepsilon_0 > 0$.

Further, we assume that the family $\mathcal{P}(\mathcal{P} \subset \mathfrak{N}_X)$ is a kernel (of a measure of noncompactness) in the sense of Definition 16.7.

Definition 16.10   We will say that the minimization problem for the functional J is *well-posed on the set D with respect to the kernel $\mathcal{P}$* if there exists a measure of noncompactness μ in the space X with the kernel $\ker \mu = \mathcal{P}$ and such that

$$\lim_{\varepsilon \to 0} \mu(D_\varepsilon) = 0. \tag{16.14}$$

The above-definition comes from [1]. Let us note that if the minimization problem for the functional J is well-posed on the set D with respect to the kernel $\mathcal{P}$ (in the sense of Definition 16.10) then the set $D_0 = \{x \in D : J(x) = m_J\}$, which consists of its minimum is compact and belongs to the kernel $\mathcal{P}$ (cf. [1]).

It is also worthwhile mentioning that in Definition 16.10, we can use an arbitrary measure of noncompactness μ_1 with $\ker \mu_1 = \mathcal{P}$ which is equivalent to the measure of noncompactness μ from Definition 16.10. From this point of view Definition 16.10 is independent on a measure of noncompactness from the equivalence class $[\mu]$ discussed earlier, which contains all measures of noncompactness with the same kernel $\mathcal{P}$ and equivalent to the measure μ (cf. Definition 16.8). On the other hand it may happen that there exist measures of noncompactness which are not equivalent to a given measure μ (cf. Section 16.3). Based on the above discussion, we will sometimes say that the minimization problem is well-posed with respect to a measure of noncompactness μ. In such a case we can use always any other measure of noncompactness μ_1 that is equivalent to the measure μ, that is, $\mu_1 \in [\mu]$.

We will not provide here the detailed properties of the well-posed minimization problems in the sense of the above accepted Definition 16.10 which were established in [1]. However, we only mention a few results obtained in [1] that seem to be essential from our point of view. First of all, let us recall that if the minimization problem for the functional J is well-posed on the set D with respect to the kernel P then the set of all accumulation points of each minimizing sequence (x_n) for the functional J on D belongs to the family P. Moreover, one can show [1] that if each minimizing sequence for J on the set D has at least one accumulation point and if $D_0 \in P$, then the minimization problem for the functional J on the set D is well-posed with respect to the kernel P.

The fundamental result obtained in [1] is contained in the following theorem. That result will be utilized in our further considerations.

Theorem 16.3 The minimization problem for the functional J on the set D is well-posed with respect to the kernel P if and only if there exists a measure of noncompactness μ with ker $\mu = P$ such that for each set $A \in \mathfrak{M}_X$ the following inequality holds:

$$m_J + \mu(A) \leqslant \sup_{x \in A} J(x). \tag{16.15}$$

16.5 Minimization Problems for Functionals Defined in Banach Sequence Spaces

In this section, we will study the minimization problem for a few functionals defined on some subsets of the classical Banach sequence spaces c_0 and l_1. Recall, that in Section 16.3 we discussed formulas expressing the Hausdorff measure of noncompactness in these sequences spaces (cf. formulas (16.4) and (16.7)). These formulas will be used in our considerations. Taking into account the fact that in the spaces c_0 and l_1 we know rather convenient formulas for the Hausdorff measure of noncompactness, we restrict ourselves in this section to the study of the minimization problem which is well-posed in the sense of Furi and Vignoli (cf. Section 16.4). Indeed, in the case when we know formulas expressing the Hausdorff measure there is no need to consider other measures of noncompactness.

We will investigate here examples of a few functionals defined on some subsets of the mentioned sequence spaces c_0 and l_1.

Example 16.1 Let us take the subset D of the sequence space c_0 defined as follows

$$D = \{x = (x_n) \in c_0 : x_n \geqslant 0 \quad \text{for } n = 1, 2, \ldots\}.$$

Obviously, the set D is closed in c_0.

Next, consider the functional J defined on the set D in the following way

$$J(x) = \frac{x_1}{1 + ||x||_{c_0}}.$$

To sake the simplicity we will drop the index c_0 in the above formula, that is, we will write

$$J(x) = J((x_1, x_2, x_3, \ldots)) = \frac{x_1}{1 + ||x||}.$$

Observe that

$$J(x) = \frac{|x_1|}{1 + ||x||} \leqslant \frac{||x||}{1 + ||x||} < 1.$$

This shows that the functional J is well defined on the set D. Obviously, J is also bounded on D.

We prove that the functional J is continuous on the set D. To this end fix a number $\varepsilon > 0$ and take arbitrary elements $x, y \in D$ such that $||x - y|| \leqslant \varepsilon$. Then we have:

$$
\begin{aligned}
|J(x) - J(y)| &= \left| \frac{x_1}{1 + ||x||} - \frac{y_1}{1 + ||y||} \right| \\
&= \frac{|x_1 + x_1||y|| - y_1 - y_1||x|||}{(1 + ||x||)(1 + ||y||)} \\
&\leqslant \frac{|x_1 - y_1| + |x_1||y|| - y_1||y|| + y_1||y|| - y_1||x|||}{(1 + ||x||)(1 + ||y||)} \\
&\leqslant \frac{|x_1 - y_1| + |x_1 - y_1|||y|| + |y_1|||y|| - ||x|||}{(1 + ||x||)(1 + ||y||)} \\
&\leqslant \frac{|x_1 - y_1|(1 + ||y||) + y_1||y - x||}{(1 + ||x||)(1 + ||y||)}.
\end{aligned}
$$

Hence, in view of the inequality $|x_1 - y_1| \leqslant ||x - y|| \leqslant \varepsilon$, we get

$$|J(x) - J(y)| \leqslant \varepsilon \frac{1 + y_1 + ||y||}{(1 + ||x||)(1 + ||y||)} \leqslant \varepsilon \frac{1 + 2||y||}{(1 + ||x||)(1 + ||y||)}.$$

The above-estimate allows us to infer that the functional J is continuous on the set D. Even more, J is uniformly continuous on bounded subsets of the set D.

In order to check if the minimization problem for the functional J is well-posed in the sense of Furi and Vignoli, we will apply Theorem 16.3. So, let us take the following subset A of the set D:

$$A = \{x = (x_n) \in D : \quad x_1 = 0, \ x_n \leqslant 1 \text{ for } n \geqslant 2\}.$$

Observe that the set A is bounded in the space c_0. Moreover, utilizing formula (16.4) for the Hausdorff measure of noncompactness in the space c_0, we can easily calculate that

$$\chi(A) = \lim_{n \to \infty} \left\{ \sup_{x=(x_n) \in A} \{\max\{|x_k| : \quad k \geqslant n\}\} \right\} = 1. \tag{16.16}$$

On the other hand, it is easy to check that

$$m_J = \inf\{J(x) : \ x \in D\} = 0. \tag{16.17}$$

Further, we obtain

$$\sup_{x \in A} J(x) = \sup \left\{ \frac{x_1}{1 + ||x||} : x = (x_n) \in A \right\} = 0. \tag{16.18}$$

Finally, combining (16.16)–(16.18), in virtue of Theorem 16.3 we conclude that the minimization problem for the functional J is not well-posed in the sense of Furi and Vignoli on the set D.

Example 16.2 Now, let us take the set D_1 in the space c_0 defined as an arbitrarily fixed subset of the set D considered in Example 16.1 that is bounded with respect to the first coordinate, that is, there exists a number $M > 0$ such that $x_1 \leqslant M$ for each $(x_1, x_2, x_3, \ldots) \in D_1$. Denote by $|D_1|$ the number $|D_1| = \sup\{x_1 : (x_1, x_2, \ldots) \in D_1\}$. Obviously, $|D_1| < \infty$. Further, let us consider the functional J on the set D_1 defined in the following way:

$$J(x) = J((x_n)) = \frac{||x||}{1 + x_1}.$$

Obviously, the functional J is well defined on the set D_1.

To check that J is continuous on the set D_1 let us fix a number $\varepsilon > 0$ and choose arbitrary elements $x, y \in D_1$ such that $||x - y|| \leqslant \varepsilon$. Then, we have

$$
\begin{aligned}
|J(x) - J(y)| &= \left| \frac{||x||}{1 + x_1} - \frac{||y||}{1 + y_1} \right| \\
&= \frac{|||x|| + y_1||x|| - ||y|| - x_1||y|||}{(1 + x_1)(1 + y_1)} \\
&\leqslant \frac{|||x|| - ||y||| + |y_1||x|| - y_1||y|| + y_1||y|| - x_1||y|||}{(1 + x_1)(1 + y_1)} \\
&\leqslant ||x - y|| + y_1|||x|| - ||y||| + ||y|||y_1 - x_1| \\
&\leqslant (1 + 2||y||)||x - y|| \leqslant (1 + 2||y||)\varepsilon.
\end{aligned}
$$

This shows the required assertion concerning the continuity of J on the set D_1.

Now, let us consider the measure of noncompactness μ in the space c_0 defined on the family $\mathfrak{M}_{c_0}$ by the formula

$$\mu(A) = \frac{1}{1 + |D_1|} \chi(A),$$

where $\chi(A)$ denotes the Hausdorff measure of noncompactness in the space c_0 (cf. formula (16.4)). Observe that μ is a regular measure of noncompactness in the space c_0 which is equivalent to the Hausdorff measure χ. Therefore, we can investigate the well-posedness of the minimization problem for the functional J on the set D_1 with respect to the measure of noncompactness μ as the well-posedness in the sense of Furi and Vignoli (cf. Section 16.4).

First, let us observe that

$$m_J = \inf_{x \in D_1} J(x) = 0. \tag{16.19}$$

Next, let us take an arbitrary nonempty and bounded subset A of the set D_1. Then, for an arbitrary element $x \in A$ and for a fixed natural number n we have

$$\frac{1}{1 + |D_1|} \max\{|x_k| : k \geq n\} \leq \max\left\{ \frac{|x_k|}{1 + x_1} : k \geq n \right\}$$

$$\leq \frac{||x||}{1 + x_1} = J(x).$$

Hence, we get

$$\frac{1}{1 + |D_1|} \sup_{x \in A}\{\max\{|x_k| : k \geq n\}\} \leq \sup_{x \in A} J(x)$$

and consequently, we derive the following inequality:

$$\frac{1}{1 + |D_1|} \lim_{n \to \infty} \left\{ \sup_{x \in A}\{\max\{|x_k| : k \geq n\}\} \right\} \leq \sup_{x \in A} J(x),$$

which can be written in the form

$$\mu(A) = \frac{1}{1 + |D_1|} \chi(A) \leq \sup_{x \in A} J(x). \tag{16.20}$$

Linking (16.19) and (16.20) and taking into account Theorem 16.3, we deduce that the minimization problem for the functional J is well-posed on the set D_1 in the sense of Furi and Vignoli.

Example 16.3 Let us take a constant α such that $0 < \alpha < 1$. Next, consider the set D_α in the space c_0 defined in the following way:

$$D_\alpha = \{x = (x_n) \in c_0 : \quad 0 \leq x_n \leq \alpha \text{ for } n = 1, 2, \ldots\}.$$

Further, let us take into account the functional J defined on the set D_α as follows:

$$J(x) = J((x_n)) = \sum_{n=1}^{\infty} x_n^n.$$

At the beginning we show that the functional J is well defined on the set D_α, that is, we show that $J(x) \in \mathbb{R}$ for any element $x \in D_\alpha$. Notice that $J(x) \geq 0$ for $x \in D_\alpha$. Next, observe that for an arbitrary element $x \in D_\alpha$, we have

$$J(x) = J((x_n)) = \sum_{n=1}^{\infty} x_n^n \leq \sum_{n=1}^{\infty} \alpha^n = \frac{\alpha}{1 - \alpha} < \infty.$$

Thus, the functional J is well defined on the set D_α.

Further, we show that the functional J is continuous on the set D_α. To this end fix arbitrarily a number ε, $0 < \varepsilon < 1$, and take $x, y \in D_\alpha$ such that $||x - y|| \leqslant \varepsilon$. Next, taking into account the fact that $x_n \to 0$ and $y_n \to 0$ as $n \to \infty$ ($x = (x_n), y = (y_n)$), we can find a natural number k such that $x_n \leqslant \varepsilon$ and $y_n \leqslant \varepsilon$ for $n \geqslant k + 1$. Hence, we obtain the following estimate:

$$|J(x) - J(y)| = \left| \sum_{n=1}^{\infty} x_n^n - \sum_{n=1}^{\infty} y_n^n \right|$$

$$\leqslant \left| \sum_{n=1}^{k} x_n^n - \sum_{n=1}^{k} y_n^n \right| + \left| \sum_{n=k+1}^{\infty} x_n^n - \sum_{n=k+1}^{\infty} y_n^n \right|$$

$$\leqslant \sum_{n=1}^{k} |x_n^n - y_n^n| + \sum_{n=k+1}^{\infty} x_n^n + \sum_{n=k+1}^{\infty} y_n^n$$

$$\leqslant \sum_{n=1}^{k} |x_n - y_n| \left(x_n^{n-1} + x_n^{n-2} y_n + x_n^{n-3} y_n^2 + \cdots + y_n^{n-1} \right)$$

$$+ 2 \sum_{n=k+1}^{\infty} \varepsilon^n \leqslant \sum_{n=1}^{k} \varepsilon \left(\alpha^{n-1} + \alpha^{n-1} + \cdots + \alpha^{n-1} \right) + 2 \frac{\varepsilon^{k+1}}{1 - \varepsilon}$$

$$\leqslant \sum_{n=1}^{k} \varepsilon n \alpha^{n-1} + 2 \frac{\varepsilon^{k+1}}{1 - \varepsilon}$$

$$\leqslant \varepsilon \left(1 + 2\alpha + 3\alpha^2 + \cdots + k\alpha^{k-1} \right) + 2 \frac{\varepsilon^{k+1}}{1 - \varepsilon}$$

$$\leqslant \frac{\varepsilon}{(1 - \alpha)^2} + 2 \frac{\varepsilon^{k+1}}{1 - \varepsilon}.$$

Hence, we deduce that the functional J is continuous on the set D_α.

In what follows, we will investigate the well-posedness of the functional J on the set D_α in the sense of Furi and Vignoli. To this end let us observe that

$$m_J = \inf_{x \in D_\alpha} J(x) = 0. \tag{16.21}$$

Next, consider the sequence (A_n) of subsets of the set D_α such that the set A_n consists of all elements $x = (x_k)$ for which $x_k = 0$ if $k = 1, 2, \ldots, n$ and $0 \leqslant x_k \leqslant \alpha$ if $k \geqslant n + 1$ ($n = 1, 2, \ldots$). Observe that for each fixed natural number n we have

$$\sup_{x \in A_n} \{ \max \{ x_k : k \geqslant n \} \} = \alpha.$$

Obviously this implies that

$$\chi(A_n) = \alpha \tag{16.22}$$

for $n = 1, 2, \ldots$. On the other hand, for an arbitrary $x \in A_n$ we get

$$J(x) = \sum_{k=n+1}^{\infty} x_k^n \leqslant \sum_{k=n+1}^{\infty} \alpha^k = \frac{\alpha^{n+1}}{1 - \alpha}.$$

Consequently, we have

$$\sup_{x \in A_n} J(x) = \frac{\alpha^{n+1}}{1 - \alpha} \tag{16.23}$$

for each $n = 1, 2, \ldots$.

Now, gathering (16.21)–(16.23) and taking into account Theorem 16.3, we see that the minimization problem for the functional J is not well-posed on the set D_α in the sense of Furi and Vignoli.

Example 16.4 Consider the same set D_α as in Example 16.3, where we assumed that α is a fixed number such that $0 < \alpha < 1$. Further, we consider the functional J_1 on the set D_α defined in the following way

$$J_1(x) = ||x||(1 + J(x))$$

$$= \max\{x_n : n = 1, 2, \ldots\}\left(1 + \sum_{n=1}^{\infty} x_n^n\right).$$

Observe that based on the fact established in Example 16.3, we deduce that the functional J_1 is well defined on the set D_α. To prove that J_1 is continuous on the set D_α let us take a number ε, $0 < \varepsilon < 1$, and choose arbitrary elements $x, y \in D_\alpha$ such that $||x - y|| \le \varepsilon$. Then, utilizing the estimates obtained in Example 16.3, we get:

$$|J_1(x) - J_1(y)|$$

$$\le |||x||(1 + J(x)) - ||y||(1 + J(x))| + |||y||(1 + J(x)) - ||y||(1 + J(y))|$$

$$\le (1 + J(x))||x - y|| + ||y|| |J(x) - J(y)|$$

$$\le \left(1 + \frac{\alpha}{1 - \alpha}\right)||x - y|| + ||y||\left(\frac{\varepsilon}{(1 - \alpha)^2} + 2\frac{\varepsilon^{k+1}}{1 - \alpha}\right)$$

$$\le \frac{\varepsilon}{1 - \alpha} + ||y||\left(\frac{\varepsilon}{(1 - \alpha)^2} + 2\frac{\varepsilon^{k+1}}{1 - \varepsilon}\right), \tag{16.24}$$

where k is a number chosen as in Example 16.3.

From estimate (16.24) it follows that the functional J_1 is continuous on the set D_α.

Now, let us fix an arbitrary nonempty subset A of the set D_α. Since D_α is bounded in c_0 this implies the boundedness of A. Then, for an arbitrary element $x \in A$ and for a fixed natural number n, we get

$$\max\{|x_k| : k \ge n\} = \max\{x_k : k \ge n\} \le \max\{x_k : k = 1, 2, \ldots\}$$

$$= ||x|| \le ||x||\left(1 + \sum_{n=1}^{\infty} x_n^n\right) = J_1(x).$$

Hence, we obtain that

$$\chi(A) \le \sup_{x \in A} J_1(x).$$

Thus, keeping in mind that

$$m_{J_1} = \inf_{x \in D_a} J_1(x) = 0$$

and applying Theorem 16.3 we deduce that the minimization problem for the functional J_1 is well-posed on the set D_a in the sense of Furi and Vignoli.

Example 16.5 Now, we will consider the classical sequence space l_1 described in Section 16.3. Recall that there is the convenient formula (16.7) expressing the Hausdorff measure of noncompactness χ in the space l_1. We will utilize that formula in our considerations.

Now, let us take into account the set D_2 in the space l_1 defined in the following way:

$$D_2 = \{x = (x_n) \in l_1 : x_n \geqslant 0 \text{ for } n = 1, 2, \ldots \text{ and there exists}$$
$$\text{a constant } M > 0 \text{ such that } \sum_{n=1}^{\infty} x_n^2 \leqslant M\}.$$

Next, let us denote by $|D_2|$ the number defined as

$$|D_2| = \sup \left\{ \sum_{n=1}^{\infty} |x_n|^2 : x = (x_n) \in D_2 \right\}.$$

Obviously, we have that $|D_2| < \infty$.

For further purposes, we will denote by $|| \cdot ||_1$ the norm of the space l_1 and by $|| \cdot ||_2$ the norm of the space l_2. Thus, if $x = (x_n)$ is an arbitrary sequence of real numbers, we define the quantities $||x||_1$ and $||x||_2$ in the following way:

$$||x||_1 = \sum_{n=1}^{\infty} |x_n|,$$

$$||x||_2 = \left(\sum_{n=1}^{\infty} x_n^2 \right)^{1/2}.$$

Obviously, $||x||_1 < \infty$ if $x = (x_n) \in l_1$ and $||x||_2 < \infty$ if $x \in l_2$.

Now, let us consider the functional $J : D_2 \to \mathbb{R}$ given by the formula

$$J(x) = J((x_n)) = \frac{x_1 + x_2 + \cdots}{1 + x_1^2 + x_2^2 + \cdots}.$$

Observe that the above formula can be written in the form

$$J(x) = \frac{||x||_1}{1 + ||x||_2^2}. \tag{16.25}$$

This formula will be useful in our further considerations.

Further, we show that the functional J is well defined on the set D_2, that is, $J(x) \in \mathbb{R}$ for any $x \in D_2$. Obviously, we have that $J(x) \geqslant 0$. Next, notice that for $x = (x_n) \in D_2$ we have that $||x||_2 \leqslant |D_2| < \infty$. Hence we infer that the functional J given by formula (16.25) is well defined on the set D_2.

In what follows, we prove that J is continuous on the set D_2. To this end let us fix arbitrarily a number $\varepsilon > 0$. We will assume that $\varepsilon < 1$. Next, fix an arbitrary element $x \in D_2$ and choose $y \in D_2$ such that $||x - y||_1 \leqslant \varepsilon < 1$. This implies that $|x_n - y_n| \leqslant \varepsilon < 1$ $(n = 1, 2, \ldots)$. Consequently, we have that $|x_n - y_n|^2 \leqslant |x_n - y_n|$ for each $n = 1, 2, \ldots$. Hence, we get

$$||x - y||_2^2 = \sum_{n=1}^{\infty} (x_n - y_n)^2 \leqslant \sum_{n=1}^{\infty} |x_n - y_n| = ||x - y||_1 \leqslant \varepsilon. \qquad (16.26)$$

Further, utilizing estimate (16.26) we obtain

$$
\begin{aligned}
|J(x) - J(y)| &= \left| \frac{||x||_1}{1 + ||x||_2^2} - \frac{||y||_1}{1 + ||y||_2^2} \right| \\
&= \frac{|\,||x||_1 + ||x||_1||y||_2^2 - ||y||_1 - ||x||_2^2||y||_1|}{(1 + ||x||_2^2)(1 + ||y||_2^2)} \\
&\leqslant |\,||x||_1 - ||y||_1| + |\,||x||_1||y||_2^2 - ||y||_1||y||_2^2 \\
&\quad + ||y||_1||y||_2^2 - ||y||_1||x||_2^2| \\
&\leqslant ||x - y||_1 + ||y||_2^2|\,||x||_1 - ||y||_1| + ||y||_1|\,||x||_2^2 - ||y||_2^2| \\
&\leqslant \varepsilon + ||y||_2^2||x - y||_1 + ||y||_1|\,||x||_2 - ||y||_2|(||x||_2 + ||y||_2) \\
&\leqslant \varepsilon + \varepsilon|D_2| + ||y||_1||x - y||_2 \cdot 2 \cdot \sqrt{|D_2|} \\
&\leqslant \varepsilon + \varepsilon|D_2| + ||y||_1 2\sqrt{|D_2|}\sqrt{\varepsilon}. \qquad (16.27)
\end{aligned}
$$

Now, notice that taking into account the assumed inequality $||x - y||_1 \leqslant \varepsilon$ we have that $||y||_1 \leqslant ||x||_1 + \varepsilon$. Linking this estimate with (16.27) we conclude that the functional J is continuous at the point x. Since x was chosen arbitrarily this implies the continuity of the functional J on the set D_2.

In what follows let us note that

$$m_J = \inf_{x \in D_2} J(x) = 0. \qquad (16.28)$$

Next, fix an arbitrary nonempty and bounded subset A of the set D_2. Further, take an element $x = (x_n) \in A$ and fix an arbitrary natural number n. Then, we have

$$\frac{1}{1 + |D_2|} \sum_{k=n}^{\infty} |x_k| = \frac{1}{1 + |D_2|} \sum_{k=n}^{\infty} x_k.$$

This implies

$$\frac{1}{1 + |D_2|} \sup_{x \in A} \sum_{k=n}^{\infty} x_k \leqslant \sup_{x \in A} \frac{\sum_{k=n}^{\infty} x_k}{1 + ||x||_2^2}.$$

Finally, keeping in mind formula (16.7) we obtain

$$\frac{1}{1 + |D_2|} \chi(A) \leqslant \sup_{x \in A} \frac{||x||_1}{1 + ||x||_2^2} = \sup_{x \in A} J(x).$$

In view of Theorem 16.3, the above inequality in conjunction with (16.28) allows us to conclude that the minimization problem for the functional J is well-posed on the set D_2 with respect to the measure of noncompactness μ defined as follows:

$$\mu(A) = \frac{1}{1 + |D_2|} \chi(A).$$

Since the measure μ is equivalent to the Hausdorff measure χ this means that the minimization problem for the functional J is well-posed on the set D_2 in the sense of Furi and Vignoli.

16.6   Minimization Problems for Functionals Defined in the Classical Space $C([a, b])$

In this section, we are going to study some functionals defined on the classical function space $C([a, b])$ which consists of functions $x(t) = x : [a, b] \to \mathbb{R}$ which are continuous on the interval $[a, b]$. This space is endowed with the classical maximum norm $||x|| = \max\{|x(t)| : t \in [a, b]\}$ (cf. Section 16.3). Let us also recall that in the case of the space $C([a, b])$ we know the convenient formula (16.3) expressing the Hausdorff measure of noncompactness χ.

We are going to discuss examples of two functionals defined on the space $C([a, b])$.

Example 16.6   Let us take the finite sequence $\{t_0, t_1, \dots, t_n\}$ of points in the interval $[a, b]$ that form an equidistance partition of the interval $[a, b]$ into n intervals, that is, $t_i = a + \frac{b-a}{n} i$ $(i = 1, 2, \dots, n)$. Next, consider the functional J defined on the space $C([a, b])$ in the following way:

$$J(x) = \sum_{i=1}^{n} |x(t_i) - x(t_{i-1})|.$$

Obviously, the functional J is well defined on the space $C([a, b])$. Moreover, we have that $J(x) \geq 0$ for every $x \in C([a, b])$. We show that the functional J is continuous on the space $C([a, b])$.

To prove this fact let us fix a number $\varepsilon > 0$ and take arbitrary functions $x, y \in C([a, b])$ such that $\max\{|x(t) - y(t)| : t \in [a, b]\} \leq \varepsilon$. Then, we get:

$$|J(x) - J(y)| = \left| \sum_{i=1}^{n} |x(t_i) - x(t_{i-1})| - \sum_{i=1}^{n} |y(t_i) - y(t_{i-1})| \right|$$

$$= \left| \sum_{i=1}^{n} [|x(t_i) - x(t_{i-1})| - |y(t_i) - y(t_{i-1})|] \right|$$

$$\leq \sum_{i=1}^{n} ||x(t_i) - x(t_{i-1})| - |y(t_i) - y(t_{i-1})||$$

$$\leqslant \sum_{i=1}^{n} |(x(t_i) - x(t_{i-1})) - (y(t_i) - y(t_{i-1}))|$$

$$\leqslant \sum_{i=1}^{n} [|x(t_i) - y(t_i)| + |x(t_{i-1}) - y(t_{i-1})|] \leqslant 2n\varepsilon.$$

From the above inequality, we deduce that the functional J is continuous on the space $C([a, b])$.

Further, let us notice that

$$m_J = \inf\{J(x) : \ x \in C([a, b])\} = 0. \tag{16.29}$$

Notice that J attains its minimum at every function $x = x(t)$ from the space $C([a, b])$ such that $x(t_0) = x(t_1) = \cdots = x(t_n)$. Particularly, if x is an arbitrary constant function then $J(x) = 0$.

Next, let us take the set $A \subset C([a, b])$ which consists of all functions $x = x(t)$ such that $0 \leqslant x(t) \leqslant 1$ for $t \in [a, b]$ and $x(t_0) = x(t_1) = \cdots = x(t_n) = 0$. Apart from this we assume that functions from the set A are arbitrarily defined on the interval $[a, b]$.

Observe that for an arbitrary number $\varepsilon > 0$, we have that $\omega(A, \varepsilon) = 1$, where the symbol $\omega(A, \varepsilon)$ denotes the modulus of continuity of the set A (cf. Section 16.3). This implies that

$$\chi(A) = \frac{1}{2}\omega_0(A) = \frac{1}{2}. \tag{16.30}$$

On the other hand, it is obvious that $J(x) = 0$ for any $x \in A$. This implies that

$$\sup_{x \in A} J(x) = 0. \tag{16.31}$$

Now, taking into account (16.29)–(16.31) and applying Theorem 16.3, we conclude that the minimum problem for the functional J on the space $C([a, b])$ is not well-posed in the sense of Furi and Vignoli.

Example 16.7 Now, we describe the functional F defined on the space $C([a, b])$ that will be the object of the study in this example. To this end, let us fix a natural number $n, n \geqslant 2$. By the symbol P_n, we will denote an arbitrary set $\{t_0, t_1, \ldots, t_n\}$ consisting of $n + 1$ points belonging to the interval $[a, b]$ such that $a = t_0 < t_1 < \cdots < t_n = b$. The set $P_n = \{t_0, t_1, \ldots, t_n\}$ forms a partition of the interval $[a, b]$ into n arbitrary intervals (not necessarily equal as in Example 16.6). By the symbol $\mathcal{P}_n$, we will denote the set of all partitions P_n of the interval $[a, b]$ into n intervals, that is,

$$\mathcal{P}_n = \{P_n = \{t_0, t_1, \ldots, t_n\} : \ a = t_0 < t_1 < \cdots < t_n = b\}$$

(cf. [37]).

Further, for a given function $x(t) = x \in C([a, b])$ we will denote by the symbol $F(x)$ the quantity defined in the following way:

$$F(x) = \sup \left\{ \sum_{i=1}^{n} |x(t_i) - x(t_{i-1})| \; : \; P_n = \{t_0, t_1, \ldots, t_n\} \in \mathcal{P}_n \right\}.$$

Let us notice that the functional F is well defined on the space $C([a, b])$. Indeed, for an arbitrarily fixed partition $P_n \in \mathcal{P}_n$, $P_n = \{t_0, t_1, \ldots, t_n\}$, we have:

$$\sum_{i=1}^{n} |x(t_i) - x(t_{i-1})| \leqslant \sum_{i=1}^{n} (|x(t_i)| + |x(t_{i-1})|) \leqslant 2n||x|| < \infty,$$

where the symbol $||\cdot||$ denotes the classical norm in the space $C([a, b])$ (see Example 16.6).

Further, we show that the functional F is continuous on the space $C([a, b])$. To do this let us take arbitrary functions $x = x(t), y = y(t) \in C([a, b])$. Next, fix an arbitrary partition $P_n = \{t_0, t_1, \ldots, t_n\}$. Then, we have:

$$\sum_{i=1}^{n} |[x(t_i) + y(t_i)] - [x(t_{i-1}) + y(t_{i-1})]|$$

$$\leqslant \sum_{i=1}^{n} |x(t_i) - x(t_{i-1})| + \sum_{i=1}^{n} |y(t_i) - y(t_{i-1})| \leqslant F(x) + F(y).$$

This implies that

$$F(x + y) \leqslant F(x) + F(y).$$

Moreover, it is easily seen that for an arbitrary function $x \in C([a, b])$ we have that $F(-x) = F(x)$. Taking into account the above established properties of the functional F, for $x, y \in C([a, b])$, we get

$$F(x) = F[(x - y) + y] \leqslant F(x - y) + F(y).$$

Consequently, we obtain

$$F(x) - F(y) \leqslant F(x - y). \tag{16.32}$$

Similarly, we can show that

$$-F(x - y) \leqslant F(x) - F(y). \tag{16.33}$$

Linking (16.32) and (16.33), we have

$$|F(x) - F(y)| \leqslant F(x - y). \tag{16.34}$$

Next, let us choose an arbitrary number $\varepsilon > 0$ and take $x, y \in C([a, b])$ such that $||x - y|| \leqslant \varepsilon$. Then, in view of (16.34) we obtain

$$|F(x) - F(y)| \leqslant F(x - y)$$

$$= \sup \left\{ \sum_{i=1}^{n} |x(t_i) - y(t_i) - x(t_{i-1}) + y(t_{i-1})| \; : \; P_n = \{t_0, t_1, \ldots, t_n\} \in \mathcal{P}_n \right\}$$

$$\leqslant \sup \left\{ \sum_{i=1}^{n} [|x(t_i) - y(t_i)| + |x(t_{i-1}) - y(t_{i-1})|] \; : \; P_n = \{t_0, t_1, \dots, t_n\} \in \mathcal{P}_n \right\}$$

$$\leqslant n\varepsilon + n\varepsilon = 2n\varepsilon.$$

Hence, it follows that the functional F is continuous on the space $C([a, b])$.

In what follows let us observe that

$$m_F = \inf\{F(x) \; : \; x \in C([a, b])\} = 0. \tag{16.35}$$

Moreover, it is not difficult to see that the functional F attains its minimum on the class of functions which are constant on the interval $[a, b]$. Further, let us take into account formula (16.3) for the Hausdorff measure of noncompactness in the space $C([a, b])$. Next, take a nonempty and bounded subset X of the space $C([a, b])$ and a number $\varepsilon > 0$. Keeping in mind the fact that the function $\varepsilon \to \omega(X, \varepsilon)$ is nondecreasing on $\mathbb{R}_+$, for a given $\delta > 0$ we can find $\varepsilon_0 > 0$ such that

$$\omega(X, \varepsilon) \leqslant \omega_0(X) + \delta \tag{16.36}$$

for $\varepsilon \leqslant \varepsilon_0$, where the quantity $\omega_0(X)$ is defined by the equality

$$\omega_0(X) = \lim_{\varepsilon \to 0} \omega(X, \varepsilon).$$

Based on inequality (16.36), we can select a function $x \in X$ such that

$$\omega(X, \varepsilon_0) - \delta \leqslant \omega(x, \varepsilon_0) \leqslant \omega(X, \varepsilon_0) \leqslant \omega_0(X) + \delta. \tag{16.37}$$

In view of the definition of the modulus of continuity we deduce that there exist numbers t, s such that $a \leqslant s < t \leqslant b, t - s \leqslant \varepsilon_0$ and

$$\omega(x, \varepsilon_0) - \delta \leqslant |x(t) - x(s)| \leqslant \omega(x, \varepsilon_0). \tag{16.38}$$

Now, let us take a partition $P_n = \{t_0, t_1, \dots, t_n\} \in \mathcal{P}_n$ such that $t_{k-1} = s, t_k = t$ for some $k \in \{1, 2, \dots, n\}$. Hence and in view of (16.38), we get

$$\sum_{i=1}^{n} |x(t_i) - x(t_{i-1})| \geqslant |x(t_k) - x(t_{k-1})| = |x(t) - x(s)| \geqslant \omega(x, \varepsilon_0) - \delta.$$

Consequently, we obtain

$$\omega(x, \varepsilon_0) \leqslant \sum_{i=1}^{n} |x(t_i) - x(t_{i-1})| + \delta$$

$$\leqslant \sup \left\{ \sum_{j=1}^{n} |x(t_j) - x(t_{j-1})| \; : \; \{t_0, t_1, \dots, t_n\} \in \mathcal{P}_n \right\} + \delta$$

$$= F(x) + \delta.$$

Next, combining (16.36)–(16.38) and the above estimate, we derive the following inequality:

$$\omega(X, \varepsilon) \leqslant \omega(X, \varepsilon_0) \leqslant \omega(x, \varepsilon_0) + \delta \leqslant F(x) + 2\delta$$

$$\leqslant \sup_{x \in X} F(x) + 2\delta$$

for $\varepsilon \leqslant \varepsilon_0$. In view of the arbitrariness of the number δ, the above inequality implies that

$$\omega_0(X) \leqslant \sup_{x \in X} F(x).$$

Hence, if we take into account equality (16.35), we infer that the minimization problem for the functional F is well-posed on the space $C([a, b])$ in the sense of Furi and Vignoli.

16.7 Minimization Problems for Functionals Defined in the Space of Functions Continuous and Bounded on the Real Half-Axis

The considerations of this section will be located in the Banach space $BC(\mathbb{R}_+)$ described in details in Section 16.3. Recall briefly that the space $BC(\mathbb{R}_+)$ consists of real functions $x = x(t)$ defined, continuous and bounded on the real half-axis $\mathbb{R}_+$ and equipped with the classical supremum norm

$$||x|| = \sup\{|x(t)| : t \in \mathbb{R}_+|\}.$$

We are going to present three examples of functionals defined on subsets of the space $BC(\mathbb{R}_+)$. We will consider the well-posedness of those functionals with respect to measures of noncompactness subsequently defined by formulas (16.10) and (16.12).

Example 16.8 For an arbitrary function $x \in BC(\mathbb{R}_+)$, let us define the following quantity:

$$O(x) = \sup\{|x(t) - x(s)| : t, s \in \mathbb{R}_+\}.$$

Observe that the functional O is well defined on $BC(\mathbb{R}_+)$, that is, $O(x) < \infty$ for $x \in BC(\mathbb{R}_+)$. Further, let us notice that the functional O is subadditive which means that

$$O(x + y) \leqslant O(x) + O(y)$$

for arbitrary $x, y \in BC(\mathbb{R}_+)$. Apart from this we have that

$$O(-x) = O(x)$$

for $x \in BC(\mathbb{R}_+)$. From the above indicated properties of the functional O we infer that

$$|O(x) - O(y)| \leqslant O(x - y) \tag{16.39}$$

for all $x, y \in BC(\mathbb{R}_+)$.

Now, for an arbitrary $x(t) = x \in BC(\mathbb{R}_+)$ let us define the quantity $J(x)$ in the following way:

$$J(x) = O(x) + \limsup_{t \to \infty} |x(t)|.$$

Further, we show that the functional J is well defined and continuous on the space $BC(\mathbb{R}_+)$. Indeed, we have

$$|J(x)| = J(x) \leqslant 2||x|| + ||x|| = 3||x||.$$

This proves that J is well defined on $BC(\mathbb{R}_+)$. To prove the continuity of J let us fix an arbitrary number $\varepsilon > 0$ and take functions $x, y \in BC(\mathbb{R}_+)$ such that $||x - y|| \leqslant \varepsilon$. Then, we have

$$|J(x) - J(y)| \leqslant |O(x) - O(y)|$$
$$+ |\limsup_{t \to \infty}|x(t)| - \limsup_{t \to \infty}|y(t)||.$$

Hence, in view of (16.39) and the properties of the limit superior, we obtain

$$|J(x) - J(y)| \leqslant O(x - y) + \limsup_{t \to \infty}||x(t)| - |y(t)||$$
$$\leqslant \sup\{|x(t) - y(t) - x(s) + y(s)| \ : \ t, s \in \mathbb{R}_+\}$$
$$+ \limsup_{t \to \infty}|x(t) - y(t)|$$
$$\leqslant \sup_{t \in \mathbb{R}_+}|x(t) - y(t)| + \sup_{s \in \mathbb{R}_+}|x(s) - y(s)| + \varepsilon \leqslant 3\varepsilon.$$

Thus, the functional J is continuous on the space $BC(\mathbb{R}_+)$.

In what follows, let us take into account the measure of noncompactness μ_a in the space $BC(\mathbb{R}_+)$ defined by equality (16.10). Our aim is to investigate the well-posedness of the functional J with respect to the measure of noncompactness μ_a on the subset D of $BC(\mathbb{R}_+)$ defined as follows:

$$D = \{x \in BC(\mathbb{R}_+) \ : \ x(t) \geqslant 0 \text{ for } t \in \mathbb{R}_+\}.$$

Obviously, we have that

$$m_J = \inf_{x \in D} J(x) = 0. \tag{16.40}$$

Further, consider the subset X of the set D containing the functions $x_n^m (m, n, = 1, 2, \dots)$ from the set D defined in the following way:

$$x_n^m(t) = \begin{cases} 0 \text{ for } t \in \left[0, m - \frac{1}{2} - \frac{1}{2n}\right] \cup \left[m - \frac{1}{2} + \frac{1}{2n}, \infty\right), \\ \text{the linear function joining points } \left(m - \frac{1}{2} - \frac{1}{2n}, 0\right) \text{ and} \\ \left(m - \frac{1}{2}, 1\right) \text{ for } t \in \left(m - \frac{1}{2} - \frac{1}{2n}, m - \frac{1}{2}\right), \\ \text{the linear function joining points } \left(m - \frac{1}{2}, 1\right) \text{ and} \\ \left(m - \frac{1}{2} + \frac{1}{2n}, 0\right) \text{ for } t \in \left(m - \frac{1}{2}, m - \frac{1}{2} + \frac{1}{2n}\right). \end{cases}$$

It is easily seen that $\omega_0(X) = 1$, where ω_0 is defined by (16.9). On the other hand, for an arbitrarily fixed $T > 0$, for $m \in \mathbb{N}$ such that $m - 1 \geqslant T$ and for an arbitrary natural number n we get

$$\sup\{|x_n^m(t)| \ : \ t \geqslant T\} = 1.$$

This implies that

$$\sup_{m,n\in\mathbb{N}} \{\sup\{|x_n^m(t)| : t \geqslant T\}\} = 1$$

and finally

$$a(X) = 1.$$

Hence, we get

$$\mu_a(X) = 2. \tag{16.41}$$

However, it can be verified that

$$O(x_n^m) = 1$$

for $m, n \in \mathbb{N}$. Moreover, we have

$$\limsup_{t\to\infty} |x_n^m(t)| = 0.$$

Hence, we get

$$J(x_n^m) = 1,$$

which yields the equality

$$\sup_{m,n\in\mathbb{N}} J(x_n^m) = \sup_{x\in X} J(x) = 1. \tag{16.42}$$

Finally, combining equalities (16.40)–(16.42) and in view of Theorem 16.3, we conclude that the minimization problem for the functional J on the set D is not well-posed with respect to the measure of noncompactness μ_a defined by formula (16.10).

Example 16.9 In the sequel, we will consider the functional J_1 defined on the function space $BC(\mathbb{R}_+)$ in the following way:

$$J_1(x) = 3||x||,$$

where $||\cdot||$ denotes the norm in $BC(\mathbb{R}_+)$. Obviously, this functional is well defined and continuous on the space $BC(\mathbb{R}_+)$. We show that the minimization problem for the functional J_1 on the set D defined in the previous example is well-posed with respect to the measure of noncompactness μ_a defined by formula (16.10). First, let us notice that

$$m_{J_1} = \inf_{x\in D} J_1(x) = 0. \tag{16.43}$$

Next, let us take an arbitrary set $X \in \mathfrak{M}_{BC(\mathbb{R}_+)}$. Fix numbers $\varepsilon > 0$ and $T > 0$. Then, for an arbitrarily fixed function $x \in X$ we have:

$$\omega^T(x, \varepsilon) = \sup\{|x(t) - x(s)| : t, s \in [0, T], |t - s| \leqslant \varepsilon\}$$
$$\leqslant \sup\{|x(t)| + |x(s)| : t, s \in [0, T]\} \leqslant 2||x||.$$

Hence, we obtain

$$\omega^T(X, \varepsilon) \leqslant 2 \sup_{x \in X} ||x||.$$

Consequently, this yields the estimate

$$\omega_0(X) \leqslant 2 \sup_{x \in X} ||x||. \tag{16.44}$$

Similarly, for a fixed number $T > 0$ and for an arbitrarily chosen $x \in X$, we get

$$\sup\{|x(t)| : t \geqslant T\} \leqslant ||x||.$$

Hence, we obtain

$$\sup_{x \in X}\{\sup\{|x(t)| : t \geqslant T\}\} \leqslant \sup_{x \in X} ||x||.$$

Finally, we derive the estimate

$$a(X) \leqslant \sup_{x \in X} ||x||. \tag{16.45}$$

Now, linking (16.43)–(16.45), we have

$$\mu_a(X) \leqslant \sup_{x \in X} J_1(x).$$

The above inequality leads us to the conclusion that the minimization problem for the functional J_1 on the set D is well-posed with respect to the measure of noncompactness μ_a defined by formula (16.10).

Remark 16.1 It is worthwhile mentioning that the minimization problem for the functional J_1 given in the above example is well-posed in the classical sense of Tikhonov. It is an immediate consequence of the definition of the functional J_1.

Example 16.10 Now, similarly as in Example 16.9, consider the functional J_1 on the set D. Next, let us take the measure of noncompactness μ defined in the space $BC(\mathbb{R}_+)$ by the formula

$$\mu(X) = \frac{1}{2}\mu_c(X), \tag{16.46}$$

where the measure μ_c is given by formula (16.12). It is easily seen that for an arbitrary nonempty and bounded subset X of the set D we have

$$\mu(X) = \frac{1}{2}\mu_c(X) = \frac{1}{2}\left[\omega_0(X) + \limsup_{t \to \infty} \operatorname{diam} X(t)\right]$$
$$\leqslant 2 \sup_{x \in X} ||x|| \leqslant \sup_{x \in X} J_1(x).$$

This implies that the minimization problem for the functional J_1 on the set D is well-posed with respect to the measure of noncompactness μ defined by formula (16.46).

References

1 Banaś, J. (2014) Measures of noncompactness and well-posed minimization problems, in *Nonlinear Analysis: Approximation Theory, Optimization and Applications* (eds Q.H. Ansari), Brikhäuser, Springer, New Delhi, pp. 109–134.

2 Tikhonov, A.N. (1966) On the stability of the functional optimization problem. *USSR J. Comput. Math. Phys.*, **6**, 631–634.

3 Levitin, E.S. and Polyak, B.T. (1966) Convergence of minimizing sequences in conditional extremum problem. *Sov. Math. Dokl.*, **7**, 764–767.

4 Furi, M. and Vignoli, A. (1970) About well-posed optimization problems for functionals in metric spaces. *J. Optim. Theory Appl.*, **5**, 225–229.

5 Barbu, V. and Precupanu, T. (2012) *Convexity and Optimization in Banach Spaces*, Springer-Verlag, Berlin.

6 Cesari, L. (1983) *Optimization: Theory and Applications: Problems in Ordinary Differential Equations*, Springer-Verlag, Berlin.

7 Clarke, F.H. (1983) *Optimization and Nonsmooth Analysis*, John Wiley & Sons, Inc., New York.

8 Donchev, A.L. and Zolezzi, T. (1993) *Well-Posed Optimization Problems*, Lecture Notes in Mathematics, vol. **1543**, Springer-Verlag, Berlin.

9 Holmes, R. (1986) *A Course on Optimization and Best Approximation*, Lecture Notes in Mathematics, vol. **1190**, Springer-Verlag, Berlin.

10 Knap, Z. and Banaś, J. (1980) Characterization of the well-posed minimum problem (in Polish). *Tow. Nauk. w Rzeszowie, Met. Numer.*, **6**, 51–62.

11 Choudhury, B.S. and Maity, P. (2016) Best proximity point results in generalized metric spaces. *Vietnam J. Math.*, **44**, 339–349.

12 Eldred, A.A. and Veeramani, P. (2006) Existence and convergence of best proximity points. *J. Math. Anal. Appl.*, **323**, 1001–1006.

13 Hussain, N., Latif, A., and Salimi, P. (2014) Best proximity point results in G-metric spaces. *Abstr. Appl. Anal.*, **2014**, Article ID 837943, 8 p.

14 Yadov, M.R., Thakur, B.S., and Sharma, A.K. (2013) Best proximity points for generalized proximity contraction in complete metric spaces. *Adv. Fixed Point Theory*, **3**, 392–405.

15 Kuratowski, K. (1930) Sur les espaces complets. *Fund. Math.*, **15**, 301–309.

16 Banaś, J. and Goebel, K. (1980) *Measures of Noncompactness in Banach Spaces*, Lecture Notes in Pure and Applied Mathematics, vol. **60**, M. Dekker, New York.

17 Darbo, G. (1955) Punti uniti in trasformazioni a condominio non compatto. *Rend. Semin. Mat. Univ. Padova*, **24**, 84–92.

18 Goldenštein, L.S., Gohberg, I.T., and Markus, A.S. (1957) Investigations of some properties of bounded linear operators with their q-norms Učen. *Zap. Kishinevsk. Univ.*, **29**, 29–36.

19 Goldenštein, L.S. and Markus, A.S. (1965) On a measure of noncompactness of bounded sets and linear operators, in *Studies in Algebra and Mathematical Analysis*, Trudy Inst. Mat. Akad. Nauk Moldav. SSR, Kishinev, pp. 45–54.

20 Akhmerov, R.R., Kamenski, M.I., Potapov, A.S., Rodkina, A.E., and Sadovski, B.N. (1992) *Measures of Noncompactness and Condensing Operators*, Birkhäuser, Basel.

21 Kuratowski, K. (1968) *Topology*, Academic Press, New York.

22 Banaś, J. and Martinon, A. (1990) Some properties of the Hausforff distance in metric spaces. *Bull. Austr. Math. Soc.*, **42**, 511–516.

23 Ayerbe Toledano, J.M., Dominguez Benavides, T., and López Acedo, G. (1997) *Measures of Noncompactness in Metric Fixed Point Theory*, Birkhäuser, Basel.

24 Banaś, J. and Krajewska, M. (2017) Existence of solutions for infinite systems of differential equations in spaces of tempered sequences. *Electron. J. Differ. Equ.*, **2017** (60), 1–28.

25 Banaś, J. and Mursaleen, M. (2014) *Sequence Spaces and Measures of Noncompactness with Applications to Differential and Integral Equations*, Springer, New Delhi, Heidelberg, New York.

26 Dronka, J. (1993) Note on the Hausdorff measure of noncompactness in L^p-spaces. *Bull. Pol. Acad. Sci. Math.*, **41**, 39–41.

27 Banaś, J. and Martinon, A. (1992) Measures of noncompactness in Banach sequence spaces. *Math. Slovaca*, **42**, 497–503.

28 Cheng, L., Shen, Q., Tu, K., and Zhang, W. A new look to measure of non-compactness of Banach spaces. *Stud. Math.*, to appear, Published online 7 July 2017 (Online First version).

29 Mallet-Paret, J. and Nussbaum, R. (2011) Inequivalent measures of noncompactness and the radius of the essential spectrum. *Proc. Am. Math. Soc.*, **139**, 917–930.

30 Mallet-Paret, J. and Nussbaum, R. (2011) Inequivalent measures of noncompactness. *Ann. Mat. Pura Appl.*, **190**, 453–488.

31 Banaś, J., Merentes, N., and Rzepka, B. (2017) Measures of noncompactness in the space of continuous and bounded functions defined on the real half-axis, in *Advances in Nonlinear Analysis via the Concept of Measure of Noncompactness* (eds J. Banas et al.), Springer, Singapore, New Delhi, pp. 1–58.

32 Bednarczuk, E. and Penot, J.P. (1992) Metrically well-set minimization problems. *Appl. Math. Optim.*, **26**, 273–285.

33 Hu, R., Fang, Y., Huang, N., and Wong, M. (2010) Well-posedness of systems of equilibrium problems. *Taiwan. J. Math.*, **14**, 2435–2446.

34 Long, X.J., Huang, N.J., and Teo, K.L. (2008) Levitin-Polyak well-posedness for equilibrium problems with functional constrains. *J. Inequal. Appl.*, **2008**, Article ID 657329.

35 Revalski, J.P. (1997) Hadamard and strong well-posedness for convex programs. *SIAM J. Optim.*, **7**, 519–526.

36 Zolezzi, T. (1996) Extended well-posedness of optimization problem. *J. Optim. Theory Appl.*, **91**, 257–266.

37 Appell, J., Banaś, J., and Merentes, N. (2014) *Bounded Variation and Around*, Series in Nonlinear Analysis and Applications, vol. **17**, De Gruyter, Berlin.

17

Some Recent Developments on Fixed Point Theory in Generalized Metric Spaces

Poom Kumam and Somayya Komal

Theoretical and Computational Science (TaCS) Centre, Department of Mathematics, Faculty of Science, King Mongkut's University of Technology Thonburi, Thung Khru Bangkok 10140, Thailand

2010 AMS Subject Classification: 47H10, 47H09

17.1   Brief Introduction

It is evident that the fixed point theory is one of the fundamental tools in nonlinear functional analysis.

In 1922, Banach contraction mapping principle (BCMP) [1] was proposed, which is the most known and crucial result in fixed point theory. It says that each contraction in a complete metric space has a unique fixed point. This theorem not only guarantees the existence and uniqueness of the fixed point but also shows how to evaluate this point. By virtue of this fact, the BCMP has been generalized in many ways over the years for example (see [75, 76, 80, 84] and references therein.).

In the last two decades, the theory of fixed points has appeared as a crucial technique in the study of nonlinear functional analysis. In fact, the techniques and the tools in fixed point theory have applications in many branches of applied mathematics and also in research field such as physics, chemistry, biology, economics, computer sciences, and many branches of engineering.

In 1965, the notion of fuzzy sets was introduced by Zadeh [2]. This new idea was really source of motivation for many mathematicians.

On the other hand, the study of existence of fixed points for non-self-mapping is also very interesting for various frameworks of different spaces. More precisely, for a given non-empty closed subsets A and B of a complete metric space (X, d), a contraction non-self-mapping $T : A \to B$ does not necessarily yields a fixed point, that is, $d(Tx, x) \neq 0$. In this case, it is quite natural to find out an element $x \in X$ such that $d(x, Tx)$ is minimum, that is, the points x and Tx are close proximity to each other. Let A and B be closed subsets of a metric space (X, d) and $T : A \to B$ be a non-self-mapping. A point x in A for which $d(x, Tx) = d(A, B)$ is called a best proximity point of T. If $A \cap B \neq \phi$, then the

Mathematical Analysis and Applications: Selected Topics, First Edition.
Edited by Michael Ruzhansky, Hemen Dutta, and Ravi P. Agarwal.

best proximity point becomes a fixed point of T. In other words, since a best proximity point reduces to a fixed point if the underlying mapping is assumed to be self-mappings, the best proximity point theorems are natural generalizations of the BCMP. In this direction, the first result was given by Fan [3] in 1969. In this pioneering work, the author in [3] introduced and established a classical best approximation theorem (Theorem 17.1).

Theorem 17.1 [3] If A is a non-empty compact convex subset of a Hausdorff locally convex topological vector space B and $T : A \to B$ is continuous mapping, then there exists an element $x \in A$ such that $d(x, Tx) = d(Tx, A)$.

Later, this result was extended in several directions by many authors. Although the best approximation theorems are adapted to present an approximate solution to the equation $Tx = x$, such results may not yield an approximate solution that is optimal. Following this initial result, a number of authors have derived extensions of *Fan's* theorem and best approximation theorem in many directions such as Eldred and Veeramani [4], Mongkolkeha and Kumam [5–7], and Basha and Veeramani [8–10] (and references therein).

Nadler [11] gave a useful lemma about Hausdorff metric. In [11], the author also characterized the celebrated Banach fixed-point theorem in the context of multivalued mappings.

Lemma 17.1 [11] If $A, B \in CB(X)$ and $a \in A$, then for each $\epsilon > 0$, there exists $b \in B$ such that $d(a, b) \le H(A, B) + \epsilon$.

The theory of multivalued mappings has applications in many areas such as in optimization problem, control theory, differential equations, economics and many branches in analysis. Due to this fact, a number of authors have focused on the topic and have published some interesting fixed-point theorems in this frame. Mizoguchi and Takahashi [12] proved a generalized result of Nadler [11], see Theorem 2 in Alesina *et al.* [13]. Theorem 2 is a partial answer of problem 9 in Reich [14].

Theorem 17.2 (Mizoguchi and Takahashi 12) Let (X, d) be a complete metric space and $T : X \to CB(X)$. Assume that

$$H(Tx, Ty) \le \alpha(d(x, y))d(x, y),$$

for all $x, y \in X$, where $\alpha : [0, \infty) \to [0, 1)$ is $\mathcal{MT}$-function (or $\mathcal{R}$-function), that is, $\lim \sup_{x \to t} \alpha(x) < 1$ for all $t \in [0, \infty)$. Then T has at least one fixed point, that is, there exists $z \in X$ such that $z \in Tz$.

Remark 17.1 [12] We obtain that if $\alpha : [0, \infty) \to [0, 1)$ is a nondecreasing function or a nonincreasing function, then α is an $\mathcal{MT}$-function. Therefore, the class of $\mathcal{MT}$-functions is a rich class, and so this class has been investigated heavily by many authors.

Very recently, Suzuki [15] gave an example, which says that *Mizoguchi −
−Takahashi's* fixed-point theorem for multivalued mappings is a real generalization of *Nadler's* result. In his remarkable paper [15], Suzuki also gave a very
simple proof of *Mizoguchi − −Takahashi's* theorem.

On the other hand, Kirk *et al.* [16] introduced the concept of a cyclic
contraction.

Let A and B be two non-empty subsets of a metric space (X, d), and let
$T : A \cup B \to A \cup B$ be a mapping. Then T is called a cyclic map if $T(A) \subseteq B$
and $T(B) \subseteq A$. In addition, if T is a contraction, then T is called cyclic
contraction.

Czerwik [17] introduced b-metric spaces the generalization of metric spaces
and provides the contraction mapping principle in b-metric spaces, which is
actually an extension of Banach contraction principle in metric spaces.

Bernfeld *et al.* [18] introduced the concept of fixed point for mappings that
have different domains and ranges, the so called past–present–future (PPF)
dependent fixed point or the fixed point with PPF dependence. Furthermore,
they gave the notion of Banach-type contraction for a non-self-mapping
and also proved the existence of PPF-dependent fixed-point theorems in the
Razumikhin class for Banach-type contraction mappings. These results are
useful for proving the solutions of nonlinear functional differential and integral
equations which may depend upon the past history, present data, and future
consideration.

In [16], the authors give a characterization of BCMP in complete metric
spaces. In [19], Kumam *et al.* introduced the concept of common limit in the
range (CLR) property for mappings $F : X \times X \times X \to X$ and $g : X \to X$, where
(X, d) is an abstract metric space, and established tripled coincidence point and
common tripled fixed-point theorems by using the CLR property. The results
in [19] generalized and extended several well-known comparable results in the
literature, in particular the results of Aydi *et al.* [20]. The concept of K-metric
spaces was reintroduced by Huang and Zhang under the name of cone metric
spaces [21], which is the generalization of a metric space. The idea of cone
metric spaces is to replace the codomain a metric from the set of real numbers
to an ordered Banach space. They reintroduced the definitions of convergent
and Cauchy sequences in the sense of an interior point of the underlying
cone. They also continued with results concerning the normal cones only.
One of the main results of Huang and Zhang [21] is fixed-point theorems for
contractive mappings in normal cone spaces. In fact, the fixed-point theorem
in cone metric spaces is appropriate only in the case when the underlying cone
is nonnormal and its interior is non-empty. Abbas *et al.* [22] introduced the
concept of w-compatible mappings and obtained a coupled coincidence point
and a coupled point of coincidence for such mappings satisfying a contractive
condition in cone metric spaces. Very recently, Aydi *et al.* [20] introduced the
concept of W-compatible mappings for mappings $F : X \times X \times X \to X$ and

$g : X \rightarrow X$, where (X, d) is an abstract metric space and established tripled coincidence point and common tripled fixed-point theorems in these spaces.

On the other hand, Kumam and Sintunavarat [23] coined the idea of common limit range property for mappings $F : X \rightarrow X$ and $g : X \rightarrow X$, where (X, d) is a metric space (and fuzzy metric spaces) and proved the common fixed-point theorems by using this property.

Starting from the background of coupled fixed points, the concept of tripled fixed points was introduced by Samet and Vetro [24] and Berinde and Borcut [25]; and it was motivated by the fact that through the coupled fixed point technique, we cannot give the solution of some problems in nonlinear analysis.

Basha [77] also obtained best proximity points for generalized contraction. Basha [26] gave necessary and sufficient conditions to claim the existence of best proximity point for proximal contraction of first and second kinds which are non-self-mapping analogues of contraction self-mappings and they also established some best proximity theorems. In [27], Basha investigated the existence of common best proximity points for pairs of non-self-mappings. In fact, in the framework of metric spaces, a common best proximity point theorem is elicited for pairs of contraction-like non-self-mappings.

The theory of modular on linear spaces and the corresponding theory of modular linear spaces were founded by Nakano [28]. The attempt to generalize the notion of a modular to avoid its restriction on a linear space or on a space with additional algebraic structure resulted in defining and developing a new modular, which works on an arbitrary set. This new modular is called a metric modular and was introduced by Chistyakov [29]. He also pointed out the connection between a modular metric space, a metric space and a modular space. He even studied one of its applications in superposition operator theory. Among the flourishing growth of fixed point theory, it is the KKM maps that attract the interests of most mathematicians.

Kumam and Sintunavarat [23] proved common fixed-point theorems for pair of weakly compatible mappings via Fuzzy metric spaces.

Chaipunya and Kumam [30] introduced and studied the circular metric space, which generalized the modular metric space, together with some elementary properties. Then, under the scope of circular metric space, they present a number of fixed-point theorems for maps satisfying the KKM property and determine the solvability of a variant of a quasi-equilibrium problem as an application of their fixed point results.

In [31], Chaipunya and Kumam describe some equivalence between the results in Sedghi *et al.* [32] and Khamsi [33]. Then, they introduced the notion of a general distance between three arbitrary points and investigated some of its properties with the derivation of some fixed point results. The literature of a distance for any triple of points in a space was first considered during the 60s by Ghler [34]. It is known as a 2-metric, the concept of which was later extended by Dhage [35] into a *D*-metric. Both notions are in no easy ways related to the classical concept of a metric. This led to the *G*-metric due to Mustafa and Sims [36].

On the other hands, Sedghi *et al.* [37] had introduced the notion of a *D**-metric, which has later been generalized by Sedghi *et al.* [32] into an *S*-metric. These achievements confirm that this kind of measurement is recently of many mathematicians' interests. One of the areas that exploited these establishments largely is the fixed point theory, especially the ones involving some generalized contractions (see, e.g., Sedghi *et al.* [32, 37], and Chaipunya and Kumam [31]). The very concept of set-valued fractional integral operator was first proposed by El-Sayed and Ibrahim [38] and this idea has opened the new universe of investigation to fractional operators equations. This type of theory can better describe nonlinear phenomena, compared to the classical theory of differential and integral equations.

In [39], they investigated the existence of fixed points for a very general class of cyclic implicit contractive set-valued operators and also proved that the class of ordered contractions is contained in our class of operators. In [39], as application for their results, they showed the solvability of delayed fractional integral inclusion problems.

On the other hand, Samet *et al.* [24] were first to introduce the concept of α-admissible self-mappings and they proved the existence of fixed point results using contractive conditions involving an α-admissible mapping in complete metric spaces. They also gave some examples and applications to ordinary differential equations of the obtained results. Subsequently, there are a number of results proved for contraction mappings via the concept of α-admissible mapping in metric spaces and other spaces.

One interesting and crucial notion is the one of coupled fixed point, introduced by Guo and Lakshmikantham [40]. Bhaskar and Laksmikantham [41] introduced the notion of mixed monotone mapping and proved some coupled fixed-point theorems for mappings satisfying the mixed monotone property. In [40], the authors observed that their theorems can be used to investigate a large class of problems and discuss the existence and uniqueness of solution for a periodic boundary value problem. Several improvements and generalizations of [41] have recently appeared. Very recently, Berinde and Borcut [25] introduced the notions of tripled fixed point. They proved existence and uniqueness results of tripled fixed point in a partially ordered complete metric space.

On the other hand, the concept of coupled best proximity point and property UC^* are first introduced by Sintunavarat and Kumam [42]. They proved existence and convergence theorems of coupled best proximity point for cyclic contraction pairs. Motivated by the interesting works [25] and [42], after that Kumam *et al.* [43] introduced the notions of tripled best proximity point and later established the existence and convergence theorems of tripled best proximity point in metric spaces. Moreover, they applied these results in uniformly convex Banach space and also studied some results on the existence and convergence of tripled fixed point in metric spaces and gave illustrative examples of their theorems.

Mongkolkeha and Kumam [44] established a new theorem for common best proximity points via proximity commuting mappings.

Sintunavarat and Kumam [45] introduced the generalized *Mizoguchi–Takahashi's* contractions and studied optimal results in uniformly convex Banach spaces.

Kumam [46] defined generalization of weak contraction named as $\mathcal{P}$-contractions in ordered metric spaces.

Sintunavarat and Kumam [47] established generalized common fixed-point theorems under the contraction conditions in complex-valued metric spaces and applied their result to unique common solution of system of Urysohn integral equation.

Sintunavarat *et al.* [48] developed the concept of finding common coupled fixed points without mixed monotone property.

Kumam *et al.* [23] defined the extension of *Mizoguchi–Takahashi's* fixed-point theorems in very fine way.

Sintunavarat and Kumam [49] established the best proximity point theorems for proximal contractions of first and second kind which are in fact the extension of Banach contraction principle in non-self-manner.

Kumam *et al.* [50] introduced the concept of C-Cauchy sequence and C-complete in complex-valued metric spaces.

Kumam *et al.* [51] introduced α_c-admissible non-self-mappings and proved the existence and convergence of the past, present, future briefly in Razumikhin class.

Kumam *et al.* proved some fixed points results about generalized Ulam–Hyers stability and well-posedness via α-admissible mappings in b-metric spaces in [52].

Kumam *et al.* [53] proved common α-Fuzzy fixed-point theorem for pair of fuzzy mappings.

Chaipunya *et al.* [54] established common fixed-point theorem in circular metric spaces.

Kumam *et al.* [55] developed fixed-point theorems for two hybrid pairs of non-self-mappings under certain property in metric spaces.

Kumam *et al.* [56] described the generalization of Ciric fixed-point theorem.

Very great work and nice results are also studied and established by Kumam *et al.* see [19, 39, 57–63, 78, 79, 81–83], and references there in.

It was a brief introduction and some historical background in fixed point theory but until yet there are too much advance developments in this field unable to capture all that progress. But many useful results and very famous, applicable notions will be defined in this chapter, which will surely not cover all the field of fixed point theory but may lead to know that the developments and event of progress in this area are going on with nice style and in detail.

17.2   Some Basic Notions and Notations

In this section, we recall some basic definitions:

For non-empty subsets A and B of a metric space (X, d), $d(A, B)$ stands for the minimum distance between A and B.

A Banach spaces X is said to be:

(1) Strictly convex if the following implication holds: for all $x, y \in X$,

$$\|x\| = \|y\| = 1$$

and $x \neq y \Rightarrow \|\frac{x+y}{2}\| < 1$.

(2) Uniformly convex if, for any ϵ with $0 < \epsilon \leq 2$, there exists $\delta > 0$ such that the following implication holds: for all $x, y \in X$,

$$\|x\| \leq 1, \|y\| \leq 1$$

and $\|x - y\| \geq \epsilon \Rightarrow \|\frac{x+y}{2}\| < 1 - \delta$.

Definition 17.1   **[27]**   A mapping $S : A \to B$ and $T : A \to B$ is said to be swapped proximally if $d(y, u) = d(y, v) = d(A, B)$ and $Su = Tv$, then these implies that $Sv = Tu$ for all $u, v \in A$ and $y \in B$.

Definition 17.2   **[27]** A is said to be approximately compact with respect to B if every sequence $\{x_n\}$ in A satisfies the condition that $d(y, x_n) \to d(y, A)$ for some $y \in B$ has a convergent subsequence.

We noticed that each set is approximately compact with respect to any set. Moreover, A_0 and B_0 are non-empty sets if A is compact and B is approximately compact with respect to A.

Definition 17.3   **[27]**   A mapping $S : A \to B$ and $T : A \to B$ is said to be commute proximally if they satisfy the following condition:

$$[d(u, Sx) = d(v, Tx) = d(A, B)] \Rightarrow Sv = Tu$$

for all $u, v, x \in A$. It is easy to see that proximal commutativity of self-mappings become commutativity of the mappings.

We denote by $CB(X)$ the class of all non-empty closed bounded subsets of a metric space (X, d). The Hausdorff metric induced by d on $CB(X)$ is given by

$$H(A, B) = \max\{\sup_{a \in A} d(a, B), \sup_{b \in B} d(b, A)\},$$

for every $A, B \in CB(X)$, where $d(a, B) = \inf\{d(a, b) : b \in B\}$ is the distance from a to $B \subseteq X$.

Remark 17.2   **[23]**   The following properties of the Hausdorff metric induced by d are well known:

(i) H is a metric on $CB(X)$.
(ii) If $A, B \in CB(X)$ and $q > 1$ is given, then for every $a \in A$, there exists $b \in B$ such that $d(a, b) \leq qH(A, B)$.

Definition 17.4 **[23]** Let A and B be non-empty subsets of a metric space (X, d) and let $T : A \to 2^B$ be a multivalued mapping. A point $x \in A$ is said to be a best proximity point of a multivalued mapping T if it satisfies the condition that

$$d(x, Tx) = d(A, B).$$

We notice that a best proximity point reduces to a fixed point for a multivalued mapping if the underlying mapping is a self-mapping.

Definition 17.5 **[64]** Let A and B be non-empty subsets of a metric space (X, d). The ordered pair (A, B) is said to satisfy the property UC if the following holds: If $\{x_n\}$ and $\{z_n\}$ are sequences in A, and $\{y_n\}$ is a sequence in B such that $d(x_n, y_n) \to d(A, B)$ and $d(z_n, y_n) \to d(A, B)$, then $d(x_n, z_n) \to 0$.

Example 17.1 **[64]** The following are examples of a pair of non-empty subsets (A, B) satisfying the property UC.

(i) Every pair of non-empty subsets A, B of a metric space (X, d) such that $d(A, B) = 0$.
(ii) Every pair of non-empty subsets A, B of a uniformly convex Banach space X such that A is convex.
(iii) Every pair of non-empty subsets A, B of a strictly convex Banach space, where A is convex and relatively compact and the closure of B is weakly compact.

Definition 17.6 **[42]** Let A and B be non-empty subsets of a metric space (X, d). The ordered pair (A, B) satisfies the property UC^* if (A, B) has property UC, and the following condition holds:
If $\{x_n\}$ and $\{z_n\}$ are sequences in A, and $\{y_n\}$ is a sequence in B satisfying:

(i) $d(z_n, y_n) \to d(A, B)$.
(ii) For every $\epsilon > 0$, there exists $N \in \mathbb{N}$ such that

$$d(x_m, y_n) \leq d(A, B) + \epsilon$$

for all $m > n \geq \mathbb{N}$,

then $d(x_n, z_n) \to 0$.

The following are examples of a pair of non-empty subsets (A, B) satisfying the property UC.

Example 17.2 **[42]** Every pair of non-empty subsets A, B of a metric space (X, d) such that $d(A, B) = 0$.

Example 17.3 **[42]** Every pair of non-empty closed subsets A, B of uniformly convex Banach space X such that A is convex (see lemma 3.7 in [4]).

Definition 17.7 **[26]** A mapping $T : A \to B$ is called a proximal contraction of the first kind if there exists $\alpha \in [0, 1)$ such that, for all $a, b, x, y \in A$,

$$d(a, Tx) = d(b, Ty) = d(A, B) \Rightarrow d(a, b) \leq \alpha d(x, y).$$

If T is self-mapping, then T is a proximal contraction of the first kind deduced to T which is a contraction mapping. But a non-self-proximal contraction is not necessarily a contraction.

Definition 17.8 **[26]** A mapping $T : A \to B$ is said to be a proximal contraction of the second kind if there exists $\alpha \in [0, 1)$ such that, for all $a, b, x, y \in A$,

$$d(a, Tx) = d(b, Ty) = d(A, B) \Rightarrow d(Ta, Tb) \leq \alpha d(Tx, Ty).$$

The necessary condition for a self-mapping T to be a proximal contraction of the second kind is that

$$d(TTx, TTy) \leq \alpha d(Tx, Ty)$$

for all x, y, in the domain of T. Therefore, every contraction self-mapping is a proximal contraction of the second kind, but the converse is not true (see [26], example 5).

Definition 17.9 **[26]** Let $S : A \to B$ and $T : B \to A$ be mappings. The pair (S, T) is said to be

(1) a cyclic contractive pair if $d(A, B) < d(x, y) \Rightarrow d(Sx, Ty) < d(x, y)$ for all $x \in A$ and $y \in B$;
(2) a cyclic expansive pair if $d(A, B), d(x, y) \Rightarrow d(Sx, Ty) > d(x, y)$ for all $x \in A$ and $y \in B$;
(3) a cyclic inequality pair if $d(A, B) < d(x, y) \Rightarrow d(Sx, Ty) \neq d(x, y)$ for all $x \in A$ and $y \in B$.

Definition 17.10 **[26]** Let $T : A \to B$ a mapping and g be an isometry. The mapping T is said to preserve isometric distance with respect to g if

$$d(Tgx, Tgy) = d(Tx, Ty)$$

for all $x, y \in A$.

17.3   Fixed Points Theorems

The applications of fixed point theory are very important and useful in diverse disciplines of mathematics. The theory can be applied to solve many problem in real world, for example: equilibrium problems, variational inequalities, and optimization problems.

The most sufficient result in this theory is BCMP is given by Banach [1], which states that every contraction self-mapping $T : X \to X$ on a complete metric spaces (X, d) has a unique fixed point, that is, $Tx = x$. It is because of its wide applied potential, this fundamental principle has been generalized in many directions over the years, we will discuss few of them here.

By considering the relationship between periodic points, fixed points, and the property $\mathcal{P}$, we will present an extended version of periodic points together with their behavior in topological spaces and cone metric spaces. It is obvious that if x_* is a fixed point of a mapping f, then it is also a fixed point of iterates f^n for all $n \in \mathbb{N}$. The converse is trivially not true in general. Whenever the converse is true for the mapping f, we say that it satisfies the property $\mathcal{P}$. The point x_* such that $x_* = f^h x_*$ for some $h \in \mathbb{N}$ is called a periodic point of f. In this case, h is called the order of x_* and if it is the least positive integer such that x_* is a fixed point of f^h, then we call it the prime order of x_*. The studies about periodic and fixed points are not limited only in metric spaces, but also in various generalizations of a metric space, for example, in G-metric spaces, probabilistic metric spaces, and cone metric spaces.

Theorem 17.3   [65]   Let X be a Hausdorff topological space and f be a self-mapping on X. Suppose that the orbit $\{f^n x\}$ converges for any $x \in X$. Then f satisfies the property $\mathcal{P}$.

Proof: Assume that there exists a point $x_* \in Fix(f^m) \backslash Fix(f)$, where $m \in \mathbb{N}$ is the least number in the sense that $x_* = f^m x_*$ and $x_* \neq f^n x_*$ for all $n \in \mathbb{N}$ with $n < m$. Thus, the sequence $\{f^n x_*\}$ may be written as follows:

$$f^n x_* = f^{n(\mathrm{mod}\ m)} x_*$$

for each $n \in \mathbb{N}$. It is obvious that $\{f^n x_*\}$ does not converge. This contradicts the hypothesis that $\{f^n x\}$ converges for any $x \in X$. Therefore, we conclude that f satisfies the property $\mathcal{P}$.

Definition 17.11   [65]   A self-mapping f on X is called a weak Picard's mapping if every orbit converges and their limits are fixed points of f. In addition, if f has exactly one fixed point, then f is called a Picard's mapping.

Corollary 17.1   [65]   Every weak Picard's mapping satisfies the property $\mathcal{P}$.

Corollary 17.2   [65]   Every Picard's mapping satisfies the property $\mathcal{P}$.

In [65] a new concept of a periodic point of order ∞ has been introduced as follows:

Definition 17.12 [65] Let f be a self-mapping on a topological space X. A point $x_* \in X$ is called a periodic point of order ∞ for f if the orbit $\{f^n x_*\}$ has at least one subsequence converging to the point x_* itself. The set of all periodic points of order ∞ for f is denoted by $Fix(f^\infty)$.

It is obvious that $\cup_{n \in \mathbb{N}} Fix(f^n) \subseteq Fix(f^\infty)$. The converse is not true in general as in the following example.

Example 17.4 [65] Let $X := \{0\} \cup \{\frac{1}{n} : n \in \mathbb{N}\}$ and $d : X \times X \to \mathbb{R}_+$ be a usual metric. Let $f : X \to X$ be a mapping given by

$$f(0) = 1,$$

and

$$f\left(\frac{1}{n}\right) = \frac{1}{n+1}$$

for all $n \in \mathbb{N}$. Then $Fix(f^n) = \phi$, for all $n \in \mathbb{N}$ while $Fix(f^\infty) = \{0\}$.

Theorem 17.4 [65] Let (X, d) be a complete cone metric space (whose underlying cone P is not necessarily normal) and let $f \in \mathcal{F}(\mathcal{X})$. Then $Fix(f^\infty)$ is non-empty.

17.3.1 Fixed Points Theorems for Monotonic and Nonmonotonic Mappings

Rhoades [66] showed weak contraction results in complete metric spaces. We state the result of Rhoades in the following. A mapping $f : X \to X$, where (X, d) is a metric space, is said to be weakly contractive if

$$d(fx, fy) \leq d(x, y)\varphi(d(x, y))$$

for all $x, y \in X$ and $\varphi : [0, +\infty) \to [0, +\infty)$ is a function satisfying:

  (i) φ is continuous and nondecreasing;
 (ii) $\varphi(t) = 0$ if and only if $t = 0$;
(iii) $\lim_{t \to +\infty} \varphi(t) = +\infty$.

We can reduce the weakly contractive to ordinary contraction by taking $\varphi(t) := kt$, where $0 \leq k < 1$.

Theorem 17.5 [66] Let (X, d) be a complete metric space and f be a weakly contractive mapping. Then f has a unique fixed point x^* in X.

A very beautiful way to generalize this theorem is to consider it in case a partial ordering is defined on the space. Recall that a relation $\sqsubseteq$ is a partial ordering on a set X if it is reflexive, antisymmetric, and transitive. By this understanding, we write $b \sqsupseteq a$ instead of $a \sqsubseteq b$ to emphasize some particular cases. Any $a, b \in X$ are said to be comparable if $a \sqsupseteq b$ or $a \sqsubseteq b$. If a set X has a partial ordering $\sqsubseteq$ defined on it, we say that it is a partially ordered set (with respect to $\sqsubseteq$) and denote it by $(X, \sqsubseteq)$. $(X, \sqsubseteq)$ is said to be a totally ordered set if any two elements in X are comparable. Moreover, it is said to be a sequentially ordered set if each element of a convergent sequence in X is comparable with its limit. Yet, if (X, d) is a metric space and $(X, \sqsubseteq)$ is a partially ordered (totally ordered or sequentially ordered) set, we say that X is a partially ordered (totally ordered or sequentially ordered, respectively) metric space, and it will be denoted by $(X, \sqsubseteq, d)$. Harjani and Sadarangani [67] carried the work of Rhoades [66] into partially ordered metric spaces. We now state the result proved in [67] as follows.

Theorem 17.6 [67] Let $(X, \sqsubseteq, d)$ be a complete partially ordered metric space and let $f : X \to X$ be a continuous and nondecreasing mapping such that

$$d(fx, fy) \leq d(x, y)\varphi(d(x, y))$$

for $x \sqsubseteq y$, where $\varphi : [0, +\infty)[0, +\infty)$ is a function satisfying:

(i) φ is continuous and nondecreasing;
(ii) $\varphi(t) = 0$ if and only if $t = 0$;
(iii) $\lim_{t \to +\infty} \varphi(t) = +\infty$.

If there exists $x_0 \in X$ such that $x_0 \sqsubseteq fx_0$, then f has a fixed point.

Harjani and Sadarangani [67] also proved fixed point theorems for noncontinuous mappings, nonincreasing mappings, and even for nonmonotonic mappings. We introduced a weak condition, which resulted in the concept called a $\mathcal{P}$-contraction. We introduced our concept of a $\mathcal{P}$-function and some of its fundamental properties. Not to be ambiguous, we assumed that $\mathbb{R}$ represents the set of all real numbers while $\mathbb{N}$ represents the set of all positive integers.

Definition 17.13 [46] Let $(X, \sqsubseteq, d)$ be a partially ordered metric space. A function $\varrho : X \times X \to \mathbb{R}$ is called a $\mathcal{P}$-function with respect to $\sqsubseteq$ in X if it satisfies the following conditions:

(i) $\varrho(x, y) \geq 0$ for every comparable $x, y \in X$;
(ii) for any sequences $\{x_n\}_{n=1}^{\infty}$, $\{y_n\}_{n=1}^{\infty}$ in X such that x_n and y_n are comparable at each $n \in \mathbb{N}$, if the limit $\lim_{n \to \infty} x_n = x$ and $\lim_{n \to \infty} y_n = y$, then $\lim_{n \to \infty} \varrho(x_n, y_n) = \varrho(x, y)$;
(iii) for any sequences $\{x_n\}_{n=1}^{\infty}$, $\{y_n\}_{n=1}^{\infty}$ in X such that x_n and y_n are comparable at each $n \in \mathbb{N}$, if the limit $\lim_{n \to \infty} \varrho(x_n, y_n) = 0$, then the limit $\lim_{n \to +\infty} d(x_n, y_n) = 0$.

If, in addition, the following condition is also satisfied:

(A) for any sequences $\{x_n\}_{n=1}^{\infty}$, $\{y_n\}_{n=1}^{\infty}$ in X such that x_n and y_n are comparable at each $n \in \mathbb{N}$, if the limit $\lim_{n \to +\infty} d(x_n, y_n)$ exists, then the limit $\lim_{n \to \infty} \varrho(x_n, y_n)$ also exists,

then ϱ is said to be a $\mathcal{P}$-function of type (A) with respect to $\sqsubseteq$ in X.

Example 17.5 **[46]** Let $(X, \sqsubseteq, d)$ be a partially ordered metric space. Suppose that the function $\varphi : [0, +\infty) \to [0, +\infty)$ is defined same as in Theorem 17.6. Then $\varphi \circ d$ is a $\mathcal{P}$-function of type (A) with respect to $\sqsubseteq$ in X.

Let $(X, \sqsubseteq, d)$ be a partially ordered metric space, a mapping $f : X \to X$ is called a $\mathcal{P}$-contraction with respect to $\sqsubseteq$ if there exists a $\mathcal{P}$-function $\varrho : X \times X \to \mathbb{R}$ with respect to $\sqsubseteq$ in X such that

$$d(fx, fy) \le d(x, y)\varrho(x, y)$$

for any comparable $x, y \in X$. Naturally, if there exists a $\mathcal{P}$-function of type (A) with respect to $\sqsubseteq$ in X such that the above inequality holds for any comparable $x, y \in X$, then f is said to be a $\mathcal{P}$-contraction of type (A) with respect to $\sqsubseteq$.

From our proposed example, it follows that in partially ordered metric spaces, a weak contraction is also a $\mathcal{P}$-contraction. The next fixed point result is for monotonic mappings.

Theorem 17.7 **[46]** Let $(X, \sqsubseteq, d)$ be a complete partially ordered metric space and $f : X \to X$ be a continuous and nondecreasing $\mathcal{P}$-contraction of type (A) with respect to $\sqsubseteq$. If there exists $x_0 \in X$ with $x_0 \sqsubseteq fx_0$, then $\{f^n x_0\}_{n=1}^{\infty}$ converges to a fixed point of f in X.

Next, if we drop the continuity of f in Theorem 17.7, then we can still guarantee a fixed point if we strengthen the condition of a partially ordered set to a sequentially ordered set.

Theorem 17.8 **[46]** Let $(X, \sqsubseteq, d)$ be a complete sequentially ordered metric space and $f : X \to X$ be a nondecreasing $\mathcal{P}$-contraction of type (A) with respect to $\sqsubseteq$. If there exists $x_0 \in X$ with $x_0 \sqsubseteq fx_0$, then $\{f^n x_0\}_{n=1}^{\infty}$ converges to a fixed point of f in X.

The next result asserts that existence of fixed points with the lack of monotonicity, as follows:

Theorem 17.9 **[46]** Let $(X, \sqsubseteq, d)$ be a complete partially ordered metric space and $f : X \to X$ be a continuous $\mathcal{P}$-contraction of type (A) with respect to $\sqsubseteq$ such that the comparability of $x, y \in X$ implies the comparability of

$fx, fy \in fX$. If there exists $x_0 \in X$ such that x_0 and fx_0 are comparable, then $\{f^n x_0\}_{n=1}^{\infty}$ converges to a fixed point of f in X.

Now, we propose an example to justify our Theorem 17.9.

Example 17.6   [46]   Let $X = [0, 1] \times [0, 1]$ and suppose that we write $x = (x_1, x_2)$ and $y = (y_1, y_2)$ for $x, y \in X$. Define $d, \varrho : X \times X \to \mathbb{R}$ by $d(x, y) = 0$, if $x = y$, and otherwise

$$d(x, y) = 2\max\{x_1 + y_1, x_2 + y_2\},$$

and $\varrho(x, y) = 0$, if $x = y$, and otherwise

$$\varrho = \max\{x_1, x_2 + y_2\}.$$

Let $\sqsubseteq$ be an ordering in X such that for $x, y \in X, x \sqsubseteq y$ if and only if $x_1 = y_1$ and $x_2 \leq y_2$. Then $(X, \sqsubseteq, d)$ is a partially ordered metric space with ϱ as a $\mathcal{P}$-function of type (A) with respect to $\sqsubseteq$ in X. Now, let f be a self-mapping on X defined by $fx = f((x_1, x_2)) = (0, \frac{x_2^2}{2})$ for all $x \in X$. It is obvious that f is continuous and nondecreasing with respect to $\sqsubseteq$. Let $x, y \in X$ be comparable with respect to $\sqsubseteq$. If $x = y$, then they clearly satisfy the condition for $\mathcal{P}$-contraction. On the other hand, if $x \neq y$, then, we note that is also satisfies the inequality $d(fx, fy) \leq d(x, y) - \varrho(x, y)$. Therefore, the condition for $\mathcal{P}$-contraction is satisfied for every comparable $x, y \in X$. Moreover, f is a continuous and nondecreasing $\mathcal{P}$-contraction of type (A) with respect to $\sqsubseteq$. Let $x_0 = (0, 0)$, so we have $x_0 \sqsubseteq fx_0$. Now, applying Theorem 17.9 of this subsection, we conclude that f has a fixed point in X which is the point $(0, 0)$.

17.3.2   PPF-Dependent Fixed-Point Theorems

Bernfeld *et al.* [18] introduced the concept of a fixed point for mappings that have different domain and range. To the best of our knowledge, there has been no discussion so far concerning the PPF-dependent fixed-point theorems via α-admissible mappings. Next, we use to mention the concept of α_c-admissible non-self-mappings and prove the existence and convergence of PPF-dependent fixed-point theorems for contraction mappings involving α_c-admissible non-self-mappings in the Razumikhin class. Furthermore, these results are extensions of PPF-dependent fixed-point theorems in [18] and also applicable for PPF-dependent coincidence points. Here, we will use E to denotes a Banach space with the norm $\|.\|_E$, I is a closed interval $[a, b]$ in $\mathbb{R}$ and $E_0 = C(I, E)$ is notation for the set of all continuous E-valued functions on I equipped with the supremum norm $\|.\|_{E_0}$ defined by

$$\|\phi\|_{E_0} = \sup_{t \in I} \|\phi(t)\|_E$$

for $\phi \in E_0$. For a fixed element $c \in I$, the Razumikhin or minimal class $\mathcal{R}_c$ of functions in E_0 contains $\phi \in E_0$ such that norm function of ϕ induced by E_0 is equal to norm of $\phi(c)$ at E. It is easy to see that the constant function is one of the mappings in $\mathcal{R}_c$. The class $\mathcal{R}_c$ is said to be algebraically closed with respect to difference if $\phi - \xi \in \mathcal{R}_c$ whenever $\phi, \xi \in \mathcal{R}_c$. In addition, we say that the class $\mathcal{R}_c$ is topologically closed if it is closed with respect to the topology on E_0 generated by the norm $\|.\|_{E_0}$. The next two notions of Bernfeld *et al.* [18] will be used in sequel.

Definition 17.14 [18] A point $\phi \in E_0$ is said to be a PPF-dependent fixed point or a fixed point with PPF dependence of the non-self-mapping $T : E_0 \to E$ if $T\phi = \phi(c)$ for some $c \in I$.

Definition 17.15 [18] The mapping $T : E_0 \to E$ is called a Banach-type contraction if there exists a real number $k \in [0, 1)$ such that the difference of $\|T\phi - T\xi\|_E \leq k\|\phi - \xi\|_{E_0}$ for all $\phi, \xi \in E_0$.

Definition 17.16 [24] Let X be a non-empty set, $f : X \to X$ and $\alpha : X \times X \to [0, \infty)$. We say that f is an α-admissible mapping if it satisfies the following condition:

$$\alpha(x, y) \geq 1 \Rightarrow \alpha(f(x), f(y)) \geq 1,$$

for $x, y \in X$.

Definition 17.17 [51] Let $c \in I$ and $T : E_0 \to E$, $\alpha : E \times E \to [0, \infty)$. We say that T is an α_c-admissible mapping if for $\phi, \xi \in E_0$, $\alpha(\phi(c), \xi(c)) \geq 1$ implies $\alpha(T\phi, T\xi) \geq 1$.

Example 17.7 [51] Let $E = \mathbb{R}$ be real Banach spaces with usual norms and $I = [0, 1]$. Define $T : E_0 \to E$ and $\alpha : E \times E \to [0, \infty)$ by $T\phi = \phi(1)$ for all $\phi \in E_0$ and $\alpha(x, y) = 1$, if $x \geq y$, otherwise 0. Then T is α_1-admissible.

The forthcoming result [51] is very interesting result for a PPF-dependent fixed point as follows:

Theorem 17.10 [51] Let $T : E_0 \to E$, $\alpha : E \times E \to [0, \infty)$ be two mappings satisfying the following conditions:

(a) There exists $c \in I$ such that $\mathcal{R}_c$ is topologically closed and algebraically closed with respect to difference.

(b) T is α_c-admissible.

(c) For all $\phi, \xi \in E_0$,

$$\alpha(\phi(c), T\phi)\alpha(\xi(c), T\xi)\|T\phi - T\xi\|_E \leq k\|\phi - \xi\|_{E_0},$$

where $k \in [0, 1)$.

(d) If $\{\phi_n\}$ is a sequence in E_0 such that $\phi_n \to \phi$ as $n \to \infty$ and $\alpha(\phi_n(c), T\phi_n) \geq 1$ for all $n \in \mathbb{N}$, then $\alpha(\phi(c), T\phi) \geq 1$.

If there exists $\phi_0 \in \mathcal{R}_c$ such that $\alpha(\phi_0(c), T\phi_0) \geq 1$, then T has a unique PPF-dependent fixed point ϕ^* in $\mathcal{R}_c$ such that $\alpha(\phi^*(c), T\phi^*) \geq 1$. Moreover, for a fixed $\phi_0 \in \mathcal{R}_c$ such that $\alpha(\phi_0(c), T\phi_0) \geq 1$, if a sequence $\{\phi_n\}$ of iterates of T in $\mathcal{R}_c$ is defined by $T\phi_{n-1} = \phi - \phi_n(c)$ for all $n \in \mathbb{N}$, then $\{\phi_n\}$ converges to a PPF-dependent fixed point of T in $\mathcal{R}_c$.

Proof: Let ϕ_0 be a point in $\mathcal{R}_c$, which is subset of E_0, such that $\alpha(\phi_0(c), T\phi_0) \geq 1$. Since $T\phi_0 \in E$, there exists $x_1 \in E$ such that

$$T\phi_0 = x_1$$

Step 1. Construct the sequence $\{\phi_n\}$ in $\mathcal{R}_c$ is subset of E_0 such that $T\phi_{n-1} = \phi_n(c)$ for all $n \in \mathbb{N}$.

Step 2. By the hypothesis (b), it is easily proved that $\alpha(\phi_{n-1}(c), T\phi_{n-1}) \geq 1$ for all $n \in \mathbb{N}$.

Step 3. Next, to prove that $\{\phi_n\}$ is a Cauchy sequence in $\mathcal{R}_c$ and by completeness property of E_0, we get that Cauchy sequence converges to a limit point $\phi^* \in E_0$ and since $\mathcal{R}_c$ is topologically closed, we have $\phi^* \in \mathcal{R}_c$.

Step 4. Now, by (d), we get $\|T\phi^* - \phi^*(c)\|_E = 0$ and so that ϕ^* is a PPF-dependent fixed point of T in $\mathcal{R}_c$. Uniqueness can be proved easily, and then finally we get unique PPF-dependent fixed point of T in $\mathcal{R}_c$, that is,

$$T\phi^* = \phi^*(c). \qquad \qquad \square$$

Remark 17.3 [51] If the Razumikhin class $\mathcal{R}_c$ is not topologically closed, then the limit of the sequence $\{\phi_n\}$ in Theorem 17.10 may be outside of $\mathcal{R}_c$, which may not be unique.

17.3.3 Fixed Points Results in *b*-Metric Spaces

In this section, we will show existence and uniqueness of some fixed point results in *b*-metric spaces.

Definition 17.18 [52] Let X be a non-empty set and let functional $d : X \times X \rightarrow [0, \infty)$ satisfy:

(b1) $d(x, y) = 0$ if and only if $x = y$;

(b2) $d(x, y) = d(y, x)$ for all $x, y \in X$;

(b3) there exists a real number $s \geq 1$ such that

$$d(x, z) \leq s[d(x, y) + d(y, z)]$$

for all $x, y, z \in X$.

Then d is called a b-metric on X and a pair (X, d) is called a b-metric space with coefficient s.

If we take $s = 1$ in Definition 17.18, then we obtain the ordinary metric spaces. Thus, the family of b-metric spaces is larger than the family of metric spaces.

Definition 17.19 [52] Let (X, d) be a b-metric space. Then a sequence $\{x_n\}$ in X is called:

(a) convergent if and only if there exists $x \in X$ such that $d(x_n, x) \to 0$ as $n \to \infty$;
(b) Cauchy if and only if $d(x_n, x_m) \to 0$ as $m, n \to \infty$.

Definition 17.20 [52] A mapping $\psi : [0, \infty) \to [0, \infty)$ is called a comparison function if it is increasing and $\psi^n(t) \to 0$ as $n \to \infty$ for any $t \in [0, \infty)$, where ψ^n is the nth iterate of ψ.

The concept of (c)-comparison function was introduced by Berinde in the following definition.

Definition 17.21 [52] A mapping $\psi : [0, \infty) \to [0, \infty)$ is called a (c)-comparison function if

(1) ψ is increasing;
(2) there exists $n_0 \in \mathbb{N}$, and $k \in (0, 1)$ and a convergent series of nonnegative terms $\sum_{n=1}^{\infty} \epsilon_n$ such that $\psi^{n+1}(t) \leq k\psi^n(t) + \epsilon_n$ for $n \geq n_0$ and any $t \in [0, \infty)$.

Definition 17.22 [52] Let $s \geq 1$ be a real number. A mapping $\psi : [0, \infty) \to [0, \infty)$ is called (b)-comparison function if

(1) ψ is increasing;
(2) there exists $n_0 \in \mathbb{N}$, and $k \in (0, 1)$ and a convergent series of nonnegative terms $\sum_{n=1}^{\infty} \epsilon_n$ such that $s^{n+1}\psi^{n+1}(t) \leq ks^n\psi^n(t) + \epsilon_n$ for $n \geq n_0$ and any $t \in [0, \infty)$.

It is clear from Definitions 17.21 and 17.22 that the idea of (b)-comparison function can be reduced to idea of (c)-comparison function by taking $s = 1$ in Definition 17.22.

Theorem 17.11 [52] Let (X, d) be a b-metric space with coefficient s, let $f : X \to X$ and $\alpha : X \times X \to [0, \infty)$ be two mappings and $\psi \in \psi_b$. Suppose that f is α-admissible and for all $x, y \in X$, we have

$$\alpha(x, f(x))\alpha(y, f(y))d(f(x), f(y)) \leq \psi(d(x, y)).$$

In addition, there exists $x_0 \in X$ such that $\alpha(x_0, f(x_0)) \geq 1$. Furthermore, if $\{x_n\}$ is a sequence in X such that $x_n \to x$ as $n \to \infty$ and $\alpha(x_n, f(x_n)) \geq 1$ for all $n \in \mathbb{N}$, then $\alpha(x, f(x)) \geq 1$. Thus, f has a unique fixed point x^* in X such that $\alpha(x^*, f(x^*)) \geq 1$.

Theorem 17.12 [52] Let (X, d) be a b-metric space with coefficient s, let $f : X \to X$ and $\alpha : X \times X \to [0, \infty)$ be two mappings and $\psi \in \psi_b$. Suppose that f is α-admissible also there exists $x_0 \in X$ such that $\alpha(x_0, f(x_0)) \geq 1$. Furthermore, if $\{x_n\}$ is a sequence in X such that $x_n \to x$ as $n \to \infty$ and $\alpha(x_n, f(x_n)) \geq 1$ for all $n \in \mathbb{N}$, then $\alpha(x, f(x)) \geq 1$ and there exists $\xi \geq 1$ such that

$$[d(f(x), f(y)) + \xi]^{\alpha(x, f(x))\alpha(y, f(y))} \leq \psi(d(x, y)) + \frac{\xi}{s}.$$

Thus, f has a unique fixed point x^* in X such that $\alpha(x^*, f(x^*)) \geq 1$.

Theorem 17.13 [52] Let (X, d) be a b-metric space with coefficient s, let $f : X \to X$ and $\alpha : X \times X \to [0, \infty)$ be two mappings and $\psi \in \psi_b$. Suppose that f is α-admissible also there exists $x_0 \in X$ such that $\alpha(x_0, f(x_0)) \geq 1$. Furthermore, if $\{x_n\}$ is a sequence in X such that $x_n \to x$ as $n \to \infty$ and $\alpha(x_n, f(x_n)) \geq 1$ for all $n \in \mathbb{N}$, then $\alpha(x, f(x)) \geq 1$ and there exists $\xi \geq 1$ such that

$$[\alpha(x, f(x))\alpha(y, f(y)) - 1 + \xi]^{d(f(x), f(y))\alpha(y, f(y))} \leq \xi^{\psi(d(x, y))}.$$

Thus, f has a unique fixed point x^* in X such that $\alpha(x^*, f(x^*)) \geq 1$.

17.3.4 The generalized Ulam–Hyers Stability in b-Metric Spaces

In this section, we will show the generalized Ulam–Hyers stability in b-metric spaces, which corresponds to Theorems 17.11, 17.12 and 17.13 respectively. Stability problems of functional analysis play the most important role in mathematics analysis and were introduced by Ulam [68], who is concerned with the stability of a group homomorphism. Afterwards, Hyers [69] gave answer for results in Banach spaces, and this type of stability is called Ulam–Hyers stability. There are many other mathematicians who found new results for generalized Ulam–Hyers stability via α-admissibility in b-metric spaces, but for this work the difference is just about some other contractions of new type with same assumptions.

Definition 17.23 [52] Let (X, d) be a b-metric space with coefficient s and let $f : X \to X$ be an operator. By definition, the fixed point equation

$$u = f(u), \ u \in X \tag{17.1}$$

is said to be generalized Ulam–Hyers stable in the framework of a b-metric space if there exists an increasing operator $\phi : [0, \infty) \to [0, \infty)$, continuous at 0 and $\phi(0) = 0$ such that for each $\epsilon > 0$ and an ϵ-solution $v^* \in X$ that is

$$d(v^*, f(v^*)) \leq \epsilon$$

there exists a solution $u^* \in X$ of the fixed point equation (17.1) such that

$$d(u^*, v^*) \leq \phi(s\epsilon).$$

If $\phi(t) := ct$ for all $t \in [0, \infty)$, where $c > 0$, then (17.1) is said to be Ulam–Hyers stable in the framework of a b-metric space.

If we take $s = 1$ in Definition 17.23, then we get generalized Ulam–Hyers stability in metric spaces. In addition, if $\phi(t) := ct$ for all $t \in [0, \infty)$, where $c > 0$, then it reduces to classical Ulam–Hyers stability.

Theorem 17.14 **[52]** Let (X, d) be a complete b-metric space with coefficients s. Suppose that all the hypothesis of Theorem 17.11 hold and also that the function $\phi : [0, \infty) \rightarrow [0, \infty)$ defined by $\phi(t) := t - s\psi(t)$ is strictly increasing and onto. If $\alpha(u^*, f(u^*)) \geq 1$ for all $u^* \in X$, which is an ϵ-solution, then the fixed point equation (17.1) is generalized Ulam–Hyers stable.

Theorem 17.15 **[52]** Let (X, d) be a complete b-metric space with coefficients s. Suppose that all the hypothesis of Theorem 17.12 hold and also that the function $\phi : [0, \infty) \rightarrow [0, \infty)$ defined by $\phi(t) := t - s\psi(t)$ is strictly increasing and onto. If $\alpha(u^*, f(u^*)) \geq 1$ for all $u^* \in X$, which is an ϵ-solution, then the fixed point equation (17.1) is generalized Ulam–Hyers stable.

Theorem 17.16 **[52]** Let (X, d) be a complete b-metric space with coefficients s. Suppose that all the hypothesis of Theorem 17.13 hold and also that the function $\phi : [0, \infty) \rightarrow [0, \infty)$ defined by $\phi(t) := t - s\psi(t)$ is strictly increasing and onto. If $\alpha(u^*, f(u^*)) \geq 1$ for all $u^* \in X$, which is an ϵ-solution, then the fixed point equation (17.1) is generalized Ulam–Hyers stable.

17.3.5   Well-Posedness of a Function with Respect to α-Admissibility in b-Metric Spaces

In this section, we will prove well-posedness of a function with respect to α-admissibility in b-metric spaces.

Definition 17.24 **[52]** Let (X, d) be a b-metric space with coefficient s, let $f : X \rightarrow X$ and $\alpha : X \times X \rightarrow [0, \infty)$ be two mappings. The fixed point problem of f is said to be well posed with respect to α if:

(a) f has unique fixed point x^* in X such that $\alpha(x^*, f(x^*)) \geq 1$;
(b) for a sequence $\{x_n\}$ in X such that $d(x_n, f(x_n)) \rightarrow 0$ as $n \rightarrow \infty$, then $x_n \rightarrow x^*$ as $n \rightarrow \infty$.

In the following theorems, we add a new condition to assure the well-posedness via α-admissibility.

(S) If $\{x_n\}$ is a sequence in X such that $d(x_n, f(x_n)) \to 0$ as $n \to \infty$, then $\alpha(x_n, f(x_n)) \geq 1$ for all $n \in \mathbb{N}$.

Theorem 17.17 [52] Let (X, d) be a complete b-metric space with coefficient s, let $f : X \to X$ and $\alpha : X \times X \to [0, \infty)$ be two mappings and $\psi \in \psi_b$. Suppose that all hypothesis of Theorem 17.11 and condition (S) hold. Then the fixed point equation (17.1) is well posed with respect to α.

Theorem 17.18 [52] Let (X, d) be a complete b-metric space with coefficient s, let $f : X \to X$ and $\alpha : X \times X \to [0, \infty)$ be two mappings and $\psi \in \psi_b$. Suppose that all hypothesis of Theorem 17.12 and condition (S) hold. Then the fixed point equation (17.1) is well posed with respect to α.

Theorem 17.19 [52] Let (X, d) be a complete b-metric space with coefficient s, let $f : X \to X$ and $\alpha : X \times X \to [0, \infty)$ be two mappings and $\psi \in \psi_b$. Suppose that all hypothesis of Theorem 17.13 and condition (S) hold. Then the fixed point equation (17.1) is well posed with respect to α.

The study of fixed point theory for multivalued mappings using Hausdroff metric spaces was initiated by Nadler [11]. He extended the Banach contraction principle to set-valued mappings.

17.3.6 Fixed Points for F-Contraction

Edelstein proved the following version of Banach contraction principle as following:

Theorem 17.20 [70] Let (X, d) be a compact metric space and $T : X \to X$ be a self-mapping. Assume that $d(Tx, Ty) < d(x, y)$ holds for all $x, y \in X$ with $x \neq y$. Then T has a unique fixed point in X.

Suzuki [71] proved generalized versions of Edelstein's results in compact metric spaces as follows:

Theorem 17.21 [71] Let (X, d) be a compact metric space and $T : X \to X$ be a self-mapping. Assume that $\frac{1}{2}d(x, Tx) < d(x, y) \quad d(Tx, Ty) < d(x, y)$ holds for all $x, y \in X$ with $x \neq y$. Then T has a unique fixed point in X.

Wardowski [72] introduced a new type of contractions called F-contraction and proved a new fixed-point theorem concerning F-contractions. He defined the fixed point results in many directions from well-known results from literature. Wardowski defined the F-contraction as follows:

Definition 17.25 [72] Let (X, d) be a metric space. A mapping $T : X \to X$ is said to be an F-contraction if there exists $\tau > 0$ such that

$$d(Tx, Ty) > 0 \Rightarrow \tau + F(d(Tx, Ty)) \leq F(d(x, y))$$

holds for all $x, y \in X$, where $F : \mathbb{R}_+ \to \mathbb{R}$ is a mapping satisfying the following conditions:

(F1) F is strictly increasing; that is, for all $x, y \in \mathbb{R}$ such that $x < y$, $F(x) < F(y)$;
(F2) For each sequence $\{\alpha_n\}_{n=1}^{\infty}$ of positive numbers, $\lim_{n \to \infty} \alpha_n = 0$ if and only if $\lim_{n \to \infty} F(\alpha_n) = -\infty$;
(F3) There exists $k \in (0, 1)$ such that $\lim_{\alpha \to 0^+} \alpha^k F(\alpha) = 0$.

We denote by $\mathcal{F}$, the set of all functions satisfying the conditions $(f1)$–$(f3)$.

From Definition 17.25 both the inequality in the F-contraction and the condition, $(F1)$ show that every F-contraction is necessarily continuous. In 2014, Secelean proved the following lemma:

Lemma 17.2 [85] Let $F : \mathbb{R}_+ \to \mathbb{R}$ be an increasing mapping and $\{\alpha_n\}_{n=1}^{\infty}$ be a sequence of positive real numbers. Then the following assertions hold:

(a) if $\lim_{n \to \infty} F(\alpha_n)$, then $\lim_{n \to \infty} \alpha_n = 0$;
(b) if $\lim_{n \to \infty} F = -\infty$ and $\lim_{n \to \infty} \alpha_n = 0$, then $\lim_{n \to \infty} F(\alpha_n) = -\infty$.

By proving Lemma 17.2, Secelean [40] showed that the condition $(f2)$, in the definition of F-contraction by Wardowski [72], can be replaced by an equivalent but a more simple condition,

$(f2')$ $\inf F = -\infty$

or, also, by

$(f2')$ there exists a sequence $\{\alpha_n\}_{n=1}^{\infty}$ of positive real numbers such that $\lim_{n \to \infty} F(\alpha_n) = -\infty$.

We denote by Υ the set of all functions satisfying the conditions $(f1), (f2')$, and $(f3')$.

Example 17.8 [85] Let $F_1(\alpha) = \frac{-1}{\alpha}$, $F_2(\alpha) = \frac{-1}{\alpha} + \alpha$, $F_3(\alpha) = \frac{1}{1 - e^{\alpha}}$, and $F_4(\alpha) = \frac{1}{e^{\alpha} - e^{-\alpha}}$. Then $F_1, F_2, F_3,$ and $F_4 \in \Upsilon$.

Definition 17.26 [73] Let (X, d) be a metric space. A mapping $T : X \to X$ is said to be an F-Suzuki contraction if there exists $\tau > 0$ such that for all $x, y \in X$ with $Tx \neq Ty$,

$$\frac{1}{2}d(x, Tx) < d(x, y) \quad \tau + F(d(Tx, Ty)) \le F(d(x, y)),$$

where $F \in \Upsilon$.

Theorem 17.22 [73] Let T be a self-mapping of a complete metric space X into itself. Suppose that $F \in \Upsilon$ and there exists $\tau > 0$ such that

$$\forall x, y \in X, \quad [d(Tx, Ty) > 0 \Rightarrow \tau + F(d(Tx, Ty))] \le F(d(x, y))$$

Then T has a unique fixed point $x^* \in X$ and for every $x_0 \in X$ the sequence $T^n x_0$ converges to x^*.

Theorem 17.23 [73] Let (X, d) be a complete metric space and $T : X \to X$ be an F-Suzuki contraction. Then T has a unique fixed point $x^* \in X$ and for every $x_0 \in X$ the sequence $\{T^n x_0\}_{n=1}^{\infty}$ converges to x^*.

17.4 Common Fixed Points Theorems

Zadeh [2] investigated the concept of a fuzzy set in his seminal paper. In the last two decades, there has been a tremendous development and growth in fuzzy mathematics. Fuzzy set theory also has applications in applied sciences such as neural network theory, stability theory, mathematical programming, modeling theory, engineering sciences, medical sciences (medical genetics, nervous system), image processing, control theory, and communication. Kumam *et al.* [39] introduced the notion of the joint common limit in the range of mappings property called (JCLR) property and proved a common fixed-point theorem for a pair of weakly compatible mappings using JCLR property in fuzzy metric space. As an application to their main result, they presented a common fixed-point theorem for two finite families of self-mappings in fuzzy metric space using the notion of pairwise commuting maps.

Definition 17.27 [39] An element x^* in A is said to be a common best proximity point of the non-self-mappings $S : A \to B$ and $T : A \to B$ if it satisfies the condition that

$$d(x^*, Sx^*) = d(x^*, Tx^*) = d(A, B).$$

It should be observed that a common best proximity point is an element at which both real-valued functions $x \to d(x, Sx)$ and $x \to d(x, Tx)$ attain the global minimum, since $d(x, Sx) \ge d(A, B)$ and $d(x, Tx) \ge d(A, B)$ for all x. Further, if the underlying mappings are self-mappings, a common best proximity point becomes a common fixed point.

Definition 17.28 [39] Let $(X, M, *)$ be a fuzzy metric space and f, g, a, b : $X \to X$. The pair (f, b) and (a, g) are said to satisfy the JCLR of b and g property (shortly, *(JCLRbg)* property) if there exists a sequence $\{x_n\}$ and $\{y_n\}$ in X such that $\lim_{n \to \infty} fx_n = \lim_{n \to \infty} bx_n = \lim_{n \to \infty} ay_n = \lim_{n \to \infty} gy_n = bu = gu$, for some $u \in X$.

Theorem 17.24 [39] Let $(X, M, *)$ be a fuzzy metric space, where $*$ is a continuous t-norm and f, g, a and b be mappings from X into itself. Further, let the pair (f, b) and (a, g) are weakly compatible and there exists a constant $k \in (0, \frac{1}{2})$ such that

$$M(fx, ay, kt) \geq \phi(M(bx, gy, t), M(fx, bx, t), M(ay, gy, t),$$
$$M(fx, gy, \alpha t), M(ay, bx, 2t - \alpha t))$$

holds for all $x, y \in X, \alpha \in (0, 2), t > 0$, and $\phi \in \Phi$. If (f, b) and (a, g) satisfy the *(JCLRbg)* property, then f, g, a and b have a unique common fixed point in X. Proof as in [39].

Azam *et al.* introduced new spaces called the complex-valued metric spaces and established the existence of fixed-point theorems under the contraction condition. Motivated by him, Kumam and Sintunavarat [50] extended and improved the condition of contraction of the results of Azam *et al.* and also applied their main results to unique common solution of system of Urysohn integral equation.

17.4.1 Common Fixed-Point Theorems for Pair of Weakly Compatible Mappings in Fuzzy Metric Spaces

In [74], we introduced the new property, which is so called a CLR for two self-mappings, and gave some examples of mappings, which satisfy this property. Moreover, we established some new existence of a common fixed-point theorem for generalized contractive mappings in fuzzy metric spaces both in the sense of Kramosil and Michalek and in the sense of George and Veeramani by using new property and gave some examples. Ours results does not require condition of closeness of range and so our theorems generalize, unify, and extend many results in literature. For further details, see [74].

Definition 17.29 [74] Suppose that (X, d) is a metric space and $f, g : X \to X$. Two mappings f and g are said to satisfy the CLR of g-property if

$$\lim_{n \to \infty} fx_n = \lim_{n \to \infty} gx_n = gx \tag{17.2}$$

for some $x \in X$.

In what follows, the common limit in the range of g property will be denoted by the *(CLRg)* property. Next, example shows about *CLRg* property of mappings f and g, which are satisfying the *CLRg* property.

Example 17.9 [74] Let $X = [0, \infty)$ be the usual metric space. Define $f, g :$ $X \to X$ by $fx = \frac{x}{4}$ and $gx = \frac{3x}{4}$ for all $x \in X$. We consider the sequence $\{x_n\} = \{\frac{1}{n}\}$. Since

$$\lim_{n \to \infty} fx_n = \lim_{n \to \infty} gx_n = 0 = g0, \tag{17.3}$$

therefore, f and g satisfy the *CLRg*-property.

In a similar mode, two self-mappings f and g of a fuzzy metric space $(X, M, *)$ satisfy the *CLRg*-property, if there exists a sequence $\{x_n\}$ in X such that fx_n and gx_n converge to gx for some $x \in X$ in the sense of Definition 17.29, which tells about convergence in KM or GM fuzzy metric spaces.

Theorem 17.25 [74] Let $(X, M, *)$ be a *KM*-fuzzy metric space satisfying the following property:

$$\forall x, y \in X, x \neq y, \exists t > 0 : 0 < M(x, y, t) < 1, \tag{17.4}$$

and let f, g be weakly compatible self-mappings of X such that, for some $\phi \in \Phi$,

$$M(fx, fy, t) \geq \phi(\min\{M(gx, gy, t), M(fx, gx, t),$$
$$M(fy, gy, t), M(fy, gx, t), M(fx, gy, t)\}),$$

for all $x, y \in X$, where $t > 0$. If f and g satisfy the *CLRg*-property, then f and g have a unique common fixed point.

Proof: Since f and g satisfy the *CLRg*-property, there exists a sequence $\{x_n\}$ in X such that

$$\lim_{n \to \infty} fx_n = \lim_{n \to \infty} gx_n = gx, \tag{17.5}$$

for some $x \in X$. Let t be a continuity point of $(X, M, *)$. Then

$$M(fx_n, fx, t) \geq \phi \left(\min \left\{ M(gx_n, gx, t), M(fx_n, gx_n, t), M(fx, gx, t), \right. \right.$$
$$\left. \left. M(fx, gx_n, t), M(fx_n, gx, t) \right\} \right),$$

for all $n \in \mathbb{N}$. By making $n \to \infty$, we have

$$M(gx, fx, t) \geq \phi(M(gx, fx, t))$$

for every $t > 0$). We claim that $gx = fx$. To prove this, we suppose the contradiction. Next, we let $z := fx := gx$. Since f and g are weakly compatible mappings, $fgx = gfx$, which implies that $fz = fgx = gfx = gz$. We claim that $fz = z$, if we suppose the contradiction, then it is obtain by some elementary calculations as:

$$M(fz, z, t) \geq \phi(M(fz, z, t))$$

for all $t > 0$, which is a contradiction. Hence $fz = z$, that is, $z = fz = gz$. Therefore, z is a common fixed point of f and g. After proving the criteria for uniqueness, we obtain that f and g have a unique a common fixed point. This finishes the proof. $\qquad \square$

17.5   Best Proximity Points

Investigation of the existence and uniqueness of a fixed point of non-self-mappings is one of the interesting subjects in fixed point theory. In fact, given non-empty closed subsets A and B of a complete metric space (X, d), a contraction non-self-mapping $T : A \to B$ does not necessarily yield a fixed point $Tx = x$. In this case, it is very natural to investigate whether there is an element x such that $d(x, Tx)$ is minimum. A notion of best proximity point appears at this point. A point x is called best proximity point of $T : A \to B$ if

$$d(x, Tx) = d(A, B) = \inf\{d(x, y) : x \in A, y \in B\},$$

where (X, d) is a metric space, and A, B are subsets of X. A best proximity point represents an optimal approximate solution to the equation $Tx = x$ whenever a non-self-mapping T has no fixed point. It is clear that a fixed point coincides with a best proximity point if $d(A, B) = 0$. Since a best proximity point reduces to a fixed point if the underlying mapping is assumed to be self-mappings, the best proximity point theorems are natural generalizations of the Banachs contraction principle. Best proximity point theorems provide an approximate solution that is optimal. In fact, a best proximity point theorem specifies sufficient conditions for the existence of an element x such that the error $d(x, Tx)$ is minimum. A best proximity point theorem is primarily dedicated to global minimization of the real-valued function, which furnishes a yardstick for the error involved for an approximate solution of the equation $Tx = x$. In addition, it is interesting to see that best proximity point theorems emerge as a natural generalization of fixed-point theorems, because a best proximity point reduces to a fixed point if the mapping under consideration turns out to be a self-mapping. In [23], Kumam *et al.* introduced the notion of a generalized multivalued cyclic contraction pair, which is an extension of Mizoguchi–Takahashis contraction mappings for non-self-version and establish a best proximity point of such mappings in metric spaces via property UC^* due to Sintunavarat and Kumam [42]. Further, they investigated best proximity point theorems in a uniformly convex Banach space. We begin this, by introducing the notion of multivalued cyclic contraction.

Definition 17.30   **[23]**   Let A and B be non-empty subsets of a metric space (X, d), $T : A \to 2^B$ and $S : B \to 2^A$. The ordered pair (T, S) is said to be a generalized multivalued cyclic contraction if there exists a function $\alpha : [d(A, B), \infty) \to [0, 1)$ with

$$\limsup_{x \to t^+} \alpha(x) < 1$$

for each $t \in [d(A, B), \infty)$ such that

$$H(Tx, Sy) \leq \alpha(d(x, y))d(x, y) + (1 - \alpha(d(x, y)))d(A, B)$$

for all $x \in A$ and $y \in B$.

Note that if (T, S) is a generalized multivalued cyclic contraction, then (S, T) is also a generalized multivalued cyclic contraction. Here, we state the some result on the existence of best proximity points for a generalized multivalued cyclic contraction pair, which satisfies the property UC^* in metric spaces.

Theorem 17.26 [23] Let A, B be non-empty closed subsets of a complete metric space X such that (A, B) and (B, A) satisfy the property UC^*. Let $T : A \to CB(B)$ and $S : B \to CB(A)$. If (T, S) is a generalized multivalued cyclic contraction pair, then T has a best proximity point in A, or S has a best proximity point in B.

If $d(A, B) = 0$, then Theorem 17.26 yields existence of a fixed point in $A \cap B$ of two multivalued non-self-mappings S and T. Moreover, if $A = B = X$ and $T = S$, then this theorem reduces to Mizoguchi–Takahashi fixed-point theorem. Every pair of non-empty closed subsets A, B of a uniformly convex Banach space such that A is convex satisfies the property UC^*. Then, we obtained the forthcoming corollary.

Corollary 17.3 [23] Let A, B be non-empty closed subsets of a uniformly convex Banach space X. Let $T : A \to CB(B)$ and $S : B \to CB(A)$. If (T, S) is a generalized multivalued cyclic contraction pair, then T has a best proximity point in A, or S has a best proximity point in B.

Now, for this event, we are going to propose a useful example.

Example 17.10 [23] Consider the uniformly convex Banach space $X = \mathbb{R}$ with Euclidean norm. Let $A = [1, 2]$ and $B = \{x : -2 \leq x \leq -1\}$. Then A and B are non-empty closed and convex subsets of X and $d(A, B) = 2$. Since A and B are convex, we have (A, B) and (B, A) satisfy the property UC^*. Let $T : A \to CB(B)$ and $S : B \to CB(A)$ be defined as in [23] for all $x \in A$ and for all $y \in B$. Let $\alpha : [d(A, B), \infty) \to [0, 1)$ be defined by $\alpha(t) = \frac{1}{2}$ for all $t \in [d(A, B), \infty) = [2, \infty)$. Next, we show that (T, S) is a generalized multivalued cyclic contraction pair with $\alpha(t) = \frac{1}{2}$ for all $t \in [2, \infty)$.
For each $x \in A$ and $y \in B$, we have

$$H(Tx, Sy) \leq \alpha(d(x, y))d(x, y) + (1 - \alpha(d(x, y)))d(A, B).$$

Therefore, all assumptions of Corollary 17.3 are satisfied, and then T has a best proximity point in A, that is, a point $x = 17$. Moreover, S also has a best proximity point in B, that is, a point $y = 1$.

We are going to propose some extended and generalized notions of the class of proximal contraction of first and second kinds which are different from another type in the literature as in Basha [26]. For such mappings, we seek the necessary condition for these classes to have best proximity points and also will give some examples to illustrate our main results.

Definition 17.31 **[49]** Let A, B be non-empty subset of metric space (X, d). $T : A \to B$ and $\kappa : A \to [0, 1)$. A mapping is said to be a generalized proximal contraction of the first kind with respect to κ if

$$d(a, Tx) = d(A, B) = d(b, Ty),$$

implies $d(a, b) \leq \kappa(x)d(x, y)$, for all $a, b, x, y \in A$.

Remark 17.4 **[49]** If we take $\kappa(x) = \alpha$ for all $x \in A$, where $\alpha \in [0, 1)$, then a generalized proximal contraction of the first kind with respect to κ reduces to a proximal contraction of the first kind. In case of a self-mapping, it is apparent that the class of contraction mapping is contained in the class of generalized proximal contraction of the first kind with respect to κ mapping.

Now, we give an example to claim that the class of proximal contraction mapping of the first kind is a proper subclass of the class of generalized proximal contractions of the first kind with respect to κ mapping.

Example 17.11 **[49]** Consider the metric space $\mathbb{C}$ with Euclidean metric. Let $A = \{(0, y) : -1 < y < 1\}$ and $\{(1, y) : -1 < y < 1\}$. Define a mapping $T : A \to B$ as follows

$$T((0, y)) = \left(1, \frac{y^2}{2}\right)$$

for all $(0, y) \in A$. It is easy to check that there is no $\alpha \in [0, 1)$ satisfying

$$d(a, Tx) = d(b, Ty) = d(A, B) \Leftrightarrow d(a, b) \leq \alpha d(x, y)$$

for all $a, b, x, y \in A$. Therefore, T is not a proximal contraction of the first kind. Consider a function κ defined by

$$\kappa(0, y) = \frac{y |y| + 1}{2}$$

Next, we prove that T is a generalized proximal contraction of the first kind with respect to κ. If $(0, y_1), (0, y_2) \in A$ such that $d(a, T(0, y_1)) = d(A, B) = d(b, T(0, y_2)) = 1$, for all $a, b \in A$, then we can easily show that $d(a, b) \leq \kappa(0, y_1)d((0, y_1), (0, y_2))$ This implies that T is a generalized proximal contraction of the first kind with respect to κ.

Definition 17.32 **[49]** Let A, B be non-empty subsets of metric space (X, d), $T : A \to B$ and $\kappa : A \to [0, 1)$. A mapping T is said to be a generalized proximal contraction of the second kind with respect to κ if for all $a, b, x, y \in A$.

Clearly, a proximal contraction of the second kind is a generalized proximal contraction of the second kind. Next, we extend the results of Basha [26] and many results in the literature.

Theorem 17.27 [49] Let (X, d) be a complete metric space and A, B be non-empty closed subsets of X such that A_0 and B_0 are non-empty. Suppose that $T : A \to B, g : A \to A$ and $\kappa : A \to [0, 1)$ are mappings satisfying the following conditions:

(a) T is a continuous generalized proximal contraction of first kind with respect to κ;
(b) $T(A_0) \subseteq B_0$ and $A_0 \subseteq g(A_0)$;
(c) g is an isometry;
(d) $\kappa(x) \leq \kappa(y)$, whenever $d(gx, Ty) = d(A, B)$. Then there exists a unique point $x \in A$ such that $d(gx, Tx) = d(A, B)$.

Proof: Let x_0 be a fixed element in A_0. From $T(A_0) \subseteq B_0$ and $A_0 \subseteq g(A_0)$, it follows that there exists a point $x_1 \in A_0$ such that $d(gx_1, Tx_0) = d(A, B)$. Continuing this process, we can construct the sequence $\{x_n\}$ in A_0 such that $d(gx_n, Tx_{n-1}) = d(A, B)$, for all $n \in \mathbb{N}$. In the next step, it is easily proved that $d(x_{n+1}, x_n) \leq (\kappa(x_0))^n d(x_1, x_0)$ for all $n \in \mathbb{N}$. Next step is to justify that $\{x_n\}$ is a Cauchy sequence in X. Since (X, d) is complete and T, g are continuous, we get

$$d(gx^*, Tx^*) = \lim_{n \to \infty} d(gx_{n+1}, Tx_n) = d(A, B).$$

Since T is a generalized proximal contraction of the first kind with respect to κ and by using the isometry, we get that point x^* is unique. $\quad\square$

In [49] Kumam *et al.* have proved very interesting results as well as applicable and more generalized than already existing results in literature.

17.6 Common Best Proximity Points

Fixed point theory highlights and brightens, the methodologies for finding a solution to nonlinear equations of the type $Tx = x$, where T is a self-mapping defined on a subset of a metric space, a normed linear space, a topological vector space, or some appropriate space. But, the equation $Tx = x$ is unlikely to have a solution when T is not a self-mapping. Consider that A and B are non-empty subsets of a metric space, and assume the non-self-mappings $S : A \to B$ and $T : A \to B$. With this nice awareness that S and T are non-self-mappings, it may possible that the equations $Sx = x$ and $Tx = x$ have no common solution, named a common fixed point of the mappings S and T. Subsequently, in the sense that there is no common solution of the preceding equations, one speculates about finding an element x that is in close proximity to Sx and Tx in the sense that $d(x, Sx)$ and $d(x, Tx)$ are minimum. Indeed, a common best proximity point theorem evaluate the existence of such an optimal approximate solution, named a common best proximity point of the mappings S and T, to the equations $Sx = x$ and $Tx = x$ when there is no common solution. Moreover, it is emphasized that the real-valued functions $x \to d(x, Sx)$ and $x \to d(x, Tx)$ investigate the degree of the error involved for any common approximate solution of the equations $Sx = x$ and $Tx = x$. Owing to the fact that the distance between x and Sx, and the distance between x and Tx are at least the distance between A and B for all x in A, a common best proximity point theorem accomplishes the global minimum of both functions $x \to d(x, Sx)$ and $x \to d(x, Tx)$ by postulating a common approximate solution of the equations $Sx = x$ and $Tx = x$ for meeting the condition that $d(x, Sx) = d(x, Tx) = d(A, B)$. This work is devoted to an interesting common best proximity point theorem for pairs of non-self-mappings satisfying a contraction-like condition, thereby producing common optimal approximate solutions of certain simultaneous fixed-point equations in Basha [27]. If K is a non-empty compact convex subset of a Hausdorff locally convex topological vector space E and $T : K \to E$ is a non-self-continuous map, then a classical best approximation theorem, due to Fan [3], asserts that there is an element x satisfying the condition that $d(x, Tx) = d(Tx, K)$. Basha [27] proved results about common best proximity point of multiobjective functions but in same year, Kumam *et al.* [44] proved some results which guarantee the generalization of work of Basha [27].

Theorem 17.28 [27] Let A and B be non-empty closed subsets of a complete metric space X such that A is approximately compact with respect to B. In addition, let A_0 and B_0 are non-empty. Assume that the non-self-mapping $S : A \to B, T : A \to B$ satisfy the following conditions:

(a) There is a nonnegative real number $\alpha < 1$ such that

$$d(Sx, Sy) \le \alpha d(Tx, Ty),$$

for all $x, y \in A$.
(b) T is continuous;
(c) S and T commute proximally;
(d) S and T can be swapped proximally;
(e) $S(A_0) \subseteq B_0$ and $S(A_0) \subseteq T(A_0)$.

Then there exists an element $x \in A$ such that $d(x, Tx) = d(A, B)$ and $d(x, Sx) = d(A, B)$. Moreover, if x^* is another common best proximity point of the mapping S and T, then it is necessary that

$$d(x, x^*) \le 2d(A, B).$$

Here are very useful results of Kumam and Mongkolkeha [44] for proximity commuting mappings about common best proximity points. The next result is actually the generalization and extension of the condition (a) of above theorem of [27]. In fact, the class of weakly contractive maps lies between the classes of contraction mappings and contractive mappings.

Theorem 17.29 [44] Let A and B be non-empty closed subsets of a complete metric space X such that A is approximately compact with respect to B. In addition, let A_0 and B_0 are non-empty. Assume that the non-self-mapping $S : A \to B, T : A \to B$ satisfy the following conditions:

(a) For each x and y are elements in A,

$$d(Sx, Sy) \le d(Tx, Ty) - \phi(d(Tx, Ty)),$$

where, $\phi : [0, \infty) \to [0, \infty)$ is a continuous and nondecreasing function such that $\phi(t) = 0$ if and only if $t = 0$.
(b) T is continuous;
(c) S and T commute proximally;
(d) S and T can be swapped proximally;
(e) $S(A_0) \subseteq B_0$ and $S(A_0) \subseteq T(A_0)$.

Then there exists an element $x \in A$ such that $d(x, Tx) = d(A, B)$ and $d(x, Sx) = d(A, B)$. Moreover, if x^* is another common best proximity point of the mapping S and T, then it is necessary that

$$d(x, x^*) \le 2d(A, B).$$

Replacing $\phi(t)$ by $(1 - \alpha)t$ where $0 \le \alpha < 1$ in Theorem 17.29, we obtain Corollary 17.4.

Corollary 17.4 [41] Let A and B be non-empty closed subsets of a complete metric space X such that A is approximately compact with respect to B. In addition, let A_0 and B_0 are non-empty. Assume that the non-self-mapping $S : A \to B, T : A \to B$ satisfy the following conditions:

(1) There exists a nonnegative real number $\alpha < 1$ such that

$$d(Sx_1, Sx_2) \le \alpha d(Tx_1, Tx_2),$$

for all $x_1, x_2 \in A$.
(2) T is continuous;
(3) S and T commute proximally;
(4) S and T can be swapped proximally;
(5) $S(A_0) \subseteq B_0$ and $S(A_0) \subseteq T(A_0)$.

Then there exists an element $x \in A$ such that $d(x, Tx) = d(A, B)$ and $d(x, Sx) = d(A, B)$. Moreover, if x^* is another common best proximity point of the mapping S and T, then it is necessary that

$$d(x, x^*) \leq 2d(A, B).$$

Example 17.12 [41] Consider complete metric space $\mathbb{C}$ with Euclidean metric. Let $A = \{(x, 1) : 0 \leq x \leq 1\}$ and $B = \{(x, -1) : 0 \leq x \leq 1\}$. Define two mappings $S : A \to B$, $T : A \to B$ as follows: $S(x, 1) = (x - \frac{x^2}{2}, -1)$ and $T(x, 1) = (x, -1)$. It is easy to see that $d(A, B) = 2$, $A_0 = A$, and $B_0 = B$. Further, S and T are continuous and A is approximately compact with respect to B. Construct $\phi : [0, \infty) \to [0, \infty)$ as $\phi(t) = \frac{t^2}{2}$ with $x > y$ for all $t \in [0, \infty)$ and $x, y \in [0, 1]$. With the help of all such useful assumption, one can be see that each hypothesis of our Theorem 17.29 satisfies here; and there exists $(0, 1) \in A$ that is a common best proximity point of S and T because

$$d((0, 1), S(0, 1)) = d((0, 1), (0, -1)) = d((0, 1), T(0, 1)) = d(A, B).$$

17.7 Tripled Best Proximity Points

The last section of this chapter is arranged an event of describing about tripled fixed point and tripled best proximity point theorems with the help of some basic definitions and examples that are related to the main results of [43].

Inspired by the work of Sintunavarat and Kumam [42], we have defined the notions for tripled best proximity points and related results.

Definition 17.33 [43] Let A, B be non-empty subsets of a metric space (X, d) and $F : A^3 \to B$ be mapping. An ordered tripled $(x, y, z) \in A^3$ is called a tripled best proximity point of F if,

$$d(x, F(x, y, z)) = d(y, F(y, x, y)) = d(z, F(z, y, x)) = d(A, B).$$

It is easy to see that, if $A = B$, then a tripled best proximity point reduces to a tripled fixed point.

Definition 17.34 [43] Let A, B be non-empty subsets of a metric space (X, d) and $F : A^3 \to B$, $G : B^3 \to A$ be two mappings. The ordered pair (f, G) is called a cyclic contraction if there exists a nonnegative number $\alpha < 1$ such that

$$d(f(x, y, z), G(u, v, w)) \leq \frac{\alpha}{3}[d(x, u) + d(y, v) + d(z, w)] + (1 - \alpha)d(A, B),$$

for all $(x, y, z) \in A^3$ and $(u, v, w) \in B^3$.

Note that, if (f, G) is a cyclic contraction, then the pair (G, F) is also a cyclic contraction. There are some lemmas, which will be useful in main theorem to prove tripled best proximity points.

Lemma 17.3 **[43]** Let A, B be non-empty subsets of a metric space (X, d) and $F : A^3 \to B, G : B^3 \to A$ be two mappings such that the ordered pair (f, G) is a cyclic contraction. If $(x_0, y_0, z_0) \in A^3$ and we define the sequence $\{x_n\}, \{y_n\}$, and $\{z_n\}$ in X by

$$x_{2n+1} = F(x_{2n}, y_{2n}, z_{2n}), \quad x_{2n+2} = G(x_{2n+1}, y_{2n+1}, z_{2n+1}),$$
$$y_{2n+1} = F(y_{2n}, x_{2n}, y_{2n}), \quad y_{2n+2} = G(y_{2n+1}, x_{2n+1}, y_{2n+1}),$$
$$z_{2n+1} = F(z_{2n}, y_{2n}, x_{2n}), \quad z_{2n+2} = G(z_{2n+1}, y_{2n+1}, x_{2n+1}),$$

for all $n \in \{0\} \cup \mathbb{N}$, then we have

$$d(x_{2n}, x_{2n+1}) \to d(A, B), \quad d(x_{2n+1}, x_{2n+2}) \to d(A, B),$$
$$d(y_{2n}, y_{2n+1}) \to d(A, B), \quad d(y_{2n+1}, y_{2n+2}) \to d(A, B),$$
$$d(z_{2n}, z_{2n+1}) \to d(A, B), \quad d(z_{2n+1}, z_{2n+2}) \to d(A, B).$$

Lemma 17.4 **[43]** Let A, B be non-empty subsets of a metric space (X, d) such that the pairs (A, B) and (B, A) have the property UC and $F : A^3 \to B, G : B^3 \to A$ be two mappings such that the ordered pair (f, G) is a cyclic contraction. For any $(x_0, y_0, z_0) \in A^3$, we define the sequence $\{x_n\}, \{y_n\}$, and $\{z_n\}$ in X by

$$x_{2n+1} = F(x_{2n}, y_{2n}, z_{2n}), \quad x_{2n+2} = G(x_{2n+1}, y_{2n+1}, z_{2n+1}),$$
$$y_{2n+1} = F(y_{2n}, x_{2n}, y_{2n}), \quad y_{2n+2} = G(y_{2n+1}, x_{2n+1}, y_{2n+1}),$$
$$z_{2n+1} = F(z_{2n}, y_{2n}, x_{2n}), \quad z_{2n+2} = G(z_{2n+1}, y_{2n+1}, x_{2n+1}),$$

for all $n \in \{0\} \cup \mathbb{N}$. Then for any $\epsilon > 0$, there exists a positive integer N_0 such that, for all $m > n \geq N_0$, we have

$$\frac{1}{3}[d(x_{2m}, x_{2n+1}) + d(y_{2m}, y_{2n+1}) + d(z_{2m}, z_{2n+1}) < d(A, B) + \epsilon.$$

Lemma 17.5 **[43]** Let A, B be non-empty subsets of a metric space (X, d) such that the pairs (A, B) and (B, A) have the property UC^* and $F : A^3 \to B, G : B^3 \to A$ be two mappings such that the ordered pair (f, G) is a cyclic contraction. For any $(x_0, y_0, z_0) \in A^3$, we define the sequence $\{x_n\}, \{y_n\}$, and $\{z_n\}$ in X by

$$x_{2n+1} = F(x_{2n}, y_{2n}, z_{2n}), \quad x_{2n+2} = G(x_{2n+1}, y_{2n+1}, z_{2n+1}),$$
$$y_{2n+1} = F(y_{2n}, x_{2n}, y_{2n}), \quad y_{2n+2} = G(y_{2n+1}, x_{2n+1}, y_{2n+1}),$$
$$z_{2n+1} = F(z_{2n}, y_{2n}, x_{2n}), \quad z_{2n+2} = G(z_{2n+1}, y_{2n+1}, x_{2n+1}),$$

for all $n \in \{0\} \cup \mathbb{N}$. Then the sequences $\{x_{2n}\}, \{y_{2n}\}, \{z_{2n}\}, \{x_{2n+1}\}, \{y_{2n+1}\}, \{z_{2n+1}\}$ are Cauchy sequences.

The next is the main result of [43], in which Kumam *et al.* proved the existence and convergence of the tripled best proximity points for cyclic contraction pairs for non-empty subsets of metric spaces via property UC^*.

Theorem 17.30 [43] Let A, B be non-empty closed subsets of a metric space X such that the pairs (A, B) and (B, A) have the property UC^* and $F : A^3 \to B, G : B^3 \to A$ be two mappings such that the ordered pair (f, G) is a cyclic contraction. For any $(x_0, y_0, z_0) \in A^3$, we define the sequence $\{x_n\}, \{y_n\}$, and $\{z_n\}$ in X by

$$x_{2n+1} = F(x_{2n}, y_{2n}, z_{2n}), \quad x_{2n+2} = G(x_{2n+1}, y_{2n+1}, z_{2n+1}),$$
$$y_{2n+1} = F(y_{2n}, x_{2n}, y_{2n}), \quad y_{2n+2} = G(y_{2n+1}, x_{2n+1}, y_{2n+1}),$$
$$z_{2n+1} = F(z_{2n}, y_{2n}, x_{2n}), \quad z_{2n+2} = G(z_{2n+1}, y_{2n+1}, x_{2n+1}),$$

for all $n \in \mathbb{N} \cup \{0\}$. Then F has a tripled best proximity point $(p, q, r) \in A^3$ and G has a tripled best proximity point $(p', q', r') \in B^3$. Moreover, we have

$$x_{2n} \to p, \ y_{2n} \to q, \ z_{2n} \to r, \ x_{2n+1} \to p', \ y_{2n+1} \to q', \ z_{2n+1} \to r'.$$

Furthermore, if $q = r$ and $q' = r'$, then

$$d(p, p') + d(q, q') + d(r, r') = 3d(A, B).$$

Proof: By Lemma 17.3, it is obtained that $d(x_{2n}, x_{2n+1}) \to d(A, B)$. With the help of Lemma 17.5, it is observed that $\{x_{2n}\}, \{y_{2n}\}$, and $\{z_{2n}\}$ are Cauchy sequences. Thus, there exist $p, q, r \in A$ such that $x_{2n} \to p, y_{2n} \to q$, and $z_{2n} \to r$. Now, we obtain

$$d(A, B) \le d(p, x_{2n1}) \le d(p, x_{2n}) + d(x_{2n}, x_{2n1}).$$

By taking limit $n \to \infty$ in last inequality, we have $d(p, x_{2n1}) \to d(A, B)$. By the similar argument, we also have $d(q, y_{2n1}) \to d(A, B)$ and $d(r, z_{2n1}) \to d(A, B)$. It follows that

$$d(x_{2n}, F(p, q, r)) = d(G(x_{2n1}, y_{2n1}, z_{2n1}), F(p, q, r))$$
$$\le [d(x_{2n1}, p) + d(y_{2n1}, q) + d(z_{2n1}, r)] + (1\alpha)d(A, B).$$

Taking limit as $n \to \infty$, we get $d(p, F(p, q, r)) = d(A, B)$.
Similarly, we can prove that $d(q, F(q, p, q)) = d(A, B)$, $d(r, F(r, q, p)) = d(A, B)$. Therefore, (p, q, r) is a tripled best proximity point of F. Using the similar fashion, one can prove that there exist $p', q', r' \in B$ such that $x_{2n+1} \to p', y_{2n+1} \to q'$ and $z_{2n+1} \to r'$. Moreover, we have $d(p', G(p', q', r')) = d(A, B)$, $d(q', G(q', p', q')) = d(A, B)$, and $d(r', G(r', q', p')) = d(A, B)$; and

hence, (p', q', r') is tripled best proximity point of G. At last, it is assumed that $q = r$ and $q' = r'$ and then it can be show that

$$d(p, p') + d(q, q') + d(r, r') = 3d(A, B).$$

For all $n \in \mathbb{N}$, we get

$$d(x_{2n}, x_{2n+1}) = d(G(x_{2n1}, y_{2n1}, z_{2n1}), F(x_{2n}, y_{2n}, z_{2n})$$
$$\leq \frac{\alpha}{3}[d(x_{2n1}, x_{2n}) + d(y_{2n1}, y_{2n}) + d(z_{2n1}, z_{2n})] + (1\alpha)d(A, B).$$

Letting $n \to \infty$, we have $d(p, p') \leq \frac{\alpha}{3}[d(p, p') + d(q, q') + d(r, r')] + (1)d(A, B)$. Similarly, we get the others as: $d(q, q') \leq \frac{\alpha}{3}[d(q, q') + d(p, p') + d(r, r')] + (1)d(A, B)$. and $d(r, r') \leq \frac{\alpha}{3}[d(p, p') + d(q, q') + d(r, r')] + (1)d(A, B)$. Thus, it follows that

$$d(p, p') + d(q, q') + d(r, r') \leq \alpha[d(p, p') + d(q, q') + d(r, r')] + 3(1)d(A, B),$$

which implies that $d(p, p') + d(q, q') + d(r, r') \leq 3d(A, B)$. Since $d(A, B) \leq d(p, p'), d(A, B) \leq d(q, q')$ and $d(A, B) \leq d(r, r')$, we have $d(p, p') + d(q, q') + d(r, r') \geq 3d(A, B)$. Finally, we can conclude that $d(p, p') + d(q, q') + d(r, r') = 3d(A, B)$. This completes the proof. $\square$

Since every pair of non-empty closed subsets A, B of a uniformly convex Banach space X such that A is convex and satisfies the property UC^*. Therefore, the upcoming corollary justifies this note.

Corollary 17.5 [43] Let A, B be non-empty closed convex subsets of a uniformly convex Banach space X and $F : A^3 \to B, G : B^3 \to A$ be two mappings such that the ordered pair (f, G) is a cyclic contraction. For any $(x_0, y_0, z_0) \in A^3$, we define the sequence $\{x_n\}, \{y_n\}$, and $\{z_n\}$ in X by

$$x_{2n+1} = F(x_{2n}, y_{2n}, z_{2n}), \quad x_{2n+2} = G(x_{2n+1}, y_{2n+1}, z_{2n+1}),$$
$$y_{2n+1} = F(y_{2n}, x_{2n}, y_{2n}), \quad y_{2n+2} = G(y_{2n+1}, x_{2n+1}, y_{2n+1}),$$
$$z_{2n+1} = F(z_{2n}, y_{2n}, x_{2n}), \quad z_{2n+2} = G(z_{2n+1}, y_{2n+1}, x_{2n+1}),$$

for all $n \in \mathbb{N} \cup \{0\}$. Then F has a tripled best proximity point $(p, q, r) \in A^3$ and G has a tripled best proximity point $(p', q', r') \in B^3$. Moreover, we have

$$x_{2n} \to p, \ y_{2n} \to q, \ z_{2n} \to r, \ x_{2n+1} \to p', \ y_{2n+1} \to q', \ z_{2n+1} \to r'.$$

Furthermore, if $q = r$ and $q' = r'$, then

$$d(p, p') + d(q, q') + d(r, r') = 3d(A, B).$$

In the favor of the above mentioned corollary of [43], we are going to describe the next example.

Example 17.13 **[43]** Consider a uniformly convex Banach space $X = \mathbb{R}$ with the usual norm and let $A = [1, 3]$ and $B = [-3, -1]$. Thus, $d(A, B) = 2$. Define two mappings $F^3 : A \to B$ and $G^3 : B \to A$ by

$$F(x, y, z) = \frac{-x - y - z - 3}{6}, G(u, v, w) = \frac{-u - v - w + 3}{6}$$

for all $(x, y, z) \in A^3$ and $(u, v, w) \in B^3$, respectively. For any $(x, y, z) \in A^3$ and $(u, v, w) \in B^3$ and fixed $\alpha = \frac{1}{2}$, we get (f, G) as a cyclic contraction.

This implies that (f, G) is a cyclic contraction with $\alpha = \frac{1}{2}$. Since A and B are closed convex, the pairs (A, B) and (B, A) satisfy the property UC^*. Therefore, all the hypothesis of Corollary 17.5 hold. Therefore, F has tripled best proximity point and G has a tripled best proximity point. We note that the point $(-1, -1, -1) \in B^3$ is a unique tripled best proximity point of G. Furthermore, we get $d(1, -1) + d(1, -1) + d(1, -1) = 6 = 3d(A, B)$.

In [43], we also investigated the tripled best proximity point result for compact subsets of metric spaces, as follows:

Theorem 17.31 **[43]** Let A, B be non-empty compact subsets of a metric space X and $F : A^3 \to B, G : B^3 \to A$ be two mappings such that the ordered pair (f, G) is a cyclic contraction. For any $(x_0, y_0, z_0) \in A^3$, we define the sequence $\{x_n\}, \{y_n\}$, and $\{z_n\}$ in X by

$$x_{2n+1} = F(x_{2n}, y_{2n}, z_{2n}), \ x_{2n+2} = G(x_{2n+1}, y_{2n+1}, z_{2n+1}),$$
$$y_{2n+1} = F(y_{2n}, x_{2n}, y_{2n}), \ y_{2n+2} = G(y_{2n+1}, x_{2n+1}, y_{2n+1}),$$
$$z_{2n+1} = F(z_{2n}, y_{2n}, x_{2n}), \ z_{2n+2} = G(z_{2n+1}, y_{2n+1}, x_{2n+1}),$$

for all $n \in \{0\} \cup \mathbb{N}$. Then F has a tripled best proximity point $(p, q, r) \in A^3$ and G has a tripled best proximity point $(p', q', r') \in B^3$. Moreover, we have

$$x_{2n} \to p, \ y_{2n} \to q, \ z_{2n} \to r, \ x_{2n+1} \to p', \ y_{2n+1} \to q', \ z_{2n+1} \to r'.$$

Furthermore, if $q = r$ and $q' = r'$, then

$$d(p, p') + d(q, q') + d(r, r') = 3d(A, B).$$

Proof: Since A is compact, the sequences $\{x_{2n}\}, \{y_{2n}\}$, and $\{z_{2n}\}$ have the convergent subsequences $\{x_{2n_k}\}, \{y_{2n_k}\}$, and , $\{z_{2n_k}\}$, respectively, such that

$$x_{2n_k} \to p \in A, \ y_{2n_k} \to q \in A, \ z_{2n_k} \to r \in A.$$

Now, we have

$$d(A, B) \leq d(p, x_{2n_k}) + d(x_{2n_k}, x_{2n_k - 1}).$$

By using Lemma 17.3, we have $d(x_{2n_k}, x_{2n_k1}) \to d(A, B)$. Taking limit as $n \to \infty$, we get $d(p, x_{2n_k1}) \to d(A, B)$. By the similar argument, we observe that $d(q, x_{2n_k1}) \to d(A, B), d(r, x_{2n_k1}) \to d(A, B)$. Note that

$$d(A, B) \leq d(x_{2n_k}, F(p, q, r)) = d(G(x_{2n_k1}, y_{2n_k1}, z_{2n_k1}), F(p, q, r))$$
$$\leq \frac{\alpha}{3}[d(x_{2n_k1}, p) + d(y_{2n_k1}, q) + d(z_{2n_k1}, r)]$$
$$+ (1\alpha)d(A, B).$$

Taking $k \to \infty$, we get $d(p, F(p, q, r)) = d(A, B)$. Similarly, we can prove that

$$d(q, F(q, p, q)) = d(A, B), d(r, F(r, q, p)) = d(A, B).$$

Thus, F has a tripled best proximity $(p, q, r) \in A^3$. In a similar way, since B is compact, we can also prove that G has a tripled best proximity point $(p', q', r') \in B^3$. Now, for next and last need to prove $d(p, p') + d(q, q') + d(r, r') = 3d(A, B)$, for this, we follow the steps of the proof of Theorem 17.30. This completes the proof. □

The next theorem is a new result for tripled fixed point via cyclic contraction.

Theorem 17.32 [43] Let A, B be non-empty closed subsets of a metric space (X, d) and $F : A^3 \to B, G : B^3 \to A$ be two mappings such that the ordered pair (f, G) is a cyclic contraction. For any $(x_0, y_0, z_0) \in A^3$, we define the sequence $\{x_n\}, \{y_n\}$, and $\{z_n\}$ in X by

$$x_{2n+1} = F(x_{2n}, y_{2n}, z_{2n}), \ x_{2n+2} = G(x_{2n+1}, y_{2n+1}, z_{2n+1}),$$
$$y_{2n+1} = F(y_{2n}, x_{2n}, y_{2n}), \ y_{2n+2} = G(y_{2n+1}, x_{2n+1}, y_{2n+1}),$$
$$z_{2n+1} = F(z_{2n}, y_{2n}, x_{2n}), \ z_{2n+2} = G(z_{2n+1}, y_{2n+1}, x_{2n+1}),$$

for all $n \in \{0\} \cup \mathbb{N}$. Then F has a tripled fixed point $(p, q, r) \in A^3$ and G has a tripled fixed point $(p', q', r') \in B^3$. Moreover, we have

$$x_{2n} \to p, \ y_{2n} \to q, \ z_{2n} \to r, \ x_{2n+1} \to p', \ y_{2n+1} \to q', \ z_{2n+1} \to r'.$$

Furthermore, if $q = r$ and $q' = r'$, then F and G have a common tripled fixed point in $(A \cap B)^3$.

Proof: Since $d(A, B) = 0$, it follows that the pairs (A, B) and (B, A) satisfy the property UC^*. Therefore, by Theorem 17.30, we claim that F has a tripled best proximity point $(p, q, r) \in A^3$, that is,

$$d(p, F(p, q, r)) = d(q, F(q, p, q)) = d(r, F(r, q, p)) = d(A, B)$$

and G has a tripled best proximity point $(p', q', r') \in B^3$, that is,

$$d(p', G(p', q', r')) = d(q', G(q', p', q')) = d(r', G(r', q', p')) = d(A, B).$$

Since $d(A, B) = 0$, we conclude that $p = F(p, q, r), q = F(q, p, q), r = F(r, q, p)$, that is, (p, q, r) is a tripled fixed point of F and similarly (p', q', r') is tripled fixed point of G. Next, we assume that $q = r$ and $q' = r'$ and then it is seen that F and G have a unique common tripled fixed point in $(A \cap B)^3$. From Theorem 17.30, we get $d(p, p') + d(q, q') + d(r, r') = 3d(A, B)$. Since $d(A, B) = 0$, we get $d(p, p') + d(q, q') + d(r, r') = 0$, which implies that $p = p'$, $q = q'$, and $'r = r$. Therefore, we conclude that $(p, q, r) \in (A \cap B)^3$ is common tripled fixed point of F and G. Hence proved. □

Next, we give one example to illustrate Theorem 17.32.

Example 17.14 [43] Consider a space $X = \mathbb{R}$ with the usual metric and let $A = [-2, 0]$ and $B = [0, 2]$. Define two mappings $F : A^3 \to B$ and $G : B^3 \to A$ by $F(x, y, z) = \frac{x+y+z}{6}$, $G(x, y, z) = \frac{u+v+w}{6}$, for all $(x, y, z) \in A^3$ and $(u, v, w) \in B^3$, respectively. Then $d(A, B) = 0$ and the ordered pair (f, G) is a cyclic contraction with $\alpha = \frac{1}{2}$. With these all assumptions all the hypothesis of Theorem 17.32 hold. Thus, F and G have common fixed point and this point is $(0, 0, 0) \in (A \cup B)^3$.

We take $A = B$ in Theorem 17.32, then we get the following results.

Corollary 17.6 [43] Let A be a non-empty closed subset of a complete metric space (X, d) and $F : A^3 \to A, G : A^3 \to A$ be two mappings such that the ordered pair (f, G) be a cyclic contraction. For any $(x_0, y_0, z_0) \in A^3$, we define the sequences $\{x_n\}$, $\{y_n\}$, and $\{z_n\}$ in X by

$$x_{2n+1} = F(x_{2n}, y_{2n}, z_{2n}), \quad x_{2n+2} = G(x_{2n+1}, y_{2n+1}, z_{2n+1}),$$
$$y_{2n+1} = F(y_{2n}, x_{2n}, y_{2n}), \quad y_{2n+2} = G(y_{2n+1}, x_{2n+1}, y_{2n+1}),$$
$$z_{2n+1} = F(z_{2n}, y_{2n}, x_{2n}), \quad z_{2n+2} = G(z_{2n+1}, y_{2n+1}, x_{2n+1}),$$

for all $n \in \{0\} \cup \mathbb{N}$. Then F has a tripled fixed point $(p, q, r) \in A^3$ and G has a tripled fixed point $(p', q', r') \in A^3$. Moreover, we have $x_{2n} \to p$, $y_{2n} \to q$, $z_{2n} \to r$, $x_{2n+1} \to p'$, $y_{2n+1} \to q'$, $z_{2n+1} \to r'$. Furthermore, if $q = r$ and $q' = r'$, then F and G have a common tripled fixed point in A^3.

If we take $F = G$ in Corollary 17.6, then we get the results in Corollary 17.7.

Corollary 17.7 [43] Let A be non-empty closed subsets of a complete metric space (X, d) and $F : A^3 \to A$ be a mapping satisfying

$$d(f(x, y, z), F(u, v, w)) \leq \frac{\alpha}{3}[d(x, u) + d(y, v) + d(z, w)]$$

for all $(x, y, z), (u, v, w) \in A^3$. Then F has a tripled fixed point $(p, q, r) \in A^3$.

17.8 Future Works

By the research experience in field of metric fixed point theory, we have many idea to extend the new thing in metric fixed point theory such as:

- To extend and study existence of other nonlinear mapping and apply it for solve a problem in a real world.
- To study properties of fixed point theory that can be apply for real world problem such as in economics and transportation.
- To study existence of generalized game equilibrium and apply to network problems.
- To study properties and application of fuzzy nonlinear in spaces.
- To study stability of fixed point, common fixed point, coincidence point, and best proximity in spaces.
- To apply metric fixed point theory for solving differential equation, integral equation, and equilibrium problem and solving method.
- To study method to find minimize of optimization problem and image processing.
- To study fuzzy optimization problem.

References

1 Banach, S. (1922) Sur les operations dans les ensembles abstraits et leurs applications aux equations integrales. *Fund. Math.*, **3**, 133–181.

2 Zadeh, L.A. (1965) Fuzzy sets. *Infor. Control*, **8**, 338–353.

3 Fan, K. (1969) Extensions of two fixed point theorems of F.E. Browder. *Math. Z.*, **112**, 234–240.

4 Eldred, A.A. and Veeramani, P. (2006) Existence and convergence of best proximity points. *J. Math. Anal. Appl.*, **323** (2), 1001–1006.

5 Mongkolkeha, C. and Kumam, P. (2012) Best proximity point theorems for generalized cyclic contractions in ordered metric spaces. *J. Optim. Theory Appl.*, **155** (1), 215–226.

6 Mongkolkeha, C., Cho, Y.J., and Kumam, P. (2013) Best proximity points for generalized proximal *C*-contraction mappings in metric spaces with partial orders. *J. Inequal. Appl.*, **2013** (1), 94.

7 Mongkolkeha, C. and Kumam, P. (2013) Best proximity points for asymptotic proximal pointwise weaker Meir-Keeler-type ψ -contraction mappings. *J. Egypt. Math. Soc.*, **21** (2), 87–90.

8 Basha, S.S. and Veeramani, P. (1997) Best approximations and best proximity pairs. *Acta. Sci. Math. (Szeged)*, **63**, 289–300.

9 Basha, S.S. and Veeramani, P. (2000) Best proximity pair theorems for multifunctions with open fibres. *J. Approx. Theory*, **103**, 119–129.

10 Basha, S.S., Veeramani, P., and Pai, D.V. (2001) Best proximity pair theorems. *Indian J. Pure Appl. Math.*, **32**, 1237–1246.

11 Nadler, S.B. Jr. (1969) Multivalued contraction mappings. *Pac. J. Math.*, **30**, 475–488.

12 Mizoguchi, N. and Takahashi, W. (1989) Fixed point theorems for multi-valued mappings on complete metric spaces. *J. Math. Anal. Appl.*, **141**, 177–188.

13 Alisina, A., Massa, S., and Roux, D. (1973) Punti uniti di multifunzzioni con conndizioni di tipo Bioyd-Wong. *Boll. Unione Mat. Ital.*, **4** (8), 29–34.

14 Reich, S. (1983) Some problems and results in fixed point theory. *Contemp. Math.*, **21**, 179–187.

15 Suzuki, T. (2008) Mizoguchi-Takahashi fixed point theorem is a real generalization of Nadler's. *J. Math. Anal. Appl.*, **340** (1), 752–755.

16 Kirk, W.A., Srinavasan, P.S., and Veeramani, P. (2003) Fixed points for mapping satisfying cyclical contractive conditions. *Fixed Point Theory*, **4** (1), 79–89.

17 Czerwik, S. (1993) Contraction mappings in *b*-metric spaces. *Acta Math. Inform. Uni. Ostrav.*, **1**, 5–11.

18 Bernfeld, S.R., Lakshmikantham, V., and Reddy, Y.M. (1977) Fixed point theorems of operators with PPF dependence in Banach spaces. *Appl. Anal.*, **6**, 271–280.

19 Wairojjana, N., Sintunavarat, W., and Kumam, P. (2014) Common tripled fixed point theorems for W-compatible mappings along with the CLR_g property in abstract metric spaces. *Journal of Inequal. Appl.*, **2014**, 133.

20 Aydi, H., Abbas, M., Sintunavarat, W., and Kumam, P. (2012) Tripled fixed point for W-compatible mappings in abstract metric spaces. *Fixed Point Theory Appl.*, **2012**, 134.

21 Huang, L.G. and Zhang, X. (2007) Cone metric spaces and fixed point theorems of contractive mappings. *J. Math. Anal.*, **332**, 1468–1476.

22 Abbas, M., Khan, A.M., and Radenoovic, S. (2010) Common coupled fixed point theorems in cone metric spaces for w-compatible mappings. *Appl. Math. Comput.*, **217** (1), 195–202.

23 Kumam, P., Aydi, H., Karapinar, E., and Sintunavarat, W. (2013) Best proximity points and extension of Mizoguchi-Takahashi's fixed point theorems. *Fixed Point Theory Appl.*, **2013**, 242.

24 Samet, B., Vetro, C., and Vetro, P. (2012) Fixed point theorem for $\alpha - \psi$-contractive type mappings. *NonLinear Anal.*, **75**, 2154–2165.

25 Berinde, V. and Borcut, M. (2011) Tripled fixed point theorems for contractive type mappings in partially ordered metric spaces. *Nonlinear Anal.*, **74** (15), 4889–4897.

26 Basha, S.S. (2011) Best proximity point theorems. *J. Approx. Theory*, **163** (11), 1772–1781.

27 Basha, S.S. (2012) Common best proximity points: global minimization of multi-objective functions. *J. Global Optim.*, **54**, 367–373.

28 Nakano, H. (1951) *Topology and Linear Topological Spaces*, Tokyo Mathematical Book Series, vol. **3**, Maruzen, Tokyo.

29 Chistyakov, W. (2010) Modular metric spaces I: basic concepts. *Nonlinear Anal.*, **72** (1), 1–14.

30 Chaipunya, P. and Kumam, P. (2013) Topological aspects of circular metric spaces and some observations on the KKM-property towards quasi-equilibrium problems. *J. Inequal. Appl.*, **2013**, 343.

31 Chaipunya, P. and Kumam, P. (2013) On the distance between three arbitrary points. *J. Funct. Spaces Appl.*, **2013**, Article ID 194631, 7 p.

32 Sedghi, S., Shobe, N., and Aliouche, A. (2012) A generalization of fixed point theorems in S-metric spaces. *Mat. Vesnik*, **64** (3), 258–266.

33 Khamsi, M.A. (2010) Remarks on cone metric space and fixed point theorems on contractive mappings. *Fixed Point Theory Appl.*, 315398. doi: 10.1155/2010/315398.

34 Gahler, S. (1963) 2-metrische Raume und ihre topologische Struktur. *Math. Nachr.*, **26** (14), 115–148.

35 Dhage, B.C. (1992) Generalized metric spaces and mappings with fixed point. *Bull. Calcutta Math. Soc.*, **84**, 329–336.

36 Muustafa, Z. and Sims, B. (2006) A new approach to generalized metric space. *J. NonLinear Convex Anal.*, **7** (2), 289–297.

37 Sedghi, S., Shobe, N., and Zhou, H. (2007) A common fixed point theorem in D^*-metric spaces. *Fixed Point Theory Appl.*, **2007**, Article ID 27906, 13 p.

38 EL-Sayed, A.M. and Ibrahin, A.G. (2001) Set valued integral equations of fractional orders. *Appl. Math. Comput.*, **11**, 81–87.

39 Chaipunya, P. and Kumam, P. (2015) Fixed Point Theorems for Cyclic Operators with Application in Fractional Integral Inclusions with Delays. Dynamical Systems, Differential Equations and Applications AIMS Proceedings, pp. 248–257.

40 Guo, D. and Lakshmikantham, V. (1987) Coupled fixed points of nonlinear operators with applications. *Nonlinear Anal. Theory Methods Appl.*, **11**, 623–632.

41 Bhaskar, T.G. and Lakshmikantham, V. (2006) Fixed point theorems in partially ordered metric spaces and applications. *Nonlinear Anal. Theory Methods Appl.*, **65** (7), 1379–1393.

42 Sintunavarat, W. and Kumam, P. (2012) Coupled best proximity point theorem in metric spaces. *Fixed Point Theory Appl.*, **2012** (1), 93.

43 Cho, Y.J., Gupta, A., Karapinar, E., Kumam, P., and Sintunavarat, W. (2013) Tripled best proximity point theorem in metric spaces. *Math. Inequal. Appl.*, **16** (4), 1197–1216.

44 Mongkolkeha, C. and Kumam, P. (2013) Some common best proximity points for proximity commuting mappings. *Optim. Lett.*, **7**, 1825–1836.

45 Sintunavarat, W. and Kumam, P. (2016) Best proximity point theorems for generalized Mizoguchi-*Takahashi's* contraction pairs. *J. Nonlinear Convex Anal.*, **17** (7), 1345–1361.

46 Chaipunya, P., Sintunavarat, W., and Kumam, P. (2012) On $\mathcal{P}$-contractions in ordered metric spaces. *Fixed Point Theory Appl.*, **2012**, 219.

47 Sintunavarat, W. and Kumam, P. (2012) Generalized common fixed point theorems in complex valued metric spaces and applications. *J. Ineq. Appl.*, **2012**, 84.

48 Sintunavarat, W., Petrusel, A., and Kumam, P. (2012) Common coupled fixed point theorems for w^*-compatible mappings without mixed monotone property. *Rend. Circolo Mat. Palermo*, **61** (3), 361–383.

49 Sintunavarat, W. and Kumam, P. (2013) The existence theorems of an optimal approximate solution for generalized proximal contraction mappings. *Abstr. Appl. Anal.*, **2013**, Article ID. 375604, 8 p.

50 Sintunavarat, W., Cho, Y.J., and Kumam, P. (2013) Urysohn integral equations approached by common fixed points in complex-valued metric spaces. *Adv. Differ. Equ.*, **49** (1), 1197–1216.

51 Agarwal, R.P., Kumam, P., and Sintunavarat, W. (2013) PPF dependent fixed point theorems for an α_c-admissible non-self mapping in the Razumikhin class. *Fixed Point Theory Appl.*, **2013**, 280.

52 Phiangsungnoen, S., Sintunavarat, W., and Kumam, P. (2014) Fixed point results, generalizzed Ulam-Hyers Stability and well-posedness via α-admissible mappings in b-metric spaces. *Fixed Point Theory Appl.*, **2014**, 188.

53 Phiangsungnoen, S., Sintunavarat, W., and Kumam, P. (2014) Common α-fuzzy fixed point theorems for fuzzy mappings via $\beta_\mathcal{F}$-admissible pair. *J. Intell. Fuzzy Syst.*, **27**, 2463–2472.

54 Chaipunya, P., Cho, Y.J., and Kumam, P. (2014) On circular metric spaces and common fixed points for an infinite family of set-valued operators. *Vietnam J. Math.*, **42** (2), 205–218.

55 Imdad, M., Chauhan, S., and Kumam, P. (2015) Fixed point theorems for two hybrid pairs of non-self mappings under joint common limit range property in metric spaces. *J. Nonlinear Convex Anal.*, **16** (2), 243–254.

56 Kumam, P., Dung, N.V., and Sitthithakerngkiet, K. (2015) A generalization of Ciric fixed point theorems. *Filomat*, **29** (7), 1549–1556.

57 Chaipunya, P., Cho, Y.J., Sintunavarat, W., and Kumam, P. (2012) Fixed point and common fixed point theorems for cyclic quasi-contractions in metric and ultrametric spaces. *Adv. Pure Math.*, **2**, 401–407.

58 Chaipunya, P. and Kumam, P. (2015) Common fixed points for an uncountable family of weakly contractive operators. *Carpathian J. Math.*, **31** (3), 307–312.

59 Phiangsungnoen, S. (2016) Ulam-Hyers Stability and well-posedness of the fixed point problems for contractive multi-valued operator in b-metric spaces. *Commun. Math. Appl.*, **7** (3), 241–262.

60 Komal, S. and Kumam, P. (2016) A new class of $\mathcal{S}$-contractions in complete metric spaces and $\mathcal{G}_p$-contractions in ordered spaces. *Fixed Point Theory Appl.*, **2016**, 76.

61 Komal, S. and Kumam, P. (2016) Global optimization using α-ordered proximal contractions in metric spaces with partial orders. *Appl. Gen. Topol.*, **17** (2), 173–183.

62 Komal, S., Kumam, P., and Gopal, D. (2016) Best Proximity point for $\mathcal{Z}$-contraction and Suzuki type $\mathcal{Z}$-Contraction mappings with an application to fractional calculus. *Appl. Gen. Topol.*, **17** (2), 185–198.

63 Komal, S., Sultana, N., Hussain, A., and Kumam, P. (2016) Optimal approximate solution for generalized contraction in complete metric space. *Commun. Math. Appl.*, **7** (1), 23–36.

64 Suzuki, T., Kikkawa, M., and Vetro, C. (2009) The existence of best proximity points in metric spaces with the property UC. *Nonlinear Anal. Theory Methods Appl.*, **71** (7–8), 2918–2926.

65 Chaipunya, P., Cho, Y.J., and Kumam, P. (2014) A remark on the property $\mathcal{P}$ and periodic points of order ∞. *Mat. Bech.*, **66** (4), 357–363.

66 Rhoades, B.E. (2001) Some theorems on weakly contractive maps. *Nonlinear Anal.*, **47**, 2683–2693.

67 Harjani, J. and Sadarangani, K. (2009) Fixed point theorems for weakly contractive mappings in partially ordered sets. *Nonlinear Anal.*, **71**, 3403–3410.

68 Ulam, S.M. (1964) *Problems in Modern Mathematics*, John Wiley & Sons, Inc., New York.

69 Hyers, D.H. (1941) On the stability of the linear functional equation. *Proc. Natl. Acad. Sci. U.S.A.*, **27** (4), 222–224.

70 Edelstein, M. (1962) On fixed and periodic points under contractive mappings. *J. Lond. Math.*, **37**, 74–79.

71 Suzuki, T. (2009) A new type of fixed point theorem in metric spaces. *Nonlinear Anal.*, **71** (11), 5313–5317.

72 Wardowski, D. (2012) Fixed point theory of a new type of contractive mappings in complete metric spaces. *Fixed Point Theory Appl.*, 94. doi: 10.1186/1687-1812-2012-94.

73 Piri, H. and Kumam, P. (2014) Some fixed point theorems concerning F-contraction in complete metric spaces. *Fixed Point Theory Appl.*, **2014**, 210.

74 Sintunavarat, W. and Kumam, P. (2011) Common fixed point theorems for pair of weakly compatible mappings in fuzzy metric spaces. *J. Appl. Math.*, **2011**, Article ID 637958, 14 p.

75 Arvanitakis, A.D. (2003) A proof of the generalized Banach contraction conjecture. *Proc. Am. Math. Soc.*, **131** (12), 3647–3656.

76 Choudhury, B.S. and Das, K.P. (2008) A new contraction principle in Menger spaces. *Acta Math. Sin. Engl. Ser.*, **24** (8), 1379–1386.

77 Basha, S.S. (2011) Best proximity point theorems generalizing the contraction principle. *Nonlinear Anal.*, **74** (17), 5844–5850.

78 Sintunavarat, W., Cho, Y.J., and Kumam, P. (2011) Coupled coincidence point theorems for contractions without commutative condition in intuitionistic fuzzy normed spaces. *Fixed Point Theory Appl.*, **54**, 1897–1906.

79 Abbas, M., Sintunavarat, W., and Kumam, P. (2012) Coupled fixed point of generalized contractive mappings on partially ordered G-metric spaces. *Fixed Point Theory Appl.*, **2012**, 31.

80 Agarwal, R.P., Alghamdi, M.A., and Shahzad, N. (2012) Fixed point theory for cyclic generalized contractions in partial metric spaces. *Fixed Point Theory Appl.*, **2012**, 40.

81 Sintunavarat, W., Cho, Y.J., and Kumam, P. (2012) Coupled fixed point theorems for weak contraction mapping under F-invariant set. *Abstr. Appl. Anal.*, **2012**, Article ID. 324874, 15.

82 Chauhan, S., Sintunavarat, W., and Kumam, P. (2013) Common fixed point theorems for weakly compatible mappings in fuzzy metric spaces using (JCLR) property. *Appl. Math.*, **3**, 976–982.

83 Mongkolkeha, C. and Kumam, P. (2012) Best proximity points theorems for generalized cyclic contractions in ordered metric spaces. *J. Optim. Theory Appl.*, **155** (1), 215–226.

84 Boyd, D.W. and Wong, J.S.W. (1969) On nonlinear contractions. *Proc. Am. Math. Soc.*, **2**, 458–464.

85 Secelean, N.A. (2013) Iterated function systems consisting of F-contractions. *Fixed Point Theory Appl.*, Article ID 277 10.1186/1687-1812-2013-277.

18

The Basel Problem with an Extension

Anthony Sofo

Victoria University, Melbourne City, Victoria 8001, Australia

2010 Mathematics Subject Classification Primary 05A10, 05A19, 33C20; Secondary 11B65, 11B83, 11M06

18.1 The Basel Problem

Pietro Mengoli (1626–1686), an Italian mathematician, posed the following problem in 1644, find the numerical value of

$$\frac{1}{1^2} + \frac{1}{2^2} + \frac{1}{3^2} + \cdots + \frac{1}{n^2} + \cdots = \sum_{n \geq 1} \frac{1}{n^2} = \zeta(2).$$

No doubt Mengoli had an approximate value, since for instance by adding the first 10 terms approximates

$$\zeta(2) \approx \frac{1968329}{1270080} = 1.54977 \cdots$$

Many other notable mathematicians including Jacob (1654–1705), Johann (1667–1748) and Daniel Bernoulli (1700–1782), Gottfried Leibniz (1646–1716), James Stirling (1682–1770), Abraham de Moivre (1667–1754), John Wallis (1616–1703) and Christian Goldbach (1690–1764) attempted to solve the problem but were only able to obtain some partial results. Wallis claimed he had evaluated $\zeta(2)$ to three decimal places, Jacob Bernoulli argued as follows, since

$$2n^2 \geq n(n+1), \text{ then } \frac{1}{n^2} \leq \frac{2}{n(n+1)}$$

also $\sum_{n \geq 1} \frac{2}{n(n+1)} = 2$, hence $\sum_{n \geq 1} \frac{1}{n^2} \leq 2$ follows.

Johann and Daniel said it's about $\frac{8}{5}$, Goldbach claimed

$$\frac{41}{25} < \zeta(2) < \frac{5}{3}$$

Dedicated to: Bella Mugeruza Lucianna

Mathematical Analysis and Applications: Selected Topics, First Edition.
Edited by Michael Ruzhansky, Hemen Dutta, and Ravi P. Agarwal.
© 2018 John Wiley & Sons, Inc. Published 2018 by John Wiley & Sons, Inc.

and Stirling had numerically calculated $\zeta(2) = 1.644934066$ to nine decimal places. It took no less than the genius Euler (1707–1783) to solve the problem. In 1735, Euler announced that $\zeta(2) = \frac{\pi^2}{6} \approx 1.6449$. Consequently this problem, which resisted a closed form solution for 90 years, has been named the Basel problem, a town in Switzerland where Euler was born. Euler's method uses the Taylor series expansion for the $\sin x$ function, which is a polynomial whose roots are $x = k\pi$, where k is an integer. Hence with the fundamental theorem of algebra the polynomial $\frac{\sin x}{x}$ can be written in terms of its roots:

$$\frac{\sin x}{x} = \sum_{n \geq 0} \frac{(-1)^n x^{2n}}{(2n+1)!} = A \prod_{n \geq 1} (x^2 - n^2 \pi^2),$$

where $A = 1$ is a proportionality constant, because $\lim_{x \to 0} \frac{\sin x}{x} = 1$. Sum each of the factors $(x^2 - n^2 \pi^2) = 0$, and then

$$\frac{\sin x}{x} = \prod_{n \geq 1} \left(1 - \frac{x^2}{n^2 \pi^2}\right),$$

now multiplying all the factors and collecting the coefficients belonging to x^2 results in

$$-\frac{1}{\pi^2} \sum_{n \geq 1} \frac{1}{n^2}.$$

Equating the coefficient of x^2 in the Taylor series gives us the closed form

$$\zeta(2) = \sum_{n \geq 1} \frac{1}{n^2} = \frac{\pi^2}{6}.$$

In a similar fashion Euler reasoned

$$\zeta(4) = \sum_{n \geq 1} \frac{1}{n^4} = \frac{\pi^4}{90}, \quad \zeta(6) = \sum_{n \geq 1} \frac{1}{n^6} = \frac{\pi^6}{945}, \dots$$

and finally discovered the celebrated result

$$\zeta(2k) = \sum_{n \geq 1} \frac{1}{n^{2k}} = \frac{(-1)^{k+1} 2^{2k-1} B_{2k}}{(2k)!} \pi^{2k}$$

for k a positive integer, B_j is the jth Bernoulli number defined by

$$\frac{z}{e^z - 1} = \sum_{j \geq 0} \frac{B_j}{j!} e^j \quad \text{for } |z| \leq 2\pi,$$

which is a Taylor expansion about $z = 0$ of the holomorphic function $h(z) = \frac{z}{e^z - 1}$, which implies $\lim_{z \to 0} \frac{z}{e^z - 1} = 1$. Since $e^z - 1$ vanishes if and only if $z = 2\pi i n$, the restriction $|z| \leq 2\pi$ means that the denominator $e^z - 1$ vanishes only for $z = 0$. Euler computed B_j up to $j = 30$, some values of B_j are

$$\begin{bmatrix} B_0 & B_1 & B_2 & B_4 & B_6 & B_8 & B_{10} \\ 1 & -\frac{1}{2} & \frac{1}{6} & -\frac{1}{30} & \frac{1}{42} & -\frac{1}{30} & \frac{5}{66} \end{bmatrix},$$

where $B_{2j+1} = 0$. Euler's fascination with prime numbers lasted throughout his life and in 1737 discovered the beautiful identity

$$\zeta(s) = \prod_{j \in P} \left(\frac{1}{1 - j^{-s}} \right)$$

for $s > 1$ and P is the set of prime numbers. In the last 50 years there has been an explosion of results, see [1–49], related to Euler sums. Two other recent representations for $\zeta(2)$ are

$$\zeta(2) = 3 \sum_{n \geq 1} \frac{1}{n^2 \binom{2n}{n}}$$

and the BBP type identities

$$\zeta(2) = \frac{27}{4} \sum_{n \geq 1} \frac{1}{2^{5n}} \left(\begin{array}{c} \dfrac{16}{(6n+1)^2} - \dfrac{24}{(6n+2)^2} - \dfrac{8}{(6n+3)^2} \\[2mm] - \dfrac{6}{(6n+4)^2} + \dfrac{1}{(6n+5)^2} \end{array} \right).$$

Interestingly, at negative integer values we have

$$\zeta(-n) = \begin{cases} -\dfrac{1}{2}, & n = 0, \\[3mm] -\dfrac{B_{n+1}}{n+1}, \end{cases} \quad \text{with } \zeta(-2n) = 0$$

while the derivative is

$$\zeta'(-2n) = \begin{cases} -\dfrac{1}{2} \ln 2\pi, & n = 0, \\[3mm] (-1)^n \dfrac{(2n)! \, \zeta(2n+1)}{2^{2n+1} \pi^{2n}}. \end{cases}$$

The summation of the reciprocal natural numbers

$$\zeta(1) = \frac{1}{1} + \frac{1}{2} + \frac{1}{3} + \cdots + \frac{1}{n} + \cdots = \sum_{n \geq 1} \frac{1}{n}$$

was first, shown to diverge by N. d'Oresme (c1320–1382). The partial sums of the zeta function are called harmonic numbers, so that

$$H_n^{(p)} = \zeta_n(p) = \sum_{j=1}^{n} \frac{1}{j^p}.$$

Interestingly, Euler conjectures that

$$\zeta(3) = \sum_{n \geq 1} \frac{1}{n^3} = \alpha \pi^2 \ln 2 + \beta \ln^2 2$$

with rational numbers α and β. A series that comes close to Euler's conjecture is

$$\zeta(3) = \frac{4}{7} \pi^2 \ln 2 - \frac{4}{21} \ln^3 2 + \frac{8}{7} \sum_{n \geq 1} \frac{1}{2^n n^3}$$

$$= \frac{2}{7} \pi^2 \ln 2 - \frac{1}{7} \int_0^\pi x^2 \cot \left(\frac{x}{2} \right) dx$$

but to date no rational numbers α and β have been found. Similarly, Euler obtained, see [50–54]

$$\sum_{n \geq 0} \frac{1}{(2n+1)^3} = \frac{3}{2} \zeta(2) \ln 2 + 2 \int_0^{\pi/2} x \ln(\sin x) \, dx.$$

The following series representation for $\zeta(3)$ is known as Apery's constant

$$\zeta(3) = \frac{5}{2} \sum_{n \geq 1} \frac{(-1)^{n+1}}{n^3 \binom{2n}{n}}.$$

There are many other series representations for $\zeta(3)$ and the odd zeta functions $\zeta(2n+1)$, and harmonic number identities [55, 56]. In a recent paper [57] the authors obtained, for $n \in \mathbb{N}$

$$\zeta(2n+1) = \frac{\pi^{2n+1} E_n}{2(2^{2n+1} - 1)(2n)!} + \frac{2^{2n+2}}{2^{2n+1} - 1} \sum_{k \geq 1} \frac{1}{(4k-1)^{2n+1}},$$

where the Euler numbers E_n are defined by

$$\frac{2e^t}{e^{2t} + 1} = \sum_{k \geq 1} \frac{t^k}{k!} E_k, \quad |t| \leq \pi.$$

Surprisingly, little is known about $\zeta(2n+1)$. For example, none of these values are known to be transcendental, although it is conjectured that they are algebraically independent over $\mathbb{Q}(\pi)$. In 1979, Apery [58] demonstrated the irrationality of $\zeta(3)$. Rivoal [59] proved that for any $\epsilon > 0$, there exist an integer $m > 0$ such that for all odd integers $(2n+1) > m$, the dimension of the $\mathbb{Q}$ vector space spanned by $\zeta(3), \zeta(5), \zeta(7), \ldots, \zeta(2n+1)$ is at least $\frac{1-\epsilon}{1+\ln 2} \ln(2n+1)$. This implies that infinitely many odd zeta values are irrational. In 2001, Zudilin [60] established that at least one element of the set $\{\zeta(5), \zeta(7), \zeta(9), \zeta(11)\}$ is also irrational. For k a positive integer the multiple zeta function of depth k is a function of k complex variables s_k defined by the series

$$\zeta(s_1, s_2, \ldots, s_k) = \sum_{n_1 > n_2 > \cdots > n_k \geq 1} \frac{1}{n_1^{s_1} n_2^{s_2} n_3^{s_3} \cdots n_k^{s_k}},$$

the series converges provided that $\Re(s_1)$ and $\Re(s_i) > 0$ for $i = 2, 3, \ldots, k$. Many interesting integral representations and algebraic relations among multiple zeta values are known [61]. Euler evaluated $\zeta(2, 3) = \zeta(3)$. Euler, further considered sums of the form

$$T(m, n) := \sum_{k \geq 1} \frac{H_k^{(m)}}{(k + 1)^n} \tag{18.1}$$

and found some explicit representations, including

$$T(1, n) := \sum_{k \geq 1} \frac{H_k}{(k + 1)^n} = (n + 2)\zeta(2n + 1) - \sum_{k=1}^{n-2} \zeta(n - k)\zeta(k + 1).$$

For odd weight $(p + q)$ Borwein (see [62–65]) recently obtained the identity

$$\sum_{n \geq 1} \frac{H_n^{(p)}}{n^q} = \left(\frac{1}{2} + \frac{(-1)^{p+1}}{2} \binom{p + q - 1}{p} + \frac{(-1)^{p+1}}{2} \binom{p + q - 1}{q} \right) \zeta(p + q)$$

$$+ \left(\frac{1 + (-1)^{p+1}}{2} \right) \zeta(p)\zeta(q) + (-1)^p \sum_{k=1}^{\left[\frac{p}{2} \right]} \binom{p + q - 2k - 1}{q - 1} \zeta(2k)\zeta(p + q - 2k)$$

$$+ (-1)^p \sum_{k=1}^{\left[\frac{q}{2} \right]} \binom{p + q - 2k - 1}{p - 1} \zeta(2k)\zeta(p + q - 2k).$$

Many other relations have also been obtained for

$$S^{++}(p, q) := \sum_{n \geq 1} \frac{H_n^{(p)}}{n^q}, \quad S^{+-}(p, q) := \sum_{n \geq 1} \frac{(-1)^{n+1} H_n^{(p)}}{n^q},$$

$$S^{-+}(p, q) := \sum_{n \geq 1} \frac{\overline{H}_n^{(p)}}{n^q}, \quad S^{--}(p, q) := \sum_{n \geq 1} \frac{(-1)^{n+1} \overline{H}_n^{(p)}}{n^q},$$

where $\overline{H}_n^{(p)} = \sum_{j=1}^{n} \frac{(-1)^{j+1}}{j^p}$. Let us define the alternating zeta function

$$\overline{\zeta}(z) = \sum_{n=1}^{\infty} \frac{(-1)^{n+1}}{n^z} = (1 - 2^{1-z})\zeta(z)$$

For first order powers $(p = 1)$, of harmonic numbers, Sitaramachandrarao [48] gave, for $1 + q$ an odd integer,

$$2S^{+-}(1, q) = (1 + q)\overline{\zeta}(1 + q) - \zeta(1 + q) - 2 \sum_{j=1}^{\frac{q}{2}-1} \overline{\zeta}(2j)\zeta(1 + q - 2j)$$

and in another special case, gave the integral

$$S^{+-}(1, 1 + 2q) = \int_0^1 \frac{\ln^{2q}(x) \ln(1 + x)}{x(1 + x)} \, dx.$$

For higher order powers ($p \geq 2$), of harmonic numbers, in the case where p and q are both positive integers and $p + q$ is an odd integer, Flajolet and Salvy [66] gave the identity:

$$2S^{+-}(p,q) = (1 - (-1)^p)\zeta(p)\overline{\zeta}(q) + \overline{\zeta}(p+q)$$
$$+2 \sum_{i+2k=q} \binom{p+i-1}{p-1} \zeta(p+i)\,\overline{\zeta}(2k)$$
$$-2 \sum_{j+2k=p} \binom{q+j-1}{q-1} (-1)^j \overline{\zeta}(q+j)\,\overline{\zeta}(2k),$$

where $\overline{\zeta}(0) = \frac{1}{2}$, $\overline{\zeta}(1) = \ln(2)$, $\zeta(1) = 0$, and $\zeta(0) = -\frac{1}{2}$ in accordance with the analytic continuation of the Riemann zeta function. We may note that

$$S^{--}(1,q) + (-1)^q S^{+-}(1,q) = \overline{\zeta}(q) \ln 2 - \sum_{j=1}^{q-1} (-1)^j \zeta(q+1-j)\,\overline{\zeta}(j)$$

and many shuffle relations are evident, such as

$$\zeta(q)\overline{\zeta}(p) + \overline{\zeta}(p+q) = S^{-+}(p,q) + S^{+-}(q,p),$$

$$\overline{\zeta}(p)\overline{\zeta}(q) + \zeta(p+q) = S^{--}(p,q) + S^{--}(q,p).$$

Much work has also recently been done on Euler sums of the form $\sum_{n\geq 1} \frac{(H_n)^m}{n^q}$ [67] and products of index (M, q)

$$\Lambda_{M,q} = \sum_{n\geq 1} \frac{H_n^{(m_1)} H_n^{(m_2)} \cdots H_n^{(m_k)}}{n^q},$$

where $M = (m_1, m_2, \ldots, m_k)$ is a partition of an integer M into k summands so that $M = m_1 + m_2 + \cdots + m_k$ and $m_1 \leq m_2 \leq \cdots \leq m_k$ where $q + m_1 + m_2 + \cdots + m_k$ is called the weight and k is the degree. Many other works may be seen in [35, 60, 61, 68–83]. Riemann (1859) extended the zeta function to complex numbers s where $\zeta(s) = \sum_{n\geq 1} \frac{1}{n^s}$ with functional equation

$$\pi^{-(1-s)/2}\Gamma\left(\frac{1-s}{2}\right) \zeta(1-s) = \pi^{-s/2}\Gamma\left(\frac{s}{2}\right) \zeta(s).$$

The Riemann hypothesis [84, 85], which still remains unsolved states, if $\zeta(s) = 0$ and s is not a negative even integer then $s = \frac{1}{2} + it$ for some real number t. If this hypothesis is true it would imply the estimate of the Gauss prime number theorem, which states

$$\pi(x) \backsim \frac{x}{\ln x} \text{ as } x \to \infty,$$

where $\pi(x)$ is the number of primes less than or equal to x (here 1 is not considered to be prime). Gauss later improved the estimate [86] to

$$\pi(x) \backsim Li(x) = \int_2^x \frac{dt}{\ln t} \text{ as } x \to \infty.$$

In the eighteenth century it became clear that the existing elementary functions were not sufficient to describe a number of existing and emerging problems in various branches of mathematics. Functions that described the new results were generally presented in the form of infinite series, integrals, and as solutions of differential equations. Some useful special functions that are required in the following analysis are now listed. Let $\mathbb{N} := \{1, 2, 3, \ldots\}$, be the set of natural numbers, $\mathbb{N}_0 := \mathbb{N} \cup \{0\}$, $\mathbb{Z} := \{\ldots, -3, -2, -1, 0, 1, 2, 3, \ldots\}$ is the set of all integers, $\mathbb{Z}_0^- := \{\ldots, -3, -2, -1, 0\}$, $\mathbb{Z}_0^+ := \{0, 1, 2, 3, \ldots\}$, $\mathbb{R}, \mathbb{R}_+$ and $\mathbb{C}$ denote the set of real, positive real, and complex numbers, respectively. The Gamma function, for $z \in \mathbb{C}$, as given by Euler in the integral form is

$$\Gamma(z) = \int_0^\infty e^{-t} t^{s-1} \, dt, \quad \mathfrak{R}(z) > 0.$$

The special case for $z \in \mathbb{N}$ reduces to, from the recurrence relation, $\Gamma(n + 1) = n\Gamma(n) = n!$. The reflection formula is an important property and

$$\Gamma(z)\Gamma(1 - z) = \frac{\pi}{\sin \pi z}.$$

the Pochhammer or shifted factorial is defined by $(\lambda)_\nu$:

$$(\lambda)_\mu := \frac{\Gamma(\lambda + \mu)}{\Gamma(\mu)} = \begin{cases} 1, & (\mu = 0; \ \lambda \in \mathbb{C}\backslash\{0\}), \\ \lambda(\lambda + 1) \cdots (\lambda + \mu - 1) & (\mu = n \in \mathbb{N}; \ \lambda \in \mathbb{C}), \end{cases}$$

it being understood conventionally that $(0)_0 := 1$ and assumed that the Γ-quotient exists. Also $(-\lambda)_\mu = (-1)^\mu (\lambda - \mu + 1)_\mu$; $\mu \in \mathbb{N}_0$. The Beta function, or Euler integral of the first kind, is

$$B(z, w) = \int_0^1 t^{z-1} (1 - t)^{w-1} \, dt$$

$$= \frac{\Gamma(z)\Gamma(w)}{\Gamma(z + w)}, \quad \mathfrak{R}(z) > 0, \mathfrak{R}(w) > 0.$$

Let

$$H_n = \sum_{r=1}^n \frac{1}{r} = \int_0^1 \frac{1 - t^n}{1 - t} \, dt = \gamma + \psi(n + 1) = \sum_{j=1}^\infty \frac{n}{j(j + n)}, \qquad H_0 := 0$$

$$(18.2)$$

be the nth harmonic number, where γ denotes the Euler–Mascheroni constant and $\psi(z)$ is the Psi (or Digamma) function defined by

$$\psi(z) := \frac{d}{dz}\{\log \Gamma(z)\} = \frac{\Gamma'(z)}{\Gamma(z)} \quad \text{or} \quad \log \Gamma(z) = \int_1^z \psi(t) \, dt.$$

Agoh [87] has given the unusual following expression for H_n. Let $z \in \mathbb{C}$, then

$$H_n = \sum_{j=1}^{n} \left(\frac{z^j}{j} - \binom{n}{j} \frac{(z-1)^j}{j} \right),$$

which for $z = 1$ reduces to (18.2). Also recently an intriguing representation for H_n has been given by Ciaurri *et al.* [88] as

$$H_n = \pi \int_0^1 \left(x - \frac{1}{2} \right) \left(\frac{\cos\left(\frac{(4n+1)\pi x}{2} \right) - \cos\left(\frac{\pi x}{2} \right)}{\sin\left(\frac{\pi x}{2} \right)} \right) dx.$$

In the case of noninteger values of n such as a value $\rho \in \mathbb{R}$, the generalized harmonic numbers $H_\rho^{(m+1)}$ may be defined, in terms of the polygamma functions

$$\psi^{(n)}(z) := \frac{d^n}{dz^n} \{\psi(z)\} = \frac{d^{n+1}}{dz^{n+1}} \{\log \Gamma(z)\} \quad (n \in \mathbb{N}_0),$$

$$= (-1)^n n! \sum_{k \geq 0} \frac{1}{(z+k)^{n+1}}$$

by

$$H_\rho^{(m+1)} = \zeta(m+1) + \frac{(-1)^m}{m!} \psi^{(m)}(\rho + 1),$$

$$(\rho \in \mathbb{R} \setminus \{-1, -2, -3, \dots\}; \ m \in \mathbb{N}),$$

where $\zeta(z)$ is the Riemann zeta function. The recurrence relation is

$$\psi^{(n)}(z+1) = \psi^{(n)}(z) + \frac{(-1)^n n!}{z^{n+1}}$$

and reflection formula

$$\psi^{(m)}(1-z) + (-1)^{m+1} \psi^{(m)}(z) = (-1)^m \pi \frac{d^m}{dz^m} (\cot(\pi z)). \tag{18.3}$$

The Hurwitz–Lerch zeta function, or transcendent

$$\Phi(z, s, a) = \sum_{n=0}^{\infty} \frac{z^n}{(n+a)^s}$$

is defined when $|z| < 1, s \in \mathbb{C}$ and $a \in \mathbb{Z} \setminus \mathbb{Z}_0^-$, $\Re(s) > 1$ when $|z| = 1$ and satisfies the recurrence

$$\Phi(z, s, a) = z\, \Phi(z, s, a+1) + a^{-s}.$$

It can be continued meromorphically to the whole complex s-plane, except for a simple pole at $s = 1$ with its residue equal to 1. The Lerch transcendent generalizes the Hurwitz zeta function at $z = 1$,

$$\Phi(1, s, a) = \zeta(s, a) = \sum_{n=0}^{\infty} \frac{1}{(n+a)^s}; \quad \Re(s) > 1, \ a \in \mathbb{C} \setminus \mathbb{Z}_0^-$$

and when $a = 1$, we have the Riemann zeta function

$$\Phi(1, s, 1) = \zeta(s, 1) = \zeta(s) = \sum_{n=0}^{\infty} \frac{1}{(n+1)^s}; \quad \Re(s) > 1.$$

When $a = 1$, we have the Polylogarithm, or de Jonquières function,

$$\Phi(z, s, 1) = Li_s(z) := \sum_{n=1}^{\infty} \frac{z^n}{n^s}, \quad s \in \mathbb{C} \text{ when } |z| < 1; \quad \Re(s) > 1 \text{ when } |z| = 1,$$

moreover,

$$\int_0^1 \frac{Li_s(px)}{x} \, dx = \begin{cases} \zeta(1+s), & \text{for } p = 1 \\ (2^{-s} - 1)\zeta(1+s), & \text{for } p = -1 \end{cases}.$$

A generalized hypergeometric function is defined by

$$_pF_q[z] = {}_pF_q \begin{bmatrix} a_1, a_2, \ldots, a_p \\ b_1, b_2, \ldots, b_q \end{bmatrix} z = {}_pF_q[(a_p); (b_q)|z]$$

$$= \sum_{n \geq 0} \frac{(a_1)_n \cdots (a_p)_n}{(b_1)_n \cdots (b_q)_n} \frac{z^n}{n!} = \sum_{n \geq 0} \frac{\prod_{j=1}^{p} (a_j)_n}{\prod_{j=1}^{q} (b_j)_n} \frac{z^n}{n!}$$

for b_j nonnegative integers or zero. When $p \leq q$; $_pF_q[z]$ converges for all complex values of z, $_pF_q[z]$ is an entire function. When $p > q + 1$; $_pF_q[z]$ converge for $z = 0$, unless it terminates, which it does when one of the parameters a_j is a negative integer, hence $_pF_q[z]$ is a polynomial in z. When $p = q + 1$ the series converges in the unit disc $|z| < 1$, and also for $|z| = 1$ provided that $\Re\left(\sum_{j=1}^{q} b_j - \sum_{j=1}^{p} a_j\right) > 0$. When $p = 2, q = 1$, we have the familiar Gauss hypergometric function

$$_2F_1 \begin{bmatrix} a, b \\ c \end{bmatrix} z = \frac{\Gamma(c)}{\Gamma(b)\Gamma(c-b)} \int_0^1 \frac{t^{b-1}(1-t)^{c-b-1}}{(1-zt)^a} \, dt,$$

where $|z| < 1$, $\Re(c - b) > 0$ and $\Re(b) > 0$.

In the next section, we consider an extension of the Basel problem, a sum of the form (18.1), and express it in closed form. Closed form expressions in mathematics are important and a good account on closed forms: what they are and why we care can be seen in [89].

18.2 An Euler Type Sum

We define a harmonic number with multiple argument as H_{pn} for $p \in \mathbb{N}\backslash\{1\}$. For $p = 1$, we write H_n as the nth harmonic number with unitary argument. In the following we will develop analytical representations for Euler type sums with inverse binomial coefficients of the type,

$$\sum_{n \geq 1} \frac{H_{pn}^{(4)}}{n \binom{n+k}{k}}. \tag{18.4}$$

Furthermore, we discuss analytical representations of the integral

$$\int_0^1 \frac{x^p \ln^3 x}{1-x} \,{}_2F_1 \left[\begin{matrix} 1, 1 \\ 2+k \end{matrix} \,\middle|\, x^p \right] dx \tag{18.5}$$

for (k, p) the set of positive integers and where ${}_2F_1 \left[\begin{matrix} \cdot, \cdot \\ \cdot \end{matrix} \,\middle|\, z \right]$ is the classical Gauss hypergeometric function. Some results, which are published, on Euler sums with multiple argument of the type (18.4) are, from [90]

$$\sum_{n \geq 1} \frac{\binom{2n}{n} H_{2n}}{4^n (2n+1)} = G,$$

where $G = 0.91596 \cdots$ is the Catalan constant, and from [91]

$$\sum_{n \geq 1} \left(\frac{\binom{2n}{n}}{4^n (2n-1)} \right)^2 H_{2n} = \frac{1}{\pi} (6 - 12 \ln 2 + 4G).$$

In the case of noninteger values of the argument $z = \frac{r}{q}$, we may write the generalized harmonic numbers, $H_z^{(\alpha+1)}$, in terms of polygamma functions

$$H_{\frac{r}{q}}^{(\alpha+1)} = \zeta(\alpha+1) + \frac{(-1)^\alpha}{\alpha!} \psi^{(\alpha)} \left(\frac{r}{q} + 1 \right), \quad \frac{r}{q} \neq \{-1, -2, -3, \ldots\},$$

where $\zeta(z)$ is the zeta function. In the following analysis, we encounter harmonic numbers at possible rational values of the argument, of the form $H_{\frac{r}{q}}^{(\alpha)}$ they maybe evaluated by an available relation in terms of the polygamma function $\psi^{(\alpha)}(z)$ or, for rational arguments $z = \frac{r}{q}$, and we also define

$$H_{\frac{r}{q}}^{(1)} = \gamma + \psi \left(\frac{r}{q} + 1 \right) \quad \text{and} \quad H_0^{(\alpha)} = 0.$$

The evaluation of the polygamma function $\psi^{(\alpha)}\left(\frac{r}{a}\right)$ at rational values of the argument can be explicitly done via a formula as given by Kölbig [92], or Choi and Cvijovic [17] in terms of the polylogarithmic or other special functions. Polygamma functions at negative rational values of the argument can also be explicitly evaluated, for example

$$H^{(4)}_{-\frac{3}{2}} = -16 - 14\zeta(4), \; H^{(3)}_{-\frac{5}{4}} = 64 + \pi^3 - 27\zeta(3).$$

Some specific values are listed in the books [93, 94]. Some recent results for sums of harmonic numbers may be seen in the works of [18, 67, 95–115] and references therein.

The following lemma is proved in [116].

Lemma 18.1 Let k be a positive integer. Then

$$A(k) = \sum_{n\geq 1} \frac{H^{(4)}_n}{n\left(\begin{array}{c} n+k \\ k \end{array}\right)}$$

$$= \frac{1}{k}\zeta(4) + \frac{1}{6(1+k)} \int_0^1 \frac{x\ln^3 x}{1-x} \, {}_2F_1\left[\begin{array}{cc} 1,1 \\ 2+k \end{array}\middle| x\right] dx \qquad (18.6)$$

$$= \zeta(5) - H_{k-1}\zeta(4) + \sum_{r=1}^{k} (-1)^{r+1}\left(\begin{array}{c} k \\ r \end{array}\right)\left(\begin{array}{c} H^{(3)}_{r-1}\zeta(2) - H^{(2)}_{r-1}\zeta(3) \\ -\sum_{j=1}^{r-1} \frac{H_j}{j^4} \end{array}\right). $$

$$(18.7)$$

We now prove the following lemma that will be required in the proof of the main theorem.

Lemma 18.2 Let $p \in \mathbb{N}$ and $j = 1, 2, 3, \ldots, p - 1$. Then,

$$B(j,k,p) = \frac{1}{k}\zeta(4) + \frac{1}{6(1+k)} \int_0^1 \frac{x^{1-\frac{j}{p}}\ln^3 x}{1-x} \, {}_2F_1\left[\begin{array}{cc} 1,1 \\ 2+k \end{array}\middle| x\right] dx \qquad (18.8)$$

$$= \sum_{n\geq 1} \frac{H^{(4)}_{n-\frac{j}{p}}}{n\left(\begin{array}{c} n+k \\ k \end{array}\right)}$$

$$= \frac{1}{k}H^{(4)}_{-\frac{j}{p}} - H_{-\frac{j}{p}}H^{(4)}_{\frac{j}{p}-1} - \left(\zeta(2) - H^{(2)}_{-\frac{j}{p}}\right)H^{(3)}_{\frac{j}{p}-1} \qquad (18.9)$$

$$+ \left(\zeta(3) - H^{(3)}_{-\frac{j}{p}}\right)H^{(2)}_{\frac{j}{p}-1} - \left(\zeta(4) - H^{(4)}_{-\frac{j}{p}}\right)H_{\frac{j}{p}-1}$$

$$+ \sum_{r=1}^{k} (-1)^{r+1} \binom{k}{r} \left(\begin{array}{c} H_{-\frac{j}{p}} H^{(4)}_{\frac{j}{p}+r-1} + \left(\zeta(2) - H^{(2)}_{-\frac{j}{p}} \right) H^{(3)}_{\frac{j}{p}+r-1} \\ - \left(\zeta(3) - H^{(3)}_{-\frac{j}{p}} \right) H^{(2)}_{\frac{j}{p}+r-1} + \left(\zeta(4) - H^{(4)}_{-\frac{j}{p}} \right) H_{\frac{j}{p}+r-1} \\ - \sum_{m=1}^{r-1} \frac{H_m}{\left(m+\frac{j}{p} \right)^4} \end{array} \right).$$

In the case when $j = 0$, (18.8) reduces to (18.6).

Proof: Let $h_n^{(4)} = H_{n-a}^{((4))} - H_{-a}^{(4)}$ and consider the following expansion:

$$\sum_{n=1}^{\infty} \frac{h_n^{(4)}}{n \binom{n+k}{k}} = \sum_{n=1}^{\infty} \frac{k! \, h_n^{(4)}}{n \prod_{r=1}^{k}(n+r)} = \sum_{n=1}^{\infty} \frac{k! \, h_n^{(4)}}{n \, (n+1)_{k+1}}.$$

Now

$$\sum_{n=1}^{\infty} \frac{h_n^{(4)}}{n \binom{n+k}{k}} = \sum_{n=1}^{\infty} \frac{k! \, h_n^{(4)}}{n} \sum_{r=1}^{k} \left(\frac{\Lambda_r}{n+r} \right), \tag{18.10}$$

where

$$\Lambda_r = \lim_{n \to -r} \frac{n+r}{\prod_{r=1}^{k} n+r} = \frac{(-1)^{r+1} r}{k!} \binom{k}{r}. \tag{18.11}$$

For an arbitrary positive sequence $X_{k,p}$, the following identity holds

$$\sum_{k=0}^{\infty} \sum_{p=0}^{n} X_{p,k} = \sum_{k=0}^{\infty} \sum_{p=0}^{\infty} X_{p,k+p}$$

hence, from (18.10)

$$\sum_{n=1}^{\infty} \frac{k! \, h_n^{(4)}}{n} \sum_{r=1}^{k} \left(\frac{\Lambda_r}{n+r} \right) = \sum_{r=1}^{k} (-1)^{r+1} r \binom{k}{r} \sum_{n=1}^{\infty} \frac{1}{n(n+r)}$$

$$\times \sum_{\lambda=1}^{n} \frac{1}{(\lambda - a)^4}$$

where

$$h_n^{(4)} = H_{n-a}^{(4)} - H_{-a}^{(4)} = \sum_{\lambda=1}^{n} \frac{1}{(\lambda - a)^4}.$$

$$\sum_{n=1}^{\infty} \frac{h_n^{(4)}}{n \binom{n+k}{k}} = \sum_{r=1}^{k} (-1)^{r+1} r \binom{k}{r} \sum_{\lambda=1}^{\infty} \frac{1}{(\lambda - a)^4}$$

$$\times \sum_{n=0}^{\infty} \frac{1}{(n+\lambda)(n+\lambda+r)}$$

$$= \sum_{r=1}^{k} (-1)^{r+1} r \binom{k}{r} \sum_{\lambda=1}^{\infty} \frac{1}{(\lambda - a)^4} \left[\frac{\psi(\lambda+r) - \psi(\lambda)}{r} \right].$$

Since we notice that

$$\psi(\lambda + r) - \psi(\lambda) = \sum_{m=0}^{r-1} \frac{1}{m + \lambda}$$

then

$$\sum_{n=1}^{\infty} \frac{h_n^{(4)}}{n \binom{n+k}{k}} = \sum_{r=1}^{k} (-1)^{r+1} \binom{k}{r} \sum_{m=0}^{r-1} \sum_{\lambda=1}^{\infty} \frac{1}{(\lambda - a)^4 (m + \lambda)}$$

$$= \sum_{r=1}^{k} (-1)^{r+1} \binom{k}{r} \left(\begin{array}{c} \displaystyle\sum_{\lambda=1}^{\infty} \frac{1}{\lambda(\lambda - a)^4} \\ + \displaystyle\sum_{m=1}^{r-1} \sum_{\lambda=1}^{\infty} \frac{1}{(\lambda - a)^4 (m + \lambda)} \end{array} \right).$$

Simplifying

$$\sum_{n=1}^{\infty} \frac{h_n^{(4)}}{n \binom{n+k}{k}} = \sum_{r=1}^{k} (-1)^{r+1} \binom{k}{r} \left[\begin{array}{c} \dfrac{H_{-a}}{a^4} + \dfrac{1}{a^3}(\zeta(2) - H_{-a}^{(2)}) \\ -\dfrac{1}{a^2}(\zeta(3) - H_{-a}^{(3)}) \\ +\dfrac{1}{a}(\zeta(4) - H_{-a}^{(4)}) \end{array} \right]$$

$$+ \sum_{r=1}^{k} (-1)^{r+1} \binom{k}{r} \sum_{m=1}^{r-1} \left(\begin{array}{c} \dfrac{H_{-a} - H_m}{(m+a)^4} + \dfrac{\zeta(2) - H_{-a}^{(2)}}{(m+a)^3} - \dfrac{\zeta(3) - H_{-a}^{(3)}}{(m+a)^2} \\ +\dfrac{\zeta(4) - H_{-a}^{(4)}}{m+a} \end{array} \right)$$

$$= \frac{H_{-a}}{a^4} + \frac{1}{a^3}(\zeta(2) - H_{-a}^{(2)}) - \frac{1}{a^2}(\zeta(3) - H_{-a}^{(3)}) + \frac{1}{a}(\zeta(4) - H_{-a}^{(4)})$$

$$+ \sum_{r=1}^{k} (-1)^{r+1} \binom{k}{r} \sum_{m=1}^{r-1} \left(\frac{H_{-a} - H_m}{(m+a)^4} + \frac{\zeta(2) - H_{-a}^{(2)}}{(m+a)^3} - \frac{\zeta(3) - H_{-a}^{(3)}}{(m+a)^2} \right.$$
$$\left. + \frac{\zeta(4) - H_{-a}^{(4)}}{m+a} \right).$$

Now

$$\sum_{n=1}^{\infty} \frac{h_n^{(4)}}{n \binom{n+k}{k}} = G(a) - H_{-a}H_a^{(4)} - (\zeta(2) - H_{-a}^{(2)})H_a^{(3)}$$

$$+ (\zeta(3) - H_{-a}^{(3)})H_a^{(2)} - (\zeta(4) - H_{-a}^{(4)})H_a$$

$$+ \sum_{r=1}^{k} (-1)^{r+1} \binom{k}{r} \begin{bmatrix} - \sum_{m=1}^{r-1} \frac{H_m}{(m+a)^4} + H_{-a}H_{r+a-1}^{(4)} \\ + (\zeta(2) - H_{-a}^{(2)})H_{r+a-1}^{(3)} \\ - (\zeta(3) - H_{-a}^{(3)})H_{r+a-1}^{(2)} \\ + (\zeta(4) - H_{-a}^{(4)})H_{r+a-1}, \end{bmatrix}$$

where

$$G(a) = \frac{H_{-a}}{a^4} + \frac{1}{a^3}(\zeta(2) - H_{-a}^{(2)}) - \frac{1}{a^2}(\zeta(3) - H_{-a}^{(3)}) + \frac{1}{a}(\zeta(4) - H_{-a}^{(4)})$$

and since, from

$$\sum_{n=1}^{\infty} \frac{h_n^{(4)}}{n \binom{n+k}{k}} = \sum_{n=1}^{\infty} \frac{H_{n-a}^{(4)} - H_{-a}^{(4)}}{n \binom{n+k}{k}} = \sum_{n=1}^{\infty} \frac{H_{n-a}^{(4)}}{n \binom{n+k}{k}} - \frac{H_{-a}^{(4)}}{k}$$

then

$$\sum_{n=1}^{\infty} \frac{H_{n-\frac{1}{q}}^{(4)}}{n \binom{n+k}{k}} = G(a) - H_{-a}H_a^{(4)} - (\zeta(2) - H_{-a}^{(2)})H_a^{(3)} + \frac{H_{-a}^{(4)}}{k}$$

$$+ (\zeta(3) - H_{-a}^{(3)})H_a^{(2)} - (\zeta(4) - H_{-a}^{(4)})H_a$$

$$+ \sum_{r=1}^{k} (-1)^{r+1} \begin{pmatrix} k \\ r \end{pmatrix} \begin{bmatrix} -\sum_{m=1}^{r-1} \frac{H_m}{(m+a)^4} + H_{-a}H_{r+a-1}^{(4)} \\ +(\zeta(2) - H_{-a}^{(2)})H_{r+a-1}^{(3)} \\ -(\zeta(3) - H_{-a}^{(3)})H_{r+a-1}^{(2)} \\ +(\zeta(4) - H_{-a}^{(4)})H_{r+a-1} \end{bmatrix},$$

and putting $a = \frac{i}{p}$ the identity (18.9) follows. $\qquad\square$

The next few theorems relate to the main results of this investigation, namely the closed form representation of the sum (18.4) and integral (18.5).

18.3 The Main Theorem

The following main theorem is proved

Theorem 18.1   Let $(k, p) \in \mathbb{N}$, then

$$F(k, p) = \sum_{n=1}^{\infty} \frac{H_{pn}^{(4)}}{n \begin{pmatrix} n+k \\ k \end{pmatrix}}$$

$$= \frac{1}{k}\zeta(4) + \frac{1}{6(1+k)} \int_0^1 \frac{x^p \ln^3 x}{1-x} \, {}_2F_1 \begin{bmatrix} 1, 1 \\ 2+k \end{bmatrix} x^p \end{bmatrix} dx \qquad (18.12)$$

$$= \left(\frac{p^3 - 1}{kp^3} \right) \zeta(4) + \frac{1}{p^4} A(k) + \frac{1}{p^4} \sum_{j=1}^{p-1} B(j, k, p), \qquad (18.13)$$

where $A(k)$ is given by (18.7) and $B(j, k, p)$ is given by (18.9).

Proof: For the integral representation (18.12), we recall that for $m \in \mathbb{N}$

$$H_n^{(m+1)} = \frac{(-1)^m}{m!} \int_0^1 \frac{(1 - x^n)\ln^m x}{1 - x} \, dx.$$

We can now write

$$\sum_{n=1}^{\infty} \frac{H_{pn}^{(4)}}{n\binom{n+k}{k}} = -\frac{1}{6}\int_0^1 \frac{\ln^3 x}{1-x}\sum_{n=1}^{\infty}\frac{(1-x^{pn})}{n\binom{n+k}{k}}\,dx$$

$$= -\frac{1}{6}\int_0^1 \frac{\ln^3 x}{1-x}\left(\frac{1}{k} - \frac{x^p}{1+k}\,{}_2F_1\left[\begin{array}{c}1,1\\2+k\end{array}\Big|\,x^p\right]\right)dx$$

$$= \frac{1}{k}\zeta(4) + \frac{1}{6(1+k)}\int_0^1 \frac{x^p\ln^2 x}{1-x}\,{}_2F_1\left[\begin{array}{c}1,1\\2+k\end{array}\Big|\,x^p\right]dx,$$

hence (18.12) follows. Now for $p \in \mathbb{N}$ and from the properties of the polygamma function with multiple argument

$$\psi^{(n)}(pz) = \delta_{n,0} + \frac{1}{p^{n+1}}\sum_{r=0}^{p-1}\psi^{(n)}\left(z + \frac{r}{p}\right),$$

where $\delta_{n,0}$ is the Kronecker delta, we are able to rewrite, in terms of harmonic numbers, and using the properties of the polygamma function, as

$$H_{pn}^{(4)} = \left(\frac{p^3-1}{p^3}\right)\zeta(4) + \frac{1}{p^4}H_n^{(4)} + \frac{1}{p^4}\sum_{j=1}^{p-1}H_{n-\frac{L}{p}}^{(4)}.$$

The fourth-order harmonic numbers $H_{n-\frac{L}{p}}^{(4)}$ may be thought of as shifted harmonic numbers, other results on summing shifted harmonic numbers are published in [117–129]. Now, summing over the integers n

$$\sum_{n=1}^{\infty}\frac{H_{pn}^{(4)}}{n\binom{n+k}{k}} = \sum_{n=1}^{\infty}\frac{1}{n\binom{n+k}{k}}\left[\begin{array}{c}\left(\frac{p^3-1}{p^3}\right)\zeta(4) + \frac{1}{p^4}H_n^{(4)}\\ +\frac{1}{p^4}\sum_{j=1}^{p-1}H_{n-\frac{L}{p}}^{(4)}\end{array}\right]$$

$$= \left(\frac{p^3-1}{p^3}\right)\zeta(4)\sum_{n=1}^{\infty}\frac{1}{n\binom{n+k}{k}} + \frac{1}{p^4}\sum_{n=1}^{\infty}\frac{H_n^{(4)}}{n\binom{n+k}{k}}$$

$$+\frac{1}{p^4}\sum_{j=1}^{p-1}\sum_{n=1}^{\infty}\frac{H_{n-\frac{L}{p}}^{(4)}}{n\binom{n+k}{k}}$$

$$= \left(\frac{p^3-1}{kp^3}\right)\zeta(4) + \frac{1}{p^4}A(k) + \frac{1}{p^4}\sum_{j=1}^{p-1}B(j,k,p),$$

which is the result (18.13). $\qquad\square$

We give an example to demonstrate the power of the above theorem.

Example 18.1

$$F(2,4) = \frac{H_{4n}^{(4)}}{n\binom{n+2}{2}} = \frac{1}{2}\zeta(4) + \frac{1}{18}\int_0^1 \frac{x^4\ln^3 x}{1-x}\, {}_2F_1\left[\begin{array}{c}1,1\\4\end{array}\Big|\,x^4\right]dx$$

$$= \frac{1}{265}\zeta(5) + \frac{19}{256}\zeta(4) - \frac{259667}{134400}\zeta(3) + \frac{53526939}{32928000}\zeta(2) + \frac{599\pi^3}{11025}$$

$$\frac{1}{2016}\psi'''\left(\frac{1}{4}\right) + \frac{1}{480}\psi'''\left(\frac{3}{4}\right) + \frac{59953072\pi}{121550625}$$

$$- \frac{2217728}{1157625}G - \frac{383424682}{121550625}\ln 2 + \frac{208307761}{31116960000}$$

$$F(2,2) = \frac{H_{2n}^{(4)}}{n\binom{n+2}{2}} = \frac{1}{2}\zeta(4) + \frac{1}{18}\int_0^1 \frac{x^2\ln^3 x}{1-x}\, {}_2F_1\left[\begin{array}{c}1,1\\4\end{array}\Big|\,x^2\right]dx$$

$$= \frac{1}{16}\zeta(5) + \frac{19}{16}\zeta(4) - \frac{215}{144}\zeta(3) + \frac{199}{144}\zeta(2)$$

$$- \frac{160}{81}\ln 2 + \frac{97}{1296}.$$

The following proposition follows directly from Theorem 18.1 and is a comment on the evaluation of the integral in (18.12).

Proposition 18.1   For $(k,p) \in \mathbb{N}$,

$$I(k,p) = \frac{1}{6(1+k)}\int_0^1 \frac{x^p\ln^3 x}{1-x}\, {}_2F_1\left[\begin{array}{c}1,1\\2+k\end{array}\Big|\,x^p\right]dx$$

$$= \frac{1}{p^4}A(k) + \frac{1}{p^4}\sum_{j=1}^{p-1}B(j,k,p) - \frac{1}{kp^3}\zeta(4),$$

where $A(k)$ is given by (18.7) and $B(j,k,p)$ is given by (18.9).

The case $k = 1$ is interesting in its own right and, therefore, we have the following result.

Corollary 18.1   Under the assumptions of Theorem 18.1, with $k = 1$, we have,

$$F(1,p) = \sum_{n=1}^{\infty} \frac{H_{pn}^{(4)}}{n(n+1)}$$

$$= \frac{1}{6}\int_0^1 \frac{(1-x^p)\ln^3 x \ln(1-x^p)}{x^p(1-x)}\,dx \tag{18.14}$$

$$= \frac{1}{p^4}\zeta(5) + \frac{1}{p^3}H_{p-1}\zeta(4) - \frac{1}{p^2}H_{p-1}^{(2)}\zeta(3) + \frac{1}{p}H_{p-1}^{(3)}\zeta(2) \quad (18.15)$$

$$+ \frac{1}{p^4}\sum_{j=1}^{p-1}\left(\left(\frac{p}{j}\right)^4 H_{-\frac{j}{p}} - \left(\frac{p}{j}\right)^3 H_{-\frac{j}{p}}^{(2)} + \left(\frac{p}{j}\right)^2 H_{-\frac{j}{p}}^{(3)} - \left(\frac{p}{j}\right)H_{-\frac{j}{p}}^{(4)}\right)$$

$$= \frac{1}{p^4}\zeta(5) + \frac{1}{p^3}H_{p-1}\zeta(4) - \frac{1}{p^2}H_{p-1}^{(2)}\zeta(3) + \frac{1}{p}H_{p-1}^{(3)}\zeta(2) \quad (18.16)$$

$$+ \frac{1}{p^4}\sum_{j=1}^{p-1}\left(\begin{array}{c}\left(\frac{p}{j}\right)^4\left(H_{\frac{j}{p}-1} + \pi\cot\left(\frac{\pi j}{p}\right)\right) \\[2mm] -\left(\frac{p}{j}\right)^3\left(2\zeta(2) - H_{\frac{j}{p}-1}^{(2)} - \pi^2\csc^2\left(\frac{\pi j}{p}\right)\right) \\[2mm] +\left(\frac{p}{j}\right)^2\left(H_{\frac{j}{p}-1}^{(3)} - \pi^3\cot\left(\frac{\pi j}{p}\right)\csc^2\left(\frac{\pi j}{p}\right)\right) \\[2mm] -\left(\frac{p}{j}\right)\left(2\zeta(4) - H_{\frac{j}{p}-1}^{(4)} - \frac{\pi^4}{3}\csc^2\left(\frac{\pi j}{p}\right)\left\{2\cot^2\left(\frac{\pi j}{p}\right) - \csc^2\left(\frac{\pi j}{p}\right)\right\}\right)\end{array}\right).$$

Proof: From (18.12)

$$F(1,p) = \sum_{n=1}^{\infty}\frac{H_{pn}^{(4)}}{n(n+1)} = \zeta(4) + \frac{1}{12}\int_0^1\frac{x^p\ln^3 x}{1-x}\,{}_2F_1\left[\begin{array}{cc}1,1 \\ 3\end{array}\middle| x^p\right]\,dx$$

$$= \zeta(4) + \frac{1}{6}\int_0^1\frac{\ln^3 x}{1-x}\left(\frac{(1-x^p)}{x^p}\ln(1-x^p) + 1\right)\,dx,$$

and since $\int_0^1\frac{\ln^3 x}{1-x}\,dx = -6\zeta(4)$ then (18.14) follows. From (18.13)

$$F(1,p) = \left(\frac{p^3-1}{p^3}\right)\zeta(4) + \frac{1}{p^4}A(1) + \frac{1}{p^4}\sum_{j=1}^{p-1}B(j,1,p)$$

$$= \frac{1}{p^4}\zeta(5) + \left(\frac{p^3-1}{p^3}\right)\zeta(4)$$

$$+ \frac{1}{p^4}\sum_{j=1}^{p-1}\left(\begin{array}{c}H_{-\frac{j}{p}}^{(4)} + \left(\frac{p}{j}\right)^4 H_{-\frac{j}{p}} - \left(\frac{p}{j}\right)^3\left(H_{-\frac{j}{p}}^{(2)} - \zeta(2)\right) \\[2mm] +\left(\frac{p}{j}\right)^2\left(H_{-\frac{j}{p}}^{(3)} - \zeta(3)\right) - \left(\frac{p}{j}\right)\left(H_{-\frac{j}{p}}^{(4)} - \zeta(4)\right)\end{array}\right),$$

since

$$\sum_{j=1}^{p-1}H_{-\frac{j}{p}}^{(4)} = -p(p-1)(p^2+p+1)\zeta(4),$$

then

$$F(1,p) = \frac{1}{p^4}\zeta(5) + \frac{1}{p^3}H_{p-1}\zeta(4) - \frac{1}{p^2}H^{(2)}_{p-1}\zeta(3) + \frac{1}{p}H^{(3)}_{p-1}\zeta(2)$$

$$+\frac{1}{p^4}\sum_{j=1}^{p-1}\left(\left(\frac{p}{j}\right)^4 H_{-\frac{\ell}{p}} - \left(\frac{p}{j}\right)^3 H^{(2)}_{-\frac{\ell}{p}} + \left(\frac{p}{j}\right)^2 H^{(3)}_{-\frac{\ell}{p}} - \left(\frac{p}{j}\right) H^{(4)}_{-\frac{\ell}{p}}\right)$$

which is the result (18.15). From the reflection relation of the polygamma function (18.3), we have, in terms of harmonic numbers,

$$H_{-\frac{\ell}{p}} = H_{\frac{\ell}{p}-1} + \pi\cot\left(\frac{\pi j}{p}\right),$$

$$H^{(2)}_{-\frac{\ell}{p}} = 2\zeta(2) - H^{(2)}_{\frac{\ell}{p}-1} - \pi^2\operatorname{cosec}^2\left(\frac{\pi j}{p}\right),$$

$$H^{(3)}_{-\frac{\ell}{p}} = H^{(3)}_{\frac{\ell}{p}-1} - \pi^3\cot\left(\frac{\pi j}{p}\right)\operatorname{cosec}^2\left(\frac{\pi j}{p}\right)$$

and

$$H^{(4)}_{-\frac{\ell}{p}} = 2\zeta(4) - H^{(4)}_{\frac{\ell}{p}-1} - \frac{\pi^4}{3}\operatorname{cosec}^2\left(\frac{\pi j}{p}\right)\left\{2\cot^2\left(\frac{\pi j}{p}\right) - \operatorname{cosec}^2\left(\frac{\pi j}{p}\right)\right\},$$

then

$$F(1,p) = \frac{1}{p^4}\zeta(5) + \frac{1}{p^3}H_{p-1}\zeta(4) - \frac{1}{p^2}H^{(2)}_{p-1}\zeta(3) + \frac{1}{p}H^{(3)}_{p-1}\zeta(2)$$

$$+\frac{1}{p^4}\sum_{j=1}^{p-1}\left[\begin{array}{c}\left(\frac{p}{j}\right)^4\left(H_{\frac{\ell}{p}-1} + \pi\cot\left(\frac{\pi j}{p}\right)\right)\\[4pt] -\left(\frac{p}{j}\right)^3\left(2\zeta(2) - H^{(2)}_{\frac{\ell}{p}-1} - \pi^2\operatorname{cosec}^2\left(\frac{\pi j}{p}\right)\right)\\[4pt] +\left(\frac{p}{j}\right)^2\left(H^{(3)}_{\frac{\ell}{p}-1} - \pi^3\cot\left(\frac{\pi j}{p}\right)\operatorname{cosec}^2\left(\frac{\pi j}{p}\right)\right)\\[4pt] -\left(\frac{p}{j}\right)\left(2\zeta(4) - H^{(4)}_{\frac{\ell}{p}-1} - \frac{\pi^4}{3}\operatorname{cosec}^2\left(\frac{\pi j}{p}\right)\left\{2\cot^2\left(\frac{\pi j}{p}\right) - \operatorname{cosec}^2\left(\frac{\pi j}{p}\right)\right\}\right)\end{array}\right].$$

The integral (18.14) follows from (18.12). The identity (18.16) is noteworthy because it introduces finite cotangent and cosecant sums, which is a separate field of study in itself. Finite cotangent and cosecant sums of the form

$$\sum_{r=1}^{p-1}\cot^m\left(\frac{\pi r}{p}\right) \text{ and } \sum_{r=1}^{p-1}\operatorname{cosec}^m\left(\frac{\pi r}{p}\right),$$

and their variations have been investigated, see [2, 5, 30, 34, 130–132]. Bettin and Conrey [133] prove a certain reciprocity formula for the cotangent sum

$$\sum_{r=1}^{p-1}\frac{r}{p}\cot\left(\frac{\pi rh}{p}\right).$$

The sum arises in connection with the Nyman–Beurling approach to the Riemann hypothesis. The author has not seen an investigation of

$$\sum_{r=1}^{p-1} r^q \cot^m \left(\frac{\pi r}{p} \right) \quad \text{and} \quad \sum_{r=1}^{p-1} r^q \csc^m \left(\frac{\pi r}{p} \right),$$

$q \in \mathbb{Z} \setminus \{0\}$, $m \in \mathbb{N}$, in the published literature. None of the integrals (18.12) and (18.14) and their generalizations can be evaluated with mathematical packages such as *Mathematica*. $\quad\square$

In the next remark, we utilize the integral (18.14) to obtain the following:

Remark 18.1 From (18.14), we have

$$F(1, 1) = \sum_{n=1}^{\infty} \frac{H_n^{(4)}}{n(n + 1)} = \zeta(5),$$

similarly utilizing

$$\sum_{n=1}^{\infty} \frac{(-1)^{n+1} H_n^{(4)}}{n(n + 1)} = \frac{49}{16}\zeta(5) - \frac{7}{4}\zeta(4) \ln 2 - \frac{3}{4}\zeta(3)\zeta(2),$$

hence after some simplification we obtain

$$\sum_{n=1}^{\infty} \frac{H_{2n}^{(4)}}{2n(2n - 1)}$$
$$= 2\zeta(5) - \frac{7}{8}\zeta(4) \ln 2 - \frac{3}{8}\zeta(3)\zeta(2) + \ln 2 - \frac{1}{4}\zeta(2) - \frac{1}{8}\zeta(3) - \frac{1}{16}\zeta(4),$$

and the integral representation results in

$$\int_0^1 \frac{\ln^3 x}{(1 - x)} \left(\frac{1}{2} \ln(1 - x^2) + \frac{x}{2} \ln \left(\frac{1 + x}{1 - x} \right) \right) \, dx$$
$$= 12\zeta(5) - \frac{45}{4}\zeta(4) \ln 2 - \frac{9}{4}\zeta(3)\zeta(2) + 6 \ln 2 - \frac{3}{2}\zeta(2) - \frac{3}{4}\zeta(3) - \frac{3}{8}\zeta(4).$$

Again this integral cannot be evaluated by *Mathematica*. A similar calculation yields

$$\sum_{n=1}^{\infty} \frac{H_{2n}^{(8)}}{(2n - 1)(2n + 1)} = -\frac{1}{7!} \int_0^1 \frac{(1 + x) \ln^7 x}{4x} \ln \left(\frac{1 + x}{1 - x} \right) \, dx$$
$$= \frac{511}{1024}\zeta(9) - \frac{1}{512}\zeta(8) - \frac{1}{256}\zeta(7) - \frac{1}{128}\zeta(6) - \frac{1}{64}\zeta(5)$$
$$- \frac{1}{32}\zeta(4) - \frac{1}{16}\zeta(3) - \frac{1}{8}\zeta(2) + \frac{1}{2} \ln 2.$$

The closed form (18.13) of the integral (18.12) is an exact identity that is expressed in finite sums of harmonic numbers and special functions. The following theorem gives a bound on the integral (18.12).

Theorem 18.2 Let $k, p \in \mathbb{N}$, then,

$$\frac{H_p^{(4)}}{1+k} < F(k,p) \le \frac{1}{k}\zeta(4) + \frac{\alpha}{6p}\psi'(p+1) \tag{18.17}$$

where

$$F(k,p) = \frac{1}{k}\zeta(4) + \frac{1}{6(1+k)} \int_0^1 \frac{x^p \ln^3 x}{1-x} \, {}_2F_1 \left[\begin{array}{c} 1,1 \\ 2+k \end{array} \middle| x^p \right] dx$$

$$= \sum_{n=1}^{\infty} \frac{H_{pn}^{(4)}}{n \left(\begin{array}{c} n+k \\ k \end{array} \right)}$$

and $\alpha = 1.4214 \cdots$.

Proof: The infinite sum $F(k,p)$ is one of positive terms, monotonic increasing and therefore,

$$F(k,p) > \frac{H_p^{(4)}}{1+k}.$$

Consider the integral inequality

$$\int_{x_0}^{x_1} |f(x)g(x)| \, dx \le \sup_{x \in [x_0, x_1]} |f(x)| \int_{x_0}^{x_1} |g(x)| \, dx$$

for integrable functions $f(x)$ and $g(x)$ and $0 \le x_0 < x_1 \in \mathbb{R}$. Now

$$\sup_{x \in [x_0, x_1]} |f(x)| = \sup_{x \in [0,1]} \left| \frac{x \ln^3 x}{1-x} \right| = 1.4214 \cdots = \alpha.$$

Also

$$\int_{x_0}^{x_1} |g(x)| \, dx = \int_0^1 \left| x^{p-1} {}_2F_1 \left[\begin{array}{c} 1,1 \\ 2+k \end{array} \middle| x^p \right] \right| dx = \frac{(1+k)\psi'(1+k)}{p},$$

therefore

$$\frac{H_p^{(4)}}{1+k} < F(k,p) \le \frac{1}{k}\zeta(4) + \frac{(k+1)\alpha}{6p(k+1)} \cdot \psi'(1+k)$$

and (18.17) follows. $\qquad\square$

18.4 Conclusion

We have given a glimpse into the history of one particular aspect of the genius of Euler in relation to the Basel problem. Alongside Archimedes, Newton, and Gauss, there is no doubt that Euler is a giant of mathematical analysis and his legacy continues to have a major impact. Furthermore we have considered a particular Euler sum and expressed its representation in closed form in terms of special functions.

References

1 Ablinger, J., Blümlein, J., and Schneider, C. (2013) Analytic and algorithmic aspects of generalized harmonic sums and polylogarithms. *J. Math. Phys.*, **54** (8), 082301, 74 p.

2 Alzer, H. and Chu, W. (2014) Two trigonometric identities. *Irish Math. Soc. Bull.*, **73**, 21–28.

3 Alzer, H. and Sondow, J. (2016) A parameterized series representation for Apéry's constant $\zeta(3)$. *J. Comput. Anal. Appl.*, **20** (7), 1380–1386.

4 Basu, A. (2008) A new method in the study of Euler sums. *Ramanujan J.*, **16** (1), 7–24.

5 Berndt, B.C. and Boon, P.Y. (2002) Explicit evaluations and reciprocity theorems for finite trigonometric sums. *Adv. Appl. Math.*, **29**, 358–385.

6 Bhatnagar, G. (2011) In praise of an elementary identity of Euler. *Electron. J. Comb.*, **18** (2), Paper 13, 44 p.

7 Blümlein, J. (2009) Structural relations of harmonic sums and Mellin transforms up to weight $w = 5$. *Comput. Phys. Commun.*, **180** (11), 2218–2249.

8 Boyadzhiev, K.N. (2012) Series with central binomial coefficients. Catalan numbers, and harmonic numbers. *J. Integer. Seq.*, **15** (1), Article 12.1.7, 11 p.

9 Boyadzhiev, K.N. and Dil, A. (2016) Geometric polynomials: properties and applications to series with zeta values. *Anal. Math.*, **42** (3), 203–224.

10 Cheon, G.-S. and El-Mikkawy, M.E.A. (2007) Generalized harmonic number identities and a related matrix representation. *J. Korean Math. Soc.*, **44** (2), 487–498.

11 Chen, K.-W. (2017) Generalized harmonic numbers and Euler sums. *Int. J. Number Theory*, **13** (2), 513–528.

12 Chen, X. and Chu, W. (2010) Dixon's 3F2(1)-series and identities involving harmonic numbers and the Riemann zeta function. *Discrete Math.*, **310** (1), 83–91.

13 Choi, J. (2014) Summation formulas involving binomial coefficients, harmonic numbers, and generalized harmonic numbers. *Abstr. Appl. Anal.*, Articel ID 501906, 10 p.

14 Choi, J. (2013) Finite summation formulas involving binomial coefficients, harmonic numbers and generalized harmonic numbers. *J. Inequal. Appl.*, **49**, 11 p.

15 Choi, J. and Srivastava, H.M. (2014) Series involving the Zeta functions and a family of generalized Goldbach-Euler series. *Am. Math. Mon.*, **121** (3), 229–236.

16 Choi, J. and Srivastava, H.M. (2013) An experimental conjecture involving closed-form evaluation of series associated with the zeta functions. *Math. Inequal. Appl.*, **16** (4), 971–979.

17 Choi, J. and Cvijović, D. (2007) Values of the polygamma functions at rational arguments. *J. Phys. A: Math. Theor.*, **40**, 15019–15028; Corrigendum, ibidem, **43** (2010), 239801, 1 p.

18 Choi, J. and Srivastava, H.M. (2011) Some summation formulas involving harmonic numbers and generalized harmonic numbers. *Math. Comput. Modell.*, **54**, 2220–2234.

19 Cohen, G.L. and Moujie, D. (1998) On a generalisation of Ore's harmonic numbers. *Nieuw Arch. Wisk. (4)*, **16** (3), 161–172.

20 Chu, W. (2012) Summation formulae involving harmonic numbers. *Filomat*, **26** (1), 143–152.

21 Chu, W. (2012) Infinite series identities on harmonic numbers. *Results Math.*, **61** (3–4), 209–221.

22 Chu, W. and Zheng, D. (2009) Infinite series with harmonic numbers and central binomial coefficients. *Int. J. Number Theory*, **5** (3), 429–448.

23 Coffey, M.W. (2005) On one-dimensional digamma and polygamma series related to the evaluation of Feynman diagrams. *J. Comput. Appl. Math.*, **183** (1), 84–100.

24 Crandall, R.E. and Buhler, J.P. (1994) On the evaluation of Euler sums. *Exp. Math.*, **3**, 275–285.

25 Dil, A. and Kurt, V. (2012) Polynomials related to harmonic numbers and evaluation of harmonic number series I. *Integers*, **12**, Paper No. A38, 18 p.

26 Dil, A. and Kurt, V. (2011) Polynomials related to harmonic numbers and evaluation of harmonic number series II. *Appl. Anal. Discrete Math.*, **5** (2), 212–229.

27 Eie, M. and Wei, C.-S. (2012) Evaluations of some quadruple Euler sums of even weight. *Funct. Approx. Comment. Math.*, **46**, Part 1, 63–77.

28 Espinosa, O. and Moll, V.H. (2010) The evaluation of Tornheim double sums. II. *Ramanujan J.*, **22** (1), 55–99.

29 Freitas, P. (2005) Integrals of polylogarithmic functions, recurrence relations, and associated Euler sums. *Math. Comput.*, **74** (251), 1425–1440.

30 Fukuhara, S. (2003) New trigonometric identities and generalized Dedekind sums. *Tokyo J. Math.*, **26** (1), 1–14.

31 Furdui, O. (2016) Harmonic series with polygamma functions. *J. Class. Anal.*, **8** (2), 123–130.

32 Furdui, O. and Vălean, C. (2016) Evaluation of series involving the product of the tail of $\zeta(k)$ and $\zeta(k+1)$. *Mediterr. J. Math.*, **13** (2), 517–526.

33 Genčev, M. (2011) Binomial sums involving harmonic numbers. *Math. Slovaca*, **61** (2), 215–226.

34 Grabner, P. and Prodinger, H. (2007) Secant and cosecant sums and Bernoulli-Nörlund polynomials. *Quaest. Math.*, **30** (2), 159–165.

35 Hoffman, M.E. (2017) Harmonic-number summation identities, symmetric functions, and multiple zeta values. *Ramanujan J.*, **42** (2), 501–526.

36 Jang, L.-C. and Kim, B.M. (2016) On identities between sums of Euler numbers and Genocchi numbers of higher order. *J. Comput. Anal. Appl.*, **20** (7), 1240–1247.

37 Jeong, Y. and Kim, D.S. (2015) Identities involving generalized harmonic numbers. *Proc. Jangjeon Math. Soc.*, **18** (2), 189–199.

38 Jung, M., Cho, Y.J., and Choi, J. (2004) Euler sums evaluatable from integrals. *Commun. Korean Math. Soc.*, **19** (3), 545–555.

39 Kalman, D. and McKinzie, M. (2012) Another way to sum a series: generating functions, Euler, and the dilog function. *Am. Math. Mon.*, **119** (1), 42–51.

40 Kitaev, S. and Liese, J. (2013) Harmonic numbers, Catalan's triangle and mesh patterns. *Discrete Math.*, **313** (14), 1515–1531.

41 Kuba, M. (2007/08) On evaluations of infinite double sums and Tornheim's double series. *Sém. Lothar. Combin.*, **58**, Article B58d, 13 p.

42 Liaw, W.-C. (2007) New relations among Euler sums of even weight. *Tamkang J. Math.*, **38** (1), 21–36.

43 Mező, I. (2014) Nonlinear Euler sums. *Pac. J. Math.*, **272** (1), 201–226.

44 Ong, Y.L., Eie, M., and Wei, C.-S. (2010) Explicit evaluations of quadruple Euler sums. *Acta Arith.*, **144** (3), 213–230.

45 Qin, H., Li, A., and Shang, N. (2014) On representation problems of Euler sums with multi-parameters. *Integr. Transf. Spec. Funct.*, **25** (5), 384–397.

46 Ram Murty, M. and Weatherby, C. (2016) A generalization of Euler's theorem for (2k). *Am. Math. Mon.*, **123** (1), 53–65.

47 Si, X., Xu, C., and Zhang, M. (2017) Quadratic and cubic harmonic number sums. *J. Math. Anal. Appl.*, **447** (1), 419–434.

48 Sitaramachandrarao, R. (1987) A formula of S. Ramanujan. *J. Number Theory*, **25**, 1–19.

49 Sittinger, B.D. (2016) Computing $\zeta(2m)$ by using telescoping sums. *Am. Math. Mon.*, **123** (7), 710–715.

50 Euler, L. (1740) De progressionibus harmonicis observationes. *Commun. Acad. Sci. Petrop.*, **7** (1734/35), 150–161; *Opera Omnia*, **14** (1), 87–100.

51 Euler, L. (1760) De seriebus divergentibus. *Novi Commun. Acad. Sci. Petrop.*, **5** (1754/55), 205–237; *Opera Omnia*, (1) **14**, 585–617; An English translation by Barbeau, E.J. and Leah, P.J. is in (1976) *Historia Math.*, **3**, 141–160.

52 Euler, L. (1750) De seriebus quibusdam considerationes. *Commun. Acad. Sci. Petrop.*, **12** (1740), 53–96; *Opera Omnia*, **14** (1), 407–462.

53 Euler, L. (1740) De summis serierum reciprocarum. *Commun. Acad. Sci. Petrop.*, **7** (1734/35), 123–134; *Opera Omnia*, **14** (1), 73–86.

54 Euler, L. (1734/35) Insttutiones calculi differentialis cum lius usu in analysi finitorum ac doctrina serierum. *Acad. Imp. Sci. Petrop. Opera Omnia*, **10** (1), 309–336.

55 Yakubovich, S. (2015) Certain identities, connection and explicit formulas for the Bernoulli and Euler numbers and the Riemann zeta-values. *Analysis (Berlin)*, **35** (1), 59–71.

56 Ito, T. (2006) On an integral representation of special values of the zeta function at odd integers. *J. Math. Soc. Jpn.*, **58** (3), 681–691.

57 Luo, Q.-M., Guo, B.-N., and Qi, F. (2003) On evaluation of Riemann zeta function (s). *Adv. Stud. Contemp. Math. (Kyungshang)*, **7** (2), 135–144.

58 Apery, R. (1979) Irrationalite de $\zeta(2)$ et $\zeta(3)$. *Asterisque*, **61**, 11–13.

59 Rivoal, T. (2000) La fonction zeta de Riemann prend une infinite de valeurs irrationnelles aux entiers impairs. *C. R. Acad. Sci. Paris*, **331**, 267–270.

60 Zudilin, V.V. (2001) One of the numbers $\zeta(5)$, $\zeta(7)$, $\zeta(9)$, $\zeta(11)$ is irrational. (Russian). *Uspekhi Mat. Nauk*, **56** 4(340), 149–150; translation in *Russian Math. Surv.*, **56** (2001) (4), 774–776.

61 Okamoto, T. (2013) On alternating analogues of the Mordell-Tornheim triple zeta values. *J. Ramanujan Math. Soc.*, **28**, 247–269.

62 Borwein, D. and Borwein, J.M. (1995) On an intriguing integral and some series related to $\zeta(4)$. *Proc. Am. Math. Soc.*, **123**, 1191–1198.

63 Borwein, J.M., Zucker, I.J., and Boersma, J. (2008) The evaluation of character Euler double sums. *Ramanujan J.*, **15** (3), 377–405.

64 Borwein, D., Borwein, J.M., and Bradley, D.M. (2006) Parametric Euler sum identities. *J. Math. Anal. Appl.*, **316** (1), 328–338.

65 Borwein, D., Borwein, J.M., and Girgensohn, R. (1995) Explicit evaluation of Euler sums. *Proc. Edinburgh Math. Soc. (2)*, **38** (2), 277–294.

66 Flajolet, P. and Salvy, B. (1998) Euler sums and contour integral representations. *Exp. Math.*, **7** (1), 15–35.

67 Sofo, A. (2015) Quadratic alternating harmonic number sums. *J. Number Theory*, **154**, 144–159.

68 Bachmann, H. and Kühn, U. (2016) The algebra of generating functions for multiple divisor sums and applications to multiple zeta values. *Ramanujan J.*, **40**, 605–648.

69 Bailey, D.H. and Borwein, J.M. (2016) Computation and structure of character polylogarithms with applications to character Mordell-Tornheim-Witten sums. *Math. Comput.*, **85** (297), 295–324.

70 Borwein, J.M. and Straub, A. (2015) Relations for Nielsen polylogarithms. *J. Approx. Theory*, **193**, 74–88.

71 Chang, C.-Y. (2016) Shuffle product formulas of two multiples of height-one multiple zeta values. *Taiwanese J. Math.*, **20** (1), 13–24.

72 Chen, K.-W. (2015) Applications of stuffle product of multiple zeta values. *J. Number Theory*, **153**, 107–116.

73 Dilcher, K., Hessami Pilehrood, Kh., and Hessami Pilehrood, T. (2014) On q-analogues of double Euler sums. *J. Math. Anal. Appl.*, **410** (2), 979–988.

74 Hessami Pilehrood, Kh., Hessami Pilehrood, T., and Tauraso, R. (2014) New properties of multiple harmonic sums modulo p and p-analogues of Leshchiner's series. *Trans. Am. Math. Soc.*, **366** (6), 3131–3159.

75 Hoffman, M.E. (2017) On multiple zeta values of even arguments. *Int. J. Number Theory*, **13** (3), 705–716.

76 Machide, T. (2017) Identities involving cyclic and symmetric sums of regularized multiple zeta values. *Pac. J. Math.*, **286**, 307–359.

77 Matsumoto, K., Nakamura, T., and Tsumura, H. (2008) Functional relations and special values of Mordell-Tornheim triple zeta and L-functions. *Proc. Am. Math. Soc.*, **136**, 2135–2145.

78 Miyagawa, T. (2016) Analytic properties of generalized Mordell-Tornheim type of multiple zeta-functions and L-functions. *Tsukuba J. Math.*, **40**, 81–100.

79 Panholzer, A. and Prodinger, H. (2005) Computer-free evaluation of an infinite double sum via Euler sums. *Sém. Lothar. Combin.*, **55**, Article B55a, 3 p.

80 Shen, Z.Y. and Cai, T.X. (2016) Some weighted sum identities for double zeta values. *Acta Math. Sin. (Engl. Ser.)*, **32** (7), 797–806.

81 Tasaka, K. (2016) On linear relations among totally odd multiple zeta values related to period polynomials. *Kyushu J. Math.*, **70**, 1–28.

82 Xu, Ce. (2017) Some evaluation of parametric Euler sums. *J. Math. Anal. Appl.*, **451** (2), 954–975.

83 Zhao, J. (2016) Identity families of multiple harmonic sums and multiple zeta star values. *J. Math. Soc. Jpn.*, **68** (4), 1669–1694.

84 Cloitre, Benoit. (2016) Good variation theory: a Tauberian approach to the Riemann hypothesis. *Int. J. Math. Comput. Sci.*, **11** (2), 133–149.

85 Connes, A. (2016) An essay on the Riemann hypothesis, in *Open Problems in Mathematics*, Springer, Cham, pp. 225–257.

86 Granville, A. (1997) *A Decomposition of Riemann's Zeta-Function. Analytic Number Theory (Kyoto, 1996)*, London Mathematical Society Lecture Note Series, vol. **247**, Cambridge University Press, Cambridge, pp. 95–101.

87 Agoh, T. (2016) On Miki's identity for Bernoulli numbers. *Integers*, **16**, Paper No. A73, 12 p.

88 Ciaurri, O., Navas, L.M., Ruiz, F.J., and Varona, J.L. (2015) A simple computation of ∎(2k). *Am. Math. Mon.*, **122** (5), 444–451.

89 Borwein, J.M. and Crandall, R.E. (2013) Closed forms: what they are and why we care. *Not. Am. Math. Soc.*, **60** (1), 50–65.

90 Hongwei, C. (2016) Interesting series associated with central binomial coefficients, Catalan numbers and harmonic numbers. *J. Integer Seq.*, **19** (1), Article 16.1.5, 11 p.

91 Campbell, J.M. and Sofo, A. (2017) An integral transform related to series involving alternating harmonic numbers. *Integr. Transf. Spec. Funct.*, **28** (7), 547–559.

92 Kölbig, K. (1996) The polygamma function $\psi(x)$ for $x = 1/4$ and $x = 3/4$. *J. Comput. Appl. Math.*, **75**, 43–46.

93 Sofo, A. (2003) *Computational Techniques for the Summation of Series*, Kluwer Academic/Plenum Publishers, New York.

94 Srivastava, H.M. and Choi, J. (2001) *Series Associated with the Zeta and Related Functions*, Kluwer Academic Publishers, London.

95 Liu, H. and Wang, W. (2012) Harmonic number identities via hypergeometric series and Bell polynomials. *Integr. Transf. Spec. Funct.*, **23** (1), 49–68.

96 Sofo, A. (2016) New results containing quadratic harmonic numbers. *J. Class. Anal.*, **9** (2), 117–125.

97 Sofo, A. (2016) Half integer values of order-two harmonic numbers sums. *Ukraïn. Mat. Zh.*, **68** (10), 1418–1429.

98 Sofo, A. (2016) Sums of quadratic half integer harmonic numbers of alternating type. *J. Class. Anal.*, **8** (2), 99–111.

99 Sofo, A. (2016) Polylogarithmic connections with Euler sums. *Sarajevo J. Math.*, **12** (1, 24), 17–32.

100 Sofo, A. and Cvijović, D. (2012) Extensions of Euler harmonic sums. *Appl. Anal. Discrete Math.*, **6** (2), 317–328.

101 Sofo, A. (2011) Summation formula involving harmonic numbers. *Anal. Math.*, **37** (1), 51–64.

102 Sofo, A. and Cerone, P. (1998) Generalisation of Euler's identity. *Bull. Austr. Math. Soc.*, **58** (3), 359–371.

103 Srivastava, H.M. (2015) Some properties and results involving the zeta and associated functions. *Funct. Anal. Approx. Comput.*, **7** (2), 89–133.

104 Sun, Z.-W. and Zhao, L.-L. (2013) Arithmetic theory of harmonic numbers (II). *Colloq. Math.*, **130** (1), 67–78.

105 Vălean, C.I. (2016) A new proof for a classical quadratic harmonic series. *J. Class. Anal.*, **8** (2), 155–161.

106 Wang, J. and Wei, C. (2016) Derivative operator and summation formulae involving generalized harmonic numbers. *J. Math. Anal. Appl.*, **434** (1), 315–341.

107 Wei, C. and Wang, Q. (2017) A Saalschütz-type identity and summation formulae involving generalized harmonic numbers. *J. Math. Anal. Appl.*, **449** (2), 1036–1052.

108 Wei, C. and Wang, X. (2016) Summation formulas involving generalized harmonic numbers. *J. Differ. Equ. Appl.*, **22** (10), 1554–1567.

109 Wei, C. (2016) Minton-Karlsson identities and summation formulae involving generalized harmonic numbers. *Integr. Transf. Spec. Funct.*, **27** (7), 592–598.

110 Wu, B.-L. and Chen, Y.-G. (2017) On certain properties of harmonic numbers. *J. Number Theory*, **175**, 66–86.

111 Xu, C., Zhang, M., and Zhu, W. (2017) Some evaluation of q-analogues of Euler sums. *Monatsh. Math.*, **182** (4), 957–975.

112 Xu, C., Yan, Y., and Shi, Z. (2016) Euler sums and integrals of polylogarithmic functions. *J. Number Theory*, **165**, 84–108.

113 Yang, J. and Wang, Y. (2017) Summation formulae in relation to Euler sums. *Integr. Transf. Spec. Funct.*, **28** (5), 336–349.

114 Zhao, J. (2015) Restricted sum formula of alternating Euler sums. *Ramanujan J.*, **36** (3), 375–401.

115 Zudilin, V.V. (2003) Algebraic relations for multiple zeta values. (Russian). *Uspekhi Mat. Nauk*, **58** (1 (349)), 3–32; translation in (2003) *Russian Math. Surv.*, **58** (1), 1–29.

116 Sofo, A. (2011) Harmonic number sums in higher powers. *J. Math. Anal.*, **2**, 15–22.

117 Sofo, A. (2016) Harmonic numbers at half integer values. *Integr. Transf. Spec. Funct.*, **27** (6), 430–442.

118 Sofo, A. and Srivastava, H.M. (2015) A family of shifted harmonic sums. *Ramanujan J.*, **37** (1), 89–108.

119 Sofo, A. (2017) A master integral in four parameters. *J. Math. Anal. Appl.*, **448**, 81–92.

120 Sofo, A. (2015) New families of alternating harmonic number sums. *Tbilisi Math. J.*, **8** (2), 195–209.

121 Sofo, A. (2016) Second order alternating harmonic number sums. *Filomat*, **30** (13), 3511–3524.

122 Sofo, A. (2016) Harmonic numbers at half integer and binomial squared sums. *Honam Math. J.*, **38** (2), 279–294.

123 Sofo, A. (2016) Integrals of logarithmic and hypergeometric functions. *Commun. Math.*, **24** (1), 7–22.

124 Sofo, A. (2016) Identities for alternating inverse squared binomial and harmonic number sums. *Mediterr. J. Math.*, **13** (4), 1407–1418.

125 Sofo, A. (2012) New classes of harmonic number identities. *J. Integer Seq.*, **15** (7), Article 12.7.4, 12 p.

126 Sofo, A. (2010) Harmonic sums and integral representations. *J. Appl. Anal.*, **16** (2), 265–277.

127 Sofo, A. (2009) Harmonic numbers and double binomial coefficients. *Integr. Transf. Spec. Funct.*, **20** (11–12), 847–857.

128 Sofo, A. (2009) Integral forms associated with harmonic numbers. *Appl. Math. Comput.*, **207**, 365–372.

129 Sofo, A. (2009) Sums of derivatives of binomial coefficients. *Adv. Appl. Math.*, **42** (1), 123–134.

130 Cvijović, D. (2009) Summation formulae for finite cotangent sums. *Appl. Math. Comput.*, **215**, 1135–1140.

131 Cvijović, D. and Srivastava, H.M. (2012) Closed-form summations of Dowker's and related trigonometric sums. *J. Phys. A*, **45** (37), 374015, 10 p.

132 da Fonseca, C.M., Glasser, L., and Kowalenko, V. (2017) Basic trigonometric power sums with applications. *Ramanujan J.*, **42**, 401–428.

133 Bettin, S. and Conrey, J.B. (2013) A reciprocity formula for a cotangent sum. *Int. Math. Res. Not. IMRN*, (24), 5709–5726.

19

Coupled Fixed Points and Coupled Coincidence Points via Fixed Point Theory

Adrian Petruşel[1] and Gabriela Petruşel[2]

[1] *Faculty of Mathematics and Computer Science, Babeş-Bolyai University, 400084 Cluj-Napoca, Romania*
[2] *Faculty of Business, Babeş-Bolyai University, 400084 Cluj-Napoca, Romania*

2010 AMS Subject Classification 47H10, 54H25

19.1 Introduction and Preliminaries

The fixed point problem for a single-valued operator $f : X \to X$ consists in the study of the following equation:

$$x = f(x), \quad x \in X, \tag{19.1}$$

where X is a given nonempty set. A solution $x \in X$ of the above equation is called a fixed point of f and the set of all fixed points of f will be denoted by $Fix(f)$. We will also denote by f^n the iterates operators of f, that is, $f^n :=$ $f \circ f \circ \cdots \circ f$ (the n-times composition).

The most important and most often applied metrical fixed point theorem is the well-known contraction mapping principle (CMP). This principle was proved in 1922 by St. Banach for the case of single-valued contractions on a Banach space and extended, in 1930, by R. Caccioppoli to the framework of a complete metric space. Then, it was successively extended to various types of generalized metric spaces and various generalized contraction mappings.

An extended version of the CMP was recently proposed by Rus [1]. We present now a slight modified variant of it, as follows.

Theorem 19.1 (see Rus [1] – The saturated principle of contraction) Let (X, d) be a complete metric space and $f : X \to X$ be an α-contraction, that is, $\alpha \in (0, 1)$ and

$$d(f(x), f(y)) \le \alpha d(x, y) \quad \text{for all } x, y \in X.$$

Then, we have the following properties:

(i) There exists $x^* \in X$ such that $Fix(f) = \{x^*\}$.

Mathematical Analysis and Applications: Selected Topics, First Edition.
Edited by Michael Ruzhansky, Hemen Dutta, and Ravi P. Agarwal.
© 2018 John Wiley & Sons, Inc. Published 2018 by John Wiley & Sons, Inc.

(ii) $Fix(f^n) = \{x^*\}$ for all $n \in \mathbb{N}$, $n \geq 2$.

(iii) For all $x \in X$, $f^n(x) \to x^*$ as $n \to \infty$ and $d(f^n(x), x^*) \leq \frac{\alpha^n}{1-\alpha} d(x, f(x))$, for each $n \in \mathbb{N}$.

(iv) $d(x, x^*) \leq \psi(d(x, f(x)))$, for all $x \in X$, where $\psi(t) = \frac{t}{1-\alpha}$, $t \geq 0$.

(v) If $\{y_n\}_{n \in \mathbb{N}}$ is a sequence in X such that

$$d(y_n, f(y_n)) \to 0 \quad \text{as} \quad n \to \infty,$$

then, $y_n \to x^*$ as $n \to \infty$.

(vi) If $\{y_n\}_{n \in \mathbb{N}}$ is a sequence in X such that

$$d(y_{n+1}, f(y_n)) \to 0 \quad \text{as} \quad n \to \infty,$$

then, $y_n \to x^*$ as $n \to \infty$.

(vii) for every $\epsilon > 0$ and every $\tilde{x} \in X$ with $d(\tilde{x}, f(\tilde{x})) \leq \epsilon$, we have that

$$d(\tilde{x}, x^*) \leq \frac{\epsilon}{1-\alpha}.$$

Notice that an operator f satisfying the condition (i) and the first part of (iii) is called a Picard operator, while a Picard operator for which the relation (iv) also holds will be called a ψ-Picard operator. The relation given in (iv) is called the retraction–displacement estimation. In this definition, $\psi : \mathbb{R}_+ \to \mathbb{R}_+$ must be a function which is increasing and continuous in 0 with $\psi(0) = 0$ and the framework could be a more general one, for example, an L-space in the sense of Fréchet [2] (a space endowed with a convergence structure).

Moreover, if the operator $f : X \to X$ satisfy the above conditions (i) and (v), then, by definition, the fixed point problem for (19.1) is said to be well posed. If $f : X \to X$ satisfies (i) and (vi) in the above theorem, then, by definition, the operator f has the Ostrowski property.

Concerning the assertion (vii), it is also easy to see that, by (i) and (iv), it follows that for every $\epsilon > 0$, if $\tilde{x}$ is an ϵ-solution of the fixed point equation (19.1), then

$$d(\tilde{x}, x^*) \leq \frac{\epsilon}{1-\alpha}.$$

By definition, if the assertions (i) and (vii) take place, then the fixed point equation (19.1) is said to be Ulam–Hyers stable.

As one can see from the above theorem, the study of the fixed point problem (19.1) behaves several mathematical phenomena: existence, uniqueness, approximation, apriori estimation, retraction–displacement estimation, well-posedness, Ostrowski property, and Ulam–Hyers stability.

In this chapter, we will focus our attention on the study of the coupled fixed point and coupled coincidence point problems for single- and multi-valued operators, via the fixed point theory approach. A study of the above mentioned mathematical phenomena with respect to the solutions of the coupled fixed point and of the coupled coincidence point problems will be presented.

With this brief introduction, we recall the following concepts and results that are needed in the main part of this chapter.

Throughout this paper $\mathbb{N}$ stands for the set of natural numbers, $\mathbb{N}^*$ is the set of natural numbers except 0, while $\mathbb{R}$ is the set of all real numbers. We will also use the same symbol $\leq$ on $\mathbb{R}^m$ for the component-wise ordering and we will make an identification between rows and columns in $\mathbb{R}^m$.

Let X be a nonempty set. A mapping $d : X \times X \to \mathbb{R}_+^m$ is called a vector-valued metric on X if all the classical axioms of the metric are fulfilled, with respect to the above-mentioned partial ordering. A nonempty set X endowed with a vector-valued metric $d : X \times X \to \mathbb{R}_+^m$ is called a generalized metric space in the sense of Perov. The notions of convergent sequence, Cauchy sequence, completeness, open and closed set, open and closed ball and so long are defined in a similar way to the metric spaces.

We denote by $M_{mm}(\mathbb{R}_+)$ the set of all $m \times m$ matrices with positive elements, by I_m the identity $m \times m$ matrix and by O_m the null $m \times m$ matrix.

Definition 19.1 A square matrix $A \in M_{mm}(\mathbb{R}_+)$ is said to be convergent towards zero if its spectral radius $\rho(A)$ is strictly less than 1. In other words, this means that all the eigenvalues of A are in the open unit disc, that is, $|\lambda| < 1$, for every $\lambda \in \mathbb{C}$ with $\det(A - \lambda I_m) = 0$ (see, e.g., [3]).

A classical result in matrix analysis is the following theorem [3].

Theorem 19.2 Let $A \in M_{mm}(\mathbb{R}_+)$. The following assertions are equivalent:

(i) *A is convergent towards zero;*
(ii) *$A^n \to O_m$ as $n \to \infty$;*
(iii) *The matrix $(I_m - A)$ is non-singular and*

$$(I_m - A)^{-1} = I_m + A + \cdots + A^n + \cdots ; \qquad (19.2)$$

(iv) *The matrix $(I_m - A)$ is non-singular and $(I_m - A)^{-1}$ has nonnegative elements;*

Notice also that if $A, B \in M_{mm}(\mathbb{R}_+)$ with $A \leq B$ (in the sense that $a_{ij} \leq b_{ij}$, for all $i, j \in \{1, 2, \dots, m\}$), then $\rho(B) < 1$ implies $\rho(A) < 1$.

Moreover, the following auxiliary result will be useful in the main sections.

Lemma 19.1 Let $A \in \mathcal{M}_{m,m}(\mathbb{R}_+)$ be a matrix convergent to zero. Then, there exists $Q > 1$ such that for any $q \in (1, Q)$ the matrix qA is convergent to 0. In particular, Q could be chosen as $\frac{1}{\rho(A)}$.

For examples and related discussion about the advantages of the vector-valued metric approach in fixed point theory, see [4, 5].

Definition 19.2 **(Bakhtin [6] and Czerwik [7])** Let X be a nonempty set and let $s \geq 1$ be a given real number. A functional $d : X \times X \to \mathbb{R}_+$ is said to be a b-metric with constant s if the classical axioms of the metric are satisfied with the following modification of the triangle inequality axiom

$$d(x, z) \leq s[d(x, y) + d(y, z)] \quad \text{for all} \quad x, y, z \in X.$$

A pair (X, d) with the above properties is called a b-metric space (also called, in some papers, quasimetric spaces or metric type spaces) with constant s.

Some examples of b-metric spaces are given, for example, in [7–9]. For example, the following generic examples are given in [9] (see also [10]) and [11].

Example 19.1 Let (X, d) be a metric space. Then, for any $\beta > 1$, $\lambda \geq 0$ and $\mu > 0$, the functional

$$J(x, y) := \lambda d(x, y) + \mu d(x, y)^\beta, \quad x, y \in X$$

defines a b-metric on X with constant $s = 2^{\beta-1}$, but J is not a metric on X.

Example 19.2 Let E be a Banach space and P a normal cone in E with $int(P) \neq \emptyset$. Denote by "$\leq$" the partially order relation generated by P.

If X is a nonempty set, then a mapping $d : X \times X \to E$ is called a cone metric on X if the usual axioms of the metric take place with respect to "$\leq$."

The cone P is called normal if there is a number $K \geq 1$ such that, for all $x, y \in E$, the following implication holds:

$$0 \leq x \leq y \Rightarrow \|x\| \leq K\|y\|.$$

If the cone P is normal with the coefficient of normality $K \geq 1$, then the functional $\hat{d} : X \times X \to \mathbb{R}_+$ defined by

$$\hat{d}(x, y) := \|d(x, y)\|$$

is a b-metric on X with constant $s = K$.

If (X, d) is a b-metric space, $x_0 \in X$ and $r > 0$, then we denote

$$B(x_0, r) := \{x \in X : d(x_0, x) < r\}, \quad \tilde{B}(x_0, r) := \{x \in X : d(x_0, x) \leq r\}.$$

It is worth to mention that the b-metric on a nonempty set X need not be continuous. Moreover, open balls $B(x_0, r)$ in such spaces need not be open sets. In this context, a set $Y \subset X$ is said to be closed if for any sequence (x_n) in Y which is convergent to some x, we have that $x \in Y$. Moreover, the well-known continuity concepts of continuous operator and operator with closed graph will be also considered in the sequential meaning. For example, $f : X \to X$ is said to have closed graph if for every sequence $\{x_n\}$ in X which converges to x such that the sequence $\{f(x_n)\}$ converges to y as $n \to \infty$, we have that $x \in X$ and $y = f(x)$. Similarly, $f : X \to X$ is said to be orbitally continuous if

$$(\forall\, x \in X)\, f^{n(i)}(x) \to y \in X, \quad i \to \infty \;\Rightarrow\; f^{n(i)+1}(x) \to f(y), \quad i \to \infty.$$

We will recall now the concept of comparison function. A function $\varphi : \mathbb{R}_+ \to \mathbb{R}_+$ is said to be a comparison function [12] if it is increasing and $\varphi^n(t) \to 0$ as $n \to \infty$, for all $t \geq 0$.

Lemma 19.2 ([12]) If $\varphi : \mathbb{R}_+ \to \mathbb{R}_+$ is a comparison function, then $\varphi(t) < t$, for any $t > 0$, $\varphi(0) = 0$ and φ is continuous at 0.

Example 19.3 The following functions $\varphi : \mathbb{R}_+ \to \mathbb{R}_+$ are comparison functions:

(1) $\varphi(t) = \alpha t$, where $\alpha \in (0, 1)$;
(2) $\varphi(t) = \frac{t}{1+t}$;
(3)

$$\varphi(t) = \begin{cases} \frac{1}{2}t, & t \in [0, 1]; \\ t - \frac{1}{2}, & t > 1. \end{cases}$$

For other examples and related results see, for example, [9, 12, 13].

19.2 Fixed Point Results

19.2.1 The Single-Valued Case

Two nice extensions of the CMP were given by Perov [14] (for the case of spaces endowed with a vector-valued metric) and Czerwik [7] (for the case of b-metric spaces). Their results are as follows.

Theorem 19.3 (Perov) Let (X, d) be a complete generalized metric space in the sense of Perov and let $f : X \to X$ be a contraction with matrix A, that is, $A \in M_{mm}(\mathbb{R}_+)$ converges towards zero and

$$d(f(x), f(y)) \leq Ad(x, y) \quad \text{for all}\;\; x, y \in X.$$

Then,

(1) $Fix(f) = \{x^*\}$;
(2) For all $x \in X$, the sequence $\{f^n(x)\}$ converges in (X, d) to x^* as $n \to \infty$;
(3) For all $x \in X$, the following estimation holds

$$d(f^n(x), x^*) \leq (I - A)^{-1} A^n d(x, f(x)), \quad n \in \mathbb{N}.$$

Remark 19.1   If we take $m = 1$ in the above theorem, then we obtain the CMP with $A := \alpha \in (0, 1)$.

Theorem 19.4   **(Czerwik)**   Let (X, d) be a complete b-metric space with constant $s \geq 1$ and let $f : X \to X$ be a nonlinear φ-contraction, that is, there exists a comparison function $\varphi : \mathbb{R}_+ \to \mathbb{R}_+$ such that

$$d(f(x), f(y)) \leq \varphi(d(x, y)) \quad \text{for all } x, y \in X.$$

Then,

(1) $Fix(f) = \{x^*\}$;
(2) For all $x \in X$, the sequence $\{f^n(x)\}$ converges in (X, d) to x^* as $n \to \infty$.

Remark 19.2   If we take $\varphi(t) = \alpha t$, $t \in \mathbb{R}_+$ (where $\alpha \in (0, 1)$), then we obtain a generalization of the CMP.

In terms of Picard operator theory, the conclusions (1) and (2) of the above theorems means that f is a Picard operator.

In the next part of this section, we will present Ran–Reurings type theorems in the context of the above generalizations of metric spaces.

We start by presenting an extension of Perov's theorem in the context of a complete vector-valued metric space endowed with a partial order.

Theorem 19.5   Let X be a nonempty set endowed with a partial order "$\preceq$," such that for every $x, y \in X$ there exists $z \in X$ which is comparable to x and y. Let $d : X \times X \to \mathbb{R}_+^m$ be a complete vector-valued metric on X and $f : X \to X$ be an operator which has closed graph with respect to d and it is increasing with respect to "$\preceq$." Suppose that there exist a matrix $A \in M_{mm}(\mathbb{R}_+)$ convergent towards zero and an element $x_0 \in X$ such that:

(i) $d(f(x), f(y)) \leq Ad(x, y)$, for all $x, y \in X$ with $x \preceq y$.
(ii) $x_0 \preceq f(x_0)$.

Then f is a Picard operator. Moreover, for each $n \in \mathbb{N}$, we have

$$d(f^n(x_0), x^*) \leq (I - A)^{-1} A^n d(x_0, f(x_0)).$$

Proof: Let $x_0 \in X$ such that $x_0 \preceq f(x_0)$. By the monotonicity assumption on f we get that $x_0 \preceq f(x_0) \preceq \cdots \preceq f^n(x_0) \preceq \cdots$. Denote $x_n := f^n(x_0)$, $n \in \mathbb{N}^*$. Then, we have the following results:

(a) $x_{n+1} = f(x_n)$, $n \in \mathbb{N}$;
(b) the sequence $\{x_n\}$ is increasing with respect to $\preceq$.
   By (b) and the contraction condition on f, we obtain
(c) $d(x_n, x_{n+1}) \leq A^n d(x_0, f(x_0))$, for each $n \in \mathbb{N}$.

By (c) and Theorem 19.2 we obtain

$$d(x_n, x_{n+p}) \le d(x_n, x_{n+1}) + \cdots + d(x_{n+p-1}, x_{n+p})$$
$$\le (A^n + \cdots + A^{n+p-1})d(x_0, f(x_0))$$
$$\le (I_m - A)^{-1}A^n d(x_0, f(x_0)) \to 0 \quad \text{as} \quad n \to \infty.$$

Thus, the sequence $\{x_n\}$ converges with respect to d to a certain element $x^* \in X$. By (a) and the closed graph assumption on f we obtain that $x^* \in Fix(f)$. Let $x \in X$ be an arbitrary element. We will show that $\{f^n(x)\}$ converges to x^* as $n \to \infty$, for every $x \in X$. We have two cases:

Case 1. If x is comparable to x_0.
Suppose, for example, $x_0 \preceq x$ (the approach for the reverse case is similar). Then, by the monotonicity assumption we obtain that $f^n(x_0) \preceq f^n(x)$, for every $n \in \mathbb{N}^*$. By the contraction condition, we obtain

$$d(f^n(x_0), f^n(x)) \le A^n d(x_0, x) \to 0 \quad \text{as} \quad n \to \infty.$$

As a consequence, $\{f^n(x)\}$ converges to x^* as $n \to \infty$, for every x comparable to x_0.

Case 2. If x is not comparable to x_0.
Then, by our hypothesis, there exists $z \in X$ which is comparable to x and to x_0. Suppose, for example, that $x_0 \preceq z$ and $x \preceq z$ (the rest of the cases can be treated in a similar way). Then, since $x_0 \preceq z$ we obtain (see Case 1) that $\{f^n(z)\}$ converges to x^* as $n \to \infty$. Now, by $x \preceq z$, the monotonicity assumption on f and the contraction condition on f, we obtain

$$d(f^n(x), f^n(z)) \le A^n d(x, z) \to 0 \quad \text{as} \quad n \to \infty.$$

Thus, $\{f^n(x)\}$ converges to x^* as $n \to \infty$.
Concerning the uniqueness of the fixed point, let us suppose that there exists $u \in Fix(f)$. Then, if u is comparable to x^*, then we obtain the following contradiction

$$0 \le d(u, x^*) = d(f^n(u), f^n(x^*)) \le A^n d(u, x^*) \to 0 \quad \text{as} \quad n \to \infty.$$

If u is not comparable to x^*, then there exists $v \in X$ which is comparable with u and with x^*. Then, we get again a contradiction by the following relations

$$0 \le d(u, x^*) \le d(u, f^n(v)) + d(f^n(v), x^*)$$
$$\le A^n d(u, v) + d(f^n(v), x^*) \to 0 \quad \text{as} \quad n \to \infty.$$

The last conclusion follows, letting $p \to \infty$, in the relation

$$d(x_n, x_{n+p}) \le (I_m - A)^{-1}A^n d(x_0, f(x_0)). \qquad \square$$

Remark 19.3 By the above theorem it follows that, if additionally one suppose that $(X, d, \preceq)$ is an ordered metric space (in the sense that, for any convergent sequences $\{y_n\} \to y^*$ and $\{z_n\} \to z^*$ in (X, d), such that $y_n \preceq z_n$ for each

$n \in \mathbb{N}$, we have $y^* \preceq z^*$), then the following partial retraction–displacement estimation holds

$$d(x, x^*) \leq (I_m - A)^{-1} d(x, f(x)) \quad \text{for each } x \in X \text{ with } x \preceq x_0.$$

In other terms, we say that f is a left Ψ-Picard operator with $\Psi : \mathbb{R}_+^m \to \mathbb{R}_+^m$ given by $\Psi(t) := (I_m - A)^{-1} t$.

In the case of a complete b-metric space, we have the following theorem, proved in [15].

Theorem 19.6 Let X be a nonempty set endowed with a partial order "$\preceq$," such that for every $x, y \in X$ there exists $z \in X$, which is comparable to x and y. Let $d : X \times X \to \mathbb{R}_+$ be a complete b-metric with constant $s \geq 1$ and $f : X \to X$ be an operator which has closed graph with respect to d and it is increasing with respect to "$\preceq$." Suppose that there exist a comparison function $\varphi : \mathbb{R}_+ \to \mathbb{R}_+$ and an element $x_0 \in X$ such that:

(i) $d(f(x), f(y)) \leq \varphi(d(x, y))$, for all $x, y \in X$ with $x \preceq y$;
(ii) $x_0 \preceq f(x_0)$.

Then f is a Picard operator.

Proof: Let $x_0 \in X$ such that $x_0 \preceq f(x_0)$. Let us denote $x_n := f^n(x_0)$, $n \in \mathbb{N}^*$. Then, we have the following results:

(a) $x_{n+1} = f(x_n)$, $n \in \mathbb{N}$;
(b) the sequence (x_n) is increasing with respect to $\preceq$;
(c) for each $n \in \mathbb{N}^*$ we have $d(x_n, x_{n+1}) \leq \varphi^n(d(x_0, f(x_0))) \to 0$ as $n \to \infty$.

Let $\epsilon > 0$. Since $\varphi^n(\epsilon) \to 0$ as $n \to \infty$, there exists $n(\epsilon) > 0$ such that $\varphi^n(\epsilon) < \dfrac{\epsilon}{4s^2}$, for each $n \geq n(\epsilon)$. Let $g := f^{n(\epsilon)}$ and $y_m := g^m(x_0)$, $m \in \mathbb{N}$. Then we have

$$d(y_m, y_{m+1}) = d(f^{n(\epsilon)m}(x_0), f^{n(\epsilon)m}(g(x_0))) \leq \varphi^{n(\epsilon)m}(d(x_0, g(x_0))) \to 0, \quad m \to \infty.$$

Hence, for $\epsilon > 0$ there exists $m(\epsilon) > 0$ such that $d(y_m, y_{m+1}) < \dfrac{\epsilon}{2s}$, for each $m \geq m(\epsilon)$. Let $\tilde{B}(y_{m(\epsilon)}; \epsilon) := \{y \in X \mid d(y, y_{m(\epsilon)}) \leq \epsilon\}$. We will show that $g : \tilde{B}(y_{m(\epsilon)}; \epsilon) \to \tilde{B}(y_{m(\epsilon)}; \epsilon)$. Indeed, let $u \in \tilde{B}(y_{m(\epsilon)}; \epsilon)$. Then

$$d(g(u), y_{m(\epsilon)}) \leq s(d(g(u), g(y_{m(\epsilon)})) + d(g(y_{m(\epsilon)}), y_{m(\epsilon)}))$$
$$= s(d(g(u), g(y_{m(\epsilon)})) + d(y_{m(\epsilon)+1}, y_{m(\epsilon)})).$$

If $u, y_{m(\epsilon)} \in X$ are comparable, then we can write directly

$$d(g(u), g(y_{m(\epsilon)})) \leq \varphi^{n(\epsilon)}(d(u, y_{m(\epsilon)})),$$

if not then there exists $z \in X$, which is comparable with $u, y_{m(\epsilon)}$. Then

$$d(g(u), g(y_{m(\epsilon)})) \leq s(d(g(u), g(z)) + d(g(z), g(y_{m(\epsilon)})))$$
$$\leq s(\varphi^{n(\epsilon)}(d(u, z)) + \varphi^{n(\epsilon)}(d(z, y_{m(\epsilon)}))).$$

Hence,

$$d(g(u), y_{m(\epsilon)}) \le s[s(\varphi^{n(\epsilon)}(d(u, z)) + \varphi^{n(\epsilon)}(d(z, y_{m(\epsilon)}))) + d(y_{m(\epsilon)+1}, y_{m(\epsilon)})] \le \epsilon.$$

As a consequence, for every $i, j \in \mathbb{N}$ with $i, j \ge m(\epsilon)$, we get

$$d(y_i, y_j) \le s(d(y_i, y_{m(\epsilon)}) + d(y_j, y_{m(\epsilon)})) \le 2s\epsilon,$$

which proves that the sequence (y_m) is Cauchy. By the completeness of the space there exists $x^* \in X$ such that $(y_m) \to x^*$ as $m \to \infty$. Since f has closed graph, it follows that g has closed graph too and thus $x^* \in Fix(g)$. Moreover,

$$y_m = g^m(x_0) \to x^* \quad \text{as } m \to \infty.$$

We will show now that for each $x \in X$ we have that $g^m(x) \to x^*$ as $m \to \infty$. Let $x \in X$. We have two cases:

(1) If x and x_0 are comparable, then

$$d(g^m(x), g^m(x_0)) = d(f^{n(\epsilon)m}(x), f^{n(\epsilon)m}(x_0)) \le \varphi^{n(\epsilon)m}(d(x, x_0)) \to 0, \ m \to \infty.$$

(2) If x and x_0 are not comparable, then there exists $w \in X$ which is comparable to x and x_0. Then, we have

$$d(g^m(x), g^m(x_0)) \le s(d(g^m(x), g^m(w)) + d(g^m(w), g^m(x_0)))$$
$$\le s(\varphi^{n(\epsilon)m}(d(x, w)) + \varphi^{n(\epsilon)m}(d(w, x_0))) \to 0, \ m \to \infty.$$

In both cases, we get that $g^m(x) \to x^*$ as $m \to \infty$, for each $x \in X$.

We will show now that x^* is a fixed point for f too. For each $x \in X$, we have

$$\lim_{m \to \infty} f(g^m(x)) = \lim_{m \to \infty} g^m(f(x)) = x^* \quad \text{and} \quad g^m(x) \to x^* \text{ as } n \to \infty.$$

Since f has closed graph, we get that $x^* \in Fix(f)$. The uniqueness of the fixed point follows in a similar way to Theorem 19.5. $\qquad\square$

Remark 19.4 If, in the above theorems, we drop the assumption that for every elements $x, y \in X$ there exists $z \in X$, which is comparable to x and y, then we lose the uniqueness of the fixed point and we can only get that the sequence $\{f^n(x)\}$ converges to x^* as $n \to \infty$, for elements $x \in X$ which are comparable to x_0.

Remark 19.5 The above theorems take place if we replace "increasing" by "decreasing" and we replace the sense of the inequalities between elements.

Remark 19.6 Notice that the approach of the proof in Theorem 19.6 does not implies an apriori estimation for the fixed point nor a retraction–displacement condition for the operator f.

In particular, if we impose the linear contraction condition on f, we get the following result, see also [6].

Theorem 19.7 Let X be a nonempty set endowed with a partial order "$\preceq$," such that for every $x, y \in X$ there exists $z \in X$ which is comparable to x and y. Let $d : X \times X \to \mathbb{R}_+$ be a complete b-metric with constant $s \geq 1$ and $f : X \to X$ be an operator which has closed graph with respect to d and is increasing with respect to "$\preceq$." Suppose that there exist $\alpha \in (0, 1)$ and an element $x_0 \in X$ such that:

(i) $d(f(x), f(y)) \leq \alpha d(x, y)$, for all $x, y \in X$ with $x \preceq y$;
(ii) $x_0 \preceq f(x_0)$.

Then,

(i) f is a Picard operator and $x^* \in X$ denotes the unique fixed point of f.
(ii) if, additionally $\alpha < \frac{1}{s}$ and the b-metric d is continuous, then we have that

$$d(f^n(x_0), x^*) \leq \frac{\alpha^n s}{1 - \alpha s} d(x_0, f(x_0)), \quad \text{for each } n \in \mathbb{N}.$$

In particular, the following retraction–displacement condition at the point x_0 holds

$$d(x_0, x^*) \leq \frac{s}{1 - \alpha s} d(x_0, f(x_0)).$$

(iii) If, additionally, $\alpha < \frac{1}{s}$ and the space X is an ordered metric space with respect to d and $\preceq$ (in the sense that for any convergent sequences $\{y_n\} \to y^*$ and $\{z_n\} \to z^*$ in (X, d), such that $y_n \preceq z_n$ for each $n \in \mathbb{N}$, we have $y^* \preceq z^*$), then f is a left ψ-Picard operator with $\psi : \mathbb{R}_+ \to \mathbb{R}_+$, $\psi(t) := \frac{s}{1-\alpha s} t$, that is,

$$d(x, x^*) \leq \frac{s}{1 - \alpha s} d(x, f(x)), \quad \text{for every } x \in X \text{ with } x \preceq x_0.$$

Proof: Conclusion (i) follows by Theorem 19.6.

For the second conclusion, notice first that if we denote, as before, by $x_n := f^n(x_0)$ for $n \in \mathbb{N}$, then $d(x_n, x_{n+1}) \leq \alpha^n d(x_0, f(x_0))$, for each $n \in \mathbb{N}$. Then, by the triangle inequality, we obtain, for $n \in \mathbb{N}$ and $p \in \mathbb{N}^*$, that

$$d(x_n, x_{n+p}) \leq \alpha^n s \frac{1 - (\alpha s)^p}{1 - \alpha s} d(x_0, f(x_0)).$$

Since $x_n \to x^*$ as $n \to \infty$, the conclusions of (ii) follow by passing to the limit in the above estimation.

(iii) Let $x \in X$ with $x \preceq x_0$. Then $x \preceq f^n(x_0)$, for every $n \in \mathbb{N}$. By the fact that $(X, d, \preceq)$ is an ordered metric space we obtain that $x \preceq x^*$. Now the conclusion follows by the following relation

$$d(x, x^*) \leq s(d(x, f(x)) + d(f(x), f(x^*))) \leq s(d(x, f(x)) + \alpha d(x, x^*)).$$

Remark 19.7 If we replace the assumption $x_0 \preceq f(x_0)$ by $f(x_0) \preceq x_0$, then, in the conditions of the above theorem and of the additional assumptions in (iii), we obtain that f is a right ψ-Picard operator with $\psi : \mathbb{R}_+ \to \mathbb{R}_+, \psi(t) := \frac{s}{1-\alpha s}t$, that is,

$$d(x, x^*) \le \frac{s}{1 - \alpha s} d(x, f(x)), \quad \text{for every } x \in X \text{ with } x_0 \le x.$$

Remark 19.8 Theorem 19.7 extends, to the case of b-metric spaces, Ran–Reurings theorem in [16], as well as Theorem 3.1 in [17], where a restriction on the constant α is imposed.

Remark 19.9 Notice that a Ran–Reurings type theorem in generalized b-metric spaces in the sense of Perov can be proved by imposing, in a similar way to Theorem 19.5, the assumption that the matrix sA is convergent toward zero.

A more general approach can be given in terms of the attraction basis of the fixed points. For this purpose, we briefly recall the notion of L-space, see [2].

Let X be a nonempty set and $s(X) := \{(x_n)_{n\in N} | x_n \in X, \quad n \in N\}$. Let $c(X) \subset s(X)$ and $Lim : c(X) \to X$ be an operator. By definition, the triple $(X, c(X), Lim)$ is called an L-space (briefly denoted by $(X, \to)$) if the following axioms are satisfied:

(i) If $x_n = x$, for all $n \in N$, then $(x_n)_{n\in N} \in c(X)$ and $Lim(x_n)_{n\in N} = x$.
(ii) If $(x_n)_{n\in N} \in c(X)$ and $Lim(x_n)_{n\in N} = x$, then for all subsequences, $(x_{n_i})_{i\in N}$, of $(x_n)_{n\in N}$ we have that $(x_{n_i})_{i\in N} \in c(X)$ and $Lim(x_{n_i})_{i\in N} = x$.

By definition, an element of $c(X)$ is a convergent sequence, while $x := Lim(x_n)_{n\in N}$ is the limit of this sequence. We will denote this fact by $x_n \to x$ as $n \to +\infty$. Actually, an L-space is a nonempty set endowed with a convergence structure. For examples and related results in fixed point theory see [18].

Recall that, if $(X, \to)$ is an L-space and $f : X \to X$ is an operator, then we denote by

$$A_f(x^*) := \{x \in X : f^n(x) \to x^* \text{as } n \to \infty\}$$

the attraction basin of a point $x^* \in X$.

Then, the following general result holds, see [19].

Theorem 19.8 Let $(X, \to)$ be an L-space, $U \subset X \times X$ be a symmetric set, such that the diagonal $\Delta(X)$ of $X \times X$ is a subset of U. Let $f : X \to X$ be an operator. We suppose:

(i) for every $x, y \in X$ there exists $z \in X$ such that $(x, z), (z, y) \in U$;
(ii) there exists $x_0, x^* \in X$ such that $x_0 \in A_f(x^*)$;
(iii) if $(x, y) \in U$ and $x \in A_f(x^*)$, then $y \in A_f(x^*)$.
(iv) f is orbitally continuous.

Then f is a Picard operator.

In particular, if we endow the L-space $(X, \rightarrow)$ by a partial order relation $\preceq$, then we will denote by

$$X_{\preceq} := \{(x, y) \in X \times X : x \preceq y \text{ or } y \preceq x\}$$

the set of all pairs in $X \times X$ with comparable elements.

Then, by Theorem 19.8, the following abstract result takes place.

Theorem 19.9 Let $(X, \rightarrow)$ be an L-space endowed with a partial order relation $\preceq$ and let $f : X \rightarrow X$ be an operator. We suppose:

(i) for every $x, y \in X$ there exists $z \in X$ such that z is comparable to x and to y;

(ii) there exists $x_0, x^* \in X$ such that $x_0 \in A_f(x^*)$;

(iii) if $(x, y) \in X_{\preceq}$ and $x \in A_f(x^*)$, then $y \in A_f(x^*)$.

(iv) f is orbitally continuous.

Then f is a Picard operator.

Proof: Take $U := X_{\preceq}$ in Theorem 19.8. $\qquad\qquad\square$

Moreover, the following general fixed point theorems in a complete generalized metric spaces in the sense of Perov, respectively in complete b-metric spaces, endowed with a partial order relation $\preceq$ can be proved. For the convenience of the reader we will only prove the second one.

Theorem 19.10 Let (X, d) be a complete generalized metric space in the sense of Perov endowed with a partial order relation $\preceq$ and let $f : X \rightarrow X$ be an operator. We suppose:

(i) for every $x, y \in X$ there exists $z \in X$ such that z is comparable to x and to y;

(ii) $X_{\preceq}$ is invariant with respect to $f \times f$, that is, $(f \times f)(X_{\preceq}) \subseteq X_{\preceq}$;

(iii) there exists $x_0 \in X$ such that $(x_0, f(x_0)) \in X_{\preceq}$;

(iv) there exists a matrix $A \in M_{mm}(\mathbb{R}_+)$ convergent to zero, such that

$$d(f(x), f(y)) \leq Ad(x, y), \quad \text{for all } (x, y) \in X_{\preceq}.$$

(v) f is orbitally continuous.

Then f is a Picard operator.

Theorem 19.11 Let (X, d) be a complete b-metric space endowed with a partial order relation $\preceq$ and let $f : X \rightarrow X$ be an operator. We suppose:

(i) for every $x, y \in X$ there exists $z \in X$ such that z is comparable to x and to y;

(ii) $X_{\preceq}$ is invariant with respect to $f \times f$, that is, $(f(x), f(y)) \in X_{\preceq}$, for each $(x, y) \in X_{\preceq}$;

(iii) there exists $x_0 \in X$ such that $(x_0, f(x_0)) \in X_{\preceq}$;

(iv) there exists a comparison function $\varphi : \mathbb{R}_+ \to \mathbb{R}_+$ such that
$$d(f(x), f(y)) \le \varphi(d(x, y)), \quad \text{for every } (x, y) \in X_{\le};$$

(v) f is orbitally continuous.

Then f is a Picard operator.

Proof: By (iii) and (ii), we obtain that $(f^n(x_0), f^{n+1}(x_0)) \in X_{\le}$, for each $n \in \mathbb{N}$. By this fact and the assumption (iv) we can prove, in a similar way to the proof of Theorem 19.6, that the sequence $x_n := f^n(x_0)$ is Cauchy. Thus, by the completeness of the space (X, d), there exists $x^* \in X$ such that $x_0 \in A_f(x^*)$. Notice also that if $(x, y) \in X_{\le}$, then, by (ii), we get that $(f^n(x), f^n(y)) \in X_{\le}$, for each $n \in \mathbb{N}$. In this case, if $x \in A_f(x^*)$ (i.e., $f^n(x) \to x^*$ as $n \to \infty$), then, by (iii), we obtain that $f^n(y) \to x^*$ as $n \to \infty$. Thus, all the assumption in Theorem 19.9 are satisfied and our theorem follows by Theorem 19.9. □

Remark 19.10 The above theorems are extensions of several results in the literature, including Theorems 19.5 and 19.6 from above. For example, if f is monotone (increasing or decreasing), then $X_{\le}$ is invariant with respect to $(f \times f)$, but the reverse implication does not always holds.

Remark 19.11 For related results concerning monotony arguments in operator equation theory see [20–22]. For the fixed point theory for mappings defined on a Cartesian product see [23].

19.2.2 The Multi-Valued Case

In this section, we will present some fixed point results for multi-valued operators, which are then applied to coupled fixed point problems.

Let (X, d) be a b-metric space and $P(X)$ be the family of all nonempty subsets of X. By $P_{cl}(X)$, we denote the family of all nonempty closed subsets of X, while $P_{cp}(X)$ denotes the family of all nonempty compact subsets of X.

We recall now some useful notions and results.

We consider, for $A, B \in P(X)$, the following generalized functionals:

(a) The gap functional generated by d:
$$D_d(A, B) := \inf\{d(a, b) \mid a \in A, \ b \in B\}. \tag{19.3}$$

(b) The excess functional of A over B generated by d:
$$e_d(A, B) := \sup\{D_d(a, B) \mid a \in A\}. \tag{19.4}$$

(c) The Hausdorff–Pompeiu functional generated by d:
$$H_d(A, B) := \max\{e_d(A, B), e_d(B, A)\}. \tag{19.5}$$

Some useful properties of these functionals are recalled [7] in the next lemma.

Lemma 19.3 If (X, d) is b-metric space with constant $s \geq 1$, then we have the following properties:

(a)
$$d(x_0, x_n) \leq sd(x_0, x_1) + \cdots + s^{n-1}d(x_{n-2}, x_{n-1}) + s^{n-1}d(x_{n-1}, x_n), \quad \forall n \in \mathbb{N}^*;$$

(b)
$$D_d(x, A) \leq s[d(x, y) + D_d(y, A)], \quad \text{for all } x, y \in X \text{and } A \in P(X);$$

(c)
$$\text{if } A \in P_{cl}(X) \text{and } x \in X \text{are such that } D_d(x, A) = 0, \text{then } x \in A.$$

(d)
$$\text{if } A, B \in P(X), \text{then for every } a \in A \text{there exists } b \in B$$

$$\text{such that } d(a, b) \leq sH_d(A, B).$$

Remark 19.12 By Czerwik [7] we also notice:

(1) If (X, d) is a b-metric space with constant $s \geq 1$, then $(P_{cl}(X), H_d)$ is a generalized b-metric space in the sense of Luxemburg–Jung (i.e., $H(A, B) \in [0, \infty]$) with the same constant $s \geq 1$, while $(P_{cp}(X), H_d)$ is a b-metric space with the same constant $s \geq 1$.

(2) If (X, d) is a complete b-metric space with constant $s \geq 1$, then $(P_{cl}(X), H_d)$ is a complete generalized b-metric space in the sense of Luxemburg–Jung with the same constant $s \geq 1$, while $(P_{cp}(X), H_d)$ is a complete b-metric space with the same constant $s \geq 1$.

If X is a nonempty set, then a fixed point for a multi-valued operator $F : X \to P(X)$ is an element $x^* \in X$ with the property $x^* \in F(x^*)$. We will denote by $Fix(F)$ the fixed point set of F.

Let us recall now a fixed point result for a multi-valued contraction in a complete generalized metric space in the sense of Perov.

Notice that if $d : X \times X \to \mathbb{R}^m$ is called a vector-valued metric on a nonempty set X and

$$d(x, y) := \begin{pmatrix} d_1(x, y) \\ \cdots \\ d_m(x, y) \end{pmatrix},$$

then the vector-valued Hausdorff–Pompeiu generalized metric on $P_{cl}(X)$ is given by

$$H_d(A, B) := \begin{pmatrix} H_{d_1}(A, B) \\ \cdots \\ H_{d_m}(A, B) \end{pmatrix}.$$

If $F : X \rightarrow P(X)$ is a multi-valued operator, then $Fix(F) := \{x \in X : x \in F(x)\}$ will denote the fixed point set for F.

In the above notations, we have the following result, which follows by Theorem 3.13 in [24].

Theorem 19.12 Let (X, d) be a complete generalized metric space in the sense of Perov and let $F : X \rightarrow P_{cl}(X)$ be a multivalued A-contraction, that is, $A \in M_{mm}(\mathbb{R}_+)$ converges to zero and

$$H_d(F(x), F(y)) \leq Ad(x, y), \quad \text{for all } x, y \in X. \tag{19.6}$$

Then,

(i) $Fix(F) \neq \emptyset$;
(ii) for each $(x, y) \in Graph(F)$ there exists a sequence $\{x_n\}_{n\in\mathbb{N}}$ of successive approximations for F (i.e., $x_0 = x, x_1 = y$ and $x_{n+1} \in F(x_n)$, for each $n \in \mathbb{N}^*$) such that $\{x_n\}_{n\in\mathbb{N}}$ is convergent to a fixed point $x^* := x^*(x, y)$ of F;
(iii) for the above defined sequence $\{x_n\}_{n\in\mathbb{N}}$ and its limit $x^*(x, y)$, the following estimations hold

$$d(x, x^*) \leq (I - A)^{-1}d(x, y)$$

and

$$d(x_n, x^*) \leq A^n(I - A)^{-1}d(x_0, x_1), \quad \text{for each } n \in \mathbb{N}^*.$$

Remark 19.13 In terms of the multi-valued weakly Picard operator theory, the conclusions (i)+(ii) of the above theorem is that F is a multi-valued weakly Picard operator. Moreover, in the same terms, the conclusions (i)–(iii) of the above theorems mean that F is a multi-valued weakly ψ-Picard operator with $\psi : \mathbb{R}_+^m \rightarrow \mathbb{R}_+^m$ given by $\psi(t) := (I - A)^{-1}t$.

We will recall now some fixed point theorems for multi-valued contractions in b-metric spaces. The following two theorems, proved by Czerwik [25], are generalizations, to the case of complete b-metric spaces, of the well-known fixed point theorems of Nadler [26] and Węgrzyk [27].

Theorem 19.13 **(Czerwik)** Let (X, d) be a complete b-metric space with constant $s \geq 1$ and $F : X \rightarrow P_{cl}(X)$ be a multivalued operator. Suppose there exists $k \in (0, \frac{1}{s})$ such that

$$H_d(F(x), F(y)) \leq kd(x, y), \quad \text{for all } x, y \in X.$$

Then, F is a multivalued weakly Picard operator, that is, $Fix(F) \neq \emptyset$ and, for every $(x, y) \in Graph(F)$, there is a sequence of successive approximations for F starting from (x, y), which converges to $x^*(x, y) \in Fix(F)$.

Theorem 19.14 **(Czerwik)** Let (X, d) be a complete b-metric space with constant $s \geq 1$ and $F : X \to P_{cp}(X)$ be a multivalued operator. Suppose that d is continuous and there exists a comparison function $\varphi : \mathbb{R}_+ \to \mathbb{R}_+$ such that

$$H_d(F(x), F(y)) \leq \varphi(d(x, y)), \quad \text{for all } x, y \in X.$$

Then, $Fix(F) \neq \emptyset$.

In the next part of this section, we will prove some Ran–Reurings type fixed point theorems for multi-valued operators in the context of a complete generalized metric space in the sense of Perov and respectively in ordered and complete b-metric spaces.

If X is a nonempty set and $F : X \to P(X)$ is a multi-valued operator, then the Cartesian product of $F \times F$ is defined by

$$F \times F : X \times X \to P(X \times X), \quad \text{given by } (F \times F)(x_1, x_2) := F(x_1) \times F(x_2).$$

The first main results of this section are some Perov type theorems for multi-valued operators in a vector-valued metric space.

Theorem 19.15 Let X be a nonempty set endowed with a partial order "$\preceq$" and let $d : X \times X \to \mathbb{R}_+^m$ be a complete vector-valued metric on X. Suppose that the multi-valued operator $F : X \to P(X)$ satisfy the following assumptions:

(1) $X_{\preceq} \in I(F \times F)$;
(2) F has closed graph with respect to d;
(3) there exists $(x_0, x_1) \in Graph(F)$ such that $(x_0, x_1) \in X_{\preceq}$;
(4) there exists a matrix $A \in M_{mm}(\mathbb{R}_+)$ convergent to zero such that

$$H_d(F(x), F(y)) \leq Ad(x, y), \quad \text{for each } (x, y) \in X_{\preceq}.$$

Then, F has at least one fixed point $x^* \in X$ and there exists a sequence $\{x_n\}_{n \in \mathbb{N}}$ of successive approximations for F starting from $(x_0, x_1) \in Graph(F)$, which converges to x^*. Moreover, the following estimation holds:

$$d(x_n, x^*) \leq A^n (I_m - A)^{-1} d(x_0, x_1), \quad \text{for all } n \in \mathbb{N}.$$

Proof: Let $x_0 \in X$ and $x_1 \in F(x_0)$ such that $(x_0, x_1) \in X_{\preceq}$. Let Q as in Lemma 19.1. For $1 < q < Q$, by Lemma 19.3 (d), there exists $x_2 \in F(x_1)$ such that

$$d(x_1, x_2) \leq q H_d(F(x_0), F(x_1)) \leq q Ad(x_0, x_1).$$

Since $(x_0, x_1) \in X_{\preceq}$, by (2), we have that $F(x_0) \times F(x_1) \subset X_{\preceq}$. It follows that $(x_1, x_2) \in X_{\preceq}$ and so we obtain (by (2)) that $F(x_1) \times F(x_2) \subset X_{\preceq}$. Let $x_3 \in F(x_2)$ such that

$$d(x_2, x_3) \leq q H_d(F(x_1), F(x_2)).$$

Thus, $(x_2, x_3) \in X_{\leq}$ and so

$$d(x_2, x_3) \leq qAd(x_1, x_2) \leq (qA)^2 d(x_0, x_1).$$

We construct the sequence $\{x_n\}_{n \in \mathbb{N}}$ of successive approximations of F starting from $x_0 \in X$ such that:

(a) $x_{n+1} \in F(x_n)$, for $n \in \mathbb{N}$;
(b) $(x_n, x_{n+1}) \in X_{\leq}$, for $n \in \mathbb{N}$;
(c) $d(x_n, x_{n+1}) \leq (qA)^n d(x_0, x_1)$, for $n \in \mathbb{N}^*$.

We prove next that the sequence $\{x_n\}$ is Cauchy. We have

$$d(x_n, x_{n+p}) \leq d(x_n, x_{n+1}) + d(x_{n+1}, x_{n+2}) + \cdots + d(x_{n+p-1}, x_{n+p}) \leq$$
$$\leq (qA)^n d(x_0, x_1) + (qA)^{n+1} d(x_0, x_1) + \cdots + (qA)^{n+p-1} d(x_0, x_1).$$

Since qA is convergent to zero, we obtain that $(I_m - qA)$ is nonsingular and

$$(I_m - qA)^{-1} = I_m + qA + \cdots + (qA)^p + \cdots .$$

Thus, by the above relation, for $n \in \mathbb{N}$ and $p \in \mathbb{N}^*$, we get that

$$d(x_n, x_{n+p}) \leq (qA)^n (I_m - qA)^{-1} d(x_0, x_1). \tag{19.7}$$

Again by the fact that the matrix qA is convergent to zero, the sequence (x_n) is Cauchy. Since the metric space (X, d) is complete, there exists $x^* \in X$ such that $x_n \to x^*$ as $n \to \infty$.

Using the assumption (2) from the hypothesis, we obtain that $x^* \in F(x^*)$ and thus $Fix(F) \neq \emptyset$. The second conclusion follows by letting $p \to \infty$ and $q \searrow 1$ in (19.7). $\qquad \square$

Following an idea from Dinevari and Frigon [28], we can obtain the following more general version of the above theorem.

Theorem 19.16 Let X be a nonempty set endowed with a partial order "$\preceq$" and let $d : X \times X \to \mathbb{R}^m_+$ be a complete vector-valued metric on X. Suppose that the multi-valued operator $F : X \to P(X)$ satisfy the following assumptions:

(1) There exists $(x_0, x_1) \in Graph(F) \cap X_{\leq}$.
(2) There exists a matrix $A \in M_{mm}(\mathbb{R}_+)$ convergent to zero such that for every $(x, y) \in X_{\leq}$ and each $u \in F(x)$ there exists $v \in F(y)$ such that $(u, v) \in X_{\leq}$ and $d(u, v) \leq Ad(x, y)$.
(3) F is $X_{\leq}$-Picard continuous from x_0, that is, the limit of any convergent sequence $(x_n)_{n \in \mathbb{N}}$, with $(x_n, x_{n+1}) \in Graph(F) \cap X_{\leq}$ for $n \in \mathbb{N}$, is a fixed point of F.

Then, F has at least one fixed point $x^* \in X$ and there exists a sequence $\{x_n\}_{n \in \mathbb{N}}$ of successive approximations for F starting from $(x_0, x_1) \in Graph(F)$, which converges to x^*. Moreover, the following estimation holds:

$$d(x_n, x^*) \leq A^n (I_m - A)^{-1} d(x_0, x_1), \quad \text{for all } n \in \mathbb{N}.$$

Proof: Since $(x_0, x_1) \in Graph(F) \cap X_\le$, by (2), there exists $x_2 \in F(x_1)$ such that $(x_1, x_2) \in X_\le$ and $d(x_1, x_2) \le Ad(x_0, x_1)$. Similarly, since $(x_1, x_2) \in Graph(F) \cap X_\le$ there exists $x_3 \in F(x_2)$ such that $(x_2, x_3) \in X_\le$ and $d(x_2, x_3) \le Ad(x_1, x_2) \le A^2 d(x_0, x_1)$. By this procedure, we inductively obtain a sequence $\{x_n\}_{n \in \mathbb{N}}$ of successive approximations of F starting from $(x_0, x_1) \in Graph(F)$ such that:

(a) $x_{n+1} \in F(x_n)$, for $n \in \mathbb{N}$;
(b) $(x_n, x_{n+1}) \in X_\le$, for $n \in \mathbb{N}$;
(c) $d(x_n, x_{n+1}) \le A^n d(x_0, x_1)$, for $n \in \mathbb{N}^*$.

As in the proof of the previous theorem, we immediately obtain that the sequence $\{x_n\}$ is Cauchy. Since the metric space (X, d) is complete, there exists $x^* \in X$ such that $x_n \to x^*$ as $n \to \infty$. Using the assumption (3) from the hypotheses, we obtain that $x^* \in F(x^*)$ and thus $Fix(F) \ne \emptyset$.

Moreover, for $n \in \mathbb{N}$ and $p \in \mathbb{N}^*$, we also get that

$$d(x_n, x_{n+p}) \le A^n (I_m - A)^{-1} d(x_0, x_1). \tag{19.8}$$

The second conclusion follows by letting $p \to \infty$ in (19.8). $\quad\square$

We will discuss now the case of Ran–Reurings type theorems for multi-valued operators in complete b-metric spaces. Following the same pattern as before we can prove the following results.

Theorem 19.17 Let X be a nonempty set endowed with a partial order "$\le$" and let $d : X \times X \to \mathbb{R}_+$ be a complete b-metric with constant $s \ge 1$ on X. Suppose that the multi-valued operator $F : X \to P(X)$ satisfy the following assumptions:

(1) $X_\le \in I(F \times F)$;
(2) F has closed graph with respect to d;
(3) there exists $(x_0, x_1) \in Graph(F) \cap X_\le$;
(4) there exists $\alpha \in (0, \frac{1}{s})$ such that

$$H_d(F(x), F(y)) \le \alpha d(x, y), \quad \text{for each } (x, y) \in X_\le.$$

Then, F has at least one fixed point $x^* \in X$ and there exists a sequence $\{x_n\}_{n \in \mathbb{N}}$ of successive approximations for F starting from $(x_0, x_1) \in Graph(F)$ which converges to x^*. Moreover, the following estimation holds:

$$d(x_n, x^*) \le \frac{s\alpha^n}{1 - \alpha s} d(x_0, x_1), \quad \text{for all } n \in \mathbb{N}.$$

Proof: Let $x_0 \in X$ and $x_1 \in F(x_0)$ such that $(x_0, x_1) \in X_\le$. For $1 < q < \frac{1}{\alpha s}$, by Lemma 19.3 (d), there exists $x_2 \in F(x_1)$ such that

$$d(x_1, x_2) \le q H_d(F(x_0), F(x_1)) \le q \alpha d(x_0, x_1).$$

Since $(x_0, x_1) \in X_{\leq}$, by (2), we have that $F(x_0) \times F(x_1) \subset X_{\leq}$. It follows that $(x_1, x_2) \in X_{\leq}$ and so we obtain (by (2)) that $F(x_1) \times F(x_2) \subset X_{\leq}$. Let $x_3 \in F(x_2)$ such that

$$d(x_2, x_3) \leq q H_d(F(x_1), F(x_2)).$$

Thus, $(x_2, x_3) \in X_{\leq}$ and so

$$d(x_2, x_3) \leq q\alpha d(x_1, x_2) \leq (q\alpha)^2 d(x_0, x_1).$$

We construct the sequence $\{x_n\}_{n \in \mathbb{N}}$ of successive approximations of F starting from $x_0 \in X$ such that:

(a) $x_{n+1} \in F(x_n)$, for $n \in \mathbb{N}$;
(b) $(x_n, x_{n+1}) \in X_{\leq}$, for $n \in \mathbb{N}$;
(c) $d(x_n, x_{n+1}) \leq (q\alpha)^n d(x_0, x_1)$, for $n \in \mathbb{N}^*$.

We prove next that the sequence $\{x_n\}$ is Cauchy. Indeed, by Lemma 19.3, for each $n \in \mathbb{N}$ and $p \in \mathbb{N}^*$, we obtain

$$d(x_n, x_{n+p}) \leq s d(x_n, x_{n+1}) + s^2 d(x_{n+1}, x_{n+2}) + \cdots + s^{p-1} d(x_{n+p-1}, x_{n+p}) \leq$$
$$\leq s(q\alpha)^n d(x_0, x_1) + \cdots + s^{p-1}(q\alpha)^{n+p-1} d(x_0, x_1) \leq$$
$$\leq s(q\alpha)^n \frac{1}{1 - sq\alpha} d(x_0, x_1).$$

This shows that the sequence $\{x_n\}$ is Cauchy. Denote by $x^* \in X$ its limit. By the closed graph assumption on F we get that $x^* \in Fix(F)$. Letting $p \to \infty$ and $q \searrow 1$ in the above relation we obtain the second conclusion. $\qquad \square$

Following again the idea from Dinevari and Frigon [28], we can obtain the following more general version of the above theorem.

Theorem 19.18 Let X be a nonempty set endowed with a partial order "$\leq$" and let $d : X \times X \to \mathbb{R}_+$ be a complete b-metric with constant $s \geq 1$ on X. Suppose that the multi-valued operator $F : X \to P(X)$ satisfy the following assumptions:

(1) there exists $(x_0, x_1) \in Graph(F) \cap X_{\leq}$;
(2) there exists $\alpha \in (0, \frac{1}{s})$ such that for every $(x, y) \in X_{\leq}$ and each $u \in F(x)$ there exists $v \in F(y)$ such that $(u, v) \in X_{\leq}$ and $d(u, v) \leq \alpha d(x, y)$;
(3) F is $X_{\leq}$-Picard continuous from x_0, that is, the limit of any convergent sequence $(x_n)_{n \in \mathbb{N}}$, with $(x_n, x_{n+1}) \in Graph(F) \cap X_{\leq}$ for $n \in \mathbb{N}$, is a fixed point of F.

Then, F has at least one fixed point $x^* \in X$ and there exists a sequence $\{x_n\}_{n \in \mathbb{N}}$ of successive approximations for F starting from $(x_0, x_1) \in Graph(F)$ which converges to x^*. Moreover, the following estimation holds:

$$d(x_n, x^*) \leq \frac{s\alpha^n}{1 - \alpha s} d(x_0, x_1), \quad \text{for all } n \in \mathbb{N}.$$

Remark 19.14 We may observe that a Ran–Reurings type theorem in generalized b-metric spaces in the sense of Perov for multi-valued operators can be proved by imposing, in a similar way to Theorems 19.15 and 19.16, the assumption that the matrix sA is convergent toward zero.

Remark 19.15 We notice that an extended study of the fixed point problem (data dependence of the fixed points, well-posedness, Ulam–Hyers stability, and Ostrowski stability) can be considered as well. For other details and results on this topic see [29–31].

Remark 19.16 In the next section, the results of this section will be applied for the study of so-called coupled fixed point problems. The fixed point results given above are also an appropriate tool for the study of the following more general system of operator equations (or inclusions):

$$\begin{cases} x = V_1(x,y) \\ y = V_2(x,y). \end{cases}$$

19.3 Coupled Fixed Point Results

19.3.1 The Single-Valued Case

The study of the coupled fixed point problems goes back in 1970s with some papers of Amann [32], Opoitsev [33–35] and Ziebur [36, 37], while the topic get a strong development starting to the works of Guo and Lakshmikantham [38] and Bhaskar and Lakshmikantham [39]. For related results concerning coupled fixed point theory see [39–46] and many other papers.

The coupled fixed point problem can be enunciated as follows: if X is a nonempty set and $V : X \times X \to X$ is a single-valued operator, find $(x,y) \in X \times X$ satisfying

$$\begin{cases} x = V(x,y) \\ y = V(y,x). \end{cases} \tag{19.9}$$

An important related problem is to find the fixed points of V, that is, the elements $x \in X$ with the property

$$x = V(x,x) \tag{19.10}$$

Notice that $V(x,x)$ is also called a diagonal operator, see [47] for a detailed study.

The following examples illustrate this approach.

Example 19.4 (**Krasnoselskii, see [43]**) Let X be a Banach space, Y be a nonempty, bounded, closed and convex subset of X and $f, g : Y \to Y$ be two operators. We suppose that:

Then, for all $(x, y), (u, v) \in X \times X$, we have that

$$\max\{d(V(x, y), V(u, v)), d(V(y, x), V(v, u))\} \leq \varphi(\max\{d(x, u), d(y, v)\}).$$

As before, if we denote $Z := X \times X$ and $F : Z \to Z, F(x, y) := (V(x, y), V(y, x))$, then Z and F satisfy all the assumptions in Theorem 19.4. Hence, the conclusion follows applying Theorem 19.4. □

Using the vector-valued metric approach, we can prove the following result.

Theorem 19.21 Let $(X, \preceq)$ be a partially ordered set and let $d : X \times X \to \mathbb{R}_+$ be a complete metric on X. Let $V : X \times X \to X$ be an orbitally continuous operator. Assume that the following conditions are satisfied:

(i) there exist $\alpha, \beta \in \mathbb{R}_+$ such that $\alpha + \beta < 1$ and for all $(x, y), (u, v) \in X \times X$ with $(x \preceq u, y \succeq v)$ or $(x \succeq u, y \preceq v)$ we have that

$$d(V(x, y), V(u, v)) \leq \alpha d(x, u) + \beta d(y, v);$$

(ii) there exist $x_0, y_0 \in X$ such that $x_0 \preceq V(x_0, y_0)$ and $y_0 \succeq V(y_0, x_0)$;
(iii) for every $(x, y), (u, v) \in X \times X$ with $x \preceq u$ and $y \succeq v$ (or reversely) we have that $V(x, y) \preceq V(u, v)$ and $V(y, x) \succeq V(v, u)$ (or reversely);
(iv) for every $(x, y), (u, v) \in X \times X$ there exists $(p, q) \in X \times X$ such that one have $(x \preceq p$ and $y \succeq q$ (or reversely)) and $(u \preceq p$ and $v \succeq q$ (or reversely)).

Then, there exists a unique solution $(x^*, y^*) \in X \times X$ of the coupled fixed point problem (19.9), such that the sequences $\{x_n\}_{n \in \mathbb{N}}, \{y_n\}_{n \in \mathbb{N}}$ in X defined, for $n \in \mathbb{N}$, by

$$\begin{cases} x_{n+1} = V(x_n, y_n), \\ y_{n+1} = V(y_n, x_n), \end{cases} \tag{19.15}$$

have the property that $\{x_n\} \to x^*$, $\{y_n\} \to y^*$ as $n \to \infty$.

Proof: We denote $Z := X \times X$ and consider on Z the following vector-valued metric $\tilde{d} : Z \times Z \to \mathbb{R}^2$ given by

$$\tilde{d}((x, y), (u, v)) := \begin{pmatrix} d(x, u) \\ d(y, v) \end{pmatrix}.$$

Then the pair $(Z, \tilde{d})$ is a generalized metric space in the sense of Perov.

Let $F : Z \to Z$ be defined as above by $F(x, y) := (V(x, y), V(y, x))$. Then, by (i), if we denote $z = (x, y)$ and $w = (u, v)$, we have that

$$\tilde{d}(F(z), F(w)) \leq A\tilde{d}(z, w), \quad \text{for all } (z, w) \in Z_{\preceq},$$

where $A := \begin{pmatrix} \alpha & \beta \\ \beta & \alpha \end{pmatrix}$ and $Z_{\preceq} := \{(z, w) \in Z \times Z : z \preceq_p w \text{ or } w \preceq_p z\}$. Moreover, by (i), the matrix A is convergent to zero.

By (ii) we obtain that $z_0 := (x_0, y_0) \in Z$ has the property that $z_0 \preceq_p F(z_0)$.

By (iii) we obtain that $(F \times F)(Z_{\le}) \subset Z_{\le}$. Thus, our conclusion follows by applying Theorem 19.10. $\qquad\square$

Remark 19.17 In particular, the mixed monotone property for V implies that the assumption (iii) of the above theorem takes place.

Remark 19.18 It is not difficult to prove a similar theorem in the context of a b-metric space (X, d) with constant $s \ge 1$. The only differences are the following:

- the space $Z : X \times X$ will be a generalized b-metric space in the sense of Perov.
- the result will follow by applying a Ran–Reurings type theorem in generalized b-metric spaces in the sense of Perov, see Remark 19.9.

19.3.2 The Multi-Valued Case

In this section, we will prove some coupled fixed point theorems in the multi-valued setting, by applying the fixed points results given in Section 1.2.2. More precisely, we will study the following problem.

If (X, d) is a metric (or a b-metric) space and $G : X \times X \to P(X)$ is a multi-valued operator, then the coupled fixed point problem for G means to find $(x, y) \in X \times X$ satisfying

$$\begin{cases} x \in G(x, y) \\ y \in G(y, x). \end{cases} \tag{19.16}$$

For related results and some extensions of this problem see [30, 31, 49].

The first result of this section is the following existence and approximation result for the above problem.

Theorem 19.22 Let X be a nonempty set endowed with a partial order relation $\le$ and let d be a complete metric on X. Consider $G : X \times X \to P(X)$ a multi-valued operator for which we suppose:

(i) $G(\cdot, \cdot)$ has the generalized strict mixed monotone property, that is, for all $z = (x, y)$ and $w = (s, t) \in Z$ and, for each $u = (u_1, u_2) \in G(x, y) \times G(y, x)$ and $v = (v_1, v_2) \in G(s, t) \times G(t, s)$ we have

$$\begin{cases} u_1 \le v_1 \\ u_2 \ge v_2 \end{cases} \quad \text{or} \quad \begin{cases} v_1 \le u_1 \\ v_2 \ge u_2 \end{cases}$$

(ii) $G : X \times X \to P(X)$ has closed graph;

(iii) there exist $z_0 := (x_0, y_0) \in X \times X$ and $z_1 := (x_1, y_1) \in G(x_0, y_0) \times G(y_0, x_0)$ such that

$$\begin{cases} x_0 \le x_1 \\ y_0 \ge y_1 \end{cases} \quad \text{or} \quad \begin{cases} x_1 \le x_0 \\ y_1 \ge y_0 \end{cases}$$

(iv) there exist $k_1, k_2 \in \mathbb{R}_+$ with $k_1 + k_2 < 1$ such that

$$H_d(G(x, y), G(u, v)) \leq k_1 d(x, u) + k_2 d(y, v),$$

for all $(x \leq u$ and $y \geq v)$ or $(u \leq x$ and $v \geq y)$.

Then, there exists $(x^*, y^*) \in X \times X$ such that

$$\begin{cases} x^* \in G(x^*, y^*), \\ y^* \in G(y^*, x^*). \end{cases} \tag{19.17}$$

Moreover, there exist two sequences $\{x_n\}_{n \in \mathbb{N}}$ and $\{y_n\}_{n \in \mathbb{N}}$ in X, with $x_{n+1} \in G(x_n, y_n)$ and $y_{n+1} \in G(y_n, x_n)$ for all $n \in \mathbb{N}$, such that $\{x_n\}_{n \in \mathbb{N}} \to x^*$, $\{y_n\}_{n \in \mathbb{N}} \to y^*$ as $n \to \infty$ and the following estimation hold

$$\begin{pmatrix} d(x_n, x^*) \\ d(y_n, y^*) \end{pmatrix} \leq A^n (I_2 - A)^{-1} \begin{pmatrix} d(x_0, x_1) \\ d(y_0, y_1) \end{pmatrix},$$

where $A := \begin{pmatrix} k_1 & k_2 \\ k_2 & k_1 \end{pmatrix}$.

Proof: Denote $Z := X \times X$ and consider on Z, for $z := (x, y) \in Z$, $w := (u, v) \in Z$, the partial order relation

$$z \preceq_P w \quad \text{if and only if} \quad (x \leq u \text{ and } y \geq v).$$

We denote

$$Z_{\preceq_P} = \{(z, w) := ((x, y), (u, v)) \in Z \times Z \quad : \quad z \preceq_P w \quad \text{or} \quad w \preceq_P z\}.$$

Let $F : Z \to P(Z)$ be an operator defined by $F(x, y) := G(x, y) \times G(y, x)$.

We endow the space Z with the vector-valued metric $\tilde{d} : Z \times Z \to \mathbb{R}_+^2$ given by

$$\tilde{d}((x, y), (u, v)) := \begin{pmatrix} d(x, u) \\ d(y, v) \end{pmatrix},$$

while on $P_{cl}(Z)$ we consider the following vector-valued Hausdorff–Pompeiu type metric generated by $\tilde{d}$:

$$\tilde{H}_{\tilde{d}}(A_1 \times A_2, B_1 \times B_2)) := \begin{pmatrix} H_d(A_1, B_1) \\ H_d(A_2, B_2) \end{pmatrix}.$$

By our hypotheses, it follows that Theorem 19.15 is applicable for the operator F. Indeed, F satisfies to a contraction type condition with a matrix A given by

$$A := \begin{pmatrix} k_1 & k_2 \\ k_2 & k_1 \end{pmatrix},$$

that is, we have that

$$\tilde{H}_{\tilde{d}}(F(x, y), F(u, v)) \leq A\tilde{d}((x, y), (u, v)), \quad \text{for all } ((x, y), (u, v)) \in Z_{\leq}.$$

Because of our assumption $k_1 + k_2 < 1$, the matrix A converges to zero.

We also observe that assumption (i) implies that $Z_{\leq_p}$ is invariant with respect to $F \times F$, that is, $(F \times F)(Z_{\leq_p}) \subseteq Z_{\leq_p}$. Moreover, since G has closed graph, it follows that F has closed graph too. The conclusion follows by Theorem 19.15.

□

Our next result is an existence and approximation result for the coupled fixed point inclusion (19.9) in the setting of a complete b-metric space.

Theorem 19.23 Let X be a nonempty set endowed with a partial order relation $\leq$ and let d be a complete b-metric on X. Let $G : X \times X \to P(X)$ be a multi-valued operator such that:

(i) $G(\cdot, \cdot)$ has the generalized strict mixed monotone property, that is, for all $z = (x, y)$ and $w = (s, t) \in Z$ and, for each $u = (u_1, u_2) \in G(x, y) \times G(y, x)$ and $v = (v_1, v_2) \in G(s, t) \times G(t, s)$ we have

$$\begin{cases} u_1 \leq v_1 \\ u_2 \geq v_2 \end{cases} \text{ or } \begin{cases} v_1 \leq u_1 \\ v_2 \geq u_2 \end{cases} ;$$

(ii) $G : X \times X \to P(X)$ has closed graph;

(iii) there exist $z_0 := (x_0, y_0) \in X \times X$ and $z_1 := (x_1, y_1) \in G(x_0, y_0) \times G(y_0, x_0)$ such that

$$\begin{cases} x_0 \leq x_1 \\ y_0 \geq y_1 \end{cases} \text{ or } \begin{cases} x_1 \leq x_0 \\ y_1 \geq y_0 \end{cases} ;$$

(iv) there exist $k \in (0, 1)$ such that

$$H_d(G(x, y), G(u, v)) \leq k \max\{d(x, u), d(y, v)\},$$

$$\forall (x \leq u, y \geq v) \text{ or } (u \leq x, v \geq y).$$

Then, there exists $(x^*, y^*) \in X \times X$ such that

$$\begin{cases} x^* \in G(x^*, y^*), \\ y^* \in G(y^*, x^*). \end{cases} \tag{19.18}$$

Moreover, there exist two sequences $\{x_n\}_{n \in \mathbb{N}}$ and $\{y_n\}_{n \in \mathbb{N}}$ in X, with $x_{n+1} \in G(x_n, y_n)$ and $y_{n+1} \in G(y_n, x_n)$ for all $n \in \mathbb{N}$, such that $\{x_n\}_{n \in \mathbb{N}} \to x^*$, $\{y_n\}_{n \in \mathbb{N}} \to y^*$ as $n \to \infty$ and the following estimation holds

$$\max\{d(x_n, x^*), d(y_n, y^*)\} \leq \frac{sk^n}{1 - ks} \max\{d(x_0, x_1), d(y_0, y_1)\}, \quad \text{for all } n \in \mathbb{N}.$$

Proof: Denote $Z := X \times X$ and consider on Z, for $z := (x, y) \in Z, w := (u, v) \in Z$, the partial order relation

$$z \leq_p w \quad \text{if and only if} \quad (x \leq u \text{ and } y \geq v).$$

We denote

$$Z_{\leq_p} = \{(z, w) := ((x, y), (u, v)) \in Z \times Z : z \leq_p w \text{ or } w \leq_p z\}.$$

Let $F : Z \to P(Z)$ be an operator defined by $F(x,y) := G(x,y) \times G(y,x)$.
We endow the space Z with the b-metric $\tilde{d} : Z \times Z \to \mathbb{R}_+$ given by

$$\tilde{d}((x,y),(u,v)) := \max\{d(x,u), d(y,v)\}.$$

By the definition of the Hausdorff–Pompeiu metric it follows that

$$H_{\tilde{d}}(A \times B, C \times D) \le \max\{H_d(A,C), H_d(B,D)\}, \forall A, B, C, D \in P(X).$$

Then, by the above relation and our assumption (iv), we have

$$H_{\tilde{d}}(F(x,y), F(u,v)) \le \max\{H_d(G(x,y), G(u,v)), H_d(G(y,x), G(v,u))\} \le$$

$$k \max\{d(x,u), d(y,v)\} = k\tilde{d}((x,y),(u,v)), \quad \text{for all } (x,y),(u,v) \in Z_{\le_p}.$$

It is easy to check that, by our hypotheses, all the conditions in Theorem 19.17 are satisfied for the multi-operator F. Thus, there exists at least one fixed point $z^* = (x^*, y^*) \in Z$ of F (which is a coupled fixed point for G) and there exists a sequence $z_n := (x_n, y_n)$ in Z, such that $z_0 = (x_0, y_0) \in Z, z_1 = (x_1, y_1) \in F(z_0)$ are arbitrary and $z_{n+1} \in F(z_n)$, for which the following estimation holds:

$$\tilde{d}(z_n, z^*) \le \frac{sk^n}{1 - ks} \tilde{d}(z_0, z_1), \quad \text{for all } n \in \mathbb{N}.$$

Remark 19.19 For related and complementary results see also [29]. □

Remark 19.20 Some applications of the above coupled fixed point theorems can be given for different systems of integral and differential inclusions, such as

$$\begin{cases} x(t) \in g(t) + \int_0^T K(s,t,x(s),y(s)) \, ds \\ y(t) \in g(t) + \int_0^T K(s,t,y(s),x(s)) \, ds, \end{cases}$$

or

$$\begin{cases} x(t) \in g(t) + \int_0^t K(s,t,x(s),y(s)) \, ds, \\ y(t) \in g(t) + \int_0^t K(s,t,y(s),x(s)) \, ds, \end{cases}$$

where $g : [0,T] \to \mathbb{R}^n$ and $K : [0,T] \times [0,T] \times \mathbb{R}^{2n} \to P(\mathbb{R}^n)$ are given operators satisfying some appropriate conditions.

19.4   Coincidence Point Results

In this section, by applying some fixed point theorems given in the previous section, we will present some coincidence point theorems in complete b-metric spaces. Some properties of the coincidence point problem will be also established. We follow the approach given in [15].

Let (X, d) and (Y, ρ) be two metric spaces and $g, t : X \to Y$ be two operators. The coincidence point problem for t and g means to find $x^* \in X$ such that $t(x^*) = g(x^*)$. We will denote by $CP(g, t)$ the coincidence point set for g and t.

The following auxiliary result, given in the context of b-metric spaces, follow in a similar way to the case of classical metric spaces, see [15].

Lemma 19.4 Let (X, d) and (Y, ρ) be two complete b-metric spaces. Let $f : X \to Y$ be an injective and continuous mapping such that $f^{-1} : f(X) \to X$ is uniformly continuous. Then $f(X)$ is a closed subset of Y.

First main result of this section is a coincidence point theorem in a complete b-metric space.

Theorem 19.24 Let (X, d) be a b-metric space with constant $s_1 \geq 1$ and Y be a nonempty set. Let ρ be a complete b-metric on Y with constant $s_2 \geq 1$ and $g, t : X \to Y$ be two operators. Suppose that the following assumptions take place:

(i) $g(X) \subset t(X)$;
(ii) $g : X \to Y$ is a φ-contraction, that is, $\varphi : \mathbb{R} \to \mathbb{R}_+$ is a comparison function and

$$d(g(x_1), g(x_2)) \leq \varphi(d(x_1, x_2)), \quad \text{for all } x_1, x_2 \in X;$$

(iii) $t : X \to Y$ is expansive, that is,

$$\rho(t(x_1), t(x_2)) \geq d(x_1, x_2), \quad \text{for all } x_1, x_2 \in X;$$

(iv) one of the following conditions hold:
    (a) $t : X \to Y$ is continuous;
    (b) $t(X)$ is closed with respect to the b-metric ρ;
    (c) the b-metrics d and ρ are continuous.
    Then $C(g, t) = \{x^*\}$.

Proof: By (iii) the operator t is an injection. Thus, $t : X \to t(X)$ is a bijection. Let $t^{-1} : t(X) \to X$. By (iii), using the notation $x_1 := t^{-1}(y_1)$ and $x_2 := t^{-1}(y_2)$, we have

$$d(t^{-1}(y_1), t^{-1}(y_2)) \leq \rho(t(t^{-1}(y_1)), t(t^{-1}(y_2))) = \rho(y_1, y_2), \text{ for all } y_1, y_2 \in t(X).$$

Thus, t^{-1} is a nonexpansive mapping and hence t^{-1} is also uniformly continuous.

(a) We suppose first that $t : X \to Y$ is continuous. Then, by Lemma 19.4 we obtain that $t(X)$ is closed in (Y, ρ) and hence $(t(X), \rho)$ is complete too. Consider now the function $h : t(X) \to t(X)$ defined by $h := g \circ t^{-1}$. Notice that h is a single-valued operator by the above remarks and it is a self-operator by condition (i). Moreover, h is a φ-contraction since, for $y_1, y_2 \in t(X)$, we have

$$\rho(h(y_1), h(y_2)) = \rho(g(t^{-1}(y_1)), g(t^{-1}(y_2))) \leq \varphi(t^{-1}(y_1), t^{-1}(y_2)) \leq \varphi(\rho(y_1, y_2)).$$

Thus, by Czerwik's theorem (see Theorem 19.4) there is a unique $y^* \in t(X)$ with $y^* = h(y^*)$. If we denote $x^* = t^{-1}(y^*)$, then we get $y^* = g(x^*) = t(x^*)$. Hence $x^* \in C(g, t)$. Uniqueness of the coincidence point follows by the uniqueness of the fixed point of h.

(b) The case when $t(X)$ is closed with respect to the b-metric ρ follows in a similar way.

(c) If $t : X \to Y$ is not necessarily continuous, suppose that the b-metrics d and ρ are continuous. Notice that, in this case, the pair $(\overline{t(X)}, \rho)$ is complete in (Y, ρ). Since t^{-1} is uniformly continuous we may define an operator $\tilde{t}^{-1}$: $\overline{t(X)} \to X$ by

$$\tilde{t}^{-1}(y) = \begin{cases} t^{-1}(y) & \text{if } y \in t(X), \\ \lim_{n \to \infty} t^{-1}(y_n) & \text{if } y \in \overline{t(X)} \setminus t(X), \end{cases}$$

where $(y_n) \subset t(X)$ is such that $y_n \to y$ as $n \to \infty$. It is easy to see (by the continuity of the b-metrics d and ρ) that $\tilde{t}^{-1}$ is nonexpansive. Consider now the operator $\tilde{h}$ defined by $\tilde{h} := g \circ \tilde{t}^{-1}$. Then as before we can prove that $\tilde{h} : \overline{t(X)} \to \overline{t(X)}$ and it is a φ-contraction. Hence, by Czerwik's fixed point theorem (Theorem 19.4) we get that there exists a unique $y^* \in \overline{t(X)}$ such that $\tilde{h}(y^*) = y^*$. Let us show that $y^* \in t(X)$. Since $y^* = \tilde{h}(y^*)$ we get that $y^* = (g \circ \tilde{t}^{-1})(y^*) \in g(X) \subset t(X)$. Next, if we denote $x^* = t^{-1}(y^*)$, then we obtain that $y^* = g(x^*) = t(x^*)$. Uniqueness of the coincidence point follows as before by the uniqueness of the fixed point of h. □

The second main result of this section is a coincidence point theorem in an ordered complete b-metric space.

Theorem 19.25 Let (X, d) be a b-metric space with constant $s_1 \geq 1$, Y be a nonempty set and "$\preceq$" be a partial order relation on X and on Y. Let ρ be a complete b-metric on Y with constant $s_2 \geq 1$ and $g, t : X \to Y$ be two operators. Suppose that the following assumptions take place:

(i) $g(X) \subset t(X)$;

(ii) there exists a comparison function $\varphi : \mathbb{R} \to \mathbb{R}_+$ such that

$$\rho(g(x_1), g(x_2)) \leq \varphi(d(x_1, x_2)), \quad \text{for all } x_1, x_2 \in X \text{ with } x_1 \preceq x_2;$$

(iii) $t : X \to Y$ is increasing with respect to $\preceq$ and expansive, that is,

$$\rho(t(x_1), t(x_2)) \geq d(x_1, x_2), \quad \text{for all } x_1, x_2 \in X;$$

(iv) g has closed graph with respect to d and ρ and it is increasing with respect to $\preceq$;

(v) one of the following conditions hold:
   (a) $t : X \to Y$ is continuous;
   (b) $t(X)$ is closed with respect to the b-metric ρ;
   (c) the b-metrics d and ρ are continuous;
(vi) there exists $x_0 \in X$ such that $t(x_0) \leq g(x_0)$;
(vii) for every $y, w \in Y$ there exists $z \in Y$ which is comparable to y and w.
Then $C(g, t) = \{x^*\}$.

Proof: By (iii) the operator t is an injection. Thus $t : X \to t(X)$ is a bijection. Hence, using again (iii) for $t^{-1} : t(X) \to X$, we have

$$d(t^{-1}(y_1), t^{-1}(y_2)) \leq \rho(t(t^{-1}(y_1)), t(t^{-1}(y_2))) = \rho(y_1, y_2), \quad \text{for all } y_1, y_2 \in t(X).$$

Thus t^{-1} is a nonexpansive mapping and hence t^{-1} is uniformly continuous. Moreover, t^{-1} is also increasing.

(a) We suppose first that $t : X \to Y$ is continuous. Then, by Lemma 19.4 we obtain again that $t(X)$ is closed in (Y, ρ) and hence $(t(X), \rho)$ is complete too. Consider now the function $h : t(X) \to t(X)$ defined by $h := g \circ t^{-1}$. Notice that h is single-valued and increasing by the above remarks and it is a self-operator by condition (i). Additionally, if we denote $y_0 := t(x_0)$, then we have $y_0 \leq h(y_0)$. Moreover, for $y_1, y_2 \in t(X)$ with $y_1 \leq y_2$, we can prove that

$$\rho(h(y_1), h(y_2)) \leq \varphi(\rho(y_1, y_2)).$$

Indeed, let $y_1, y_2 \in t(X)$ such that $y_1 \leq y_2$. Then, there exist $x_1, x_2 \in X$ such that $y_1 = t(x_1)$ and $y_2 = t(x_2)$. Since t^{-1} is increasing we get that $t^{-1}(y_1) \leq t^{-1}(y_2)$. Then, by (ii) and (iii), we get

$$\rho(h(y_1), h(y_2)) \leq \varphi(d(t^{-1}(y_1), t^{-1}(y_2))) \leq \varphi(\rho(y_1, y_2)).$$

Then, by Theorem 19.6, there exists a unique $y^* \in t(X)$ such that $y^* = h(y^*)$. As a consequence, if we denote $x^* := t^{-1}(y^*)$, then we obtain $y^* = g(x^*) = t(x^*)$.
(b) The case when $t(X)$ is closed with respect to the b-metric ρ follows in a similar way.
(c) If $t : X \to Y$ is not necessarily continuous, suppose that the b-metrics d and ρ are continuous. Notice that the pair $(\overline{t(X)}, \rho)$ is complete in (Y, ρ). Since t^{-1} is uniformly continuous we may define an operator $\tilde{t}^{-1} : \overline{t(X)} \to X$ by

$$\tilde{t}^{-1}(y) = \begin{cases} t^{-1}(y), & \text{if } y \in t(X), \\ \lim_{n \to \infty} t^{-1}(y_n) & \text{if } y \in \overline{t(X)} \backslash t(X), \end{cases}$$

where $(y_n) \subset t(X)$ is such that $y_n \to y$ as $n \to \infty$. It is easy to see (by the continuity of the b-metrics d and ρ) that $\tilde{t}^{-1}$ is nonexpansive. Consider now the operator $\tilde{h}$ defined by $\tilde{h} := g \circ t^{-1}$. Then, as before, we can prove that $\tilde{h} : \overline{t(X)} \to \overline{t(X)}$ and it satisfies the following relation:

$$\rho(\tilde{h}(y_1), \tilde{h}(y_2)) \leq \varphi(\rho(y_1, y_2)), \quad \text{for all } y_1, y_2 \in t(X) \text{ with } y_1 \leq y_2.$$

Hence, again by Theorem 19.6 there exists a unique $y^* \in \overline{t(X)}$ such that $\tilde{h}(y^*) = y^*$. Let us show that $y^* \in t(X)$. Since $y^* = \tilde{h}(y^*)$ we get that

$$y^* = (g \circ \tilde{t}^{-1})(y^*) \in g(X) \subset t(X).$$

If we denote $x^* = t^{-1}(y^*)$, then we obtain that $y^* = g(x^*) = t(x^*)$. The uniqueness of the element x^* follows by the assumption (vii). □

Our next result is a data dependence theorem for the solution set of the coincidence point problem.

Theorem 19.26 Let (X, d) be a b-metric space with constant $s_1 \geq 1$, Y be a nonempty set and "$\leq$" be a partial order relation on Y. Let ρ be a complete b-metric on Y with constant $s_2 \geq 1$ and $g, t : X \to Y$ be two operators satisfying all the assumptions of Theorem 19.25. Denote by x^* the unique coincidence point of g and t. Let $g_1, t_1 : X \to Y$ be two operators having at least one coincidence point $x_1^* \in X$. We also suppose that:

(i) $t_1 : X \to Y$ is injective, $t_1(X) \subset t(X)$ and $t(X)$ is a closed subset of (Y, ρ); item there exist $\eta_1, \eta_2, \eta_3 > 0$ such that

$$\rho(g(x), g_1(x)) \leq \eta_1, \quad \text{for all } x \in X;$$

$$\rho(t(x), t_1(x)) \leq \eta_2, \quad \text{for all } x \in X;$$

$$d(t^{-1}(y), t_1^{-1}(y)) \leq \eta_3, \quad \text{for all } y \in t_1(X);$$

(ii) the function $\gamma : \mathbb{R}_+ \to \mathbb{R}_+$, $\gamma(t) := t - s_2\varphi(t)$ satisfies the condition $\lim_{t \to \infty} \gamma(t) = \infty$.

Then, the following estimation holds

$$d(x^*, x_1^*) \leq s_2(\psi(\eta_1, \eta_3) + \eta_2),$$

where $\psi : \mathbb{R}_+^{\not{=}} \to \mathbb{R}_+$ is given by

$$\psi(\eta_1, \eta_3) := \sup\{t \geq 0 \mid t - s_2\varphi(t) \leq s_2^2(\varphi(\eta_3) + \eta_1)\}.$$

Proof: Let us consider $h : t(X) \to t(X)$ defined by $h := g \circ t^{-1}$ and $h_1 : t_1(X) \to t_1(X)$ defined by $h_1 := g_1 \circ t_1^{-1}$. Denote $y^* = t(x^*) = g(x^*)$ and $y_1^* = t_1(x_1^*) = g_1(x_1^*)$. Then y^* and y_1^* are fixed points for h and respectively

h_1 and, by the proof of Theorem 4.2, the operator h is a φ-contraction. Then, we have the following estimation:

$$\rho(y^*, y_1^*) = \rho(h(y^*), h_1(y_1^*)) \leq s_2(\rho(h(y^*), h(y_1^*)) + \rho(h(y_1^*), h_1(y_1^*)))$$
$$\leq s_2(\varphi(\rho(y^*, y_1^*)) + \eta),$$

where $\eta > 0$ is given by the following relation

$$\rho(h(y_1^*), h_1(y_1^*)) = \rho((g \circ t^{-1})(y_1^*), (g_1 \circ t_1^{-1})(y_1^*))$$

$$\leq s_2(\rho((g \circ t^{-1})(y_1^*), (g \circ t_1^{-1})(y_1^*)) + \rho((g \circ t_1^{-1})(y_1^*), (g_1 \circ t_1^{-1})(y_1^*)))$$

$$\leq s_2(\varphi(d(t^{-1}(y_1^*), t_1^{-1}(y_1^*))) + \eta_1) \leq s_2(\varphi(\eta_3) + \eta_1) := \eta.$$

Hence,

$$\rho(y^*, y_1^*) - s_2\varphi(\rho(y^*, y_1^*)) \leq s_2^2(\varphi(\eta_3) + \eta_1).$$

We conclude that

$$\rho(y^*, y_1^*) \leq \psi(\eta_1, \eta_3) := \sup\{t \geq 0 \mid t - s_2\varphi(t) \leq s_2^2(\varphi(\eta_3) + \eta_1)\}.$$

Since t is expansive, we get that

$$d(x^*, x_1^*) \leq \rho(t(x^*), t(x_1^*))$$
$$\leq s_2(\rho(t(x^*), t_1(x_1^*)) + \rho(t_1(x_1^*), t(x_1^*)))$$
$$\leq s_2(\rho(y^*, y_1^*) + \eta_2).$$

As a conclusion

$$d(x^*, x_1^*) \leq s_2(\psi(\eta_1, \eta_3) + \eta_2).$$

We recall now the concept of well-posed coincidence problem. □

Definition 19.4 Let (X, d) and (Y, ρ) be two b-metric spaces with constants $s_1 \geq 1$ and respectively $s_2 \geq 1$. Let $g, t : X \to Y$ be two operators. By definition, the coincidence problem for g and t is well-posed if:

(i) $C(g, t) = \{x^*\}$;
(ii) for any sequence $\{x_n\}_{n \in \mathbb{N}}$ in X for which $\rho(g(x_n), t(x_n)) \to 0$ as $n \to \infty$, we have that $x_n \to x^*$ as $n \to \infty$.

The following theorem is a well-posedness result for the coincidence point problem in b-metric spaces.

Theorem 19.27 Let (X, d) be a b-metric space with constant $s_1 \geq 1$, Y be a nonempty set and "$\preceq$" be a partial order relation on Y. Let ρ be a complete b-metric on Y with constant $s_2 \geq 1$ and $g, t : X \to Y$ be two operators satisfying all the assumptions of Theorem 19.25. Additionally, suppose that the mapping $\psi : \mathbb{R}_+ \to \mathbb{R}_+, \psi(t) = t - s_2^2 \varphi(t)$ is a bijection such that $\psi^{-1}(u_n) \to 0$ as $u_n \to 0$, for $n \to \infty$. Then, the coincidence problem for g and t is well-posed.

Proof. By Theorem 19.25, we have that $C(g, t) = \{x^*\}$. Let $\{x_n\}_{n \in \mathbb{N}}$ be a sequence in X such that $\rho(g(x_n), t(x_n)) \to 0$ as $n \to \infty$. Then, we have

$$
\begin{aligned}
d(x_n, x^*) &\leq \rho(t(x_n), t(x^*)) \\
&\leq s_2(\rho(t(x_n), g(x_n)) + \rho(g(x_n), t(x^*))) \\
&\leq s_2\rho(t(x_n), g(x_n)) + s_2^2(\rho(g(x_n), g(x^*)) + \rho(g(x^*), t(x^*))) \\
&\leq s_2\rho(t(x_n), g(x_n)) + s_2^2\varphi(d(x_n, x^*)).
\end{aligned}
$$

Thus

$$
d(x_n, x^*) - s_2^2\varphi(d(x_n, x^*)) \leq s_2\rho(t(x_n), g(x_n))
$$

and so

$$
d(x_n, x^*) \leq \psi^{-1}(s_2\rho(t(x_n), g(x_n))) \to 0 \quad \text{as } n \to \infty. \qquad \square
$$

We will study now the Ulam–Hyers stability of the coincidence point problem. For a general study of this problem in generalized metric spaces see Rus [50].

Definition 19.5 Let (X, d) and (Y, ρ) be two b-metric spaces with constants $s_1 \geq 1$ and respectively $s_2 \geq 1$. Let $g, t : X \to Y$ be two operators. By definition, the coincidence problem for g and t is Ulam–Hyers stable if there exists an increasing function $\psi : \mathbb{R}_+ \to \mathbb{R}_+$ continuous in 0 with $\psi(0) = 0$ such that for each $\epsilon > 0$ and each ϵ-solution $\tilde{x} \in X$ of the coincidence problem for g and t (i.e., $\rho(t(\tilde{x}), g(\tilde{x})) \leq \epsilon$), there exists a solution $x^* \in X$ of the coincidence problem for g and t such that $d(x^*, \tilde{x}) \leq \psi(\epsilon)$.

Theorem 19.28 Let (X, d) be a b-metric space with constant $s_1 \geq 1$, Y be a nonempty set and "$\leq$" be a partial order relation on Y. Let ρ be a complete b-metric on Y with constant $s_2 \geq 1$ and $g, t : X \to Y$ be two operators satisfying all the assumptions of Theorem 19.25. Additionally, suppose that the mapping $\gamma : \mathbb{R}_+ \to \mathbb{R}_+, \gamma(t) = t - s_2^2\varphi(t)$ is strictly increasing and onto. Then the coincidence problem for g and t is Ulam–Hyers stable.

Proof: By Theorem 19.25, we have that $C(g, t) = \{x^*\}$. Let $\epsilon > 0$ and $\tilde{x} \in X$ such that $\rho(t(\tilde{x}), g(\tilde{x})) \leq \epsilon$. Then we have

$$
\begin{aligned}
d(x^*, \tilde{x}) &\leq \rho(t(x^*), t(\tilde{x})) \\
&\leq s_2(\rho(t(x^*), g(\tilde{x})) + \rho(g(\tilde{x}), t(\tilde{x}))) \\
&\leq s_2^2(\rho(t(x^*), g(x^*)) + \rho(g(x^*), g(\tilde{x}))) + s_2\epsilon \\
&\leq s_2^2\varphi(d(x^*, \tilde{x})) + s_2\epsilon.
\end{aligned}
$$

Hence, $d(x^*, \tilde{x}) - s_2^2 \varphi(d(x^*, \tilde{x})) \le s_2 \epsilon$ and so $d(x^*, \tilde{x}) \le \gamma^{-1}(s_2 \epsilon)$. This completes the proof. □

The last result of this section is another coincidence point theorem of Ran–Reurings type. The result is a slight extension of Theorem 1 in [51] and a generalization of Theorem 3 in [52].

Theorem 19.29 Let (X, d) be a b-metric space with constant $\lambda \ge 1$, Y be a nonempty set and "$\le$" be a partial order relation on Y. Let ρ be a b-metric on Y with constant $s \ge 1$ and $g, t : X \to Y$ be two operators with closed graph. Suppose that the following assumptions take place:

(i) $t(X) \subset g(X)$;
(ii) $(t(X), \rho)$ is a complete subset of Y;
(iii) there exists a comparison function $\varphi : \mathbb{R}_+ \to \mathbb{R}_+$ such that

$$\rho(t(x), t(y)) \le \varphi(\rho(g(x), g(y))), \quad \text{for all } x, y \in X \text{ with } g(x) \le g(y);$$

(iv) there exists $x_0 \in X$ such that $g(x_0) \in t(X)$ and $g(x_0) \le t(x_0)$;
(v) t is increasing with respect to g, that is,

$$x_1, x_2 \in X \quad \text{and} \quad g(x_1) \le g(x_2) \Rightarrow t(x_1) \le t(x_2).$$

Then, there exists $x^* \in X$ such that $g(x^*) = t(x^*)$ and the sequence $\{z_n\}$ defined by $g(z_{n+1}) = t(z_n)$ (where $n \in \mathbb{N}$ and $z_0 := x_0 \in X$) converges to x^* as $n \to \infty$.
If, in addition:

(vi) for every $y, w \in Y$ there exists $z \in Y$ which is comparable to y and w;
(vii) g is an injection, then $C(t, g) = \{x^*\}$ and the sequence $\{z_n\}_{n \in \mathbb{N}}$ defined by $g(z_{n+1}) = t(z_n)$, starting from any point $z_0 \in X$ converges to x^*.

Proof: Let $x_0 \in X$ such that $g(x_0) \le t(x_0)$. Let us define $f := t \circ g^{-1}$. Then, we have for f the following properties:

(1) f is a single-valued operator on $t(X)$;
(2) $f : t(X) \to t(X)$;
(3) f has closed graph;
(4) $\rho(f(y_1), f(y_2)) \le \varphi(\rho(y_1, y_2))$, for all $y_1, y_2 \in t(X)$ with $y_1 \le y_2$;
(5) f is increasing on $t(X)$;
(6) If $y_0 := g(x_0)$, then $y_0 \le (t \circ g^{-1})(y_0) = f(y_0)$.

By Theorem 19.6, we obtain that f is a Picard operator. Thus, $Fix(f) = \{y^*\}$. Then, $(t \circ g^{-1})(y^*) = y^*$. Thus, if we denote $x^* := g^{-1}(y^*)$, then we have $t(x^*) = g(x^*) = y^*$, showing that x^* is a coincidence point for t and g. Moreover, the sequence $y_{n+1} := f(y_n)$ (where $n \in \mathbb{N}$), starting from $y_0 := g(x_0) \in t(X)$ converges to y^* as $n \to \infty$, while the sequence z_n defined by $g(z_{n+1}) = t(z_n)$ (where $n \in \mathbb{N}$ and $z_0 := x_0 \in X$) converges to x^* as $n \to \infty$.

The uniqueness of the coincidence point follows by (vi) and (vii). Indeed, by the first part of this theorem there exist $x^* \in X$ and $y^* \in t(X)$ such that $t(x^*) = g(x^*) = y^*$. Suppose that there exist $u^* \in X$ and $v^* \in t(X)$ such that $t(u^*) = g(u^*) = v^*$. We have two cases:

Case 1. If $g(x^*)$ and $g(u^*)$ are comparable, that is, $g(x^*) \preceq g(u^*)$ or reversely. Let $f : t(X) \to t(X), f(y) := t \circ g^{-1}(y)$ with the above six properties. Suppose, for example, that $g(x^*) \preceq g(u^*)$. Then $t(x^*) \preceq t(u^*)$ and so

$$\rho(y^*, v^*) = \rho(f(y^*), f(v^*)) \leq \varphi(\rho(y^*, v^*)).$$

By the properties of the comparison function φ we get that $\rho(y^*, v^*) = 0$. Thus, $y^* = v^*$ which implies $g(x^*) = g(u^*)$. By the injectivity of g we get that $x^* = u^*$.

Case 2. Suppose that $g(x^*)$ and $g(u^*)$ are not comparable. Then, there exists $z \in Y$ such that z is comparable to $g(x^*)$ and $g(u^*)$. Suppose $g(x^*) \preceq z \preceq g(u^*)$. Thus, $y^* \preceq z \preceq v^*$. Consider $f : t(X) \to t(X), f(y) := t \circ g^{-1}(y)$. Then, since f is increasing, we get that $f^n(y^*) \preceq f^n(z) \preceq f^n(v^*)$, for all $n \in \mathbb{N}^*$.

Let $y_1, y_2 \in t(X)$ with $y_1 \preceq y_2$. By the monotonicity of f we obtain that $f^n(y_1) \preceq f^n(y_2)$, for all $n \in \mathbb{N}^*$. Now, by induction, we get

$$\rho(f^n(y_1), f^n(y_2)) \leq \varphi^n(\rho(y_1, y_2)), \quad \text{for all } n \in \mathbb{N}^*.$$

Applying the above relation we get

$$\begin{aligned}
\rho(y^*, v^*) &= \rho(f^n(y^*), f^n(v^*)) \\
&\leq s(\rho(f^n(y^*), f^n(z)) + \rho(f^n(z), f^n(v^*))) \\
&\leq s(\varphi^n(\rho(y^*, z)) + \varphi^n(\rho(z, v^*))) \to 0 \quad \text{as } n \to \infty.
\end{aligned}$$

Thus, $y^* = v^*$. As before, by the injectivity of g we obtain $x^* = u^*$. The rest of the cases can be treated similarly. $\qquad\square$

Remark 19.21 Our results are in connection with several previous theorems given in Buică [53], Falset and Mleşniţe [54], Mleşniţe [52], and Rus [55].

Let us consider, as an illustration of the previous results, an integral equation of the following form:

$$\begin{cases} T(x(s)) = \displaystyle\int_0^s g(p, x(p)) \, dp, & \text{for } s \in [0, \alpha] \quad (\text{with } \alpha > 0), \\ x(0) = 0, \end{cases} \tag{19.19}$$

where

(i) $T : \mathbb{R}_+ \to \mathbb{R}_+$ and $T(0) = 0$;
(ii) $g : [0, \alpha] \times \mathbb{R}_+ \to \mathbb{R}_+$

are two continuous mappings.

If we consider

$$X := C_+([0, \alpha]) = \{x : [0, \alpha] \to \mathbb{R}_+ : x \text{ is continuous and } x(0)=0\}$$

and the operators $t, G : X \to X$ given by

$$tx(s) := T(x(s)) \quad \text{and} \quad Gx(s) := \int_0^s g(p, x(p)) \, \mathrm{d}p,$$

then our problem can be rewritten as a coincidence point problem of the following form:

$$tx = Gx, \quad x \in X.$$

Notice that on $C_+([0, \alpha])$ we can define a partial order relation by

$$x \leq_C y \quad \text{if and only if} \quad x(s) \leq y(s), \quad \forall \, s \in [0, \alpha]$$

and a Bielecki type norm given by

$$\|x\|_B := \max_{s \in [0,\alpha]} (|x(s)| e^{-\tau s} k), \quad \text{where } \tau > 0,$$

with respect to which the space X is complete.

We have the following existence and uniqueness result for (19.19).

Theorem 19.30 Consider the functional-integral equation (19.19). We suppose that:

(i) $T : \mathbb{R}_+ \to \mathbb{R}_+$ and $g : [0, \alpha] \times \mathbb{R}_+ \to \mathbb{R}_+$ are two continuous mappings;

(ii) T is onto, increasing and expansive;

(iii) there exist $\tau > 0$ and a function $\psi : \mathbb{R}_+ \to \mathbb{R}_+$ such that for arbitrary $q > 1$ we have $\psi(qt) \leq q\psi(t)$, for all $t \in \mathbb{R}_+$, the function $\varphi(t) = \frac{1}{\tau}\psi(t)$ is a comparison function and

$$|f(s, u) - f(s, v)| \leq \psi(|u - v|) \; \forall \; s \in [0, \alpha] \text{ and } \forall u, v \in \mathbb{R}_+ \text{with } u \leq v;$$

(iv) $f(s, \cdot) : \mathbb{R}_+ \to \mathbb{R}_+$ is increasing, for all $s \in [0, \alpha]$;

(v) there exists $x_0 \in C_+([0, \alpha])$ a lower solution of (19.19), that is,

$$T(x_0(s)) \leq \int_0^s g(p, x_0(p)) \, \mathrm{d}p, \quad \text{for } s \in [0, \alpha].$$

Then, the functional-integral equation (19.19) has a unique solution in $C_+([0, \alpha])$.

Proof: Consider the space X endowed with the partial order $\leq_C$ and the Bielecki type norm $\|x\|_B$ and the operators t, G defined as above. Then, we have the following results:

(a) $G(X) \subset t(X) = X$;

(f) there exists $z_0 := (x_0, y_0) \in Z$ such that $S(z_0) \preceq_P T(z_0)$;
(g) for all $z, \tilde{z} \in Z$ there exists $w \in Z$ which is comparable (with respect to $\preceq_P$) to z and $\tilde{z}$.

Hence, our conclusion follows by Theorem 19.25.

As before, if the contraction condition (ii) is satisfied on $X \times X$, then no monotonicity assumptions are needed and we obtain the following result.

Theorem 19.34   Let (X, d) be a b-metric space with constant $s_1 \geq 1$, (Y, ρ) be a complete b-metric space with constant $s_2 \geq 1$ and $G : X \times X \to Y, s : X \to Y$ be two operators. We suppose:

(i) $G(X \times X) \subset s(X)$;
(ii) there exists a comparison function $\varphi : \mathbb{R}_+ \to \mathbb{R}_+$ such that

$$\rho(G(x, y), G(u, v)) + \rho(G(y, x), G(v, u)) \leq \varphi(d(x, u) + d(y, v)),$$

for all $(x, y), (u, v) \in X \times X$;
(iii) s is expansive, that is, $\rho(s(x_1), s(x_2)) \geq d(x_1, x_2)$, for all $x_1, x_2 \in X$;
(iv) s is continuous or $s(X)$ is closed or the b-metrics d and ρ are continuous.

Then, there exists a unique $z^* = (x^*, y^*) \in X \times X$ solution for the coupled coincidence system (19.21).

It is easy to see that a similar result takes place under a slight modification of the contraction condition, as follows.

Theorem 19.35   Let (X, d) be a b-metric space with constant $s_1 \geq 1$, (Y, ρ) be a complete b-metric space with constant $s_2 \geq 1$ and $G : X \times X \to Y, s : X \to Y$ be two operators. We suppose:

(i) $G(X \times X) \subset s(X)$;
(ii) there exists a comparison function $\varphi : \mathbb{R}_+ \to \mathbb{R}_+$ such that

$$\max\{\rho(G(x, y), G(u, v)), \rho(G(y, x), G(v, u))\} \leq \varphi(\max\{(d(x, u), d(y, v))\}),$$

for all $(x, y), (u, v) \in X \times X$;
(iii) s is expansive, that is, $\rho(s(x_1), s(x_2)) \geq d(x_1, x_2)$, for all $x_1, x_2 \in X$;
(iv) s is continuous or $s(X)$ is closed or the b-metrics d and ρ are continuous.

Then, there exists a unique $z^* = (x^*, y^*) \in X \times X$ solution for the coupled coincidence system (19.21).

Proof: The proof runs in a similar way to the proof of Theorem 5.4 working with the the b-metric

$$\tilde{d}((x, y), (u, v)) := \max\{d(x, u), d(y, v)\}$$

on $Z := X \times X$ and with the b-metric

$$\tilde{\rho}((x, y), (u, v)) := \max\{\rho(x, u), \rho(y, v)\}$$

on $W := Y \times Y$. The conclusion follows by Theorem 3.12 in [54]. $\qquad\square$

References

1 Rus, I.A. (2016) Some variants of contraction principle, generalizations and applications. *Stud. Univ. Babeş-Bolyai Math.*, **61**, 343–358.

2 Fréchet, M. (1928) *Les espaces abstraits*, Gauthier-Villars, Paris.

3 Varga, R.S. (2000) *Matrix Iterative Analysis*, Springer Series in Computational Mathematics, vol. 27, Springer-Verlag, Berlin.

4 Precup, R. (2009) The role of matrices that are convergent to zero in the study of semilinear operator systems. *Math. Comput. Model.*, **49**, 703–708.

5 Petruşel, A., Petruşel, G., and Urs, C. (2013) Vector-valued metrics, fixed points and coupled fixed points for nonlinear operators. *Fixed Point Theory Appl.*, **2013**, 218.

6 Bakhtin, I.A. (1989) The contraction mapping principle in almost metric spaces. *Funct. Anal. Unianowsk, Gos. Ped. Inst.*, **30**, 26–37.

7 Czerwik, S. (1993) Contraction mappings in b-metric spaces. *Acta Math. Inform. Univ. Ostraviensis*, **1**, 5–11.

8 Berinde, V. (1993) Generalized contractions in quasimetric spaces. *Semin. Fixed Point Theory*, **3**, 3–9.

9 Kirk, W.A. and Shahzad, N. (2014) *Fixed Point Theory in Distance Spaces*, Springer-Verlag, Heidelberg.

10 Xia, Q. (2009) The geodesic problem in quasimetric spaces. *J. Geom. Anal.*, **19**, 452–479.

11 Bota, M., Molnar, A., and Varga, C. (2011) On Ekeland's variational principle in b-metric spaces. *Fixed Point Theory*, **12**, 21–28.

12 Rus, I.A. (2001) *Generalized Contractions and Applications*, Transilvania Press, Cluj-Napoca.

13 Van Luong, N. and Thuan, N.X. (2011) Coupled fixed points in partially ordered metric spaces and application. *Nonlinear Anal.*, **74**, 983–992.

14 Perov, A.I. (1964) On the Cauchy problem for a system of ordinary differential equations. *Pviblizhen. Met. Reshen. Differ. Uravn.*, **2**, 115–134.

15 Petruşel, A., Petruşel, G., and Yao, J.-C. (2017) Fixed point and coincidence point theorems in b-metric spaces with applications. *Appl. Anal. Discrete Math.*, **11**, 199–215.

16 Ran, A.C.M. and Reurings, M.C.B. (2004) A fixed point theorem in partially ordered sets and some applications to matrix equations. *Proc. Am. Math. Soc.*, **132**, 1435–1443.

17 Petruşel, A., Petruşel, G., Samet, B., and Yao, J.-C. (2016) Coupled fixed point theorems for symmetric contractions in b-metric spaces with applications to operator equation systems. *Fixed Point Theory*, **17**, 459–478.

18 Rus, I.A. (2003) Picard operators and applications. *Sci. Math. Jpn.*, **58**, 191–219.

19 Petruşel, A. and Rus, I.A. (2005) Fixed point theorems in ordered *L*-spaces. *Proc. Am. Math. Soc.*, **134**, 411–418.

20 Guo, D., Cho, Y.J., and Zhu, J. (2004) *Partial Ordering Methods in Nonlinear Problems*, Nova Science Publishers Inc., Hauppauge.

21 Heikkilá, S. and Lakshmikantham, V. (1991) *Monotone Iterative Techniques for Discountinuous Nonlinear Differential Equations*, Marcel Dekker.

22 Krasnoselskii, M.A. and Zabreiko, P.P. (1984) *Geometrical Methods of Nonlinear Analysis*, Springer-Verlag, Berlin.

23 Şerban, M.A. (2002) *Fixed Point Theory for Operators on Cartesian Product Spaces*, Cluj University Press, (in Romanian).

24 Petruşel, A., Urs, C., and Mleşniţe, O. (2015) Vector-valued metrics in fixed point theory. *Contemp. Math.*, **636**, 149–165.

25 Czerwik, S. (1998) Nonlinear set-valued contraction mappings in b-metric spaces. *Atti Sem. Mat. Univ. Modena*, **46**, 263–276.

26 Nadler, S.B. (1969) Multi-valued contraction mappings. *Pac. J. Math.*, **30**, 475–488.

27 Węgrzyk, R. (1982) Fixed point theorems for multifunctions and their applications to functional equations. *Disser. Math. (Rozprawy Mat.)*, **201**, 1–28.

28 Dinevari, T. and Frigon, M. (2013) Fixed point results for multivalued contractions on a metric space with a graph. *J. Math. Anal. Appl.*, **405**, 507–517.

29 Petruşel, A., Petruşel, G., Samet, B., and Yao, J.-C. (2016) Coupled fixed point theorems for symmetric multi-valued contractions in b-metric space with applications to systems of integral inclusions. *J. Nonlinear Convex Anal.*, **17**, 1265–1282.

30 Petruşel, A., Petruşel, G., and Samet, B. (2016) Scalar and vectorial approaches for multi-valued fixed point and multi-valued coupled fixed point problems in *b*-metric spaces. *J. Nonlinear Convex Anal.*, **17**, 2049–2061.

31 Petruşel, A., Petruşel, G., and Yao, J.-C. (2016) A study of a system of operator inclusions via a fixed point approach and applications to functional-differential inclusions. *Carpathian J. Math.*, **32**, 349–361.

32 Amann, H. (1977) *Order Structures and Fixed Points*, SAFA 2 University of Calabria, 1–51.

33 Opoitsev, V.I. (1975) Heterogenic and combined-concave operators. *Syber. Math. J.*, **16**, 781–792 (in Russian).

34 Opoitsev, V.I. (1979) A generalization of the theory of monotone and concave operators. *Trans. Moscow Math. Soc.*, **36**, 243–280.

35 Opoitsev, V.I. and Khurodze, T.A. (1984) Nonlinear operators in spaces with a cone. *Tbilis. Gos. Univ. Tbilisi*, 271 (in Russian).

36 Ziebur, A.D. (1962) Uniqueness and the convergence of successive approximations. *Proc. Am. Math. Soc.*, **13**, 899–903.

37 Ziebur, A.D. (1965) Uniqueness and the convergence of successive approximations. II. *Proc. Am. Math. Soc.*, **16**, 335–340.

38 Guo, D. and Lakshmikantham, V. (1987) Coupled fixed points of nonlinear operators with applications. *Nonlinear Anal.*, **11**, 623–632.

39 Bhaskar, T.G. and Lakshmikantham, V. (2006) Fixed point theorems in partially ordered metric spaces and applications. *Nonlinear Anal.*, **65**, 1379–1393.

40 Berinde, V. (2011) Generalized coupled fixed point theorems for mixed monotone mappings in partially ordered metric spaces. *Nonlinear Anal.*, **74**, 7347–7355.

41 Lakshmikantham, V. and Ćirić, L. (2009) Coupled fixed point theorems for nonlinear contractions in partially ordered metric spaces. *Nonlinear Anal.*, **70**, 4341–4349.

42 Bota, M.-F., Petruşel, A., Petruşel, G., and Samet, B. (2015) Coupled fixed point theorems for singlevalued operators in *b*-metric spaces. *Fixed Point Theory Appl.*, **2015**, 231. doi: 10.1186/s13663-015-0482-3.

43 Chen, Y.-Z. (1991) Existence theorems of coupled fixed points. *J. Math. Anal. Appl.*, **154**, 142–150.

44 Guo, D. and Lakshmikantham, V. (1988) *Nonlinear Problems in Abstract Cones*, Academic Press.

45 Rus, M.D. (2014) The method of monotone iterations for mixed monotone operators in partially ordered sets and order-attractive fixed points. *Fixed Point Theory*, **15**, 579–594.

46 Xiao, J.-Z., Zhu, X.-H., and Shen, Z.-M. (2013) Common coupled fixed point results for hybrid nonlinear contractions in metric spaces. *Fixed Point Theory*, **14**, 235–250.

47 Petruşel, A. and Rus, I.A. (2016) Contributions to the fixed point theory of diagonal operators. *Fixed Point Theory Appl.*, **2016**, 95. doi: 10.1186/s13663-016-0589-1.

48 Krasnoselskii, M.A. (1955) Two remarks on the method of successive approximations. *Uspehi Mat. Nauk*, **10**, 123–127.

49 Diamond, P. and Opoitsev, V.I. (2001) Stability of nonlinear difference inclusions. *Dyn. Cont. Discrete Impuls. Syst. Ser. A: Math. Anal.*, **8**, 353–371.

50 Rus, I.A. (2011) Ulam stability of the operatorial equations, in *Functional Equations in Mathematical Analysis*, Springer Optimization and Its Applications, vol. **52** (eds T. Rassias and J. Brzdek), Springer-Verlag, New York, 287–305.

51 Petruşel, A., Petruşel, G., and Samet, B. (2016) A study of the coupled fixed point problem for operators satisfying a max-symmetric condition in *b*-metric spaces with applications to a boundary value problem. *Miskolc Math. Notes*, **17**, 501–516.

52 Mleşniţe, O. (2013) Existence and Hyers-Ulam stability results for a coincidence problem with applications. *Miskolc Math. Notes*, **14**, 183–189.

53 Buică, A. (2001) *Coincidence Principles and Applications*, Cluj University Press, Cluj-Napoca.

54 Falset, J.G. and Mleşniţe, O. (2014) Coincidence problems for generalized contractions. *Appl. Anal. Discrete Math.*, **8**, 1–15.

55 Rus, I.A. (1990/91) Some remarks on coincidence theory. *Pure Math. Manuscr.*, **9**, 137–148.

56 Petruşel, A., Petruşel, G., and Yao, J.-C. (2017) Contributions to the coupled coincidence point problem in *b*-metric spaces with applications. *Filomat*, **31**, 3173–3180.

20

The Corona Problem, Carleson Measures, and Applications

Alberto Saracco

Dipartimento di Scienze Matematiche, Fisiche e Informatiche, Università degli Studi di Parma, 43124 Parma, Italy

20.1 The Corona Problem

20.1.1 Banach Algebras: Spectrum

In order to state the corona problem, a few notions on Banach algebras are needed. In this section, we will recall the needed definitions and results, without proofs. The interested reader can find the theorems proven in [1, Section 12.1].

A commutative algebra B over $\mathbb{C}$ with unit $\mathbf{1}$ is said to be a *Banach algebra* if it is a Banach space with a norm $\| \cdot \|$ such that

$$\|\mathbf{1}\| = 1, \quad \|xy\| \leq \|x\| \ \|y\|$$

for all $x, y \in B$.

A Banach algebra B is called an *uniform algebra* if for all $b \in B$

$$\|b^2\| = \|b\|^2.$$

This condition is very restrictive and the only uniform algebras are algebras of $\mathbb{C}$-valued functions.

From now on, B will always denote a Banach algebra.

The *spectrum* of an element $b \in B$ is the subset of $\mathbb{C}$

$$\sigma(b) = \{\lambda \in \mathbb{C} : \lambda \cdot \mathbf{1} - \mathbf{b} \text{ is not invertible}\}.$$

For every $b \in B$, the spectrum $\sigma(b)$ is a compact subset of $\mathbb{C}$.

A *character* of B is a homomorphism of algebras $\chi : B \to \mathbb{C}$. Every character χ of a Banach algebra is continuous. In particular the norm of every character is bounded by 1. If the character χ is not trivial (i.e., does not send B in 0) than since $\chi(\mathbf{1}) = 1 \ \|\chi\| = 1$.

Mathematical Analysis and Applications: Selected Topics, First Edition.
Edited by Michael Ruzhansky, Hemen Dutta, and Ravi P. Agarwal.
© 2018 John Wiley & Sons, Inc. Published 2018 by John Wiley & Sons, Inc.

The *spectrum* of B is the set $\mathfrak{M}(B)$ of all nontrivial characters of B. If B^* denotes the dual space of B than

$$\mathfrak{M}(B) \subset \{b^* \in B^* : \|b^*\| \leq 1\},$$

thus the spectrum, with the topology induced by the weak-* topology on B^* (called the *Gelfand topology*), is compact, thanks to Banach–Alaoglu.

$\mathfrak{M}(B)$ equipped with the Gelfand topology is a compact Hausdorff space. For every $b \in B$, the map $\hat{b} : \mathfrak{M}(B) \to \mathbb{C}$ defined by

$$\hat{b}(\chi) = \chi(b)$$

is called the *Gelfand transform* of b. By $\hat{B}$ we denote the set of all Gelfand transforms of B.

The Gelfand topology is the weakest topology that makes every Gelfand transform a continuous map.

The map $\Gamma : B \to C^0(\mathfrak{M}(B))$ sending an element of B to its Gelfand transform is continuous (the space of continuous $\mathbb{C}$-valued functions on the spectrum being endowed with the sup norm).

20.1.2 Banach Algebras: Maximal Spectrum

The closure of any proper ideal of a Banach algebra B is a proper ideal. Hence maximal ideals are closed. The set of all maximal ideals of B, $\Omega(B)$, is called the *maximal spectrum* of B.

There is a natural bijection between the spectrum and the maximal spectrum of algebra, sending a nontrivial character to its kernel:

$$T : \mathfrak{M}(B) \to \Omega(B), \quad \varphi \mapsto \text{Ker } \varphi.$$

The spectrum of a Banach algebra is a nonempty compact space.

20.1.3 The Algebra of Bounded Holomorphic Functions and the Corona Problem

Let D be a domain (i.e., an open connected set) in $\mathbb{C}^n$ or more generally in a complex manifold or a complex space X.

By $\mathcal{O}(D)$, we denote the algebra of holomorphic functions on D and by $H^\infty(D)$ the algebra of bounded holomorphic functions on D. $H^\infty(D)$ is a Banach algebra when endowed with the supnorm:

$$\|f\|_\infty = \sup_{z \in D} |f(z)|.$$

Assume $H^\infty(D)$ separates the points of D, that is, given any two points $z \neq w \in D$ there is a function $f \in H^\infty(D)$ such that $f(z) \neq f(w)$.

Then there is a natural embedding ι of the domain D into the (maximal) spectrum $\mathfrak{M}(H^\infty(D))$ given by

$$\iota(z) = \{f \in H^\infty(D) | f(z) = 0\},$$

that is, z is sent to the ideals of functions vanishing in z.

Obviously, $\iota(D)$ is not all of the spectrum (e.g., the ideals of functions vanishing at a point of the boundary is not in the image), also for the compactness of the spectrum.

Consider the set

$$\mathfrak{M}(H^\infty(D))\backslash\overline{\iota(D)},$$

called the *corona*.

The *corona conjecture* states that the corona is empty, that is, the domain D naturally embeds densely in the spectrum $\mathfrak{M}(H^\infty(D))$.

We remark that the corona conjecture has an analytic equivalent form:

Theorem 20.1 Assume bounded holomorphic functions separate points of D.

$\iota(D)$ is dense in $\mathfrak{M}(H^\infty(D))$ if and only if for all $f_1, \dots, f_n \in H^\infty(D)$ such that

$$\sum_{i=1}^{n} |f_i(z)|^2 \geq \delta > 0$$

for all $z \in D$, there exists $g_1, \dots, g_n \in H^\infty(D)$ such that

$$\sum_{i=1}^{n} f_i(z)g_i(z) \equiv 1 \quad \forall z \in D.$$

For a proof, see [1, p. 342].

20.2   Carleson's Proof and Carleson Measures

The conjecture (which is false in all generality) was first posed by Kakutani in 1941 for the unit disk $\Delta \subset \mathbb{C}$.

A first step towards the solution of the corona conjecture for the disk Δ was done by Newman in 1959 [2], and a positive answer was given by Carleson [3].

Carleson's original proof was based on Carleson measures.

Carleson measures were defined by Carleson for all $H^p(\Delta)$ spaces. $H^p(\Delta)$ for $p \geq 1$ is the Banach space of all holomorphic functions with finite norm

$$\|f\|_p^p = \lim_{r\to 1} \frac{1}{2\pi} \int_0^{2\pi} |f(re^{i\theta})|^p \, d\theta.$$

Definition 20.1 Let μ be a positive Borel measure on the disk Δ. μ is said to be a *Carleson measure* iff there exists a constant C such that for all $f \in H^p(\Delta)$

$$\int_\Delta |f(z)|^p \, d\mu \leq C\|f\|_p^p.$$

Observe that μ is a Carleson measure for $H^p(\Delta)$ iff the identity maps $H^p(\Delta)$ continuously into $L^p(\Delta, \mu)$.

If this inclusion is compact, we say that μ is a *vanishing Carleson measure*.

Using Carleson measures and interpolations methods for Blaschke products [4], Carleson was able to prove the corona conjecture for the disk.

Both the corona theorem and the methods of Carleson's proof had a huge impact on mathematics in the subsequent decades. In the next sections, we will follow the developments of the corona conjecture and of the theory of Carleson measures from 1962 to today.

20.2.1 Wolff's Proof

Later, in 1980, Wolff gave in a seminar a more elementary proof (written down by Gamelin [5]) based on $\bar{\partial}$-estimates – following a strategy outlined by Hörmander [6] – in which he proved a stronger result obtaining an explicit bound on the norms of the functions g_j of Theorem 20.1:

Theorem 20.2 For all $n \in \mathbb{N}$ and $\delta > 0$ there exists a constant $C_1(n, \delta)$ such that if $f_1, \dots, f_n \in H^\infty(\Delta)$ satisfy $\|f_j\|_\infty \le 1$ and

$$\sum_{i=1}^n |f_i(z)|^2 \ge \delta$$

for all $z \in \Delta$, there exists $g_1, \dots, g_n \in H^\infty(\Delta)$ such that

$$\sum_{i=1}^n f_i(z)g_i(z) \equiv 1 \quad \forall z \in D$$

and $\|g_j\|_\infty \le C_1(n, \delta)$.

The constant $C_1(n, \delta)$ can be explicitly computed (see, e.g., [1, p. 347]) and is a polynomial in n.

20.3 The Corona Problem in Higher Henerality

In the last 50 years, the corona problem has been treated in several different situations. In this chapter, we cover only a few of those results. There are also matrix-valued (both in the complex and the quaternionic case) and operator-valued versions of the corona problem which we do not treat here. The interested reader may read a more complete story in [7].

20.3.1 The Corona Problem in $\mathbb{C}$

In view of the Riemann mapping theorem, the corona problem has a positive answer for all simply connected domains $D \subsetneq \mathbb{C}$. In $\mathbb{C}$ the problem has a trivial positive answer due to the fact that the only bounded holomorphic entire functions are constants.

In 1970, Gamelin [8] proved that if $D \subset \mathbb{CP}^1$ is such that there exists an integer m such that the complement of each component of D has at most m components (i.e., each connected component of D is $(m + 1)$-connected), then the corona problem has a positive answer.

Actually Gamelin proves the following stronger version of the theorem, similar to Theorem 20.2:

Theorem 20.3 Let $D \subset \mathbb{CP}^1$ be an $(m + 1)$-connected domain. For all $n \in \mathbb{N}$ and $\delta > 0$ there exists a constant $C_m(n, \delta)$ such that if $f_1, \dots, f_n \in H^\infty(D)$ satisfy $\|f_j\|_\infty \leq 1$ and

$$\sum_{i=1}^n |f_i(z)|^2 \geq \delta$$

for all $z \in D$, there exists $g_1, \dots, g_n \in H^\infty(D)$ such that

$$\sum_{i=1}^n f_i(z)g_i(z) \equiv 1 \quad \forall z \in D$$

and $\|g_j\|_\infty \leq C_m(n, \delta)$.

Open problem. The corona conjecture is still an open problem for an arbitrary open subset D of $\mathbb{C}$.

Remark 20.1 In the cited paper, Gamelin shows that an affirmative answer to this question is equivalent to the boundedness for each integer $n \geq 1$ and $\delta > 0$ of the best possible constants $C_m(n, \delta)$ of Theorem 20.3.

In case the best possible constants are not bounded, a counterexample to the corona conjecture for $D \subset \mathbb{C}$ is easy to produce.

20.3.2 The Corona Problem in Riemann Surfaces: A Positive and a Negative Result

In 1963, Alling [9] proved that if X is a finite open Riemann surface, then the corona conjecture has a positive answer: X is dense in the spectrum of bounded analytic functions on X.

A finite open Riemann surface X is a proper open connected surface of a compact Riemann surface W whose boundary coincides the boundary of the complement of X in W and is consists of a finite number of closed analytic arcs. Since $W \setminus X$ has nonempty interior, Riemann–Roch theorem shows that bounded holomorphic functions separate points of X and the corona conjecture is meaningful.

This theorem of Alling implies the corona theorem for finitely connected domains of the projective plane $\mathbb{CP}^1$ bounded by analytic curves, that is, a weaker version of Theorem 20.3 by Gamelin, which was to be proved 7 years later.

It is known that the corona problem for Riemann surfaces has in general a negative answer. The first counterexample was due to B. Cole (see Chapter 4 in [10]).

The construction of the example runs along these lines, using the idea of Gamelin in $\mathbb{C}$ (see Remark 20.1).

Fix $\delta > 0$. First are constructed a sequence $\{R_m\}_{m \in \mathbb{N}}$ of finite bordered Riemann surfaces and bounded holomorphic functions $f_{1,m}, f_{2,m}$ of norm at most 1 on R_m such that $|f_{1,m}(z)| + |f_{2,m}(z)| \geq \delta$ on R_m and if $g_{1,m}, g_{2,m} \in H^\infty(R_m)$ satisfy

$$f_{1,m} g_{1,m} + f_{2,m} g_{2,m} \equiv 1$$

on R_m then either $\|g_{1,m}\|_\infty$ or $\|g_{2,m}\|_\infty$ are greater than m. This means that $\{R_m\}_{m \in \mathbb{N}}$ is a sequence of finite bordered Riemann surfaces where, if the corona problem has positive answer, the best possible constants are unbounded: $C_{R_m}(2, \delta) \geq m$.

The unboundedness of these constants allows the construction of a connected Riemann surface R containing all the Riemann surfaces R_m and their borders as disjoint subsets and of two holomorphic functions f_1, f_2 on R whose norms are bounded by 1 such that $|f_1| + |f_2| \geq \delta$ on R and there are no holomorphic bounded functions g_1, g_2 on R such that $f_1 g_1 + f_2 g_2 \equiv 1$ on R.

Cole's counterexample can be easily fattened in order to provide a counterexample in any dimension.

Cole's example was later modified by Nakai [11] in order to show that in Riemann surfaces of Parreau–Widom type, whose algebra of bounded analytic functions shares many properties with $H^\infty(\Delta)$, the corona problem does not have a positive answer. The counterexample is of course a Riemann surface with infinite genus and infinite connectivity.

20.3.3 The Corona Problem in Domains of $\mathbb{C}^n$

While for domains in $\mathbb{C}$ it is still unknown whether the corona conjecture has always a positive answer, in $\mathbb{C}^n$ several counterexamples are known.

In several complex variables, the corona conjecture fails even for domains of holomorphy, as the following example, due to Sibony [12], shows.

Example 20.1 Let $\{z_j\}_{j \in \mathbb{N}}$, be a sequence of points in the unit disk Δ without limit points in Δ and such that each point in $b\Delta$ is a nontangential limit of a subsequence of $\{z_j\}$. Let $\{\lambda_j\}_{j \in \mathbb{N}}$ be a sequence of positive numbers such that $\sum_j \lambda_j$ is finite. Defining the function

$$V(z) = \prod_{j=1}^\infty \left| \frac{z - z_j}{2} \right|^{\lambda_j}$$

on the disk Δ and

$$M(\Delta, V) = \{(z, w) \in \Delta \times \mathbb{C} : |w| < e^{-V(z)}\}$$

is a bounded domain of holomorphy such that all bounded holomorphic functions $f \in H^\infty(M(\Delta, V))$ are restriction of bounded holomorphic functions on the bidisk Δ^2 (for details we refer to [1, pp. 340, 347–348]). Thus, the corona problem has a negative answer for this domain, which is a Runge domain.

Using the previous example, Sibony [13] was able to construct a domain in $\mathbb{C}^3$ that shows that the corona conjecture has a negative answer even for a pseudo-convex bounded domain whose boundary is strictly Levi-convex at all points but one of its boundary. This is extremely interesting since the related problem for the algebra $A(\overline{\Omega})$ of holomorphic functions continuous up to the boundary has a positive answer for pseudoconvex bounded domains Ω, that is, the spectrum of the algebra coincides with $\overline{\Omega}$, as proved by Sibony himself with Hakim [14].

Open problem. The corona problem is still open in the simple domains as the unit ball and the unit polidisk in $\mathbb{C}^n$, for $n > 1$.

20.3.4 The Corona Problem for Quaternionic Slice-Regular Functions

20.3.4.1 Slice-Regular Functions $f : D \to \mathbb{H}$

Let $\mathbb{H}$ be the space of quaternions, that is, of elements $q = x_0 + ix_1 + jx_2 + kx_3$, where $x_l \in \mathbb{R}$ $(l = 0, \dots, 3)$ and $i^2 = j^2 = k^2 = ijk = -1$. Quaternions are a skew field.

Many attempts have been done in the last century to define a notion similar to that of holomorphic functions in $\mathbb{C}$ for functions $f : \mathbb{H} \to \mathbb{H}$. The main problem in doing so in a satisfying manner (i.e., being able to reproduce some of the nice properties and theorems that can be obtained in the complex case) lies in the lack of commutativity of quaternions, which has many nasty consequences. Just to explicitly name a few, it makes necessary a distinction between right and left differentiation; and monomials are not only the ones written in a nice way aq^n or $q^n a$ but also things of the sort $a_0 q a_1 q \cdots a_n q a_{n+1}$. Thus, differential conditions are pretty difficult to use (and any differential condition in $\mathbb{C}$ has two natural generalizations in $\mathbb{H}$) and even simple functions as polynomials are rather complicated.

Several definitions given in early 1900 had the problem of having too few regular functions (for example not even the identity was regular) and of the notion not being preserved by pointwise multiplication of functions. To make a long story short, a satisfying definition was given in 2006 by Gentili and Struppa [15], where a notion of slice-regularity was introduced.

We briefly recall some basic definitions. For details and proofs, the interested reader should refer to [16] or [17] for an even more general theory on alternating algebras.

Let $\mathbb{S} \subset \mathbb{H}$ be the sphere of imaginary units of $\mathbb{H}$, that is, the set of quaternions q of modulus $|q| = \sqrt{x_0^2 + x_1^2 + x_2^2 + x_3^2} = 1$ and such that $q^2 = -1$. Notice that

$$\mathbb{S} = \{ ix_1 + jx_2 + kx_3 \in \mathbb{H} \mid x_1^2 + x_2^2 + x_3^2 = 1 \}.$$

For any fixed $I \in \mathbb{S}$ the set $\mathbb{C}_I = \{ x + Iy \mid x, y \in \mathbb{R} \}$ is a complex plane.

For $q = x_0 + Iy_0 \in \mathbb{H}$, $y \neq 0$, we denote by $[q]$ the two-dimensional sphere of quaternions with same real part and same modulus as q; if $q \in \mathbb{R}$, $[q] = \{q\}$. A domain D is said to be *axially symmetric* if for each $q \in D$ the set $[q]$ is contained in D. The *axially symmetric completion* of a domain D is the domain

$$\hat{D} = \bigcup_{q \in D} [q].$$

A domain $D \subset \mathbb{H}$ is said to be *s-regular* if its slices $D_I = D \cap \mathbb{C}_I$ are connected for every $I \in \mathbb{S}$.

Definition 20.2 Let $D \subset \mathbb{H}$ be a domain and $f : D \to \mathbb{H}$ be $\mathbb{R}$-differentiable. The function f is said to be (*left*) *slice-regular* if for every $I \in \mathbb{S}$ its restriction f_I to $D_I = D \cap \mathbb{C}_I$ satisfies

$$\overline{\partial}_I f(x + Iy) := \frac{1}{2} \left(\frac{\partial}{\partial x} + I \frac{\partial}{\partial y} \right) f_I(x + Iy) = 0.$$

We will call $\mathcal{SR}(D)$ the set of (left) slice-regular functions on D.

Notice that polynomials and power series with coefficients on the right, that is, of the special form $f(q) = \sum q^n a_n$ are left slice-regular. A similar definition of right slice-regularity can be given as to make right slice regular all polynomials and power series with coefficients to the left.

Any (left) slice-regular function on D can be extended to a (left) slice-regular function defined on the axially symmetric completion of D. Thus, it is mostly interesting to study (left) slice-regular functions in axially symmetric domains.

The notion of (left) slice-regularity is mostly interesting for domains intersecting the real axis, as if $D \cap \mathbb{R} = \emptyset$, slice-regularity does not even imply continuity. It can however be given a different definition that agrees with the one above on domains intersecting the real axis and is meaningful and interesting also in the case the domain does not intersect $\mathbb{R}$. We are not interested in such a definition here, as we will treat just the unit ball.

The sum of two slice-regular functions is slice-regular, but their pointwise product is not. Therefore, a new product has been introduced on slice-regular functions, to turn them into an algebra. We do not present here the complete definition of $*$-multiplication, but just how it works for power series. Let $f(q) = \sum q^n a_n$ and $g(q) = \sum q^n b_n$ be two power series converging on $B(0, r) \subset \mathbb{H}$ for some $r > 0$. Then

$$(f * g)(q) = \sum q^n c_n, \quad c_n = \sum_{k=0}^{n} a_k b_{n-k},$$

that is, their $*$-multiplication is defined as if the coefficients commuted with the variable q.

A big problem when dealing with slice-regular functions is that the composition of two slice-regular functions is not necessarily slice-regular, so something in the spirit of Riemann uniformization theorem is actually either impossible or useless.

20.3.4.2 The Corona Theorem in the Quaternions

The results in this section are due to Shelah [18].

Slice-regular functions (and their zeros) behave quite weirdly. While for the disk $\Delta \subset \mathbb{C}$ the condition $\sum_{k=1}^{m} |f_k|^2 \geq \delta > 0$ for functions $f_k \in H^\infty(\Delta)$ is both necessary and sufficient for the existence of $g_k \in H^\infty(\Delta)$ such that $\sum_k f_k g_k \equiv 1$, things are quite different in the ball $\mathbb{B} \subset \mathbb{H}$.

First of all, since $*$-multiplication is not commutative we have to consider two different problems (H^∞ denotes the space of bounded left slice-regular functions):

1. The left inverse problem: given $f_1, \dots, f_n \in H^\infty(\mathbb{B})$ under which condition there exist $g_1, \dots, g_n \in H^\infty(\mathbb{B})$ such that

$$\sum_{k=1}^{n} g_k * f_k \equiv 1?$$

2. The right inverse problem: given $f_1, \dots, f_n \in H^\infty(\mathbb{B})$ under which condition there exist $g_1, \dots, g_n \in H^\infty(\mathbb{B})$ such that

$$\sum_{k=1}^{n} f_k * g_k \equiv 1?$$

While – as in the complex case – for the right inverse problem the condition

$$\sum_{k=1}^{m} |f_k|^2 \geq \delta > 0 \tag{20.1}$$

is necessary, for the left inverse problem it is not, as the following example shows.

Let $f_1(q) = q - i$ and $f_2(q) = (q - i)j$. Then both functions have a zero in $q = i$, yet defined $g_1(q) = \frac{i}{2}$ and $g_2(q) = \frac{k}{2}$,

$$(g_1 * f_1)(q) + (g_2 * f_2)(q) = \frac{i}{2} * (q - i) + \frac{k}{2} * (q - i)j$$

$$= q(i + kij) + \frac{-i^2 - ijk}{2} = 1.$$

Open problem. It is unknown whether condition (20.1) is sufficient for the left or right inverse problem.

In [18] also a quaternionic matrix-valued corona theorem is proved.

20.4   Results on Carleson Measures

Carleson measure, as already noted at the beginning of this chapter, were initially introduced by Carleson in the setting of Hardy spaces. However, their definition makes sense in a much bigger generality.

Let D be any complex domain (of an Euclidean complex space or of some complex manifold or even complex space). Let A be a Banach algebra of holomorphic functions on D and assume $A \subset L^p(D)$ for some $p > 0$.

Definition 20.3   Let μ be a positive Borel measure on D. μ is said to be a *Carleson measure* of $A \subset L^p(D)$ iff there exists a constant C such that for all $f \in H^p(\Delta)$

$$\int_D |f(z)|^p \, d\mu \leq C \|f\|_p^p.$$

Observe that μ is a Carleson measure for $H^p(D)$ iff the identity maps A continuously into $L^p(D, \mu)$.

If this inclusion is compact, we say that μ is a *vanishing Carleson measure* of A.

20.4.1   Carleson Measures of Hardy Spaces of the Disk

The first algebras for which Carleson measure were studied are Hardy spaces. While there seems not to be a priory reasons for Carleson measures relative to different Hardy spaces $H^p(\Delta)$ to have relations with each other, the notion of Carleson measure is actually not dependent on $p > 0$, as proved by Duren in 1959 [19].

Theorem 20.4   A finite positive Borel measure μ is Carleson of $H^p(\Delta)$ if and only if the μ-measure of the Carleson boxes

$$S_{\theta_0,h} = \{re^{i\theta} \in \Delta | 1 - h \leq r < 1, |\theta - \theta_0| \leq h\}$$

is controlled by a fixed multiple of h, that is, there exists $C > 0$ such that $\mu(S_{\theta_0,h}) \leq Ch$ for all h and all θ_0.

This means that a measure μ is a Carleson of a Hardy space of the disk if and only if the μ-measure of any Carleson box $S_{\theta_0,h}$ is bounded by (a fixed multiple of) its side.

It is sufficient to prove the theorem for $p = 2$, after factoring Blaschke products out of the bounded holomorphic function f, considering $f^{2/p}$ instead of f.

The functions $f_\alpha(z) = (1 - \alpha z)^{-1}$, $\alpha \in \Delta$ show the necessity of the condition, while the sufficiency is the difficult part to prove.

The Carleson boxes are not invariant under biholomorphisms, so the characterization given by Theorem 20.4 is not an intrinsic characterization. It is a characterization based on the extrinsic Euclidean metric (i.e., based on how the disk lies inside $\mathbb{C}$) rather than based on the intrinsic hyperbolic Poincaré metric of the disk.

20.4.2 Carleson Measures of Bergman Spaces of the Disk

If $D \subset \mathbb{C}^n$ is a bounded domain then we define the *Bergman* space of functions $A^p(D)$ as the space of holomorphic $L^p(D)$ functions endowed with the L^p norm.

The Carleson measures of the Bergman spaces of the disk $\Delta \subset \mathbb{C}$ were investigated by Hastings [20], who found a characterization in terms of the measures of Carleson boxes. Again, thanks to this result, the set of Carleson measures of $A^p(\Delta)$ does not depend on p (but it is different from the set of Carleson measures for the Hardy spaces).

Theorem 20.5 A finite positive Borel measure μ is Carleson of $A^p(\Delta)$ if and only if the μ-measure of the Carleson boxes

$$S_{\theta_0,h} = \{re^{i\theta} \in \Delta | 1 - h \le r < 1, |\theta - \theta_0| \le h\}$$

is controlled by a fixed multiple of h^2, that is, there exists $C > 0$ such that $\mu(S_{\theta_0,h}) \le Ch^2$ for all h and all θ_0.

This means that a measure μ is a Carleson of a Bergman space of the disk if and only if the μ-measure of any Carleson box $S_{\theta_0,h}$ is bounded by (a fixed multiple of) its Lebesgue area.

As already noticed, the Carleson boxes are not invariant under biholomorphisms and this is not an intrinsic characterization.

A characterization of Carleson measure of Bergman spaces of the disk in terms of hyperbolic balls was given by Luecking [21].

Consider in Δ the Poincaré distance ω or its strictly related pseudohyperbolic distance ρ (the relation between them being $\rho = \tanh(\omega)$). Let

$$B_\Delta(z_0, r) = \left\{ z \in \Delta \Big| \left| \frac{z - z_0}{1 - \bar{z}_0 z} \right| < r \right\}$$

be the Poincaré disk of center $z_0 \in \Delta$ and pseudohyperbolic radius $0 < r < 1$. Then the following theorem holds.

Theorem 20.6 A finite positive Borel measure μ on Δ is Carleson of $A^p(\Delta)$ if and only if for some (for all) $0 < r < 1$ there is a constant $C_r > 0$ such that

$$\mu(B_\Delta(z_0, r)) \le C_r \lambda(B_\Delta(z_0, r)),$$

λ being the Lebesgue measure, that is, if and only if the μ-measure of Poincaré disks are uniformly controlled by their Lebesgue measure.

The Poincaré disks are invariant under automorphisms of the disk, thus this characterization – which is always a characterization based on a bound on the μ-measure in terms of the Lebesgue measure – is way better than the previous one.

20.4.3   Carleson Measures in the Unit Ball of $\mathbb{C}^n$

The first results on Carleson measures in several complex variables has been obtained for the unit ball $\mathbb{B}^n \subset \mathbb{C}^n$ by Cima and Wogen [22]. Again the characterization is given – similarly to the one by Hastings (or the original one by Carleson for $H^p(\Delta)$ – in terms of Carleson boxes. A *Carleson box* in $\mathbb{B}^n$ of location ζ and dimension t is the set

$$S(\zeta, t) = \{z \in \mathbb{B}^n \,||1 - z \cdot \zeta| < t\},$$

that is, the set of points whose scalar product with z is sufficiently near to 1. As in the case on the unit disk, Carleson boxes are defined in terms of the extrinsic euclidean geometry.

Theorem 20.7   A positive Borel measure μ on $\mathbb{B}^n$ is Carleson of the Bergman space $A^p(\mathbb{B}^n$ if and only if there is a constant $C > 0$ such that, for all Carleson boxes

$$\mu(S(\zeta, t)) \leq C\lambda(S(\zeta, t)),$$

where λ is the Lebesgue measure, that is, the μ-measure of all Carleson boxes is uniformly bounded by their Lebesgue measure.

Again this theorem shows that the set of Carleson measure is the same for all Bergman spaces of the unit ball.

To provide a characterization that only uses the intrinsic geometry of the unit ball, a distance with properties similar to those of the Poincaré distance in the disk is needed. There are actually several possibilities.

In great generality, we will define two pseudodistances on a connected complex manifold (or even a complex space) X, that is, functions

$$d : X \times X \to \mathbb{R}_+$$

symmetric, satisfying the triangular inequality $(d(x, z) \leq d(x, y) + d(y, z))$ and such that $d(x, y) = 0$ if $x = y \in X$. We remark that if the last *if* is actually an *iff*, d is a distance.

Both pseudodistances we define are based on the Poincaré distance ω on the disk Δ. The Carathéodory pseudodistance c_X is based on holomorphic functions from X to Δ, while the Kobayashi pseudodistance k_X is based on holomorphic maps form Δ to X. Let $z, w \in X$, then we define:

- (Carathéodory) $c_X(z, w) = \sup\{\omega(h(z), h(w)) \mid h : X \to \Delta$ holomorphic$\}$;
- (Kobayashi) $k_X(z, w) = \inf\{\sum_{j=0}^{m} \omega(\zeta_j, \eta_j)\}$, where $\zeta_j, \eta_j \in \Delta$ are such that there are holomorphic disks $\phi_j : \Delta \to X$ connecting z and w: $z = \phi_0(\zeta_0)$, $\phi_j(\eta_j) = \phi_{j+1}(\zeta_{j+1})$, $\phi_m(\eta_m) = w$.

Both c_X and k_X are pseudodistances (in general not coinciding) that are decreasing under holomorphic maps:

Proposition 20.1 Let $f : X \to Y$ be holomorphic maps. Then for all $z, w \in X$

$$c_Y(f(z), f(w)) \leq c_X(z, w),$$
$$k_Y(f(z), f(w)) \leq k_X(z, w).$$

Hence, these pseudodistances are invariant under biholomorphisms and actually reflect the complex structure of the manifold (space) X.

If k_X is a (complete) distance, X is said to be (complete) Kobayashi hyperbolic; if c_X is a (complete) distance, X is said to be (complete) Carathéodory hyperbolic. While this notions are usually different, they coincide (and coincide with several other notions of hyperbolicity, as Brody hyperbolicity) in case of domains of $\mathbb{C}^n$ with convexity properties (as convexity [23] or $\mathbb{C}$-convexity [24]).

Due to Proposition 20.1 and the fact that $k_{\mathbb{B}^n}$ and $c_{\mathbb{B}^n}$ are explicitly known (and are distances), bounded domains of $\mathbb{C}^n$ are Kobayashi (and Carathéodory) hyperbolic.

For more details about this distances we refer the reader to [25, Section 2.3] or [26, 27].

An intrinsic characterization of Carleson measures of Bergman spaces of the unit ball similar to the one provided by Luecking in dimension one has been proved by Duren and Weir [28], using the Kobayashi distance, or more precisely the pseudohyperbolic distance ρ: $\rho = \tanh(k_{\mathbb{B}^n})$.

Let the Kobayashi ball of center z_0 and pseudohyperbolic radius r be

$$B_\rho(z_0, r) = \{z \in \mathbb{B}^n | \rho(z, z_0) < r\}.$$

Theorem 20.8 A finite positive Borel measure μ on Δ is Carleson of $A^p(\mathbb{B}^n)$ if and only if for some (for all) $0 < r < 1$ there is a constant $C_r > 0$ such that

$$\mu(B_\rho(z_0, r)) \leq C_r \lambda(B_\rho(z_0, r)),$$

λ being the Lebesgue measure, that is, if and only if the μ-measure of Kobayashi balls are uniformly controlled by their Lebesgue measure.

In the very same paper [28], Duren and Weir obtained yet another characterization of Bergman spaces of the unit ball, based on the Bergman kernel.

If $D \subset \mathbb{C}^n$, the space $A^2(D)$ of L^2 holomorphic functions is a separable Hilbert space endowed with the inner product

$$\langle f, g \rangle = \int_D f \bar{g} \, d\lambda.$$

Given a point $z \in D$, consider the continuous functional e_z of evaluation at z: $e_z(f) = f(z)$ for all $f \in A^2(D)$. Thanks to Riesz representation theorem, e_z can be represented as a scalar product, that is, there is $K_D(z, \cdot) \in A^2(D)$ such that

$$f(z) = e_z(f) = \langle f, K_D(z, \cdot) \rangle = \int_D f \overline{K_D}(z, \cdot) \, d\lambda,$$

for every $f \in A^2(D)$. $K_D(z, \cdot)$ is a reproducing kernel, known as the Bergman kernel.

We recall that the Bergman kernel on the diagonal is real and strictly positive, more precisely $K_D(z,z) = \|K_D(z,\cdot)\|_2^2$. For details and properties of the Bergman kernel, refer, for example, to [1, Section 1.6].

For any finite positive Borel measure μ one can define its Berezin transform as the function $B\mu : D \to \mathbb{R}_+$ given by

$$B\mu(z) = \int_D \frac{|K_D(\zeta,z)|}{K_D(z,z)}\, \mathrm{d}\mu(\zeta).$$

Duren and Weir [28] proved the following theorem.

Theorem 20.9 Let μ be a finite positive Borel measure on $\mathbb{B}^n$. μ is a Carleson measure of $A^p(\mathbb{B}^n)$ if and only if its Berezin transform $B\mu$ is bounded.

20.4.4 Carleson Measures in Strongly Pseudoconvex Bounded Domains of $\mathbb{C}^n$

Proving results in the disk Δ or in the ball $\mathbb{B}^n$ is usually made easier by the abundance of tools we have at our disposal working there, for example, the huge automorphism groups, which are transitive. Usually this approach cannot be generalized to other domains, as even strongly pseudoconvex domains which are not biholomorphic to the ball admit very few automorphisms.

But one should expect most results obtained in the ball to still be valid for strictly convex (or more generally strongly pseudoconvex) bounded domains of $\mathbb{C}^n$.

Cima and Mercer [29], while studying the properties of the composition operator between holomorphic functions, characterized Carleson measures of $A^p(D)$ for D bounded strongly pseudoconvex domain of $\mathbb{C}^n$. In doing this, Carleson measures with a given weight were introduced.

Their characterization depends on the extrinsic Euclidean geometry of $\mathbb{C}^n$ and shows that also in this setting the set of Carleson measures of $A^p(D)$ is the same for all p.

In order to get an intrinsic characterization of Carleson measures of bounded strongly pseudoconvex domains, it is necessary to have an understanding on the geometry of Kobayashi balls and precise estimates on the Bergman kernel, both on and out of the diagonal.

In this way, for bounded strongly pseudoconvex domains, the very same characterizations as in the ball case of Carleson measures (via Berezin transform or via estimates on the μ-volume of Kobayashi balls) hold, as Abate and Saracco [30] proved in 2011.

Theorem 20.10 Let μ be a finite positive Borel measure on a strongly pseudoconvex bounded domain $D \subset \mathbb{C}^n$. Then the following statements are equivalent:

1. μ is a Carleson measure of $A^p(D)$ for some (and hence all) $p \in (0, +\infty)$;

2. the Berezin transform of μ is a bounded function;
3. for every $r \in (0, 1)$ there exists a positive constant C_r such that for all $z_0 \in D$
$$\mu(B_D(z_0, r)) \le C_r \lambda(B_D(z_0, r));$$
4. for some $r \in (0, 1)$ there exists a positive constant C_r such that for all $z_0 \in D$
$$\mu(B_D(z_0, r)) \le C_r \lambda(B_D(z_0, r)).$$

20.4.5 Generalizations of Carleson Measures and Applications to Toeplitz Operators

As observed in Definition 20.3, the definition of Carleson measure can be restated as the continuity of the identity map between A and $L^p(D, \mu)$ and so it makes sense to ask for something more, as, for example, the compactness of this inclusion.

If the inclusion is compact, the Carleson measure μ is said to be vanishing. The results we stated in the previous sections have their corresponding counterparts in the setting of vanishing Carleson measure. The idea of vanishing Carleson measure is mostly useful in the study of Toeplitz operators.

Let $D \subset \mathbb{C}^n$ be a domain. The reproducing Bergman kernel $K_D(z, \zeta)$ is very interesting because not only it is a means to reproduce the values of a function $f \in A^p(D)$ by an integral:

$$f(z) = \int_D K_D(z, \zeta) f(\zeta) \, d\lambda(\zeta), \qquad (20.2)$$

(λ as always denotes the Lebesgue measure) but also because the very same integral expression gives a projection B of $\mathcal{O}(D)$ onto $A^2(D)$, called the *Bergman projection*:

$$Bf(z) = \int_D K_D(z, \zeta) f(\zeta) \, d\lambda(\zeta), \quad f \in \mathcal{O}(D). \qquad (20.3)$$

Since for each $p \ge 1$ there are functions in $A^p(D)$ but not in $A^q(D)$ for any $q > p$, one cannot in general expect the Bergman projection to send functions in $A^p(D)$ in $A^q(D)$ if $q > \max(2, p)$. However, it might be useful to have an operator which regularizes holomorphic functions giving better growth conditions.

With this idea in mind Čučković and McNeal [31] in 2006 introduced the Toeplitz operators T_{δ^η} of the following form:

$$T_{\delta^\eta} f(z) = B(\delta^\eta f) = \int_D K_D(z, \zeta) \delta(\zeta)^\eta f(\zeta) \, \lambda(\zeta),$$

where $\delta(\zeta) = d(\zeta, \partial D)$ is the Euclidean distance of ζ from the boundary of D and $\eta > 0$, thus $\mu = \delta^\eta \cdot \lambda$ is a positive Borel measure. If D is a bounded domain, $\mu = \delta^\eta \cdot \lambda$ is a finite positive Borel measure.

Čučković and McNeal proved that these Toeplitz operators actually give a gain in growth in the case D is a bounded strongly pseudoconvex domain.

More generally, if $D \subset \mathbb{C}^n$ is a domain and μ a finite positive Borel measure on D, we can define the Toeplitz operator associated to μ as

$$T_\mu f(z) = \int_D K_D(z, \zeta) \mu(\zeta).$$

The first to consider such operators was Kaptanoğlu [32] in 2007 for the unit ball of $\mathbb{C}^n$.

In the most general setting of bounded strongly pseudoconvex domains, Abate *et al.* [33] provided an interesting link between θ-Carleson measures and properties of Toeplitz operators.

As we have already observed, a finite positive Borel measure is Carleson of $A^p(D)$ ($D \subset \mathbb{C}^n$ being bounded strongly pseudoconvex) iff μ is (geometric) Carleson, that is, for some (for all) $0 < r < 1$ there is $C_r > 0$ such that

$$\mu(B_D(z_0, r)) \leq C_r \lambda(B_D(z_0, r))$$

for all Kobayashi balls.

Thus, given $\theta > 0$, we can define a finite positive Borel measure to be a *(geometric) θ-Carleson measure* iff for some (for all) $0 < r < 1$ there is $C_r > 0$ such that

$$\mu(B_D(z_0, r)) \leq C_r \lambda(B_D(z_0, r))^\theta$$

for all Kobayashi balls.

If, moreover,

$$\lim_{z_0 \to \partial D} \mu(B_D(z_0, r)) \lambda(B_D(z_0, r))^\theta = 0,$$

μ is said to be a *vanishing (geometric) θ-Carleson measure*.

If $\theta = 1$ we get again the notions of Carleson measure or vanishing Carleson measure.

Theorem 20.11 Let $D \subset \mathbb{C}^n$ be a bounded strongly pseudoconvex domain, μ a finite positive Borel measure on D and $1 < p < r < \infty$. The following are equivalent:

1. $T_\mu : A^p(D) \to A^r(D)$ continuously (resp. compactly);
2. μ is a (resp. vanishing) $\left(1 + \frac{1}{p} - \frac{1}{r}\right)$-Carleson measure.

Actually the notion of (resp. vanishing) geometric Carleson measure is linked to the continuity (resp. compacticity of the inclusion of a Bergman space with weight in $L^p(D)$). For details, see [33].

20.4.6 Explicit Examples of Carleson Measures of Bergman Spaces

For Bergman spaces, as we have seen, the property of being Carleson for a finite positive Borel measure μ is a property of having the μ-measure of sets (as Carleson boxes or Kobayashi balls) bounded by (a multiple of) the Lebesgue measure.

Thus, some Carleson measures are easily constructed: If $\mu = C\lambda$ or even $\mu = f\lambda$, with f bounded on the domain D, then μ is Carleson.

On the other side of the spectrum lie measures that are sum of Dirac's deltas. Obviously any finite positive Borel measure with compact support is a Carleson measure, and so the property of being Carleson in this setting is interesting to explore for converging sums of Dirac's deltas.

To construct an important class of examples it is necessary to introduce the notion of uniformly discrete sequence. Let (X, d) be a metric space. A sequence of points $\{x_k\}_{k\in}$ is said to be *uniformly discrete* if there exists $\delta > 0$ such that any two points of the sequence are at least δ-apart from each other.

Carleson measures and uniformly discrete sequences are linked by the following theorem, proved in the unit ball by Duren and Weir [28], Jevtič *et al.* [34], Massaneda [35], and in bounded strongly pseudoconvex domains by Abate and Saracco [30].

Theorem 20.12 Let $D \subset \mathbb{C}^n$ be a bounded strongly pseudoconvex domain and $\Gamma = \{z_k\}_{k\in\mathbb{N}} \subset D$. Then the following are equivalent:

- Γ is a finite union of uniformly discrete (with respect to the Kobayashi distance) sequence;
- $\mu = \sum_{z_j\in\Gamma} d(z_j, \partial D)\delta_{z_j}$ is a Carleson measure of $A^p(D)$.

(δ_{z_j} is the Dirac measure at z_j and $d(\cdot, \partial D)$ is the euclidean distance from the boundary.)

20.4.7 Carleson Measures in the Quaternionic Setting

20.4.7.1 Carleson Measures on Hardy Spaces of $\mathbb{B} \subset \mathbb{H}$

As we have seen, $*$-multiplication turns $\mathcal{SR}(D)$ into a (right) algebra, the definition of Carleson measure makes sense also in the quaternionic setting. As the only results known so far are in the ball $\mathbb{B} \subset \mathbb{H}$, we will give the definitions there.

Let $p > 0$, $I \in \mathbb{S}$ and $\mathbb{B}_I = \mathbb{B} \cap \mathbb{C}_I$. We define the following subalgebra of $\mathcal{SR}(\mathbb{B})$:

$$H^p(\mathbb{B}) = \{f \in \mathcal{SR}(\mathbb{B}) | \|f\|_p < \infty\},$$

where

$$\|f\|_p = \sup_{I\in\mathbb{S}} \lim_{r\to 1^-} \left(\int_0^{2\pi} |f(re^{I\theta}|^p \, d\theta \right)^{1/p}.$$

Definition 20.4 A finite positive Borel measure μ on $\mathbb{B}$ is said to be a *Carleson measure* for $H^p(\mathbb{B})$ if there exists $C > 0$ such that for all $f \in H^p(\mathbb{B})$

$$\int_{\mathbb{B}} |f(q)|^p \, d\mu(q) \leq C\|f\|_p^p.$$

A finite positive Borel measure μ on $\mathbb{B}$ is said to be a *slice Carleson measure* for $H^p(\mathbb{B})$ if there exists $C > 0$ such that for all $f \in H^p(\mathbb{B})$ and all $I \in \mathbb{S}$

$$\int_{\mathbb{B}_I} |f(x + Iy)|^p \, d\mu_I(x + Iy) \le C\|f\|_p^p,$$

where μ_I is the restriction of μ on $\mathbb{B}_I$.

The first results on Carleson and slice Carleson measures in the quaternionic unit ball were obtained on $H^2(\mathbb{B})$ by Arcozzi and Sarfatti [36] and then extended to $H^p(\mathbb{B})$ by Sabadini and Saracco [37].

Slice Carleson measures are in particular Carleson measures. The characterization of both these classes of measures is given in terms of the quaternionic analog of Carleson boxes.

Definition 20.5 If $q = re^{J\theta_0} \in \mathbb{B}$ ($J \in \mathbb{S}$), we define the *Carleson box* $S_I(\theta_0, r)$ in the plane $\mathbb{C}_I$ as

$$S_I(\theta_0, r) = \{\rho e^{I\theta} \mid |\theta - \theta_0| \le r, \ 1 - r \le \rho < 1\},$$

and the *symmetric Carleson box* $S(\theta_0, r)$ as

$$S(\theta_0, r) = \bigcup_{I \in \mathbb{S}} S_I(\theta_0, r).$$

Then the following characterizations hold [36, 37].

Theorem 20.13 A finite positive Borel measure on $\mathbb{B}$ is

- a slice Carleson measure of $H^p(\mathbb{B})$ iff there exists $C > 0$ such that for all Carleson boxes $\mu_I(S_I(\theta_0, r)) \le Cr$;
- a Carleson measure of $H^p(\mathbb{B})$ iff there exists $C > 0$ such that for all symmetric Carleson boxes $\mu(S(\theta_0, r)) \le Cr$.

20.4.7.2 Carleson Measures on Bergman Spaces of $\mathbb{B} \subset \mathbb{H}$

As in the complex case, Bergman spaces of slice-regular functions can be defined in a broader class of domains. The first definitions in this setting were given by Colombo *et al.* [38–41].

Let $D \subset \mathbb{H}$ be an axially symmetric s-domain. For $p > 0$, let the slice-regular Bergman space $A^p(D)$ be the set of all slice regular function $f : D \to \mathbb{H}$ with finite norm

$$\|f\|_{A^p} = \left(\sup_{I \in \mathbb{S}} \int_{D_I} |f(x + Iy)|^p \lambda(x + Iy) \right)^{1/p},$$

λ being the Lebesgue measure on $\mathbb{C}_I$.

Definition 20.6   Let $D \subset \mathbb{H}$ be an axially symmetric s-domain and μ be a finite positive Borel measure on D. μ is said to be

* a *slice Carleson measure* on $A^p(D)$ if there exists $C > 0$ such that for all $f \in A^p(D)$

$$\int_{D_I} |f(x + Iy)|^p \, d\mu_I(x + Iy) \le \|f\|_{A^p}^p;$$

* a *Carleson measure* on $A^p(D)$ if there exists $C > 0$ such that for all $f \in A^p(D)$

$$\int_D |f(q)|^p \, d\mu(q) \le \|f\|_{A^p}^p.$$

The only results known so far are obtained in the unit ball $\mathbb{B} \subset \mathbb{H}$, for $A^2(\mathbb{B})$ [39–41] and $A^p(\mathbb{B})$ [37].

Open problem. So far no characterization of slice Carleson or Carleson measures is known in a general axially symmetric s-domain.

As in the H^p case, being slice Carleson of $A^p(\mathbb{B})$ implies being Carleson of $A^p(\mathbb{B})$.

Differently from what happens in the complex case, for the quaternionic case no characterization of Carleson measures of $A^p(\mathbb{B})$ is possible in terms of pseudohyperbolic balls holds [37]. The main problem in generalizing the result is the lack of a Moebius-like transformation of the unit quaternionic ball preserving slice-regular function, that is, the fact that the composition of slice-regular functions may not be slice-regular.

Instead, there is a characterization based on the hyperbolic geometry of the slices. Let $B_I(x + Iy, r)$ be a pseudohyperbolic disk of pseudohyperbolic radius r in $\mathbb{B}_I$ and $B(x + Iy, r)$ its axially symmetric completion

$$B(x + Iy, r) = \bigcup_{J \in \mathbb{S}} B_J(x + Jy, r).$$

Let us denote by $A(x + Iy, r)$ the euclidean area of $B_I(x + Iy, r)$.

Then, slice Carleson and Carleson measure were characterized by Sabadini and Saracco [37] in the following way:

Theorem 20.14   Let μ be a finite positive Borel measure on $\mathbb{B}$.

1. μ is slice Carleson of $A^p(\mathbb{B})$ if and only if for every (for some) $0 < r < 1$ there is C such that for all $x + Iy \in \mathbb{B}$

$$\mu_I(B_I(x + Iy, r)) \le CA(x + Iy, r);$$

2. μ is Carleson of $A^p(\mathbb{B})$ if and only if for every (for some) $0 < r < 1$ there is C such that for all $x + Iy \in \mathbb{B}$

$$\mu(B(x + Iy, r)) \le CA(x + Iy, r).$$

References

1 Della Sala, G., Saracco, A., Simioniuc, A., and Tomassini, G. (2006) *Lectures on Complex Analysis and Analytic Geometry*, Appunti, Scuola Normale Superiore di Pisa (Nuova Serie), vol. 3, Edizioni della Normale, Pisa, xx+430.

2 Newman, D.J. (1959) Some remarks on the maximal ideal structure of H^∞. *Ann. Math. (2)*, **70**, 438–445.

3 Carleson, L. (1962) Interpolations by bounded analytic functions and the corona problem. *Ann. Math.*, **76**, 547–559.

4 Carleson, L. (1958) An interpolation problem for bounded analytic functions. *Am. J. Math.*, **80**, 921–930.

5 Gamelin, T.W. (1980) Wolff's proof of the corona theorem. *Israel J. Math.*, **37** (1–2), 113–119.

6 Hormander, L. (1967) Generators for some rings of analytic functions. *Bull. Am. Math. Soc.*, **73**, 943–949.

7 Douglas, R.G., Krantz, S.G., Sawyer, E.T., Treil, S., and Wicks, B.D. (2014) A history of the corona problem, in *The Corona Problem. Connections Between Operator Theory, Function Theory, and Geometry*, Fields Institute Communications, vol. 72, Springer, New York, pp. 1–29; Fields Institute for Research in Mathematical Sciences, Toronto, ON, viii+231 p.

8 Gamelin, T.W. (1970) Localization of the corona problem. *Pac. J. Math.*, **34**, 73–81.

9 Alling, N.L. (1964) A proof of the corona conjecture for finite open Riemann surfaces. *Bull. Am. Math. Soc.*, **70**, 110–1112.

10 Gamelin, T.W. (1978) *Uniform Algebras and Jensen Measures*, London Mathematical Society Lecture Note Series, vol. 32, Cambridge University Press, Cambridge, New York, iii+162 p.

11 Nakai, M. (1982) Corona problem for Riemann surfaces of Parreau-Widom type. *Pac. J. Math.*, **103** (1), 103–109.

12 Sibony, N. (1975) Prolongement des fonctions holomorphes bornes et métrique de Carathéodory. *Invent. Math.*, **29** (3), 205–230.

13 Sibony, N. (1987) Probléme de la couronne pour des domaines pseudoconvexes á bord lisse. *Ann. Math. (2)*, **126** (3), 675–682.

14 Hakim, M. and Sibony, N. (1980) Spectre de $A(\overline{\Omega})$ pour des domaines bornés faiblement pseudoconvexes réguliers. *J. Funct. Anal.*, **37** (2), 127–135.

15 Gentili, G. and Struppa, D.C. (2006) A new approach to Cullen-regular functions of a quaternionic variable. *C. R. Math. Acad. Sci. Paris*, **342** (10), 741–744.

16 Gentili, G., Stoppato, C., and Struppa, D.C. (2013) *Regular Functions of a Quaternionic Variable*, Springer Monographs in Mathematics, Springer, Berlin, Heidelberg.

17 Colombo, F., Sabadini, I., and Struppa, D.C. (2011) *Noncommutative Functional Calculus. Theory and Applications of Slice Hyperholomorphic Functions*, Progress in Mathematics, vol. 289, Birkhäuser-Springer.

18 Shelah, Y. (2016) Quaternionic Wiener algebras, factorization, and the corona theorem, MS thesis. Tel Aviv University.

19 Duren, P.L. (1969) Extension of a theorem of Carleson. *Bull. Am. Math. Soc.*, **75**, 143–146.

20 Hastings, W.W. (1975) A Carleson measure theorem for Bergman spaces. *Proc. Am. Math. Soc.*, **52**, 237–241.

21 Luecking, D. (1983) A technique for characterizing Carleson measures on Bergman spaces. *Proc. Am. Math. Soc.*, **87** (4), 656–660.

22 Cima, J.A. and Wogen, W.R. (1982) A Carleson measure theorem for the Bergman space on the ball. *J. Oper. Theory*, **7** (1), 157–165.

23 Bracci, F. and Saracco, A. (2009) Hyperbolicity in unbounded convex domains. *Forum Math.*, **21** (5), 815–825.

24 Nikolov, N. and Saracco, A. (2007) Hyperbolicity of ℂ-convex domains. *C. R. Acad. Bulgare Sci.*, **60** (9), 935–938.

25 Abate, M. 1989 *Iteration Theory of Holomorphic Maps on Taut Manifolds*, Research and Lecture Notes in Mathematics. Complex Analysis and Geometry, Mediterranean Press, Rende, xvii+417 pp.

26 Jarnicki, M. and Pflug, P. (1993) *Invariant Distances and Metrics in Complex Analysis*, De Gruyter Expositions in Mathematics, vol. 9, Walter de Gruyter & Co., Berlin, xii+408 pp.

27 Kobayashi, S. (1998) *Hyperbolic Complex Spaces*, Grundlehren der Mathematischen Wissenschaften [Fundamental Principles of Mathematical Sciences], vol. 318, Springer-Verlag, Berlin, xiv+471 pp.

28 Duren, P.L. and Weir, R. (2007) The pseudohyperbolic metric and Bergman spaces in the ball. *Trans. Am. Math. Soc.*, **359** (1), 63–76.

29 Cima, J.A. and Mercer, P.R. (1995) Composition operators between Bergman spaces on convex domains in ℂⁿ. *J. Oper. Theory*, **33** (2), 363–369.

30 Abate, M. and Saracco, A. (2011) Carleson measures and uniformly discrete sequences in strongly pseudoconvex domains. *J. Lond. Math. Soc. (2)*, **83** (3), 587–605.

31 Čučković, Ž. and McNeal, J.D. (2006) Special Toeplitz operators on strongly pseudoconvex domains. *Rev. Mat. Iberoam.*, **22**, 851–866.

32 Kaptanoğlu, H.T. (2007) Carleson measures for Besov spaces on the ball with applications. *J. Funct. Anal.*, **250**, 483–520.

33 Abate, M., Raissy, J., and Saracco, A. (2012) Toeplitz operators and Carleson measures in strongly pseudoconvex domains. *J. Funct. Anal.*, **263** (11), 3449–3491.

34 Jevtić, M., Massaneda, X., and Thomas, P.J. (1996) Interpolating sequences for weighted Bergman spaces of the ball. *Michigan Math. J.*, **43**, 495–517.

35 Massaneda, X. (2006) A^{-p} interpolation in the unit ball. *J. Math. Anal. Appl.*, **318**, 37–42.

36 Arcozzi, N. and Sarfatti, G. (2017) From Hankel operators to Carleson measures in a quaternionic variable. *Proc. Edimburg Math. Soc.*, **60** (3), 565–585.

37 Sabadini, I. and Saracco, A. (2017) Carleson measures for Hardy and Bergman spaces in the quaternionic unit ball. *J. Lond. Math. Soc.* doi: 10.1112/jlms.12035.

38 Colombo, F., Gonzáles-Cervantes, J.O., Sabadini, I., and Shapiro, M. (2013) *On Two Approaches to the Bergman Theory for Slice Regular Functions*, Springer INdAM Series, vol. 1, Springer, Milano, pp. 39–54.

39 Colombo, F., Gonzáles-Cervantes, J.O., and Sabadini, I. (2012) On slice biregular functions and isomorphisms of Bergman spaces. *Complex Variab. Ellip. Equ.*, **57**, 825–839.

40 Colombo, F., Gonzáles-Cervantes, J.O., and Sabadini, I. (2013) The C-property for slice regular functions and applications to the Bergman space. *Complex Variab. Ellip. Equ.*, **58**, 1355–1372.

41 Colombo, F., Gonzáles-Cervantes, J.O., and Sabadini, I. (2015) Further properties of the Bergman spaces of slice regular functions. *Adv. Geom.*, **15**, 469–484.

Index

Mathematical Analysis and Applications: Selected Topics, First Edition.
Edited by Michael Ruzhansky, Hemen Dutta, and Ravi P. Agarwal.
© 2018 John Wiley & Sons, Inc. Published 2018 by John Wiley & Sons, Inc.